# HANDBOOK OF NEUROCHEMISTRY

## VOLUME III

## METABOLIC REACTIONS IN THE NERVOUS SYSTEM

# HANDBOOK OF NEUROCHEMISTRY

Edited by Abel Lajtha

---

# HANDBOOK OF NEUROCHEMISTRY

Edited by Abel Lajtha

*New York State Research Institute*
*for Neurochemistry and Drug Addiction*
*Ward's Island*
*New York, New York*

## VOLUME III

## METABOLIC REACTIONS IN THE NERVOUS SYSTEM

PLENUM PRESS • NEW YORK–LONDON • 1970

Library of Congress Catalog Card Number 68-28097

**SBN 306-37703-9**

ISBN 978-1-4615-7162-9 ISBN 978-1-4615-7160-5 (eBook)
DOI 10.1007/978-1-4615-7160-5



softcover reprint of the hardcover 1st edition 1970

A Division of Plenum Publishing Corporation
227 West 17th Street, New York, New York 10011

United Kingdom edition published by Plenum Press, London
A Division of Plenum Publishing Company, Ltd.
Donington House, 30 Norfolk Street, London W.C. 2, England

# Contributors to this volume:


M. H. Aprison — Institute of Psychiatric Research and Departments of Psychiatry and Biochemistry, Indiana University Medical Center, Indianapolis, Indiana (page 381)

Robert Balazs — Medical Research Council, Neuropsychiatric Research Unit, Carshalton, Surrey, England (page 1)

Claude F. Baxter — Neurochemistry Laboratories, Veterans Administration Hospital, Sepulveda, California (page 289)

H. C. Buniatian — Institute of Biochemistry, Academy of Sciences, Armenian SSR, Yerevan (page 399)

Stephen R. Cohen — New York State Research Institute for Neurochemistry and Drug Addiction, Ward's Island, New York (page 87)

R. V. Coxon — University Laboratory of Physiology, Oxford, England (page 37)

Amedeo F. D'Adamo, Jr. — Saul R. Korey Department of Neurology and Department of Biochemistry, Albert Einstein College of Medicine of Yeshiva University, New York (page 525)

Robert A. Davidoff — Institute of Psychiatric Research and Departments of Psychiatry and Biochemistry, Indiana University Medical Center, Indianapolis, Indiana. (page 381)

Alan N. Davison — Department of Biochemistry, Charing Cross Hospital Medical School, London, England (page 547)

M. K. Gaitonde — Medical Research Council, Neuropsychiatric Research Unit, Carshalton, Surrey, England (page 225)

Gordon Guroff — Laboratory of Biochemical Sciences, National Institute of Child Health and Human Development and Experimental Therapeutics Branch, National Heart Institute, National Institutes of Health, Bethesda, Maryland (page 209)

| | |
| --- | --- |
| J. N. Hawthorn | Department of Medical Biochemistry and Pharmacology, University of Birmingham, England (page 491) |
| M. Kai | Department of Medical Biochemistry and Pharmacology, University of Birmingham, England (page 491) |
| Francis N. LeBaron | Department of Biochemistry, The University of New Mexico School of Medicine, Albuquerque, New Mexico (page 561) |
| Walter Lovenberg | Laboratory of Biochemical Sciences, National Institute of Child Health and Human Development and Experimental Therapeutics Branch, National Heart Institute, National Heart Institute, Bethesda, Maryland (page 209) |
| N. Marks | New York State Research Institute for Neurochemistry and Drug Addiction, Ward's Island, New York (page 133) |
| Cyril L. Moore | Department of Neurology, Albert Einstein School of Medicine, Bronx, New York (page 53) |
| Norman S. Radin | Mental Health Research Institute, University of Michigan, Ann Arbor, Michigan (page 415) |
| Maurice M. Rapport | New York State Psychiatric Institute and College of Physicians and Surgeons, Columbia University, New York (page 509) |
| Abraham Rosenberg | Columbia University Research Service, Goldwater Memorial Hospital, New York (page 453) |
| R. J. Rossiter | Department of Biochemistry, University of Western Ontario, London, Canada (page 467) |
| Paula M. Strasberg | Department of Neurology, Albert Einstein School of Medicine, Bronx, New York (page 53) |
| Harold J. Strecker | Albert Einstein School of Medicine, Yeshiva University, New York (page 173) |

| | |
|---|---|
| K. P. Strickland | Department of Biochemistry, University of Western Ontario, London, Canada (page 467) |
| Lars Svennerholm | Department of Neurochemistry, Psychiatric Research Centre, Faculty of Medicine, University of Goteborg, Sweden (page 425) |
| C. J. van den Berg | Central Institute for Brain Research, Ijdijk, Amsterdam (page 355) |
| Robert Werman | Institute of Psychiatric Research and Departments of Psychiatry and Biochemistry, Indiana University Medical Center, Indianapolis, Indiana (page 381) |

# PREFACE

When the projected volumes of the *Handbook* are completed, most of our current knowledge of the biochemistry of nervous systems will have been touched upon. A number of the chapters will have dealt with the correlations of the biochemical findings with morphological and physiological parameters as well. Considering the abysmal lack of such attempts, even in the recent past, this is a sign of great progress. If the reader's eventual goal is to derive the "laws" that relate various aspects of animal and human behavior to underlying physiological and biochemical function, these admirable volumes will help him to establish a firm biochemical base from which to operate. It is certain that the future approaches to the various problems of the information-processing functions of the nervous system will require an integrated understanding of the essence of all of the scientific disciplines which are grouped under the general name of neurobiology. The rich feast of information offered up in this *Handbook* will enable those in the non-chemical disciplines to pick and choose those areas of chemical information pertinent to their immediate interests. Similar types of compendia by physiologists, anatomists, cyberneticists, and psychologists have been helpful to chemists and continue to be so. However, when all is said and done, it is likely that important generalizations and synthetic insights will occur in the minds of *individuals* who have assimilated enough information from the various related areas to enable them to discuss the contact points of the moving edges of the several fields. Such individuals probably will come largely from the generation of neuroscientists being trained now and from their students. Let us hope that the availability of the thoughts and data of the excellent group of contributors to the *Handbook* will aid them in their exciting search.

Eugene Roberts

*Duarte, California*
*September* 1969

# CONTENTS

*Chapter 4*



*Chapter 5*

*Chapter 10*

*Chapter 1*

# CARBOHYDRATE METABOLISM

**Robert Balázs**

*Medical Research Council*
*Neuropsychiatric Research Unit*
*Carshalton, Surrey, England*

## I. INTRODUCTION

There is a general agreement that under normal physiological conditions the main substrate utilized by the brain is glucose. The earlier evidence for this was obtained mainly by investigating the overall metabolism of the brain *in vivo*. Further evidence which has more recently become available relates to the intermediate steps and enzymic mechanisms underlying the metabolism of glucose in the brain. The recent advances can be attributed to several factors: (1) techniques for preventing postmortem change have been improved,[1,2] (2) more sensitive and specific methods have been developed for determining the intermediates of glucose metabolism,[1,2] and (3) the study of the kinetic properties of the individual enzymes has increased our understanding of the mechanisms concerned in glucose metabolism.[3-7] As a result of extensive studies in several laboratories, and especially by the Bücher, Krebs, and Lowry groups, it is now possible to construct a fairly detailed map of the different pathways of glucose metabolism in the brain. The pattern of relative activities of the enzymes concerned in the different pathways, combined with new information on the regulation of glucose metabolism, and the results of tracer studies have provided a firm biochemical basis for the unique role assigned to glucose in the energy metabolism of the brain.

The brain is cytologically heterogeneous and there is experimental evidence of metabolic compartmentation in the tissue [*cf.* Berl *et al.*[8]]. However, the present review deals mainly with the broad outlines of glucose metabolism in the brain as a whole. Compartmentation is discussed in other chapters of the *Handbook* (*cf.* Volume II, Berl and Clarke and van den Berg).

Earlier literature on carbohydrate metabolism in nervous tissues is given in a number of reviews and books.[9-16]*

* Abbreviations used in this chapter for metabolites and enzymes are listed in Tables III and IV and Fig. 2.

## II. GLUCOSE AS THE MAIN SUBSTRATE FOR OXIDATION IN THE BRAIN

### A. Evidence from Overall Brain Metabolism

The cerebral respiration rate is relatively high (Table I) and under basal conditions the brain accounts for approximately 20% of the total $O_2$ requirement in man, although the weight of the brain is only about 2% of that of the body.[(10)] The main evidence on the utilization of substrates by the brain can be summarized as follows[(9,10)]: (1) The respiratory quotient (R.Q.) is approximately 1.0 under various conditions: this indicates that the substrate oxidized is mainly carbohydrate. (2) Glucose is the only metabolite taken up from the blood by the brain in a sufficient amount to account for the observed respiration rate. The theoretical ratio of $O_2$ uptake to glucose consumed is 6, the measured value is 5.5, and the difference can be accounted for by lactate formation (Table I). (3) When the supply of glucose to the brain is reduced, the respiration rate decreases and cerebral function is impaired. Complete restoration of function is apparently effected only by glucose or by substances providing glucose either directly or indirectly.

### B. Evidence from Tracer Studies

The central role of glucose in the energy metabolism of brain has been further investigated by tracer methods. Some of the results reported were apparently at variance with the earlier conclusions.

#### 1. *Respiratory* $^{14}CO_2$

It has been reported that in perfused brain preparations or in cerebral cortex slices only a fraction of the respiratory $CO_2$ produced was derived from $^{14}$C-glucose [*cf.* Geiger[(11)]]. However, these experiments have certain limitations. During the course of these studies it came to be realized that the small pools of intermediates produced by the immediate breakdown of glucose are in equilibrium with relatively large metabolite pools, such as those of amino acids associated with the tricarboxylic acid cycle; this results in isotopic dilution.[(17–19)] The amino acid pools in the brain are not homogeneous; this may result in pools which equilibrate at different rates with the corresponding cycle intermediates and with each other. Furthermore, at certain steps isotopic exchange of a kind not necessarily involving net metabolic reactions can occur. Thus, measurement of the incorporation of label into the final product ($CO_2$) without simultaneous chemical determination of the intermediates does not give a true estimate of the contribution of different possible substrates to the respiration. For example, the specific activity of $^{14}CO_2$ produced by slices of brain cortex respiring in the presence of $^{14}$C-glucose is reduced when unlabeled glutamate is also provided. This effect, however, does not mean that glucose utilization is inhibited by glutamate. On the contrary, quantitative determinations showed that glutamate removal was suppressed by glucose in different brain preparations.[12,20,21] The effect of glutamate on the conversion of $^{14}C$-glucose to $^{14}CO_2$ is probably related to the retention of the label in the glutamate pool after isotopic exchange between α-oxoglutarate and glutamate (*cf.* Section II,B,2). Finally, some of the preparations used were probably not "intact" physiologically, resulting in increased catabolism of macromolecules whose breakdown products were continuously fed into the main path of glucose combustion.* Thus, by using better experimental techniques, the specific activity of $^{14}CO_2$ relative to glucose (RSA) was raised *in vitro* from 0.6 to 0.8[22] and from 0.3 observed in perfused brain preparations[17] to values approximating 1.0 in the intact dog brain.[23]† The substrate requirement of excised sympathetic ganglion was also thoroughly investigated and it was found that approximately 90% of the $CO_2$ formed could be accounted for by the oxidation of $^{14}C$-glucose.[15]

### 2. *Fate of $^{14}C$-Glucose in Brain*

The distribution of $^{14}C$ in brain is different from that of most other tissues after $^{14}C$-glucose has been metabolized.[18,25–32] The results have shown that a rapid and extensive conversion of $^{14}C$-glucose into amino acids is quite characteristic for the adult brain; in liver and immature brain most of the $^{14}C$ was recovered in glucose when $\frac{2}{3}$ of the $^{14}C$ was in amino acids in adult brain (Table II). Furthermore, the rapid flux of glucose carbon into amino acids develops sharply at the so-called "critical period" during which the cortex becomes functionally mature.[33]

These observations have raised some controversy as far as the major substrate and pathway of cerebral oxidative metabolism are concerned. On closer examination,[20,21] however, the results have supported the previous conclusion that the main substrate for cerebral oxidation is glucose, and the results have contributed to the better understanding of the operation of a number of factors characteristic for the pattern of oxidative metabolism in brain. It has been argued that the rapid and extensive incorporation of $^{14}C$ into brain glutamate reflects a combination of special features of brain metabolism largely unrelated to that of glutamate.[20,21] It has been shown that the net metabolism of glutamate in the presence of glucose is relatively low (*cf.* Section II,B,1), but an exchange of $^{14}C$ catalyzed by the amino acid transferases occurs between α-oxoglutarate and glutamate; the rate of this

* It has recently been shown that the conversion of glucosecarbon into metabolites on the path between glucose and $CO_2$ correlates with the functional state of the perfused brain preparation as indicated by the EEG recording. When function is well preserved most of the intermediates are derived from glucose, but as brain function is impaired there is a progressive increase in the proportion of metabolites formed from non-glucose carbon.[135]

† In contrast to the results found in the dog, it has been reported that the RSA of $^{14}CO_2$ is low even in the intact cat brain (0.45).[24] The discrepancy may be due to differences in the method used for collecting venous blood.

**TABLE I**

**Rate of Overall Glucose Metabolism in Brain Tissue (mmole/kg/min)**

| | | | $O_2$[a] | $CO_2$[a] | Glucose | Lactate | ~P[b] |
|---|---|---|---|---|---|---|---|
| *In vivo* | Normal | Man[(10)] | 1.6 | 1.6 | 0.3 | 0.002–0.1 | |
| | | Cat[(17)c] | | 1.9–2.2 | 0.4–0.8 | 0.1–0.7 | |
| | | Mouse[(1)] | | | 0.76[d] | | 25 |
| | | Rat[(34)] | | | 0.96[e] | | |
| | Convulsion | Cat | | 3–5[(17)] | +10[(51)f] | +12[(51)f] | |
| | | Mouse[(48)] | | | | 10 | +11–18; +40–80[g] |
| | Anesthesia | Man[(10)] | 0.9 | | | | |
| *In vitro* | Normal | Cerebral cortex slices[(12,14)h] | 1–1.7 | | 0.3–0.4[i] | 0.1–0.2[i] | |
| | Stimulated[j] | Cerebral cortex slices[(12,14)h] | 2–2.5 | | | 1.2–1.7;7[k] | |
| | Anaerobic | Cerebral cortex slices[(12,14)h] | | | | 1.7 | |
| | | Mitochondria (rat)[(80)l] | 3.5–4.5 | | | | |
| | | Ground tissue anaerobic (rat)[(14)] | | | | 9.3 | |

*Footnotes to Table I*

[a] The R.Q. of intact brain and of different brain preparations was found to be near to 1.0.[9,10,12]

[b] Utilization rate of available sources of ~P in meq/kg/min; for normal conditions it was calculated from the initial change which takes place during ischemia in the contents of CrP, ATP, glucose, glycogen, and lactate.[1]

[c] Perfused cat brain preparations.

[d] Calculated from the value for ~P utilization, assuming that 33 eq of ~P are generated for each mole of glucose consumed.

[e] Calculated from the values of total radioactivity and the specific activity of glucose in brain at short times after the administration of [U-$^{14}$C] glucose.

[f] Calculated from the change in metabolites during 10-sec experimental period, i.e., increase in rate above the normal metabolic activity.

[g] Increase above the normal rate of use of ~P; the first and second range of values are calculated for 50 sec and for the initial 3-sec period, respectively.

[h] Different species including man, cat, guinea pig, and rat.

[i] Wide range of values are usually obtained due to the high rate of aerobic lactate formation at the beginning of the incubation. The values given refer to the second ½-hr incubation period.

[j] Means of stimulation: electrical pulses, incubation in media containing high concentration of $K^+$ ions, or uncouplers of phosphorylation.

[k] Calculated for the initial 20 sec after electrical stimulation.

[l] Calculated from maximal respiration rate obtained with mitochondrial preparations ($q_{O_2}$ about 120), assuming that 1 g brain contains 40–50 mg protein in the crude mitochondrial fraction.

## TABLE II

### Fate of $^{14}$-C Glucose in Brain and Liver

A. Distribution of $^{14}C$ in acid-soluble fraction of organs at 10–30 min after the administration of $^{14}C$-glucose (%)[a]

| | Brain | | | Liver |
|---|---|---|---|---|
| | Adult | | Immature | Adult |
| | Rat[40] | Rat[100]; Cat[32] | Rat[33] (7-day-old) | Cat[32] |
| "Glucose" fraction | 12[b] | 3–8(1.5–4)[c] | 6.5 | 79(70)[c] |
| Amino acids | 56 | 70 | 5 | 8 |
| Glutamate | 37 | 40 | | 5 |
| Lactate | 21 | 10[d] | 15[e] | 2[d] |

B. Conversion of $^{14}C$ glucose in brain (in percentage of $^{14}C$ glucose metabolized in 1 hr)

| | Perfused cat brain[17,18] | Cerebral cortex slices [25,26,28,38] |
|---|---|---|
| Respiratory $^{14}CO_2$ | 30 | 20 |
| Amino acids | 22 | 7–16 |
| Glutamate | 13 | 3–11 |
| Lactate | 46 | 60–70 |

[a] 95–99% of the total $^{14}C$ in the organs was recovered in the acid-soluble fraction.
[b] Probably contains compounds other than glucose as well.
[c] Cat; the "glucose" fraction contains neutral compounds including glucose, inositol, glycerol, etc.; values in parentheses represent the $^{14}C$ recovered in glucose.
[d] Also includes succinate.
[e] Also includes carboxylic acids other than lactate.

reaction is several times more rapid than the rate at which the intermediates of the tricarboxylic acid cycle are oxidized.[20,21] The incorporation of $^{14}C$ into glutamate from labeled glucose by isotopic exchange must depend on the activity of the enzymes catalyzing the exchange, the size of the glutamate pool, and the rate of formation of labeled α-oxoglutarate. The exchange rate is apparently high in almost every type of tissue; therefore, the differences in the rate of conversion of $^{14}C$ into glutamate must be related to factors other than exchange between α-oxoglutarate and glutamate.

*a. Rate of Formation of Labeled α-Oxoglutarate.* Evidence was obtained for the operation of several factors that result in more $^{14}C$ from labeled glucose entering the tricarboxylic acid cycle in brain than in other tissues. The specific activity of glucose in brain approached that in blood soon after the injection.[34] Pyruvate becomes rapidly labeled,[35] and the specific activities of lactate and alanine are close to that of glucose in the brain even at short times after the administration of $^{14}C$-glucose.[34] Furthermore, the specific activity of $^{14}CO_2$ produced by the brain rapidly approaches that of [3-$^{14}C$] glucose in the blood.[19]* These results are consistent with relatively little dilution of the labeled glucose which is metabolized in the brain by glycolysis. Factors contributing to this are as follows: (1) the glucose and glycogen pool sizes are relatively low (Table IV); (2) the rate of glucose breakdown to pyruvate is high (Table I) and the back reaction (glucose formation from metabolites via pyruvate), which is important under many conditions in other tissues, is negligible in brain (*cf.* Section III,A); (3) the steady-state level of glycolytic intermediates is low (Table IV); and (4) uptake of lactate from the blood is limited.[9]

Dilution of material at the level of acetyl-CoA as a result of fatty acid oxidation is also low in brain. Fatty acids can be oxidized by brain, but their uptake from the blood/or the rate of oxidation, or both, are low when compared with glucose.[36] The overall R.Q. of the brain under conditions which result in substantial fatty acid oxidation in other tissues has been found to be near 1.0.[9]

These considerations indicate that the specific activity of α-oxoglutarate will be relatively high in brain owing to the relatively small dilution in the glycolytic pathway and in the initial parts of the tricarboxylic acid cycle. In other words, glucose is utilized rapidly in the glycolytic pathway and the contribution of reactions other than glycolysis or oxidation of pyruvate is limited in brain.

*b. Effect of Glutamate Pool Size.* The content of glutamate is higher in brain than in other tissues (Table IV) and factors which contribute to this effect have been discussed.[37] It has been shown that the incorporation of $^{14}C$ into glutamate from $^{14}C$-pyruvate or α-oxoglutarate is related to the glutamate pool size *in vitro* [*cf.* Balázs and Haslam[21]]. Thus, when isotopic equilibrium is established, the total $^{14}C$ incorporated into glutamate in brain can be greater than in other tissues. However, glutamate is compartmented in brain.[8] The various factors concerned in the incorporation of $^{14}C$ into glutamate will affect each pool separately. It is of interest that the metabolic compartmentation of amino acids is affected in $K^+$-stimulated brain slices.[38]

* The metabolism of [3-$^{14}C$] glucose in glycolysis results in the formation of [1-$^{14}C$] pyruvate and C-1 of pyruvate is eliminated as $CO_2$ when pyruvate enters the tricarboxylic acid cycle (*cf.* Fig. 2). Thus the specific activity of $^{14}CO_2$ relative to that of [3-$^{14}C$] glucose indicates the dilution of metabolites formed from $^{14}C$-glucose up to the stage of acetyl-CoA.

## TABLE III

### Pathways of Glucose Metabolism in Brain: Enzyme Activities and Comparison with Liver

| Enzyme[a] | Reaction[b] | $K_{equ}$[c] | Cerebral enzyme activity[d] (mmole/kg wet wt./min) | Activity ratio[e] (Brain/Liver) | $K_A$ ($\mu$M)[f] | $K_B$ ($\mu$M)[f] |
|---|---|---|---|---|---|---|
| | | *(1) Glycolysis* | | | | |
| Hexokinase (Hk) | Glucose + ATP = G6P + ADP | 3800[4] | 15–27[3,56,71] | 17[41] | 40[3]<br>1.6–3.1 × $10^3$ (fructose)[54,60] | 130[3][g] |
| Phosphoglucoisomerase (PGI) | G6P = F6P | 0.45[4] | 90–330[3,87] | 1.1[41] | 210[3] | |
| Phosphofructokinase (PFK) | F6P + ATP = FDP + ADP | 910[4] | 15–27[3,87] | 4[41] | 270[3] | 25[3][h] |
| Aldolase | FDP = GAP + DAP | 1.4 × $10^{-4}$[4]<br>0.9 × $10^{-4}$[3] | 10–35[3,71,87] | 1.4[41,105] | 12[3] | |
| Triosephosphate isomerase (TPI) | DAP = GAP | 0.05[4]<br>0.036[3] | 4000[106] | 1.8[41] | 120(GAP)[106] | |
| Glyceraldehyde-phosphate dehydrogenase (GAPD) | GAP + $NAD^+$ + $P_i$ = 1,3-DPG + NADH + $H^+$ | 0.09[4] | 90–190[3,71,105] | 0.7[41,105] | 44[3] | 22[3]<br>0.4–2.2 × $10^3$ ($P_i$)[3] |
| Phosphoglycerate kinase (PGK) | 1,3-DPG + ADP = 3PG + ATP | 2400[4] | 170–330[3,71] | 9[42] | 9[3] | 70[3] |
| Phosphoglyceromutase (PGlyM) | 3PG = 2PG | 0.1[3] | 70–210[3,71] | 1.7[107] | 240[3] | |
| Enolase | 2PG = PEP + $H_2O$ | 2.6[4]<br>4.6[3] | 30–50[3,71] | 3[42] | 33[3] | |

**Table III**—*continued*

| Enzyme[a] | Reaction[b] | $K_{equ}$[c] | Cerebral enzyme activity[d] (mmole/kg wet wt./min) | Activity ratio[e] (Brain/Liver) | $K_A$ ($\mu$M)[f] | $K_B$ ($\mu$M)[f] |
|---|---|---|---|---|---|---|
| Pyruvate kinase (PK) | PEP + ADP = Pyr + ATP | 19,500[4] | 150[3] | 12[42] | 55[3] | 180[3] |
| Lactate dehydrogenase (LDH) | Pyr + NADH + $H^+$ = lactate + $NAD^+$ | 19,000[2] | 90–180[3,71,108] | 0.3[41,105] | 140[3] | 2.8[3] |
| | | *(2) Pentose phosphate cycle* | | | | |
| Glucose 6-phosphate dehydrogenase (G6PD) | G6P + $NADP^+$ = 6PG(lactone) + NADPH + $H^+$ | $6 \times 10^{-7}$[103][i] | 0.6–2.5[89,109] | 0.3[41] | 240[109] | 3[109] |
| 6-Phosphogluconate dehydrogenase (6PGD) | 6PG + $NADP^+$ = Ru5P + $CO_2$ + NADPH + $H^+$ | | 0.4–2.5[88,109] | 0.2[41] | <130[109] | 2[109] |
| Transketolase (TK) | R5P + xylulose 5P = S7P + GAP | 1.2[104] | 0.1–0.5[88,89] | 0.2[89] | | |
| Transaldolase (TA) | S7P + GAP = E4P + F6P | 0.95[123] | 0.6[89] | 0.3[89] | | |
| | | *(3) Glycogen metabolism*[j] | | | | |
| Glycogen-UDP glucosyl-transferase | $(Glucosyl)_n$ + UDPG = $(glucosyl)_{n+1}$ + UDP | 6.9[102] | 0.5[110] | 0.2[110] | | 2000(UDPG)[111] |
| Phosphorylase | $(Glucosyl)_n + P_i$ = $(glucosyl)_{n-1}$ + G1P | 0.4[4] | 12[112] | 0.4[41] | 7200(G1P)[112] | |
| Phosphoglucomutase (PGM) | G1P = G6P | 15.7[4] | 17[112] | 0.1[113] | 1000[112] | 40($Mg^{2+}$)[112] |
| | | *(4) Gluconeogenesis* | | | | |
| Glucose 6-phosphatase (G6Pase) | G6P + $H_2O$ = glucose + $P_i$ | 650[101] | 0.07[46] | 0.005[41,46] | | |
| Fructose diphosphatase (FDPase) | FDP + $H_2O$ = F6P + $P_i$ | 650[102] | [k] | [k] | | |
| Malic enzyme (ME) | Pyr + $CO_2$ + NADPH + $H^+$ = malate + $NADP^+$ | 0.56[101] | 1.57[133][g] | 1[133][g] | 1200(Pyr)[44] $3.5 \times 10^4$($CO_2$)[44] | 10(NADPH)[44] |

**Table III**—*continued*

| Enzyme[a] | Reaction[b] | $K_{equ}$[c] | Cerebral enzyme activity[d] (mmole/kg wet wt./min) | Activity ratio[e] (Brain/Liver) | $K_A$ ($\mu$M)[f] | $K_B$ ($\mu$M)[f] |
|---|---|---|---|---|---|---|
| Phosphopyruvate carboxylase (PEPC) | PEP + GDP + $CO_2$ = OAA + GTP | 0.1[101] | 2[43][m] | 0.1[43][m] | 30(OAA)[44] | 150(GTP)[44] |
| Pyruvate carboxylase (PC) | Pyr + $CO_2$ + ATP + $H_2O$ = OAA + ADP + $P_i$ | 5.9[101] | 0.38[133] | 0.05[134] | 400(Pyr)[44]<br>1000($CO_2$)[44] | 30[44] |
| *(5) Polyols and glucuronate pathway* | | | | | | |
| Aldose reductase | Glucose + NADPH + $H^+$ = sorbitol + $NADP^+$ | $40 \times 10^8$[92][n] | 0.002[92][n] | | 37000[92] | |
| L-Hexonate dehydrogenase | Glucuronate + NADPH + $H^+$ = gulonate + $NADP^+$ | $16 \times 10^8$[92][n] | 0.026[92][n] | | 1700[92] | |
| *(6) Some auxiliary enzymes of glucose metabolism* | | | | | | |
| α-Glycerolphosphate dehydrogenase (αGPD)[o] | αGP + $NAD^+$ = DAP + NADH + $H^+$ | $8.9 \times 10^{-5}$[2] | 2–8[3,71,105,106] | 0.04[41,105] | 37(DAP)[3]<br>100 (DAP)[106] | 2.2(NADH)[3]<br>10(NADH)[106] |
| α-Glycerolphosphate oxidase (αGPO)[p] | αGP + (acceptor)→DAP + (reduced acceptor) | | 2.5–7.0[76,106] | | 3000[106]<br>9500[76] | |
| Malate dehydrogenase (MDH) | Malate + $NAD^+$ = OAA + NADH + $H^+$ | $9.8 \times 10^{-6}$[2] | 90–670[105,109] | 0.7[105] | 46(OAA)[109] | 23(NADH)[109] |
| Aspartate aminotransferase (AspAT) | Glu + OAA = αOG + Asp | 6.7[4] | 170[21,37] | 0.7[21,37] | 590(αOG)[109]<br>2600(Asp)[109] | |
| Alanine aminotransferase (AlaAT) | Glu + Pyr = αOG + alanine | 1.5[4] | 5[37] | 0.1[37] | | |
| Isocitrate dehydrogenase (ICD) | IC + $NADP^+$ = αOG + NADPH + $H^+$ | 15.9[101] | 3–17[109,114] | 0.03[114] | <20[109] | |

*Footnotes to Table III*

[a] The conventional name of enzyme is given; the abbreviations used in this chapter are indicated in parentheses.
[b] The abbreviations used for the metabolites are given in Table IV and Fig. 2 (for intermediates in pentose phosphate cycle).
[c] The apparent equilibrium constants ($K_{equ}$) are calculated from the values given for $\Delta G'$ in the following Refs.: 4, 101, and 102. $\Delta G' = -RT \ln K_{equ}$ where $\Delta G'$ is the change in the free energy under standard conditions at *p*H 7.0, $R$ is the gas constant (1987 cal/mole/deg.), and $T$ is the absolute temperature (the values were calculated for 37°C). Other values are taken from the references indicated.
[d] Usually the range of activities given by the different authors is indicated. The data were corrected for 38°C: the temperature coefficients reported by Lowry and Passonneau[3] were used for the glycolytic enzymes, and $Q_{10} = 1.7$ in case of the other enzymes.
[e] For the calculation of the activity ratios, the mean values from the previous column[d] were used for brain; the references from which the hepatic enzyme activities were taken are indicated.
[f] Apparent Michaelis constant ($K_m$): $K_A$ denotes the substrates other than nucleotides and $K_B$ the nucleotide substrates. The $K_m$ values relate to the reactants given in the chemical equations, unless indicated. Some of the enzymes listed are known to be present in multiple forms in the brain which is a possibility in case of other enzymes, as well. Furthermore, in some of the enzyme reactions the $K_m$ for one substrate is influenced by the concentration of the other substrate. Therefore, the kinetic constants may be regarded as attributes of the tissue as a whole under a given set of conditions rather than as properties of the pure enzyme species.
[g] For the properties of the purified Hk and Hk isoenzymes (*cf.* Refs. 56, 59 and 60).
[h] For more detailed kinetic properties *cf.* Ref. 63.
[i] $K_{equ}$ determined at *p*H 6.4 and 28°C.
[j] More detailed analysis of the enzymes involved in glycogen metabolism is given by Coxon in this *Handbook* (Chapter 2 of this volume).
[k] The activity of FDPase was not measurable in brain.[47]
[l] The reaction was assayed in the direction malate → Pyr + $CO_2$. The values for liver were quoted as approximately the same as for brain.[133]
[m] Approximate values calculated from the data given by Utter[43] for the enzymic activity observed in extracts of acetone powder of tissues; the activity of PEPC was related to that of PK since absolute values for the activity of this enzyme in tissue homogenates were available.
[n] The $K_{equ}$ of aldose reductase was determined with xylose and xylitol as substrates (*p*H 8), and that of L-hexonate dehydrogenase with the gulconate-glucuronate substrate pair (*p*H 8.3). The enzyme activities were calculated from the data of Moonsammy and Stewart[92] (ox brain). It was assumed that the total activity of both enzymes was in the supernatant fraction. The combined activity of the two enzymes with glucose as substrate was calculated from the data of Stewart *et al.*[91] for rat brain: $V_{max}$ 0.03 (38°C).
[o] Cytoplasmic enzyme.
[p] Mitochondrial enzyme: the enzyme reaction was found to be irreversible.

*c. Conversion of Glucose Carbon into Proteins, Lipids, and Nucleic Acids.* Until now, the role of glucose in the energy metabolism of brain was mainly considered, but glucose also provides the precursors in the tissue for the formation of essential cell constituents, lipids, nucleic acids, and proteins. At 5 hr after the injection of $^{14}C$-glucose, about one-third of the $^{14}C$ remaining in the brain was recovered in lipids and proteins, respectively, and less than one-tenth in nucleic acids.[30] The reverse reaction, the formation and consecutive oxidation of metabolites as a result of the catabolism of these cell constituents, will also occur. Under certain conditions the contribution of these reactions to the total oxidation in the tissue can be quite high, e.g., in the perfused cat brain preparations,[11] but under normal conditions it is low and is related to the turnover of these substances, which is very low compared with that of the intermediates of glycolysis of the tricarboxylic acid cycle. (Nevertheless, the possibility of high turnover of proteins and other macromolecules in special sites of neurons and under conditions of stimulation has to be considered).

It appears, therefore, that studies with radioisotopes have confirmed the earlier observations showing that under physiological conditions glucose is the main substrate for the energy metabolism of nervous tissues. There are indications, however, that during insulin hypoglycemia[39] or under conditions of stimulation[11] the contribution of substances other than glucose to cerebral energy metabolism is increased. Thus, during convulsions induced by drugs or electrical stimulation, the cerebral respiration rate increased (Table I) but the incorporation of $^{14}C$ from glucose into $CO_2$ decreased.[17,23] Furthermore, an apparent block in the metabolism of [2-$^{14}C$] glucose was observed during periodic convulsions in rat brain.[40] On the other hand, in stimulated sympathetic ganglia there was no indication of accelerated oxidation of substrates other than glucose.[15]

## III. PATHWAYS OF CARBOHYDRATE METABOLISM

### A. Enzyme Profile and Steady-State Level of Intermediates of Carbohydrate Metabolism in Brain

Several pathways for the metabolism of glucose are found in mammalian cells and have also been detected in nervous tissue. Table III lists some of the available kinetic data for the relevant enzymes in brain, and Table IV gives the steady-state levels of intermediates. It has to be emphasized that when the activity of an enzyme is measured under optimal conditions the information obtained is not necessarily relevant to the actual activity of this enzyme under the conditions prevailing in the cell. For example, the substrate levels found under physiological conditions are usually well below those required for maximum enzyme activities (compare the Michaelis constants given in Table III with the steady-state concentration of metabolites in Table IV). However, these data are important as they

## TABLE IV
## Levels of Intermediates Related to Glucose Metabolism

| | Blood[a] (μmole/liter) | Brain (μmole/kg wet wt.) | Brain/liver[b] |
|---|---|---|---|
| *(1) Glycogen and glycolytic intermediates* | | | |
| Glycogen (acid-soluble) | | 2200[1,117] | 0.1[116] |
| (residual) | | 1900[117] | 0.03[116] |
| UDP-glucose (UDPG) | | 118[98] | 0.1[98] |
| Glucose | 6–8[115,116] | $1\text{–}1.5 \times 10^3$ [1,115] | 0.18[116] |
| Glucose 6-phosphate (G6P) | 18[2] | 80–160[1,118] | 0.32[2] |
| Fructose 6-phosphate (F6P) | | 16; 50[1][c] | 0.14[2]; 0.45[2][c] |
| Fructose disphosphate (FDP) | 12[2] | 120; 27[1][c] | 5.4[2]; 1.2[2][c] |
| Dihydroxyacetone phosphate (DAP) | 25[2] | 46; 13[1][c] | 1.2[2]; 0.3[2][c] |
| Glyceraldehyde phosphate (GAP) | | 1.0[1][c] | 0.3[119][d] |
| 3-Phosphoglycerate (3PG) | 41[2] | 25[1][c] | 0.07[2][c] |
| 2-Phosphoglycerate | | 3[1][c] | 0.1[119][c] |
| Phosphoenolpyruvate (PEP) | | 4[1][c] | 0.03[2][c] |
| Pyruvate (Pyr) | 190[2] | 91[118]; 39[1][c] | 0.6[2]; 0.25[2][c] |
| Lactate | 2200[2] | 2000[1,50]; 770[1][c] | 1.3[2]; 0.5[2][c] |
| α-Glycerol phosphate (αGP) | | 48[1][c] | 0.19[2][c] |
| *(2) Metabolites related to pathways of glucose metabolism other than glycolysis* | | | |
| Fructose | | 40[91]; 390[e] | |
| Sorbitol | | 20[91]; 150[e] | |
| Myo-inositol | 33–110[95] | 5700[91]<br>$11\text{–}15 \times 10^3$ [95] | 11[95] |
| Ascorbate | 90–130[120] | $2.9\text{–}3.4 \times 10^3$ [120] | 1.5[120] |
| UDP-glucuronic acid | | 9[98] | 0.03[98] |
| *(3) Substances closely associated with glucose metabolism* | | | |
| Citrate | | 330[118] | 1.2[119] |
| α-Oxoglutarate (αOG) | | 130[118] | |
| Oxaloacetate (OAA) | | 4[118] | 0.5–1[2,119] |
| Malate | 31[2] | 438[118] | 1[2] |
| Aspartate (Asp) | | 2500[100] | 2.5[37] |
| Glutamate | | 9700[100] | 3.7[37] |
| Alanine (Ala) | | 600[100] | 0.5[37] |
| *(4) ATP, CrP, and nicotinamide nucleotide coenzymes* | | | |
| ATP | 440[2] | 3000[14] | 1[2] |
| Creatine phosphate (CrP) | | 3000[14] | [f] |
| $NAD^+$ | | 322[121] | 0.5[121] (NAD⁺ and NADH) |
| NADH | | 95[121] | |
| $NADP^+$ | | 5[122] | 0.06[122] (NADP⁺ and NADPH) |
| NADPH | | 27[122] | |

[a] Rat (whole blood).
[b] Ratio of metabolite levels in brain to liver; the references for the liver values are indicated.
[c] These values are for anesthetized 10-day-old mice.
[d] Perfused liver preparation.
[e] Rat sciatic nerve.[91]
[f] Very low values in liver.

may indicate which enzymes could be, on the basis of their activity or kinetic properties, most susceptible to metabolic control. Furthermore, the enzyme profiles are apparently quite characteristic for each tissue. Therefore, the conclusion is almost inescapable that they are intimately connected with the physiological function of the organ. It is felt, therefore, that the consequences of the cerebral enzyme profile in the carbohydrate metabolism of the tissue can be better understood when comparison is made with the enzyme pattern of another organ, liver, which is characterized by a high capacity for glucose formation and storage, and biosynthetic reactions in general.

The comparison of the enzyme pattern of the two organs shows the following differences (Table III):

1. The potential activity of the glycolytic pathway is apparently higher in brain than in liver. The activity of Hk which is the enzyme initiating intercellular glucose metabolism is about 20 times higher in brain than in liver, and the activities of all the other kinases of glycolysis (PFK, PGK, and PK) are also much higher in brain. These enzymic steps are probable control points since they are linked through the involvement of adenine nucleotides in the reactions to the "energy status" of the cell.[6] To show that the enzymic makeup of brain is in favor of the breakdown of glucose in glycolysis, the following ratios for brain and liver, respectively, can be calculated: for the branching of pathways at G6P, PFK/G6PD = 11 and 1 and PFK/PGM = 1 and 0.04; for the direction of glycolytic flux, PK/PEPC = 75 and 1 and Hk/G6Pase = 300 and 0.1.

2. The pathway for glucose synthesis from pyruvate is negligible in brain. This pathway involves some reactions of glycolysis in reversal and some additional reactions which overcome the energy barriers preventing a direct reversal of glycolysis.[4,43,44] These reactions are: (1) pyruvate → PEP, (2) FDP → F6P, and (3) G6P → glucose. Reaction (1) involves $CO_2$ fixation presumably catalyzed in liver by pyruvate carboxylase. The activity of this enzyme is apparently very low in brain.[44,134] However, $CO_2$ fixation has been observed in nervous tissue,[45] which may partly occur by a reaction catalyzed by the malic enzyme. Consecutive enzymic reactions, involving MDH and PEPC, could result in PEP formation from pyruvate. These enzymes have been detected in brain[44] (*cf.* Table III). The activity of the other two critical enzymes of gluconeogenesis, FDPase and G6Pase, was found to be negligible in brain[46,47]; thus, glucose formation *in situ* from pyruvate, which is an important pathway in liver, can be ruled out in this tissue.

3. The potential activity for glycogen synthesis is much smaller in brain than in liver.

4. The activities of enzymes in the pentose phosphate cycle are apparently low in brain. The activity of another important enzyme reaction forming NADPH in the cytoplasm, ICD, is also relatively low in brain.*

* Only 30–40% of the activity of ICD in whole homogenates of brain[114] compared with about 90% of that in liver[107] is recovered in the cell sap.

5. The activity of αGPD in brain is only 4% of that in liver, but the activities of AspAT and MDH determined in whole homogenates of brain are near the values obtained in liver (*cf.* Section III.B.4). AspAT and MDH are present in both tissues in the form of isoenzymes found in mitochondria and cell sap, respectively.

The comparison of the enzymic makeup of the brain with that of the liver clearly demonstrates that glucose metabolism in the brain is programmed for the rapid breakdown of glucose in the energy-yielding pathway of glucose metabolism, which is glycolysis. It is of interest in this respect that the ready-made energy content (ATP and CrP) is relatively high in the brain (Table IV).

The steady-state level of metabolites in conjunction with knowledge concerning the enzyme pattern of a tissue may provide important information on the control of enzyme reactions in metabolic pathways, when comparison is made with another tissue or with the level of metabolites under conditions of altered flux.* Additional support for the pattern of carbohydrate metabolism suggested by the enzyme profile of brain is obtained from the data concerning the steady-state level of metabolites (Table IV). The content of glucose in brain is less than 20% of that in the plasma; this indicates that the rate of utilization is relatively high compared with that of the transport. However, the transport of glucose from blood to brain is not a static situation; during convulsions it is apparently increased severalfold.[48] Furthermore, in anesthesia the cerebral glucose level is more

* One of the methods in the search for enzymes involved in the regulation of metabolic pathways is the identification of enzymes which in the steady-state catalyze reactions which are far displaced from equilibrium [*cf.* Newsholme and Gevers[6]]. The "nonequilibrium" reactions can be specified: the mass action ratios (the product of the concentration of products divided by the product of concentration of reactants, Γ) of these reactions are greatly different from the values of the apparent equilibrium constants ($K_{equ}$). Nonequilibrium enzymes may be considered as potential control sites since they provide bottle-necks to substrate flow. In order to specify which of the nonequilibrium enzymes may be subject to metabolic control (regulatory enzyme), the steady-state levels of intermediates of the pathway are compared under conditions in which the overall rates of the pathway are different. This information may suggest which of the nonequilibrium steps are involved in the regulation of the pathway. If the substrate concentration changes in a direction opposite to the change in the rate of the pathway, this is considered as evidence for a regulatory enzyme.[6,79] However, the failure of observing this change does not necessarily mean that a nonequilibrium enzyme is not regulatory. Additional information for the identification of regulatory enzymes is provided by the study of the kinetic properties of the nonequilibrium enzymes.

The following example may elucidate the means of identification of regulatory enzymes of a metabolic pathway. Lowry *et al.*[1,3] studied the regulation of cerebral glycolysis *in vivo* by comparing the steady-state levels of intermediates under normal conditions (low rate of glycolysis) with those in anoxia (high rate of glycolysis). They have observed that the content of F6P (the substrate for PFK) decreased and that of FDP (the product of the reaction) increased when the glycolytic rate was stimulated. Thus, PFK can be considered as a regulatory enzyme in cerebral glycolysis. The investigation of the kinetic properties of PFK has provided the necessary information for the better understanding of the control mechanism at this enzymic step [*cf.* Lowry and Passonneau[63]] (*cf.* Section III.B.2).

than doubled and the increase cannot be accounted for exclusively by the reduced rate of metabolism.[49] The glycogen stores in the brain are only about 1% of that observed in the liver of fed rats, and the level of UDP sugars and sugar derivatives is much higher in liver. Except for the relatively high level of FDP in the brain, the contents of the other glycolytic intermediates, especially those between GAP and PEP, are relatively low: these observations may be related to the one-directional glycolytic flow of glucose in brain *versus* the two-way traffic in liver. On the other hand, the levels of intermediates of the tricarboxylic acid cycle are comparable in the two tissues, but the amount of amino acids, which are closely associated with glucose metabolism, is relatively high in brain. The content of NAD is about 50%, but that of NADP is only 6% of the values observed in liver. NAD is the hydrogen carrier in reactions mainly leading to energy formation, whereas NADP is the cofactor usually in biosynthetic reactions. These data give, therefore, additional emphasis to the general metabolic pattern of the brain which favors energy production *versus* energy conservation.

## B. Glycolysis

The rate of glucose utilization is 0.3–1 mmole/kg/min in the brain of different species (Table I). More than 90% of the glucose consumed is apparently metabolized in the glycolytic pathway.[19] Under normal conditions *in situ*, only a small fraction of the pyruvate formed is converted to lactate, the rest is oxidized through the tricarboxylic acid cycle in the mitochondria to $CO_2$ and water (Table I). The complete oxidation of glucose would liberate about 690 kcal/mole, whereas the conversion of glycogen or glucose to lactate yields only 52 and 47 kcal/mole, respectively.[4] The efficiency of glycolysis and pyruvate oxidation is relatively high: the energy is conserved in the form of ATP [$\Delta G'$, about 8.0 kcal/mole[4]], and glycolysis results in the formation of 3 or 2 mole of ATP depending on whether glycogen or glucose is metabolized, whereas the complete combustion of pyruvate yields about 30 mole of ATP per glucose equivalent. The energy gain when pyruvate formed in glycolysis is oxidized instead of reduced to lactate is, therefore, substantial. However, the conversion of pyruvate to lactate is greatly increased in brain under conditions of stimulation. Even "emotional excitation" results in a sudden severalfold increase in the lactate content of the brain.[50] Lactate formation increased 30–100-fold in the initial period of convulsive activity *in vivo*[48,51] or after electrical stimulation *in vitro*[14] (Table I). In the same time the rate of oxidation of glucose also increased. In Table I the rates of cerebral glycolysis and respiration, as observed in cell-free preparations under optimal conditions, are also listed. It is of interest that the maximal rates of glucose utilization, lactate formation, and respiration observed during the initial period of convulsive activity are very near the values determined for the apparent metabolic potential of the tissue. These data indicate that compared with resting condition the rate of glucose consumption can increase severalfold but that of oxygen uptake can only treble. It is suggested that the discrepancy between the maximal activity of oxidative reactions in mitochondria and that of pyruvate formation in the glycolytic pathway might account for the observed increase in lactate production during excitation.* Thus, the increase in cerebral lactate formation observed under these conditions is not necessarily a consequence of inadequate $O_2$ supply to the cells. It has indeed been observed that during convulsive activity the $O_2$ saturation of the venous blood draining the brain of paralyzed dogs and monkeys does not decrease, owing to a significant increase in cerebral blood flow.[(52)]

The reactions of glycolysis are similar in brain to those in other tissues. Glucose is brought into the pathway by phosphorylation. A series of preparatory reactions follows in which the free-energy change is small. The chemical changes are so organized that a package of energy can be released in two reactions, at the GAPD and PK steps (Table III). Since good reviews are available,[(4,12–14)] this presentation is limited to the tabulation of some kinetic data of the glycolytic enzymes and to a description of some of the enzymes which have been recently studied in the brain in more detail.

### 1. *Hexokinase (Hk)*

It has long been known that among mammalian tissues brain contains the highest Hk activity.[(53)] In contrast to the other glycolytic enzymes, a high proportion of the total Hk activity in the brain, and varying degrees also in other tissues, appears to be associated with mitochondria [*cf.* Crane[(54)]]. The variation in the fraction of the Hk recovered in the particulate residue of the tissue (35–90%) reported from different laboratories is probably related to the preparation of the homogenate. The binding of the enzyme to the particles is influenced by the experimental conditions: both reversible and irreversible solubilizations of the pigeon heart Hk have been reported.[(55)]

In brain the particle-bound Hk can be activated by several means, which result only in partial solubilization, but the enzyme can be almost completely released from mitochondria by a combination of experimental conditions.[(56–59)] The binding of Hk to the mitochondrial membrane is probably by ionic bonds, which can be broken by raising the ionic strength of the medium, but not by urea.[(59)]

Electrophoresis in starch gel showed that Hk is present in multiple forms in different tissues.[(60)] Type I (the slowest moving band under the experimental conditions) is the major constituent in brain tissue[(56–60)] and has been obtained in purified form (mol. wt. = 96,000).[(60)]

Brain Hk reacts with a number of sugar substrates.[(54)] The reaction is rather specific for ATP; ITP can substitute for ATP but the enzymic

* An additional factor resulting in elevated lactate levels may be a limitation in the rate of disposal of cytoplasmic NADH by reactions other than the reduction of pyruvate to lactate (*cf.* Section III.B.4).

activity is very low.[60] Detailed kinetic analyses have indicated that the two substrates (glucose and ATP) are not present on the enzyme simultaneously but an enzyme–P or enzyme–glucose intermediate participates in the reaction.[61] Most reports agree that the kinetic properties of the Hk obtained from the soluble and the particulate fraction of brain are similar,[57,59,60] but some differences are also noted.[56] The observation that the uptake of fructose by the brain is small compared with that of glucose is probably related to the much lower affinity of the Hk to fructose than to glucose (Table III).[9–11,60] Brain Hk activity is inhibited by both products of the reaction (G6P and ADP)[54,62]; this negative feedback might be an important factor in the control of glucose metabolism. Another possibility of control might be provided by a reversible binding of Hk to intracellular particles[55]; this effect has not yet been demonstrated in brain preparations.

## 2. *Phosphofructokinase* (*PFK*)

PFK is kinetically one of the most complicated of enzymes, and the kinetic properties can probably account, at least partly, for the control of glucose utilization in brain.[63,64] The enzyme is inhibited by one of its substrates (ATP), and it is activated by the other substrate (F6P) and by both products (ADP and FDP). Furthermore, activation and inhibition have been observed by substances whose concentration changes according to the metabolic state of the tissue (e.g., AMP, $P_i$, $NH_4^+$, citrate).

Lowry and Passonneau[63] have summarized their extensive kinetic studies on brain PFK by proposing a model for the enzyme (Fig. 1). According to this model there are two separate substrate sites on the enzyme. This was indicated by the observation that with noninhibiting levels of ATP the $K_m$ for each substrate was independent of the concentration of the other substrate. There are separate binding sites for the inhibitors and activators which are indicated in Fig. 1.

The kinetic evidence is that ATP increases the apparent affinity for another inhibitor, citrate, and *vice versa*; similarly $NH_4^+$, AMP, $P_i$, and F6P are synergistic stimulators. Conversely, an activator (e.g., F6P) decreases the affinity for an inhibitor (e.g., citrate), or an inhibitor (e.g.,ATP) decreases the apparent affinity for the stimulators. These results suggest a model (Fig. 1) in which inhibitor sites are so situated that the occupation of the sites by ATP or citrate tends to separate the two substrate sites from each other and also tends to separate the points of attachment of the activators. On the other hand, when the deinhibitor sites are occupied, the substrate sites are closer to each other, and the points of attachment of the inhibitors are more separated.

The properties of PFK show that the activity of this enzyme is regulated by many substances of importance in cellular metabolism. It can be expected that under normal conditions the enzyme in the brain will be mainly in the inhibited form as a result of high ATP and low F6P, AMP, and $P_i$ concentrations. During convulsive activity or anoxia, the internal environment

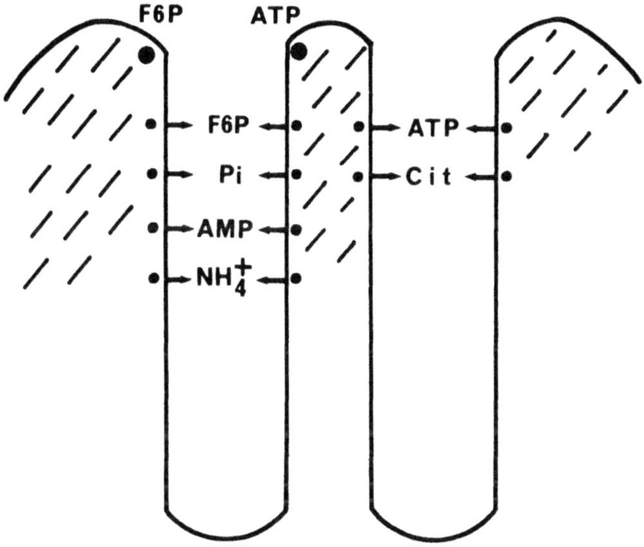


Fig. 1. Model for PFK [Lowry and Passonneau[63]]. The large solid circles represent substrate sites. Inhibitor (right) and deinhibitor sites (left) are each represented as having two points for attachment. The inhibitors and activators and the number of binding sites on each enzyme molecule are as follows (other compounds which probably bind to the same sites are given in parentheses): Inhibitors—ATP, 2; citrate, 1. Activators—F6P (FDP), 3; $P_i$, 1; AMP (ADP), 1; $NH_4^+(K^+)$, 1.

changes and ATP content decreases, whereas there is an increase in the levels of AMP, $P_i$, F6P, and $NH_4^+$, resulting in the activation of the enzyme.[1,3,48] From the point of view of transient fine adjustment to the requirement of the cells, the large number of activator or inactivator molecules involved and the fact that their effects are synergistic might be of major importance. They can provide relatively sharp activation and cutoff points; a moderate change in conditions may result in great change in enzymic activity.[63,64]

### 3. *Lactate Dehydrogenase* (*LDH*)

It has been shown that LDH in brain is present in several different molecular forms with the same substrate specificity.[65-67] The LDH isoenzymes have been characterized in brain as in other tissues on the basis of electrophoretic mobilities, reaction rates with different NAD analogues, and enzyme kinetic determinations. The generally accepted nomenclature of the LDH isoenzymes is based on the electrophoretic mobilities: type I (or heart type, H) is the most negatively charged form and type V (or muscle type, M) is the most positive form. In adult mammalian brain, type I is the dominant form (about 40% of the total LDH activity). Cahn *et al.*[68] have postulated that LDH is a tetramer (mol./wt. = 150,000) built up essentially of two different types of subunits (H and M) probably elaborated by different genes. If both H and M genes are active in the same cell, then the monomers recombine in a random fashion in groups of four, yielding five different molecular species. Excess pyruvate inhibits H-LDH at lower concentration than M-LDH and the $K_{Pyr}$ is about five times higher for the M than the H form. It has been suggested that the differences in kinetic properties have functional significance, but some observations are at variance with this hypothesis.

There is a progressive change in the LDH isoenzyme pattern during brain maturation.[67] In the brain of the newborn rat only 10% of the LDH is present in the H form, but in the adult brain 40% of the activity is in this form. The changes in the isoenzyme pattern of cerebral LDH apparently correlate with the resistance of the animal to anoxia.

The LDH isoenzymes have been obtained in purified form from human and beef brain.[65,66]

### 4. *Subcellular Distribution of Glycolytic Enzymes: Functional Consequences*

The subcellular distribution of glycolytic enzymes has been studied in brain.[69–72] Initial observations that substantial glycolytic activity is associated with the crude mitochondrial fraction in brain are related to the presence of nerve-ending particles in this fraction which contain the glycolytic enzymes in the entrapped cytoplasm.[71–73] However, a great part of the hexokinase activity appears to be associated with the mitochondria (*cf.* Section III.B.1).

One of the important consequences of the cytoplasmic localization of the glycolytic enzymes is the extramitochondrial formation of NADH in the GAPD reaction.

More than 80% of the pyruvate formed from glucose is oxidized in the mitochondria under normal conditions. The cell must therefore dispose of the 80% of NADH produced in glycolysis by means other than lactate formation, to enable the continuous oxidation of GAP. This question has recently been raised by experiments which showed that the metabolism of pyruvate was affected by some inhibitors in a different way when pyruvate was formed in glycolysis or was offered as an exogenous substrate to brain slices.[74] Cytoplasmic NADH is impermeable to mitochondria; therefore, the reducing equivalents have to be carried to the mitochondria by indirect routes. Reactions which may also be of importance in brain have been suggested [*cf.* Klingerberg and Bücher[7] and Krebs[75]].

The reducing equivalents generated in glycolysis can be transferred to DAP by the action of αGPD in the cell sap. The equilibrium of this reaction is in favor of αGP formation. αGP can enter the mitochondria and is there oxidized to DAP in the irreversible reaction catalyzed by αGPO. DAP can then pass into the cytoplasm and in turn participate in NADH oxidation. The αGP shuttle,[7,77] therefore, can operate as an oxidation pathway for the cytoplasmic NADH, even if the ratio of NADH to $NAD^+$ is much higher in the mitochondria than in the cell sap. This cycle, however, is probably of small significance in brain; the activity of the cytoplasmic enzyme is relatively low (the activity ratio of αGPD to LDH is 0.04; Table III), and the concentration of αGP is well below the $K_m$ of the αGPO. In accordance with that it was observed that the rate of formation of αGP after transition from aerobiosis to anaerobiosis was only about 6% of that of lactate.[3] Similarly in synaptosome-containing mitochondrial preparations, approximately 7% of the NADH formed in glycolysis was oxidized by αGPD.[78] The transport of cytoplasmic NADH to the mitochondria can be effected by the combined operation of MDH and AspAT, and this is probably the major route in brain. These enzymes have relatively high activity in brain (Table III), and they are localized both in the cell sap and the mitochondria, which is a requirement of such a "carrier" system. However, the ratio of NADH to $NAD^+$ in the mitochondria is probably much higher than that in the cell sap; thus, transfer of NADH by the malate system in the direction of the mitochondria can occur only if special conditions are met. For example, the ratios of the concentrations of mitochondrial to cytoplasmic OAA and $H^+$ are low and that of malate high; alternatively, the flow of reducing equivalents may be facilitated by active transport of malate into mitochondria.[75] It is of interest, however, that the constituents of the MDH and AspAT systems determined in the whole tissue appear to be in equilibrium, whereas the steady-state levels of the αGP–DAP couple are displaced from equilibrium.*

* It has been pointed out by Hohorst *et al.*[2] that the NAD-linked dehydrogenases in the extramitochondrial compartment are in equilibrium with each other through a common pool of NAD. This apparently applies to the LDH and MDH systems in whole brain. From the chemical equations in Table III.

$$\frac{[NADH]\,[H^+]}{[NAD^+]} = \frac{[\text{lactate}]}{[\text{Pyr}]} K_L = \frac{[\text{malate}]}{[\text{OAA}]} K_M$$

Thus,

$$\frac{[\text{lactate}]}{[\text{Pyr}]} : \frac{[\text{malate}]}{[\text{OAA}]} = K_M : K_L$$

From Table III $K_M = 9.8 \times 10^{-6}$ and $K_L = 53 \times 10^{-6}$ (the $K_{equ}$ relates to the reverse direction, lactate → Pyr); hence, $K_M : K_L = 0.18$. The ratio of the levels of intermediates given in Table IV is 0.22. However, similar calculations show that the αGPD system in the brain is not in equilibrium with the LDH and MDH systems.[1,79]

The constituents of the AspAT reaction are near equilibrium in the steadystate. The $K_{equ}$ is 6.7 and the value for the mass action ratio (Γ) obtained by substituting the data given in Table IV is 8.2.

An important consequence of the observations of Hohorst *et al.*[2] is that the true ratio of $NAD^+/NADH$ can be calculated for that compartment of the cell in which the dehydrogenases are in equilibrium through the $NAD^+$–NADH couple. It is necessary to calculate the $NAD^+/NADH$ ratio by indirect means since the free contents of $NAD^+$ and NADH are probably not identical with the amounts determined in tissue extracts;

[*continued on page 22*

5. *Pasteur Effect*

The aerobic control of glucose metabolism, however, means more than the prevention of the reduction of pyruvate to lactate. If the effect of respiration were solely to oxidize pyruvate and NADH formed in glycolysis, the lactate formation observed under anaerobic conditions would be reduced only by the amount of lactate which can be accounted for by $O_2$ uptake (1 mole of $O_2 = \frac{1}{3}$ mole of lactate). In brain slices the $O_2$ uptake (approximately 1.3 mmole/kg/min) would account for the oxidation of about 0.4 mmole lactate. Since the anaerobic lactate formation is about 1.5 mmole/kg/min, the rate of aerobic lactate production would be about 1 mmole/kg/min. But it is only about 0.1–0.2 mmole/kg/min. It follows that respiration inhibits glycolysis; this phenomenon is called the Pasteur effect.

The suppression of glycolysis is not dependent on respiration in general. Respiration can be maintained, even stimulated, under conditions when the glycolytic flux is increased; this can be effected by uncouplers of phosphorylation or by means that increase the cellular expenditure of ATP, such as stimulated functional activity (convulsions *in situ*, or stimulation by electrical pulses or by high extracellular $K^+$ion concentration *in vitro*) [*cf.* Greville[(13)]]. Such conditions have one factor in common: they result in an alteration of the "energy status" of the cell, i.e., the contents of ADP, AMP, and $P_i$ increase and the ATP level decreases. Since these substances are reactants in some reversible glycolytic reactions (GAPD, PGK, and PK, Table III), the resulting change in their concentration will stimulate glycolysis. However, the first steps of glucose utilization are dependent on ATP rather than on $P_i$ and ADP. Additional control mechanisms are therefore required. They can operate at the level of glucose transport or at the level of the Hk and PFK activities, and the properties of these two enzymes are compatible with such control mechanisms (*cf.* Sections III.B.1, 2). The control has recently been studied *in situ* and in cell-containing preparations,[(1,3,6,79)] and the results are discussed in this *Handbook* (Volume IV, Chapter 1).

The Pasteur effect was also studied in model systems containing mitochondria and cell sap or purified enzymes of glycolysis [*cf.* Racker[(5)]].

---

*Footnote continued from page 21*

a considerable proportion of the nucleotides is bound to cell constituents. The thermodynamic situation in the cell, however, is related to the free levels of $NAD^+$ and NADH. The calculation can be made, e.g., by determining the Pyr and lactate levels:

$$\frac{[NAD^+]}{[NADH]} = \frac{[Pyr] \times [H^+]}{[lactate] \times K_L}$$

$$K_L \text{ (at } p\text{H 7.0)} = 5.3 \times 10^{-5} \times 10^{-7} = 5.3 \times 10^{-12}$$

Thus,
$$\frac{[NAD^+]}{[NADH]} \sim 1000$$

The value calculated from the estimated amounts of $NAD^+$ and NADH is about 3.5.

The crude mitochondrial fraction of brain constitutes a similar system due to the presence of entrapped cytoplasm in the synaptosomes.(73) A reversible inhibition of glycolytic flux under aerobic conditions has been observed in this system.(78,80) Two different mechanisms were identified: one related to restricted availability of mitochondrial ATP for the phosphorylation of glucose, and another involving a reversible inhibition of GAPD. The first mechanism, which was observed when the system operated with endogenous adenine nucleotides, behaves very much like the Pasteur effect in the whole tissue. The main similarities are: (1) both glucose consumption and lactate formation are inhibited under aerobic conditions; (2) the aerobic inhibition of glycolysis is removed by respiratory inhibitors and by uncouplers of phosphorylation; and (3) the inhibition is completely reversible. The results indicated that the NADH-flavoprotein site of oxidative phosphorylation is of major importance in the aerobic control of glycolysis in this system. The other control, which involves the inhibition of GAPD under aerobic conditions, could be removed by dimercaptopropanol (BAL) but not by monothiols. The latter observation suggests that contrary to the properties of GAPD from other sources dithiol groups may be involved in the active site of the cerebral enzyme.

## C. Pentose Phosphate Cycle (PPC)

An alternative pathway of glucose utilization involves the NADP-linked oxidation of G6P, followed by oxidative decarboxylation at the C-1 position. The resulting pentose phospate is converted back to hexosemonophosphate by a series of nonoxidative reactions (Fig. 2). This pathway operates also in nervous tissues, and the enzyme activities, at least at the initial steps, are adequate for the observed resting glucose consumption in the brain (Table III). However, the contribution of the PPC to glucose utilization in the brain is apparently very low(26,69); the value calculated for human brain is only about 1% of the metabolic flux of glucose.(19) Even the increased glucose utilization in electrically stimulated cerebral cortical slices proceeds apparently by way of the glycolysis.(81a) It has been reported, however, that the utilization of glucose through the PPC is relatively high in medulla(81) and in frog brain.(82)

In most of these studies the contribution of PPC to glucose oxidation was investigated only qualitatively. The incorporation of $^{14}C$ from [6-$^{14}C$] glucose into respiratory $CO_2$ or into a triose-P derivative, such as lactate has usually been compared with that from [1-$^{14}C$] glucose. Figure 2 shows that C-1 of G6P is lost as $CO_2$ by oxidation in the PPC. On the other hand, in glycolysis both C-1 and C-6 of glucose are converted to C-3 of pyruvate; thus the decarboxylation which follows in the tricarboxylic acid cycle does not distinguish between C-1 or C-6 of G6P. $^{14}C$-Triozephosphates (lactate) arise from both [1-$^{14}C$] and [6-$^{14}C$] glucose metabolized via glycolysis, but only from [6-$^{14}C$] glucose utilized via the PPC. Therefore, equal yields of $^{14}CO_2$ or $^{14}C$-lactate from [1-$^{14}C$]- and [6-$^{14}C$]glucose have been taken to indicate a

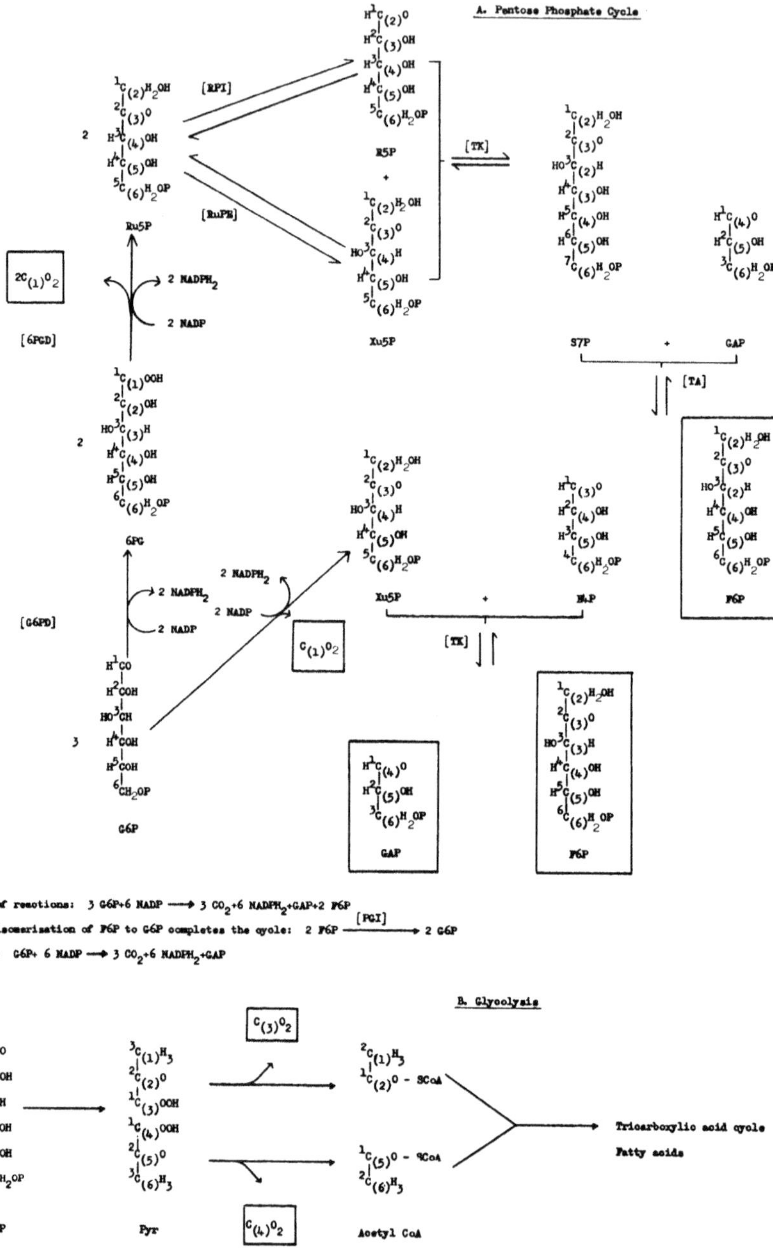


Fig. 2.(A) Changes in the distribution of carbon of hexosemonophosphate after one passage through the pentose phosphate cycle. (B) Distribution of carbon in pyruvate formed in glycolysis. The superscript numerals on the left of C indicate the carbon atom positions of each intermediate; the subscript numerals in parentheses refer to the carbon position in the parent compound, G6P. The enzymes are given in brackets and the products in boxes. 6PG, 6Pgluconate; Ru5P, ribulose5- P; R5P, ribose 5-P; Xu5P, xylulose 5-P; S7P, sedeheptulose 7-P; E4P, erythrose 4-P; [RPI], ribose P isomerase; [RuPE], ribuloseP 3-epimerase.

predominantly glycolytic process, whereas high relative yield of $^{14}CO_2$ from [1-$^{14}$C]glucose or of $^{14}$C-lactate from [6-$^{14}$C]glucose to suggest a significant contribution of PPC to glucose utilization. However, Katz and Wood[83] have pointed out that at least the following factors must be considered for the evaluation of pathways of glucose metabolism by these methods: (1) F6P formed in the pentose cycle may re-enter the G6P pool after isomerization. The specific activity of G6P is affected by recycling with [1-$^{14}$C]glucose as the substrate but not with [6-$^{14}$C]glucose (Fig. 2). (2) Some glucose may be metabolized in pathways other than glycolysis and PPC. (3) The oxidation of triose-P to $CO_2$ is usually incomplete. Katz and Wood have shown that when glucose is metabolized only by glycolysis and PPC the pathways can be evaluated from the ratio of $^{14}$C in triose-P derived from [1-$^{14}$C]- and [6-$^{14}$C]glucose. Thus, in the case of mammalian cerebral cortex the evidence pointing to the low contribution of PPC to glucose utilization is apparently acceptable.[26,69] On the other hand, these authors have shown that the pattern of glucose metabolism cannot be assessed even qualitatively from the ratios of $^{14}$C in $CO_2$ derived from [1-$^{14}$C]- and [6-$^{14}$C]glucose, unless additional information is obtained on the above factors (1)–(3), and they have outlined methods whereby this might be done. However, these techniques have not yet been applied to the study of nervous tissues[81,82] in which there might be a significant participation of PPC in glucose metabolism.

The physiological role of the cycle is suggested by the sum of the reactions of the PPC given in Fig. 2. First of all, the oxidation of glucose in the PPC does not lead to the formation of ATP; thus, the role of the cycle is not primarily concerned with energy metabolism. Two products are formed, however, in the PPC which are of importance in biosynthetic reactions, namely, pentose phosphates and NADPH. The importance of local nucleotide synthesis in adult brain is indicated by the restricted entry of nucleotides in the tissue from the blood. Furthermore, it has been shown that in the brain the incorporation of $^{14}$C into RNA from $^{14}$C-glucose is appreciable.[30] The significance of the PPC during early development is probably greater than in the adult brain. In chick embryo the activity of G6PD at 3-days incubation is about four times as high as at hatching. Furthermore, a correlation has been observed between G6PD activity, mitotic rate, and RNA content.[84] The greater involvement of the PPC in the metabolism of mammalian brain during early development is suggested by the observation that the activities of G6PD and 6PGD are relatively high in the immature brain.[85] On the other hand, the yields of $^{14}CO_2$ from glucose labeled in C-1 and C-6 respectively, are similar in the fetal and adult cerebral cortex *in vitro*, and this observation has been taken to suggest that the relative rate of the PPC is as low in the immature as in the adult brain.[81]

NADPH is the coenzyme in a number of synthetic reactions, such as fatty acid synthesis, reductive carboxylation of pyruvate to malate,[44] and hydroxylation of aromatic amino acids.[86] The initial steps in the pentose phosphate cycle (G6PD and 6PGD) are among the important enzyme reactions supplying NADPH to the different tissues. In brain, in particular, G6PD and 6PGD are important, because NADPH formation by the ICD-catalyzed oxidation of isocitrate in the cell sap is relatively small (Table III). A connection between lipid metabolism and the activity of the PPC has been suggested by the observation that heavily myelinated tracts in the white matter of the rabbit brain have much higher levels of G6PD activity than lightly myelinated ones.[87] Furthermore, the activity of TK (expressed on protein basis) is also higher in the white matter and spinal cord than in the cerebral cortex.[88] The cofactor of TK is thiamine pyrophosphate and it has been observed that the enzyme activity is decreased in the brain of thiamine-deficient rats.[88]

The reaction catalyzed by TK is the rate-limiting step of the PPC, when the enzymes are assayed under optimal conditions *in vitro*.[89] However, the overall operation of the cycle is probably limited by the supply of $NADP^+$. It has been shown that the rate of NADPH oxidation is an important factor in the control of the cycle: glucose utilization in the PPC was increased when NADPH oxidation was stimulated by providing either artificial electron acceptors[81a] or oxidized glutathione (GSSG).[89a] The enzymic reduction of GSSG, in turn, may affect the maintenance of the redox potential of the cells.[89a]

## D. Formation and Breakdown of Glycogen

Althought the amount of cerebral glycogen is relatively small (Table IV), the rate of incorporation of $^{14}C$ from $^{14}C$-glucose is comparable with that observed in other tissues containing much larger glycogen stores.[90] The regulation of glycogen metabolism by the contribution of separate routes for the synthesis (through glycogen synthetase) and breakdown (phosphorylase) with the appropriate kinases and phosphatases is similar to that in other tissues and is discussed in detail in this *Handbook* (Volume III, Chapter 2).

## E. Polyols and the Glucuronate Pathway

It has recently been observed that in case there is an overflow of carbohydrates in the body, such as in diabetes or galactosemia, an alternative pathway of hexose metabolism, involving reduction to polyols, is detectable also in nervous tissues. In diabetes the levels of sorbitol and fructose are high in the brain and especially in peripheral nerve.[91] The latter tissue contains appreciable amounts of these compounds even under normal conditions (Table IV). Glucose is reduced to sorbitol in a NADPH-linked reaction, which is catalyzed in brain either by aldose reductase or by L-hexonate dehydrogenase[91,92] (Table III). Sorbitol is oxidized, in turn, to fructose by the NAD-linked sorbitol dehydrogenase.

In response to insulin, the glucose level falls rapidly in the nervous tissues of alloxan diabetic rats, but the contents of sorbitol and fructose decrease together and at a slow rate. These observations indicate that sorbitol is removed via fructose. The initial step of fructose metabolism in brain is probably exclusively that catalyzed by Hk, and the rate of fructose phosphorylation will be low relative to that of glucose at the corresponding substrate concentrations (Table IV). The difference in the fructose content of brain and peripheral nerve has also been attributed to the difference in the Hk activity and in the $K_m$ for fructose in the two tissues.[91]

Galactitol accumulation has been observed in the brain of galactosemic patients or animals kept on galactose-rich diets.[91,93] The rate of galactitol disappearance from the nervous tissues after the introduction of normal diet was slow, suggesting that galactitol is removed from the brain by a slow diffusion process.

It has been observed that glucose is converted to galactose in brain and all the enzymic steps leading to the formation of galactolipids from glucose have been detected in the tissue.[94]

Among the different tissues investigated (except lens and some endocrine glands), brain appears to have the highest content of myo-inositol[95] (Table IV). In view of the very low permeability of brain to polyols, inositol is probably synthesized *in situ*. It has been shown that $^{14}$C-glucose is converted into free inositol in brain, and the kinetics of labeling suggests that free inositol is the precursor of the different inositol-containing compounds in the tissue.[96] The enzymic steps in the formation of inositol are not yet fully understood; it seems that the synthesis is not a direct cyclization of the carbon skeleton of glucose, but probably involves the cleavage of the glucose molecule.[97] However, $^{14}$C-acetate is apparently not incorporated into inositol.[96] The catabolism of inositol probably proceeds through glucuronic acid.[97] This substance can also be formed from glucose *via* UDP-glucuronic acid which has been detected in brain[98] (Table IV). The reduction of glucuronic acid to gulonic acid in the brain is catalyzed by L-hexonate dehydrogenase.[92] Gulonic acid is the precursor of ascorbate, but no evidence is yet available for the formation of ascorbate in rat brain.[99] However, the cerebral ascorbate concentration is quite high (about 3 mmoles/kg; Table IV). Ascorbic acid is known to be a cofactor in the hydroxylation of aromatic amino acids[86]; this is probably the rate-limiting step in the formation of 5HT, one of the important biogenic amines involved in cerebral transmission processes.

Aldose reductase and hexonate dehydrogenase have been purified from brain.[92] It was observed that the enzymes from brain, as from other tissues, catalyze the reduction of a wide range of substrates with NADPH. It has not yet been investigated, however, whether or not all the enzymic steps of the glucuronate–xylulose pathway[97] are present in the brain.

## IV. FACTORS AFFECTING GLUCOSE METABOLISM

The adult pattern of cerebral energy metabolism develops during the maturation of the central nervous system. The developmental changes involve more than a quantitative increase in enzyme activities. A unique organization of metabolic pathways is brought about resulting in predominant utilization of glucose via glycolysis in the tricarboxylic acid cycle. This effect is indicated by the utilization of glucose carbon which changes to the pattern characteristic for the adult during the "critical period"[33] (*cf.* Section II,B,2).

These biochemical changes coincide with the electrophysiological manifestations of cerebral maturation. The electrical activity of neurons is associated with transmembrane ionic movements down electrochemical gradients. Restoration and maintenance of these gradients require energy, and additional energy is probably also needed for the neurosecretory processes involved in synaptic transmitter action. Proper control mechanisms must operate therefore in order to provide energy which is ultimately supplied in the brain by the combustion of glucose, according to the functional need. The great flexibility of carbohydrate metabolism in brain is indicated by the observations that glucose utilization can vary by a factor of 100 according to the functional state [rates (mmole/kg/min) in anesthesia about 0.15 and in the initial phase of convulsions more than 10; Table I]. The regulation of glucose metabolism is affected at almost every level of biological organization, such as cerebral circulation,[10] transport of glucose from blood to brain,[48] and the rates of individual enzyme reactions as a result of environmental changes on regulatory enzymes.[3,6,48,79] The effects of neurotropic agents on cerebral carbohydrate metabolism have also been studied in isolated nervous tissues.[124] It is apparently possible to simulate some effects of depressant agents under *in vitro* conditions. The respiration of brain cortex slices with glucose as substrate can be stimulated by alteration of the cation concentration of the medium[13,124] or by electrical pulses.[14] The stimulated respiration is markedly inhibited by low concentrations of depressants, which have little or no effect on the unstimulated oxygen uptake.[14,124] The selective sensitivity of stimulated respiration has led to the discovery that polyacidic and polybasic substances, such as gangliosides and histones, affect the excitability of cerebral cortex slices through combination with tissue constituents.[125]

Since glucose plays such an essential role in cerebral energy metabolism, it is not surprising that carbohydrate metabolism in brain is relatively insensitive to a number of factors that have powerful effects on other organs. The effect of insulin in brain is still controversial: it has been reported that insulin facilitates the entry of glucose in nervous tissues,[126,127] but other studies failed to find an effect either on glucose transport[128] or on glucose utilization.[25] Thyroid hormones have apparently no effect on cerebral respiration rate in the adult man.[10] Neonatal thyroidectomy results in a decrease of some enzyme activities related to glucose metabolism, but the activities of the glycolytic enzymes which have been studied are apparently not affected in the brain [*cf.* Balázs *et al.*[108] and Hamburgh and Flexener[129]].*

* It has recently been shown that the development of the adult pattern of cerebral glucose metabolism is retarded after neonatal thyroidectomy and is advanced by thyroid treatment in infancy.[136]

The important investigations of Peters and co-workers [*cf.* Peters[130] on the effects of thiamine deficiency in pigeon brain led to the suggestion that a biochemical lesion is responsible for the neurological disorder. Thiamine deficiency was found to be associated with a severe defect in carbohydrate metabolism characterized by an accumulation of lactate in the region of the lower brainstem. Thiamine pyrophosphate is the cofactor of a number of enzymes involved in carbohydrate metabolism, such as Pyr and αOG dehydrogenases and transketolase. The oxidative decarboxylation of pyruvate, however, was apparently unaffected in the brains of thiamine-deficient rats or chicks *in vivo*,[131] and the cerebral activities of Pyr and αOG dehydrogenases were not reduced when assayed *in vitro*.[132] On the other hand, a significant decrease in the activity of TK has been observed in rat brain.[88] During progressive depletion, the initial decrease in TK activity appeared to be the greatest in the lateral pontine tegmentum, which is the site of the most pronounced pathological changes in the rat brain. It has been suggested that the pentose phosphate cycle might be involved in the metabolism of oligodendroglia and through that in the maintenance of the integrity of myelin in the central nervous system (*cf.* Section III,C). These observations further emphasize the need to study the enzymes and metabolic pathways in specific regions of the brain and even in individual cell types. Recent advances in this direction are treated in other chapters of this *Handbook*.

## V. REFERENCES


1. O. H. Lowry, J. V. Passonneau, F. X. Hasselberger, and D. W. Schulz, Effect of ischemia on known substrates and cofactors of the glycolytic pathway of the brain, *J. Biol. Chem.* **239**:18–30 (1964).
2. H. J. Hohorst, F. H. Kreutz, and Th. Bücher, Über Metabolitgehalte und Metabolit-Konzentrationen in der Leber der Ratte, *Biochem. Z.* **332**:18–46 (1959).
3. O. H. Lowry and J. V. Passonneau, The relationships between substrates and enzymes of glycolysis in brain, *J. Biol. Chem.* **239**:31–42 (1964).
4. H. A. Krebs and H. L. Kornberg, *Energy Transformations in Living Matter*, pp. 213–285, Springer Verlag, Berlin (1957).
5. E. Racker, *Mechanisms in Bioenergetics*, Academic Press, New York (1965).
6. E. A. Newsholme and W. Gevers, Control of glycolysis and Glucogenesis in liver and kidney cortex, *in Vitamins and Hormones* (R. S. Harris, I. G. Wool, and J. A. Loraine, eds.), Vol. 25, 1–87, Academic Press, New York (1967).
7. M. Klingerberg and Th. Bücher, Biological oxidations, *Ann. Rev. Biochem.* **29**: 669–708 (1960).
8. S. Berl, A. Lajtha, and H. Waelsch, Amino acid and protein metabolism. VI. Cerebral compartments of glutamic acid metabolism, *J. Neurochem.* **7**:186–197 (1961).
9. H. E. Himwich, *Brain Metabolism and Cerebral Disorder*, Williams & Wilkins Co., Baltimore (1951).
10. S. S. Kety, General metabolism of the brain *in vivo*, *in Metabolism of the Nervous System* (D. Richter, ed.), pp. 221–236, Pergamon Press, New York (1957).

11. A. Geiger, Correlation of brain metabolism and function by the use of a brain perfusion method *in situ*, *Physiol. Rev.* **38**:1–20 (1958).
12. K. A. C. Elliot and L. S. Wolfe, Brain tissue respiration and glycolysis, *in Neurochemistry* (K. A. C. Elliot, I. H. Page, and J. H. Quastel, eds.), pp. 177–211, Charles C. Thomas, Springfield, Illinois (1962).
13. G. D. Greville, Mechanisms of carbohydrate metabolism in the brain, *in Neurochemistry* (K. A. C. Elliot, I. H. Page, and J. H. Quastel, eds.), pp. 238–266, Charles C. Thomas, Springfield, Illinois (1962).
14. H. McIlwain, *Biochemistry of the Central Nervous System*, 3rd edition, Churchill Ltd., London (1966).
15. M. G. Larrabee and J. D. Klingman, Metabolism of glucose and oxygen in mammalian sympathetic ganglia at rest and in action, *in Neurochemistry* (K. A. C. Elliot, I. H. Page, and J. H. Quastel, eds.), pp. 150–176, Charles C. Thomas, Springfield, Illinois (1962).
16. J. H. Quastel, Respiration in the central nervous system, *Physiol. Rev.* **19**:135–183 (1939).
17. A. Geiger, Y. Kawakita, and S. S. Barkulis, Major pathways of glucose utilization in the brain in perfusion experiments *in vivo* and *in situ*, *J. Neurochem.* **5**:323–338 (1960).
18. S. S. Barkulis, A. Geiger, Y. Kawakita, and V. Aguilar, A study on the incorporation of $^{14}C$ derived from glucose into the free amino acids of the brain cortex, *J. Neurochem.* **5**:339–348 (1960).
19. W. Sachs, Cerebral metabolism of doubly labelled glucose in humans *in vivo*, *J. Appl. Physiol.* **20**:117–130 (1965).
20. R. J. Haslam and H. A. Krebs, The metabolism of glutamate in homogenates and slices of brain cortex, *Biochem. J.* **88**:566–578 (1963).
21. R. Balázs and R. J. Haslam, Exchange transamination and the metabolism of glutamate in brain, *Biochem. J.* **94**:131–142 (1965).
22. H. Gainer, C. L. Allweis, and I. L. Chaikoff, Precursors of $CO_2$ produced by rat brain slices stimulated electrically and by 2,4-dinitrophenol, *J. Neurochem.* **9**:433–442 (1962).
23. H. Gainer, C. L. Allweis, and I. L. Chaikoff, Precursors of metabolic $CO_2$ produced by the brain of the anaesthetized intact dog: The effect of electrical stimulation, *J. Neurochem.* **10**:903–908 (1963).
24. A. Barkai and C. L. Allweis, The contribution of blood glucose to the $CO_2$ produced by the narcotised brain of the intact cat, *J. Neurochem.* **13**:23–33 (1966).
25. A. Beloff-Chain, R. Catanzaro, E. B. Chain, I. Masi, and F. Pocchiari, Fate of [U-$^{14}C$]glucose in brain slices, *Proc. Roy. Soc. (London) Ser. B* **144**:22–28 (1955).
26. D. B. Tower, The effects of 2-deoxy-D-glucose on metabolism of slices of cerebral cortex incubated *in vitro*, *J. Neurochem.* **3**:185–205 (1958).
27. R. B. Roberts, J. B. Flexner, and L. B. Flexner, Biochemical and physiological differentiation during morphogenesis—XXIII. Further observations relating to the synthesis of amino acids and proteins by the cerebral cortex and liver of the mouse, *J. Neurochem.* **4**:78–90 (1959).
28. M. Kini and J. H. Quastel, Carbohydrate–amino acid interrelations in brain cortex *in vitro*, *Nature* **184**:252–256 (1959).
29. H. Busch, E. Fujiwara, and L. M. Keer, Metabolic pattern for [1-$^{14}C$]glucose in tissues of tumor-bearing rats, *Cancer Res.* **20**:50–57 (1960).
30. R. Vrba, M. K. Gaitonde, and D. Richter, The conversion of glucose carbon into protein in the brain and other organs of the rat, *J. Neurochem.* **9**:465–475 (1962).

31. J. E. Cremer, Amino acid metabolism in rat brain studied with $^{14}C$-labelled glucose, *J. Neurochem.* **11**:165–185 (1964).
32. M. K. Gaitonde, S. A. Marchi, and D. Richter, The utilization of glucose in the brain and other organs of the cat, *Proc. Roy. Soc. (London) Ser. B* **160**:124–136 (1964).
33. M. K. Gaitonde and D. Richter, Changes with age in the utilization of glucose carbon in liver and brain, *J. Neurochem.* **13**:1309–1318 (1966).
34. M. K. Gaitonde, Rate of utilization of glucose and compartmentation of α-oxoglutarate and glutamate in rat brain, *Biochem. J.* **95**:803–810 (1965).
35. J. R. Lindsay and H. S. Bachelard, Incorporation of $^{14}C$ from glucose into α-keto acids and amino acids in rat brain and liver *in vivo*, *Biochem. Pharmacol.* **15**:1045–1052 (1966).
36. C. L. Allweis, T. Landau, M. Abeles, and J. Magnes, The oxidation of albumin bound [U-$^{14}C$] palmitate to $^{14}CO_2$ by the perfused cat brain, *J. Neurochem.* **13**:795–804 (1966).
37. R. Balázs, Control of glutamate metabolism. The effect of pyruvate, *J. Neurochem.* **12**:63–76 (1965).
38. R. Balázs, Y. Machiyama, and D. Richter, Metabolism of GABA in cerebral cortex, *in First International Meeting of the International Society for Neurochemistry, Strasbourg, July*, p.13 (1967).
39. J. K. Tews, S. H. Carter, and W. E. Stone, Chemical changes in the brain during insulin hypoglycemia and recovery, *J. Neurochem.* **12**:679 (1965).
40. F. N. Minard and I. K. Mushahwar, The effect of periodic convulsions induced by 1,1-dimethylhydrazine (UDMH) on the synthesis of rat brain metabolites from [2-$^{14}C$] glucose, *J. Neurochem.* **13**:1–11 (1966).
41. H. B. Burch, O. H. Lowry, A. M. Kuhlman, J. Skerjance, E. J. Diamant, S. R. Lowry, and P. Von Dippe, Changes in patterns of enzymes of carbohydrate metabolism in the developing rat liver, *J. Biol. Chem.* **238**:2267–2273 (1963).
42. H. B. Burch, Substrates of carbohydrate metabolism and their relation to enzyme levels in liver from rats of various ages, *in Advances in Enzyme Regulation* (G. Weber, ed.), Vol. 3, pp. 185–197 Pergamon Press, Oxford (1965).
43. M. F. Utter, The role of $CO_2$ fixation in carbohydrate utilization and synthesis, *Ann. N. Y. Acad. Sci.* **72**:451–461 (1959).
44. M. F. Utter, Pathways of phosphoenolpyruvate synthesis in glycogenesis, *Iowa State J. Sci.* **38**:97–113 (1963).
45. H. Waelsch, S.-C. Cheng, L. J. Coté, and H. Naruse, $CO_2$ fixation in the nervous system, *Proc. Natl. Acad. Sci.* **54**:1249–1253 (1965).
46. H. G. Hers and Ch. De Duve, Le systéme hexose phosphatique. II. Repartition de l'activité G6Pase dans les tissues, *Bull. Soc. Chim. Biol.* **32**:20–29 (1950).
47. H. A. Krebs and M. Woodford, Fructose 1,6-diphosphatase in striated muscle, *Biochem. J.* **94**:436–445 (1965).
48. L. J. King, O. H. Lowry, J. V. Passonneau, and V. Venson, Effects of convulsants on energy reserves in the cerebral cortex, *J. Neurochem.* **14**:599–611 (1967).
49. C. I. Mayman, P. D. Gatfield, and B. McL. Beckenbridge, The glucose content of brain in anaesthesia, *J. Neurochem.* **11**:483–487 (1964).
50. D. Richter and R. M. C. Dawson, Brain metabolism in emotional excitement and in sleep, *Am. J. Physiol.* **154**:73–79 (1948).
51. J. R. Klein and N. S. Olsen, Effect of convulsive activity upon the concentration of brain glucose, glycogen, lactate and $P_i$, *J. Biol. Chem.* **167**:747–756 (1947).
52. F. Plum, J. B. Posner, and B. Troy, Cerebral metabolic and circulatory responses to induced convulsions in animals, *Arch. Neurol.* **18**:1–13 (1968).

53. C. Long, Studies involving enzymic phosphorylation. I. The hexokinase activity of rat tissues, *Biochem. J.* **50**:407–415 (1952).
54. R. K. Crane, Hexokinases and Pentokinases *in The Enzymes* (P. D. Boyer, H. Lardy, and K. Myrbäck, eds.), Vol. 6, pp. 47–66, Academic Press, New York (1962).
55. A. Hernandez and R. K. Crane, Association of heart hexokinase with subcellular structure, *Arch. Biochem. Biophys.* **113**:223–229 (1966).
56. H. S. Bachelard, The subcellular distribution and properties of hexokinases in the guinea pig cerebral cortex, *Biochem. J.* **104**:286–292 (1967).
57. J. E. Wilson, The latent hexokinase activity of rat brain mitochondria, *Biochem. Biophys. Res. Commun.* **16**:123–127 (1967).
58. D. Biesold and P. Teichgräber, Activation and solubilization of particle bound hexokinase in rat brain, *Biochem, J.* **103**:13C—14C (1967).
59. D. Biesold and B. E. Leonard, Some properties of hexokinase in the developing rat brain, *in Ontogenesis of the Brain*, Proc. Int. Symp. Neuroontogeneticum (Prague) (L. Ljílek and S. Trojan, eds.) pp. 259–268, Universita Karolina Pragensis, Prague (1968).
60. L. Grossbard and R. T. Schimke, Multiple hexokinases of rat tissues. Purification and comparison of soluble forms. *J. Biol. Chem.* **241**:3546–3560 (1966).
61. H. J. Fromm and V. Zewe, Kinetic studies of the brain hexokinase reaction, *J. Biol. Chem.* **237**:1661–1667 (1962).
62. H. Weil-Malherbe and A. D. Bone, Studies on hexokinase. 1. The hexokinase activity of rat brain extracts, *Biochem. J.* **49**:339–347 (1951).
63. O. H. Lowry and J. V. Passonneau, Kinetic evidence for multiple binding sites on phosphofructokinase. *J. Biol. Chem.* **241**:2268–2279 (1966).
64. D. Garfinkel, A simulation study of mammalian phosphofructokinase, *J. Biol. Chem.* **241**:286–294 (1966).
65. E. D. Wachsmuth and G. Pfleiderer, Biochemische Untersuchungen an Kristallinen Isoenzymen der Lactatdehydrogenase aus menschlichen Organen. *Biochem. Z.* **336**: 545–556 (1963).
66. L. B. Flexner, G. De la Haba, and J. B. Flexner, Further studies on the components of lactic dehydrogenase of cerebral cortex, *J. Neurochem.* **9**:313–320 (1963).
67. V. Bonavita, F. Ponte, and G. Amore, Lactate dehydrogenase isoenzymes in the nervous tissue—IV. An ontogenetic study on the rat brain, *J. Neurochem.* **11**:39–47 (1964).
68. R. D. Cahn, N. O. Kaplan, L. Levine, and E. Zwilling, Nature and development of lactate dehydrogenases, *Science* **136**:962–969 (1962).
69. D. Di Pietro and S. Weinhouse, Glucose oxidation in rat brain slices and homogenates, *Arch. Biochem. Biophys.* **80**:268–282 (1959).
70. R. Balázs and J. R. Lagnado, Glycolytic activity associated with rat brain mitochondria, *J. Neurochem.* **5**:1–17 (1959).
71. M. K. Johnson, The intracellular distribution of glycolytic and other enzymes in rat brain homogenates and mitochondrial preparations, *Biochem. J.* **77**:610–618 (1960).
72. R. Tanaka and L. G. Abood, Isolation from rat brain of mitochondria devoid of glycolytic activity, *J. Neurochem.* **10**:571–576 (1963).
73. V. P. Whittaker, The application of subcellular fractionation techniques to the study of brain function, *Progr. Biophys. Mol. Biol.* **15**:39–96 (1965).
74. J. E. Cremer, Studies on brain cortex slices. Differences in the oxidation of $^{14}C$-labelled glucose and pyruvate revealed by the action of triethyltin and other toxic agents, *Biochem. J.* **104**:212–222 (1967).

75. H. A. Krebs, Mitochondrial generation of reducing power, *in Biochemistry of Mitochondria* (E. C. Slater, Z. Kaniuga, and L. Wojtczak, eds.), pp. 104–113, Academic Press, New York (1967).
76. R. L. Ringler and T. P. Singer, Studies on the mitochondrial α-glycerophosphate dehydrogenase. 1. Reaction on the dehydrogenase of electron acceptors and the respiratory chain, *J. Biol. Chem.* **234**:2211–2217 (1959).
77. B. Sacktor, L. Packer, and R. W. Estabrook, Respiratory activity of brain mitochondria, *Arch. Biochem. Biophys.* **80**:68–71 (1959).
78. R. Balázs, The aerobic inhibition of glycolysis associated with brain mitochondria, *Biochem. J.* **86**:494–509 (1963).
79. F. S. Rolleston and E. A. Newsholme, Control of glycolysis in cerebral cortex slices, *Biochem. J.* **104**:524–533 (1967).
80. R. Balázs, K. Magyar, and D. Richter, The operation of the tricarboxylic acid cycle in brain mitochondrial preparations, *in Comparative Neurochemistry* (D. Richter, ed.), pp. 225–248, Pergamon Press, Oxford (1964).
81. C. A. Villee and J. M. Loring, Alternative pathways of carbohydrate metabolism in foetal and adult tissue, *Biochem. J.* **81**:488–494 (1961).
81a. J. J. O'Neill, S. H. Simon, and W. W. Shreeve, Alternate glycolytic pathways in brain. A comparison between the action of artificial electron acceptors and electrical stimulation, *J. Neurochem.* **12**:797–802 (1965).
82. M. M. De Piras and J. A. Zaduinsky, Effect of potassium and ouabain on glucose metabolism by frog brain, *J. Neurochem.* **12**:657–661 (1965).
83. J. Katz and H. G. Wood, The use of $^{14}CO_2$ yields from glucose-1- and -6-$^{14}C$ for the evaluation of the pathways of glucose metabolism, *J. Biol. Chem.* **238**:517–523 (1963).
84. A. M. Burt and B. S. Wenger, Glucose-6-phosphate dehydrogenase activity in the brain of the developing chick, *Develop. Biol.* **3**:84–95 (1961).
85. R. E. Kuhlman and O. H. Lowry, Quantitative histochemical changes during the development of rat cerebral cortex, *J. Neurochem.* **1**:173–180 (1956).
86. E. M. Gal, J. C. Armstrong, and B. Ginsberg, The nature of *in vitro* hydroxylation of L-tryptophan by brain tissue, *J. Neurochem.* **13**:643–654 (1966).
87. D. B. McDougal, D. W. Schultz, J. V. Passonneau, J. R. Clark, M. A. Reynolds and O. H. Lowry, Quantitative studies of white matter. 1. The enzymes involved in glucose-6-phosphate metabolism, *J. Gen. Physiol.* **44**:487–498 (1961).
88. P. M. Dreyfus, Transketolase activity in the nervous sytem, *in Ciba Foundation Study Group No. 28, Thiamine Deficiency* (G. E. W. Wolstenholme and M. O'Connor, eds.), pp. 103–111, Churchill Ltd., London (1967).
89. F. Novello and P. McLean, The pentose phosphate pathway of glucose metabolism. Measurement of the non-oxidative reactions of the cycle, *Biochem. J.* **107**:775–791 (1968).
89a. S. S. Hotta, Glucose metabolism in brain tissue; the hexosemonophosphate shunt and its role in glutathione reduction, *J. Neurochem.* **9**:43–51 (1962).
90. R. V. Coxon, E. C. Gordon-Smith, and J. R. Henderson, The incorporation of isotopic carbon into the cerebral glycogen of rabbits, *Biochem. J.* **97**:776–781 (1965).
91. M. A. Stewart, W. R. Sherman, M. M. Kurien, G. I. Moonsammy, and M. Wisgerholf, Polyol accumulations in nervous tissue of rats with experimental diabetes and galactosaemia, *J. Neurochem.* **14**:1057–1066 (1967).
92. G. I. Moonsammy and M. A. Stewart, Purification and properties of brain aldose reductase and L-hexonate dehydrogenase, *J. Neurochem.* **14**:1187–1193 (1967).
93. M. W. Wells, T. A. Pittman, H. J. Wells, and T. J. Egan, The isolation of galactitol from the brains of galactosemia patients, *J. Biol. Chem.* **240**:1002–1004 (1965).

94. R. M. Burton, M. A. Sodd, and R. O. Bradley, The incorporation of galactose into galactolipids, *J. Biol. Chem.* **233**:1053–1060 (1958).
95. R. M. C. Dawson and N. Freinkel, The distribution of free mesoinositol in mammalian tissues, including some observations on the lactating rat, *Biochem. J.* **78**: 606–610 (1961).
96. G. Hauser, The formation of free and lipid myo-inositol in the intact rat, *Biochem. Biophys. Acta.* **70**:278–289 (1963).
97. S. Hollmann, *Non-glycolytic Pathways of Metabolism of Glucose*, pp. 83–114, Academic Press, New York (1964).
98. K. P. Wong and T. L. Sourkes, Determination of UDPG and UDPGA in tissues, *Anal. Biochem.* **21**:444–453 (1967).
99. J. J. Burns and G. Ashwell, L-Ascorbic acid, *in The Enzymes* (P. D. Boyer, H. Lardy, and K. Myrbäck, eds.), Vol. 3, pp. 387–406, Academic Press, New York (1960).
100. M. K. Gaitonde, D. R. Dahl, and K. A. C. Elliott, Entry of glucose carbon into amino acids of rat brain and liver *in vivo* after injection of uniformly $^{14}$C-labelled glucose, *Biochem. J.* **94**:345–352 (1965).
101. M. J. Johnson, Enzymic equilibria and thermodynamics, *in The Enzymes* (P. D. Boyer, H. Lardy, and K. Myrbäck, eds.), Vol. 3, pp. 407–441, Academic Press, New York (1960).
102. H. A. Lardy, D. O. Foster, E. Shrago, and P. D. Ray, Metabolic and hormonal regulation of phosphopyruvate synthesis, *in Advances in Enzyme Regulation* (G. Weber, ed.), Vol. 2, pp. 39–47, Pergamon Press Oxford (1964).
103. L. Glaser and D. H. Brown, Purification and properties of D-glucose-6-phosphate dehydrogenase, *J. Biol. Chem.* **216**:67–79 (1955).
104. A. G. Datta and E. Racker, Mechanism of action of transketolase. 1. Properties of the crystalline yeast enzyme, *J. Biol. Chem.* **236**:617–623 (1961).
105. A. Delbrück, H. Schinemassek, K. Bartsch, and Th. Bucher, Enzym-Verteilungsmuster in einigen Organen und in experimentellen Tumoren der Ratte und der Maus, *Biochem. Z.* **331**:297–311 (1959).
106. D. B. McDougal, Jr., R. T. Shimke, E. M. Jones, and E. Touhill, Quantitative studies of white matter. II. Enzymes involved in triose phosphate metabolism, *J. Gen. Physiol.* **47**:419–432 (1964).
107. D. Pette, W. Luh, and Th. Bucher, A constant-proportion group in the enzyme pattern of the Embden–Meyerhof chain, *Biochem. Biophys. Res. Commun.* **7**:419–424 (1962).
108. R. Balázs, S. Kovács, P. Teichgräber, W. A. Cocks, and J. T. Eayrs, Effects of neonatal thyroid deficiency on the developing brain, *J. Neurochem.* **15**:1335–1349 (1968).
109. N. R. Roberts, R. R. Coehlo, O. H. Lowry, and E. J. Crawford, Enzyme activities of giant squid axoplasm and axon sheath, *J. Neurochem.* **3**:109–115 (1958).
110. L. F. Leloir, J. M. Olavarria, S. H. Goldenberg, and H. Carminatti, Biosynthesis of glycogen from uridine diphosphate glucose, *Arch. Biochem. Biophys.* **81**:508–520 (1959).
111. B. M. Breckenbridge and E. T. Crawford, Glycogen synthesis from uridine diphosphate glucose in brain, *J. Biol. Chem.* **235**:3054–3057 (1960).
112. M. V. Buell, O. H. Lowry, N. R. Roberts, M. L. W. Chang, and J. I. Kapphahn, The quantitative histochemistry of the brain. V. Enzymes of glucose metabolism, *J. Biol. Chem.* **232**:979–993 (1958).
113. G. Weber, M. C. Henry, S. R. Wagle, and D. S. Wagle, Correlation of enzyme activities and metabolic pathways with growth rate of hepatomas, *in Advances in Enzyme Regulation* (G. Weber, ed.), Vol. 2, pp. 335–346, Pergamon Press, Oxford (1964).

114. S. S. Hotta, R. H. Laatsch, and P. V. Myron, Jr., Isocitrate dehydrogenase activity in isolated guinea pig mitochondria, *J. Neurochem.* **10**:841–847 (1963).
115. J. Mark, Y. Godin, and P. Mandel, Glucose and lactic acid content of the rat brain, *J. Neurochem.* **15**:141–143 (1968).
116. H. B. Stoner, Studies on the mechanism of shock, the quantitative aspects of glycogen metabolism after limb ischaemia in the rat. *Brit. J. Exptl. Pathol.* **39**:635–651 (1958).
117. H. Balzer and D. Palm, Über den Mechanismus der Wirkung des Reserpins auf den Glykogengehalt der Organe, *Arch. Exptl. Pathol. Pharmakol.* **243**:65–84 (1962).
118. N. D. Goldberg, J. V. Passonneau, and O. H. Lowry, Effects of changes in brain metabolism on the levels of citric acid cycle intermediates, *J. Biol. Chem.* **241**:3997–4003 (1966).
119. J. R. Williamson, Mechanism for the stimulation *in vivo* of hepatic gluconeogenesis by glucagon, *Biochem. J.* **101**:11C–14C (1966).
120. R. Rajalakshmi, J. Malathy, and C. V. Ramakrishnan, Effect of dietary protein content on regional distribution of ascorbic acid in the rat brain, *J. Neurochem.* **14**: 161–167 (1967).
121. O. H. Lowry, J. V. Passonneau, D. W. Schulz, and M. K. Rock, The measurement of pyridine nucleotides by enzymatic cycling, *J. Biol. Chem.* **236**:2746–2755 (1961).
122. H. B. Burch, M. E. Bradley, and O. H. Lowry, The measurement of triphosphopyridine nucleotides and reduced triphosphopyridine nucleotide and the role of haemoglobin in producing erroneous triphosphopyridine nucleotide values, *J. Biol. Chem.* **242**:4546–4554 (1967).
123. E. Racker, Transaldolase, *in The Enzymes* (P. D. Boyer, H. Lardy, and K. Myrbäck, eds.), Vol. 5, pp. 407–412, Academic Press, New York (1961).
124. J. H. Quastel, Effects of drugs on metabolism of the brain *in vitro*, *Brit. Med. Bull.* **21**:49–56 (1965).
125. H. McIlwain, Polybasic and polyacidic substances or aggregates and the excitability of cerebral tissues, electrically stimulated *in vitro*, *Biochem. J.* **90**:442–448 (1964).
126. O. J. Rafaelson, Action of insulin on carbohydrate uptake of isolated rat spinal cord, *J. Neurochem.* **7**:33–44 (1961).
127. W. J. H. Butterfield, M. E. Abrams, R. A. Sells, G. Sterky, and M. J. Wichelow, Insulin sensitivity of the human brain, *Lancet* **9**:557–560 (1966).
128. C. R. Park, L. H. Johnson, J. H. Wright, Jr., and H. Batsel, Effect of insulin on transport of several hexoses and pentoses into cells of muscle and brain, *Am. J. Physiol.* **191**:13–18 (1957).
129. M. Hamburgh and L. B. Flexner, Biochemical and physiological differentiation during morphogenesis. XXI. Effect of hypothyroidism and hormone therapy on enzyme activities of the developing cerebral cortex of the rat, *J. Neurochem.* **1**:279–288 (1957).
130. R. A. Peters, Significance of thiamine in the metabolism and function of the brain, *in Neurochemistry* (K. A. C. Elliott, I. H. Page, and J. H. Quastel, eds.), pp. 267–275 Charles C. Thomas, Springfield, Illinois (1962).
131. R. E. Koeppe, R. M. O'Neal, and C. H. Hahn, Pyruvate decarboxylation in thiamine deficient brain, *J. Neurochem.* **11**:695–699 (1964).
132. C. J. Gabler, Studies on the physiological functions of thiamine. 1. The effects of thiamine deficiency and thiamine antagonists on the oxidation of α-keto acids by rat tissues, *J. Biol. Chem.* **236**:3112–3120 (1961).
133 L. Salganicoff and R. E. Koeppe, Subcellular distribution of pyruvate carboxylase, diphosphopyridine nucleotide and triphosphopyridine nucleotide isocitrate dehydrogenases, and malate enzyme in rat brain, *J. Biol. Chem.* **243**:3416–3420 (1968).

134. H. V. Henning and W. Seubert, Zum Mechanismus der Gluconegesese und ihre Steigerung. I. Quantitative Bestimmung der Pyruvate Carboxylase in Rohextracten der Rattenleber, *Biochem. Z.* **340**:160–170 (1964).
135. S. Otsuki, S. Watanabe, K. Ninomiya, T. Hoaki and N. Okumura, Correlation between [U-$^{14}$C] glucose metabolism and function in perfused cat brain, *J. Neurochem.* **15**:859–865 (1968).
136. J. A. Cocks, R. Balázs, and J. T. Eayrs, The effect of thyroid hormones on the biochemical maturation of the rat brain, *Biochem. J.* **111**:187 P (1969).

*Chapter 2*

# GLYCOGEN METABOLISM


## R. V. Coxon

*University Laboratory of Physiology*
*Oxford, England*


## I. INTRODUCTION

The isolation of glycogen from dog brain was reported in 1938 by Kerr[1] who established its identity from elementary analysis, optical rotation, reaction with iodine, and the products of hydrolysis. Kerr[2] had estimated the quantity present in dog brain as in the region of 80–100 mg/100 g of tissue. He also examined a few other species and since then "glycogen" has been measured in various ways in the brains of other animals, but only rarely has it been isolated and fully characterized.

## II. CHEMISTRY OF GLYCOGEN

### A. Quantitative Determination in Nervous Tissue

Glycogen is generally determined in nervous tissue by digestion with potassium hydroxide, to which glycogen is resistant, followed or accompanied by precipitation with alcohol. The resulting precipitate is contaminated with cerebrosides which have to be removed by repeated washing with a chloroform–methanol mixture or some similar solvent. The washed precipitate is finally hydrolyzed and its quantity determined as glucose by one of the conventional methods, either chemical or enzymatic, but may, if there is a sufficient amount, be collected for structural studies or positive identification as was done by Kerr.[1]

### B. Molecular Weight

Glycogen samples isolated from a variety of tissues, especially muscle and liver, have shown a high degree of polydispersity, and, according to Stetten and Stetten,[3] their average molecular weights may vary almost continuously from 1 to 100 million, depending on the method of isolation. A sample of glycogen from rabbit brain prepared in the writer's laboratory[4] by alcoholic precipitation following digestion of the tissue in aqueous potassium hydroxide for 20 min had a molecular weight of 2 million as determined in the ultracentrifuge. However, Khaikina[5] reports molecular weights between 280,000 and 610,000 determined by light-scattering measurements on samples of glycogen from brains subjected to treatment with potassium hydroxide for some hours.

## C. Structure of Glycogen

The generally accepted structure of the glycogen molecule (Fig. 1) is that of a branched homopolymer of glucose, as proposed by Meyer and Fuld.[6] While there is chemical supporting evidence, the main justification for this model is derived from studies involving stepwise degradation with two enzymes which specifically attack the α-1-4-glucosidic linkages and the 1-6 linkages, respectively. These enzymes are phosphorylase, which brings about a phosphorolytic rupture of the 1-4 linkage, and amylo-1-6-glucosidase which attacks the 1-6 linkages hydrolytically. Digestion with phosphorylase alone gives an indication of the proportion of glucose residues located in the outer side chains. In the case of cerebral glycogen from rats this was found by Minard *et al.*[7] to be 33% as compared with 38% reported for glycogen from rat liver by Illingworth *et al.*[8]

Fig. 1. Schematic representation of the structure of a segment of a glycogen molecule. Branch points are indicated by filled hexagons. R denotes reducing end group.

By measuring the proportion of free and phosphorylated glucose liberated by phosphorylase and 1-6-glucosidase acting in combination, the frequency of branch points and hence average lengths of the linear glucose chains can be calculated. Another method of determining structure is by end group analysis using periodate; this method has been applied to brain glycogen from rabbits by Khaikina and Goncharova,[9] who estimate the average chain length as 12, whereas similar determinations on liver glycogen give a figure of 16.

A complete enzymatic study of the structure of cerebral glycogen has not to the writer's knowledge been carried out.

### D. Binding of Glycogen to Other Cell Constituents

Extraction with cold trichloracetic acid yields only a fraction (40% in the case of muscle) of the total glycogen present in many tissues and this is sometimes referred to as lyoglycogen in contradistinction to the residue which is recoverable by treatment with strong alkali and is called desmoglycogen.[3] Khaikina and Krachko[10] claim that 20% of the total cerebral glycogen is in the free state, the remainder being either protein-bound (60%) or lipid-bound (20%). Similar distinctions between free and total glycogen have been recognized in other tissues and the metabolism of the two fractions has been found to differ in certain circumstances.

## III. DISTRIBUTION AND QUANTITY OF GLYCOGEN IN NERVOUS TISSUE

### A. Brain

Glycogen has been demonstrated histochemically in the brains of many vertebrates including mammals, birds, amphibia, and some fish. Analytical data also are available for a number of common animals and some of these data are assembled in Table 1.

It has long been held [see Kerr[2]] that unless the brain is rapidly frozen the glycogen content will be seriously underestimated due to the rapid destruction (presumably by enzymes) at higher temperatures. Freezing is generally carried out in liquid air or nitrogen.

Glycogen is found in the forebrain, midbrain, and hindbrain.[11]

### B. Spinal Cord

Histochemical evidence for the presence of glycogen in both neurons and neuropil in the spinal cord of the monkey has been produced by Manocha *et al.*,[12] but little or no information seems to be available on the quantities present. In certain birds, however, a so-called "glycogen body"

TABLE I

Glycogen Content of Brain (mg/100g)

| Species | Content | Authors |
|---|---|---|
| Dog | 77 - 130 | Kerr[2] |
| Cat | 77 - 101 | Kerr[2] |
| Rabbit | 70 - 99 | Kerr[2] |
| Rat | 95 — 124 | Křivánek[58] |
| Mouse | 46 — 68 | Carter and Stone[56] |
| Guinea pig[a] | 21 — 32 | Kleinzeller and Rybova[35] |
| Monkey | 97 — 115 | Coxon and Larkins[b] |
| Man | 112 (±36) | Kirsch and Leitner[59] |
| Sea turtle | 306 | Kerr and Ghantus[39] |

[a] Tissue not quick-frozen.
[b] Unpublished.

is found in the dorsal columns of the spinal cord. The biological function of this organ is unknown[13] but it is composed of glycogen to the extent of some 75% of its dry weight in the newly hatched chick [see Doyle and Watterson[14]]. These authors also found glycogen to be present in the cord cephalad to the glycogen body in quantities of the order of 100 mg/100 g.

## C. Ganglia

### 1. *Spinal*

Berthold[15] has reported the histochemical demonstration of glycogen in the neurons of these ganglia in the frog (see below).

### 2. *Sympathetic*

Glycogen has been found in the sympathetic ganglia of rats by Ribaupierre *et al.*,[16] while a comparative histochemical study of the sympathetic ganglia of a number of different animals by Ershova[17] revealed the glycogen present in these ganglia taken from many fish, reptiles, birds, and mammals, but absent from the ganglia of bats, rats, and horses.

### 3. *Peripheral Nerves*

Lowry *et al.*[18] report the occurrence of glycogen in a concentration of some 18 mg/100 g in the tibial nerve of the rabbit.

## IV. ENZYME SYNTHESIS AND DEGRADATION OF GLYCOGEN

### A. General

Since the discovery by Leloir and Cardini[19] in 1957 of an enzyme catalyzing the addition of glucose units derived from uridine diphosphate-glucose (UDPG) to a polysaccharide primer in 1-4 glucosidic linkage, it has been recognized that the synthesis of glycogen in living tissues is predominantly accomplished by a pathway different from that involved in its degradation. Clearly such an arrangement, apart from any other consideration, offers very important advantages to the cell in regard to metabolic control. The scheme depicted in Fig. 2 assumes that formation and destruction of glycogen take place by different mechanisms and it is clear that, starting from glucose 1-phosphate, five enzymes are required for the main cycle, *viz.*: (1) UDPG pyrophosphorylase, (2) UDPG glycogen transglucosylase (also known as glycogen synthetase), (3) amylo-1-4 → 1-6 transglucosidase (branching enzyme), (4) glycogen phosphorylase, and (5) amylo-1-6-glucosidase (debranching enzyme).

The subsidiary cycle required for the regeneration of UTP involves also UDP-kinase [(6) in Fig. 2]. It may be noted that *in vitro* the action of UDPG glycogen transglucosylase only proceeds in the presence of preexisting polysaccharide: if the same is true *in vivo* the implication is that a small quantity of primer must be present in the fertilized ovum from which an animal grows.

### B. In Nervous Tissue

As far as the nervous system is concerned the enzymes numbered (1), (2), and (4) above have been demonstrated in brain extracts, while there is

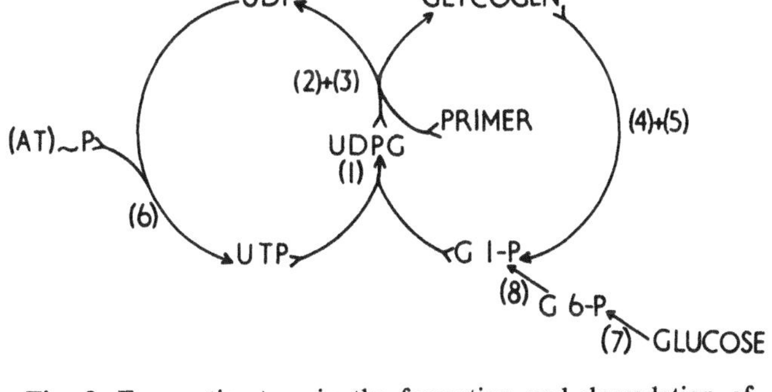


Fig. 2. Enzymatic steps in the formation and degradation of glycogen. UDP, uridine diphosphate; UTP, uridine triphosphate; G 1–P, glucose 1-phosphate; G 6-P, glucose 6-phosphate; UDPG, uridine-diphosphate glucose; ~P, high-energy phosphate; (AT), probably adenosine–tri. Numbers in parentheses refer to enzymes mentioned in the text.

histochemical evidence suggestive of the presence of (3); also a paper (which has not been seen by the present writer) is said to exist reporting the presence of (5) in rabbit brain. References to the literature relating to the above enzymes in nervous tissue are given by McIlwain[20] and by Greville.[21]

In the living brain where the starting material for glycogen synthesis is likely to be glucose derived from the blood stream, hexokinase [(7) in Fig. 2] and phosphoglucomase [(8) in Fig. 2] would also be required and both have in fact been found in cerebral tissue.[20]

### C. Adenyl Cyclase

In liver it has been found that phosphorylase is activated by a cyclic form of adenosine monophosphate (cyclic AMP). The same type of activation has been found to occur with brain phosphorylase and, since the brain contains adenyl cyclase[21] which leads to the production of cyclic AMP from ATP, it seems likely that this type of control may also operate in the brain. Whether, as in other tissues, hormonal effects are exerted through AMP is uncertain (see Section VIII B on adrenalin below) but this is clearly a possibility to be borne in mind.

## V. CELLULAR AND SUBCELLULAR LOCALIZATION

The evidence for the localization of glycogen in the nervous system, both as regards the cell types in which it is found and also its intracellular distribution, is inevitably largely dependent on the interpretation of histochemical data. Both the light microscope and the electron microscope have been used in collecting such data and the usual reservations regarding specificity and technical artifacts must apply in their evaluation. In the case of the light microscope a much favored reaction for demonstrating glycogen in tissues generally is the periodic acid Schiff (PAS) reaction which is generally combined with a test for susceptibility to digestion with either ptyalin or malt diastase.[22] Using the PAS method, Manocha *et al.*[12] have found positively reacting material in neurons and also in the neuropil of monkey spinal cord, but the strength of the reaction in the neurons was only reduced and not abolished by exposure to diastase and may not, therefore, have been exclusively attributable to glycogen.

The lead tetra-acetate Schiff reaction is a modification of the PAS reaction in which the lead salt replaces periodic acid but according to Pearse[22] it offers no special advantages over the original procedure. Using the lead tetra-acetate reaction, Shimizu and Kumamoto[23] made an extensive study of the distribution of glycogen in the brains of mice, rats, guinea pigs, and rabbits and reported its presence in both neurons and glia. However, some of the material designated as glycogen by Shimizu and Kumamoto[23] was found extracellularly and this casts some doubt on the reliability of the localization [see Crome *et al.*[24]].

The time-honored Best carmine stain for glycogen which has been widely employed in studies on the liver has been applied also to nervous tissue. The chemical basis of this staining reaction is not fully understood, but its specificity seems to be fairly high. By carmine staining glycogen has been identified in the axon of degenerating peripheral nerve.[24, 26]

In general, the results of histochemical studies with the light microscope have suggested a cytoplasmic location for glycogen granules without going much further, although some preferential deposition in the region of the Golgi apparatus has been suggested by Shimizu and Kumamoto.[23]

Among cerebral tumors, according to Tiraspolskaya,[27] glycogen is prominent in astrocytomata and spongioblastomata but less so in oligodindiogliomata and medulloblastomata.

### A. Electron Microscope

The visualization of glycogen with the electron microscope depends upon achieving some sort of contrast. The Best carmine stain has been employed for this purpose[28] relying upon the calcium and aluminum contained in the dye. However, a very effective method for visualizing glycogen is one depending upon treatment of the section with a lead salt. Although the underlying chemistry does not appear to have been worked out in detail, the method has received powerful vindication from some careful studies by Revel,[29] who has demonstrated a good correlation between the ultrastructural characteristics of glycogen deposits stained with lead and their constitution as revealed by chemical methods. Using this technique, Berthold[15] showed the presence of putative glycogen granules in the perikarya of neurons in the frog's spinal ganglia and obtained some evidence suggestive of their presence also within mitochondria. Glycogen deposits identified by lead staining have also been detected in the astrocytes of the rat cerebrum following laser irradiation,[30] and similar granules have been described by Konishi[31] in the endings of cerebellar mossy fibers. These were found in mice under the age of 3 weeks but disappeared thereafter.

Curiously enough no systematic studies of glycogen in the adult brain using the electron microscope have come to the writer's notice, although Résibois-Grégoire and Dourov[32] have given an elegant description of the findings in glycogenosis where they observed extensive deposits in the astroglia.

## VI. TURNOVER OF GLYCOGEN IN THE NERVOUS SYSTEM

### A. Chemical Studies

The lability of glycogen in brain was recognized by Kerr[2] in his early work, its rapid disappearance following excision of the brain being accompanied by the appearance of lactic acid. More recently, Lowry *et al.*[33]

have reported on the rate of disappearance following decapitation in the mouse. They find that glycogen in the brain within the severed head of a mouse maintained at 38°C may persist at its original concentration for 6 sec, after which degradation takes place at a rate of about 40 mg/100 g brain/min. The rate of synthesis has been studied in brain slices and figures of 18 mg/100 g/hr have been reported by LeBaron[(34)] and of 8 mg/100 g/hr by Kleinzeller and Rybova[(35)]; in both reports the guinea pig was the source of the tissue. Some incorporation of $^{14}C$ from glucose has also been observed in such slices.[(35a)]

## B. Studies with $^{14}C$

Studies by Russian workers, in particular Prokhorova,[(36)] demonstrated that the readiness with which the brain incorporated $^{14}C$ from glucose into its glycogen was comparable with that of the liver. Detailed comparisons carried out by Coxon *et al.*[(4)] showed that the specific activity of the glycogen of brain substantially exceeded that of both cardiac and skeletal muscle over a period of 8 hr following an intravenous injection of glucose-$^{14}C$ in the rabbit. The same group of workers also demonstrated some incorporation into cerebral glycogen of carbon derived from glutamate and bicarbonate, while Khaikina and Goncharova[(9)] have reported that $^{14}C$ administered subcutaneously as carboxyl-labeled acetic acid is also incorporated into cerebral glycogen. Stetten and Stetten,[(3)] on the basis of their own studies of liver and muscle glycogen, have drawn attention to the inhomogeneous nature of glycogen molecules in relation to the incorporation of radioactive glucose. They found that this inhomogeneity was evident both with respect to the position in the molecule where the radioactive glucose was located (which varied with time) and with respect to the size of molecule in which the labeled glucose was preferentially incorporated. However, judged by these criteria, muscle glycogen and liver glycogen behaved differently. Such information is not available for cerebral glycogen, probably largely because of the paucity of material available for such investigations. However, so long as one is aware of this complication straightforward comparisons of the overall speed of incorporation of $^{14}C$ between one organ and another can clearly afford a general indication of the metabolic stability or instability of the glycogen in a particular organ *in vivo*. Interpreted in this way, the results of experiments with $^{14}C$ labeling leave no doubt that cerebral glycogen undergoes relatively rapid renewal in the normal adult animal, be it rat or rabbit.

Khaikina and Goncharova[(9)] in their studies with $^{14}C$-labeled precursors carried out a fractionation of the labeled glycogen from rat brain into three portions: the free, the protein-bound, and the lipid-bound (see p. 39). Of these, the free glycogen showed the highest specific activity, the protein-bound the lowest, while the lipid-bound fraction attained an intermediate value between the other two at 3 hr after administration of the precursor. In their experiments with acetate-$^{14}C$ on guinea pigs, Khaikina and Goncharova[(9)] found that the free glycogen was more strongly labeled at

$1\frac{1}{2}$ hr post injection than at 3 hr, while the reverse held for both bound fractions, suggesting a possible precursor–product relationship. As in the glucose-$^{14}C$ experiments of Coxon *et al.*,[4] the total glycogen underwent a progressive rise in specific activity over the period from $1\frac{1}{2}$ to 3 hr after the dose of precursor.

## VII. ENVIRONMENTAL INFLUENCES ON CEREBRAL GLYCOGEN

### A. Diet

The effects of fasting on the level of cerebral glycogen has been investigated by Prasannan *et al.*[37] by comparing a group of rats which had fasted for 24 hr with a second group which had fed *ad libitum* and been injected intraperitoneally with glucose. The mean glycogen level in the fed group was 51 mg ± 11 mg/100 g, while in the fasted group it was 38 mg ± 8 mg/100 g, a difference which was significant by the *t* test at the 0.01 level. The effect of an intraperitoneal injection of 50 mg of glutamic acid[38] was to raise the level in fasting rats by about 70% (41–73), but this effect was not additive with cortisone (see below). These results stand in contrast to those of Kerr and Ghantus[39] who found no change in the cerebral glycogen of dogs after 3-day fasting on the one hand or glucose loading on the other.

### B. Hypoxia

Chance and Yaxley[40] reported that a few minutes exposure to pure nitrogen resulted in no change in cerebral glycogen in mice, while mice exposed to 12% $CO_2$ in $N_2$ suffered from convulsions and underwent a 30% increase in their cerebral glycogen. The estimations in these experiments appear to have been carried out without quick freezing.

On the other hand, a few minutes exposure to a simulated altitude of 12,000 m ($p0_2 \simeq 35$ torr) led to a reduction in cerebral glycogen in rats according to Jilek *et al.*,[41] while exposure to an altitude of 12,500 ft for longer periods led to a fall between 1 and 4 weeks with a return to normal levels after 9 weeks in a study reported by Woolley and Timiras.[42]

Following carotid ligation Krulich *et al.*[43] noted a fall in cerebral glycogen in 12-day-old rats, which, however, was much less conspicuous in adult animals.

### C. Temperature

Fujita[44] has reported a fall in cerebral glycogen in rats when the body temperature is maintained at 22°C for some hours. He reports also a similar fall when the body temperature is maintained at 39°C. However, he employed an unusual method for measuring glycogen which led to control values of 500 mg/100 g, which is much higher than that found by most other workers (*cf.* Table I). Petschek and Timiras,[11] who estimated glycogen in several areas of the brain in rats, found some fall in concentration (from 85–67 mg/100 g) when the animals were maintained in an environmental temperature at 2.5°C for 60 hr, although the body temperature in the animals remained normal.

### D. Injury

Changes in cerebral glycogen have been reported in association with many types of local injury. Shimizu and Hamuro[45] from histochemical examination reported the appearance of glycogen deposits in the brain adjacent to stab wounds.

Lierse *et al.*[46] have reported PAS-positive material in both glial and neuronal cells following X-irradiation of the brain, while Shabadash[47] reported, also using histochemical methods, that in the spinal cord of frogs gamma-irradiation leads to a decrease of glycogen in the motor neurons. Lampert *et al.*,[30] using electron microscopy, noted the appearance of glycogen in the astrocytes following exposure of the brain to a laser beam. An electron microscopic study of the degenerative changes following tract-section by Laatsch and Cowan[48] revealed what appeared to be massive deposits of glycogen in the astrocytes in the neighbourhood of degenerating axon terminals in the dentate gyrus of the rat.

### E. Drugs

A variety of drugs (including hormones which will be discussed separately later) have been tested for their effects on cerebral glycogen. Most of these are known excitants or depressants. Among the convulsant group dimethylhydrazine has been carefully studied by Minard *et al.*,[7] who found that administration of the drug was followed by a fall in cerebral glycogen which, however, was not proportional to the number of convulsions produced when repeated doses of the provoking agent were given.

Klein and Olsen[49] in a study embracing a wide variety of convulsant drugs found in cats an invariable depression of glycogen content in the brain. This depression generally amounted to 30%.

Cyanide has also been shown by Albaum *et al.*[50] to cause some depression of the glycogen content of rat brain. The glycogen estimated in this instance appears to have been what other authors would have called "bound glycogen" since it was not extracted by cold trichloracetic acid, and the rats did not convulse because they were under pentobarbitone.

The effect of caffeine as an example of a central nervous stimulant was investigated by Prokhorova and Tupikova,[51] who found that it produced some depression of the glycogen content, though not an impressive one (72 ± 10 to 56 ± 7 mg/100 g). Conversely, the same authors found that the induction of mild narcosis with chloral hydrate or morphia led to a small rise in cerebral glycogen (72 ± 10 to 84 ± 12 mg/100 g). Metamphetamine, another stimulant, also causes a fall in cerebral glycogen according to

Estler and Ammon,[52] but a curious feature of these experiments is that the control level of glycogen (6 mg/100 g) appears to be very low in comparison with other reported figures, possibly due to an unusual analytical method.

The effect of ethyl alcohol was examined by Estler and Ammon[53] who found a substantial fall in cerebral glycogen following its intravenous administration to mice. The effect, however, was not seen in adrenalectomized mice and may therefore be an indirect one (see Sect. VIII. B on adrenalin).

Heim *et al.*[54] also studied the effect of methanol, which unlike ethanol induced a slight rise of cerebral glycogen in mice, as does propanolol.[52]

Chlorpromazine is said to be without effect on the level of cerebral glycogen,[44] while "ipraside" was found to produce a fall in rabbits but not in cats.[55]

### F. Electroshock

The evidence with regard to electrically provoked convulsions is contradictory. Chance and Yaxley[40] report a rise in cerebral glycogen following such treatment in rats, while Klein and Olsen[49] report a fall in cats which were under anaesthesia.

### G. Enzyme Defects

Glycogen storage disease or glycogenosis is a condition in which exceptionally heavy deposits of glycogen may appear in various organs including the brain in some cases. Several different forms of the disease have been distinguished, some of which can be attributed to defective functioning of a particular enzyme involved in the glycogen cycle.[3] One type (called Pompes disease) affects the nervous system in a rather specific manner and is attributed to a deficiency of lysosomal α-1-4-glucosidase. However, other forms of glycogenosis may also result in deposits of glycogen in the CNS.[32]

## VIII. HORMONAL INFLUENCES ON CEREBRAL GLYCOGEN

### A. Insulin

Kerr and Ghantus[39] as far back as 1936 discovered that insulin-induced hypoglycemia was accompanied by a marked depletion of the cerebral glycogen in rabbits. This has since been confirmed in mice by Carter and Stone[56] and in rats by Timiras *et al.*[57]

### B. Adrenalin (Epinephrine)

Kerr and Ghantus[39] noted that adrenalin did not affect the cerebral glycogen in dogs but their experiments were complicated by the fact that the adrenalin was given to phlorinized animals. However, Estler and Ammon[53] found that adrenalin given in combination with ethyl alcohol led to a fall in cerebral glycogen which they attribute to an effect of the alcohol on the blood–brain barrier, permitting adrenalin to reach the nervous system more readily than usual. In this connection it will be recalled that the phosphorylase of brain is sensitive to cyclic AMP which is involved in the activation of phosphorylase by adrenalin in muscle and liver.

Coxon *et al.*[4] found that the rate of incorporation of glucose-$^{14}$C into cerebral glycogen was unaffected by adrenalin in rabbits but the possible effect of alcohol on this finding has not yet been tested.

### C. Adrenocortical Steroids

Timiras *et al.*[57] reported a substantial rise in cerebral glycogen following treatment of rats with hydrocortisone. Fujita[44] reports a similar rise following treatment with a synthetic corticoid analog, while Coxon *et al.*[4] found that treatment with hydrocortisone considerably enhanced the incorporation of glucose-$^{14}$C into the cerebral glycogen of rabbits.

The rise in glycogen level noted by Timiras *et al.*[57] following administration of hydrocortisone was prevented by concomitant administration of insulin.

### D. Adrenocorticotrophin (ACTH)

In view of the findings with hydrocortisone it is surprising to learn that Fujita[44] noted a fall in cerebral glycogen following treatment with ACTH. Moreover he found that when both ACTH and a synthetic glucocorticoid were administered together the effect of the former predominated. Presumably, therefore, it represents an extra-adrenal action on the trophic hormone.

## IX FUNCTIONAL IMPORTANCE OF GLYCOGEN IN THE NERVOUS SYSTEM

It is generally assumed that the glycogen found in normal nervous structures constitutes a reserve of oxidizible or fermentable carbohydrate. This idea is supported by the rapidity with which the glycogen is degraded under conditions of circulatory arrest or hypoglycemia, but of itself it hardly affords a sufficient explanation of the continuous turnover of the glycogen under normal circumstances. Presumably the glycogen forms part of a small pool of carbohydrate exchanging all the time with the free glucose of the brain and spinal cord.

Perhaps the most direct evidence for the participation of glycogen in the functional activity of nervous tissue comes from a recent report by Ribaupierre *et al.*[16] This is concerned with experiments on the isolated sympathetic ganglion of the rat. This structure is capable of synthesizing glycogen when incubated *in vitro* and the glycogen so synthesized is capable of being utilized in the provision of energy for impulse transmission. This role of glycogen is indicated by the observation that a ganglion which is rich in glycogen will continue to transmit impulses when deprived of glucose for a longer period than will a similar ganglion which is poor in glycogen.

## X. CONCLUSION

Although the quantities found are much smaller than in muscle, glycogen is very widely distributed in nervous structures. In view of its turnover as revealed by isotopic incorporation it is to be regarded as a dynamic rather than a static constituent of such tissues. It is found in both nervous and glial elements and its concentration is subject to variation in response to a number of endogenous and exogenous influences. Thus it may play an important part in metabolism both in normal adult life and possibly more particularly during development and in the reaction to injuries of many types.

## XI. REFERENCES


1. S. E. Kerr, The carbohydrate metabolism of brain. VI. Isolation of glycogen, *J. Biol. Chem.* **123**:443–449 (1938).
2. S. E. Kerr, The carbohydrate metabolism of brain. I. The determination of glycogen in nerve tissue, *J. Biol. Chem.* **116**:1–7 (1936).
3. DeWitt Stetten, Jr., and Marjorie Stetten, Glycogen metabolism, *Physiol. Rev.* **40**: 505–537 (1960).
4. R. V. Coxon, E. C. Gordon-Smith, and J. R. Henderson, The incorporation of isotopic carbon ($^{14}C$) into the cerebral glycogen of rabbits, *Biochem. J.* **97**:776–781 (1965).
5. B. I. Khaikina and V. E. Yakushko, Polyglucosides of the brain, *Federation Proc.* **24**:T991–T994 (1965).
6. K. H. Meyer and M. Fuld, L'arrangement des restes de glucose dans le glycogène, *Helv. Chim. Acta* **24**:375–378 (1941).
7. F. N. Minard, C. H. Kang, and I. K. Mushahwar, The effect of periodic convulsions induced by 1,1-dimethylhydrazine on the glycogen of rat brain, *J. Neurochem.* **12**: 279–286 (1965).
8. B. Illingworth, J. Larner, and G. T. Cori, Structure of glycogen and amylopectins. I. Enzymatic determination of chain length, *J. Biol. Chem.* **199**:631–640 (1952).
9. B. I. Khaikina and Ye. Ye. Goncharova, *in Problems of the Biochemistry of the Nervous System* (A. V. Palladin, ed.), pp. 87–95, Pergamon Press, Oxford (1964).
10. B. I. Khaikina and L. S. Krachko, *Ukr. Biokhim. Zh.* **29**:1 (1957), cited by B. I. Khaikina and Ye. Ye. Goncharova, *in Problems of the Biochemisrty of the Nervous System* (A. V. Palladin, ed.), pp. 87–95, Pergamon Press, Oxford (1964).
11. R. Petschek and P. S. Timiras, Electroshock seizures and brain chemistry after acute to moderate cold, *Am. J. Physiol.* **205**:1163–1166 (1963).
12. S. L. Manocha, T. R. Shantha, and G. H. Bourne, Histochemical studies on the spinal cord of squirrel monkey (*Saimiri sciureus*), *Exptl. Brain Res.* (*Berlin*) **3**:25–39 (1967).
13. R. L. Friede and A. E. Vossler, Histochemistry of the glycogen body of the turkey spinal cord, *Histochemie* **4**:330–335 (1964).
14. W. L. Doyle and R. L. Watterson, The accumulation of glycogen in the "glycogen body" of the nerve cord of the developing chick, *J. Morphol.* **85**:391–403 (1949).
15. C.-H. Berthold, Ultrastructural appearance of glycogen in the $\beta$-neurons of the lumbar spinal ganglia of the frog, *J. Ultrastruct. Res.* **14**:254–267 (1966).

16. F. de Ribaupierre, G. Siegrist, M. Dolivo, and Ch. Rouiller, Synthèse et utilisation in vitro du glycogène dans le ganglion sympathique cervical du rat, *Helv. Physiol. Pharmacol. Acta* **24**:C48–49 (1966).
17. V. P. Ershova, Comparative cytochemical study of the sympathetic neurons of the thoracic section of the marginal trunk in vertebrates, *Ark. Anat. Gistol. i Embriol.* **48**:24–32 (1965).
18. M. A. Stewart, J. V. Passonneau, and O. H. Lowry, Substrate changes in peripheral nerve during ischaemia and Wallerian degeneration, *J. Neurochem.* **12**:719–727 (1965).
19. L. F. Leloir and C. E. Cardini, Biosynthesis of glycogen from uridine diphosphate glucose, *J. Am. Chem. Soc.* 79:6340–6341 (1957).
20. H. McIlwain, *Biochemistry and the Central Nervous System*, 3rd ed., Churchill, London (1966).
21. G. D. Greville, *in Neurochemistry* (K. A. C. Elliott, I. H. Page, and J. H. Quastel, eds.), 2nd ed., pp. 238–266, Thomas, Springfield (1962).
22. A. G. E. Pearse, *Histochemistry*, Churchill, London (1960).
23. N. Shimizu and T. Kumamoto, Histochemical studies on the glycogen of the mammalian brain, *Anat. Record* **114**:479–497 (1952).
24. L. Crome, J. N. Cumings, and S. Duckett, Neuropathological and Neurochemical aspects of generalized glycogen storage disease, *J. Neurol. Neurosurg. Psychiat.* **26**: 422–430 (1963).
25. S. Blumcke, H. Themann, and H. R. Niedorf, Disposition of glycogen during the degeneration and regeneration of the sciatic nerves of rabbits (light and electron microscope studies), *Acta Neuropathol.* **5**:69–81 (1965).
26. R. Schnabel, Über die Speicherung mucopolysaccharidartiger Substanzen im Gehirn bei generalisierter Glykogenose (Typ II), *Acta Neuropathol.* **4**:646–658 (1965).
27. M. M. Tiraspolskaya and M. N. Toropova, Glycogen content of C.N.S. tumors, *Federation Proc.* **23**:T343–344 (1964).
28. H. Themann, Zur elektronenmikroskopischen Darstellung von Glykogen mit Best's Carmin, *J. Ultrastruct. Res.* **4**:401–412 (1960).
29. J. P. Revel, Electron microscopy of glycogen, *J. Histochem. Cytochem.* **12**:104–114 (1964).
30. P. W. Lampert, J. L. Fox, and K. M. Earle, Cerebral edema after laser radiation. An electron microscopic study, *J. Neuropathol. Exptl. Neurol.* **25**:531–541 (1966).
31. A. Konishi, Occurrence of glycogen in developing cerebellar mossy fiber endings: An electron microscopy study. *Arch. Histol. Japan.* **27**:451–464 (1966).
32. A. Resibois-Gregoire and N. Dourov, Electron microscopic study of a case of cerebral glycogenosis, *Acta Neuropathol.* **6**:70–79 (1966).
33. O. H. Lowrey, J. V. Passonneau, F. X. Hasselberger, and D. W. Schulz, Effect of ischaemia on known substrates and cofactors of the glycolytic pathway in brain, *J. Biol. Chem.* **239**:18–30 (1964).
34. F. N. LeBaron, The resynthesis of glycogen by guinea-pig cerebral cortex slices, *Biochem. J.* **61**:80–85 (1955).
35. A. Kleinzeller and R. Rybova, Glycogen synthesis in brain cortex slices and some factors affecting it, *J. Neurochem.* **2**:45–57 (1957).
35a. M. E. Phillips and R. V. Coxon, unpublished observations.
36. M. I. Prokhorova, *in Biochemistry of Nervous System*, Coll. Papers, Acad. Sci. Ukrain. SSR Kiev (1954), cited by B. I. Khaikina and Ye. Ye. Goncharova, *in Problems of the Biochemistry of the Nervous System* (A. V. Palladin, ed.) pp. 87–95, Pergamon Press, Oxford (1964).
37. K. G. Prasannan, R. Rajan, and K. Subrahmanyam, Brain glycogen in the fed and fasting state, *Indian J. Med. Res.* **51**:703–707 (1963).

38. K. G. Prasannan, R. Rajan, and K. Subrahmanyam, Effect of glutamic acid on the brain and liver glycogen of normal and cortisone-treated rats, *Indian J. Med. Res.* **52**:208–212 (1964).
39. S. E. Kerr and M. Ghantus, The carbohydrate metabolism of brain. II. The effect of varying the carbohydrate and insulin supply on the glycogen, free sugar, and lactic acid in mammalian brain, *J. Biol. Chem.* **116**:9–20 (1936).
40. M. R. A. Chance and D. C. Yaxley, Central nervous function and changes in brain metabolite concentration, *J. Exptl. Biol.* **27**:311–323 (1950).
41. L. Jilek, S. Trojan, and E. Travnickova, Lactic acid and glycogen changes in the rat brain due to aerogenic (altitude) hypoxia during ontogenesis, *Physiol. Bohemoslov.* **15**:532–537 (1966).
42. D. E. Woolley and P. S. Timiras, Changes in brain glycogen concentration in rats during high altitude (12,470 ft) exposure, *Proc. Soc. Exptl. Biol. Med.* **114**:571–574 (1963).
43. L. Krulich, L. Jilek, and S. Trojan, The effect of oligaemia on the content of glycogen and lactic acid in the brain of the rat during ontogeny, *Physiol. Bohemoslov.* **11**:58–63 (1962).
44. T. Fujita, Studies on glycogen contents of cerebral cortex in hyperthermia, hypothermia and administration of several drugs, *Kobe J. Med. Sci.* **13**:39–45 (1967).
45. N. Shimizu and Y. Hamuro, Deposition of glycogen and changes in some enzymes in brain wounds, *Nature* **181**:781–782 (1958).
46. W. Lierse, K. Gritz, and H. Franke, Histochemical demonstration of glycogen and mucopolysaccharides in the guinea-pig brain following Roentgen irradiation, *Fortschr. Gebiete Roentgenstrahlen Nuklearmed.* **103**:612–618 (1965).
47. A. L. Shabadash, Cytochemical examination of the glycogen of the central nervous system of frogs after gamma irradiation, *Arkh. Anat. Gistol. i Embriol.* **44**:26–36 (1963).
48. R. H. Laatsch and W. M. Cowan, Electron microscopic studies of the dentate gyrus of the rat. II. Degeneration of commissural afferents, *J. Comp. Neurol.* **130**:241–262 (1967).
49. J. R. Klein and N. S. Olsen, Effect of convulsive activity upon the concentration of brain glucose, glycogen, lactate and phosphates, *J. Biol. Chem.* **167**:747–756 (1947).
50. H. G. Albaum, J. Tepperman, and O. Bodansky, The *in vitro* inactivation by cyanide of brain cytochrome oxidase and its effect on glycolysis and on the high energy phosphorus compounds in brain. *J. Biol. Chem.* **164**:45–51 (1946).
51. M. I. Prokhorova and Z. N. Tupikova, The rate of turnover of carbohydrates and lipids in the brain and liver during excitation and drug-induced sleep, *in Problems of the Biochemistry of the Nervous System* (A. V. Palladin, ed.), pp. 96–104, Pergamon Press, Oxford (1964).
52. C.-J. Estler and H. P. T. Ammon, The influence of propanolol on the metamphetamine-induced changes of cerebral function and metabolism, *J. Neurochem.* **14**:799–805 (1967).
53. C.-J. Estler and H. P. T. Ammon, The effect of ethanol and adrenaline on the phosphorylase activity and glycogen content of the brain, *J. Neurochem.* **12**:871–876 (1965).
54. F. Heim, C-J. Estler, H. P. T. Ammon, and U-D. Wiedmann, The effect of methanol on brain metabolism, *J. Neurochem.* **13**:1495–1500 (1966).
55. E. E. Goncharova and A. A. Musyalkovskaya, Effect of ipraside on the glycogen A content in the brain of animals, *Ukr. Biokhim. Zh.* **36**:829 (1964).
56. S. H. Carter and W. E. Stone, Effect of convulsants on brain glycogen in the mouse, *J. Neurochem.* **7**:16–19 (1961).

57. P. S. Timiras, D. M. Woodbury, and D. H. Baker, Effect of hydrocortisone acetate, desoxycorticosterone acetate, insulin, glucagon and dextrose, alone or in combination, on experimental convulsions and carbohydrate metabolism, *Arch. Intern. Pharmacodyn.* **105**:450–467 (1956).
58. J. Křivánek, Changes of brain glycogen in the spreading EEG-depression of LEAO, *J. Neurochem.* **2**:337–343 (1958).
59. W. M. Kirsch and J. W. Leitner, Glycolytic metabolites and co-factors in human cerebral cortex and white matter during complete ischaemia, *Brain Res.* **4**:358–368 (1967).

*Chapter 3*

# CYTOCHROMES AND OXIDATIVE PHOSPHORYLATION

**Cyril L. Moore and Paula M. Strasberg**

*Department of Neurology*
*Albert Einstein College of Medicine*
*Bronx, New York*

## I. INTRODUCTION

The electron transport systems of brain include mitochondrial and non-mitochondrial bound cytochromes. The nonmitochondrial (microsomal) systems are involved in the oxidation of NAD(P)H, with eventual transfer of electrons to some acceptor either natural or artificial.

The mitochondrial system involves structurally organized units of cytochromes, flavoprotein dehydrogenases, nonheme iron, and coenzyme $Q_{10}$. This composite of enzyme systems is involved in the oxidation of both $NAD^+$ and non-$NAD^+$ linked substrates, with the eventual reduction of oxygen. The energy derived from such oxidations can be utilized in the synthesis of ATP or utilized in several energy-linked functions. Before getting to the cytochromes and the mechanism(s) of oxidative phosphorylation, however, it is necessary to have relatively pure mitochondrial preparations, thereby minimizing the problem of contaminant artifact.

## II. MITOCHONDRIAL PREPARATIONS

The preparation of pure tightly coupled brain mitochondria has been the torment of neurobiochemists for several years. Recently, however, several methods have been used with varying degrees of success. Those which would be of interest to the reader are discussed here.

For those who are interested in the subcellular localization of enzymes, the methods developed in the laboratories of Whittaker[1,2] and De Robertis[3,4] are generally acceptable with the reservation that the hypotonic treatment of nerve-ending particles used to liberate entrapped mitochondria does result in mitochondrial swelling and possible release of intramitochondrial substances.

The above preparative methods are inadequate for those who are involved in the study of mitochondrial functions such as electron and ion transport, oxidative phosphorylation, and the partial reactions involved in high-energy bond synthesis. Alternative procedures must therefore be used.

Without discussing the older methods which are in essence duplications of the Schneider–Hogeboom procedure,[5] let us take a brief look at the best available methods for the students of electron transport and oxidative phosphorylation.

The method of Ozawa *et al.*[6,7] advocates an initial suspending medium composed of 0.3 M mannitol and 0.1 mM EDTA, *p*H 7.4, in the absence of any buffer. The decapitation of the rat and removal of the brain is rapid (15 sec) and the ensuing homogenization is classical, involving first the use of a loose-fitting pestle followed by a tight-fitting pestle in a Potter–Elvehjem homogenizer. Each homogenization requires approximately 30 sec. The mitochondrial fraction obtained by differential centrifugation is devoid of massive amounts of myelin, but is in no way characterized as pure. Nonetheless these mitochondria are tightly coupled, having respiratory control ratios* (RCRs) as high as 16.5 even in the absence of bovine serum albumin (BSA).

The method of Jobsis which originally appeared as an addendum to one of his publications includes the use of bacillus proteinase in the homogenizing medium.[9] This allows a maximal yield of mitochondria with low contamination and a high RCR. This procedure calls for mincing the tissue in a solution consisting of 225 mM mannitol, 50 mM sucrose, 20 mM tris–HCl, *p*H 7.6, and 0.5 mM EDTA. One milligram bacillus proteinase, 5 mg BSA, and 10 mg $KHCO_3$ are added per gram of tissue. After loose homogenization the suspension is kept at 0°C and *p*H 7.4 for 10 min. The mitochondria are collected by differential centrifugation. In some of Jobsis' unpublished work infinitely high RCRs, dependent upon the balance between EDTA and $Mg^{2+}$ ions, can be obtained.[9,10] This is represented in Fig. 1(a).

In the laboratory of Abood[11,12] mitochondria are separated from other subcellular particles on a ficoll density gradient. This does not result in the lack of morphological preservation as in the case of other procedures. The RCR values are in the area of 4–5, and we would presume that they could be even higher if the $Mg^{2+}$ ion concentration were properly controlled.

Our own method encompasses both the Jobsis' procedure[9] and the ficoll density gradient procedure of Tanaka and Abood.[11] Rat, rabbit,

* Respiratory control ratio (RCR) is the ratio of the rates of oxygen uptake by mitochondria oxidizing substrate in the presence (state 3) and the absence (state 4) of ADP.[8]

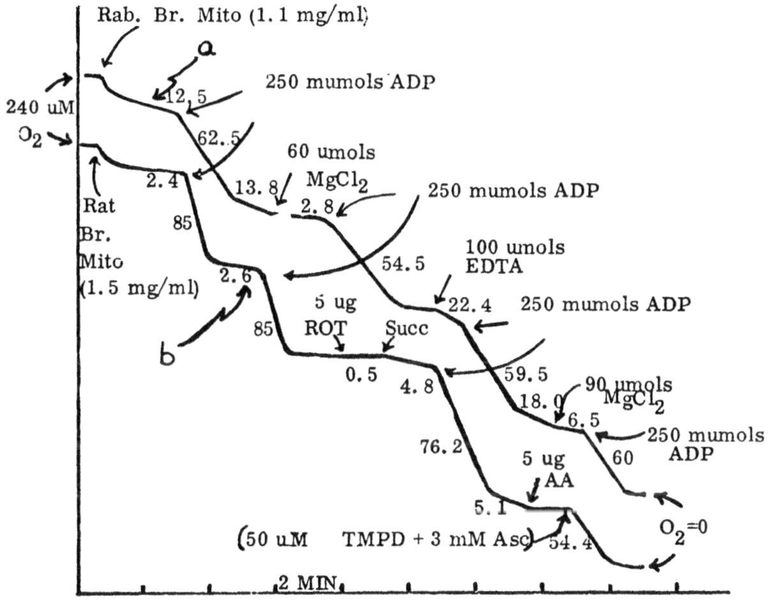


Fig. 1. Tracing (a): Rabbit brain mitochondria, 1.1 mg/ml, in the absence and presence of $MgCl_2$, showing the effect of $MgCl_2$ on state 4 respiration and thus on the RCR.
Tracing (b): Rat brain mitochondria, 1.5 mg/ml, prepared according to the method of Moore, showing the RCR with glutamate+malate (5 mM), succinate (5 mM), and TMPD+ascorbate.

or beef brain cortex is homogenized in a Potter–Elvehjem homogenizer (0.007-inch clearance) in a solution of 225 mM mannitol, 75 mM sucrose, 5 mM tris–HC1, *p*H 7.4, and 0.2 mM EDTA (MSET), and then 50 mg BSA, 10 mg bacillus proteinase, and 10 mg $KHCO_3$ are added per 10 g of cortical tissue. The loosely homogenized tissue is kept at 0°C for 10 min. Tight homogenization (0.002-inch clearance) is then applied, and this latter homogenate centrifuged at $3000 \times g$ for 30 sec. The supernatant solution is then centrifuged at $23{,}000 \times g$ for 3 min, and the crude pellet is shaken to remove the white fluffy layer. After washing once in MSET and once in mannitol–sucrose–tris–HCl (MST), the pellet is placed on a density gradient of 5–31% ficoll* in MST and centrifuged at $32{,}000 \times g$ for 45 min. The major fraction of the mitochondria does not sediment through 20% ficoll as is stated elsewhere.

The resultant fractions (Fig. 2) are washed once with 20 volumes of MST and the difference spectra of fraction 4 are shown in Figs. 3 and 4. The electron micrographs are shown in Figs. 5a, 5b, 5c, and 5d.

* The ficoll is dialyzed for 6 hr against two changes of 500 volumes of glass-distilled water and then overnight against 500 volumes of water, and dried before use.

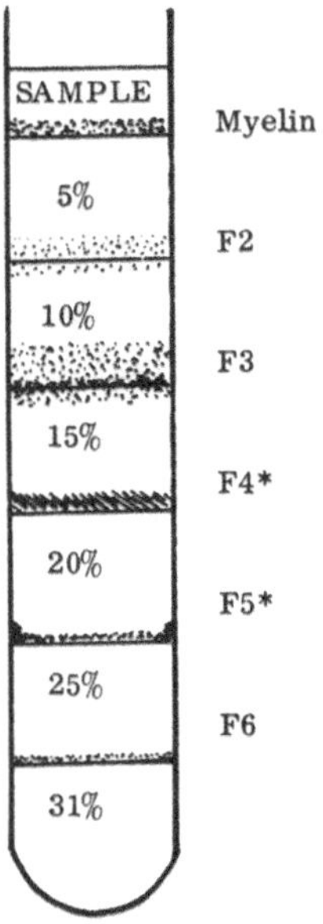


Fig. 2. Density gradient centrifugation of beef brain through ficoll-MST. The ficoll concentration is increased from 5 to 31 % in MST solution. The numbers are shown in the figure. Electron-micrographic pictures of fractions $F_2$–$F_5$ are shown in Figs. 5a–5d.

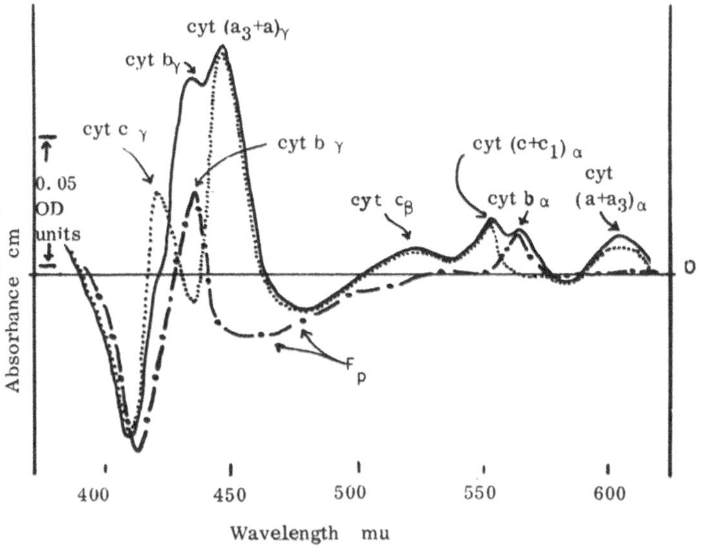


Fig. 3. Difference spectra of beef brain mitochondria $F_4$. (a)—·— Succinate and antimycin A reduced vs. oxidized. (b) —— Dithionite reduced vs. oxidized. (c) · · · · · Dithionite reduced vs. succinate and antimycin A reduced. Note: In some preparations the cvtechrome *b* peak in the soret region is higher than shown here.

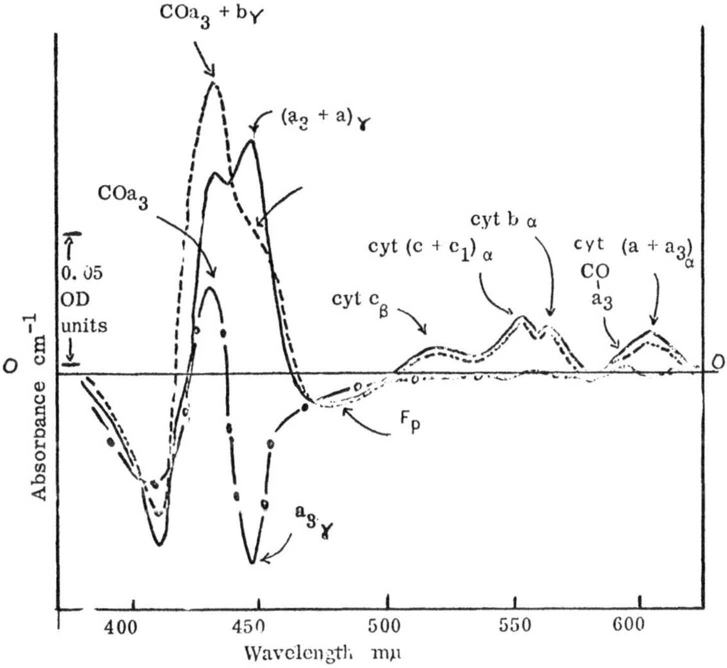


Fig. 4. Difference spectra of beef brain mitochondria $F_4$. (a) —— Dithionite reduced *versus* oxidized. (b) - - - - - Dithionite reduced + CO *versus* oxidized. Note the decrease of the cyt *a* peak and increase in the CO $a_3$ peak swamping the cytochrome *b* peak. (c) — · — Dithionite reduced + CO *versus* dithionite reduced. Note the peak of CO $a_3$ at 428.6 mμ and the valley for cytochrome $a_3$ at 447 mμ.

The RCRs of the ($F_4$) mitochondria in the presence of 60 μM $MgCl_2$ and 10 mM KCl range from 9 to < 30 [Fig. 1(b)], with glutamate and malate as substrate.

## III. ELECTRON TRANSPORT SYSTEMS OF BRAIN I

The cytochromes of brain include the mitochondrial and microsomal enzymes which are similar to those found in other cells. Cytochrome $P_{450}$ which is found in microsomes of the liver, kidney, adrenal cortex, and medulla and in mitochondria of the adrenal cortex, placenta, and testis, is absent from brain.(13,14) Carbon monoxide combines with the cytochrome to produce a characteristic reduced *versus* oxidized difference spectrum with a broad absorption maximum at 450 mμ as described by Klingenberg,(15) Garfinkel,(16) and Omura and Sato,(17) who also demonstrated the participation of $P_{450}$ in drug and steroid hydroxylation.(18)

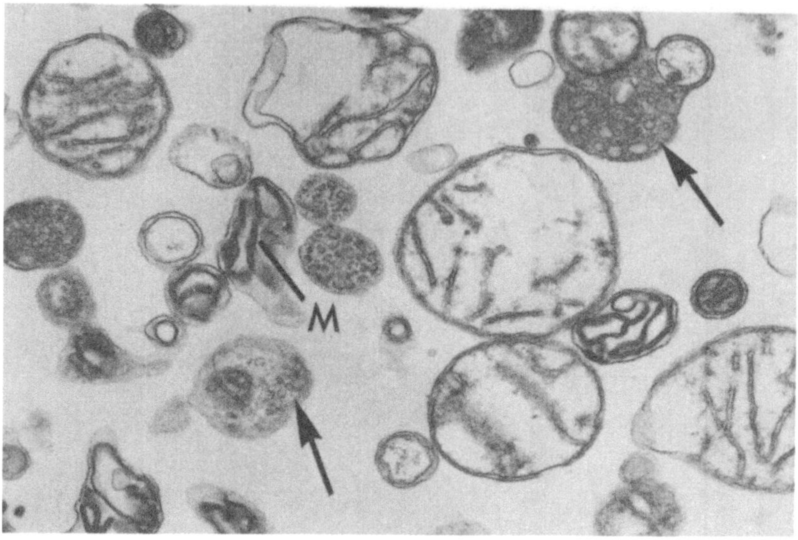


Fig. 5a. Figures 5a–5d are Electron Micrographs of Rat Brain Mitochondrial Fractions $F_2$–$F_5$, respectively. $F_2$ fraction showing nerve endings with synaptic vesicles (arrow) and contracted mitochondria (M) (Green's energized twisted). ×27,000.

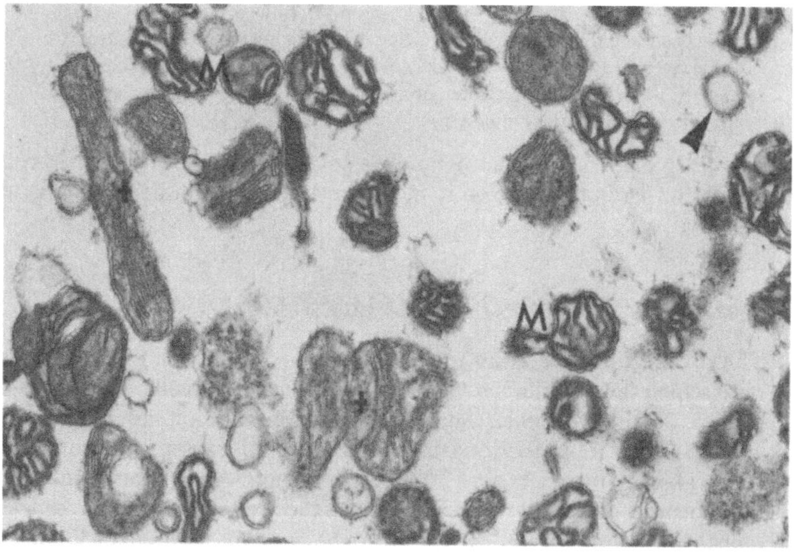


Fig. 5b. $F_3$ fraction showing mitochondria with contracted cristae (M) (Green's energized twisted), some normal mitochondria (+), and some smooth vesicles (arrow). ×21,500.

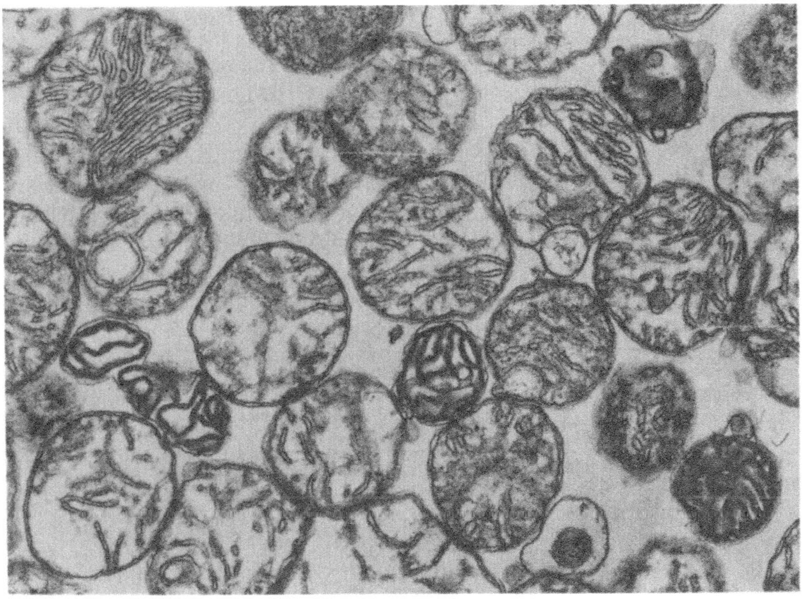

Fig. 5c. $F_4$ fraction containing normal mitochondria. ×23,500.

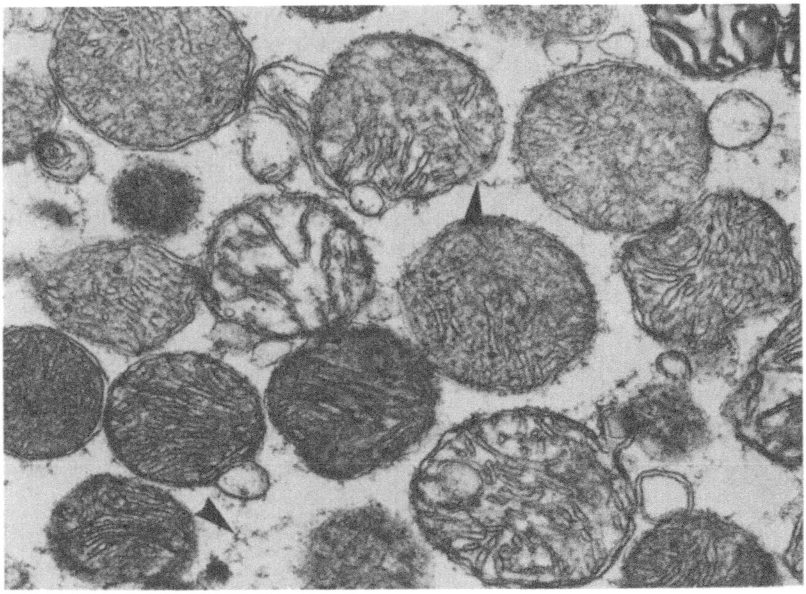

Fig. 5d. $F_5$ fraction clumped slightly on the gradient. Note: Intact mitochondria with filamentous "protein" material (arrow) are adherent to outer membrane and dispersed between mitochondria. × 27,900.

### A. Microsomal *b*

Inouye and co-workers have demonstrated the presence of cytochrome $b_5$ in brain microsomes of several species.[13,14] The spectral characteristics are identical to those of $b_5$ of liver microsomes,[13,14,19] with absorption maxima of 554, 525, and 424 m$\mu$. Like that of liver, brain microsomal $b_5$ is fully reducible by NADH and NADPH and is slowly autoxidizable.[13]

Giuditta and Strecker reported that brain microsomal preparations catalyze reduction of cytochrome *c* by NADH and NADPH.[20] Since Strittmatter found that reduced cytochrome $b_5$ can rapidly reduce cytochrome *c*,[21] it seems likely that the reductase activity of Giuditta and Strecker is mediated through cytochrome $b_5$, and this is discussed below.

The redox potential ($E'_0$) of microsomally bound cytochrome $b_5$ was estimated to be +7 mV at *p*H 7.0 and 20°C. The similar enzyme in liver has an $E'_0$ of −100 mV in the bound form and +20 mV in the soluble form. This discrepancy is quite significant and is blamed on the difficulty of the method used.[13]

The cytochrome $b_5$ content of bovine brain microsomes varies from 12 m$\mu$moles/g microsomal protein in the cortical layer to 4–5 m$\mu$moles in the cerebellum, brainstem, and spinal cord. Rabbit brain microsomes have 2–4 times (20–40 m$\mu$moles/g) as much cytochrome $b_5$,[13] and rabbit liver over 100 times (1 $\mu$mole/g) as much as bovine brain, with a disproportionately higher amount in the lighter microsomal fraction.[13]

In brain microsomes, several electron transfer pathways have been described: the NADH and NADPH cytochrome *c* reductase activities, operating both with and without the involvement of cytochrome $b_5$; the NADH oxidase activity; and the naphthoquinone-dependent NADH oxidase.

These pathways have been studied by Inouye and co-workers,[13,14] who have presented the scheme shown in Fig. 6.

The NADH and NADPH cytochrome *c* reductases and the naphthoquinone-requiring reductase, like those of liver,[22–24] are flavoproteins. The NADPH oxidase requiring naphthoquinone as cofactor has been

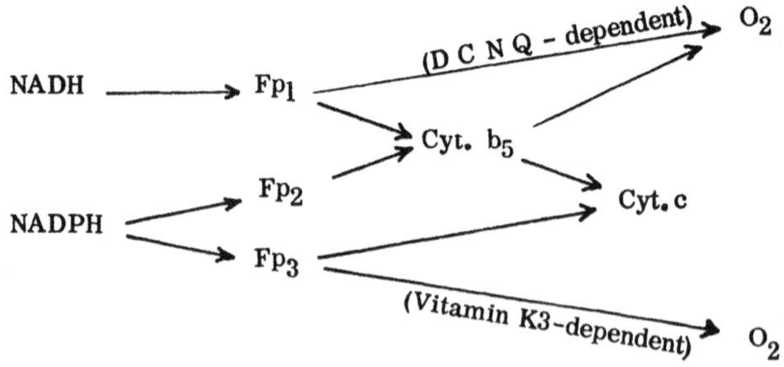


Fig. 6. Scheme of electron transfer pathways in brain microsomes.[14]

purified from liver and shown to be identical with microsomal NADPH cytochrome *c* reductase.[25] The flavin nucleotide concentrations of rabbit brain microsomes are for FAD 23 mμmoles/g protein and for FMN 14 mμmoles/g protein.

## B. Mitochondrial $b_5$

The presence of cytochrome $b_5$ in the outer membrane of liver and heart mitochondria has been reported by Parsons[26] and by Sottocasa *et al.*[27] This has been contested by Allman and Tan,[28] who claim that the activity observed in the outer membrane preparations of mitochondria is due to microsomal contamination of the preparations. In our laboratory, we have been able to prepare liver mitochondrial outer membrane fractions free of microsomal contamination without cytochrome $P_{450}$ or NADPH–cytochrome *c* reductase activity, but with kynurenine hydroxylase activity.[29] The cytochrome $b_5$ of mitochondria is involved in the oxidation of extramitochondrial NADH by cytochrome *c*. This pathway is insensitive to rotenone and antimycin A.

The oxidation of exogenous NADH by intact brain mitochondria by a rotenone- and antimycin A-insensitive pathway has been observed in fractions 4 and 5 of our purified brain preparations. The reaction requires cytochrome *c* and, by inference, is most likely mediated through cytochrome $b_5$. The inner membrane fraction of disrupted brain mitochondria contains NADH–cytochrome *c* reductase which is rotenone-, amytal-, and antimycin A-sensitive, and, if the analogy to liver is correct, this pathway does not involve cytochrome $b_5$ which is absent from inner membrane fractions of mitochondria.

No quantitative data are as yet available on brain mitochondrial cytochrome $b_5$.

## C. Diaphorases

Some other enzymes of the brain involved in the reoxidation of reduced $NAD^+$ are diaphorases I and II isolated in the laboratory of Strecker.[30–33] Adler *et al.*[34] introduced the name *diaphorase* for the enzymes originally described by Warburg, Straub, Hogness, and others. These flavoprotein enzymes catalyze the reoxidation of NADH or NADPH by methylene blue, ferricyanide, or other artificial electron acceptors.

These enzymes differ in their extractability from brain tissue. Diaphorase I is easily solubilized with 0.25 M sucrose solution or with water,[32] whereas diaphorase II is extractable with phosphate buffer from the residue obtained after repeated water extraction of the rat brain homogenates.[33]

These two diaphorases catalyze the oxidation of NADPH by benzoquinone, naphthoquinone, thiazine indophenols, ferricyanide, and methylene blue. Equations (1) and (2) exemplify the reaction with menadione as acceptor. Strecker and co-workers were able to identify peroxide production as shown in equation (2):[30]

$$NADH + H^+ + \text{Menadione (Ox.)} \rightleftharpoons NAD^+ + \text{Menadione (Red.)} \quad (1)$$

$$\text{Menadione (Red.)} + O_2 \rightleftharpoons \text{Menadione (Ox.)} + H_2O_2 \quad (2)$$

These enzymes show a high sensitivity to the uncouplers of oxidative phosphorylation (amytal, chlorpromazine, dicumarol, and DNP), as well as to inhibitors of mitochondrial electron transport including *o*-phenanthroline and α,α′-dipyridil, both of which react with nonheme iron.

## IV. ELECTRON TRANSPORT SYSTEMS OF BRAIN II

The components of the electron transport chain are similar in liver, heart, skeletal muscle, kidney, and brain mitochondria. They include the cytochromes, flavoproteins, nonheme iron, and coenzyme $Q_{10}$ (ubiquinone). The links between substrates and the chain involve $NAD^+$-linked dehydrogenases and flavin-linked dehydrogenases.

For the basic methods involved in the study of these components, the reader is referred to Chance[35] and Chance and Baltscheffsky[36,37]; fluorescence determinations of the nicotinamide nucleotides are described by Maitra and Estabrook.[38] For other methods see Estabrook and Pullman.[39]

Sacktor and Packer[40] have presented data for the relative concentrations of cytochromes, flavoproteins, and the nicotinamide dinucleotides in brain mitochondria. Their data are compared in Table I with our own

**TABLE I**

**Concentrations of Respiratory Enzymes in Brain Preparations**[a]

| Component | Wavelength (mμ) | $\Delta E_{(red-ox)}$ ($mM^{-1}cm^{-1}$) | Concentration (moles × $10^{-10}mg^{-1}$protein) | | |
|---|---|---|---|---|---|
| | | | $Br_1$ | $Br_2$ | Liver |
| Cyt *a* | 605–630 | 16.0 | 2.8 | 2.04 | 2.0 |
| Cyt *b* | 564–575 | 21.4 | 1.8 | 1.21 | 1.8 |
| Cyt $c+c_1$ | 550–540 | 19.6 | 3.2 | 2.87 | 3.4 |
| Fp | 465–500 | 11.0 | (3.4) | 1.97 | 7.2 |
| Cyt $a_3$ | 445–455 | (89)91.0 | 3.5 | 2.48 | 2.2 |
| PN | 340–375 | 6.0 | — | 8.93 | — |
| | 366–450 | (std.) | 12.5 | — | 38.0 |

[a] The concentrations of the respiratory enzymes determined from the OD changes of the reduced–oxidized component, measured at the pairs of wavelengths listed. $\Delta E_{(red-ox)}$ is the millimolar extinction coefficient for a 1-cm light path. $Br_1$, Moore (unpublished); $Br_2$, Sacktor and Packer[40]; Liver, Chance and Hess.[57]

results and those of liver preparations. Our data indicate higher values than those of Sacktor and Packer, which is to be expected since we have eliminated myelin and other nonmitochondrial contaminations.

## A. Cytochromes

Figure 3, tracing a, shows the oxidized *versus* dithionite reduced spectrum of our brain mitochondrial preparation. Tracing b shows the $\alpha$ and $\gamma$ peaks of cytochrome *b* at 564 and 432 m$\mu$, respectively, obtained by reduction with succinate in the presence of antimycin A. The $\beta$ peak at 530 m$\mu$ is not outstanding in spectra obtained at room temperature and is seen at temperatures of liquid nitrogen. Antimycin A inhibits the transfer of electrons from cytochrome *b* to *c*, causing reduction of cytochrome *b* by electrons derived from oxidation of either succinate or NADH-linked substrates. This fact can be used to show more clearly the soret peak of cytochrome *c*. Thus, by subtracting the peak of reduced cytochrome *b* from the fully reduced spectrum as in Fig. 3, the $\gamma$ peak of cytochrome *c* is seen clearly at 417 m$\mu$; the $\alpha$ and $\beta$ peaks of cytochrome *c* at 550 and 512 m$\mu$ can also be seen. Low-temperature difference spectra have revealed peaks for cytochrome *c* at 554 m$\mu$ and for $c_{\alpha 1}$ and $c_{\alpha 2}$ at 548 and 545 m$\mu$, respectively. Low-temperature difference spectra of cytochrome *b* reveal absorption maxima in the $\alpha$, $\beta$, and soret regions at 561, 530, and 431 m$\mu$, respectively.

Cytochrome oxidase has recently and almost exhaustively been discussed at the International Symposium on Oxidases and Related Oxidation Reduction Systems.[(41)] It was originally described by Ehrlich[(42)] and Rohman and Spitzer[(43)] and was first known as indophenol oxidase and later as Atmungsferment.[(44)] This oxidase has shrouded itself in a mask of secrecy, denying scientists the knowledge of its true identity.

To date, experiments and discussions of experiments are still going on in an attempt to determine if cytochrome oxidase is composed of two discrete chemical entities or whether one is dealing with a unique protein, a part of which has the capacity to react with substances such as $CN^-$ or CO, thereby altering part of its chemistry and, with it, part of its spectrophotometric characteristics.

The chemical data on cytochrome oxidase indicate that there is 1 heme per 72,000 mol. wt. and a copper–heme ratio of 1 : 1. The possibility of there being two individual heme proteins, *a* and $a_3$, each with its own heme prosthetic group in cytochrome oxidase has been proposed by Morrison.[(45)] He has been able to destroy 16–18% of the heme associated with the $a_3$ portion of the complex, and with it the CO binding capacity of the oxidase.

Jacobs has found that the reduced form of molybdenum octacyanide, having a redox potential ($E'_0$ = + 0.73 mV) close to that of the oxygen–water couple, is capable of selectively reducing "cytochrome $a_3$" of the oxidase complex.[(46)] The resultant difference spectrum shows a band between 585–595 m$\mu$ with approximately 25% of the absorbance of the totally reduced oxidase at 605 mμ. A trough at 630–650 mμ was reported but no data were presented on the absorption peaks in the soret region of the spectrum.

Here, we shall often refer to cytochrome oxidase as *cytochrome* $a + a_3$, as we attempt to describe its presence and characteristics in brain mitochondria.

Cytochrome oxidase is inhibited by azide, carbon monoxide, sodium sulfide, and cyanide. The action of azide has recently been studied by Wilson and Chance, and it was shown to act between the $a$ and $a_3$ components of cytochrome oxidase, as well as acting as an uncoupler of phosphorylation at site I.(47)

The reduced versus oxidized difference spectra seen in Figs. 3 and 4 show α and soret peaks of cytochrome $a + a_3$ appearing at 605 and 443 mμ respectively. In the presence of CO (Fig. 4), there is a small decrease of approximately 10% in absorbancy at the 605-mμ peak and a shift to a lower wavelength at about 590 mμ. The soret peak decreases substantially (about 40%) and a new peak appears at 428.6 mμ. The ratio of $a$ to $a_3$ activity is approximately 1 : 1.23.(40)

Administration of very low concentrations of cyanide *in vivo* results in convulsive activity, 80–90% reduction of cerebral arterio-venous difference in oxygen, and approximately 50% inhibition of brain cytochrome oxidase activity.(48) At the concentrations of cyanide used (5 mg/kg body weight), there is very little combination with hemoglobin. Nevertheless, the isolated cytochrome oxidase is inhibited.(49) The *in vitro* inhibition of cytochrome oxidase is relieved by methemoglobin, known to reduce the clinical effects of cyanide poisoning.

Potassium cyanide, when added to mitochrondria, also combines with either oxidized or reduced cytochrome oxidase.(50,51) It has been reported for rat liver mitochondria that in the presence of $CN^-$ the absorption bands of cytochrome oxidase in the reduced preparation shift from 443 to 439 mμ in the soret region and from 605 to 599 mμ in the α region of the spectrum. Our own data indicate that in brain mitochrondria the shifts are similar. There is also a broadening of the α peak.

Without adding to the controversy as to the individuality of the components of cytochrome oxidase, let it suffice to say that the oxidase from brain is now being purified in our laboratory for further study, since we have discovered a lethal genetic abnormality of the CNS (Kinky Hair Disease) in which there appears to be an 80–90% decrease in cytochrome oxidase, and the emergence of a new pigment in the soret region of cytochrome *b*.

## B. Flavoproteins

Mitochrondrial flavoproteins can be estimated either spectrophotometrically or fluorometrically.(52,53) The spectrophotometric method is subject to interference from nonheme iron(54) and from cytochromes.(52) The fluorometric methods(53) are more specific but give results compatible with the spectrophotometric procedures.

Recent studies of Chance *et al.* demonstrate the interaction of several flavoproteins in the electron transport chain, The location of these flavoproteins in the chain has been brought about by the use of stop flow and fluorescence measuring techniques.[55] The crossover point technique used by Ohnishi *et al.* demonstrated a crossover point between the flavoproteins $FPD_1$ and $FPD_2$ in *Streptomyces cerevisiae*.[56] These data are in agreement with the scheme presented in Fig. 7.

While these sophisticated techniques have not as yet been applied to brain mitochondrial preparations, the concentration of flavoproteins as calculated spectrophotometrically is found to be in the range of 0.25 to 0.30 μmole/g mitochondrial protein, compared to 0.72 μmole/g liver mitochondrial protein.[57]

The flavoprotein dehydrogenases shown in Fig. 7 include lipoamide dehydrogenase (FPL), NADH dehydrogenase ($FPD_1$), succinic dehydrogenase ($FP_S$), and another protein $FDP_2$ that is reducible by succinate via $FP_S$ and an *n*-bromosuccinimide-sensitive step. $FPD_2$ is also reducible by complex 1 via a rotenone- and amytal-sensitive pathway.

Other flavoproteins include α-glycerophosphate dehydrogenase and fatty acyl-CoA dehydrogenase. All of these flavoproteins are present in brain mitochondria.

## C. Coenzyme $Q_{10}$ (Ubiquinone)

According to the available evidence, the possible locations of $CoQ_{10}$ and nonheme iron in the electron transport chain of mitochondria are as shown in Fig. 7. The studies in Green's laboratory have shown that mitochondrial CoQ is reduced to the quinol in anaerobic mitochondria, and the quinol reoxidized under aerobic conditions.[58]

Szarkowska[59] has shown that pentane extraction of lyophilized beef heart mitochondria yields a preparation that is unable to oxidize NADH

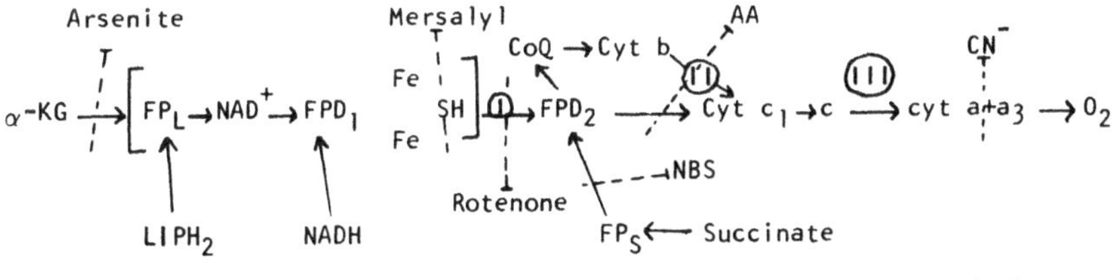


Fig. 7. A representation of the components of the electron transport chain brought up to date by the studies of Chance and co-workers.[55] $FP_L$, lipoamide dehydrogenase; $FPD_1$, NADH dehydrogenase; $FPD_2$, flavoprotein dehydrogenase between succinate dehydrogenase ($FP_S$) and CoQ; NBS *n*-bromosuccinimide which inhibits between $FP_S$ and $FPD_2$. Encircled numerals I, II, and III represent the three probable sites of phosphorylation. Rotenone and amytal inhibit at the same segment, while mersalyl, a mercurial, inhibits the SH function as shown in the first pair of brackets enclosing complex I. The dashed lines indicate sites of inhibition.

or succinate unless $CoQ_{10}$ and cytochrome *c* are added. The restored activity is sensitive to rotenone and antimycin A.

Data of Redfearn and Burgos[60] have indicated that certain preparations of beef heart particles devoid of $CoQ_{10}$ are able to oxidize succinate if cytochrome *c* is added. The rate, however, is very low, and the possible contamination by low concentrations of CoQ could not be excluded. Kroger and Klingenberg[61] have obtained data indicative of redox changes of CoQ in mitochondria during state 4-to-state 3 transitions and during ATP-dependent reversed electron flow. There is still some doubt as to whether the time constants of the oxidation–reduction cycles of CoQ as reported are small enough to justify its presence on a direct path between substrates and oxygen. Thus, Chance would rather believe that CoQ could act as an electron sink which could be depleted during state 3 respiration,[56] but which is not obligatory to the transport system.

The presence of $CoQ_{10}$ in brain and its depletion caused by vitamin E deficiency has been established by Edwin and co-workers.[62,63] The studies of di Prisco *et al.*,[64] showing that the NADH oxidase system of treated brain mitochondrial preparations devoid of $CoQ_{10}$ is capable of using $CoQ_1$ and $CoQ_2$ as electron acceptors, could be indicative of $CoQ_{10}$ being a natural electron acceptor in brain mitochondria as it is in mitochondria from other sources. Since no detailed studies have been carried out on the exact position of CoQ in the electron transport system of brain mitochondria its tentative position can be no better localized than in other sytems.

## D. Nonheme Iron

Only a very small percentage of brain iron can be accounted for in the iron porphyrin cytochromes of the mitochondria. Nonheme iron occurs as part of several complexes in mammalian mitochondria: (1) α-glycerophosphate dehydrogenase, (2) complex 1, and (3) succinic dehydrogenase, which has four atoms of nonheme iron for each FAD per $2 \times 10^5$ g molecular weight.

The inhibition of the brain mitochondrial electron transport from $NAD^+$-linked substrates by iron chelators, such as α,α′-dipyridyl and *o*-phenanthroline (0.5 mM), definitely implicates nonheme iron in the electron transport chain. The spectrophotometric data obtained in the presence of these iron chelators indicate an increase in the steady-state oxidation levels of cytochromes *b*, *c*, and *a*. This would position nonheme iron on the substrate side of the cytochromes. The studies of Sanadi *et al.* suggest that the nonheme iron of complex 1 (*cf.* Fig. 7) is on the oxygen side of the flavoproteins.[65] Nonheme iron has a characteristic ESR signal at $g = 1.94$ which can be induced by substrate oxidation. The studies of Beinert and Palmer indicate the kinetics of oxidation of nonheme iron are of the same order of magnitude as are the kinetics of the cytochromes.[66]

Theonyl trifluoroacetone (TTA) has been found to inhibit succinate oxidation by chelating nonheme iron.[67] The action of TTA has not been tested in brain.

## E. Substrate Oxidation

Several biological and nonbiological compounds have been shown to be substrates for the complex of enzymes of the electron transport system of brain mitochondria. Oxidation of these substrates is linked to the cytochrome system through $NAD^+$ or flavoproteins or by direct reduction of one of the components of the electron transport chain. *In vitro*, oxidation of several components of the citric acid cycle which are capable of penetrating mitochondria is demonstrable by measuring oxygen consumption either manometrically or polarographically using the platinum electrode.[68]

$NAD^+$-linked substrates shown to work *in vitro* include malate, glutamate, α-ketoglutarate, β-hydroxybutyrate, and pyruvate. Succinate, α-glycerophosphate (not of the citric acid cycle), and CoA derivatives of fatty acids (fatty acyl CoA) are oxidized by various flavoproteins.

The scheme of oxidation shown in Fig. 8 is a simpler representation of that given by Chance and co-workers (Fig. 7), but is more appropriate here.

With γ-aminobutyric acid as substrate in the presence of α-ketoglutarate, brain mitochondria respire at a rate of approximately 60 mμmoles $O_2$/min/mg protein. Ascorbate, which is capable of nonenzymatic reduction of cytochrome *c*, results in a very high rate of respiration. The data on the oxidation of various substrates by mitochondria are compiled in Table II. The reason for high rates of respiration obtainable with some substrate pairs, for example, glutamate plus malate, is not always evident since with the single substrate (malate) in the short-term experiments product inhibition may not be substantial. In other instances the inhibition of malate dehydrogenase by oxaloacetate is substantial, and, in this circumstance, transamination of the oxaloacetate with glutamate or condensation with acetyl CoA formed from pyruvate relieves the inhibition and results in a higher rate of respiration.

Saktor *et al.* have demonstrated the stimulation of α-glycerophosphate oxidation by $Mg^{2+}$ ions,[69] an observation confirmed by Voss *et al.*[70,71]; however, it should be remembered that $Mg^{2+}$ ions can cause an increase in state 4 respiration with all substrates.[10]

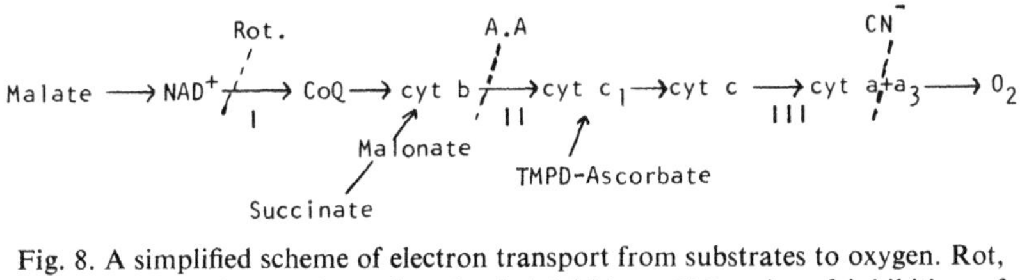


Fig. 8. A simplified scheme of electron transport from substrates to oxygen. Rot, rotenone inhibition; A.A, antimycin A inhibition; $CN^-$, site of inhibition of cyanide as well as azide and carbon monoxide. The numbers I, II, and III refer to the probable sites of phosphorylation.

TABLE II

**Rates of Substrate Oxidation, Respiratory Control, and ADP : O Ratios of Brain Mitochondrial Preparations**[a]

| Substrate | *CM* $-MgCl_2$ Ox | *CM* $+MgCl_2$ Ox | *CM* $+MgCl_2$ RCR | *CM* ADP:O | *SP* Ox | *O* Ox | *O* RCR | *O* ADP:O |
|---|---|---|---|---|---|---|---|---|
| Succ. | 12.6 | 4.8 | 15.0 | 1.6 | 7.2 | 21.5 | 3.8 | 1.97 |
| α-GP | 7.8 | — | (3.2) | 1.3 | 6.0 | | 1.4[SP] | — |
| α-GP+succ. | 19.8 | 5.9 | 13.2 | 1.8 | 14.2 | — | — | — |
| Malate | 5.4 | — | — | 2.2 | — | — | — | 2.10[T] |
| Glutam. | 7.6 | 2.1 | 38.1 | 2.6 | 5.0 | 5.0 | 11.2 | 2.62 |
| Glutam. + Malate | 8.8 | 2.4 | 31.2 | 2.8 | — | 5.0 | 10.2 | 2.99 |
| Pyr. | 3.0 | — | — | 1.6 | — | — | — | 1.40[T] |
| Pyr. + Mal. | 9.8 | — | — | 2.7 | — | 8.2 | 7.3 | 3.00 |
| Glutamine | — | — | — | 1.7 | 3.6 | 6.2 | 2.7 | 1.94 |
| Gline + Mal. | — | — | — | 2.4 | — | 6.8 | 3.1 | 2.48 |
| α-KG | 9.0 | 3.4 | 20.3 | 2.7 | 4.8 | 8.4 | 7.4 | 2.70 |
| α-KG+Mal. | 11.8 | — | — | 2.5 | — | 10.2 | 5.8 | 2.48 |
| Asc.+TMPD | 33.2 | — | — | 0.7 | — | — | — | — |
| GABA | — | — | — | — | 2.5 | | 2.2[SP] | 3.0[SP] |

[a] Rate of substrate oxidation by brain mitochondria. Data compiled from: *CM*, Moore unpublished; *SP*, Sacktor and Packer[40]; *O*, Ozawa *et al.*[6]; *T*, Tanaka and Abood.[11] Respiratory control (RCR) and ADP : O ratios are presented.

Evidence for the sites of entrance of electrons into the cytochrome system during substrate oxidation and the relationship of the component cytochromes to one another is based on the redox potentials of various couples as seen in Table III and the kinetic and spectrophotometric data obtained with specific inhibitors.[39] The inhibition of electron transfer between the $NAD^+$-linked substrates and cytochrome *b* is brought about by rotenone, piericidin A, amytal, TTA, and *o*-phenanthroline on the oxygen side of $NAD^+$ and [in most instances[72,73]] on the substrate side of cytochrome *b*, resulting in an increase in the steady-state level of NADH and an oxidation of the cytochromes.[74] During glutamate oxidation, malonate inhibition results in a decreased level of NADH and an oxidation of the cytochromes. This inhibition by malonate is most likely a reflection of the inhibition of the oxidation of succinate formed from glutamate, with a resultant decrease in α-ketoglutarate available for oxidation. This is in agreement with the finding of Tower that glutamate and GABA accumulate in cerebral cortical slices in the presence of malonate.[75]

The oxidation of succinate is inhibited by malonate and antimycin A, but not by rotenone. With antimycin A present (0.1–0.3 μg/mg protein)

TABLE III

**Thermodynamic Relationships**[a]

| | $E_0'$(mV) | $\Delta E_0'$(mV) | $\Delta G$(kcal) | Phosphorylation sites—thermodynamics and crossover points |
|---|---|---|---|---|
| $NAD^+$ | −320 | | | |
| | | 270 | 12.20 | + |
| Flavoprotein | − 50 | | | |
| | | 10 | 4.05 | 0 |
| Cyt *b* | + 40 | | | |
| | | 220 | 9.90 | + |
| Cyt *c* | +260 | | | |
| | | 30 | 1.35 | 0 |
| Cyt *a* | +290 | | | |
| | | 530 | 23.50 | + |
| Oxygen | +820 | | | |

[a] Oxidation–reduction potentials $E'_0$ of the components of the electron transport chain between $NAD^+$ and oxygen. The energy differences between the couples are calculated from the $E'_0$ values. $\Delta G$ for ADP + Pi $\rightleftharpoons$ ATP + $H_2O$ = + 9.0 kcal, and so on this basis and on the basis of the crossover point data of Chance and Williams[8] the + signs indicate probable regions of phosphorylation.

cytochrome *b* becomes reduced, while cytochromes *c*, *a*, and $a_3$ become more oxidized. With $CN^-$, cytochromes $a + a_3$, *c* and *b* all become reduced. This indicates that succinate oxidation is linked to the electron transport chain at the level of cytochrome *b* and that antimycin A prevents the transfer of electrons from cytochrome *b* to *c*. This does not exclude the possibility of a succinate-linked reduction of $NAD^+$ with subsequent oxidation of the formed NADH by cytochrome *b*. This indeed does occur by an energy-dependent mechanism and is discussed under reversed electron transfer. This, however, is not the normal pathway, since rotenone which inhibits the oxidation of NADH does not inhibit the succinate reduction of cytochrome *b*. As would be expected, antimycin A also prevents the oxidation of NADH.

The inhibition of respiration by antimycin A is overcome by tetramethylphenylenediamine (TMPD) as first shown by Jacobs.[76] It was later shown to act with ascorbate in reducing cytochrome *c*, or alone in a bypass of the antimycin A site, resulting in oxidation of the reduced cytochrome *b*. This, however, results in a bypass of the second phosphorylation "site," and only one ATP/1/2 $O_2$ reduced is observed.

This pathway also exists in brain mitochondria as seen from Fig. 3(b). The nonenzymatic reduction of cytochrome *c* by TMPD + ascorbate occurs in the presence of rotenone, antimycin A, or $CN^-$, but its reoxidation with a P : O ratio of 1 is $CN^-$- and azide-sensitive.

### F. Reversed Electron Flow

The reduction $NAD^+$ by substrates not directly linked to this nucleotide has been shown to be energy-dependent,[74,77-79] which is to be predicted from the thermodynamic data ($E'_0$ values) in Table III. This phenomenon has been demonstrated in brain mitochondria,[40,80] where it was found that the steady-state level of NADH could be contributed to by succinate to the extent of 88%. This reduction is sensitive to amytal and to uncouplers. Glutamate can contribute over 25% of the steady-state level of NADH. The data are presented in Table IV. Ascorbate can also supply electrons for the reduction of $NAD^+$ in a similar energy-dependent mechanism.

In the brain system and in several of our liver preparations, it has not been necessary to add exogenous ATP to observe this reduction as originally observed by Chance,[74] indicating the presence and availability of the necessary energy for reversal of electron transport.

**TABLE IV**

**Steady-State Levels of $NAD^+$ Reduction by Non-$NAD^+$-Linked Substrates**[a]

| Additions | Steady-state levels of $\frac{NADH}{NAD^+ + NADH}$ |
|---|---|
| No addition | 0.05–0.10 |
| Glutamate | 0.20–0.32 |
| Succinate | 0.75–0.88 |
| Amytal | 0.05–0.12 |
| Amytal + Succinate | 0.05–0.08 |
| Oligomycin | 0.05–0.10 |
| Oligomycin + Succinate | 0.08–0.12 |
| DNP | 0 |
| DNP + Succinate | 0 |
| Amytal + ATP + Succinate | 0.05–0.08 |
| Oligomycin + ATP + Succinate | 0.15–0.22 |

[a] The reduction of $NAD^+$ by succinate in the absence and presence of inhibitors amytal and oligomycin, and the uncoupling agent DNP. Data from Sacktor and Packer[40] and our own observations made by fluorescence measurements at 366-mμ incident light and 450-mμ fluorescent emission.[38]

## V. OXIDATIVE PHOSPHORYLATION

### A. Theories

The studies of Kalckar[81] introduced the true significance of the fact that aerobic phosphorylation is coupled to respiration and is independent of glycolytic (or substrate level) phosphorylation. Belister and Tsibakowa[82] presented stoichiometric data showing the ratio of esterified phosphate to oxygen utilized (P : O) as being at least 2.0. The rapid rate of investigation that followed in the laboratories of Cori, Ochoa, Hunter, Lardy, Green, Lipmann, Lehninger, Keilin, Hartree, and others resulted in a basic but cloudy understanding of the problem of oxidative phosphorylation, which was recognized as occurring in the mitochondria.

The basic characteristics of oxidative phosphorylation can be stated for mitochondria in general: oxidation of $NAD^+$-linked substrates usually results in a P : O ratio of approximately 3.0; oxidation of succinate or α-glycerophosphate (flavoprotein-linked substrates) results in P : O ratios of approximately 2.0; and ascorbate oxidation via a nonenzymatic reduction of cytochrome *c* results in a P : O ratio of approximately 1.0.

The crossover point methods of Chance and Williams[83] were used to identify the probable sites of phosphorylation. By studying the redox state of the components of the electron transport chain, prior to and after the addition of limiting amounts of ADP to mitochondria in the presence of excess substrate, three crossover points where a deficiency of ADP caused the carrier at the substrate side of the crossover point to become more oxidized were identified. One crossover point was found to occur between $NAD^+$ and flavoprotein (site I), the second between cytochrome *b* and cytochrome *c* (site II), and the third between cytochrome *c* and oxygen (site III). Thermodynamic data ($\Delta G$ values) are in agreement with the postulation of three phosphorylation sites.

With the aid of specific inhibitors, the sites of energy conservation have been well established.

What has remained a mystery is not the thermodynamics of phosphorylation but the basic biophysical chemistry of the process. There are essentially three schools of thought, or better, three camps of concentration:

First, there are those proposing the involvement of common chemical intermediates in the process of oxidative phosphorylation. Slater[84] has pointed to the following two possible mechanisms.

(1) According to the first, some substance (C, X, or 1) combines with the electron carrier during the coupled oxidation–reduction, resulting in formation of a high-energy intermediate. This intermediate carrier ~ C, (~ X or ~ 1), can react with Pi and ADP to form ATP as shown in equations (3) and (4) as follows:

$$AH_2 + B + C \rightleftharpoons A \sim C + BH_2 \qquad (3)$$

$$\frac{A \sim C + Pi + ADP \rightleftharpoons A + C + ATP}{\text{Net } AH_2 + B + Pi + ADP \rightleftharpoons A + BH_2 + ATP} \qquad (4)$$

Lehninger and Wadkins elaborated on this mechanism for $NAD^+$-linked phosphorylation,[85] stating essentially that Pi does not react directly with the electron carriers.

(2) In Slater's second case, Pi may combine directly with the electron carrier either before or during the oxidation–reduction reactions as shown in equations (3a) and (4a) as follows:

$$AH_2 + B + Pi \rightleftharpoons A \sim P + BH_2 \qquad (3a)$$

$$A \sim P + ADP \rightleftharpoons ATP \qquad (4a)$$

$$\text{Net} \quad AH_2 + B + Pi + ADP \rightleftharpoons A + BH_2 + ATP$$

The net results of the two mechanisms are identical.

Taking as an example of phosphorylation a known substrate level phosphorylation, e.g., the glyceraldehyde-3-phosphate dehydrogenase reaction, it is found that a high-energy thioester serves as the common intermediate. This is used to exemplify the possibility of the Slater-type mechanism involving a high-energy intermediate. While this is indeed probable, there is still no clear-cut evidence in the form of an isolated phosphorylated or other high-energy compound.

When oxidation is uncoupled from phosphorylation by DNP, Pi is not required for electron transport (which now proceeds at a faster rate). The implication is that there is increased interaction of the electron transport carrier in the absence of Pi.

In spite of the fact that there is no satisfactory proof for either of the two mechanisms proposed by Slater, there is a division of thought on the more acceptable one. The disagreement relates to whether the energized form of the electron transport carrier is in the reduced or oxidized form, and this debate continues with dwindling enthusiasm.[85a]

Second, Penniston and co-workers have postulated energized complexes of the respiratory carriers which undergo conformational changes (as seen in electron micrographs) as being involved as intermediaries in oxidative phosphorylation.[86a,b]

These states are nonenergized, energized, and energized twisted, and they describe three conditions of the mitochondrial system under observation. According to this hypothesis,[87] no common high-energy intermediates have to be postulated, but instead the transfer of electrons results in energized membranes whose energy can be transduced into the work of ion transport, energy-dependent transhydrogenase reaction, or ATP synthesis. While these studies have been carried out with beef heart mitochondria, our electron micrographs of brain mitochondria shown in Figs. 5a–5d also contain examples of these three states,. The universality of these observations and their pertinence as to cause or effect of oxidative phosphorylation have been questioned.

Third, the chemiosmotic coupling theory of oxidative phosphorylation proposed by Mitchell and co-workers[88–91] states that the mitochondrial membrane is anisotropic and that there is a *p*H gradient developed across this membrane during electron transport. This gradient (a result of the separation of $H^+$ and $OH^-$ charges) across the mitochondrial membrane is used as an energy supply in the formation of ATP from ADP and Pi by the reversal of the anisotropic ATPase in the membrane.

Some of the experiments used by Mitchell to promote his theory work satisfactorily in that they do not refute his thesis. He has shown that when mitochondria are suspended in an alkaline environment and the *p*H is dramatically decreased, there is a rapid synthesis of ATP within the first few seconds. It has also been shown by Mitchell that DNP, arsenate, and dicumarol, when added to intact mitochondria, stimulate $H^+$ ion uptake by the mitochondria as his theory demands. Jagendorf and Uribe have also demonstrated a *p*H-gradient-dependent synthesis of ATP by chloroplasts[(92)] in support of the theory.

Mitchell and Moyle have also demonstrated an acid jump during a transition from anaerobiosis to aerobiosis in liver mitochondria and suggest that this results from resumption of electron transport in the presence of $O_2$.[(90)] It has been shown by Chance and Mahler,[(93)] by Lehninger and co-workers[(94)] using liver mitochondria, and by Moore[(95)] using brain mitochondria that this acid shift is due greatly to the translocation of ions. There is still, however, the possibility that the small residual *p*H change unaccounted for by any measurable ion translocation could be the change pertinent to Mitchell's theory.[(91)]

The data supplied by the laboratories of Mueller *et al.*[(96)] of Finkelstein,[(97)] and of others using lipid bilayers indicate that antibiotics are capable of altering the resistance of such membranes by increasing the permeability to cations. The work of Hopfer *et al.* on the effect of the uncoupling agents DNP and carbonylcyanide *p*-trifluoromethoxyphenylhydrazone (CFCCP) on similar membrane systems[(98)] indicates that uncouplers of oxidative phosphorylation cause an increase in the conductance of the lipid bilayer as a result of the increased proton transport across the membrane. The maximum change in conductance is *p*H-dependent. The *p*H optima are 4.0 and 7.2 for DNP and CFCCP, respectively. The data of Hemker[(99)] and of Myers and Slater[(100)] using rat liver mitochondria indicate that the concentration of DNP required for maximal stimulation of ATPase decreases as the *p*H is decreased from 8.0 to 5.0, and the resultant ATPase effect is also greatly enhanced. The magnitude of the conductance changes is of the order of $6.7 \times 10^{-6}$ mho/cm$^2$ for the model membrane and compares favorably with $7.2 \times 10^{-6}$ mho/cm$^2$ for mitochondrial membrane as calculated by Mitchell and Moyle.[(101)] The direction of the proton movement is indeed compatible with Mitchell's hypothesis. This stresses as one of its supporting ideas that uncouplers work through their lipid solubility and their ability to act as proton conductors, thereby destroying the ion gradient across the mitochondrial membrane and with it the driving force for synthesis of ATP. There is no inhibition of electron transport, as is also the case with DNP, arsenate, dicumarol, and CFCCP.

With these "misunderstandings" hovering over us, we can now examine oxidative phosphorylation in brain mitochondrial preparations.

## B. Coupled Phosphorylation

The methods of determining P : O ratios of mitochondria have progressed from the use of the Warburg manometers and the glucose hexokinase trap to polarographic determinations of the amount of oxygen used during the period of phosphorylation of a known limiting amount of ADP. The resultant ratio of phosphorylation to oxygen used is referred to as the ADP : O ratio. This procedure is adequately described elsewhere.[(102)] Mitochondria in a completely controlled state should not in the absence of phosphorylation utilize any oxygen, i.e., electron transport from substrate to oxygen should be zero. This, or course, is the ideal case in which (high-energy) intermediates of oxidative phosphorylation are not being utilized (transduced) in energy-linked functions of mitochondria such as ion translocation or fatty acid activation. In this ideal case, the RCR would be infinite. Thus, in more practical terms, the RCR tells the degree to which electron transport is coupled to phosphorylation.

With brain mitochondrial preparations it has been observed as shown in Fig. 1 that in the controlled state (state 4) the rate of respiration and thus the RCR can be a function of the $Mg^{2+}$ ion concentration.

The ADP : O ratios of brain mitochondria oxidizing different substrates are indicative of one, two, or three phosphorylations per atom of oxygen reduced. Values gleaned from the literature are listed in Table IV. While several substrates enter the electron transport chain at similar loci, the ADP : O ratios and the RCR for such substrates may be quite different. A pertinent example is that of the oxidation of α-glycerophosphate and succinate.[(40)] Although these two substrates are oxidized at approximately equal rates by brain mitochondria, ADP phosphorylation with succinate results in a higher RCR than with α-glycerophosphate. A similar case exists for glutamate and glutamine.[(40)] Low respiratory control during α-glycerophosphate oxidation has also been reported for mitochondria from insect flight muscle.[(52,103–105)]

Figure 7 is descriptive of possible mechanisms of oxidative phosphorylation, including the participation of the partial reactions described above and the possible loci of action of the various inhibitors. Sacktor and Packer[(40)] as well as Ozawa *et al.*[(7)] have shown that brain mitochondria like those of kidney mitochondria are capable of oxidizing glutamine at a rate similar to glutamate but with a lower ADP : O ratio.

The low P : O ratio with malate alone as substrate is inexplicable, while its inability to function as a lone substrate in many preparations can be explained on the basis of the inhibition of MDH by oxaloacetate.

Phosphorylative activity with GABA as substrate depends on transfer of its $-NH_2$ group to α-ketoglutarate during conversion to succinic semialdehyde which is oxidized by $NAD^+$ to succinate. The respiratory rate with the two substrates is more than additive. These findings by

Bacila *et al.*[70,71] are contrary to those of Sacktor *et al.*[106] The latter investigators have stated that GABA oxidation by mitochondria can proceed in the absence of α-ketoglutarate but goes via a direct oxidative pathway, and they have provided evidence to this effect. The facts that the binding of GABA requires $Na^+$[107,108] and that the respiratory medium used by Bacila *et al.* was void of $Na^+$ might be the answer to this discrepancy.

### 1. *Coupling Factors*

Studies carried on in the laboratories of Racker,[109–112] of Sanadi,[113] and of others have resulted in the preparation from mitochondria of certain factors capable not only of carrying out some of the partial reactions of oxidative phosphorylation, but also of restoring phosphorylative capability to certain submitochondrial preparations. The isolation of such factors has not been achieved with brain mitochondria, and so the interested reader is referred to a review by Pullman and Schatz.[113a]

It should be mentioned here that Strecker and co-workers isolated a factor from brain mitochondria which is capable of stimulating the respiratory rate of mitochondria from several sources.[114] Exactly what the nature of this factor is has not been determined.

## C. Some Partial Reactions of Oxidative Phosphorylation

### 1. *ATPases*

Working on the effect of dinitrophenol (DNP) on the phosphorylative activity of mitochondria, Loomis and Lipmann were the first to demonstrate the effect of 2,4-dinitrophenol on mitochondria.[115] This compound, when added to intact liver mitochondria, was found to inhibit phosphorylation with a resultant increase in the rate of respiration. This compound, however, inhibited some of the other partial reactions of oxidative phosphorylation [ATP–Pi exchange, the oxygen ($^{18}O$) exchange, ATP–ADP exchange].

Tanaka and Abood[116] have presented data showing an inhibition of brain mitochondrial respiration with $10^{-6}$ M DNP or higher. There was, however, only a slight stimulation of ATPase activity as determined by the release of Pi from ATP. Wechsler[117] found a 20% stimulation of respiration by $2.5 \times 10^{-4}$ M DNP. On the other hand, Ozawa *et al.*[6] have observed an 11-fold stimulation of ATPase activity at $5 \times 10^{-5}$ M DNP. The difference between the systems of Ozawa *et al.* and Tanaka and Abood was the presence of millimolar amounts of $Mg^{2+}$ ion in the system used by Tanaka and Abood. As was shown by Ozawa *et al.*, there is a $Mg^{2+}$ ion stimulated ATPase being maximal at 2 mM $MgCl_2$. Tuena *et al.* have reported a 30–75% increase in DNP-stimulated ATPase activity of brain mitochondria.[118] Here the concentration of DNP was $10^{-4}$ M, which resulted in a four-fold stimulation of ATPase of liver mitochondria.

In our own experiments we have found that increasing the concentration of $Mg^{2+}$ ion above 0.1 mM in the absence of EDTA decreases the RCR by increasing the state 4 rate. This might be a demonstration of the utilization of either ATP or some high-energy intermediates of oxidative phosphorylation for $Mg^{2+}$ ion transport. The uptake of $Mg^{2+}$ ions by beef heart mitochondria has been shown by Brierley *et al.* to utilize the energy derived from electron transport.[(119)] $Mg^{2+}$ ion uptake, like $Ca^{2+}$ ion uptake by liver mitochondria, is not inhibited by oligomycin when it is activated by electron transport, but is inhibited when supported by ATP. DNP inhibits the uptake regardless of the source of energy. Thus, it would seem that part of the ATPase activity described by Ozawa could be accounted for by $Mg^{2+}$ ion transport energized by ATP.

It is noteworthy that ischemia[(6)] seems to result in production of an inhibitor to the 2,4-DNP stimulated ATPase of brain mitochondria. Aging of brain mitochondria at 22°C results in a rapid loss of DNP-stimulated ATPase but very little loss of $Mg^{2+}$-stimulated ATPase. An inhibitor has been isolated from these aged preparations, and the inhibition is reversed by serum albumin. This inhibitor also inhibits state 3 respiration but stimulates the state 4 rate. The phosphorylating competence of the mitochondria is constant for the first 20 min but is rapidly lost thereafter.[(118)] The indication is, therefore, that the $Mg^{2+}$–ATPase and DNP–ATPase have different sensitivities and possibly involve different loci or enzymes and that the ATPase enzyme stimulated by DNP is indeed involved in ATP synthesis. That this is indeed the case is attested to by the finding that a coupling factor ($F_1$) isolated by Racker and co-workers possesses ATPase activity, as well as the ability to return phosphorylative activity to certain nonphosphorylating submitochondrial preparations.[(109–113)]

$Na^+$ ions (30 mM) have been found to inhibit DNP-stimulated ATPase. Higher concentrations of $K^+$ ions are required for inhibition.[(120)]

Under the influence of the antibiotics gramicidin and valinomycin, brain mitochondria are stimulated to accumulate monovalent cations. Gramicidin induces the uptake of $Li^+$, $Na^+$, $Rb^+$, $Cs^+$, and $K^+$ ions, while valinomycin induces the transport of $Rb^+$, $Cs^+$, and $K^+$ ions. The accumulation of these ions is energy-dependent and can be supported either from substrates of the electron transport system or by ATP. With the latter as the source of energy, ATP is cleaved with the resultant ATPase activity which is oligomycin-sensitive. Some aspects of induced ion transport will be presented later.

### 2. *Exchange Reactions*

The ADP–ATP exchange reaction of fresh brain mitochondria is depressed as much as 33% by DNP.[(118)] This exchange is similar to that described by Wadkins and Lehninger[(121a,b)] in its high degree of insensitivity to aging for up to 72 hr. The ADP–ATP exchange enzyme promotes phosphorylation coupled to the oxidation of reduced cytochrome *c*. The exchange reaction is sensitive to oligomycin, but DNP is ineffective. The sensitivity to oligomycin enhances the probability that the exchange reaction is an integral part of ATP synthesis.

The $^{32}P$–ATP exchange reaction of fresh brain mitochondria is almost completely inhibited by $10^{-4}$ M DNP. Aging also causes an inhibition of exchange by as much as 90% within 24 hr. Azide ($10^{-3}$ M) causes 85% inhibition and oligomycin 90% inhibition of the ATP–$^{32}P$ exchange reaction.

It would appear, however, that the distinction between the $P^{32}$–ATP exchange reaction of liver and brain mitochondria is more apparent than real. According to Tuena *et al.*,[118] the ATPase activity of brain and liver in the absence of DNP is inhibited by oligomycin to a greater extent with brain than with liver mitochondria. It would seem that the uninhibited level of activity (0.6 $\mu$mole phosphate/15 min) could be attributable to nonenzymatic cleavage of ATP. This would mean that the inhibition by oligomycin (or $10^{-3}$ M azide) is the same from both liver and brain with the exception that there is more ATPase activity elicited in the absence of DNP with brain than liver mitochondrial preparations. The data with KCN indicate a lack of inhibition of ATPase by $10^{-3}$ M KCN in the absence or presence of DNP. However, at $10^{-2}$ M KCN concentration (two orders of magnitude higher than required to inhibit electron transport) there is a 30% inhibition of ATPase in the absence of DNP in the brain preparations while with liver mitochondria inhibition only occurs in the presence of DNP ($10^{-7}$ M). Wadkins and Lehninger[121a,b] and Low *et al.*[122] have shown that the ATP–$^{32}P$ exchange reaction decreases significantly when the respiratory carriers of liver mitochondria are maintained in the reduced state. Thus, the high concentration of $CN^-$ and azide used in the experiments of Tuena *et al.* is enough to block completely electron transport, thereby forcing the carriers into a reduced state. This would result in an inhibition of the ATP–$^{32}P$ exchange reaction.

Thus as with liver mitochondria[123] the inhibition by oligomycin of phosphorylating respiration is overcome by DNP, and the DNP-stimulated ATPase is inhibited by oligomycin but not by ouabain. The roles of azide and $CN^-$ await further clarification.

No data so far have been obtained using the $O^{18}$ exchange reaction in the elucidation of brain mitochondrial phosphorylation. If, however, it is similar to that of liver, then it would be safe to say that during ATP synthesis the bridge oxygen of the terminal phosphate is furnished by the ADP molecule.

Some of these partial reactions, while consistent with the theory of oxidative phosphorylation, involve some chemical intermediate common to both.

## VI. SOME ENERGY-LINKED FUNCTIONS OF BRAIN MITOCHONDRIA

### A. Ion Transport

Brain mitochondria have been shown to accumulate mono- and divalent cations in energy-dependent reactions.[80,95] The uptake of

$K^+$ ions by brain mitochondria depends upon the age of the preparation and the availability of energy. During the transition to anaerobiosis, we have observed an uptake of $K^+$ as well as a release of $K^+$ by mitochondria even when the extramitochondrial $K^+$ level is in the millimolar range. This leakage can be reversed aerobically in the presence of valinomycin or gramicidin. In the presence of valinomycin or gramicidin, the mitochondria release their $K^+$ when electron transport is inhibited or the system becomes anaerobic. ATP can, however, reverse both these conditions, and $O_2$ the latter.

The existence of $Na^+$-, $K^+$-, and $Mg^{2+}$-stimulated ATPases of brain mitochondrial preparations(6,10,120,124) is a reflection of the ability of these organelles to accumulate these cations with ATP as the source of energy. $Ca^{2+}$ ions are also accumulated by brain mitochondria and, as with liver mitochondria, one observes respiratory stimulation upon addition of $Ca^{2+}$ ions, as well as oxidation–reduction cycles of cytochrome *b* measured spectrophotometrically at 430–410 mμ. The activation of these ATPases, unlike that of the plasma membrane, is insensitive to ouabain.

## B. Fatty Acid Activation

Beattie and Basford(125) have reported the activation of fatty acids by a high-energy intermediate of oxidative phosphorylation and have also pointed to the probable interaction of $K^+$ ions at the same site.

## C. Hexokinase Reaction

Brain mitochondria contain as much as 80% of the hexokinase activity of the tissue.(126) ATP synthesized by brain mitochondria can be utilized by the bound hexokinase in the phosphorylation of glucose.(10) Thus, glucose added to state 3 mitochondria causes a prolongation of this state until the hexokinase becomes inhibited by the formed glucose 6-phosphate. If glucose is added after a state 3–4 transition, a pseudostate 3 rate is obtained for a brief period. This has been postulated as one possible physiological purpose of the intracellular localization of this enzyme in brain mitochondria.(127)

## D. Reversed Electron Transfer

The energy-linked reduction of $NAD^+$ by succinate has been described above. This reaction is intimately related to the energy-linked reduction of acetoacetate. The anaerobic reduction of acetoacetate to β-hydroxybutyrate is ATP-dependent, clearly demonstrating the dependence of the reduction on a source of energy. The aerobic reduction of acetoacetate by succinate is inhibited by amytal, rotenone, DNP, and oligomycin.(128,129a,b)

# VII. ACKNOWLEDGMENT

We wish to thank Dr. Joseph Osinchak, Dr. Doris Murray, and Mr. Lawrence Gonsalez for their assistance in preparing the electron micrographs.

## VIII. REFERENCES


1. E. G. Grey and V. P. Whittaker. The isolation of nerve endings from brain: An electron microscopic study of cell fragments derived by homogenization and centrifugation, *J. Anat.* **96**:79–88 (1962).
2. V. P. Whittaker. The application of subcellular fractionation techniques to the study of brain function, *Progr. Biophys. Mol. Biol.* **15**:39–96 (1965).
3. E. De Robertis, G. Rodriguez de Lores Arnaiz, L. Salganicoff, A. Pellegrino de Iraldi, and L. M. Zieher. Isolation of synaptic vesicles and structural organization of the acetylcholine system within brain nerve endings, *J. Neurochem.* **10**:225–235 (1963).
4. G. Rodriguez de Lores Arnaiz and E. De Robertis. Cholinergic and non-cholinergic nerve endings in the rat brain. II. Subcellular localization of monoamine oxidase and succinate dehydrogenase, *J. Neurochem.* **9**:503–508 (1962).
5. W. C. Schneider and G. H. Hogeboom. Intracellular distribution of enzymes. V. Further studies on the distribution of cytochrome *c* in rat liver homogenates, *J. Biol. Chem.* **183**:123–128 (1950).
6. K. Ozawa, K. Seta, H. Takeda, K. Ando, H. Handa, and C. Araki. On the isolation of mitochondria with high respiratory control from rat brain, *J. Biochem.* **59**:501–510 (1966).
7. K. Ozawa, K. Seta, and H. Handa, The effect of magnesium on brain mitochondrial metabolism, *J. Biochem.* **60**:268–273 (1966).
8. B. Chance and G. R. Williams. The respiratory chain and oxidative phosphorylation, *Advan. Enzymol.* **17**:65–134 (1956).

9a. F. F. Jobsis. A study of preparative procedures for brain mitochondria, *Biochim. Biophys. Acta* **74**:60–68 (1963).

9b. F. F. Jobsis and H. Vreman. Unpublished observation.

10. C. L. Moore and F. F. Jobsis, submitted for publication.
11. R. Tanaka and L. G. Abood, Isolation from rat brain of mitochondria devoid of glycolytic activity, *J. Neurochem.* **10**:571–576 (1963).
12. A. A. Abdel-Latif and L. G. Abood, Biochemical studies of mitochondria and other cytoplasmic fractions of developing rat brain, *J. Neurochem.* **11**:9–15 (1964).
13. A. Inouye and Y. Shinagawa, Cytochrome $b_5$ and related oxidative activities in mammalian brain microsomes, *J. Neurochem.* **12**:803–813 (1965).
14. A. Inouye, Y. Shinagawa, and J. Shinagawa, Cytochrome $b_5$ and related oxidative activities in brain microsomes of fowl, toad and carp, *J. Neurochem.* **13**:385–390 (1966).
15. M. Klingenberg, Pigments of rat liver microsomes, *Arch. Biochem. Biophys.* **75**:376–386 (1958).
16. D. Garfinkel, Studies on pig liver microsomes. I. Enzymic and pigment composition of different microsomal fractions, *Arch. Biochem. Biophys.* **77**:493–509 (1958).
17. T. Omura and R. Sato, A new cytochrome in liver microsomes, *J. Biol. Chem.* **237**:PC 1375–PC 1376 (1962).
18. T. Omura and R. Sato, The CO-binding pigment of liver microsomes. I. Evidence for its hemoprotein nature, *J. Biol. Chem.* **239**:2370–2378 (1964).
19. K. Hirota, T. Omura, H. Nishibayashi, and R. Sato, Reprints of 15th Koso Kagaku Symposium, Osaka, (1963), p. 55 (*cf.* Ref. 18, p. 2377).
20. A. Giuditta and H. J. Strecker, Alternate pathways of pyridine nucleotide oxidation in cerebral tissue, *J. Neurochem.* **5**:50–61 (1959).

21. P. Strittmatter and E. G. Ball, The intracellular distribution of cytochrome components and of oxidative enzyme activity in rat liver, *J. Cellular Comp. Physiol.* **43**:57–78 (1954).
22. B. L. Horecker, TPN-cytochrome reductase in liver, *J. Biol. Chem.* **183**:593–605 (1950).
23. G. H. Williams and H. Kamin, Microsomal TPN-cytochrome *c* reductase of liver, *J. Biol. Chem.* **237**:587–595 (1962).
24. A. H. Phillips and R. G. Langdon, Hepatic TPN-cytochrome *c* reductase. Isolation, characterization and kinetic studies, *J. Biol. Chem.* **237**:2652–2660 (1962).
25. H. Nishibayashi, T. Omura, and R. Sato, A flavoprotein oxidizing NADPH isolated from liver microsomes, *Biochim. Biophys. Acta* **67**:520–522 (1963).
26. D. F. Parsons, Mitochondrial structure: Two types of subunits on negatively stained mitochondrial membranes, *Science* **140**:985–987 (1963).
27. G. L. Sottocasa, B. Kuylenstierna, L. Ernster, and A. Bergstrand, An electron transport system associated with the outer membrane of liver mitochondria. A biochemical and morphological study, *J. Cell Biol.* **32**:415–438 (1967).
28. D. W. Allman and W. C. Tan, Localization of enzymes in the mitochondrial outer membranes, *Federation Proc.* **27**:461 (1968).
29. W. Cammer and C. L. Moore, in press.
30. S. Englard and H. J. Strecker, Oxidation of reduced pyridine nucleotide in brain, *Federation Proc.* **15**:248 (1956).
31. W. Levine, A. Giuditta, and H. J. Strecker, Brain diaphorases, *J. Neurochem.* **6**:28–36 (1960).
32. A. Giuditta and H. J. Strecker, Purification and some properties of a brain diaphorase, *Biochem. Biophys. Res. Commun.* **2**:159–163 (1960).
33. E. Harper and H. J. Strecker, Purification of some properties of a second brain diaphorase, *J. Neurochem.* **9**:125–134 (1960).
34. E. Adler, H. von Euler, and G. Gunther, Diaphorase I and II, *Nature* **143**:641–642 (1939).
35. B. Chance, Techniques for the assay of the respiratory enzymes, *in Methods in Enzymology* (S. P. Colowick and N. O. Kaplan, eds.), Vol. 4, pp. 273–329, Academic Press, New York (1957).
36. B. Chance and H. Baltcheffsky, Respiratory enzymes in oxidative phosphorylation. VII. Binding of intramitochondrial reduced pyridine nucleotide, *J. Biol. Chem.* **233**:736–739 (1958).
37. B. Chance and H. Baltcheffsky, Spectroscopic effects of *ADP* upon the respiratory pigments of rat–heart–muscle sarcosomes, *Biochem. J.* **68**:283–295 (1958).
38. P. K. Maitra and R. W. Estabrook, Fluorimetric method for the enzymic determination of glycolytic intermediates, *Anal. Biochem.* **7**:472–484 (1964).
39. R. W. Estabrook and M. Pullman (eds.), *Methods in Enzymology*, Vol. 10, Academic Press, New York (1967).
40. B. Sacktor and L. Packer, Reactions of the respiratory chain in brain mitochondrial preparations, *J. Neurochem.* **9**:371–382 (1962).
41. T. E. King, H. S. Mason, and M. Morrison, (eds.), *Oxidases and Related Redox Systems*, Vol. 2, Wiley, New York (1965).
42. P. Ehrlich, *in das Sauerstoff-Bedurfnis des Organismus. Eine farbeanalytische Studie*, Hirschwald, Berlin (1885).
43. F. Rohman and W. Spitzer, Ueber Oxydations–wirkungen thierischer Gewebe, *Ber.* **28**:567–572 (1895).
44. O. Warburg, Uber Eisen, den sauerstoffubertragenden Bestandteil des Atmungsferments, *Biochem. Z.* **152**:479–497 (1924).

45. M. Morrison, The cytochrome *c* oxidase system: Its components and their reactions, *in Oxidases and Related Redox Systems* (T. E. King *et al.*, eds.), pp. 639–666, Wiley, New York (1965).
46. E. E. Jacobs, A general discusssion of the distinction of cytochromes *a* and $a_3$, *in Oxidases and Related Redox Systems* (T. E. King, *et al.*, eds.), p. 542, Wiley, New York (1965.)
47. D. F. Wilson and B. Chance, Azide inhibition of mitochondrial electron transport. I. The aerobic steady state of succinate oxidation, *Biochim. Biophys. Acta* **131**:421–430 (1967).
48. H. E. Himwich, *in Brain Metabolism and Cerebral Disorders*, Williams and Wilkins, Baltimore (1951).
49. H. G. Albaum, J. Tepperman, and O. Bodansky. The *in vivo* inactivation by CN of brain cytochrome oxidase and its effect on glycolysis and on the high energy P compounds in brain, *J. Biol. Chem.* **164**:45–51 (1946).
50. W. W. Wainio, Reactions of cytochrome oxidase, *J. Biol. Chem.* **212**:723–733 (1955).
51. I. Sekuzu, S. Takemori, T. Yonetani, and K. Okunuki, Studies on cytochrome *a*. II. Spectral properties of cytochrome *a*, *J. Biol. Chem.* **46**:43–49 (1959).
52. M. Klingenberg and T. Bucher, Flugmustelmitochondrien aus Locusta migratoria mit Atmungskontrolle. Aufban and Fusammensetzung der Atmungskette, *Biochem. Z.* **331**:312–333 (1959).
53. B. Chance and B. Schoner, Fluorometric studies of flavin components of the respiratory chain, *in Flavins and Flavoproteins* (E. C. Slater, ed.), Vol. 8, pp. 510–519 (*Biochim. Biophys. Acta* Library), Elsevier, Amsterdam (1966).
54. S. Minakami, T. Cremona, R. L. Ringler, and T. P. Singer, Studies on the respiratory chain-linked reduced NAD dehydrogenase. III. Catalytic properties of the enzyme from beef heart, *J. Biol. Chem.* **238**:1529–1537 (1963).
55. B. Chance, L. Ernster, P. B. Garland, C. P. Lee, P. A. Light, T. Ohnishi, C. I. Ragan, and D. Wong, Flavoproteins of the mitochondrial respiratory chain, *Proc. Natl. Acad. Sci.* (*U.S.*) **57**:1498–1505 (1967).
56. T. Ohnishi, Flavoprotein and non-heme iron in yeast mitochondria and submitochondrial particles, *Federation Proc.* **27**:298 (1968).
57. B. Chance and B. Hess, Metabolic control mechanisms. I. Electron transfer in the mammalian cell, *J. Biol. Chem.* **234**:2404–2412 (1962).
58. R. S. Criddle, R. M. Bock, D. E. Green, and H. Tisdale, Physical characteristics of proteins of the electron transfer system and interpretation of the structure of the mitochondrion, *Biochemistry* **1**:827–842 (1962).
59. L. Szarkowska, The restoration of DPNH oxidase activity by coenzyme Q (ubiquinone), *Arch. Biochem. Biophys.* **113**:519–525 (1966).
60. E. R. Redfearn and J. Burgos, Ubiquinone (coenzyme Q) and the respiratory chain, *Nature* **209**:711–713 (1966).
61. A. Kroger and M. Klingenberg, On the role of ubiquinone in mitochondria. II. Redox reactions of ubiquinone under the control of oxidative phosphorylation, *Biochem. Z.* **344**:317–336 (1966).
62. E. E. Edwin, A. T. Diplock, J. Bunyan, and J. Green, Studies on vitamin E. 6. The distribution of vitamin E in the rat and the effect of α-tocopherol and dietary selenium on ubiquinone and ubichromenol in tissues, *Biochem. J.* **79**:91–105 (1961).
63. A. T. Diplock, J. Bunyan, J. Green, and E. E. Edwin, Studies on vitamin E. 7. The effect of thiamine, riboflavin, and pantothenic acid on ubiquinone and ubichromenol in the rat, *Biochem. J.* **79**:105–108 (1961).
64. G. di Prisco, M. Banay-Schwartz, and H. J. Strecker, Stimulation of mitochondrial

NADH oxidase by brain mitochondrial extracts, *J. Neurochem.* **12**:113–121 (1965).
65. D. R. Sanadi, R. L. Pharo, and L. A. Sordahl, Non-heme iron in NADH–coenzyme $Q_{10}$ reductase, *in Non-heme Iron Proteins* (A. San Pietro, ed.), Yellow Springs, Ohio, pp. 429–438 (1965).
66. H. Beinert and G. Palmer, On the function and disposition of copper in cytochrome oxidase, *in Oxidases and related Redox Systems* (T. E. King *et al.*, eds.), pp. 567–590, Wiley, New York (1965).
67. A. L. Tappel, Inhibition of electron transport by antimycin A, alkyl hydroxy naphthoquinones and metal coordination compounds, *Biochem. Pharmacol.* **3**:289–296 (1960).
68. J. B. Chappell, Malate oxidation, *in Biological Structure and Functions*, Vol. 2, IUB/IUBS Proc. 1st Intern. Symp., Stockholm, pp. 81–83. Academic Press, New York (1961).
69. B. Sacktor, L. Packer, and R. W. Estabrook, Respiratory activity of brain mitochondria, *Arch. Biochem. Biophys.* **80**:68–71 (1959).
70. D. O. Voss, A. P. Campello, and M. Bacila, The respiratory chain and the oxidative phosphorylation of rat brain mitochondria, *Biochem. Biophys. Res. Commun.* **4**:48–51 (1961).
71. M. Bacila, R. Campello, C. H. M. Vianna, and D. O. Voss, The respiratory chain of rat cerebrum and cerebellum mitochondria: Respiration and oxidative phosphorylation, *J. Neurochem.* **11**:231–242 (1964).
72. G. Palmer, D. J. Horgan, H. Tisdale, P. Singer, and H. Beinert, Studies on the respiratory chain-linked reduced NAD dehydrogenase. XIV. Location of the sites of inhibition of rotenone, barbiturates, and piericidin by means of EPR spectroscopy, *J. Biol. Chem.* **243**:844–847 (1968).
73. D. J. Horgan and T. P. Singer, Characteristics of the binding of rotenone in the respiratory chain and the inhibition sites of amytal and piericidin A, *Biochem. J.* **104**:50c–52c (1967).
74. B. Chance and G. Hollunger, The interaction of energy and electron transfer reactions in mitochondria; 1. General properties and nature of the products of succinate-linked reduction of pyridine nucleotide (1534–1543); III. Substrate requirements for pyridine nucleotide reduction in mitochondria (1555–1561); IV. The pathway of electron transfer (1562–1568); *J. Biol. Chem.* **236**:1534, 1555, 1562 (1961).
75. D. B. Tower, The effects of 2-deoxy-D-glucose on metabolism of slices of cerebral cortex incubated *in vitro*, *J. Neurochem.* **3**:185–205 (1958).
76. E. E. Jacobs, Phosphorylation coupled to electron transport initiated by substituted phenylenediamines, *Biochem. Biophys. Res. Commun.* **3**:536–539 (1960).
77. B. Chance and G. Hollunger, Succinate-linked pyridine nucleotide reduction in mitochondria, *Federation Proc.* **16**:163 (1957).
78. T. Bucher and M. Klingenberg, Wege des Wasserstoffs in der Lebendigen Organisation, *Angew. Chem.* **70**:552–570 (1958).
79. H. Low, H. Krueger, and D. M. Ziegler, On the reduction of externally added DPN by succinate in submitochondrial particles, *Biochem. Biophys. Res. Commun.* **5**:231–237 (1961).
80. C. L. Moore, Gramicidin induced ion transport in Brain mitochondria preparations, *J. Neurochem.* **15**:883–902 (1968).
81. H. Kalckar, CXXVIII. The nature of phosphoric esters formed in kidney extracts, *Biochem. J.* **33**:631–641 (1939).
82. V. A. Belister and E. T. Tsibakowa, Sur le mécanisme des phosphorylations couplées avec la respiration, *Biokhimiya* **4**:516–535 (1939) (*cf.* Ref. 84c, p. 329).

83. B. Chance and G. R. Williams, Respiratory enzymes in oxidative phosphorylation. 1. Kinetics of oxygen utilization, *J. Biol. Chem.* **217**:383–393 (1955).
84a. E. C. Slater, Mechanism of phosphorylation in the respiratory chain, *Nature* **172**:975–978 (1953).
84b. E. C. Slater, Mechanism of energy conservation in mitochondrial oxido-reductions (166–179); Is there a ~P intermediate? (550–552), *in Regulation of Metabolic Processes in Mitochondria*, Vol. 7, pp. 166 and 550 (*Biochim. Biophys. Acta Library*) (1966).
84c. E. C. Slater, Oxidative phosphorylation, *in Comprehensive Biochemistry* (M. Florkin and E. H. Stotz, eds.), Vol. 14, pp. 327–396, Elsevier, Amsterdam (1966).
85. A. L. Lehninger and C. L. Wadkins, Oxidative phosphorylation, *Ann. Rev. Biochem.* **31**:47–78 (1962).
85a. A. L. Lehninger, ed., *The Mitochondrion*, Benjamin, New York, pp. 118–121 (1964).
86a. J. T. Penniston, R. A. Harris, J. Asai, and D. E. Green, The conformational basis of energy transformations in membrane systems. I. Conformation changes in mitochondria, *Proc. Natl. Acad. Sci.* (*U.S.*) **59**:624–631 (1968).
86b. R. A. Harris, J. T. Penniston, J. Asai, and D. E. Green, The conformation basis of energy transformations in membrane systems. II. Correlation between conformational change and functional states, *Proc. Natl. Acad. Sci.* (*U.S.*) **59**:830–837 (1968).
87. P. D. Boyer, Oxidative phosphorylation, *in Biological Oxidations* (T. P. Singer, ed.), Wiley, New York, pp. 193–235 (1967).
88. P. Mitchell, Coupling of phosphorylation to electron and hydrogen transfer by a chemiosmotic type of mechanism, *Nature* **191**:144–148 (1961).
89. P. Mitchell, Molecule, group, and electron translocation through natural membranes, *Biochem. Soc. Symp.* **22**:142–169 (1962).
90. P. Mitchell and J. Moyle, Proton transport phosphorylation. Some experimental tests, *Prod. Federation European Biochem. Soc., April*, pp. 53–74, (1966). E. C. Slater, Z. Kaniuga, and L. Wojtczak, *Biochemistry of Mitochondria*, Academic Press, New York (1967).
91. R. A. Reid, J. Moyle, and P. Mitchell, Synthesis of ATP by a protonmotive force, in rat liver mitochondria, *Nature* **212**:257–258 (1966).
92. A. T. Jagendorf and E. Uribe, ATP formation caused by acid–base transition of spinach chloroplasts, *Proc. Natl. Acad. Sci.* (*U.S.*) **55**:170–177 (1966).
93. B. Chance and L. Mahler, Proton movements in mitochondrial membranes, *Nature* **212**:372–376 (1966).
94. C. S. Rossi, N. Siliprandi, E. Carafoli, J. Bielawski, and A. L. Lehninger, Proton movement across the mitochondrial membrane supported by hydrolysis of ATP, *European J. Biochem.* **2**:332–340 (1967).
95. C. L. Moore, Cation accumulation during the anaerobic–aerobic transition in brain mitochondria, *Biochemistry* **7**:300–304 (1968).
96. P. Mueller, D. O. Rudin, H. Ti Tien, and W. C. Westcott, Formation and properties of bimolecular lipid membranes, *in Recent Progress in Surface Science* (J. F. Danielli, K. G. A. Parkhurst, and A. C. Riddiford, eds.), Academic Press, New York, pp. 379–393 (1964).
97. A. Finkelstein and A. Cass, Effect of cholesterol on the water permeability of thin lipid membranes, *Nature* **216**:717–718 (1967).
98. U. Hopfer, A. L. Lehninger, and T. E. Thompson, Protonic conductance across phospholipid bilayer membranes induced by uncoupling agents for oxidative phosphorylation, *Proc. Natl. Acad. Sci.* (*U.S.*) **59**:484–490 (1968).
99. H. C. Hemker, Lipid solubility as a factor influencing the activity of uncoupling phenols, *Biochim. Biophys. Acta* **63**:46–54 (1962).

100. D. K. Myers and E. C. Slater, The enzymic hydrolysis of ATP by liver mitochondria, *Biochem. J.* **67**:572–579 (1957).
101. P. Mitchell and J. Moyle, Acid–base titration across the membrane system of rat liver mitochondria, *Biochem. J.* **104**:588–600 (1967).
102. B. Hagihara, Techniques for the application of polarography to mitochondrial respiration, *Biochim. Biophys. Acta* **46**:134–142 (1961).
103. B. Sacktor and D. G. Cochran, The respiratory metabolism of insect flight muscle. I. Manometric studies of oxidation and concomitant phosphorylation with sarcosomes, *Arch. Biochem. Biophys.* **74**:266–276 (1958).
104. C. T. Gregg, C. R. Heisler, and L. F. Remmert, Pyruvate and α-GP oxidation in insect tissue, *Biochim. Biophys. Acta* **31**:593–595 (1959).
105. B. Chance and B. Sacktor, Respiratory metabolism of insect flight muscle. II. Kinetics of respiratory enzymes in flight muscle sarcosomes. *Arch. Biochem. Biophys.* **76**:509–531 (1958).
106. B. Sacktor, L. Packer, J. Cummins, and B. E. Hackley, Jr., Oxidation of GABA by brain mitochondria, *in Inhibition in the Nervous System and GABA* (E. Roberts, ed.), pp. 182–188 (1960).
107. P. Strasberg and K. A. C. Elliott, Further studies on the binding of γ-aminobutyric acid by brain, *Can. J. Biochem.* **45**:1795–1807 (1967).
108. H. Weinstein, S. Varon, D. R. Muhleman, and E. Roberts, A carrier mediated transfer model for the accumulation of $^{14}C$ γ-aminobutyric acid by subcellular brain particles, *Biochem. Pharmacol.* **14**:273–288 (1965).
109. M. E. Pullman and E. Racker, Spectrophotometric studies of oxidative phosphorylation, *Science* **123**:1105–1107 (1957).
110. M. E. Pullman, H. S. Pinefsky, A. Datta, and E. Racker, Partial resolution of the enzymes catalyzing oxidative phosphorylation. I. Purification and properties of soluble DNP-stimulated ATPase, *J. Biol. Chem.* **235**:3322–3329 (1960).
111. E. Racker, Studies of factors involved in oxidative phosphorylation, *Proc. Natl. Acad. Sci.* (*U.S.*) **48**:1659–1663 (1962).
112. T. E. Conover, R. L. Prairie, and E. Racker, Partial resolution of the enzymes catalyzing oxidative phosphorylation. III. A new coupling factor required by submitochondrial particles extracted with phosphatides, *J. Biol. Chem.* **238**:2831–2837 (1963).
113. R. Sanadi, Energy requiring reduction of pyridine nucleotide by ascorbate in the presence of coenzyme Q or menadione, *J. Biol. Chem.* **238**:PC482–PC483 (1963).
113a. M. E. Pullman and G. Schatz, Mitochondrial oxidations and energy coupling, *Ann. Rev. Biochem.* **36**:539–610 (1967).
114. G. di Prisco, M. Banay-Schwartz, and H. J. Strecker, A stimulatory factor for mitochondrial NADH oxidase, *Biochem. Biophys. Res. Commun.* **15**:116–120 (1964).
115. W. F. Loomis and F. Lipmann, Inhibition of phosphorylation by azide in kidney homogenate, *J. Biol. Chem.* **179**:503–504 (1949).
116. R. Tanaka and L. G. Abood, Relationship of structure to phosphorylative activity in rat brain mitochondria, *J. Neurochem.* **10**:7–16 (1963).
117. M. B. Wechsler, Studies on oxidative phosphorylation and ATPase activity of fresh and frozen brain mitochondria, *Arch. Biochem. Biophys.* **95**:494–498 (1961).
118. M. Tuena, A. Gomez-Puyou, A. Pena-Diaz, and G. H. Massieu, Studies on ATPase and ATP exchange reaction in brain mitochondria and the effect of 2,4-DNP, *J. Neurochem.* **11**:527–536 (1964).
119. G. B. Brierley, E. Bachmann, and D. E. Green, Active transport of Pi and magnesium ions by beef heart mitochondria, *Proc. Natl. Acad. Sci.* (*U.S.*) **48**:1928–1935 (1962).

120. D. S. Beattie and R. E. Basford, Sodium stimulated ATPase activity of rat brain mitochondria, *J. Neurochem.* **15**:325–333 (1968).
121a. C. L. Wadkins and A. L. Lehninger, The ATP–ADP exchange reaction of oxidative phosphorylation, *J. Biol. Chem.* **233**:1589–1597 (1958).
121b. C. L. Wadkins, The ATP–ADP exchange reaction of intact rat liver mitochondria, *J. Biol. Chem.* **236**:221–224 (1961).
122. H. Low, P. Siekevitz, L. Ernster, and O. Lindberg, Studies on the relation of the ATP–Pi exchange reaction of mitochondria to oxidative phosphorylation, *Biochim. Biophys. Acta* **29**:392–405 (1958).
123. H. A. Lardy, D. Johnson, and W. C. McMurray, Antibiotics as tools for metabolic studies. I. A survey of toxic antibiotics in respiratory, phosphorylative and glycolytic systems, *Arch. Biochem. Biophys.* **78**:587–597 (1958).
124. S. Lovtrup, A comparative study of the influence of chlorpromazine and imipramine on mitochondrial activity, *J. Neurochem.* **12**:743–749 (1965).
125. D. S. Beattie and R. D. Basford, Brain mitochondria. IV. The activation of fatty acids in bovine brain mitochondria, *J. Biol. Chem.* **241**:1412–1418 (1966).
126. R. K. Crane and A. Sols, The non competitive inhibition of brain hexokinase by G6P and related compounds, *J. Biol. Chem.* **210**:597–606 (1954).
127. C. L. Moore, Purification and properties of two hexokinases from beef brain, *Arch. Biochem. Biophys.* **128**:734–744 (1968).
128. M. Klingenberg and H. von Hafen, Wege des wasserstoffs in Mitochondrien. I. Die wasserstoffubertragung von Succinat zu Acetoacetat, *Biochem. Z.* **337**:120–145 (1963).
129a. M. Klingenberg and P. Schollmeyer, Zur Reversibilitat der Oxydativen Phosphorylierung. ATP-Abbangige Atmungskontrolle und Reduction von DPN in Mitochondrien, *Biochem. Z.* **333**:335–350 (1960).
129b. M. Klingenberg and P. Schollmeyer, Zur Reversibilitat der Oxydativen Phosphorylierung. II. Der Einfluss von ATP auf die Atmungskette in Atmenden Mitochondrien, *Biochem. Z.* **335**:231–242 (1961).

*Chapter 4*

# PHOSPHATASES


## Stephen R. Cohen

*New York State Research Institute for Neurochemistry and Drug Addiction*
*Ward's Island*
*New York, New York*


---

## I. INTRODUCTION[2]*

Phosphatases are a diverse and ubiquitous group of enzymes which occur in most organisms and tissues. To date all nervous systems have been found to have at least nonspecific alkaline and acid phosphatases. No observed activity probably means no activity under experimental conditions, not the absence of phosphatases. The nonspecific alkaline and acid phosphatases have been extensively studied. Besides these, many specific phosphatases have been found in nervous tissue, but studies of these enzymes have been limited. The occurrence and gross histological localization of nonspecific alkaline and acid phosphatase in nervous tissue of mammals and birds are well known. Although no one species has been studied in detail through its complete life cycle, the general pattern of changes in activity and distribution of these nonspecific phosphatases during development is also fairly clear. Much less is known about specific or nonspecific phosphatases in other vertebrates, and practically nothing is known about phosphatases in invertebrate nervous systems. Most of the investigations have relied largely on histochemical methods, where applicable. As a result, the properties of phosphatases in nervous tissue are poorly known and must be extrapolated from the properties of similar enzymes isolated from other sources. For some phosphomonoesterases, even the demonstration of a specific phosphatase rather than nonspecific phosphatase activity is doubtful, and at most only the general *p*H range, that is, acid, neutral (physiological), or alkaline, is clear. Almost without exception, the function of these enzymes is not known.

* Citations appended to a section heading are to texts, monographs, or reviews covering some broad aspect of the subject. Such references should be consulted for further detail.

This chapter is a survey of phosphatases in nervous tissue with references to more detailed reviews of certain aspects. The topics covered include the types of phosphatases present, their occurrence and localization, presence in invertebrate nervous systems, changes during development, responses to injury, and speculations on the role of these enzymes. The properties of nonspecific alkaline and acid phosphatases from various sources are summarized. Because of the numerous sources of error in the experimental methods, a critical sketch of the principal methods has been included. ATPases and related membrane-bound phosphatases associated with facilitated transport have been omitted from this review, but are treated by Albers and Siegel, in *Handbook of Neurochemistry Volume* 4, and in Arvy's review of phosphatases in nervous tissue.[2] The treatment of pathology is limited to the effect of certain types of injury; other conditions including tumors, hemorrhage, and disease have been excluded. Finally, neither neurosecretory cells *per se*, the pituitary gland, nor the pineal gland are considered.

## II. METHODS [3,4,6,8]

The two principal ways to locate and study tissue phosphatases are *biochemical methods* and *histochemical methods*. These two approaches complement each other. Both should be used if possible.

Biochemical methods have the advantages of good characterization of an individual enzyme, quantitative measurement of enzyme activity, and facilitation of study of the properties of an individual enzyme. Characterization is especially important with phosphatases, with their overlapping specificities and ranges of activity. These advantages come from the precise and reproducible control of reaction conditions that is possible and from the possibility of purifying individual enzymes; isozymes can be demonstrated, for example, by acrylamide gel electrophoresis. Variation of different enzymes or isozymes with genetic factors, age, and physiological or pathological conditions can be studied. Quantitative biochemical methods depend on good recovery—or at least consistent recovery—of active enzyme, preferably in a condition approximating its intracellular state. They are subject to artifacts from denaturation and from addition or removal of inhibitors and activators, to interactions with other tissue components in the preparation, and with particulate, membrane-segregated or membrane-bound enzymes, with gross disruption of their normal environment. Although Lowry,[89,90] Hydén,[75] and others have measured enzymes from single neuron cell bodies and small fragments of tissue prepared by microdissection, biochemical methods are ordinarily limited to measuring the average activity of a macroscopic region. Crude intracellular localization has been achieved by studying the supernatant and the different particulate fractions from homogenized tissue.[42,49]

Histochemical techniques have several advantages compared to biochemical methods. Ideally, they locate enzyme activity precisely *in situ* in tissue type and even within the cell organelle. They are inherently very sensitive, permitting small amounts of tissue to be studied. They avoid the difficulties and artifacts produced by extracting, isolating, and purifying an enzyme. They are better suited for surveys and comparison studies of many specimens, systems, and species. Unfortunately, phosphatase stains have severe limitations, the most serious being their susceptibility to artifacts. Both fixing, embedding, and sectioning before staining[32,137] and the staining mixture itself may inactivate enzymes, remove soluble enzymes, or grossly alter enzyme properties, producing either a false negative reaction, or, if activity remains, a misleading picture of its distribution. There is no completely satisfactory method for treating tissue. Those methods which distort tissue structure the least affect phosphatases the most, and *vice versa*. Artifacts are also produced by diffusion and by multiple enzyme systems where the product of one reaction is the substrate for the next. Artifacts become more of a problem the more closely enzyme activity is to be defined. Most phosphatase techniques are satisfactory for gross qualitative localization of activity in an organ; none may be suitable for accurate localization for electron microscopy.

Histochemical techniques are restricted to phosphatases that liberate inorganic phosphate.[3] There are no satisfactory staining methods for nucleases, phosphodiesterases, phospholipase C or D, phosphoproteinases, or phosphatidic acid phosphatases.[4] Although stains for specific phosphatases have been developed, their specificity is mediocre for the most part, especially when nonspecific alkaline or acid phosphatases are also present. Most staining methods are only semiquantitative at best. Because only a restricted range of activity can be studied in any one preparation, the presence of enzyme may be missed in regions of relatively low activity, unless regions of relatively high activity in the same preparation are intentionally overstained. Enzymes have been assayed by extracting and measuring the dye produced by the staining reaction,[58] but this method cannot be used for fine-grained localization of activity.

The two principal types of phosphatase stains are heavy-metal stains, which were developed independently by Gomori[63,64] and Takamatsu,[149] and azo-dye coupling stains, which have been developed extensively by Burstone and by Pearse. In the Gomori method, inorganic phosphate is released from a phosphate ester substrate and precipitated as an insoluble, heavy-metal phosphate which is then converted to an insoluble, dark-colored sulfide, usually CoS for alkaline phosphatase and PbS for acid phosphatase. This heavy-metal precipitate makes these stains especially suitable for electron microscopy. By selecting the phosphate ester, adjusting the *p*H, and adding inhibitors for conflicting phosphatases, these stains can be made more or less specific for various phosphomonoesterases. Because Gomori stains detect local concentrations of inorganic phosphate rather than phosphatase activity *per se*, artifacts are produced by inorganic phosphate unless it is removed from tissue before staining. Errors are also produced by other substances or structures that bind the heavy-metal ions, especially $Pb^{2+}$, used for staining, and by precipitation of heavy-metal phosphate at a pre-existing center for nucleation rather than at the site of enzyme activity. These precipitates may be misleading because they often appear to be sharp and clearly localized with none of the diffuse blurring at the boundaries that is typical of diffusion artifacts. Diffusion artifacts are especially troublesome with the alkaline phosphatase stain where they prevent localization of the enzyme in nervous tissue by electron microscopy. Gomori stains require meticulous technique. It is difficult to get consistent, uniform staining, and often successive serial sections on the same slide stain differently.[(55)]

Azo-dye coupling stains for phosphatases consist of a stabilized diazonium salt and a substrate which is the phosphate ester of a diazonium coupling agent, usually a naphthol or a naphthol AS (amidosäure) derivative.* The phosphatase hydrolyzes the ester, freeing the coupling agent which reacts with the diazonium salt to form an azo dye. Because the color is produced by the liberated coupling agent instead of by inorganic phosphate, these stains are much more specific for phosphatase. In simultaneous coupling stains, diazonium salt reacts with the coupling agent as it is formed. In post coupling stains, diazonium salt is added to react with the free coupling agent only after the enzymatic hydrolysis is ended. Because the dye is formed as the reaction proceeds, diffusion of liberated coupling agent causes less difficulty in simultaneous coupling stains, permitting a greater choice of substrates. Against this, the necessity of keeping both substrate and diazonium salt in the same mixture under conditions where non-enzymatic hydrolysis of the substrate is minimal and the reaction to form dye is rapid places severe restrictions on the buffers, activators, or inhibitors that can be used in the staining mixture. In addition, the stabilizing agent for the diazonium salt (e.g., zinc chloride) may inhibit activity or activate competing enzymes. With either variant, staining is slower than with Gomori stains, and therefore erratic and blurred staining and artifacts from migration are more of a problem. Fortunately, such artifacts can often be recognized by their diffuse boundaries. Some of the dyes formed, especially when a naphthol phosphate ester rather than a naphthol AS phosphate ester is used, stain adjacent tissue or other regions within the cell. The specialized substrates needed limit azo-dye coupling stains to non-specific alkaline and acid phosphatase.

Comparative studies of Gomori and azo-coupling stains for non-specific phosphatases have revealed significant differences.[(95,108,127,136,147)] Many of these may be due to artifacts, which are a more serious problem with Gomori stains. In general, Gomori stains stain structures, especially cell nuclei, that are not stained by azo-coupling stains in addition to most structures that are. For example, Rogers[(127)] found that the Gomori alkaline phosphatase technique stained cell nuclei more heavily than

* There are also azo-dye coupling stains for esterases, sulfatases, peptidases, etc., which act on the same principle as the stains for phosphatases.

cytoplasm in whole cells, but that it stained only cytoplasm in preparations of isolated nuclei or cytoplasm. Many workers agree with Deane[44] that Gomori staining of cell nuclei for either acid or alkaline phosphatase is an artifact; some disregard any activity revealed by Gomori stains but not by azo-coupling stains (where these stains can be used). This may not always be correct. Applying starch gel electrophoresis to human serum, Sandler and Bourne[139] found that $\alpha$-naphthol acid phosphate, a substrate for azo-coupling stains, and sodium $\beta$-glycerophosphate, a common substrate for Gomori stains, gave a different location for the nonspecific alkaline phosphatase band.

Because of the broad activity of nonspecific alkaline and acid phosphatases, it is important to differentiate clearly any specific phosphatase activity from nonspecific activity. This is best done biochemically. The ideal method is to separate the specific phosphatase from other phosphatases. A satisfactory method without separation is to demonstrate specific activity under conditions where other phosphatases are either inactive or have been inactivated. If other phosphatases cannot be eliminated, the effect of *p*H, of activators, or of inhibitors on activity with various substrates can be investigated and the kinetic data analyzed. This method, unless carefully used, is unreliable because carefully purified enzymes have been shown to have different kinetic behavior and to be affected differently by activators and inhibitors depending on the substrate. The least satisfactory biochemical method is to infer the existence of more than one phosphatase from the differences in the ratios of activity toward several substrates that is shown by preparations from related but distinct tissues, such as cerebral cortex and white matter. Histochemically, specific phosphatases can best be demonstrated by using specific enzyme stains (which may not be very specific) and comparing the pattern of staining to the pattern for other phosphatases. If no specific stain is available, the treatment of tissue before staining can be varied or the stain modified in various ways (e.g., by changing *p*H or substrate, or by adding activators or inhibitors) and changes in the distribution of activity observed. Because the apparent distribution of even nonspecific acid and alkaline phosphatase can depend on the techniques used, histochemical methods are untrustworthy since they depend on differences in the pattern of enzyme activity.

## III. TYPES AND PROPERTIES OF PHOSPHATASES IN NERVOUS TISSUE

The phosphatases considered in this review may be classified for convenience as nonspecific *alkaline phosphatases*, nonspecific *acid phosphatases*, and *other phosphatases*. The phosphatases of the nervous system have not been well characterized; therefore, unless a specific reference is made to nervous tissue, the following descriptions of the properties of nonspecific alkaline phosphatase and nonspecific acid phosphatase are based on studies of enzymes from various other sources.

## A. Alkaline Phosphatase[7,10]

Nonspecific alkaline phosphatases (3.1.3.1) are widely distributed in many types of plant and animal tissue and in microorganisms. They are often associated with membranes or other cellular structures and must then be solubilized with acetone or *n*-butanol before they can be extracted from tissue homogenates.[39,99] The enzymatic properties of alkaline phosphatases from the most diverse sources are similar; therefore, the properties of those that have been characterized probably represent the properties of these enzymes in the nervous system.

These enzymes are phosphomonoesterases that hydrolyze a wide variety of substrates,[41,61,71] including phosphate esters of aliphatic alcohols (e.g., $\alpha$- and $\beta$-glycerophosphate and choline phosphate), phenols (e.g., phenyl phosphate), cyclic alcohols (e.g., inositol phosphate), nucleoside phosphates [e.g., AMP, ADP, and ATP[71]], inorganic pyrophosphate,[50,102] amides (e.g., creatine phosphate), and acyl phosphates. Activity against several substrates depends on the source of the enzyme; for example, alkaline phosphatase from *Escherichia coli* does not hydrolyze creatine phosphate.[61] Alkaline phosphatases show no phosphodiesterase,[71] sulfatase, or carboxylesterase activity.

Purified phosphatases have maximum activity at about *p*H 9, depending on source, degree of purification, and salts in the medium. The $K_m$ and $V_{max}$ often vary little with substrate.[61,71] Besides hydrolyzing phosphate esters, they can also phosphorylate suitable acceptors,[100,163] e.g., ethanolamine to form ethanolamine phosphate. The P–O bond apparently is cleaved to produce a hydroxy compound (e.g., the alcohol of the ester) and a phosphoryl enzyme intermediate which, if the reaction is hydrolysis, then reacts with water to release inorganic phosphate and regenerate enzyme, or, if the reaction is transphosphorylation, then transfers the phosphate group to an acceptor with the production of free enzyme.[12,51,52] Alkaline phosphatases are activated *in vitro* by low concentrations of a number of divalent cations, especially $Mg^{2+}$ and $Zn^{2+}$; at higher concentrations, $Zn^{2+}$ ions are inhibitory.[41,114] They are competitively inhibited by inorganic phosphate ions, by arsenate ions, and by phosphonate ions presumably by competition for the active site.[41,50,86] They are strongly inhibited by low concentrations of $Be^{2+}$ ions.[68,80] Many other divalent and trivalent metal ions also markedly affect activity in low concentrations[41,114] (S. R. Cohen, unpublished). Unlike nonspecific acid phosphatases, they are not inhibited by low concentrations of fluoride ions.[71] They are markedly inhibited by metal-complexing agents, which is evidence that they are metalloenzymes.[36,118] Purified preparations from swine kidneys[91] and *E. coli*[118] show the essential metal to be zinc. Presumably, because zinc-deficient diets decrease the phosphatase activity in a number of other species,[161] this is also true for enzymes from other sources.

To the extent to which it has been characterized, alkaline phosphatase from goat brain cortex ribosomes has the properties of a typical nonspecific phosphatase. It differs from many of these enzymes in its unusually high activity against nucleoside mono-, di-, and triphosphates and its lack of activity against most sugar monophosphates.(41)

Alkaline phosphatases frequently consist of several isozymes which may differ from tissue to tissue in the same species. In some cases these isozymes have essentially identical enzymatic properties,(114) while in others they differ noticeably.(26) Saraswathi and Bachhawat(140) have separated two isozymes with somewhat different substrate specificities and distribution from sheep brain, and Cunningham and Field(39) have shown the presence of two isozymes in extracts of guinea pig brain.

## B. Acid Phosphatase(7,9)

Nonspecific acid phosphatases (3.1.3.2) are widely distributed in many organisms and types of tissue. Although they occur in solution within cells and in biological fluids, their presence in lysosomes is so characteristic that some workers consider all intracellular particulate bodies with acid phosphatase activity to be lysosomes.(112) The enzymatic properties of different nonspecific acid phosphatases vary more than those of alkaline phosphatases. Future research may show that the category "acid phosphatase" includes several classes of nonspecific acid phosphatases with distinctly different properties. The following description is based primarily on studies of two well characterized enzymes, human prostate acid phosphatase(110,111,160) and bovine milk acid phosphatase.(28) Extrapolation from the properties of these soluble enzymes to the particulate enzymes of nervous tissue must be made with some caution.

Acid phosphatases are phosphomonoesterases that hydrolyze a wide variety of substrates including phosphate esters of alcohols and phenols (e.g., α- and β-glycerophosphate and phenyl phosphate), polyphosphates (e.g., inorganic pyrophosphate, ADP, and ATP), nucleotides, phosphoamides (e.g., creatine phosphate), and phosphoproteins. There are marked variations in specificity. Human prostate enzyme attacks aliphatic esters (e.g., α- and β-glycerophosphate),(160) but bovine milk enzyme does not.(28) These enzymes show no phosphodiesterase, sulfatase, or carboxylesterase activity.

The optimum *p*H is 4 to 6, but both optimum *p*H and other enzymatic properties including $V_{max}$ and $K_m$ depend on the substrate and medium(110,160) as well as the enzyme. Besides hydrolyzing phosphate monoesters, they also transphosphorylate suitable acceptors.(101,160) The mechanism is apparently the same as that of alkaline phosphatase (*q.v.*). The effect of activators and inhibitors on acid phosphatases varies considerably. Some, like alkaline phosphatase, are activated, or stabilized, or both, by $Mg^{2+}$ ions, while others are not. Acid phosphatases are competitively inhibited by inorganic phosphate. Unlike alkaline phosphatases, they are inhibited competitively by hydroxy acids, especially L-tartaric acid,[110,160] and many are inhibited by $F^-$ ions. One acid phosphatase from brain is only slightly inhibited by $F^-$ ions, however.[162] Unlike alkaline phosphatases, they are not inhibited by metal-complexing agents; this is evidence that they are not metalloenzymes. These enzymes are very sensitive to surface denaturation.[160] Such denaturation may be the cause of apparent deviations from Michaelis–Menten kinetics and of conflicting reports on enzymatic properties.

Nonspecific acid phosphatase from brain resembles acid phosphatase from other sources in being inhibited by L-tartrate, by arsenate, and by fluoride ions.[15,72] Three forms of acid phosphatase have been found in neural tissue from rats,[15] and four major forms have been found in human brain tissue by gel electrophoresis.[17,27] In this latter case, all bands were seen with the substrate α-naphthyl phosphate, while only the slowest band was seen with β-glyceryl phosphate. At least two of these bands contained subbands. From histological studies of the difference in appearance and staining properties of lysosomes, Blank has concluded that there are at least eight types of acid phosphatase in cat retina[29] and at least nine types in spinal cord of chicken.[30] Isozymes of acid phosphatase from neural tissue may differ significantly in several properties, including *p*H dependence and effect of inhibitors.[162]

## C. Other Phosphatases

Over one dozen specific phosphatases have been reported in nervous tissue. Some were observed histochemically, usually by some variant of the Gomori phosphate precipitation technique. Others, especially those for which there are no adequate staining methods, were studied biochemically. Irrespective of method, the possibility of mistaking nonspecific alkaline or acid phosphatase for a specific phosphatase is always present in such research. Except for the observations on neurokeratinal phosphatases, the following list is restricted to studies, where, from the criteria in the Methods section, I believe that a specific phosphatase was probably observed. Because these studies are fragmentary and negative results have rarely been reported, a description of specific phosphatase activity in certain species or organs does not mean that the enzyme is limited to these, but means only that reliable studies have not been made on other species or organs. Most likely, most of the specific phosphatases discussed here and others not yet detected are widely distributed in both vertebrate and invertebrate nervous systems.

### 1. *Triphosphoinositide Phosphatases*

The complexity that phosphatase enzyme systems may have is illustrated by the hydrolysis of triphosphoinositides by extracts of beef brain. This hydrolysis occurs by the following mechanism[60,158]:

$$\text{Triphosphoinositide} \xrightarrow{+H_2O} \text{Inositol triphosphate} + \text{Diglyceride}$$

$$\downarrow (-Pi) \qquad\qquad\qquad\qquad \downarrow (-Pi)$$

$$\text{Diphosphoinositide} \xrightarrow{+H_2O} \text{Inositol diphosphate} + \text{Diglyceride}$$

$$\downarrow (-Pi)$$

$$\text{Monophosphoinositide (Phosphatidyl inositol)} \xrightarrow{+H_2O} \text{Inositol phosphate} + \text{Diglyceride}$$

Dawson and Thompson isolated two specific phosphatases which are required for this process: *triphosphoinositide phosphodiesterase*,[159] and *triphosphoinositide phosphomonoesterase*.[43] Triphosphoinositide phosphodiesterase converts a triphosphoinositide or a diphosphoinositide to triphosphoinositol or diphosphoinositol, respectively, and the corresponding 1,2-diglyceride. Activity is apparently limited to these two substrates. It is inactive toward phosphatidyl inositol and other phosphate diesters. It is activated by $Mg^{2+}$ and $Ca^{2+}$ under certain conditions, but does not require metal ions. Triphosphoinositide phosphomonoesterase dephosphorylates triphosphoinositides to diphosphoinositides and inorganic phosphate, and to a lesser extent dephosphorylates triphosphoinositol to diphosphoinositol. It is inactive against all other phosphate monoesters tested; in particular it does not dephosphorylate diphosphoinositol or monophosphoinositol. This enzyme has an absolute requirement of $Mg^{2+}$ or $Mn^{2+}$ ions and is most active near neutral *p*H. A third enzyme for this process, *monophosphoinositide phosphodiesterase*, which converts a monophosphoinositide to inositol phosphate and the corresponding diglyceride, has been obtained from guinea pig brain.[60]

Triphosphoinositide phosphomonoesterase and diphosphoinositide phosphomonoesterase occur in the neuronal cytoplasmic fraction of rat brain, with little activity in the myelin or the synaptosomal fractions. Triphosphoinositide phosphomonoesterase is not present in glia. In contrast, phosphatidic acid phosphatase (3.1.3.4) (*q.v.*) occurs only in particulate fractions.[134]

## 2. *Inorganic Pyrophosphatase*

Inorganic pyrophosphatase (3.6.1.1) that is clearly distinct from both nonspecific acid and nonspecific alkaline phosphatase has been found biochemically in several species including human, rat, rabbit, sheep[65,66] and pig.[142] It has maximum activity at physiological *p*H [7 to 7.8,[65,142]] requires $Mg^{2+}$ ions, and shows no phosphomonoesterase activity. The level of activity varies with the region of the brain. Comparisons of enzymatic properties from different regions of sheep brain show that there may be several isozymes. Using a Gomori-type stain, Fischer demonstrated inorganic pyrophosphatase in mouse, rat, pigeon, frog, and carp brain.[54] He found it to be present in some cells but not others, with no obvious rule for the intercellular distribution of activity. It is in dendrites of Purkinje cells but not in the cell bodies. This observed erratic distribution may be largely an artifact.

### 3. *Glucose 6-Phosphatase*

Glucose 6-phosphatase (3.1.3.9) has been found both biochemically, in beef brain, and histochemically, with a modified Gomori stain, in rat tissue. The biochemical and histochemical activities may be due to different enzymes. Beef brain enzyme also hydrolyzes fructose 6-phosphate, phosphohydroxypyruvate, and α- and β-glycerophosphate, but not phosphoserine, *p*-nitrophenylphosphate, phosphoethanolamine, 3-phosphoglycerate ($COOHCHOHCOPO_3H_2$), and 5′-mononucleotides; it does not catalyze phosphate exchange between phosphohydroxypyruvate and hydroxypyruvate. Maximum activity is at physiological *p*H (7.5). The enzyme is activated by $Mg^{2+}$ and inhibited by $Mn^{2+}$, $F^-$, and inorganic phosphate. Arsenate has no effect.[81]

In the rat olfactory bulb, the layer of olfactory fibers stains intensely for glucose 6-phosphatase, but not all cells in this layer show activity. Glomerula, neurons, and dendrites stain weakly to moderately in the olfactory bulb, while synapses stain heavily.[144] Cerebral and trigeminal ganglion neurons stain, but the staining is variable; in some cells it appears as an intensely stained region surrounding the nucleus and in others it is spread more or less diffusely throughout the perikaryon.[153,155] The observed distribution of activity in these histochemical studies may reflect both artifacts and some nonspecific acid phosphatase activity.

### 4. *Acyl Phosphatase*

Grisolia and co-workers have isolated an enzyme from beef brain, acylphosphatase (3.6.1.7), which hydrolyzes the mixed acid anhydrides, acyl phosphate and carbamyl phosphate.[69,122] The optimum *p*H ranges from 3 to 5 depending on substrate and conditions. Activity is inhibited by inorganic phosphate. With a molecular weight of only 13,000, it is one of the smallest enzymes known. This enzyme is distinct from non-specific acid phosphatase which also attacks acyl phosphates.

### 5. *Phosphodiesterases*

Two distinct phosphodiesterases (3.1.4.1) have been characterized in brain preparations. One is an endonuclease from lamb brain which is especially active against thermally denatured DNA and has been used to measure denaturation.[70] Hydrolysis is on the 3′-hydroxyl side of the diester bond. It differs from both spleen and venom enzymes, which are exonucleases. This enzyme is activated by low concentrations of both $Mg^{2+}$ and $Mn^{2+}$ ions, but is inhibited by $Mg^{2+}$ at concentrations greater than $10^{-3}$ M. The optimum *p*H range is 7 to 9. The other is found in the ribosomal fraction from goat brain cortex.[40] With bis(*p*-nitrophenyl) phosphate, the only substrate studied, it has a broad *p*H optimum in the alkaline range. It is moderately inhibited by inorganic phosphate, but, unlike nonspecific alkaline phosphatase, is not inhibited by arsenate.

### 6. *Thiamine Pyrophosphatase and Nucleoside Diphosphatase—Golgi Apparatus Markers*

Using modified Gomori stains, thiamine pyrophosphatase and nucleoside diphosphatase (3.6.1.6) activities have been found to be reliable enzymatic markers for the Golgi apparatus in vertebrate tissues, including nervous tissue, and in onion root tips.[62,113] As with many histochemical studies, it is not known whether this activity represents one or a complex of enzymes. Electron microscopy shows that the pattern of staining is the same in position and structure as the staining produced by the classical stains for Golgi apparatus, osmium tetraoxide and silver carbonate.[21] There is no activity in either the endoplasmic reticulum (Nissl substance)[62] or in lysosomes[113] (revealed by acid phosphatase staining) adjacent to the Golgi apparatus. The thiamine pyrophosphatase staining technique has shown that the Golgi apparatus in neurons in the brain, spinal ganglia, and sympathetic ganglia occurs in a large variety of forms which can be used to classify these cells.[143] The larger cerebral neurons are frequently more heavily stained and usually contain a more elaborate Golgi apparatus than smaller cerebral neurons.[25]

Thiamine phosphatase also occurs in invertebrate neurons, but with a different distribution (*q.v.*).

In addition to neurons, this enzyme is present in vessels, including capillaries and arterioles, and microglia.[25,143,148] Evidence of its presence in astrocytes is conflicting.[25,148] It is absent from axons and myelin sheaths, but traces may occur in Schwann cells that are external to the myelin sheaths.[143]

The response of microglia and the Golgi apparatus of neurons to injury has been studied (*q.v.*) using this enzyme as a marker.

### 7. *5′-Nucleotidase*

5′-Nucleotidase (3.1.3.5) occurs in the nervous system of mammals. Studies of the localization of this enzyme are contradictory, reflecting shortcomings of available techniques. [Reis' biochemical data[123] for relative activity in white and in gray matter are suspect, as there is no indication of proper precautions being taken to prevent loss of enzyme.] The bulk of evidence is that, unlike nonspecific alkaline and acid phosphatase, it is more concentrated in white matter than in gray, at least in rats.[104,107,154] Here, it occurs in fiber tracts where it is associated with the myelin sheath[18,104,106,107]; it is also concentrated in the pia and arachnoid membranes,[104,107] in the choroid plexus,[106] and in ependymal cells.[77] Unlike alkaline phosphatase, it is not present in capillaries. In the hypothalamus of the squirrel monkey (*Saimiri sciureus*), nuclei containing neurosecretory neurons (e.g., the supraoptic and paraventricular nuclei) stain more deeply for this enzyme than other nuclei do. In these nuclei both neurons and neuropil stain moderately, while glia stain moderately to strongly.[77] Barron and Boshes[18] observed 5′-nucleotidase staining in the cytoplasm of astroglia but not in that of neurons. Tewari and Bourne,[153,156] on the other hand, found staining in the endoplasmic reticulum of neurons.

Scott[141] observed a peculiar banded pattern of high activity in the molecular layer of the mouse cerebellum. There is no evident correlation between this pattern and the structure of function or this organ; similar patterns have not been reported for other species.

At physiological *p*H (7.5) this enzyme is more active in human tissue than nonspecific alkaline phosphatase, but its activity falls and alkaline phosphatase is more active at higher *p*H.[124]

Changes in activity and distribution of this enzyme during development (*q.v.*) and in response to injury (*q.v.*) have been studied.

## 8. *Cyclic Ribonucleoside 2′,3′-Phosphodiesterase*

A specific phosphatase which cleaves the ring of cyclic ribonucleoside 2′,3′-phosphates to produce ribonucleoside 2′-phosphates has been obtained from brain and other nervous tissues of mammals[48,83] and other vertebrates.[83] This enzyme has an optimum *p*H range of 5.5 to 7.5 and is inactive against cyclic ribonucleoside 3′,5′-diphosphates.[46] It is firmly bound to myelin and, consequently, is richest in brain regions (e.g., corpus callosum and cerebral white matter) which contain myelinated fibers.[83] It is either absent or present in low concentrations in non-nervous tissue.[46]

## 9. *Cyclic 3′,5′-Nucleotide Phosphodiesterase*

Cyclic 3′,5′-nucleotide phosphodiesterase has been found in nerve-ending particles from rat brain homogenates.[33,34] An appreciable amount of enzyme, which may have escaped from ruptured particles was observed dissolved in the supernatant. The activity of enzyme associated with nerve endings is increased by the detergent "Triton X-100," while the activity of dissolved enzyme is not affected. This suggests that much of the activity in normal tissue is latent, being masked by the structures containing the enzyme.[34] Because this enzyme is inhibited by physiological concentrations of inorganic pyrophosphate, citrate, and all nucleotide triphosphates, Cheung[33] suggests that it may help control adenosine 3′,5′-phosphate (cyclic AMP) concentration.

## 10. *Phosphatidic Acid Phosphatase*

Chicken brain microsomes contain water-insoluble phosphatidic acid phosphatase (3.1.3.4) which can be solubilized by deoxycholate.[119,120] This enzyme hydrolyzes α- and β-glyceryl phosphate at about one-seventh the rate for phosphatidic acid, but does not hydrolyze lecithin. Unlike many phosphatases, its activity is not affected by $Mg^{2+}$ ions. Changes during development have been studied (*q.v.*). In rat and pig brain, this enzyme is also entirely in particulate fractions.[11,134] The enzyme from pig brain also hydrolyzes long-chain alkyl phosphates at a rate comparable to or greater than the rate of hydrolysis of phosphatidic acid.[11]

## 11. *Phospholipase C*

Roitman and Glatt[129] have isolated phospholipase C (3.1.4.3) from rat brain, an enzyme which hydrolyzes sphingomyelin to *N*-acyl sphingosine (ceramide) and phosphoryl choline, and lecithin to 1,2-diglyceride and phosphoryl choline. Interestingly, although both these phospholipids are important constituents of nervous tissue, this is the first report of phospholipase C in this tissue.

## 12. *Phosphoprotein Phosphatase*

The enzyme phosphoprotein phosphatase (3.1.3.16) was isolated from beef brain and partially purified by Rose and Heald,[131] who showed the activity to be distinct from nonspecific acid or alkaline phosphatase. It hydrolyzes α-casein, but releases only a small part of the protein phosphate. It does not hydrolyze simple phosphate monoesters or inorganic pyrophosphate. It is activated by $Mg^{2+}$ and $Mn^{2+}$ ions. The optimum *p*H is 5.4. In guinea pig brain, activity is divided approximately equally between soluble and bound enzyme.[130]

## 13. *Neurokeratinal Phosphatases*

Tewari and Bourne[151,152] studied the phosphatase activity of myelinated nerves by applying Gomori-type stains with several substrates to fresh-frozen and fixed sciatic nerves from a large variety of mammals and other vertebrates, including man. The myelin of nerves from all species showed a similar pattern of staining for the following enzymes: nonspecific alkaline phosphatase (using the substrate glycerol phosphate), nucleoside 5′-monophosphatase, nucleoside diphosphatase, nucleoside triphosphatase, and glucose 6-phosphatase. The stain revealed a honeycomb pattern of more or less hexagonal tubes extending radially from the central core of axon. The authors believe that this pattern coincides with the network of neurokeratin. Unfortunately, their techniques could not reveal whether the activity was due to a single phosphatase or, as is more likely, to several phosphatases associated with the neurokeratin.

Unlike the other phosphatases, acid phosphatase activity did not appear in the myelin, but was restricted to the axon and periaxon.

## IV. DISTRIBUTION AND LOCALIZATION OF ALKALINE PHOSPHATASE[5]

The histochemical distribution of alkaline phosphatase (3.1.3.1) has been studied for several vertebrate species.[84,145,146,164] Both the activity and the distribution within different regions of the vertebrate nervous system, and even within individual cells, vary greatly, each species having its own characteristic pattern.[146] Activity is higher in gray matter than in white. Representative ratios are 2 to 1 for dog brain[94] and 4.8 to 1 for rhesus monkey (*Macaca mulatta*).[57] Using both biochemical assays and a quantitative azo-coupling stain, Friede prepared a detailed map of alkaline phosphatase in the brain of the rhesus monkey[57] with the results listed in Table I.

The predominant concentration of alkaline phosphatase in brain tissue is in the blood vessels.[78,84,108,138,146,154] As Fig. 1 shows, activity can be

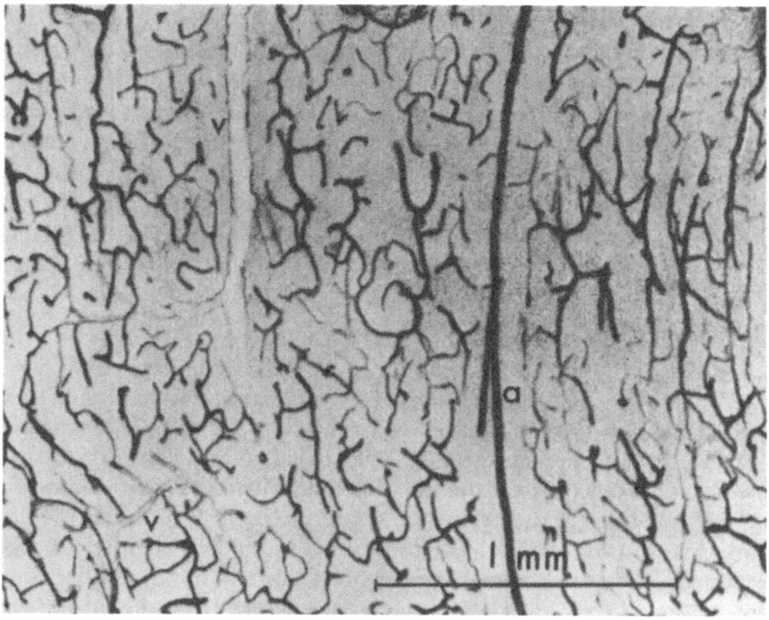


Fig. 1. Alkaline phosphatase activity in the cerebral cortex of a child. Staining occurs in blood vessels only, decreasing gradually from the arteries (a) toward the veins. Large veins (v) show little staining. Simultaneous azo-dye coupling method was used. From Friede.[5]

so concentrated that this is the only site of histochemical staining. Brain is not unique in this. In most other tissues, including those with other characteristic sites of high alkaline phosphatase activity (e.g., brush border of gut), capillaries have high concentrations of this enzyme. Characteristically, some alkaline phosphatase staining occurs at the surface of brain tissue and diffusely in the neuropil of gray matter. In brain, activity in blood vessels is high in small arteries, decreases gradually in the capillary network, and is low or absent in veins.(16) This activity may be located in the cytoplasm of endothelial cells.(16) In addition to being more vascular than white matter, the level of activity associated with blood vessels is higher in gray matter. A zone of perivascular activity which has been attributed to pericytes and glia is frequently observed.(146)

The staining of central nervous tissue, especially nuclei and tracts, is extremely diverse. In some species the neuropil of nuclei shows only a slight diffuse staining; in others (e.g., guinea pig) certain nuclei and tracts stain intensely while others stain only slightly.(145) In the squirrel monkey (*Saimiri sciureus*), hypothalamic nuclei containing neurosecretory cells (i.e., the supraoptic and paraventricular nuclei) stain more darkly than surrounding tissue. This staining is largely confined to the capillaries. The neuropil and neuronal cell bodies stain only slightly, and the glia do not stain.(77)

With some exceptions, there is a more or less continuous sheath of alkaline phosphatase activity covering all internal and external surfaces of the central nervous system, including the spinal cord. This layer includes the superficial portion of the molecular layer of the cerebral cortex, the surface of the brainstem and the cerebral ventricles, and the attachment of the cranial nerves to the brain. Activity is also found in the choroid plexus,(78) where it may be concentrated in the epithelial layer.(146) In general there is little activity in the ependymal cells.(84,145) The sheet of phosphatase activity is apparently located instead in the underlying endothelial cells and the capillaries.(84) In some species strong activity is observed in the pia-arachnoid membrane(78,108,145); other species, however, including man, have little activity in the meninges.(84) Cellular elements of the cerebrospinal fluid lack activity.(84)

In many species alkaline phosphatase is found in the sheaths and blood vessels of cranial and peripheral nerves. Characteristic sharp concentrations of enzyme occur at the nodes of Ranvier and the Schmidt-Lanterman incisures. Because the staining shows a different sensitivity to *p*H, this "nodal" enzyme may not be the same as capillary alkaline phosphatase.(116,117)

Histochemical studies of the microdistribution of alkaline phosphatase have given inconsistent results even when made by the same workers. Although much of this was undoubtedly due to technical factors, some must have been caused by species variation and by differences in different cells of the same class, for example, different types of spinal neurons (see below). There is little or no enzyme in peripheral axons, or in most other structures

## TABLE I

**Distribution of Acid Phosphatase Activity and Alkaline Phosphatase Activity in the Brain of the Rhesus Monkey[a]**

| Region | Type of region[b] | Acid phosphatase[a] | Alkaline phosphatase[a] |
|---|---|---|---|
| Medulla oblongata | | | |
| Gracile nucleus | — | — | 80 |
| Nucleus of descending trigeminal nerve tract | — | 83 | 104 |
| Nucleus of solitary tract | G | 110 | 105 |
| Reticular nucleus | M | 66 | 57 |
| Lateral vestibular nucleus | G | 87 | 66 |
| Medial vestibular nucleus | G | — | 116 |
| Pyramidal tract | W | 27 | 21 |
| Nucleus of facial nerve | — | — | 108 |
| Cerebellum | | | |
| Cortex | — | 114 | 116 |
| Fastigial nucleus | M | — | 59 |
| Dentate nucleus | M | — | 17 |
| White matter | W | 28 | 17 |
| Pons | | | |
| Nucleus of abducent nerve | G | 91 | 119 |
| Superior vestibular nucleus | G | 91 | 86 |
| Nucleus pontis (compact pts.) | M | 73 | 79 |
| Brachia pontis | W | 32 | — |
| Pyramidal tract | — | — | 40 |
| Superior pedunculus | W | — | 21 |
| Midbrain (inferior colliculus) | | | |
| Inferior colliculus | G | 86 | 150 |
| Central gray matter (dorsal portion) | G | 99 | 93 |
| Nucleus of median raphe | M | 84 | 67 |
| Reticular nucleus | M | 73 | 43 |
| Midbrain (superior colliculus) | | | |
| Superior colliculus (superficial layer) | G | 92 | 97 |
| Superior colliculus (deep layers) | G | 68 | 49 |
| Central gray matter | G | 103 | 62 |
| Nucleus of oculomotor nerve | — | — | 124 |
| Nucleus niger | G | 93 | 104 |
| Cerebral pedunculi | W | 41 | 13 |
| Reticular nucleus | M | 66 | — |
| Diencephalon | | | |
| Lateral thalamic nucleus | G | 83 | 109 |
| Region of dorsomedial nucleus | G | 89 | 93 |
| Lateral geniculate nucleus | — | 78 | 140 |
| Internal capsule | W | 34 | 23 |

TABLE I—*continued*

| Region | Type of region[b] | Acid phosphatase[a] | Alkaline phosphatase[a] |
|---|---|---|---|
| Diencephalon—*cont.* | | | |
| Optic tract | W | 33 | 28 |
| Tuber cinereum | G | 97 | 112 |
| Subthalamic nucleus | — | 107 | 177 |
| Basal telencephalic centers | | | |
| Caudate nucleus (caput) | G | 100 | 83 |
| Putamen (anterior portion) | G | 97 | 100 |
| Putamen (posterior portion) | G | 85 | |
| External pallidum | M | 75 | 44 |
| Internal pallidum | M | 83 | |
| Internal capsule | W | 43 | — |
| Frontal cortex | | | |
| Cortex | G | 99 | 132 |
| White matter | W | 42 | 24 |
| Parietal cortex | | | |
| Cortex | G | 101 | 177 |
| White matter | W | 35 | 32 |
| Occipital cortex | | | |
| Cortex (area 18) | G | 102 | 115 |
| Cortex (area 17) | G | 90 | |
| White matter | W | 50 | 19 |
| Temporal cortex | | | |
| Cortex | G | 125 | 160 |
| White matter | — | 48 | 35 |
| Hippocampus (selective sampling of pyramidal layer of fascia dentata and hippocampus) | — | 127 | — |
| Hippocampus (general) | — | — | 98 |
| Average for pure samples of white matter | W | 36 | 22 |
| Average for pure samples of gray matter | G | 95 | 106 (all regions) 96 (all except cortex) 146 (cortex only) |
| Average for samples of gray matter that invariably include transfixing fiber bundles | M | 73 | 69 |

[a] All activities are expressed as a percentage of that of the caudate nucleus or putamen, respectively. The values are averages from measurements of the histochemical simultaneous azo coupling method in three monkeys each, with an average of 10 measurements for each region in each monkey [acid phosphatase, Friede and Knoller(59); alkaline phosphatase, Friede(57)].

[b] Notation: W, regions of pure white matter; G, pure samples of gray matter; and M, (mixed) samples of gray matter with transfixing fiber bundles [from Friede(5)].

## TABLE II
## Nonspecific Alkaline and Acid Phosphatase Activity in Selected Regions of the Mammalian Central Nervous System[a]

| Region | Mammal | Alkaline phosphatase | Acid phosphatase | Reference |
|---|---|---|---|---|
| Cerebellum | Rhesus monkey | | | 125[b] |
| Hemisphere | | | | |
| Molecular layer | | 0.37 ± 0.04 | 1.7 ± 0.2 | |
| Granular layer | | 0.27 ± 0.04 | 1.2 ± 0.1 | |
| Subjacent white matter | | 0.087 ± 0.006 | 0.32 ± 0.02 | |
| Vermis | | | | |
| Molecular layer | | 0.28 ± 0.02 | 1.8 ± 0.2 | |
| Granular layer | | 0.31 ± 0.04 | 1.18 ± 0.05 | |
| Subjacent white matter | | 0.084 ± 0.01 | 0.31 ± 0.04 | |
| Cerebellum | Rhesus monkey | | | 72[c] |
| Molecular layer | | | 0.11 ± 0.01 | |
| Granular layer | | | 0.29 ± 0.03 | |
| White matter | | | 0.036 ± 0.002 | |
| Purkinje cell layer | | | 0.36 ± 0.05 | |
| Hippocampus | Rabbit | | | 90[d] |
| Alveus | | 0.24 ± 0.02 | 1.67 ± 0.08 | |
| Oriens | | 0.25 ± 0.01 | 4.58 ± 0.05 | |
| Pyramidial layer | | 0.23 ± 0.01 | 3.2 ± 0.1 | |
| Radiata | | 0.29 ± 0.01 | 3.9 ± 0.1 | |
| Lacunosum | | 0.32 ± 0.01 | 3.0 ± 0.2 | |
| Molecular layer | | 0.34 ± 0.01 | 3.8 ± 0.1 | |
| Average | | 0.60 | 2.6 | |
| Anterior horn | Rabbit | | | 89[e] |
| Motor neurons | | 0.08 ± 0.03 | | |
| Neuropil | | 0.14 ± 0.06 | | |
| Dorsal root ganglion | Rabbit | | | 89[e] |
| Cell bodies | | 1.3 ± 0.4 | | |
| Myelinated fibers | | 0.03 ± 0.01 | | |
| Anterior horn | Man | | | 72[c] |
| Motor neurons | | | 3.7 ± 0.4 | |
| Neuropil | | | 0.16 ± 0.02 | |
| Anterior horn | Rhesus monkey | | | 72[c] |
| Motor neurons | | | 1.5 ± 0.4 | |
| Neuropil | | | 0.15 ± 0.01 | |

[a] All activities were determined by microchemical methods using unfixed, freeze-dried tissue. All activities were given in moles substrate consumed per kilogram dry weight per hour, except for reference 90 where activity was given per kilogram protein. Where necessary, average values and errors were computed and values were rounded off to the proper number of significant figures. Activities within any one study may be compared, but activities from different studies are only roughly comparable because of differences in technique.

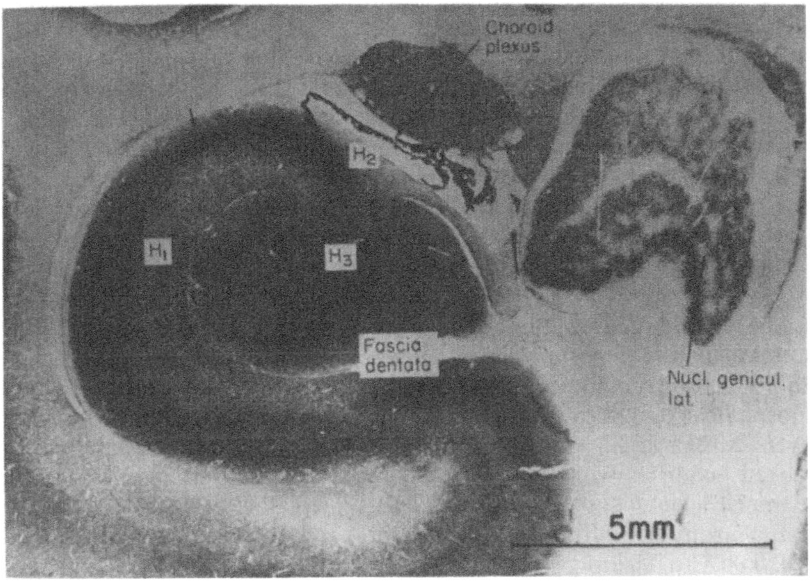


Fig. 2. Acid phosphatase activity in the hippocampus of a rhesus monkey. Note the great differences in the staining in adjacent regions. Stain is intense in segment $H_2$ and in the cell band of the fascia dentata. The lateral geniculate nucleus shows much less stain than the hippocampus. Note the intense stain in the choroid plexus, which is not continuous with the ependyma. From Friede.[5]

---

*Footnotes to Table II continued*

[b] Incubated at 38°C with *p*-nitrophenyl phosphate substrate and $MgCl_2$ activator. Succinate buffer at *p*H 5.3 was used for acid phosphatase, and glycine buffer at *p*H 10.2 was used for alkaline phosphatase. Tissue samples were 0.7 to 5 μg dry weight (3.5 to 25 μg wet weight).

[c] Incubated at 38°C with α-naphthyl acid phosphate substrate and $MgCl_2$ activator in acetate buffer at *p*H 5.3 containing bovine serum albumin to minimize surface inactivation. Tissue samples were 0.3 to 0.8 μg dry weight for cerebellum, 0.005 to 0.015 μg dry weight for motor neurons, and 0.04 to 0.08 μg dry weight for neuropil.

[d] Incubated at 38°C with *p*-nitrophenyl phosphate substrate and $MgCl_2$ activator. Succinate buffer at *p*H 5.3 was used for acid phosphatase, and glycine buffer at *p*H 10.0 for alkaline phosphatase. Tissue samples were 1 to 4 μg dry weight (5 to 20 μg wet weight).

[e] The assay procedure is not given, but presumably it was similar to the alkaline phosphatase assay in footnote *d* above. Tissue samples were 0.009 to 0.046 μg dry weight. The investigator believes his values to be low.

not mentioned so far. In most species, Schwann cells have little activity; such activity as has been reported may be an artifact. Activity is low in most neuronal cell bodies[108,154]; when present, it is localized at the cell membrane.[156] Cerebellar Purkinje cells stain variably, some having sparse activity while others in the same specimen do not.[154] Glial cell bodies are inactive.[154] The enzyme in the neuropil may be localized at least partly at the synapses.[108,154]

Several microchemical studies of the enzyme distribution in selected brain regions have been made. The results from three such studies are listed in Table II. In the first, Robins and Smith[125] assayed 0.7- to 5-$\mu$g dry weight samples of freeze-dried cerebellar tissue from the rhesus monkey; in the second, Lowry and co-workers assayed 1- to 4-$\mu$g dry weight samples of rabbit hippocampus.[90] The experimental uncertainty in these studies is too great for conclusions to be drawn from differences in alkaline phosphatase activity of different layers in gray matter, even though some layers consist largely of neuronal cell bodies, while others contain closely packed dendrites with few cell bodies. In the third study, Lowry, using ultramicrochemical methods, measured the alkaline phosphatase in individual neurons and in fragments of neuropil and myelinated fiber of the same size, 0.009- to 0.046-$\mu$g dry weight, from rabbit spinal cord.[89] As Table II shows, the activity in dorsal root neurons is over 15 times that in anterior horn cells. Such great differences may partly explain divergent histochemical studies of intracellular distribution.

## V. DISTRIBUTION AND LOCALIZATION OF ACID PHOSPHATASE[5]

The histochemical distribution of acid phosphatase (3.1.3.2) within the nervous system has been studied for several species.[14,15,106,145,146,165] The distribution is similar for most mammalian species[14]; in this respect it differs from the distribution of alkaline phosphatase. Acid phosphatase levels, however, vary greatly with species. Acid phosphatase is found in both gray and white matter. It is concentrated in gray matter but less so than alkaline phosphatase. Typical values for the ratio of activities are 2.1 to 1 for dog brain,[94] 2.6 to 1 for rhesus monkey brain,[59] and 2.5 to 1 for rabbit spinal cord.[137] Friede and Knoller, using both a quantitative azo-coupling stain and assays of homogenized tissue have mapped the acid phosphatase distribution in the brain of the rhesus monkey with the findings shown in Table I.[59]

The histological distribution of acid phosphatase frequently shows extremely sharp boundaries between adjacent regions. Activity in the hippocampus and fascia dentata is sharply banded (Fig. 2) with obvious contrasts between the deeply stained $H_2$ segment of the hippocampus and the $H_1$ and $H_3$ segments. Similarly, in the cerebellar cortex the strongly staining layer of Purkinje cells contrasts sharply with surrounding tissue (Fig. 3).[106,154]

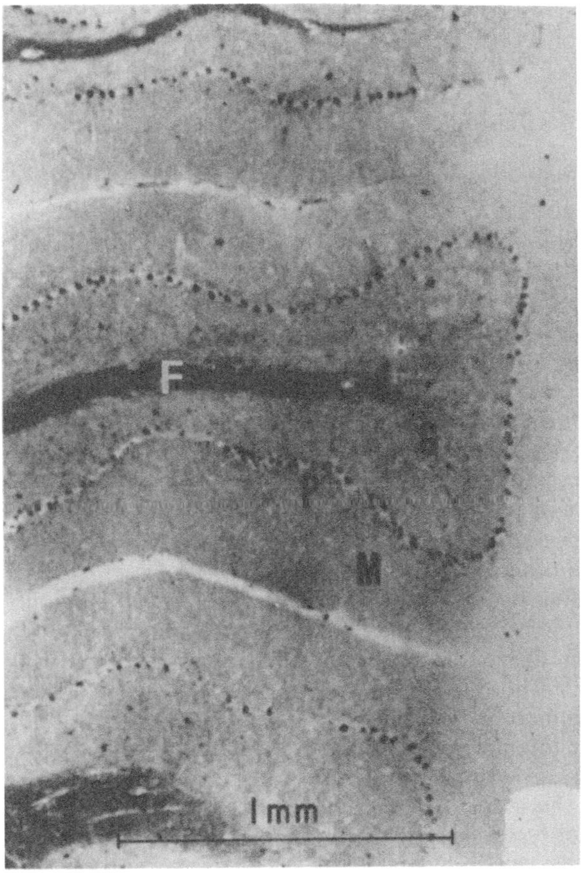


Fig. 3. Acid phosphatase activity in the cerebellum of a rat. Highest staining appears in the Purkinje cells (P). The fiber tracts (F) stain a light brown in contrast to the black color in the neuron cell bodies. At this low magnification, the molecular layer (M) and the granular layer (G) appear unstained. Section was stained for 30 min at 37°C using the Gomori method. From Becker *et al.*[24]

Hypothalamic nuclei containing neurosecretory neurons (e.g., the supraoptic nucleus and the paraventrical nucleus) also have marked activity in contrast to surrounding tissue.[37,77,106,136]

Acid phosphatase in gray matter is present in both neuropil and cell bodies. High activity in cell bodies is often accompanied by low activity in neuropil and *vice versa*. This inverse relation minimizes differences between nuclei. The sparse activity in white matter is observed in all types of glia and in thick axons. (The activity in axons may be at least partly artifactual). High concentrations of acid phosphatase occur in the epithelium of the choroid plexus and in the ependymal cells.[14,15,24,77,145,164] The activity here is mainly concentrated in the brush border of cells.[78] The meninges also show some activity.[109,145] Blood vessels show activity,[24,109] but in contrast to alkaline phosphatase it is not exceptionally high.[14,78] Axons of peripheral nerves and their myelin sheaths have little or no activity; the activity of uninjured peripheral nerves is apparently confined to the perinuclear zone of their Schwann cells.[14,15] The staining found by Wolf *et al.*[165] and by de Sibrik and O'Doherty[146] probably represents nonspecific interaction with lead ions from the Gomori stain.

On a cellular level, neuronal cell bodies show acid phosphatase staining, with larger neurons (e.g., large pyramidal cells in the cerebral cortex, Purkinje cells and large neurons in the dentate nucleus of the cerebellum, and neurons in the magnocellular nuclei of the hypothalamus) staining more heavily than smaller ones.[14,24,77,95,136,165] Studies of glia are inconsistent. Astrocytes and oligodendroglia show little activity.[15,24] Becker *et al.*[24] observed considerable acid phosphatase activity in microglia, while Anderson and Song[14] observed little activity. Nandy and Bourne[109] report that glia associated with strongly staining neurons have more activity than others.

Intracellularly, acid phosphatase is largely confined to lysosomes.[14,15,24,82,109] These are found in all types of cells, but the number present, their size, and staining properties vary, even for cells of the same type.[27,30,157] Those cells which stain most heavily (see above) have larger and more numerous lysosomes.[24] Neuroglia have few lysosomes, which reflects their low phosphatase activity. Although they are concentrated in the cell body, lysosomes are sometimes also found in the dendrites, and possibly in the axons, of neurons, especially those neurons with a high population of lysosomes.[24,82,154] They apparently migrate into these processes from the cell body where they are formed.[157] They also occur in the processes as well as cell bodies of glia[82] and in the perikaryon of endothelial cells. Reports on the distribution of lysosomes in the perikaryon of neurons differ. Becker *et al.*[24] found them to be randomly distributed. Others, however, have found striking differences in distribution within the same type of cell[153,156,157] and have proposed a theory of the function of acid phosphatase based on these variations (*q.v.*). Localized concentrations of acid phosphatase at synapses have been detected by both light microscopy and electron microscopy[67,109,154]; however, Kreutzberg and Hager, using electron microscopy, found no evidence of synaptic phosphatase.[82] The acid phosphatase content of cells sometimes increases with age. Thus, Bergmann cells in the human cerebellar cortex have little activity in infants, but are even more active than Purkinje cells in adults.[115] Lipofuscin deposits in cells are generally accompanied by significant acid phosphatase activity.[115,147] This is located in the pigment granules,[82] which may represent altered lysosomes.[14]

The results of several microchemical studies on the acid phosphatase distribution in selected nervous tissues are listed in Table II. Because of differences in method, only relative values, not absolute values, from different studies can be compared. These data confirm histochemical observations of extremely sharp gradations in enzyme activity. Although in the spinal cord the activity in nerve cell bodies is significantly greater than the activity in neuropil, there is no clear distinction between neurons and neuropil in the hippocampus or in the cerebellar cortex. It is hard to reconcile Robins and Smith's findings(125) and Hirsch's findings(72) for acid phosphatase in monkey cerebellar cortex. The differences in absolute activity and in the relative activity of the molecular layer compared to the granular layer seem too great to be due entirely to technical differences. Tewari and Bourne's histochemical studies of rat cerebellum(154) agree with Robins and Smith's results by showing greater staining in the molecular layer than in the granular layer.

## VI. INVERTEBRATES

Nonspecific alkaline and acid phosphatase have been found by staining techniques in the retina and in all cells bodies in the optic lobe of cuttlefish (*Sepia officinalis*),(46,47) squid (*Loligo vulgaris*),(46) and octopus (*Octopus vulgaris* and *Eledone moschata*).(46) Neither enzyme is found in capillaries. By contrast, the most characteristic site for alkaline phosphatase in vertabrate neural tissue is arteries and capillaries. Neither enzyme is found in nerve fibers,(46) except possibly in cuttlefish where some material in the fibers sometimes stains for alkaline phosphatase.(47) This positive reaction may be an artifact. In cuttlefish, the cytoplasm of the optic lobe cell bodies stains diffusely for alkaline phosphatase; nuclei do not stain.(47) Staining for acid phosphatase is very irregular. Some optic lobe cell bodies stain deeply, others lightly, with no correlation with either cell size or location.(47)

In snails (*Helix pomatia*), Domján and Minker(45,96) using Gomori staining techniques found acid phosphatase in cells of the central nerve ganglia and of the intramural ganglia of the intestinal canal. Staining for this enzyme was located in the cell cytoplasm and in those portions of cell processes next to the cell body. Processes forming nerve bundles were unstained. They observed staining for alkaline phosphatase in the "intracellular substance" of gangila but, surprisingly, they observed no staining in nerve cells or cell processes, or near nerve elements. This observation is most likely either an artifact or a reflection of a relatively low activity in neuronal bodies compared with intracellular substance. In the related species, *H. aspersa*, both acid phosphatase and thiamine pyrophosphatase that is distinct from nonspecific phosphatases are associated with the phospholipid lamellae which form the surface of certain lipid globules in the neuronal cytoplasm.(85) This contrasts with vertebrates where acid phosphatase is associated with lysosomes, and thiamine pyrophosphatase with the Golgi apparatus. Alkaline phosphatase is associated somewhat more diffusely with these same lipid bodies which may be the equivalent of the vertebrate Golgi apparatus.

Motor neurons in thoracic ganglia of locusts (*Locusta migratoria*) contain both acid phosphatase-staining particles and thiamine pyrophosphatase-staining particles, but the cytological localization is quite different from that in vertebrates or snails.[87] The acid phosphatase is found in bodies with the microstructure and osmium tetroxide-staining characteristic of vertebrate Golgi apparatus; these structures may represent the equivalent of the Golgi apparatus in the insect neuron. The thiamine pyrophosphatase is restricted to smaller spheroidal bodies which lack both osmium tetroxide-staining and the structure of Golgi apparatus.

## VII. CHANGES DURING DEVELOPMENT

Unlike most enzymes (e.g., ATPase, succinic dehydrogenase, cytochrome *c*) which increase steadily to adult values,[55] the activity of both alkaline and acid phosphatase in the central nervous system is greater during certain stages of development than it is in the adult. A peak of phosphatase activity also occurs in several other immature tissues, including some with little activity in the mature animal.[98]

Biochemical studies have been made of alkaline phosphatase activity in the immature brain of various species, both before and after birth. The alkaline phosphatase activity in rat brain from birth to maturity, per unit dry weight of tissue, is shown in Fig. 4. A peak of enzyme activity, $2\frac{1}{2}$ to 3 times the activity in adult brain, is observed in the hypothalamus 10 days after birth and in the cerebral cortex 15 days after birth. In the thalamus, activity decreases from the 15th day after birth (data for younger animals were not given). Adult values are reached by about the 40th day.[37] Although obscured by experimental scatter, a similar pattern was observed in guinea pigs, where alkaline phosphatase activity in the cerebral cortex is approximately constant at 2 times adult levels from the 27th day of gestation (the earliest embryo studied) to birth.[56] Because of the greater maturity of this animal at birth, the peak is prenatal. A variant of this pattern is found in chickens.[128] Enzyme activity in most parts of the brain decreases from the 6th day of incubation to a minimum at about the 10th day, then increases to a maximum at the 17th to 20th day, depending on the region, and finally decreases to adult values. The sharp rise from the 10th day occurs shortly before the rapid increase in acetylcholine esterase activity and corresponds to the period of greatest mitotic activity,[127] morphological differentiation, and synthesis of RNA, protein, and phospholipids.[128]

Comparative histochemical studies have been made of alkaline phosphatase in the developing nervous system of various species. Although they range from the earliest stages of differentiation to postnatal maturation, no one study includes all stages. Activity appears early in embryonic

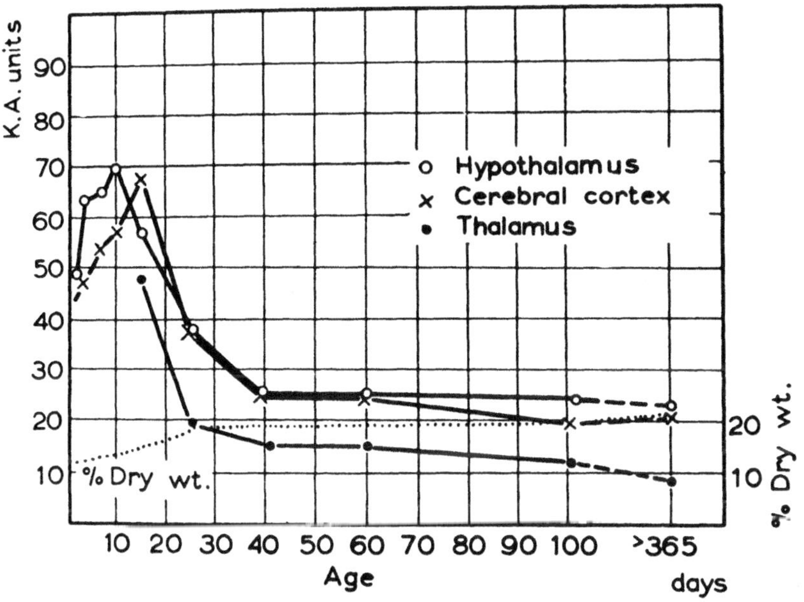


Fig. 4. Alkaline phosphatase activity and percentage dry weight in the rat brain during growth. Assays of tissue suspensions were done by the method of King and Armstrong [*Can. Med. Assoc. J.* **31**:376 (1934)] using disodium phenyl phosphate substrate for 15 min at *p*H 9.8 and 37°C in the presence of 1 mM magnesium chloride. Activities are expressed in King–Armstrong units: milligrams phenol liberated under standard conditions per gram dry tissue. From Cohn and Richter.(37)

development. Intense activity is observed in the neural tissue of 1-day-old chick embryos(97) and the brain of 2-day chick embryos(98) and in the neural tube of 18- to 26-day human embryos, where the activity is greater than anywhere else.(132) In mice, alkaline phosphatase activity increases to a maximum during the earlier stages of morphological differentiation, then decreases as differentiation progresses. Activity appears in the neural plate by the 6th day after fertilization, increases to a maximum as the neural plate differentiates into the spinal cord and the brain vescicles, and then decreases in most parts of the central nervous system after the 11th day. At the 10th and 11th days nerve fibers growing from ventral roots stain strongly, permitting them to be traced easily through adjacent tissues.(35) A similar pattern of changes has been observed in fetal rats.(92,103)

The capillaries, especially those of the choroid plexus, stain more heavily than nervous tissue in newborn rats.(37) Fiber tracts (e.g., the optic chiasma and the corpus callosum) also stain markedly; this staining of fiber tracts compared to adjacent gray matter decreases in mature animals.(37) Unhatched chickens show a similar pattern of activity.(126) Fiber tracts in the brain stain well during the later stages of embryonic development (the

14th day of incubation to hatching), a period of active myelination and of glial staining for alkaline phosphatase; some of these tracts no longer stain in adult birds. In mouse brain only the capillaries of the choroid plexus, dura mater, and medulla stain for alkaline phosphatase at birth. By the 10th day after birth, capillaries in the midbrain and hindbrain also stain and by the 20th day capillaries show the adult pattern of activity, except in the cerebellum, where they stain somewhat less heavily.[88]

Rogers found Gomori staining for alkaline phosphatase in the nuclei of chick neuroblasts.[126] This is evidence for the occurrence of this enzyme in these cells but not necessarily for its localization in nuclei.[127] At a later stage (14th day of incubation to hatching), most neurons are unstained or only lightly stained by the Gomori procedure.[126]

A comparison for various species of histochemical observations on the earliest stages of neural differentiation with histochemical and biochemical studies of the later stages of development suggests that there may be two separate peaks of alkaline phosphatase activity. If this is so, the observed drop in enzyme activity in chick embryos between the 6th and 10th days of incubation may represent the decrease following the first peak.

The postnatal changes in acid phosphatase activity of maturing rat brain per unit dry weight of tissue are shown in Fig. 5. At birth the activity is roughly at adult levels. It increases to a maximum of 1½ times adult activity at 10 to 15 days after birth, then falls to adult values by about the 40th day after birth. No difference is observed between cerebral cortex, thalamus, and hypothalamus.[35] Despite bad scatter, biochemical data for guinea pig cerebral cortex show a similar pattern. A broad maximum of twice adult activity occurs from the 33rd day of gestation (the earliest embryo studied) to birth.[55] Maximum activity occurs prenatally in the guinea pig because of its greater maturity at birth. Comparison of Fig. 4 with Fig. 5 (taken from the same study) shows that the rise and fall of acid phosphatase activity is smaller and more gradual with a broader maximum than that of alkaline phosphatase activity. The periods of maximum enzyme activity coincide, however, with the peak of acid phosphatase activity embracing the separate, sharper peaks of alkaline phosphatase activity in the cerebral cortex, the thalamus, and the hypothalamus.

The total acid phosphatase activity (soluble and bound) per unit weight of protein in low-speed supernatants from homogenates of mouse cerebral hemispheres treated with Triton X-100 also shows changes in activity with maturation similar to those in rat and guinea pig brain. In 1-day-old animals the activity is about 20% greater than in adults. The activity reaches a maximum of about 1½ times adult activity by the 5th day after birth, gradually declines to adult values by the 40th day, then remains essentially constant until early senescence at the 500th day after birth. The changes in activity of "free" (soluble) enzyme per unit weight of protein in similar supernatants from homogenates that were not treated with Triton X-100 are quite different. The activity is constant at about 70% of activity of free enzyme in adults from the 1st to the 10th day after birth, rises to adult

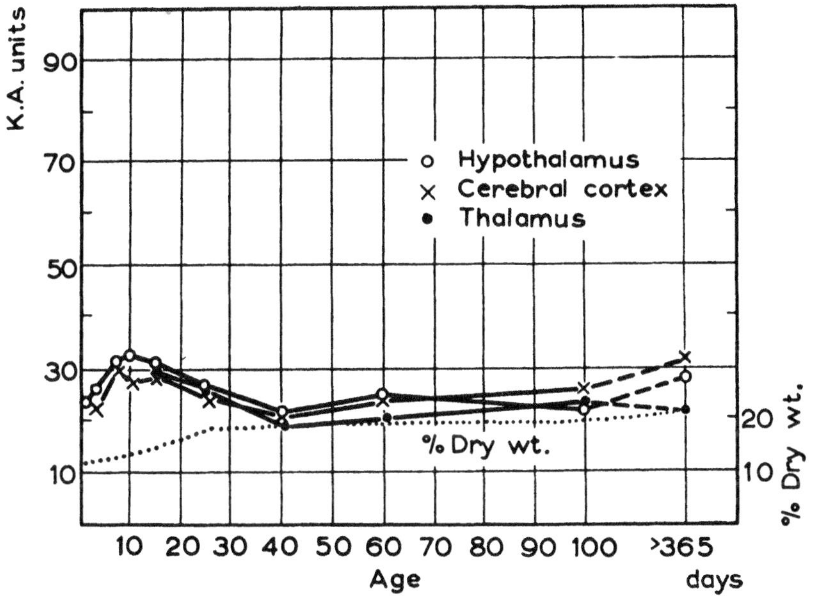


Fig. 5. Acid phosphatase activity and percentage dry weight in the rat brain during growth. Assays of tissue suspensions were done by the method of King and Armstrong [*Can. Med. Assoc. J.* **31**:376 (1934)] using disodium phenyl phosphate substrate for 60 min at *p*H 4.9 and 37°C. Activities are expressed in King–Armstrong units: milligrams phenol liberated under standard conditions per gram dry tissue. From Cohn and Richter.[37]

values between the 10th and the 20th day, then remains constant until early senescence. The comparison of free enzyme activity to total activity is interesting, even though this ratio depends on the experimental procedures. On the 1st day, free activity is about 40% of total activity; at the peak of total activity on the 5th day, free activity is 35% of total; and in the mature animal after the 40th day, it is 70 to 75% of total activity.[162] Because free acid phosphatase activity is always a significant fraction of total acid phosphatase activity, the moderate decrease in total activity with maturation must come from a great decrease in bound enzyme. (Since histochemical staining techniques are restricted almost entirely to insoluble enzyme, they exaggerate the decrease upon maturation.) In microsomes from embryonic chicken brain, acid phosphatase, which presumably consists entirely of bound enzyme, decreases rapidly between the 5th day of incubation and the 10th day, then gradually drops to adult levels.[119]

Histochemically, acid phosphatase has been demonstrated in neural tissue of 1-day-old chick embryos.[98] These same studies have furthermore shown a qualitative decrease in activity in embryonic chick nervous system during the first 8 days of incubation.[97,98] The high activity observed in these studies of early chick embryos and the high activity observed in biochemical studies on immature rats and fetal guinea pigs may represent two separate peaks of enzyme activity.

In human infants[14] and newborn rats[14,37] staining for acid phosphatase is largely limited to the choroid plexus (*cf.* alkaline phosphatase) and the ependyma. By the 10th day after birth, rats show darker staining in several hypothalamic nuclei; by the 15th day, more nuclei show appreciable staining, including those neurosecretory nuclei which have marked acid phosphatase activity in adult animals. The cerebral cortex, however, is paler than the brainstem and shows little differentiation between layers. The adult pattern for acid phosphatase staining is reached between the 15th and the 60th day.[37] The staining of neurosecretory nuclei for acid phosphatase has been observed in various animals including cow embryos,[150] and presumably it is widespread. Cytologically, neuroblasts show no acid phosphatase activity (differing from alkaline phosphatase) but immature neurons do.[14] Meyer finds that the prenatal development of acid phosphatase activity in humans parallels the maturation of neuroblasts and RNA synthesis. It is highest in cells with maximum protein synthesis, e.g., Purkinje cells and large nerve cells in the dentate nucleus.[95]

Naidoo[104,107] found that 5′-nucleotidase (3.1.3.5) activity in maturing rat brain increases very gradually, reaching adult values about 100 days after birth. The increase in activity in nerve fibers from different regions parallels the pattern of myelination in those regions. Activity arises much more slowly than for several other enzymes, e.g., ATPase, carbonic anhydrase, and cytochrome oxidase, which reach adult values 20 to 40 days after birth.

Salway *et al.* found the activity of triphosphoinositide phosphomonoesterase in rat brain increases rapidly between the 12th and the 25th day after birth, a period associated with rapid myelination.[133]

Pomazanskaya studied the development of phosphatidic acid phosphatase (3.1.3.4) activity in 1- to 20-day chick embryos, 1- to 2-day-old chicks, and adult fowl.[119,120] Enzyme activity increases rapidly from the 11th to the 16th day of incubation when it reaches adult values. At no period does the activity decrease. This contrasts strongly with the rise to a maximum and subsequent fall of alkaline and acid phosphatase activity.

## VIII. RESPONSE TO INJURY[1]

Only the effects of nerve section or crush, anoxia, and ischemia on phosphatases in the nervous system are considered in this review. Other aspects of pathology are discussed elsewhere in this *Handbook* and in Arvy's review.[2]

Of all phosphatases, the response of nonspecific acid phosphatase to injury has been studied most. As a rule, injury increases soluble acid phosphatase. As in other organs, soluble acid phosphatase increases markedly and rapidly in isolated rat brains during autolysis, although both bound and total acid phosphatase decline.[3] In the living nervous system, the total acid phosphatase increases by synthesis in undamaged cells at the site of injury (e.g., Schwann cells, glia, or pericytes, depending on the location),[15] by migration to and proliferation of phagocytes, including microglia, and other types of cells[14,15,148] (e.g., reactive Schwann cells in the peripheral nervous system), and possibly by synthesis in damaged cells. An increase in the number of microglia occurs in many pathological conditions besides trauma; the accompanying increase in phosphatase content is so characteristic that acid phosphatase staining has been used for a microglial stain.[5] The concentration of several other enzymes also increases in these reactive glia.

The increase in acid phosphatase activity in cut or crushed nerve is both large and prolonged. Typical data for the distal portion of peripheral nerve where there can be no synthesis of enzyme in the axon are shown in Fig. 6 for cut cat sciatic nerve which was prevented from regenerating and in Fig. 7 for crushed cat sciatic nerve which gradually regenerated during the course of the studies.[73] Biochemical studies of cut rabbit sciatic nerve

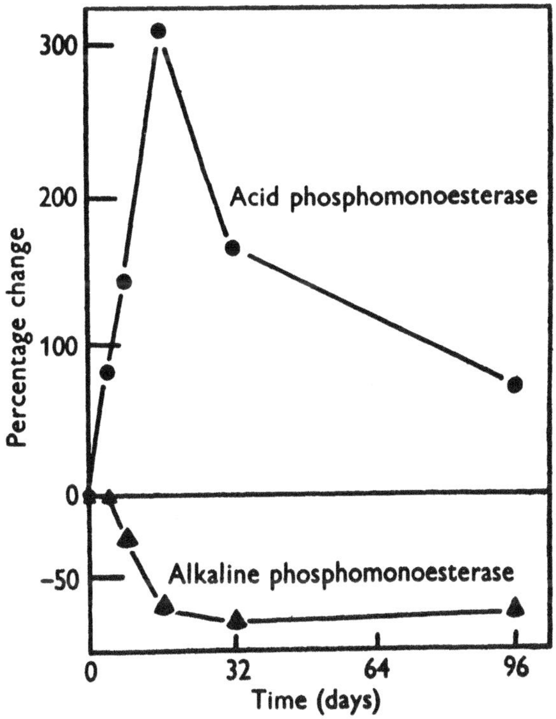


Fig. 6. Percentage change in the activity of acid phosphatase (●——●) and alkaline phosphatase (▲——▲) in the distal portion of degenerating sciatic nerve of the cat after section. Activities were measured by a modification of the method of King and Armstrong [*Can. Med. Assoc. J.* **31**:376 (1934)] using 4 mM disodium phenyl phosphate substrate with 1 mM magnesium chloride activator for 1 hr at 37°C. Citrate buffer with *p*H 4.9 was used for acid phosphatase, and carbonate–bicarbonate buffer with *p*H 9.9 was used for alkaline phosphatase. Activities were compared per unit wet weight of nerve. Each point is the average of three or more determinations. From Hollinger *et al.*[73]

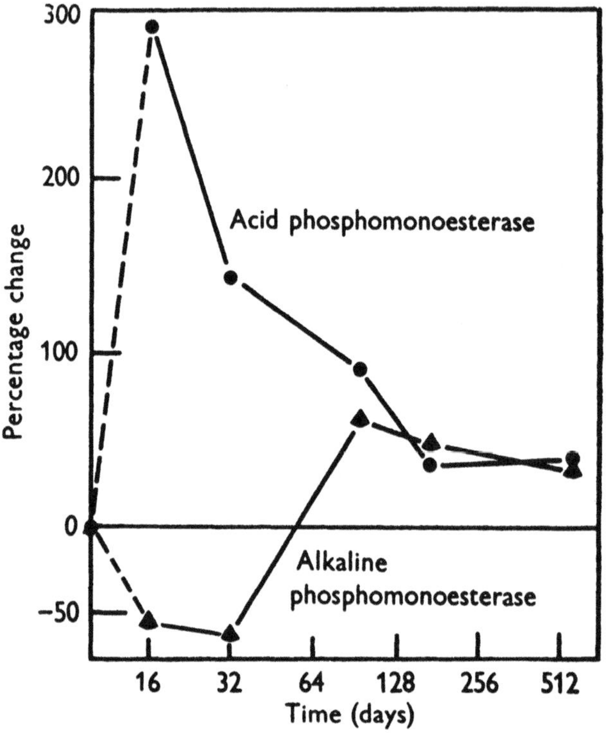


Fig. 7. Percentage change in the activity of acid phosphatase (●——●) and alkaline phosphatase (▲——▲) in the distal portion of regenerating sciatic nerve of the cat after crush (see Fig. 6 for methods). From Hollinger *et al.*[73]

gave very similar results.[93,135] A similar, but less pronounced and much slower increase occurs in rabbit optic nerve (which is distal to its missing retinal neurons), where activity reaches a maximum of $1\frac{1}{2}$ times normal from 45 to 100 days after removing the eye.[93] In the proximal portion of regenerating rabbit sciatic nerve, activity within 1 cm of the cut gradually rises until by the 13th day after the operation (when regeneration has begun to bridge the cut) it is 5 times normal. This increase is limited to the regenerating tip. Further back, in the proximal segment 1 to 2 cm from the cut, activity decreases to $\frac{1}{2}$ to $\frac{3}{4}$ of normal. In regenerating nerve, 2 months after the operation, the activity in these two proximal segments is the same, 3 times normal, compared with the 6-times-normal activity of the 1-cm distal segment immediately across the cut.[135] Much of the increase during the first 2 weeks after nerve section comes from proliferating Schwann cells[135] at and distal to the cut; however, at 2 months there is no longer any correlation between Schwann cells and activity.[135] Acid phosphatase activity in the corresponding nerve on the unoperated side has been reported to also increase markedly after peripheral nerve section.[135] Several histochemical studies have shown an increase in staining for acid phosphatase in the corresponding anterior horn motoneurons after section of a peripheral nerve, whether or not there was regeneration.[19,21,31,38] As in the severed nerve, activity increases gradually for the first 2 to 4 weeks, and increased activity may persist for 6 months.[19] Anderson and Song,[14] however, observed no difference in staining for acid phosphatase between anterior horn motoneurons of the operated and of the control sides. Unfortunately, the results of biochemical studies of the reaction of the spinal cord following peripheral nerve section are questionable. Fieschi and Soriani[53] found no change in activity of guinea pig cord. Samorajski and Fitz[137] reported that activity in both the operated and the unoperated side of rabbit cord rises threefold by 1 week after nerve section, and then returns to normal by 2 weeks after operation. In parallel histochemical studies, they observed no change in activity on either side and only slight chromatolysis in the injured motoneurons.

The effects of injury on alkaline phosphatase have been studied much less than on acid phosphatase. The patterns of change are complex, partly because the various elements of nervous tissue (e.g., blood vessels, neurons, glia, fibers, etc.) react differently and partly because of diverse species-specific patterns for this enzyme in normal tissue. Alkaline phosphatase increases in blood vessels near a lesion.[105,138,148] Capillaries may also proliferate, further increasing the amount of enzyme.[116,148] The alkaline phosphatase of microglia at a cerebral lesion rises,[105] which may be one consequence of the proliferation of these cells and of their phagocytic activity. As Figs. 6 and 7 show, the alkaline phosphatase of cut or crushed peripheral nerve decreases greatly, reaching a minimum of roughly ½ normal activity 2 weeks after injury. This depression may persist for months.[73,135] If there is regeneration (Fig. 7), the activity of the regenerating distal portion eventually rises to well above normal values.[73,135] This increased activity, which is associated with remyelination[73,116] (compare with alkaline phosphatase activity during development), may persist for over a year. It is at least partially localized at the nodes of Ranvier and the Schmidt–Lanterman incisures.[117] Like acid phosphatase, alkaline phosphatase in the corresponding nerve on the unoperated side of experimental animals has been reported to be greater than in the corresponding nerves of unoperated control animals.[135] Activity in the spinal cord also decreases for prolonged periods following injury to peripheral nerves.[53,137] Surprisingly, activity increases in the proximal stump,[116] especially in the portion within 1 cm of the cut, where it reaches a maximum of 4 times normal activity by the 12th day after operation.[135] There is a lesser increase in the proximal segment 1 to 2 cm from the cut.[135] This activity in the proximal stump may be due to the sprouting of nerve fibers and other restorative processes.

Cutting or crushing peripheral nerve increases 5′-nucleotidase activity to over twice normal by 1 month after injury. After 4 months the activity is still $1\frac{1}{2}$ times normal.[73] This increase is suggestive of the high activity of this enzyme in myelinating tracts during development (*q.v.*). When peripheral nerve is cut, 5′-nucleotidase in spinal cord drops moderately and remains below normal for at least 1 month. This change is not limited to the region containing the cell bodies of the damaged neurons, but may occur in all higher regions of the cord[53] and even in the neuropil of ipsilateral nuclei in the medulla, where decreased staining for this enzyme may persist for 6 months.[20]

Cutting a peripheral nerve increases the thiamine pyrophosphatase-nucleoside diphosphatase in the spinal cord[21,148] and in ipsilateral medullary nuclei.[20] A definite effect is observed within 1 day[148] to 3 weeks,[20] depending on the distance from the wound, and persists for at least a month. The increase comes from heightened enzyme activity in proliferating and reacting microglia[20,21,148] which react within 1 to 3 days of injury[20,148] even preceding changes in neuronal lysosomes or Golgi apparatus (see below). The activity of this enzyme associated with neuronal Golgi apparatus decreases (see below).

Because of their use as markers for lysosomes and for the Golgi apparatus, respectively, the cytochemistry of acid phosphatase and of thiamine pyrophosphatase–nucleoside diphosphatase has been studied extensively in damaged mammalian neurons.[19,21–23,25] The course of events is remarkably similar in different types of injury, although the rate of change is very different. After an initial period the lysosomes swell, becoming large spherical acid-phosphatase-positive bodies, the "cytolysosomes." In severe, irreversible damage, the cytolysosomes gradually clump and fragment. The number of lysosomes in the cell decreases and acid phosphatase activity is lost in the later stages of reaction.[23,25] In less severe reversible damage, such as from peripheral nerve section, lysosomes swell less, forming smaller cytolysosomes; the number of lysosomes and the amount of acid phosphatase increases instead of decreasing; and eventually, as the neuron recovers, the cytolysosomes are replaced by, or revert to, lysosomes.[19,23] The detailed response of neuronal lysosomes to nerve section has recently been studied by electron microscopy.[74] This response of lysosomes is not limited to mammals. Acid phosphatase staining reveals an analogous, although not identical, reaction in the neurons of squid optic lobe after ischemic anoxia.[47]

Degenerative changes in the Golgi apparatus (and chromatolysis) lag behind changes in lysosomes.[22,25] The Golgi apparatus first appears loosened; then, depending on the extent of damage, it may fragment and may even disintegrate completely. Thiamine pyrophosphatase–nucleoside diphosphatase activity decreases and in irreversible injury finally disappears.[22]

Neurons vary considerably in their sensitivity to injury. In the brain, Purkinje cells and some of the neocortical neurons are most sensitive and react fastest[23,25]; neurons in the corpus striatum, the thalamus, and the brainstem are most resistant.[22,23,25]

The most rapid response occurs after anoxia, with or without accompanying ischemia.[22,23] Here cytolysosomes are sometimes observed in brain tissue taken immediately after injury, and the process is essentially completed in the more sensitive neurons 8 hr after injury.[23] Damage to the Golgi apparatus of these neurons is apparent 45 to 90 min after treatment and is extensive 4 hr after injury.[22] Similar changes occur in motoneuron lysosomes from anoxic cord.[79] Postmortem autolytic changes in the lysosomes and Golgi apparatus of brain neurons closely resemble changes produced by anoxia,[22,23] making it difficult to use autopsy material to study the cytology of pathological conditions. The response is slower in brain damaged by injection of diphtheria toxin.[25] In the most sensitive cells, cytolysosomes form 12 to 18 hr after injection and acid phosphatase has disappeared 24 to 36 hr after injection. The Golgi apparatus starts to disintegrate after 18 hr. In chromatolytic neurons produced by peripheral nerve damage, the process is even slower and is less complete.[19,21] Definite lysosomal reaction is apparent 7 days after injury, reaches a maximum by 3 to 4 weeks and persists with little change for 2 months, then slowly reverts to normal, but with some reaction persisting for over 6 months. There is some dissolution of the Golgi apparatus, but these changes lag behind the lysosomal reaction. Recovery of the Golgi apparatus appears to precede subsidence of the lysosomal reaction. After 1 month, staining for thiamine pyrophosphatase–nucleoside diphosphatase activity is more intense on the injured side than on the unoperated control side.[148] From the appearance of the Golgi apparatus, recovery appears to be well advanced by 2 months. In some motoneurons, signs of reaction persist for 4 months.[21]

The cytological changes in damaged neurons can be explained by postulating that the lysosome plays the key role. Injury increases permeability of the lysosomal membrane releasing the acid hydrolases, including acid phosphatase, which they contain. This process may well be autocatalytic with an initial release of small amounts of hydrolases which then attack lysosomal membranes releasing additional hydrolases. Evidence for autocatalysis is the induction period between injury and observable change in lysosomes. These hydrolases also attack various other cellular structures causing the various changes, including chromatolysis by injury to the Nissl substance, and the fragmentation and ultimate breakup of the Golgi apparatus that, as predicted from this hypothesis, follow changes in the lysosomes. If injury is slight, with ultimate recovery of the neuron, there may be only partial release of hydrolases which ultimately destroy each other. With severe injury, the process may be aggravated by hydrolases from surrounding cells, which penetrate and destroy the damaged neurons.

## IX. FUNCTIONS OF PHOSPHATASES

With a few notable exceptions (e.g., the phosphoinositide dephosphorylation enzyme system, *q.v.*) little is known about the function of phosphatases in either nervous or non-nervous tissue. [Morton[7] has suggested several plausible roles.] This has not stopped speculation; if anything, it has stimulated theories, many of which depend entirely or largely on "guilt by association." Often, the alleged relation may be only an artifact. Even if present, the relation may be casual rather than causal.

From the response of tissues to injuries (*q.v.*), it appears clear that lysosomal acid phosphatases, together with other lysosomal acid hydrolases, remove damaged and dead tissue. In healthy tissue they may destroy injurious substances within the cells and remove intracellular structures and materials during the normal turnover of cellular components. The conspicuous differentiation and specialization of structure and function during development is accompanied by dedifferentiation, death, and removal of more primitive structures. The association of acid phosphatase in these latter processes is shown by its high concentration in the tail of metamorphizing tadpoles (*Rana* sp. and *Xenopus* sp.), and in regressing Mullerian ducts of chick embryos.[112] This enzyme, which appears early, may have the same function in the developing nervous system. Although alkaline phosphatase is not considered to be an enzyme for gross catabolic processes, it may function similarly, because changes in its activity parallel changes in acid phosphatase activity.

From the marked occurrence of nonspecific alkaline phosphatase in kidney, gut, placenta, and mammary glands, as well as in brain capillaries, the choroid plexus, brain investments, and nodes of Ranvier, it appears plausible that this enzyme is involved in various secretory and transport processes. It may also take part in the blood–brain barrier mechanisms. Increase in alkaline phosphatase occurs with increase in permeability of blood vessels in edema, meningioma, meningo-encephalomyelitis, etc.[138] The area postrema and neurohypophysis which, unlike the rest of the central nervous system, can be stained *in vivo* by dye injected in the bloodstream, contain alkaline phosphatase in the connective tissue surrounding blood vessels.[138] The acid phosphatase in the choroid plexus, ependyma,[15] and elsewhere may be involved in secretion.[14] These enzymes, and several specific phosphatases, may also take part in intracellular transport. If so, regions with concentrations of phosphatases would not necessarily show marked secretion or active transport. In transport processes, phosphatases presumably dephosphorylate substrates that were phosphorylated to permit transport.

Certain phosphatases may be regulatory enzymes, an obvious function that has been suggested for many enzymes that do not appear to fit into any known metabolic pathway. Cyclic 3′,5′-nucleotide phosphodiesterase (*q.v.*) may control the concentration of "cyclic AMP." The ribosomal phosphatases, nonspecific alkaline phosphatase with its high activity against nucleoside phosphates,[41] and phosphodiesterase[40] (*q.v.*) may help to regulate protein synthesis, but there is no evidence for this. Other possibilities are control of metabolic rates,[7] either through uncoupling oxidative phosphorylation, e.g., by acyl phosphatase (3.6.1.7), inorganic pyrophosphatase (3.6.1.1), etc., or by destroying intermediates, e.g., by glucose 6-phosphatase (3.1.3.9).

Dephosphorylation may also provide both organic intermediates, e.g., choline from phosphocholine, 1,2-diglycerides from phosphatidic acids, and inorganic phosphate for metabolic processes.

The properties of an isolated enzyme may give a misleading impression of its functions in tissue. Several researchers have suggested that a principal function of phosphatases is transphosphorylation. This role has been criticized on the grounds that specific phosphatases do not transphosphorylate—only nonspecific alkaline and acid phosphatases do—and that (if unnatural acceptors are excluded) the concentration of phosphate acceptor needed to compete efficiently with water *in vitro* is much higher than the concentration present in tissue. When associated with subcellular structures, as many are, and as part of an ordered complex of enzymes, the properties of phosphatases may be quite different from their properties *in vitro*. In intact cells they may be able to phosphorylate an acceptor molecule furnished by an adjacent part of the enzyme system; that is, the local concentration of the acceptor may be great enough, even though the average concentration within the cell is too low for phosphorylation *in vitro*.

Most probably, phosphatases are necessary to form, maintain, and replace the specialized phospholipid membranes and sheaths of nervous tissue, but there is only circumstantial evidence for this. The increase of 5′-nucleotidase activity which accompanies myelination (*q.v.*)[104,107] may only reflect an increase in the number of myelin sites available for this enzyme, rather than any role in myelination. The changes in phosphatidic acid phosphatase during development can be explained similarly. The proposed connection of alkaline phosphatase in the later stages of development with myelination, mitotic index, and differentiation of nervous tissue[126–128] (see Section VII) would be more convincing if: (1) there were greater uniformity from species to species in adult patterns of enzyme distribution, (2) much of the activity did not reside in the capillary system, which suggests that the observed changes reflect development of and changes in permeability of capillaries, and (3) there were some obvious correlation between activity in developing fiber tracts and the structure of these tracts.

Using both Gomori stains and azo-coupling stains, Tewari and Bourne and Tewari and Sood have observed what they regard as significant variations in the distribution of several enzymes including acid phosphatase,[153,156,157] glucose 6-phosphatase,[153,155] and 5′-nucleotidase[153,156] within neurons. In some cell bodies the enzyme or particulate bodies containing the enzyme being studied were close to the nucleus; in others the distribution was more uniform throughout the perikaryon. Using thiamine pyrophosphatase staining as a marker, Shanthaveerappa and Bourne have observed what appear to be analogous changes in the Golgi apparatus.[143] These workers postulate that phosphatases shift cyclically to different regions within the cell and that the Golgi apparatus changes cyclically either to meet metabolic and physiological requirements or because neurons function in teams, alternating periods of activity and rest, and that the observed variations in distribution of phosphatase and in appearance of the Golgi apparatus represent different phases of this shift. Even though the same enzyme was studied by several staining techniques, these observations could be dismissed as artifacts, except that other dynamic changes in nervous tissue are known. These include changes in the distribution of the Nissl substance (chromatolysis) especially after fatigue or injury, rhythmic pulsation of glia and Schwann cells, rotation of the nucleus in neurons,[121] and reversible changes in the activity of cytochrome oxidase and succinic acid oxidase of neurons upon physiological stimulation.[76]

## X. ACKNOWLEDGMENTS

I am indebted to Professor Donald H. Ford for reading the manuscript and for many helpful suggestions.

I wish to thank the following organizations and individuals for permission to use copyrighted material: Academic Press and Dr. R. L. Friede for Figs. 1 and 2 and Table I, which appeared originally as Figs. 43 and 49 and Table XI in Ref. 5; Rockefeller University Press and Drs. N. H. Becker, S. Goldfischer, W.-Y. Shin, and A. B. Novikoff for Fig. 3 which appeared originally as Fig. 5 in Ref. 24; Pergamon Press and Drs. P. Cohn and D. Richter for Figs. 4 and 5 which appeared originally as Figs. 1 and 2 in Ref. 37; *The Biochemical Journal* and Drs. D. M. Hollinger, R. J. Rossiter, and H. Upmalis for Figs. 6 and 7 which appeared originally as Figs. 1 and 2 in Ref. 73; Pergamon Press and Dr. H. E. Hirsch; Harper and Row and Dr. O. H. Lowry; The American Society of Biological Chemists and Drs. O. H. Lowry, N. R. Roberts, K. Y. Leiner, M.-L. Wu, A. L. Farr, and R. W. Albers; and the Association for Research in Nervous and Mental Disease and Drs. E. Robins and D. E. Smith for Table II, which was compiled from Tables 1–3 in Ref. 72, Table 1 in Ref. 89, Table V in Ref. 90, and Table 49A in Ref. 125.

## REFERENCES

### A. General References, Reviews, and Monographs

1. (a) C. W. M. Adams, Chap. 10, Disorders of neurones and neuroglia, pp. 403–436; (b) C. W. M. Adams, M. Z. M. Ibrahim, and S. Leibowitz, Chap. 11, Demyelination, pp. 437–487; (c) C. W. M. Adams, Chap. 13, Histochemistry of cerebrovascular degeneration, pp. 519–546; (d) R. G. Spector, Chap. 14, Enzyme chemistry of anoxic brain injury, pp. 547–557; (e) K. C. Dixon, Chap. 15, Ischaemia and the neurone, pp. 558–598, *in Neurohistochemistry* (C. W. M. Adams, ed.), Elsevier Publishing Co., New York (1965).

2. L. Arvy, *Les Phosphatases du Tissu Nerveux*, Hermann, Paris (1966). This is an excellent recent review covering much the same material as this article but from a somewhat different point of view. It should be consulted for additional details and for additional references, especially to the older literature.
3. T. Barka and P. J. Anderson, Chap. IX, Histochemical demonstration of hydrolytic enzymes, pp. 203–211; Chap. X, Phosphatases, pp. 212–256, *in Histochemistry; Theory, Practice, and Bibliography*, Harper and Row Publishers, New York (1963).
4. M. S. Burstone, Chap. 4, Sec. V, Substantivity and histochemical localizations; pp. 138–152; Chap. 5, Phosphatases, pp. 160–292, *in Enzyme Histochemistry and Its Application in the Study of Neoplasms*, Academic Press, New York (1962).
5. R. L. Friede, Chap. VIII, Alkaline and acid phosphatases and nonspecific esterases, *in Topographic Brain Chemistry*, pp. 178–223, Academic Press, New York (1966). This is an excellent survey of its field, the "chemical anatomy" of the brain, primarily at the level of histology, and regional anatomy, rather than cellular and intracellular localization.
6. G. G. Glenner, Chap. 4, Enzyme histochemistry, *in Neurohistochemistry* (C. W. M. Adams, ed.), pp. 109–160, Elsevier Publishing Co., New York (1965).
7. R. K. Morton, Chap. 2, Phosphatases, *in Comprehensive Biochemistry* (M. Florkin and E. H. Stotz, eds.), Vol. 16, pp. 55–84, Elsevier Publishing Co., New York (1965). A good brief summary of the enzymatic properties of the various phosphatases.
8. A. G. E. Pearse, Chap. XIII, The principles of hydrolytic enzyme histochemistry, pp. 363–383; Chapter XIV, Alkaline phosphatases, pp. 384–430; Chap. XV, Acid phosphatases, pp. 431–455, *in Histochemistry, Theoretical and Applied*, 2nd ed., Little, Brown, and Co., Boston (1960). This is one of the classic texts on histochemistry. Although somewhat out of date for the newer azo-dye coupling stains, it is an excellent critical source for older methods and for principles and artifacts.
9. G. Schmidt, Chap. 2, Nonspecific acid phosphomonoesterases, *in The Enzymes*, (P. D. Boyer, H. Lardy, and K. Myrbäck, eds.), 2nd ed., Vol. 5, pp. 37–47, Academic Press (1961). Although dated and somewhat sketchy, this is the best and most recent easily accessible review devoted to these enzymes.
10. T. C. Stadtman, Chap. 4, Alkaline phosphatases, *in The Enzymes* (P. D. Boyer, H. Lardy and K. Myrbäck, eds.), 2nd ed., Vol. 5, pp. 55–71, Academic Press (1961). Although somewhat dated, this is the most recent easily accessible review and is an excellent survey of the topic at the time it was written.

## B. Other References

11. B. W. Agranoff, Hydrolysis of long-chain alkyl phosphates and phosphatidic acid by an enzyme purified from pig brain, *J. Lipid Res.* **3**:190–196 (1962).
12. W. N. Aldridge, The rate of formation and decomposition of phosphoryl-phosphatase (*Escherichia coli*), *Biochem. J.* **92**:23–25 (1964).
13. P. J. Anderson, The effect of autolysis on the distribution of acid phosphatase in rat brain, *J. Neurochem.* **12**:919–925 (1965).
14. P. J. Anderson and S. K. Song, Acid phosphatase in the nervous system (I, II), *J. Neuropathol. Exptl. Neurol.* **21**:263–283 (1962).
15. P. J. Anderson, S. K. Song, and N. Christoff, The cytochemistry of acid phosphatase in neural tissue: Separation, validation, and localization, *Proc. 4th Intern. Congr. Neuropathol.*, Munich, 1961 (H. Jacob, ed.), Vol. 1, *Histochemistry and Biochemistry of the Diseases of the Central and Peripheral Nervous System*, pp.75–79, Georg Thieme Verlag, Stuttgart (1962).

16. R. G. Bannister and F. C. A. Romanul, The localization of alkaline phosphatase activity in cerebral blood vessels, *J. Neurol. Neurosurg. Psychiat.* **26**:333–340 (1963).
17. K. D. Barron, J. Bernsohn, and A. R. Hess, Zymograms of neural acid phosphatases. Implications for slide histochemistry, *J. Histochem. Cytochem.* **12**:42–44 (1964).
18. K. D. Barron and R. Boshes, Histochemical demonstration of 5-nucleotidase in the central nervous system. Effects of manganous ion and *p*H, *J. Histochem. Cytochem.* **9**:455–457 (1961).
19. K. D. Barron and S. Sklar, Response of lysosomes of bulbospinal motoneurons to axon section, *Neurology* **11**:866–875 (1961).
20. K. D. Barron and T. O. Tuncbay, Phosphatase in cuneate nuclei after brachial plexectomy, *Arch. Neurol.* **7**:203–210 (1962).
21. K. D. Barron and T. O. Tuncbay, Phosphatase histochemistry of feline cervical spinal cord after brachial plexectomy. Hydrolysis of β-glycerophosphate, thiamine pyrophosphate and nucleoside diphosphates, *J. Neuropathol. Exptl. Neurol.* **23**:368–386 (1964).
22. N. H. Becker, The cytochemistry of anoxic and anoxic-ischemic encephalopathy in rats. III. Alterations in the neuronal Golgi apparatus identified by nucleoside diphosphatase activity, *Am. J. Pathol.* **40**:243–252 (1962).
23. N. H. Becker and K. D. Barron, The cytochemistry of anoxic and anoxic-ischemic encephalopathy in rats. I. Alterations in neuronal lysosomes identified by acid phosphatase activity, *Am. J. Pathol.* **38**:161–175 (1961).
24. N. H. Becker, S. Goldfischer, W.-Y. Shin, and A. B. Novikoff, The localization of enzyme activities in the rat brain, *J. Biophys. Biochem. Cytol.* **8**:649–663 (1960).
25. N. H. Becker, A. B. Novikoff, and S. Goldfischer, A cytochemical study of the neuronal Golgi apparatus, *Arch. Neurol.* **5**:497–503 (1961).
26. F. J. Behal and M. Center, Heterogeneity of calf intestinal alkaline phosphatase, *Arch. Biochem. Biophys.* **110**:500–505 (1965).
27. J. Bernsohn and K. D. Barron, Multiple molecular forms of brain hydrolases, *Intern. Rev. Neurobiol.* **7**:297–344 (1964).
28. E. W. Bingham and C. A. Zittle, Purification and properties of acid phosphatase in bovine milk, *Arch. Biochem. Biophys.* **101**:471–477 (1963).
29. M. Blank, Enzymmuster *p*H-abhängiger saurer Phosphatasen der Katzenretina, *Acta Histochem.* **28**:8–50 (1967).
30. M. Blank and R. Oehlschlägel, *p*H-abhängige intracelluläre Verteilung der sauren Phosphatase im Rückenmark des Huhnes, *Gallus domesticus*, *Histochemie* **6**:187–208 (1966).
31. D. Bodian and R. C. Mellors, The regenerative cycle of motoneurons, with special reference to phosphatase activity, *J. Exptl. Med.* **81**:469–488 (1945).
32. M. S. Burstone, The relationship between fixation and techniques for the histochemical localization of hydrolytic enzymes, *J. Histochem. Cytochem.* **6**:322–339 (1958).
33. W. Y. Cheung, Properties of cyclic 3′,5′-nucleotide phosphodiesterase from rat brain, *Biochemistry* **6**:1079–1087 (1967).
34. W. Y. Cheung and L. Salganicoff, Cyclic 3′,5′-nucleotide phosphodiesterase: Localization and latent activity in rat brain, *Nature* **214**:90–91 (1967).
35. A. D. Chiquoine, Distribution of alkaline phosphomonoesterase in the central nervous system of the mouse embryo, *J. Comp. Neurol.* **100**:415–439 (1954).
36. S. R. Cohen and I. B. Wilson, Measurement of the zinc dissociation constants of alkaline phosphatase from *Escherichia coli* by equilibration with zinc ion buffers, *Biochemistry* **5**:904–909 (1966).

37. P. Cohn and D. Richter, Enzymic development and maturation of the hypothalamus, *J. Neurochem.* **1**:166–172 (1956).
38. H.-J. Colmant, Aktivitätsschwankungen der sauren Phosphatase im Rückenmark und den Spinalganglien der Ratte nach Durchschneidung des Nervus ischiadicus, *Arch. Psychiat. Nervenkrankh.* **199**:60–71 (1959).
39. V. R. Cunningham and E. J. Field, Alkaline phosphatase isozyme systems of the guinea pig in health and in experimental allergic encephalomyelitis, *J. Neurochem.* **11**:281–285 (1964).
40. R. K. Datta and J. J. Ghosh, Phosphodiesterase activity of ribosome preparation from goat brain, *J. Neurochem.* **10**:285–286 (1963).
41. R. K. Datta and J. J. Ghosh, Alkaline phosphomonoesterase activity of goat brain cortex ribosomes, *J. Neurochem.* **11**:779–786 (1964).
42. A. N. Davison and N. A. Gregson, Cytochemistry of the nervous system, Chap. 6, *in Neurohistochemistry* (C. W. M. Adams, ed.), pp. 189–235, Elsevier Publishing Co., New York (1965).
43. R. M. C. Dawson and W. Thompson, The triphosphoinositide phosphomonoesterase of brain tissue, *Biochem. J.* **91**:244–250 (1964).
44. H. W. Deane, Nuclear location of phosphatase activity: Fact or artifact? *J. Histochem. Cytochem.* **11**:443–444 (1963).
45. Gy. Domján and E. Minker, Untersuchung der Phosphomonoesterasen in den zentralen Nervenganglien der Weinbergschnecke (*Helix pomatia*), *Acta Biol. Acad. Sci. Hung.* **11**:219–229 (1960).
46. J. Drukker and J. P. Schadé, The localization of some enzymes in the optic system of cephalopods, *Acta Morphol. Neerl. Scand.* **5**:290–291 (1963).
47. J. Drukker and J. P. Schadé, Degeneration patterns in the optic lobe of cephalopods, *Progr. Brain Res.* **14**:122–142 (1965).
48. G. I. Drummond, N. T. Iyer, and J. Keith, Hydrolysis of ribonucleoside 2′,3′-cyclic phosphates by a diesterase from brain, *J. Biol. Chem.* **237**:3535–3539 (1962).
49. C. de Duve, General principles, Chap. 1, *in Enzyme Cytochemistry* (D. B. Roodyn, ed.), pp.1–26, Academic Press, New York (1967).
50. R. H. Eaton and D. W. Moss, Inhibition of the orthophosphatase and pyrophosphatase activities of human alkaline phosphatase, *Biochem. J.* **102**:917–921 (1967).
51. H. N. Fernley and P. G. Walker, Phosphorylation of *Escherichia coli* alkaline phosphatase by substrate, *Nature* **212**:1435–1437 (1966).
52. H. N. Fernley and P. G. Walker, Phosphorylation of calf-intestinal alkaline phosphatase by substrate, *Biochem. J.* **102**:48–49 (1967).
53. C. Fieschi and S. Soriani, Enzymic activities in the spinal cord after sciatic section. Alkaline and acid phosphatases, 5-nucleotidase and ATP-ase, *J. Neurochem.* **4**:71–77 (1959).
54. E. Fischer, Anorganische alkalische Pyrophosphatase in Nervenzellen, *Acta Histochem.* **24**:382–384 (1966).
55. J. B. Flexner and L. B. Flexner, Biochemical and physiological differentiation during morphogenesis. VII. Adenylpyrophosphatase and acid phosphatase activities in the developing cerebral cortex and liver of the fetal guinea pig, *J. Cellular Comp. Physiol.* **31**:311–320 (1948).
56. J. B. Flexner, C. L. Greenblatt, S. R. Cooperband, and L. B. Flexner, Biochemical and physiological differentiation during morphogenesis. XIX. Alkaline phosphatase and aldolase activities in the developing cerebral cortex and liver of the fetal guinea pig, *Am. J. Anat.* **98**:129–138 (1956).
57. R. L. Friede, A quantitative mapping of alkaline phosphatase in the brain of the rhesus monkey, *J. Neurochem.* **13**:197–203 (1966).

58. R. L. Friede and M. Knoller, Quantitative tests of histochemical methods for phosphomonoesterases in brain tissue, *J. Histochem. Cytochem.* **13**:125–132 (1965).
59. R. L. Friede and M. Knoller, A quantitative mapping of acid-phosphatase in the brain of the rhesus monkey, *J. Neurochem.* **12**:441–450 (1965).
60. R. O. Friedel, J. D. Brown, and J. Durrell, Monophosphatidyl inositol inositol-phosphohydrolase in guinea pig brain, *Biochim. Biophys. Acta* **144**:684–686 (1967).
61. A. Garen and C. Levinthal, A fine-structure genetic and chemical study of the enzyme alkaline phosphatase of *E. coli*. I. Purification and characterization of alkaline phosphatase, *Biochim. Biophys. Acta* **38**:470–483 (1960).
62. S. Goldfischer, The Golgi apparatus and the endoplasmic reticulum in neurons of the rabbit, *J. Neuropathol. Exptl. Neurol.* **23**:36–45 (1964).
63. G. Gomori, Microtechnical demonstration of phosphatase in tissue sections, *Proc. Soc. Exptl. Biol. Med.* **42**:23–26 (1939).
64. G. Gomori, Distribution of acid phosphatase in the tissues under normal and under pathologic conditions, *Arch. Pathol.* **32**:189–199 (1941).
65. J. J. Gordon, Properties of brain pyrophosphatase, *Biochem. J.* **46**:96–99 (1950).
66. J. J. Gordon, Observations on brain phosphatases, *Biochem. J.* **55**:812–817 (1953).
67. M. K. Gordon, K. G. Bench, G. G. Deanin, and M. W. Gordon, Histochemical and biochemical study of synaptic lysosomes, *Nature* **217**:523–527 (1968).
68. R. S. Grier, M. B. Hood, and M. B. Hoagland, Observations on the effects of beryllium on alkaline phosphatase, *J. Biol. Chem.* **180**:289–298 (1949).
69. S. Grisolia, J. Caravaca, and B. K. Joyce, Purification and properties of brain carbamyl and acyl phosphatase, *Biochim. Biophys. Acta* **29**:432–433 (1958).
70. J. W. Healy, D. Stollar, M. I. Simon, and L. Levine, Characterization of phosphodiesterase from lamb brain, *Arch. Biochem. Biophys.* **103**:461–468 (1963).
71. L. A. Heppel, D. R. Harkness, and R. J. Hilmoe, A study of the substrate specificity and other properties of the alkaline phosphatase of *Escherichia coli*, *J. Biol. Chem.* **237**:841–846 (1962).
72. H. E. Hirsch, Acid phosphatase localization in individual neurons by a quantitative histochemical method, *J. Neurochem.* **15**:123–130 (1968).
73. D. M. Hollinger, R. J. Rossiter, and H. Upmalis, Chemical studies of peripheral nerve during Wallerian degeneration. 4. Phosphatases, *Biochem. J.* **52**:652–659 (1952).
74. E. Holtzman, A. B. Novikoff, and H. Villaverde, Lysosomes and GERL in normal and chromatolytic neurons of the rat ganglion nodosum, *J. Cell. Biol.* **33**:419–435 (1967).
75. H. Hydén, The chemistry of single neurons: A study with new methods, *in Proc. 1st Intern. Neurochem. Symp.*, Oxford, 1954 (H. Waelsch, ed.), *Biochemistry of the Developing Nervous System*, pp. 358–371, Academic Press, New York (1955).
76. H. Hydén, Dynamic aspects on the neuron–glia relationship. A study with microchemical methods, Chap. 4, *in The Neuron* (H. Hydén, ed.), pp. 179–219, Elsevier Publishing Co., New York (1967).
77. K. Iijima, T. R. Shantha, and G. H. Bourne, Enzyme-histochemical studies on the hypothalamus with special reference to the supraoptic and paraventricular nuclei of squirrel monkey (*Saimiri sciureus*), *Z. Zellforsch. Mikroskop. Anat.* **79**:76–91 (1967).
78. J. S. Kaluza and M. S. Burstone, Staining patterns of phosphatases of the central nervous system with azo-dye methods, *J. Neuropathol. Exptl. Neurol.* **23**:477–485 (1964).

79. F. I. Khattab, Alterations in acid phosphatase bodies (lysosomes) in cat motoneurons after asphyxiation of the spinal cord, *Exptl. Neurol.* **18**:133–140 (1967).
80. F. W. Klemperer, J. M. Miller, and C. J. Hill, The inhibition of alkaline phosphatase by beryllium, *J. Biol. Chem.* **180**:281–288 (1949).
81. M. L. Kornguth and E. A. Stubbs, Hydrolysis of phosphohydroxypyruvate and β-glycerophosphate by a phosphatase preparation from beef brain, *Arch. Biochem. Biophys.* **109**:104–109 (1965).
82. G. W. Kreutzberg and H. Hager, Electron microscopical demonstration of acid phosphatase activity in the central nervous system, *Histochemie* **6**:254–259 (1966).
83. T. Kurihara and Y. Tsukada, The regional and subcellular distribution of 2′,3′-cyclic nucleotide 3′-phosphohydrolase in the central nervous system, *J. Neurochem.* **14**:1167–1174 (1967).
84. H. Landow, E. Kabat, and W. Newman, Distribution of alkaline phosphatase in normal and in neoplastic tissues of the nervous system. A histochemical study, *(A.M.A.) Arch. Neurol. Psychiat.* **48**:518–530 (1942).
85. N. J. Lane, Thiamine pyrophosphatase, acid phosphatase, and alkaline phosphatase in the neurones of *Helix aspersa*, *Quart. J. Microscop. Sci.* **104** (3rd series):401–412 (1963).
86. C. Lazdunski and M. Lazdunski, Etude cinétique du mécanisme d'action catalytique de la phosphatase alcaline d'*Escherichia coli*, *Biochim. Biophys. Acta* **113**:551–566 (1966).
87. R. S. Lee, Phosphatases in the neurones of *Locusta migratoria*, *Quart. J. Microscop. Sci.* **104**(3rd series):475–481 (1963).
88. W. Lierse, Die alkalische Phosphatase in den Hirngefässen der Maus (*Mus muris*) während der postnatalen Entwicklung, *Z. Mikroskop.-Anat. Forsch.* **70**:48–61 (1963).
89. O. H. Lowry, Quantitative analysis of single nerve cell bodies, Chap. IV, *in Progress in Neurobiology. II. Ultrastructure and Cellular Chemistry of Neural Tissue* (H. Waelsch, ed.), pp. 69–82, Hoeber-Harper, New York (1957).
90. O. H. Lowry, N. R. Roberts, K. Y. Leiner, M.-L. Wu, A. L. Farr, and R. W. Albers, The quantitative histochemistry of brain. III. Ammon's horn, *J. Biol. Chem.* **207**:39–49 (1954).
91. J. C. Mathies, Preparation and properties of highly purified alkaline phosphatase from swine kidneys, *J. Biol. Chem.* **223**:1121–1127 (1958).
92. R. J. McAlpine, Selected observations on the early development of the motor neurons in the brain stem and spinal cord of the white rat as revealed by the alkaline phosphatase technique, *J. Comp. Neurol.* **113**:211–243 (1959).
93. R. E. McCaman and E. Robins, Quantitative biochemical studies of Wallerian degeneration in the peripheral and central nervous systems. II. Twelve enzymes, *J. Neurochem.* **5**:32–42 (1959).
94. A. R. McNabb, Enzymes of gray matter and white matter of dog brain, *Can. J. Med. Sci.* **29**: 208–215 (1951).
95. P. Meyer, Histochemistry of the developing human brain. I. Alkaline phosphatase, acid phosphatase, and AS esterase in the cerebellum, *Acta Psychiat. Neurol. Scand.* **39**:123–138 (1963).
96. E. Minker and G. Domján, Histochemical studies of phosphatases in the intramural ganglion cells of the intestinal canal of *Helix pomatia*, *Acta Biol. Acad. Sci. Hung.* **12**:137–140 (1961).
97. F. Moog, The distribution of phosphatase in the spinal cord of chick embryos of one to eight days incubation. *Proc. Natl. Acad. Sci. (U.S.)* **29**:176–183 (1943).

98. F. Moog, Localizations of alkaline and acid phosphatases in the early embryogenesis of the chick, *Biol. Bull.* **86**:51–80 (1944).
99. R. K. Morton, The purification of alkaline phosphatases of animal tissues, *Biochem. J.* **57**:595–603 (1954).
100. R. K. Morton, The phosphotransferase activity of phosphotases. 2. Studies with alkaline phosphomonoesterases and some substrate-specific phosphatases, *Biochem. J.* **70**:139–150 (1958).
101. R. K. Morton, The phosphototransferase activity of phosphatases. 3. Comparison of enzymic catalysis by acid phosphatase with nonenzymic catalysis at acid *p*H values, *Biochem. J.* **70**: 150-155 (1958).
102. D. W. Moss, R. H. Eaton, J. K. Smith, and L. G. Whitby, Association of inorganic-pyrophosphatase activity with human alkaline-phosphatase preparations, *Biochem. J.* **102**:53–57 (1967).
103. J. Mulnard, Contribution à la connaissance des enzymes dans l'ontogénèse. Les phosphomonestérases acide et alcaline dans le développement du rat et de la souris. *Arch. Biol.* (*Liege*) **66**:525–685 (1955).
104. D. Naidoo, The activity of 5′-nucleotidase determined histochemically in the developing rat brain, *J. Histochem. Cytochem.* **10**:421-434 (1962).
105. D. Naidoo, Alkaline phosphatase at the site of cerebral injury, *Acta Histochem.* **15**:182–184 (1963).
106. D. Naidoo and O. E. Pratt, The localization of some acid phosphatases in brain tissue, *J. Neurol. Neurosurg. Psychiat.* **14**:287–294 (1951).
107. D. Naidoo and O. E. Pratt, the development of adenosine 5′-phosphatase activity with the maturation of the rat cerebral cortex, *Enzymologia* **16**:298–304 (1954).
108. K. Nandy and G. H. Bourne, Alkaline phosphatases in brain and spinal cord, *Nature* **200**:1216–1217 (1963).
109. K. Nandy and G. H. Bourne, Histochemical studies on the distribution of acid naphthol AS-phosphatase in the spinal cord of the rat, *Acta Anat.* **61**:84–91 (1965).
110. V. N. Nigam, H. M. Davidson, and W. H. Fishman, Kinetics of hydrolysis of the orthophosphate monoesters of phenol, *p*-nitrophenol, and glycerol by human prostatic acid phosphatase, *J. Biol. Chem.* **234**:1550–1554 (1959).
111. V. N. Nigam and W. H. Fishman, Catalysis of phosphoryl transfer by prostatic acid phosphatase, *J. Biol. Chem.* **234**:2394–2398 (1959).
112. A. B. Novikoff, Lysosomes and related particles, Chap. 6, *in The Cell* (J. Brachet and A. E. Mirsky, eds.), Vol. 2, pp. 423–488, Academic Press, New York (1961).
113. A. B. Novikoff and S. Goldfischer, Nucleosidediphosphatase activity in the Golgi apparatus and its usefulness for cytological studies, *Proc. Natl. Acad. Sci.* (*U.S.*) **47**:802–810 (1961).
114. R. Nüske and H. Venner, Die Isoenzymfraktionen der unspezifischen alkalischen Phosphomonoesterase, *Biochem. Z.* **346**:226–243 (1966).
115. S. Olsen and C. Petri, Histochemical localization of acid phosphatase in the human cerebellar cortex, *Acta Psychiat. Neurol. Scand.* **39**:112–122 (1963).
116. B. Pinner and J. B. Campbell, Alkaline phosphatase activity of incisures and nodes during degeneration and regeneration of peripheral nerve fibers, *Exptl. Neurol.* **12**:159–172 (1965).
117. B. Pinner, J. F. Davison, and J. B. Campbell, Alkaline phosphatase in peripheral nerves, *Science* **145**:936–938 (1964).
118. D. J. Plocke, C. Levinthal, and B. L. Vallee, Alkaline phosphatase of *Escherichia coli:* A zinc metalloenzyme, *Biochemistry* **1**:373–378 (1962).

119. L. F. Pomazanskaya, Phosphatidic acid phosphatase of chick-brain microsomes during ontogeny, *Dokl. Akad. Nauk SSSR* **155**:208–211 (1964). [Translation, *Dokl. Biochem. Sect.* **154–156**:56–58 (1964).]*
120. L. F. Pomazanskaya, Phosphatidic acid phosphatase in the subcellular fractions of the chick brain during development, *Zh. Evolyutsionnoi Biokhim. i Fiziol.* **1**:320–324 (1965). [*Biol. Abstr.* **47** (1966), #67244.]†
121. C. M. Pomerat, W. J. Hendelman, C. W. Raiborn, Jr., and J. F. Massey, Dynamic activities of nervous tissue *in vitro*, Chap. 3, *in The Neuron* (H. Hydén, ed.), pp. 119–178, Elsevier Publishing Co., New York (1967).
122. L. Raijman, S. Grisolia, and H. Edelhoch, Further purification and properties of brain acyl phosphatase, *J. Biol. Chem.* **235**:2340–2342 (1960).
123. J. L. Reis, Über die spezifische Phosphatase der Nervengewebe, *Enzymologia* **2**:110–116 (1937).
124. J. L. Reis, The specificity of phosphomonoesterases in human tissue, *Biochem. J.* **48**:548–551 (1951).
125. E. Robins and D. E. Smith, A quantitative histochemical study of eight enzymes of the cerebellar cortex and subjacent white matter in the monkey, *Res. Publ. Assoc. Res. Nervous Mental Disease* **32**: 305–327 (1953).
126. K. T. Rogers, Studies on chick brain of biochemical differentiation related to morphological differentiation and onset of function. III. Histochemical localization of alkaline phosphatase, *J. Exptl. Zool.* **145**:49–59 (1960).
127. K. T. Rogers, Cell fraction studies on nuclear vs. cytoplasmic localization of alkaline phosphatase in the very early development of the chick brain. *Exptl. Cell Res.* **34**:100–110 (1964).
128. K. T. Rogers, L. De Vries, J. A. Kepler, C. R. Kepler, and E. R. Speidel, Studies on chick brain of biochemical differentiation related to morphological differentiation and onset of function. II. Alkaline phosphatase and cholinesterase levels and onset of function. *J. Exptl. Zool.* **144**:89–103 (1960).
129. A. Roitman and S. Glatt, Isolation of phospholipase-C from rat brain, *Israel J. Chem.* **1**:190 (1963).
130. S. P. R. Rose, The localization of cerebral phosphoprotein phosphatase, *Biochem. J.* **83**: 614–622 (1962).
131. S. P. R. Rose and P. J. Heald, A phosphoprotein phosphatase from ox brain, *Biochem. J.* **81**:339–347 (1961).
132. F. Rossi and E. Reale, The somite state of human development studied with the histochemical reaction for the demonstration of alkaline glycerophosphatase, *Acta Anat.* **30**:656–681 (1957).
133. J. G. Salway, J. L. Harwood, M. Kai, G. L. White, and J. N. Hawthorne, Enzymes of phosphoinositide metabolism during rat brain development, *J. Neurochem.* **15**:221–226 (1968).
134. J. G. Salway, M. Kai, and J. N. Hawthorne, Triphosphoinositide phosphomonoesterase activity in nerve cell bodies, neuroglia and subcellular fractions from whole rat brain, *J. Neurochem.* **14**:1013–1024 (1967).
135. T. Samorajski, Changes in phosphatase activity following transection of the sciatic nerve, *J. Histochem. Cytochem.* **5**:15–27 (1957).

* In preparing this review, the English translation was used instead of the original article in Russian.

† In preparing this review, the abstract was used instead of the original article in Russian.

136. T. Samorajski and G. R. Fitz, Histochemical analysis of phosphomonoesterase in the hypothalamus and pituitary gland of the rat, *Lab. Invest*, **9**:517–534 (1960).
137. T. Samorajski and G. R. Fitz, Phosphomonoesterase changes associated with spinal cord chromatolysis, *Lab. Invest.* **10**:129–143 (1961).
138. T. Samorajski and J. McCloud, Alkaline phosphomonoesterase and blood brain permeability, *Lab. Invest.* **10**:492–501 (1961).
139. M. Sandler and G. H. Bourne, Histochemical studies of phosphatases separated by starch gel electrophoresis, *Exptl. Cell Res.* **24**:174–177 (1960).
140. S. Saraswathi and B. K. Bachhawat, Heterogeneity of alkaline phosphatase in sheep brain. *J. Neurochem.* **13**:237–246 (1966).
141. T. G. Scott, A unique pattern of localization within the cerebellum of the mouse, *J. Comp. Neurol.* **122**:1–7 (1964).
142. U. S. Seal and F. Binkley, An inorganic pyrophosphatase of swine brain, *J. Biol. Chem.* **228**:193–199 (1957).
143. T. R. Shanthaveerappa and G. H. Bourne, The thiamine pyrophosphatase technique as an indicator of the morphology of the Golgi apparatus in the neurons, *Acta Histochem.* **22**:155–178 (1965).
144. N. N. Sharma. Studies on the histochemical distribution of glucose-6-phosphatase, glucose-6-phosphate dehydrogenase, β-glucuronidase and glucan phosphorylase in olfactory bulb of rat, *Acta Histochem.* **27**:165–171 and plate X (1967).
145. N. Shimizu, Histochemical studies on the phosphatase of the nervous system, *J. Comp. Neurol.* **93**:201–217 (1950).
146. I. de Sibrik and D. S. O'Doherty, Phosphatases and phospholipases in the central nervous system, *Arch. Neurol.* **2**:537–546 (1960).
147. M. Silva-Pinto and A. Coimbra, Comparative studies of the central nervous system phosphatases employing the Gomori and azo-dye methods, *Acta Anat.* **54**:157–173 (1963).
148. J. Sjöstrand, Changes of nucleoside phosphatase activity in the hypoglossal nucleus during nerve regeneration, *Acta Physiol. Scand.* **67**:219–228 (1966).
149. H. Takamatsu, Histologische und biochemische Studien über die Phosphatase (I. Mitteilung). Histochemische Untersuchungsmethodik der Phosphatase und der Verteilung in verschiedenen Organen und Geweben, *Transactiones Societatis Pathologicae Japonicae* (*Nippon Byori Gakkai Kaishi*) **29**:492–498 (1939).
150. S. Talanti, E. Kivalo, and A Kivalo. The acid phosphatase activity in the hypothalamic magnocellular nuclei of the cow embryo, *Acta Endocrinol.* **29**:302–306 (1958).
151. H. B. Tewari and G. H. Bourne, The structure and biochemical identity of the neurokeratin network of myelinated nerve fibers, *Bibliotheca Anat.* **2**:111–127 (1961).
152. H. B. Tewari and G. H. Bourne, New morphological identity and functional significance of the neurokeratin network of myelinated nerve fibers, *Pathol. Biol. Semaine Hop.* **9**:919–924 (1961).
153. H. B. Tewari and G. H. Bourne, Histochemical evidence of metabolic cycles in spinal ganglion cells of rat, *J. Histochem. Cytochem.* **10**:42–64 (1962).
154. H. B. Tewari and G. H. Bourne, Histochemical studies on the distribution of alkaline and acid phosphatases and 5-nucleotidase in the cerebellum of rat, *J. Anat.* **97**:65–72, (1963).
155. H. B. Tewari and G. H. Bourne, On the intracellular distribution of glucose-6-phosphatase in the neurons of cerebrum and trigeminal ganglion of the rat, *J. Histochem. Cytochem.* **11**:121–122 (1963).
156. H. B. Tewari and G. H. Bourne, Histochemical studies on the distribution of

alkaline and acid phosphatases and 5-nucleotidase in the trigeminal ganglion cells of rat, *Acta Histochem.* **17**:197–207 (1964).

157. H. B. Tewari and P. P. Sood, On the distribution of acid phosphatase among the olfactory neurons of some vertebrates, *Histochemie* **11**:62–70 (1967).
158. W. Thompson and R. M. C. Dawson, The hydrolysis of triphosphoinositide by extracts of ox brain, *Biochem. J.* **91**:233–236 (1964).
159. W. Thompson and R. M. C. Dawson, The triphosphoinositide phosphodiesterase of brain tissue, *Biochem. J.* **91**:237–243 (1964).
160. K. K. Tsuboi and P. B. Hudson, Acid phosphatase. III. Specific kinetic properties of highly purified human prostatic phosphomonoesterase, *Arch. Biochem. Biophys.* **55**:191–205 (1955).
161. B. L. Vallee, Chap. 34, Zinc. *in Mineral Metabolism* (C. L. Comar and F. Bronner, eds.), Vol. IIB, pp. 443–482, Academic Press, New York (1962).
162. M. A. Verity and W. J. Brown, Structure-linked activity of lysosomal enzymes in the developing mouse brain, *J. Neurochem.* **15**:69–80 (1968).
163. I. B. Wilson, J. Dayan, and K. Cyr, Some properties of alkaline phosphatase from *Escherichia coli:* Transphosphorylation, *J. Biol. Chem.* **239**:4182–4185 (1964).
164. G. B. Wislocki and E. W. Dempsey, The chemical cytology of the chorioid [*sic*] plexus and blood brain barrier of the rhesus monkey (*Macaca mulatta*), *J. Comp. Neurol.* **88**:319–345 (1948).
165. A. Wolf, E. A. Kabat, and W. Newman, Histochemical studies on tissue enzymes. III. A study of the distribution of acid phosphatases with special reference to the nervous system. *Am. J. Pathol.* **19**:423–439 (1943).

*Chapter 5*

# PEPTIDE HYDROLASES


## N. Marks

*New York State Research Institute*
*for Neurochemistry and Drug Addiction*
*Ward's Island*
*New York, New York*


## I. INTRODUCTION

Exopeptidases constitute a group of hydrolytic enzymes capable of cleaving the $NH_2$— or COOH— terminal amino acids of peptides or polypeptides. Despite the presence in brain of peptides with important physiological properties, the role of these hydrolases in peptide and protein turnover is still obscure. It is well known that brain proteins are in dynamic equilibrium with their environment, with the processes of synthesis matching those of degradation. Protein breakdown as a consequence is an orderly process in which different batteries of endo- and exopeptidases are linked together to form a degradative pathway leading to the liberation of smaller peptides or amino acids. So far only simple schemes can be postulated for the pathways of intracellular hydrolysis of proteins and peptides.

Rapid advances in the characterization of exopeptidases occurred with the introduction of synthetic substrates by Bergmann[(1)] in the 1930's; however, many puzzling questions remain regarding their role and their true physiological substrates within the cell. The simple hydrolysis of peptides, amides, or esters *in vitro* may not correspond with the situation within the tissue. It is the purpose of this review to correlate the many factors that may be considered important to the functional role of brain exopeptidases, particularly enzyme localization, accessibility of substrate, effect of cofactors, activation of zymogen precursors and isoenzymes. The individual categories of enzymes are described, followed by a brief essay on their role in hormone turnover, disease, and transport processes. Special consideration is given to specialized brain areas closely linked to endocrine function that are known to contain relatively large pools of free peptides. Brain itself is not characterized by a high concentration of free peptides, with the exception of the tripeptide glutathione. The application of newer analytical procedures has revealed a large number of peptides present in trace quantities that do not have any known physiological role.[2,3] These products may represent intermediates in the degradative pathway or artifacts induced by the extraction procedure.

## A. Comment on the Classification

The scheme adopted by the IUB[4] follows that proposed by the classical work of Bergmann.[1] Exopeptidases are classified on the basis of those that require substrates with a free $NH_2$— terminus (aminopeptidases 3.4.1) or COOH— terminus (carboxypeptidases 3.4.2.) or those that are specific for dipeptide substrates with both terminal groups free (dipeptide hydrolases 3.4.3; Table I). With the exception of carboxypeptidases, most exopeptidases are unavailable in a satisfactory state of purity and frequently exhibit a broader range of specificity than implied by the Enzyme Commission (E.C.) classification. Indeed, the E.C. classification bears a postscript noting that other peptidases undoubtedly exist but cannot be assigned a systematic name without additional characterization studies. The use of more than one substrate and consideration of other criteria are often essential for the differentiation of exopeptidases. Some confusion exists in the literature arising from the naming of exopeptidases on the basis of complex polypeptide substrates, e.g., insulinase, oxytocinase, vasopressinase, protaminase (now termed carboxypeptidase B), and recently angiotensinase. It is probable that many of these enzymes represent examples of commonly occurring exo- or endopeptidases. These terms are best avoided in keeping with the recommendations of the E.C.[4] Although the introduction of synthetic substrates has led to great advances in the characterization of exopeptidases, some caution must be exercised since the hydrolysis of synthetic substrates does not necessarily correspond to the activity *in vivo*. As an illustration, amino-acylated naphthylamines introduced by Gomori[5] for detection of leucine aminopeptidase have led to the discovery of a different class of hydrolases, termed by the trivial name "arylamidases." These enzymes occur in high concentrations in the CNS and consequently are accorded some description in the present review. The true physiological substrates for this new class of hydrolases is unknown, and they have not yet received a classification number. There is now evidence that enzymes hydrolyzing the higher homologues such as dipeptidyl arylamides correspond in some cases to known exopeptidases[6] (see Section VI, A, 2).

## B. Early Literature

The literature survey for this review revealed exceptionally few references related to the measurements of exopeptidase in the CNS prior to the 1950's. These early studies are difficult to assess due to the complexities of measuring activity in crude extracts prepared by different procedures without regard to the broad specificity of exopeptidases, effect of metals, and other factors. Blum *et al.*[7] is credited with the first observations (1936) of "dipeptidase" activity in whole brain homogenates of several species.

## TABLE I

**Classification of Exopeptidases (E.C.3.4)**

| E.C. listing | Systematic name | Trivial name | Typical substrate (metal requirement) | |
|---|---|---|---|---|
| 3.4.1 | α-Aminopeptide aminoacidhydrolase | — | — | |
| 3.4.1.1 | L-leucyl-peptide hydrolase | Leucine-aminopeptidase | Leu-$NH_2$ | ($M^{2+}$) |
| 3.4.1.2 | Amino acyl-oligopeptide hydrolase | Aminopeptidase | | |
| 3.4.1.3 | Amino acyl-dipeptide hydrolase | Aminotripeptidase | Leu–Gly–Gly | |
| 3.4.2 | α-Carboxypeptide aminoacidhydrolase | | | |
| 3.4.2.1 | | Carboxypeptidase A | Z-Gly–Phe | ($Zn^{2+}$) |
| 3.4.2.2 | | Carboxypeptidase B | Hippuryl-arginine | ($Zn^{2+}$) |
| 3.4.3 | Dipeptide hydrolases | | | |
| 3.4.3.1 | | Glycyl-glycine dipeptidase | Gly–Gly | ($Co^{2+}$) |
| 3.4.3.2 | | Glycyl-leucine dipeptidase | Gly–Leu | ($M^{2+}$) |
| 3.4.3.3 | Aminoacyl-histidine hydrolase | Carnosinase | β-Ala–His | ($M^{2+}$) |
| 3.4.3.4 | Aminoacyl-1-methyl histidine hydrolase | Anserinase | β-Ala-methyl histidine | ($Zn^{2+}$) |
| 3.4.3.5 | | Cysteinyl-glycine dipeptidase | Cys–Gly | ($Mn^{2+}$) |
| 3.4.3.6 | | Iminodipeptidase | Pro–Gly | ($M^{2+}$) |
| 3.4.3.7 | | Imidodipeptidase (prolidase) | Gly–Pro | ($Mn^{2+}$) |

Abderhalden and Ceaser[8] reported the hydrolysis of di- and some tripeptides containing leucine in extracts prepared from fresh bovine brain and acetone powders with the highest activity is most cases in the grey matter. Kies and Schwimmer[9] also reported hydrolysis of di- and tripeptides with similar extracts with activity higher in brain compared to muscle; the substrates employed were Leu–Gly, Gly–Leu, Gly–Gly, Ala–Gly, Leu–Gly–Gly. In a survey of a large number of tissues, Price *et al.*[10] reported that brain exceeded muscle with Gly–Ala as the substrate but was less than kidney, liver, lung, and the spleen. Adams and Smith[11] were the first to demonstrate a $Mn^{2+}$ activated exopeptidase for crude pituitary extracts. Hanson and Tendis[12] using largely chromatographic procedures indicated that most exopeptidases when tested with dipeptide substrates resided in the cell sap rather than the mitochondrial and microsomal fractions. Pope and Anfinsen[13–15] undertook a detailed survey of the anatomical localization using histochemical procedures with DL–Ala–Gly as the substrate (see Section III, E). The first attempt at the classification of brain exopeptidases based on their specificity and metal requirements can be attributed to Uzman and his group[16–20] in the early 1960's. These and later studies devoted to separation and purification are discussed in detail in the appropriate subsections.

## II. AMINOPEPTIDASES (3.4.1)

### A. Leucine Aminopeptidase (LAP) (3.4.1.1)

Precise information concerning the properties of aminopeptidases is not well represented in the literature. This is due to difficulties connected with the purification, stability, enzyme multiplicity, and an apparent overlap in specificities. LAP (or L-leucyl peptide hydrolase; Table I) has attracted the most attention as a potential tool for protein sequence analysis. This property, the hydrolysis of proteins and polypeptides, may indicate an important role in brain protein turnover, but studies with brain as the source of enzyme are limited. Since the problems related to LAP are representative of the entire group 3.4.1, some description is accorded below to this enzyme.

In 1929 Linderstrøm-Lang[21] reported that erepsin preparations contained an enzyme that rapidly hydrolyzed Leu–Gly or Leu–Gly–Gly but had lower activities with Ala–Gly and Gly–Gly. Subsequently this enzyme was partially purified and shown to be $Mg^{2+}$ or $Mn^{2+}$ activated. In addition to peptides, amides are also hydrolized; substrates considered to be specific are Leu-, Norleu-, Norval-$NH_2$.[23] Purified pig kidney or lens LAP is a large protein of about 300,000 mol. wt.[24,25] and exhibits a preference for substrates with a hydrophobic side chain. All amino acids are susceptible to end-group cleavage although the rates for Pro and Cys are notably low. Most commercial preparations appear to contain endopeptidase contaminants.[26] The action of LAP is determined by the stereochemical configuration and the amino acid residues adjacent to the $NH_2$ terminus.[26] In a study of dihexapeptides of L- and D-Ala, rapid hydrolysis occurred for bonds formed from L-Ala residues but only very slow hydrolysis occurred if one residue was of the D configuration.[27] LAP exhibits a pronounced esterolytic activity, equivalent on a molar basis to about 10% of the peptide bond-splitting activity.[28,29] Lens aminopeptidase rapidly hydrolyzes ester substrates without alteration in the ratio of the esterase to peptidase activity or in metal dependency during purification.[30] The significance of these two types of hydrolase activity shown *in vitro* is not understood but appears to be a property shared by a number of proteolytic enzymes.

The only attempts to characterize LAP in the brain have been with crude extracts. Brecher[31] reported the hydrolysis of Leu-, Try-, and Phe-$NH_2$ in crude mitochondrial and microsomal fractions with activation by $Mn^{2+}$ and to a lesser extent by $Mg^{2+}$ and $Co^{2+}$. With Tyr- or Phe-$NH_2$ the activity was higher in the soluble fractions. In our own studies we observed only low activities with Leu-$NH_2$ in soluble brain extracts but very high activities with Leu–Gly and Leu–Gly–Gly.[32] The ratio for these activities evidently varies with LAP from different sources; for liver, 0.85 : 1.2 : 1.0[33]; for lens, 0.93 : 1.3 : 1.0[30]; and for muscle, 2.0 : 2.0 : 1.0.[34] There appears to be considerable heterogeneity in enzyme from different sources: Brain LAP like muscle LAP is intracellular with *p*H optima of about 7.0 compared to 8.5 for lens LAP; mucosal enzyme is extracellular and less stable than that of other tissues.[25,35–37]

## B. Aminotripeptidase (3.4.1.3)

Enzymes specific for tripeptides are widely distributed in all animal tissues (amino acyl-dipeptide hydrolase; Table I). These enzymes show a marked specificity for tripeptides with neutral amino acid residues, and unlike LAP they are normally without metal requirements.[38] The point of hydrolysis is the bond adjacent to the terminal amino group which must be both $\alpha$ and of the L-configuration; the requirements are less strict for the carboxyl end group. Hydrolysis has been reported for dipeptides with intermolecular distances between end groups approximating that of tripeptides, e.g., glycyl-$\delta$-aminovalerate and glycl-$p$-amino benzoate.[66] Tripeptidases appear to be inhibited by $Cd^{2+}$ and a number of drugs including local anesthetics.[40,41]

Uzman[16] attempted to classify tri- and dipeptidases on the basis of sensitivity to EDTA and $Co^{2+}$. Triglycine Leu–, Val–, DL–Ala–Gly–Gly were hydrolyzed by glycerol–phosphate extracts of mouse brain slices, and unlike dipeptidases were unaffected by EDTA or $Co^{2+}$. More recently a brain tripeptidase was purified with marked specificity for tripeptides containing glycine.[32,42–44] This enzyme was localized in the postmicrosomal supernatant with some 20% in the crude microsomal and mitochondrial fractions (Table II). Some tripeptidase activity was associated with microsomal membranous fractions, and the outer mitrochondrial membrane

## TABLE II

### Distribution of Aminopeptidase, Carboxypeptidase, and Arylamidase in Brain Subcellular Fractions[32]

| | Relative enzyme activities[a] | | | | |
|---|---|---|---|---|---|
| Substrate | H | N | Mt | Mc | Supt |
| Aminopeptidase | | | | | |
| Leu–Gly | 100 | 4 | 7 | 6 | 61 |
| Gly–Gly[b] | 18 | 2 | 11 | 5 | 1 |
| Leu–Gly–Gly | 100 | 3 | 10 | 8 | 55 |
| Leu–Leu–Leu | 30 | 3 | 5 | 5 | 25 |
| Carboxypeptidase | | | | | |
| Z-Leu–Tyr | 5 | 2 | 2 | 0 | 1 |
| Arylamidase | | | | | |
| α-Asp-β-NA | 3 | 0 | 0 | 0 | 1 |
| γ-Glu-β-NA | 2 | 0 | 1 | 0 | 1 |
| Lys-β-NA | 83 | 5 | 10 | 6 | 37 |
| Arg-β-NA | 70 | 3 | 6 | 4 | 35 |
| Leu-β-NA | 50 | 3 | 11 | 7 | 26 |
| Ala-β-NA | 77 | 5 | 46 | 12 | 21 |
| Met-β-NA | 63 | 4 | 15 | 6 | 20 |
| Gly-β-NA | 16 | 1 | 5 | 2 | 16 |
| Phe-β-NA | 13 | 1 | 3 | 1 | 3 |
| Ser–Tyr-β-NA | 15 | 1 | 1 | 1 | 5 |

[a] Values are relative to activity in the homogenate (H) with Leu–Gly as substrate. Assays were done at *p*H 7.6 with 2 mM peptide or 0.5 mM arylamide and incubated 30 min at 37°C. Key: H, homogenate; N, nuclei; Mt, mitochondria; Mc, microsomes; Supt, supernatant.

[b] 0.1 mM $Co^{2+}$.

was prepared by treatment with phospholipase A or by detergent treatment (Table III).[32,105] Brecher and Sobel[45] reported the presence in brain acetone powder extracts of several enzymes with tripeptidase activity; one enzyme was inhibited by $Mn^{2+}$ and showed a marked specificity for Leu–Gly–Gly and triglycine, and the second enzyme was activated by $Mn^{2+}$ and hydrolyzed dipeptides such as Leu–Gly in addition to tripeptides containing leucine.

Recent studies have shown that there is considerable heterogeneity of exopeptidases that are capable of hydrolyzing tripeptides. In addition to the specific enzyme isolated from rat brain extracts hydrolyzing Leu-Gly-Gly and similar tripeptides,[44] a new enzyme has been isolated from hog brain supernatant fractions that hydrolyzes both tripeptide substrates and substituted arylamides.[103] The favored substrates for this "aminotripeptidase" were the tripeptides Met–Met–Ala, trialanine, and trimethionine. This enzyme could be differentiated from the specific aminotripeptidase by the

TABLE III

**Distribution of Enzymes in Myelin, Nerve Ending, and Mitochondrial Subfractions** [32,106,107]

| Substrate | Per cent recovery in the subfractions[a] Mt ($P_2$) | Myelin A | Synaptosomes BCD | Mt E |
|---|---|---|---|---|
| Aminopeptidase | | | | |
| Leu–Gly–Gly | *100* | 21 | 42 | 6 |
| Leu–Leu–Leu | 75 | 9 | 24 | 10 |
| Carboxypeptidase | | | | |
| Z-Gly–Phe | 11 | 4 | 7 | 1 |
| Z-Leu–Tyr | 6 | 2 | 3 | 1 |
| Arylamidase | | | | |
| Leu·β·NA | *100* | 6 | 42 | 3 |
| Ala·β·NA | 50 | 6 | 22 | 3 |
| Arg·β·NA | 98 | 7 | 57 | 3 |

[a] Crude mitochondria ($P_2$) and subfractions prepared by the methods of Marks and Lajtha.[208] Values used for comparison are italicized. See Table II for other details.

inhibition noted in the presence of low concentrations of puromycin or puromycin analogues; the $K_i$ with Met–Met–Ala was $7 \times 10^{-6}$M and was close to that obtained in the presence of monoacyl (Arg-βNA) and dipeptidyl arylamides (Arg–Arg-βNA, Lys-Ala-βNA. See Table IV). The aminopeptidases that are puromycin sensitive and capable of splitting dipeptide substrates are probably similar to the enzymes shown to be present in the pituitary (Section VI, 2) and the kidney.[97] The presence in brain of a family of related aminopeptidases in high concentration points to a significant role in brain peptide turnover, as shown in the scheme in Fig. 1.

## III. DIPEPTIDASES (3.4.3)

Particular care must be exercised in the identification of dipeptidases (hydrolysis of dipeptides with both a free —$NH_2$ and —COOH) since many dipeptide substrates are cleaved by LAP. Dipeptidases are generally differentiated from each other and other exopeptidases on the basis of activation with different metal ions.[46-48] This method of classification is complex and must await purification of the enzymes involved for verification. Diglycine hydrolysis is inhibited by $Co^{2+}$ and $Mg^{2+}$, but there are also reports of significant hydrolysis of some dipeptides in the absence of added metal ions.[49,50] As noted in the introduction there are several early reports of dipeptidase-like activity in brain homogenates from several species (Section I, B). A recent survey utilizing electrophoretic methods also

## TABLE IV

**Summary of the Arylamidases and Aminopeptidases Present in Nerve Tissue**

| Major substrate | Nomenclature[a] | Alternative substrates -βNA | Alternative substrates Peptide(ester) | pH (activators) | Inhibitors | Tissue source references |
|---|---|---|---|---|---|---|
| Leu-$NH_2$ | LAP | Leu | Leu-Gly | 7–8<br>$Mn^{2+}$ | | Brain[31,32] |
| Leu–Gly–Gly | Aminotripeptidase | — | Met–Met–Ala<br>Ala–Gly–Gly<br>Ala–Ala–Ala | 7–8 | | Brain[32,42–45]<br>PN[179,213] |
| Gly–Gly | Diglycinase | — | — | 7–8<br>$Co^{2+}$ | Leu<br>Val<br>Asp<br>α-Aminobutyric | Brain[19,32,55] |
| α-Glu-βNA | Arylamidase A | | α-Asp<br>β-Asp | 7–8<br>$Ca^{2+}$ | — | Sera[80]<br>Brain[32,44] |
| Leu-βNA | Arylamidase N | Ala-<br>Met-<br>Ser–Tyr- | — | 7–8<br>–SH<br>$M^{2+}$<br>$M^{3+}$ | $Cu^{2+}$, $Cd^{2+}$<br>$Hg^{2+}$, TPCK<br>PCMB<br>Puromycin | Brain[32,44,105,109,110]<br>PN[109] |
| Arg-βNA | Arylamidase B | Arg–Arg-<br>Lys–Ala<br>Ala–Ala<br>Lys–Lys<br>Leu–Ala- | Met–Met–Ala<br>Ala–Ala–Ala<br>Met–Met–Met<br>Leu–Gly–Gly | 7.0, 9.0<br>$M^{2+}$<br>$M^{3+}$<br>–SH | $Cu^{2+}$, $Cd^{2+}$<br>$Hg^{2+}$, TPCK<br>PCMP<br>Puromycin<br>Actinomycin | Brain[32,44,100,105]<br>PN[179,213] |
| Arg-βNA<br>Lys-βNA | Arginyl<br>Arylamidase<br>"Dipeptidase" | Leu-<br>Met-<br>Phe- | Lys–Val-<br>$Ala_2$–$Ala_5$<br>$Lys_2$–$Lys_4$ | 6.8<br>$Mn^{2+}$<br>$Co^{2+}$<br>$Zn^{2+}$ | Dipeptides (except with Gly)<br>EDTA | Pituitary[76,115] |

**TABLE IV**—*continued*

| Major substrate | Nomenclature[a] | Alternative substrates -βNA | Alternative substrates Peptide(ester) | pH (activators) | Inhibitors | Tissue source references |
|---|---|---|---|---|---|---|
| Lys-βNA | Lysyl<br>Arylamidase<br>"Polypeptidase" | Arg-<br>Met-<br>Leu- | Lys–Val<br>$Ala_4$–$Ala_5$<br>$Lys_5$ | 6.5–8.0<br>$Mn^{2+}$<br>$Co^{2-}$<br>$Zn^{2+}$<br>–SH | Di- and tripeptides<br>Bradykinin<br>Puromycin<br>Histones<br>Bases<br>EDTA | Pituitary[6,148] |
| Ser–Tyr-βNA | Dipeptidyl<br>Aminopeptidase I<br>(Dipeptidyl transferase?) | His–Ser<br>Ala–Ala<br>Gly–Phe<br>Ser–Met | Alar→$Ala_2$<br>Gly–Phe-$NH_2$<br>Ser–Tyr–Ser–Met | | | Pituitary[147] |
| Lys–Ala-βNA | Dipeptidyl<br>Aminopeptidase II<br>(Carboxytripeptidase dipeptide esterase) | Arg–Arg<br>Leu–Ala<br>Ala–Ala | Lys–Ala–$OCH_3$<br>Ala–Ala–Ala<br>Met–Met–Ala<br>Met–Met–Met<br>Ser–Met–Glu<br>Ser–Met–Glu | 4.5 (peptide)<br>5.5 (ester) or arylamide | Puromycin<br>$Na^+$, $N^+$, $Li^+$<br>Tris cations | Pituitary[6,148]<br>(lysosomes) |
| Arg–Arg-βNA | Dipeptidyl<br>Aminopeptidase III | | $Ala_4$→2$Ala_2$<br>$Ala_5$→$Ala_2$+$Ala_3$<br>$Ala_6$→3$Ala_2$<br>$Lys_4$→2Lys<br>$Phe_4$→2Phe<br>Val–Leu–Ser–Glu–Gly | 6.5, 9.0<br>–SH | $Arg_2$<br>PCMB<br>$Lys_2$-βNA | Pituitary[6,149]<br>Brain[105] |

[a] The nomenclature adopted is that proposed by the individual authors.

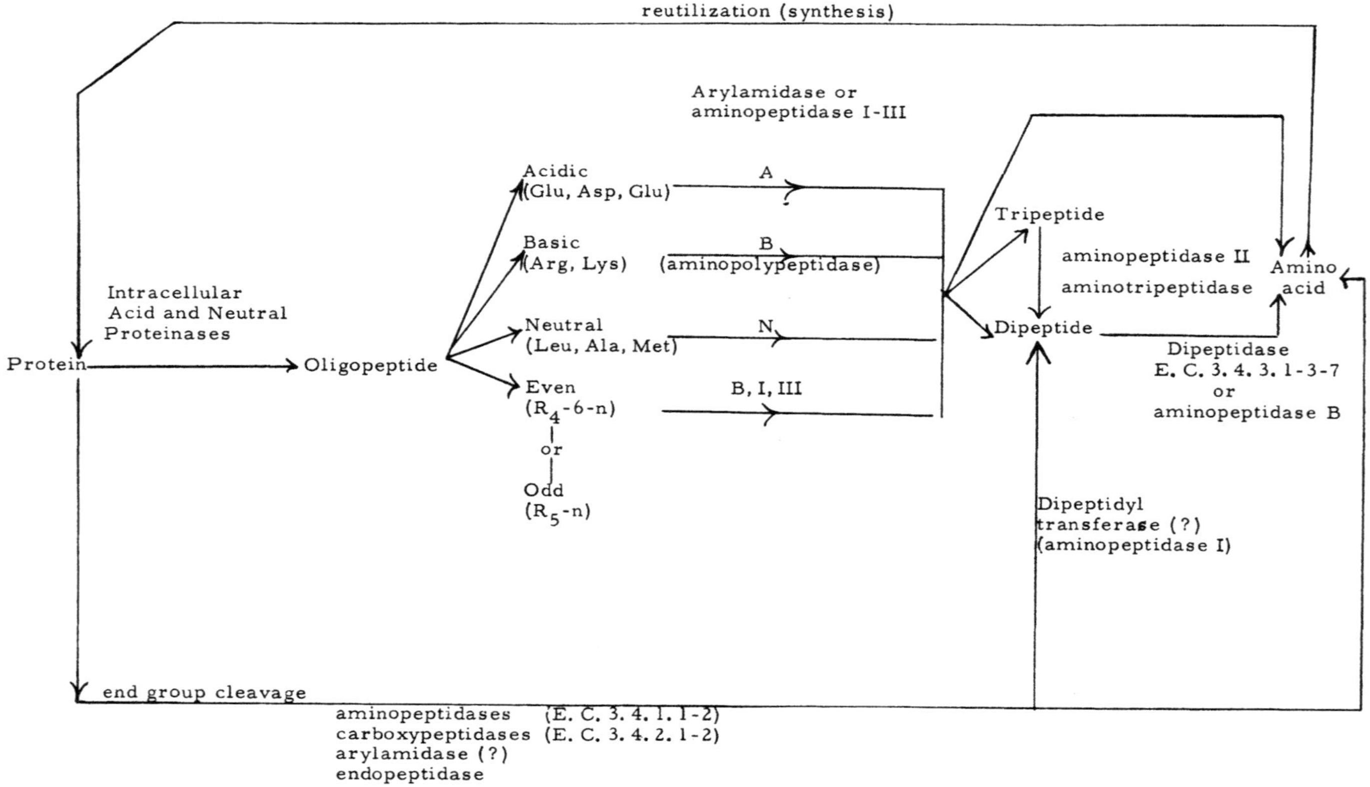


Fig. 1. Hypothetical scheme for peptide turnover in brain. Nomenclature based on that in Tables I and IV. The end products, amino acids are then available for return to the cellular pools or reutilization for protein synthesis or energy production.

reported dipeptidase activities in the blood, serum, and brain in a large number of animals.[51–53] There have been very few attempts at the characterization of brain dipeptidases.

## A. Glycine–Glycine Dipeptidase (3.4.3.1)

Glycerol phosphate extracts of brain slices contain diglycinase activity with specific activation by $Co^{2+}$.[19] This enzyme has not been fully characterized, but specific metal-activated diglycinases are known to occur in various mammalian tissues[50,140] and in bacteria.[54] Diglycine hydrolysis in brain is inhibited by various amino acids, particularly Leu, α-aminobutyric acid, valine, and aspartate.[20] Similar phenomena were reported for liver[50] and bacterial diglycinase.[54] This inhibition may represent a control mechanism for the release of glycine from peptide precursors. Glycine has aroused a great deal of interest as an inhibitor of neuronal function in spinal neurons.[55] Glycine occurs in high concentration in the ventral gray matter of cat spinal cord, which is rich in interneuron processes.[56] In our own studies on diglycinase distribution we showed a high concentration in crude brain particulates, particularly nuclei, synaptosomes, and mitochondria, with marked activation by low concentration of $Co^{2+}$[32] (Table I). In a recent report Brecher and Koski[57] reported a *p*H optimum of 8 in ox and human brain homogenates with activity largely present in soluble supernatant fractions.

## B. Carnosinase and Anserinase (3.4.3.3.–4)

Specific dipeptidases for carnosine (β-alanylhistidine) and anserine (β-alanyl-1-methylhistidine) occur in liver and kidney.[58,140] Only trace quantities of these components occur in brain (0.5 mg/100 g)[59] compared to considerably higher concentration in other excitable tissues, notably muscle (150–200 mg/100 g wet wt.).[58] Also, there is now good evidence for the presence of homocarnosine (γ-aminobutyrylhistidine, 8 mg/100 g) and homoanserine (γ-aminobutyryl-1-methylhistidine, 1 mg/100 g) in bovine brain.[60] In a study related to $^{14}C$-β-alanyhistidine uptake in rat brain slices, Abrahams *et al.*[61] reported a 30% hydrolysis in 2 hr as measured by the release of $^{14}C$-β-alanine, suggesting the presence of a specific carnosinase. Other than this observation, there have been no detailed studies of this and related enzymes in nervous tissue.

## C. Cysteinyl–Glycine Dipeptidase (3.4.3.5)

Ribosomal preparations from various tissues and bacteria readily split Cys–Gly by an enzyme claimed to be a distinct dipeptidase.[62,63] Of the many dipeptides tested only those with amino terminal Leu–, Met–, and Cys– appear to serve as substrates for ribosomal enzymes; Matheson and Murayama[64] have suggested that peptidases play a role in polypeptide formation by removal of terminal methionyl groups. Brain is exceptionally rich in $\gamma$-glutamylcysteinyl-glycine (glutathione, 80 mg/100 g), which might suggest a role for cysteinyl–glycine dipeptidase after removal of the $\gamma$-glutamyl moiety. Glutathione can be synthesized from Cyst–Glu by a reversible transpeptidation reaction.[65]

## D. Imino- and Imidodipeptidases (3.4.3.6.-7)

Despite the relatively high concentration of proline in brain tissue (0.3 $\mu$ moles/g protein), very little attention has been directed toward these enzymes. Specific dipeptidases for proline dipeptides have been purified from erythrocytes from the intestinal mucosa and the kidney[39,66,140]; two groups can be distinguished: (1) iminodipeptidase or L-prolyl-amino acid hydrolase and (2) imidodipeptidase or amino-acyl-L-proline hydrolase. Iminopeptidases are specific for dipeptides where proline supplies the $\alpha$-imino group, and imidopeptidases split substrates where proline is bound by the imino group. Examples of both substrates, Pro–Gly and Gly–Pro, are hydrolyzed by a variety of brain extracts.[25] Imino- and imidodipeptidases are $Mn^{2+}$ dependent and are specific only for proline or hydroxyproline in dipeptide linkage and can be distinguished from peptidases that split N-terminal proline from proteins or N-terminal amino acids adjacent to proline.[67,215] One such enzyme, termed "iminopeptidase", is believed to be involved in some aspects of collagen metabolism. Collagen is the only animal protein known to contain hydroxyproline; there are only trace quantities of collagen in brain and spinal cord, and this is probably derived from nonneuronal elements.[68] The slightly higher levels in the spinal cord (0.5%) are of some interest in relation to the induction of experimental allergic encephalomyelitis in animals.[69] There is evidence that hydroxylation of proline occurs at selected residues already in peptide linkage, and it is of considerable interest to the possible role of hydrolases to note that a number of oligopeptides containing the tripeptide sequence $(\text{Pro–Gly–Pro})_n$ act as competitive inhibitors to the hydroxylase activity.[70]

### 1. *ε-Peptidases*

Spinal cord also contains basic proteins rich in lysine which are involved in the induction of experimental allergic encephalomyelitis.[71] A peptidase specific for dipeptide substrates with the ε-amino bond of lysine has been reported in extracts of rat and hog kidney.[72] This enzyme remains tentative but could be involved in the degradation of some nerve tissue proteins. Peptide bonds with the ε-amino group occur in bovine growth hormone and in collagen.[72,137]

## E. Histochemical Studies

In the previous decade there were many studies devoted to the anatomical location of exopeptidases in the brain. The most frequently quoted work is that of Pope and Anfinsen,[13–15] who employed the quantitative microchemical procedures of Linderstrøm-Lang[22] for the study of DL-Ala–Gly hydrolysis. They reported high activity in the intralaminar layers II, IV, Vb, and VIb with activity in the human frontal isocortex some threefold that of the rat. It was originally suggested that the neuronal perikarya are the principal intracortical sites of this exopeptidase activity, but recalculation of the data on a dry weight or protein weight basis indicated no difference between gray and white matter.[73] Consequently it must be concluded that exopeptidase activity also occurs in the processes of neurons and neuroglia in addition to the cell bodies. It should be noted that DL-Ala–Gly is not a specific substrate for dipeptidases but could also be hydrolyzed by LAP. A similar approach utilizing quantitative microchemical techniques was adopted by Uzman *et al.*[16–20] in studying the exopeptidases in histological slices of mouse brain. In a survey of substrate requirements they showed that glycerol–phosphate extracts favored peptides with a lipophilic side chain and an aliphatic or aromatic C-terminus. Studies on the absolute stereospecific requirements were incomplete, but racemic peptides appeared to be favored compared to those with only the L-configuration.

## IV. ARYLAMIDE AMINO ACID HYDROLASES

Gomori[5] introduced Leu-β-naphthylamide for the histological detection of aminopeptidases. Hydrolysis of the substrate released β-naphthylamine (coupled to amino acids at the C-terminal position) which formed azo dyes with diazonium compounds. An excellent account of the chemistry is given by Burstone.[74] Other synthetic substrates used for the same purpose are nitroanilides[75,76] and aminonitriles.[77] There is no E.C. listing for these hydrolases but various names have been suggested: aminopeptidase A or B,[78] naphthylamidase,[35] arylamidase,[33] and aminoacylnaphthylamidase aminohydrolase.[79] Currently the term arylamidase has gained general acceptance regardless of the chromogenic substrate employed. In the early studies it was assumed that enzymes hydrolyzing Leu–βNA were identical to LAP, but Patterson *et al.*[33,35] showed that these activities represented different enzymes. Most tissues, including brain,[44] and microorganisms contain a spectrum of arylamidases; the different specificities appear to be conferred by the amino acid moieties. The classification adopted for purposes of this review must be regarded as tentative until all the enzymes have been purified and characterized.

### A. Arylamidase A

Glenner[80] first reported the presence of a $Ca^{2+}$-activated enzyme specific for α-Asp- and α-Glu-βNA in rat kidney microsomal preparations. This enzyme, termed arylamidase A, is present in sera, epithelium of pancreatic ducts, islets of Langerhans,[80–84] and brain.[44] Enzymes specific for glutamyl-containing peptides are of exceptional interest in view of the high content of free and bound acidic amino acids in the nervous system. Brain homogenates and subcellular fractions hydrolyze $\alpha,\gamma$-Glu- and $\alpha,\beta$-Asp-$\beta$NA, but activity is less than 5% compared to the activity measured with neutral and basic analogues. Brain arylamidase A activity is only slightly activated by $Ca^{2+}$; it is particulate bound, with the highest activity present in the crude mitochondrial fractions.(32)

### 1. *Inactivation of Angiotensin II*

Relatively little is known about the degradation of angiotensin II (H-Asp$^1$-Arg$^2$-Val$^3$-Tyr$^4$-Val$^5$-His$^6$-Pro$^7$-Phe$^8$-OH) by specific exo- or endopeptidases. Khairallah and Page(85) reported the presence of three enzymes in plasma: (1) a $Ca^{2+}$-activated exopeptidase specific for Asp$^1$–Val$^5$–angiotensin II and similar in properties to arylamidase A, (2) a second exopeptidase specific for Asn$^1$–Val$^5$–angiotensin II, and (3) an endopeptidase capable of degrading all analogues. A number of other studies support the view that the first enzyme is related to an arylamidase specific for acidic arylamide substrates.(82,86) This enzyme is present in serum, red cells, and kidney homogenates. There seems to be some doubt concerning the identity of the third chymotryptic-like enzyme,(87) which is present in kidney, active at *p*H 5.0, and inhibited by DFP. With an angiotensin analogue relatively resistant to aminopeptidases, [$\beta$-L-Asp$^1$]-angiotensin, Regoli *et al.*(87) reported the release by kidney homogenate of a N-terminal tetrapeptide plus four free amino acids. Johnson and Ryan(88) recently demonstrated that the release of free amino acids was probably due to the action of a carboxypeptidase that is present in high concentration in rabbit liver. Yang *et al.*(89) have also described an enzyme specific for the C-terminal end of angiotensin II, Pro$^7$-Phe$^8$, but which is not identical to known carboxypeptidases.

### 2. *Metabolism of γ-Glutamyl Peptides*

The histochemical location of $\gamma$-Glu-$\beta$NA in the brain is distinct from that of $\alpha$-Glu-$\beta$NA.(81,90) Metabolism of $\gamma$-glutamyl peptides can occur by transfer of peptide- or protein-bound $\gamma$-glutamate to suitable acceptors, and the view was once held that transpeptidation or transamidation played an important role in the synthesis of polypeptides.(91,92) Peptides containing $\gamma$-glutamate occur in high concentration in brain, glutathione being present at a level of 3–4 $\mu$moles/g in the rat.(93) More recently trace levels of $\gamma$-Glu–Glu, $\gamma$-Glu–Gly, and $\gamma$-Glu–Glu-$NH_2$ at 7 $\mu$g/g were reported present in brain extracts.(94,95) The trace activity observed in brain homogenates with $\gamma$-Glu-$\beta$NA could represent transpeptidation rather than hydrolase activity (Table II).(44) $\gamma$-Glutamyl transpeptidases are widely distributed in tissues and plants with the activity in brain some 100-fold lower than that of the kidney.(80,90,96) Cysteinylglycine is the product of transpeptidation, with GSH as one of the substrates; the reactions are reversible with reformation of GSH.(65,214) This may represent a pathway for degradation of GSH and similar tripeptides since Cys–Gly can be subsequently split by LAP and specific dipeptidases. There are a number of other enzymes involved in metabolism of γ-glutamyl peptides; one is γ-glutamyl transferase, which represents a reversal of the glutamine synthetase pathway with specific binding sites for glutamate, ATP, and $NH_3$.[98,214] This enzyme is distinct from transglutaminase that mediates the exchange of protein γ-glutamyl groups derived from glutamine with primary amines.[99]

## B. Arylamidase B

There is good evidence on the basis of a highly purified liver enzyme for the existence of a separate arylamidase specific for basic arylamides[78] termed arylamidase B in the present review. Previously, Nachlas *et al.*[100] in an evaluation of arylamide hydrolysis in different tissues reported the splitting of Arg-βNA in brain homogenates. Arylamidase B has been purified from brain[44] and displays some differences in properties compared to those of the pituitary (Section VI, A) and purified liver enzyme.[78] Hopsu *et al.*[78,101,102] succeeded in purifying liver arylamidase B several hundredfold and have studied its specificity requirements in detail. It is specific for Lys- and Arg-arylamides which contain an ε-amino group on the sixth atom from the susceptible α-amino group; other arylamides without this structure such as the histidine, ornithine, and ε-amino caproic analogues are not hydrolyzed. The liver enzyme is stereospecific and without esterolytic activity, and unlike the brain it is halide dependent but without any metal requirements.[44] The purified brain enzyme is inactivated by exposure to EDTA and reactivated by the addition of a variety of metal ions, except $Cd^{2+}$, $Hg^{2+}$, and $Cu^{2+}$, which are inhibitory. Also, the brain enzyme is inactivated by compounds such as PCMB but activated by cysteine and mercaptoethanol. Probably the most unusual feature of brain arylamidase is the potent inhibition by puromycin (Ki, $2 \times 10^{-6}$)[44] and by actinomycin.[103] The inhibition by puromycin has been successfully employed to differentiate arylamidases from other exopeptidases.[44,103] Liver arylamidases are considered to be exclusively lysosomal; their presence in soluble fractions has been attributed to leakage arising from drastic extraction methods.[104] In brain, even with mild homogenization techniques, the activity was widely distributed in all fractions, ranging from 40% in the soluble supernatant to 10% in the crude mitochondria, 6% in microsomes, and 4% in nuclei (Table II).[32] Purification of mitochondria on continuous sucrose gradients indicated the greatest association with synaptosomes and particles that resembled lysosomes when seen under the electron microscope (Table III).[105] Fragmentation of intact purified mitochondria and microsomal preparations indicated a small but significant association of arylamidases with the membrane components.[105-107]

## C. Arylamidase N

Early studies with Leu-βNA were performed solely in relation to LAP activity. Since these two activities are distinct and also differ in properties from enzymes hydrolyzing basic and acid arylamides, it has been presumed that a separate category of neutral arylamidases is present in tissues.[44,79,108,182] Adams and Glenner[109] reported the hydrolysis of Leu-$\beta$NA in corpus callosum and in frontal gray matter of adult and newborn rats. In rat brain the highest activity was present in the microsomal supernatant fraction (Leu-$\beta$NA, 43%), followed by crude mitochondria (22%) and microsomes (13%) and nuclei (6%) (Table II).[32] In further subfractionation of crude mitochondria, arylamidase N was associated with synaptosomes, and with lysosome-like granules.[105] The highest activity occurred with Ala-, followed by Met-, Leu-, Gly, Phe-$\beta$NA.[32,44] Brecher and Barefoot[110] reported the highest activity in microsomes with reduced activity with 4-methoxy-amino acid arylamides analogues. Liver and kidney arylamidase was reported to occur in microsomal and soluble cytoplasmic fractions[35,111] with an altered pattern of distribution during malignancy, but recent studies indicate a lysosomal origin.[104,112]

Attempts to purify neutral arylamidase have not met with the same success as in the case of arylamidase B. This difficulty may be attributed in part to the multiple electrophoretic forms noted in many tissues, in sera, spleen, pancreas, lymph node, and intestinal tract.[79,113] Partially purified arylamidase N from the brain was similar in properties to that described for the liver[114] and the pituitary.[115] The brain enzyme required sulfhydryl compound for maximum stability; its activity was inhibited by dialysis in the presence of EDTA and reactivated by a variety of di- and trivalent metal ions.[44] Arylamidase N, like the basic enzyme, was potently inhibited by puromycin (Ki, $1 \times 10^{-5}$).

## V. CARBOXYPEPTIDASE (3.4.2)

The availability of highly purified crystalline preparations has led to some notable achievements in the knowledge of the enzyme structures and the mechanisms of catalysis. Carboxypeptidases split C-terminal end groups of proteins and serve as an important tool in determining the structure of polypeptides. These enzymes occur in high concentration in extracellular secretions of the pancreas, which is the major source for purification purposes. The nervous system contains only low concentrations of carboxypeptidases, and the few studies available are in relation to the hydrolysis of kinin-like polypeptides. A feature of particular interest to the regulatory influences of intracellular protein breakdown is the existence of intracellular zymogen granules in the pancreas. The inactive procarboxypeptidases represent an intracellular storage site that can be activated by endopeptidases such as trypsin and by a variety of chemical treatments.

On the basis of precise information with highly purified preparations, two major groups are recognized: (1) carboxypeptidase A, which is specific for peptides with neutral amino acid C-terminal end groups except proline, and (2) carboxypeptidase B, which is specific for peptides with C-terminal basic amino groups. In this respect the classification is analagous with the tentative scheme adopted in this review for arylamidases (Section IV). Carboxypeptidases are metalloenzymes normally requiring $Zn^{2+}$ to exhibit peptide hydrolase activity.[116] The chief means for identification of these enzymes are the synthetic substrates introduced by Bergmann.[1] In general such substrates contain carbobenzoxy protected N-terminal groups, although N-halogen acylamino and ester compounds are sometimes employed. Enzymes hydrolyzing acylamino acids or peptides probably belong to the general class of carboxypeptidases, although others have considered them as a distinct group termed "acylases."[25,120]

There are several reports on the hydrolysis of acetylated compounds by brain extracts, that of *S*-acetylglutathione, *S,N*-diacetylcysteine,[117] and *N*-acetyl-aspartate.[119,120] Goldstein[121] recently separated and purified two enzymes hydrolyzing *N*-acetyl-aspartate from rat cell-sap fractions; one enzyme was similar to kidney acylase II and the second to acylase I, which is specific for *N*-acetyl and chloroacetyl- derivatives of several monocarboxylic amino acid substrates.[120]

## A. Carboxypeptidase A

There are no detailed studies of this enzyme in the brain. The purified enzyme favors peptides with C-terminal aromatic residues such as Phe-, Tyr-, or Try- or the branched aliphatic amino acids Leu- or Ile-; it is inactive toward Pro- or Hyp- as the terminal or penultimate residues. Extracts of brain are active with Z-Gly–Phe, but this activity is considerably lower than with the aminopeptidase substrates.[32] In the author's studies the highest activity occurred with those dipeptides that correspond with some of the sequences present in insulin β-chain; Z-Phe–Phe, Z-Val–Phe, and Z-Gly–Phe.[32] This activity was particulate bound, with the highest concentration present in nuclei and synaptosomes.

### 1. *Comment on Carboxypeptidase Zymogens*

Activation of zymogens has long constituted an effective approach to the elucidation of some aspects of enzyme catalysis.[116] The factors that affect activation of digestive enzymes may be involved in the intracellular regulation of protein breakdown in tissues. In the case of trypsinogen and chymotrypsinogen the release of a peptide fragment from the amino-terminus results in conformational changes accompanied by enzyme activation. The picture is more complicated in the case of bovine carboxypeptidase A since the zymogen is composed of three subunits: One subunit is the precursor of the exopeptidase, the second gives rise to an endopeptidase, and the third remains to be identified.[122] The relationship between these enzymes may serve as a guide for elucidating some of the mechanisms of brain protein turnover. For example, the endopeptidase subunit can combine with the active exopeptidase to yield a dimeric complex with carboxypeptidase A activity. The different carboxypeptidase subunits can be disaggregated by succinylation.[122] Species variation are also apparent since the spiny dogfish contains only a monomeric form of the precursor zymogen.[123] The full delineation of the complex activation process requires a knowledge of the amino acid sequence of the zymogen which is presently unavailable. Studies of the active center of carboxypeptidases have revealed a fascinating homology with other proteolytic enzymes, particularly trypsin and chymotrypsin; the amino acids serine and histidine are involved in the active centers of these enzymes, suggesting a common evolutionary origin.[124] Contrary to previous expectations, carboxypeptidase A appears to exist in a number of alloptropic variants, Aα, Aβ, and Aγ.[125]

Procarboxypeptidase is a metallozymogen with $Zn^{2+}$ as the metal bound to an —SH group and a second donor group yet to be identified. Replacement of $Zn^{2+}$ by $Cd^{2+}$ or $Hg^{2+}$ leads to loss of exopeptidase activity with a concomitant increase in esterase activity.[126,127] It is of interest that the monomeric dogfish zymogen exhibits esterase activity that disappears during the activation process.[128] It is particularly intriguing that a number of chemical and physical treatments can alter the function of bovine carboxypeptidase A to an esterase activity: tetranitromethane, acylation with mono- and dicarboxyoic acid anhydrides, photoxidation with methylene blue, hydrogen peroxide, ultraviolet radiation, etc. In contrast, N-substituted hydrolysis products of synthetic peptides and esters inhibit esterase activity but activate peptidase activity. These compounds are referred to as modifiers and have properties akin to N-terminal blocking agents. It appears that these treatments modify the tyrosyl residue, essential for exopeptidase activity; of the 19 tyrosines in carboxypeptidase A only 2 are essential for enzyme actions.[129] Peptide substrates can form stable apocarboxypeptidase complexes in the absence of metals, in contrast to the binding of ester substrates.[127,130] It appears that many peptides act as competitive inhibitors; the competition of the inhibitor for the metal binding site may be a significant factor in peptide breakdown within the cell.

## B. Carboxypeptidase B (3.4.2.2)

These exopeptidases have attracted great interest because they inactivate a group of important biologically active peptides known collectively as the kinins. These peptides have related chemical structures with C-terminal basic end groups, e.g., bradykinin, kalliden, angiotensin II; they are generally vasoactive and are involved in tissue permeability and inflammatory processes. Kinin production represents another example of precursor activation by proteolytic enzymes; kalliden is released by addition of kallikrein to serum pseudoglobulin. Krivoy and Kroeger[131] reported that the enzyme responsible for inactivation of bradykinin in brain extracts of rabbit, rat, pig, and pigeon was similar in all properties to pancreatic carboxypeptidase B. Like the pancreatic enzyme it was inhibited by phenothiazines, which are known also to potentiate *in vivo* the activity of bradykinin. Some indication of the catalytic event was supplied by the formation of a drug-enzyme complex in the presence of $Zn^{2+}$ that failed to potentiate the effect of bradykinin.

A number of other carboxypeptidase-like enzymes are known to inactivate bradykinin. Erdos and Yang[132] reported that the hydrolysis of bradykinin by homogenized kidney occurred in two steps; after removal of the terminal Arg a second carboxypeptidase cleaved the C-terminal phenylalanine. These enzymes could be distinguished from an exopeptidase that released the C-terminal $Phe^8$–$Arg^9$ dipeptide.[133] As mentioned in the case of angiotensin inactivation, it is perhaps premature to advance names for the enzymes in the absence of adequate characterization. This is especially pertinent in the case of bradykinin since it can be inactivated by endopeptidases, iminopeptidase, etc. There are several recent reviews summarizing the vast literature on this subject.[132,134,135]

## VI. ANATOMICAL LOCATION OF EXOPEPTIDASES

The association of physiologically active peptides in different brain areas suggests that exopeptidases may play a hitherto unsuspected role in the regulation of many body activities. Surprisingly, there have been only limited studies on the peptide composition of the nervous system, the subcellular localization, biosynthesis, and degradation. The bovine posterior and anterior glands contain the same amino acid spectrum as whole brain, except that taurine, phosphoethanolamine, and cystathionine are present in a slightly higher concentration; the bovine pineal gland is exceptionally rich in cystathionine, with a level (128 mg/100 g) considerably higher than that of primates.[136] Expressed in terms of dry weight the peptide composition is about 4% for the posterior pituitary and 0.02% for the pineal gland.[2]

### A. Pituitary Gland

The hormones of the pituitary gland are either proteins or polypeptides with molecular weights of 1000–50,000; the number of amino acid residues range from 8 (vasopressin) to 188 (growth hormone).[137] This scatter in molecular size indicates that degradation can be mediated by both exo- and endopeptidases. The large number of nonhormonal peptides recently reported in trace quantities may represent intermediates in the synthesis and degradation of the active components. Many pituitary hormones are located in different particulate fractions and in some cases are stored in granules. Extracts of posterior pituitary contain protein, termed neurophysin, capable of forming a complex with the peptide hormones oxytocin and vasopressin in an approximate ratio of 1 : 3.[138] Recent work demonstrated the presence of neurophysin in porcine serum and the appropriate target organs for oxytocin and vasopressin, but its absence in the brain, skeletal muscle, spleen, and liver.[139] There is now some evidence that exopeptidases are associated with alteration in the levels of hormone prior to parturition.[187] The interactions of exopeptidases with these hormones, or their protein carriers, constitutes an important area of research. Early studies showed the presence of a $Mn^{2+}$ activated aminopeptidase in crude pituitary extracts,[11] but the relationship of this and other enzymes to hormone activity is still obscure. For example, the removal of the terminal phenylalanine from the COOH-terminus of bovine growth hormone by carboxypeptidase does not cause any change in the growth-promoting activity.[141] Even the limited digestion of bovine growth hormone with chymotrypsin of 25% produces no alteration of the biological activity even though the active core (α) posesses distinct physiochemical and immunological properties.[137,142] The minimum molecular structure required for hormonal activity is of particular interest in relation to the role of exopeptidases; studies on the chemical synthesis of ACTH show that the octadecapeptide sequence $Ser^1$–$Tyr^2$ . . . $Lys^{15}$–$Lys^{16}$–$Arg^{17}$–$Arg^{18}$ possesses steroidogenic potency *in vivo* comparable with the natural hormone ($Ser^1$–Tyr . . . Lys–Lys–Arg–Arg–Pro . . . Phe.[39,143] Recent studies have placed emphasis on pituitary enzymes that are specific for the peptide sequences containing the terminal dipeptide $Ser^1$–$Tyr^2$ of ACTH and that also hydrolyze arylamide synthetic substrates[147]; there are no studies on pituitary enzymes specific for the —COOH terminal groups.

### 1. *Arylamidases*

There have been many recent efforts to characterize the spectrum of peptides in the anterior pituitary gland. For this purpose mono- and dipeptidylarylamide substrates have been employed. Ellis[76] reported the presence of thiol-activated arylamidases in nonparticulate cell sap capable of hydrolyzing *p*-nitroanilides of Lys-, Arg-, and to a lesser extent Met-, Leu-, Phe-, and Ala-. In a comparison between the anterior and posterior lobes, Jouan and Rocaboy[144] reported higher activities in the two lobes with Thr-, Ser-, and Val-βNA. Vanha-Perttulla and Hopsu[145–146] separated five arylamidases components hydrolyzing Leu-βNA from water extracts of rat anterior pituitary by DEAF–cellulose chromatography: Three components were inhibited by EDTA with different sensitivities on reactivation by $Co^{2+}$, $Mn^{2+}$, and $Mg^{2+}$; the other components were differentiated by their *p*H optima and the effect of cysteine. Ellis *et al.*[6,115,147–149] supplied evidence for five different arylamidases with different specificities when tested with monoaminoacyl and dipeptide substituted arylamides or selected peptides. The relationship of these newly described enzymes to other accepted exopeptidases is somewhat complex and can be established only when more highly purified preparations are available. A comparison of these enzymes, including substrate specificity and the effect of inhibitors and activators, is shown in Table IV; the nomenclature is that supplied by the authors. The first two enzymes bear some similarities to the arylamidases of other tissues described in Section IV. "Arginyl arylamidase" preferentially hydrolyzes

Lys- and Arg- *p*-nitroanilides but is also active to a lesser extent with neutral amino acid analogues. Its chief difference from purified basic arylamidase of brain is the lack of effect with addition of puromycin and thiols. On the basis of its specificity against peptide substrates, Ellis and Perry[115] considered this enzyme as a possible "dipeptidase" although it was capable of hydrolyzing other oligopeptides (Table IV). The lysyl enzyme was distinguished by the preference shown for Lys-*p*NA, its activation by thiol compounds, and its inhibition by puromycin. Its activity appeared to be restricted to the larger oligopeptides, and this enzyme was tentatively named "aminopolypeptidase." The authors envisaged the degradative pathway as an initial hydrolysis of oligopeptides by an "aminopolypeptidase" followed by the dipeptidase activity to form free amino acids. Arginyl arylamidase appeared to contain an active aminotripeptidase which is known to occur in a number of tissues including that of the brain.[44] This aminotripeptidase may be identical to the dipeptidyl arylamidase II described below.

## 2. *Dipeptidyl Aminopeptidase*

A second group of pituitary exopeptidases have now been described that can be differentiated from monoacyl arylamidases by their ability to hydrolyze dipeptide-substituted arylamide analogues such as Ser–Tyr-, Lys–Ala-, and Arg–Arg-βNA (Table IV). The first such enzyme ,"dipeptidyl aminopeptidase I", assayed with Ser–Tyr-βNA exhibits less specificity than the related aminopeptidases II and III.[147] It is noteworthy that this substrate fulfills the substrate requirements for cathepsin C (or dipeptidyl transferase),[92] an unsubstituted terminal residue together with a penultimate aromatic amino acid,[91] as confirmed by the ability to hydrolyze Gly–Phe-$NH_2$ and Gly–Phe-βNA, but unlike cathepsin C it can also hydrolyze $Ala_2$-βNA, His–Ser-βNA, etc. (Table IV).[147] However, this enzyme, like cathepsin C and aminopeptidase II, was located largely in pituitary lysosomes.[6] A possible physiological function was demonstrated by McDonald *et al.*,[147] since aminopeptidase I hydrolyzed the terminal dipeptidases Ser–Tyr and Ser–Met from the decapeptide corresponding to the $NH_2$-terminus of ACTH. However, aminopeptidase I was not similar to the halide-dependent enzyme obtained from beef liver, capable of cleaving the terminal His–Ser from glucogen.[151]

Aminopeptidase-II assayed with Lys–Ala-βNA showed a considerably wider specificity when tested with a variety of dipeptide arylamides, dipeptide esters, and tripeptides.[6] The authors noted no distinction of the three activities in preparations more than 1000-fold purified (molecular weight approximately 130,000). The major substrate was Lys–Ala-, followed by Arg–Arg-βNA with about one-third of the activity. The rate of ester hydrolysis with Lys–Ala-$OCH_3$ actually exceeded by a factor of 3–4 the hydrolysis of the arylamide analogue. In contrast to aminopeptidase I and III, aminopeptidase specificity was restricted to unsubstituted tripeptides;

the rate of hydrolysis of tripeptides exceeded that of arylamide but was equal to that of the dipeptidase esters. McDonald *et al.*[63,148] attributed the inhibitory effect of puromycin and amicetin to a cationic effect related to molecular size; smaller cations such as tris, $K^+$, $Na^+$, $Li^+$ caused a proportionately smaller inhibition.

Arylamidase II was widely distributed in a number of animal tissues; thyroid was the highest (100%), followed by spleen, kidney, pituitary (70–80%), thymus (50%), adrenal, liver, testis (20–30%), pancreas, whole brain (12%), and serum (4%).[6] Granules separated by sucrose gradients from freshly extracted glands and morphologically identical in appearance to lysosomes also contained a high concentration of dipeptidyl arylamidases I and II. Lysosomes were previously observed by Smith and Farquhar[152] in anterior pituitary with altered hydrolytic activity during lactation as demonstrated by the ability to assimulate and degrade hormone-bearing secretory granules. In contrast to the lysosomal origin, arylamidase III (Table IV) and the monoaminoacyl arylamidases were predominantly cytoplasmic.

Arylamidase III catalyzes cleavage of Arg–Arg-βNA at βH 9.0 and also hydrolyzes peptides with a minimum of four amino acids; tetrahexapeptides and poly-L-alanine gave dipeptide products; pentapeptides gave a di- and tripeptide end product, but tri- and dipeptides were not noticeably hydrolyzed. The enzyme was activated by 2-mercaptoethanol and inhibited by pCMB but was unaffected by the addition of EDTA.[149]

## B. Hypothalamus

There is now good evidence that oxytocin and vasopressin and related peptides are synthesized in the hypothalamic region and stored in the posterior pituitary; granules containing these peptides have been detected in the hypothalamic–hypophysial tract by histochemical methods and have been isolated by sucrose gradient techniques.[152–154] The mechanism for the release of these peptides from the storage sites is not clear but could involve proteolytic enzymes; the situation may be analogous to the release of exopeptidases from inactive pancreatic zymogen precursors. It has been established that under different physiological conditions, such as dehydration and lactation, there is an increased release of neurosecretory material [155,156] accompanied by increased levels of acid phosphatase[157] and enzymes hydrolyzing Leu-βNA.[158]

### 1. *Inactivation of Oxytocin and Vasopressin*

Oxytocin and vasopressin are rapidly inactivated by a variety of body fluids and tissues homogenates.[154] Tuppy[160] considered that enzymes hydrolyzing Leu-, Ala-, Gly-, and Cystine-di-βNA in sera were related to the exopeptidases that inactivated oxytocin and that increased severalfold during pregnancy. The enzyme from retroplacentar serum hydrolyzing both cystine-di-βNA and oxytocin has been purified 4500-fold. In a study of enzymes that inactivate oxytocin in pregnant and nonpregnant dogs, Hooper (1964)[161] reported a changed distribution in subcellular fractions obtained from the hypothalamic region. In pregnant dogs the fractions rich in myelin and synaptosomes contained a higher level of the inactivating enzyme compared to the subfractions obtained from the nonpregnant animals. Hypothalamic extracts are also capable of inactivating substance P, bradykinin, and vasopressin.[162,163] In a survey of six different brain areas, the enzymes inactivating vasopressin were the highest in the hypothalamus compared to the cerebellum, thalamus, cortex, caudate nucleus, and white matter. By way of contrast, enzymes inactivating bradykinin were equally distributed in all areas except in the white matter, where it was in slightly lower concentration.[164]

### 2. *Hypothalamic Releasing Factors*

The polypeptide nature of these factors suggests that exopeptidases may be involved in their release and activation. These factors arise from the hypothalamic region and are transported to the adenohypophysis by means of the circulation. Interruption of the blood supply has led to the identification of specific hormonal releasing factors for α- and β-corticotrophins, and thyrotropic, follicle, luteinizing, and growth hormones.[165] Extracts of the hypothalamus, especially the medial basal tuberal region,[166] were active in the release of anterior pituitary hormones both *in vitro* and *in vivo*.[167–168] As noted previously, the hypothalamus and sera contain exopeptidases that hydrolyze pituitary hormones, and these may be involved in the levels of hormone-releasing factors.

## C. Pineal Gland

Despite the known interrelationships between the pineal and pituitary glands, studies concerning the pineal peptide composition and turnover are scanty. The enzymes studied in pineal glands tisue, especially the hydrolases and transferases, differ in properties from those of normal brain tissue.[169] In a comparison of arylamidase N activities in the different brain areas, Jouan and Rocaboy[144] reported that with Ala- and Phe-*p*NA as the substrates the activity in the pineal gland equaled that in the anterior and posterior pituitary but was less than that in whole brain tissue. With Leu-βNA the activity was the same in all four tissues; with Val-βNA, only trace activities were observed in the pineal compared with high activities in the pituitary gland extracts.

## D. CSF

The attempted correlations between disease states and the enzyme levels in the CSF are obscured by the large and unspecific increase in protein composition.[170] Early studies showed that CSF could split polypeptides with increased "polypeptidase" activity in patients.[171] Abderhalden[172] also reported the hydrolysis of DL-Leu–Gly–Gly and in some cases DL-Leu–

Gly in the CSF of a large group of patients. Later studies by Stern *et al.*[173] showed that the activity with the tripeptides Leu–Gly–Gly and triglycine was activated by $Co^{2+}$. In patients with brain tumors Green and Perry[174] reported increased levels of enzymes hydrolyzing Leu-βNA, a substrate that would indicate the presence of LAP and arylamidase N activities. Chapman and Wolf[175] reported in some disease states the appearance of vasodilator polypeptides in the CSF of man. The polypeptides had biological activities similar to those of bradykinin, and their formation was attributed to the appearance of an unidentified proteolytic enzyme. Enzymes that inactivate the polypeptide were not explored, but it would be of interest to determine if specific exopeptidases such as carboxypeptidase B are involved.

In a detailed study of exopeptidase activity in the CSF Wiechert[176] reported the rapid hydrolysis within 7 hr on intracisternal injection of the following peptides: Gly–Leu, Gly-D-Leu, Ala–Gly, Pro–Gly, Gly–Phe, triglycine, Leu-$NH_2$, and DL-Leu–Leu. The hydrolysis of the peptides *in vitro* was considerably slower with the best activity with the LAP substrate Leu-$NH_2$ and only trace activities with the iminodipeptidase substrate and with the glycine–glycine dipeptidase substrate Gly–Gly. Riekkinen and Rinne[177] were able to separate several groups of exopeptidases from human CSF by column chromatography capable of hydrolyzing neutral and basic arylamides, with the greatest activity shown by Leu- and Arg-βNA. The *p*H optima of these exopeptidases were about 7.0 with a varying molecular weight of 80–140,000.

## E. Spinal Cord

The induction of acute experimental allergic encephalomyelitis (EAE) by proteins and polypeptides of the spinal cord has been the subject of considerable interest.[178] Enzymatic degradation is considered to play an important role in the formation of the active immunological agents.[179] In a comparison of several exopeptidases in the lobster ventral cord, the hydrolysis of the following substrates has been shown: Leu–Gly–Gly (aminotripeptidase), Arg-βNA (arylamidase B), Leu-βNA (arylamidase N), and Z–Leu–Tyr (carboxypeptidase B). The exopeptidase activities in the head and thorax regions of the spinal cord exceeded that in the abdomen and tail regions. This gradient of activity may be correlated with aspects associated with axonal flow. The relatively high concentration of neutral and acid proteinase present in the ventral cord material together with the high exopeptidase activities probably indicates a large protein breakdown and turnover.[213]

There have been a number of studies of the exopeptidases in the dorsal roots, ventral roots, and dorsal-root ganglia. Extracts from these different areas hydrolyzed bradykinin and substance P but not oxytocin and vasopressin.[162,164] The concentration of enzymes inactivating substance P was similar in dorsal and ventral roots but not in the case of bradykinin, where the concentration was higher in the dorsal-root ganglia followed by the dorsal and ventral roots. Hooper[164] attributed this enzyme gradient to aspects associated with axonal flow.

Importance has been ascribed to the intraspinal distribution of glycine and other amino acids.[55] Glycine, α-alanine, and serine, but not cystathionine, were higher in extracts of spinal cord compared to brain material.[56] Curtis *et al.*[56] have suggested, however, some caution in the interpretation of the transmitter function of amino acids on the basis of concentration; formation of glycine by synthetic or degradative processes at the presynaptic terminal may obviate the requirement for a large pool of stored transmitter substance. It is of some interest that lobster ventral cord contains unusually high concentrations of glycine, suggesting a metabolic role in addition to possible transmitter functions.[213]

## VII. PERIPHERAL NERVE

Most of the data available concern the complex changes accompanying Wallerian degeneration. With transection of the peripheral nerve there is considerable destruction of all cellular elements, accompanied by an increased level of amino-nitrogen compounds, notably the amino acids. In the hen some amino acids increased 15–33% at 14 days[180] and in the rabbit the increase was in the range 22–60%.[181] McCamen and Robins[181] showed in a study of 12 different enzymes that maximum degeneration occurred at 14 days, with a large increase in "dipeptidase" activity with Leu–Ala as the substrate, and also acid phosphatase, isocitric dehydrogenase, etc. Since the Wallerian degeneration is accompanied by the cellular influx of macrophages and by the proliferation of Schwann cells, it cannot be assumed that the increased levels of enzymes originated from the peripheral nerve itself. The role of exopeptidases in Wallerian degeneration is not clear; presumably they contribute to the increased levels of amino acids, and the cessation of their activities may be required for regeneration.

In rat sciatic homogenates, Adams and Glenner[109] reported high arylamidase N activity with Leu-βNA as the substrate. This enzyme differed in properties from that reported as present in brain; it was inhibited by sulfhydryl compounds, ascorbate, $CN^-$, $Mn^{2+}$, and DFP but not by metal chelating agents. Activity was present in the different anatomical areas and in the myelin sheath, neurilemma, perineurium, endonerium, and Schwann cells. Datta *et al.*[213] also observed arylamidase N activity in sciatic nerve extracts of several species, the lobster, the crab, and the rat. In the unmyelinated invertebrate nerve, the activity with Arg-βNA was 3–5-fold higher than with Leu-βNA as the substrate. As in the lobster ventral cord, there was a gradient of exopeptidases on comparison of the distal and proximal segments. The presence of peptide hydrolases along the nerve trunk suggests that degradation of axoplasmic proteins and peptides is not confined to the nerve-ending region. In all species the highest activity occurred with Leu–Gly–Gly as the substrate, but appreciable carboxypeptidase A activity was detected in the rat sciatic nerve.

## VIII. CONCLUSIONS

The very multiplicity of peptide hydrolases, their ubiquitous distribution in all brain areas and cellular fractions, and the broad specificity patterns all suggest many different roles of exopeptidases in the nervous system. The following brief essay represents an attempt to integrate the functional aspects of exopeptidases in relation to (1) hormone turnover, (2) disease processes, (3) transport, and (4) protein turnover.

### A. Hormone Turnover

Modification of proteins or peptides may result in the formation of physiologically active compounds from inactive precursors or, alternatively, in the inactivation of active proteins. A recent example supporting this concept was the formation of bradykinin-like materials from the polypeptide kalliden-10 by arylamidase B.[183] The precursor polypeptide itself is obtained from serum-α-globulin by trypsin digestion, and the production of this hormone serves as an example of the sequential action of both endo- and exopeptidases. It can be expected that peptidyl hormones contain peptide linkages susceptible to attack by hydrolytic enzymes. Numerous examples exist in the literature, some already noted, for the inactivation of angiotensin-II, bradykinin, oxytocin, vasopressin, ACTH, substance P, etc. Inactivation *in vitro* suggests a possible role *in vivo*, namely, that the level of peptide hydrolase determines the tissues or blood level of circulating hormone.[187] It is also conceivable that activation from inactive precursors plays a role in the availability of hormones or hormone-releasing factors. A plasma enzyme is known to act on the decapeptide angiotensin I, releasing the dipeptide His–Leu and producing the highly vasoactive pressor material angiotensin II.[184,185] This enzyme may be similar to the aminopeptidase recently shown to be present in the kidney that is capable of splitting the terminal dipeptide moiety from Gly–Pro–Ala, Gly–Pro-βNA, Gly–Pro–Gly–Gly, and Ala–Ala-βNA.[97,150] During pregnancy or on the injection of estradiol, conditions known to suppress ovulation, the level of enzymes degrading oxytocin in the hypothalamus is increased.[187] It is unfortunate that the terminology adopted by some investigators implies specific hormone degrading enzymes; the bewildering array of such enzymes prompts the question of whether some of these are identical to other known amino- and carboxypeptidases.

### B. Disease Processes

A large number of inborn metabolic disorders are associated with generalized aminoacidurias. No specific exopeptidase enzyme defects as such have been detected, but a number of clinical disorders are associated with increased turnover and excretion of amino acids and peptides. In cerebromacular disease (Tay–Sachs), there is an increased excretion of carnosine and anserine in addition to histidine and methylhistidine.[188] Cystathioninuria is associated with increased urinary excretion of cystathionine; hepatolenticular degeneration (Wilson's disease) the Fanconi syndrome also result in aminoaciduria and increased peptide excretion.[189,190] Some of these defects have been ascribed to faulty reabsorption in the renal tubules; aspects associated with transport are briefly described below.

## C. Transport

A number of recent studies have implicated membrane proteins, or proteins adjacent to the membranes (periplasmic)[191], as components for the transport in bacteria of sulfate,[192] amino acids,[193,194] galactose[195] and other sugars.[196] Recently one protein carrier component for sugars was described as heat stable, molecular weight 10,000; it could be phosphorylated by a membrane enzyme component in the presence of phosphoenolpyruvate.[196] Consequently, enzymes involved in the turnover of transport carriers are regarded as components of the "permease" or "kinase" transport systems in bacteria.[197] Separate transport systems for peptides may also exist, although peptides are known to penetrate tissues and bacteria at low rates compared with the passage of free amino acids.[198] In bacteria separate transport systems appear to exist for several di- and tripeptides which are distinct from those for amino acids.[199–201]: Gly–Gly, e.g., is transported unchanged in some *E. coli* mutants that are devoid of glycyl–glycyl dipeptidase (3.4.3.1).[200] In *S. foecalis*, Gly–Gly is transported unchanged but is believed to be hydrolyzed within the cell: the hydrolysis of Gly–Gly appeared to stimulate the incorporation of glycine by exchange reactions with external glycine.[201] *E. coli* mutants have been described for the transport of dilysine and oligopeptides such as tetralysine and tetraarginine.[202] There is now evidence that some exopeptidases are involved in the transport of peptides in the kidney brush-border epitheal cells[203] which contain a high concentration of enzymes hydrolyzing Leu-$\beta$NA, Leu-$NH_2$, Leu–Gly, and Leu–Gly–Gly.[204] Arylamidases also occur in the form of globular repeating units 50–60 Å on the surface of isolated liver and hepatoma plasma membranes.[205] In preliminary studies from our laboratory, arylamidases and aminopeptidases were associated with synaptosomal membrane fractions and to a lesser extent with purified mitochondrial inner and outer membranes.[105]

There have also been some studies on the uptake of dipeptides in rat brain slices. Araham *et al.*[61] reported the active transport of carnosine but not homocarnosine in rat brain slices. Both these dipeptides are constituents of brain tissue.[206,207] The uptake of carnosine was qualitatively similar to histidine but with different time sequences; maximum values for histidine uptake occurred at 30 min, compared with 4 hr for the dipeptide. Since other peptides decreased the uptake of carnosine to a greater extent than amino acids, the authors suggested a separate general mechanism for the transport of dipeptides in the brain. The significance of exopeptidases in the brain membrane fractions to the transport mechanism is not readily apparent. The increase in pool size as a result of degradation of peptides could serve to drive the transport mechanism by the increase in the rate of exchange reactions.

### D. Protein Turnover

In conclusion, it is to be emphasized that only simple schemes can be proposed for the pathways of protein and peptide turnover within cells. The pathways of brain protein breakdown are discussed in a later chapter[(208)]; a hypothetical scheme based on the enzymes described in this review is illustrated in Fig. 1. It is proposed that the initial event is the cleavage of internal peptide bonds in proteins to form oligopeptides that are susceptible to attack by the appropriate exopeptidase. The intracellular localization of exo- and endopeptidases at similar or adjacent sites within the cell permits a sequential breakdown of proteins to form oligopeptides or amino acids. For this purpose most tissues are endowed with batteries of exopeptidases capable of dealing with oligopeptides or specifically with di- or tripeptides. The concept of specific exopeptidases for larger peptides is novel and remains to be established. Such enzymes would encompass some of the common peptidyl hormones described in the present review. The availability of arylamide substrates has stimulated the search for aminopeptidases capable of cleaving a variety of peptides related to hormonal structures. Arylamidases are present in the CNS in high concentrations, but the true physiological substrates are unknown. Presumably the native substrates are naturally occurring polypeptides, with the specificity determined by the nature and number of amino acid residues.

In addition to their role in degradation, exopeptidases may be involved in some aspects of polypeptide synthesis. Exopeptidases may function like cathepsin C (dipeptidyl transferase) in the formation of some oligopeptides by the process of transamidation. Hydrolytic enzymes are now known to catalyze the release of di-Phe-*t*RNA, *N*-acetyl-amino-acids, and *N*-substituted oligopeptidyl-*t*RNA in *E. coli*.[(209–210)] The enzymes involved have not been characterized but are believed to include enzymes capable of hydrolyzing the ester linkage, such as a number of exopeptidases or acylases (Section V). Exopeptidases may play a regulatory role by alterations of the accessory protein factors recently shown to be required during polypeptide synthesis. Different protein factors are required for the binding of aminoacyl *t*RNA or of RNA to ribosomes[(211,212)] and for *N*-formyl methionine-*t*RNA at chain initiations. It is of particular interest that puromycin, which affects synthesis by early chain termination, also dramatically affects a number of behavioral factors. The fact that puromycin and its analogues are potent inhibitors of arylamide hydrolases points to a more complex effect *in vivo* than hitherto supposed.

Its effect therefore may be on a degradative and also a synthetic pathway, or on pathways common to both processes. In the pathway for breakdown distinct exopeptidases are envisaged for *even-* and *odd-*numbered peptides, as suggested by the work on purified pituitary hydrolases (Table IV). The possibility also exists that polypeptides and proteins are degraded directly to amino acids by simple end group cleavage or, as recently demonstrated, by liberation of dipeptide sequences. Brain extracts are rich in di- and tripeptidases capable of splitting polypeptide end products to amino acids. The liberated amino acids are more easily transported out of the nervous system or returned to the cellular pool for reutilization in synthesis or for energy production. In this manner peptide breakdown by exopeptidases forms an essential component in the brain economy and for protein turnover.

## IX. REFERENCES

1. M. Bergmann, A classification of proteolytic enzymes, *Advan. Enzymol.* **2**:49–68 (1942).
2. B. Shome and M. Saffran, Peptides of the hypothalamus, *J. Neurochem.* **13**:433–448 (1966).
3. S. Lande, A. B. Lerner, and V. G. Upton, Pituitary peptides. Isolation of new peptides related to β-melanocyte-stimulating hormone, *J. Biol. Chem.* **240**:4259–4263 (1965).
4. *Enzyme Nomenclature*, Elsevier, Amsterdam (1965).
5. G. Gomori, Chromogenic substrates for amino peptidases, *Proc. Soc. Exptl. Biol. Med.* **87**:559–561 (1954).
6. J. K. McDonald, F. H. Leibach, R. E. Grindeland, and S. Ellis, Purification of Dipeptidyl Aminopeptidases II (Dipeptidyl Arylamidase II) of the Anterior Pituitary Gland. Peptide and dipeptide esterase activities, *J. Biol. Chem.* **243**:4143–4150 (1968).
7. E. Blum, A. I. Yakovchuk, and A. I. Yarmoskevich, Proteolytic enzymes of the brain, *Bull. Biol. Med. Exptl. U.R.S.S.* **1**:17–18 (1936).
8. E. Abderhalden and G. Ceaser, Untersuchungen über das vorkommen von polypeptidasen in centralen und peripheren nerven system, *Fermentforschung* **16**:255–262 (1940).
9. M. W. Kies and S. Schwimmer. Observations on proteinase in brain, *J. Biol. Chem.* **145**:685–691 (1942).
10. V. E. Price, A. Meister, J. B. Gilbert, and J. P. Greenstein, Separation of dehydropeptidases and analogous L- and D-peptidases, *J. Biol. Chem.* **181**:535–547 (1947).
11. E. S. Adams and E. L. Smith, Proteolytic activity of pituitary extracts, *J. Biol. Chem.* **191**:651–664 (1951).
12. H. Hanson and N. Tendis, Darstellung von Zellbestandteil-präparationen aus Hirngewebe und ihre Peptidase–Aktivität im Vergleich zu Niere und Leber, *Z. Ges. Inn. Med. Ihre Grenzebiete* **5**:224–233 (1954).
13. A. Pope, Quantitative distribution of dipeptidase and acetylcholinesterase in architectonic layers of rat cerebral cortex, *J. Neurophysiol.* **15**:115–130 (1952).

14. A Pope, The intralaminar distribution of dipeptidase activity in human frontal isocortex, *J. Neurochem.* **4**:31–41 (1959).
15. A. Pope and C. B. Anfinsen, Histochemical distribution of peptidase activity in the central nervous system of the rat, *J. Biol. Chem.* **173**:305–311 (1948).
16. L. L. Uzman, M. K. Rumley, and S. Van Den Noort, The substrate specificity of mouse brain peptidase activity, *J. Neurochem.* **6**:299–310 (1961).
17. L. L. Uzman, M. K. Rumley, and S. Van Den Noort, Dipeptidase activity of the brain, *Nature* **186**:559–560 (1960).
18. S. Van Den Noort and L. L. Uzman, Effect of metal ions on brain peptidase activity, *Proc. Soc. Exptl. Biol. Med.* **108**:32–34 (1961).
19. L. L. Uzman, S. Van Den Noort, and M. K. Rumley, Properties and classification of some brain peptidases, *J. Neurochem.* **9**:241–252 (1962).
20. L. L. Uzman, M. K. Rumley, and S. Van Den Noort, The inhibition of cerebral diglycinase by α-amino, α-keto and α-hydroxy acids, *J. Neurochem.* **10**:795–804 (1963).
21. K. Linderstrøm-Lang, Über Darmerepsin, *Z. Physiol. Chem.* **182**:151–174 (1929).
22. K. Linderstrøm-Lang and H. Holter, Studies on enzymatic histochemistry XI. The distribution of peptidase in the gastric and duodenal mucosa of the pig, *Compt. Rend. Lab. Carlsberg Sern. Chim.* **20**:42–56 (1935).
23. E. L. Smith and P. H. Spackmann, Leucineaminopeptidase V. Activation, specificity, and mechanism of action, *J. Biol. Chem.* **212**:221–229 (1955).
24. E. D. Wachsmuth, Untersuchungen zur Struktur der Aminopeptidase aus Partikeln von Schweinenieren, *Biochem. Z.* **346**:467–473 (1967).
25. H. Hanson, *in Hoppe-Segler Thierfelder Handbuch der Physiologisch- und Pathologisch-Chemischen Analyse*, Vol. 6C, p. 1, Springer, Berlin (1966).
26. S. R. Himmelhoch and E. A. Peterson, Preparation of leucine aminopeptidase free of endopeptidase activity, *Biochemistry* **7**:2085–2092 (1968).
27. I. Schecter and A. Berger, The hydrolysis of diasteroisomers of alanine peptides by carboxypeptidase A and leucine aminopeptidase, *Biochemistry* **5**:3371–3383 (1966).
28. J. B. Wolff and R. A. Resnick, Aminopeptidase of the outer lens. I. Metal ion requirements and synergistic activation II Substrate specificity, *Biochim. Biophys. Acta* **73**:588–622 (1963).
29. S. Fittkau, D. Glässer, and H. Hanson, Zur Aktivität und Spezifität der Leucinaminopeptidase in Augenlinsen. Aminosäure-und Dipeptidamilide als Substrate, *Z. Physiol. Chem.* **322**:101–111 (1960).
30. A. Spector, Lens aminopeptidase I. Purification and properties, *J. Biol. Chem.* **238**:1353–1357 (1963).
31. A. S. Brecher, The distribution and activity of calf brain peptidases, *J. Neurochem.* **10**:1–6 (1963).
32. N. Marks, R. K. Datta, and A. Lajtha, The relationship of aminotripeptidase and arylamidase to protein breakdown in the brain, *in Macromolecules and the Nervous System* (Z. Lodin, ed.), Excerpta Medica, Amsterdam (1968).
33. E. K. Patterson, S. H. Hsiao, A. Keppel, and S. Sorot, Studies on dipeptidases and aminopeptidases. II. Zonal electrophoretic separation of rat liver peptidases, *J. Biol. Chem.* **240**:710–715 (1965).
34. R. L. Joseph and W. J. Saunders, Leucine aminopeptidase in extracts of swine muscle, *Biochem. J.* **100**:827–832 (1966).
35. E. K. Patterson, S. H. Hsiao, and A. Keppel, Studies on dipeptidases and aminopeptidases. I. Distinction between leucine aminopeptidase and enzymes that hydrolyze L-leucyl-β-naphthylamide, *J. Biol. Chem.* **238**:3611–3620 (1963).

36. G. F. Bryce and B. R. Rabin, The assay and reaction kinetics of leucine aminopeptidase from swine kidney, *Biochem. J.* **90**:509–512 (1964).
37. E. L. Smith and R. L. Hill, Leucine aminopeptidase, *in The Enzymes*, 2nd ed., Vol. 4, pp. 37–62, Academic Press, New York (1960).
38. E. L. Smith, Aminopeptidases, *in Methods in Enzymology* (S. P. Colowick and N. O. Kaplin, eds.), Vol. 2, pp. 83–114, Academic Press, New York (1955).
39. N. C. Davis and E. L. Smith, Partial purification and specificity of iminodipeptidase, *J. Biol. Chem.* **200**:375–384 (1953).
40. D. Ellis and J. S. Fruton, On the proteolytic enzymes of animal tissues. IX. Calf thymus tripeptidase, *J. Biol. Chem.* **191**:153–159 (1951).
41. M. Ziff and A. A. Smith, Inhibition of aminotripeptidase, *Proc. Soc. Exptl. Biol. Med.* **80**:761–764 (1952).
42. R. K. Datta, N. Marks, and A. Lajtha, Purification and properties of brain tripeptidase, *Federation Proc.* **26**:452 (1967).
43. N. Marks, Separation of brain peptidases and proteinases, *Biochem. J.* **103**:40–41P (1967).
44. N. Marks, R. K. Datta, and A. Lajtha, Partial resolution of brain arylamidases and aminopeptidases, *J. Biol. Chem.* **243**:2882–2889 (1968).
45. A. S. Brecher and R. E. Sobel, Studies on ox-brain aminopeptidases, *Biochem. J.* **105**.641–646 (1967).
46. S. Simmonds and N. O. Toye, The role of metal ions in the peptidase activity in E. Coli, *J. Biol. Chem.* **242**:2086–2093 (1967).
47. L. Josefsson and T. Lindberg, Intestinal dipeptidases. Spectrophotometric determination and characterization of dipeptidase activity in pig intestinal mucosa, *Biochim. Biophys. Acta* **105**:149–161 (1965).
48. A. Rosenberg, *The Role of Metal Ions in the Catalytic Action of Peptidases*, p. 14, Almquist and Wiksell, Uppsala (1960).
49. G. B. Robinson and B. Shaw, The hydrolysis of dipeptides by different regions of rat small intestine, *Biochem. J.* **77**:351–356 (1960).
50. H. G. Wilcox and M. Fried, Studies on rat-liver glycylglycine dipeptidase, *Biochem. J.* **87**:192–199 (1963).
51. H. von Euler and H. Hasselquist, Electrophoretic enzyme determinations in the blood, serum and brain of higher animals, *Arch. Kemi* **25**:129–133 (1966).
52. H. von Euler, H. Hasselquist and K. Kyyroe, Electrophoretic enzyme determinations in the blood, serum and brain of higher animals, Pts. II and III, *Arch. Kemi* **25**:151–157 and 257–262 (1966).
53. H. von Euler, K. Kyyroe, and H. Hasselquist, Electrophoretic enzyme determinations in the blood, serum and brain of higher animals, Pt. IV, *Arch. Kemi* **25**: 97–107 (1966).
54. A. Nishi, Inhibition of yeast glycylglycine dipeptidase by amino acids, *J. Biochem. Tokyo* **47**:47–59 (1959).
55. L. T. Graham, R. P. Shank, R. Werman, and M. H. Aprison, Distribution of some synaptic transmitters in cat spinal cord: Glutamic acid, aspartic acid, $\gamma$-aminobutyric acid, glycine and glutamine, *J. Neurochem.* **14**:465–472 (1967).
56. D. R. Curtis, L. Hösli, and G. A. R. Johnston, A pharmacological study of the depression of spinal neurons by glycine and related amino acids, *Exptl. Brain Res.* **6**:1–18 (1968).
57. A. S. Brecher and I. R. Koski, Studies on mammalian dipeptidases, *Arch. Intern. Physiol. Biochem.* **75**:821–834 (1968).
58. A. Meister, *Biochemistry of the Amino Acids*, 2nd ed., Vol. 1, p. 119, Academic Press, New York (1965).

59. E. A. Hosein and M. Smart, The presence of anserine and carnosine in brain tissue, *Can. J. Biochem. Physiol.* **38**:569–573 (1960).
60. T. Nakajima, F. Wolfgram, and W. G. Clark, The isolation of homoanserine from bovine brain, *J. Neurochem.* **14**:1107–1112 (1967).
61. D. Araham, J. J. Pisano, and S. Udenfriend, Uptake of carnosine and homocarnosine by rat brain slices, *Arch. Biochem. Biophys.* **104**:160–165 (1964).
62. D. J. McCorquodale, Some properties of a ribosomal cysteinylglycinase in Escherichia Coli B, *J. Biol. Chem.* **238**:3914–3920 (1963).
63. F. Binkley, Purification and properties of renal glutathionase, *J. Biol. Chem.* **236**: 1075–1082 (1961).
64. A. T. Matheson and T. Murayama, The limited release of ribosomal peptidase during formation of *Escherichia coli* spheroplasts, *Can. J. Biochem.* **44**:1407–1415 (1966).
65. P. J. Fodor, A. Mîller, A. Neidle, and H. Waelsch, Enzymatic synthesis of glutathione by a transfer reaction, *J. Biol. Chem.* **203**:991–1002 (1953).
66. E. L. Smith, Peptide bond cleavage (survey), *in The Enzymes* (P. D. Boyer, H. Lardy, and K. Myrback, eds.), Vol. 4, pp. 1–10, Academic Press, New York (1960).
67. S. Sand, A. Berger, and E. Katchalski, Proline iminopeptidase. II. Purification and comparison with iminodipeptidade (prolinase), *J. Biol. Chem.* **237**:2207–2212 (1962).
68. R. U. Margolis, Acid mucopolysaccharides and proteins of bovine whole brain, white matter and myelin, *Biochim. Biophys. Acta* **141**:91–102 (1968).
69. E. Roboz, N. Henderson, and N. W. Klies, A collagen-like compound isolated from bovine spinal cord, *J. Neurochem.* **2**:254–259 (1958).
70. J. J. Hutton, A. Marglin, B. Witkop, J. Kurz, A. Berger, and S. Udenfriend, Synthetic polypeptidases as substrates and inhibitors of collagen proline hydroxylase, *Arch. Biochim. Biophys.* **125**:779–785 (1968).
71. E. R. Einstein, J. Csejtey, W. J. Davis, A. Lajtha, and N. Marks, Enzymic degradation of the encephalitogenic protein, *Intern. Arch. Allergy Appl. Immunol.* **34** (in press).
72. J. D. Padayatty and H. V. Kley, Studies on E-Peptidase, *Biochemistry* **5**:1394–1399 (1966).
73. R. L. Friede, *Topographic Brain Chemistry*, Academic Press, New York (1966).
74. M. S. Burstone, *Enzyme Histochemistry and Its Application in the Study of Neoplasm*, Academic Press, New York (1962).
75. H. Tuppy, U. Wiesbauer, and E. Wintersberger, Aminosäure-*p*-nitroanilide als Substrate für Aminopeptidasen und andere proteolytische Fermente, *Z. Physiol. Chem.* **329**:278–288 (1962).
76. S. Ellis, A thiol activated aminopeptidase of the pituitary, *Biochim. Biophys. Res. Commun.* **12**:452–456 (1963).
77. A. Szewczuk, M. Kochman, and T. Baranowski, Dipeptide nitriles as substrates for colorimetric determination of aminopeptidases, *Acta Biochim. Polon* **12**:357–367 (1965).
78. V. K. Hcpsu, K. K. Mäkinen, and G. G. Glenner, Purification of a mammalian peptidase selective for N-terminal arginine and lysine residues: Aminopeptidase B, *Arch. Biochem. Biophys.* **114**:557–566 (1966).
79. E. E. Smith and A. M. Rutenberg, Starch gel electrophoresis of human tissue enzymes which hydrolyze L-leucyl-$\beta$-naphthylamide, *Science* **152**:1256–1257 (1966).
80. G. G. Glenner, P. J. McMillan, and J. E. Folk, A mammalian peptidase specific for the hydrolysis of N-terminal $\alpha$-1-glutamyl and aspartyl residues, *Nature* **194**:867 (1962).
81. G. G. Glenner and J. E. Folk, Glutamyl peptidases in rat and guinea pig kidney slices, *Nature* **192**:338–340 (1962).

82. I. Nagatsu, L. Gillespie, J. M. George, J. E. Folk, and G. G. Glenner, Serum amino peptidases, "angiotensinase" and hypertension. II. Amino acid β-naphthylamide hydrolysis by normal and hypertensive serum, *Biochem. Pharmacol.* **14**:853–857 (1965).
83. I. Nagatsu and J. Hara, Relationship between parathyroid function and serum aminopeptidase A. activity, *Nature* **213**:206–207 (1967).
84. I. Nagatsu, T. Nagatsu, and G. G. Glenner, Species difference of serum amino acid-β-naphthylamidases, *Enyzmologica* **34**:73–76 (1968).
85. P. A. Khairallah and I. H. Page, Plasma angiotensinases, *Biochem. Med.* **1**:1–8 (1967).
86. R. Hess, Arylamidase activity related to angiotensinase, *Biochim. Biophys. Acta* **99**:316–324 (1965).
87. D. Regoli, B. Riniker, and H. Brunner, The enzymatic degradation of various angiotensin. II. derivatives by serum, plasma or kidney homogenate, *Biochem. Pharmacol.* **12**:637–646 (1963).
88. D. C. Johnson and J. W. Ryan, Degradation of angiotensin-II by a carboxypeptidase of rabbit liver, *Biochim. Biophys. Acta* **160**:196–203 (1968).
89. H. Y. T. Yang and E. G. Erdos, T. S. Chiang, New enzymatic route for the inactivation of angiotensin, *Nature* **218**:1224–1226 (1968).
90. Z. Albert, M. Orlowski, Z. Rzucidlo, and J. Orlowska, Studies on γ-glutamyl transpeptidase activity and its histochemical location in the central nervous system of man and different animal species, *Acta Histochem.* **25**:312–320 (1966).
91. J. S. Fruton, Cathepsins, *in The Enzymes* (P. D. Boyer, H. Lardy, and K. Myrback, eds.), pp. 233–241, Academic Press, New York (1960).
92. H. Wurtz, A. Tanaka, and J. S. Fruton, Polymerisation of dipeptide anides by cathepsin C, *Biochemistry* **1**:19–28 (1962).
93. H. McIlwain and M. A. Trezize, The speed of several cerebral reactions involving the nicotinamide coenzymes, *Biochem. J.* **65**:288–296 (1957).
94. Y. Kakimoto, T. Nakajima, A. Kanazawa, M. Takesada, and I. Sano, Isolation of γ-L-glutamyl-L-glutamic acid and γ-L-glutamyl-L-glutamine from bovine brain, *Biochim. Biophys. Acta* **93**:333–338 (1964).
95. A. Kanazawa, Y. Kakimoto, T. Nakajima, H. Shimuzu, M. Takesada, and I. Sano, Isolation and identification of γ-L-glutamylglycine from bovine brain, *Biochim. Biophys. Acta* **97**:460 (1965).
96. A. Z. Orlowski and A. Meister, Isolation of γ-glutamyl transpeptidase from hog kidney, *J. Biol. Chem.* **240**:338–347 (1965).
97. V. K. Hopsu-Harvu, P. Rintola, and G. G. Glenner, A hog kidney aminopeptidase liberating N-terminal dipeptides. Partial purification and characterization, *Acta Chem. Scand.* **22**:299–308 (1968).
98. S. Berl, Glutamine synthetase. Determination of its distribution in brain during development, *Biochemistry* **5**:916–922 (1966).
99. J. H. Pincus and H. Waelsch, The specificity of transglutaminase I. Human haemoglobin as a substrate for the enzyme II. Structural requirements of the amine substrate *Arch Biochem. Biophys.* **126**:34–52 (1968).
100. M. M. Nachlas, T. P. Goldstein, and A. M. Seliqman, An evaluation of aminopeptidase specificity with seven chromogenic substrates, *Arch. Biochem. Biophys.* **97**: 223–231 (1962).
101. V. K. Hopsu, K. K. Makinen, and G. G. Glenner, Characterization of aminopeptidase B: Substrate specificity and affector studies, *Arch. Biochem. Biophys.* **114**: 567–575 (1966).
102. K. K. Mäkinen and V. K. Hopsu-Havu, A simplified method for purification of rat liver aminopeptidase B, *Arch. Biochem. Biophys.* **118**:257–258 (1967).

103. N. Marks, Purification and specificity of brain aminopeptidases-II, *J. Biol. Chem.* (to be submitted).
104. S. Mahavedan and A. L. Tappel, Arylamidases of rat liver and kidney, *J. Biol. Chem.* **242**:2369–2374 (1967).
105. N. Marks, B. D'Monte, C. Bellman, and A. Lajtha, Protein turnover in brain mitochondrial membranes, *Brain Res.* **1** (in press).
106. R. K. Datta, N. Marks, and A. Lajtha, *Exopeptidase Activities of Brain Subcellular Membrane Fractions*, p. 49, First Meeting of the International Society of Neurochemistry, Strasbourg (1967).
107. R. K. Datta, N. Marks, and A. Lajtha, Peptide breakdown in cerebral mitochondria, *Indian J. Biochem.* **4**:37 (1967).
108. G. Pfleiderer, P. G. Celliers, M. Stanulovic, E. D. Wachsmuth, H. Determan, and G. Braunitzer, Eigenschaften und analytische Anwendung der aminopeptidase aus Nierenpartikeln, *Biochem. Z.* **340**:552–564 (1964).
109. C. W. M. Adams and G. G. Glenner, Histochemistry of myelin-IV. Aminopeptidase activity in CNS and PNS, *J. Neurochem.* **9**:233–239 (1962).
110. A. S. Brecher and S. W. Barefoot, The distribution of arylamidase activity in brain, *Arch. Intern Physiol. Biochem.* **75**:816–820 (1967).
111. K. Felgenhauer and G. G. Glenner, The enzymatic hydrolysis of amino acid-β-naphthylamides. II. Partial purification and properties of a particle-bound cobalt-activated rat kidney aminopeptidase, *J. Histochem. Cytochem.* **14**:401–413 (1966).
112. B. Sylven and U. Lippi, The suggested lysosomal localization of aminoacyl naphthylamide splitting enzymes, *Exptl. Cell Res.* **40**:145–147 (1965).
113. M. Rybak, M. Petáková, and E. Simonianová, Cleavage of glycine, alanine and leucine *p*-nitroanilides by some animal arylaminopeptidases, *Collection Czech. Chem. Commun.* **32**:1051–1057 (1967).
114. F. J. Behal, R. A. Klein, and F. B. Dawson, Separation and characterization of aminopeptidase and arylamidase components of human liver, *Arch. Biochem. Biophys.* **115**:545–554 (1966).
115. S. Ellis and M. Perry, Pituitary arylamidase and peptidases, *J. Biol. Chem.* **241**: 3679–3686 (1966).
116. H. Neurath, Carboxypeptidase A and B *in The Enzymes* (P. D. Boyer, H. Lardy, and K. Myrback, eds.), Vol. 4, Pt. A, pp. 11–36, Academic Press, New York (1960).
117. H. J. Strecker, P. Mela, and H. Waelsch, Brain thioesterases, *J. Biol. Chem.* **212**: 223–233 (1955).
118. F. B. Goldstein, Biosynthesis of *N*-acetyl-L-aspartic acid, *J. Biol. Chem.* **234**:2702–2706 (1959).
119. H. C. Buniatian, V. S. Hovhannissian, and G. V. Aprikan, The participation of *N*-acetyl-L-aspartic in brain metabolism, *J. Neurochem.* **12**:695–703 (1965).
120. S. M. Birnbaum, L. Levinlow, R. B. Kinsley, and J. P. Greenstein, Specificity of amino acid acylases, *J. Biol. Chem.* **194**:455–470 (1952); and *in Methods in Enzymology* (S. P. Colowick and N. O. Kaplan, eds.), Vol. 2, pp. 109–119, Academic Press, New York (1955).
121. F. B. Goldstein (personal communication).
122. J. R. Brown, R. N. Greenshields, M. Yamasaki, and H. Neurath, The subunit structure of bovine procarboxypeptidase A-S6. Chemical properties and enzymatic activities of the products of molecular disaggregation, *Biochemistry* **2**:867–876 (1963).
123. J. W. Prahl and H. Neurath, Pancreatic enzymes of the spiny pacific dogfish-II. Procarboxypeptidase B and carboxypeptidase B, *Biochemistry* **5**:4137–4145 (1966).
124. H. Neurath, Evolution of structure and function of proteases, *Science* **158**:1638–1644 (1967).

125. K. S. V. Sampath-Kumar, J. B. Clegg, and K. A. Walsh, The N-terminal sequence of bovine carboxypeptidase A and its relation to zymogen activation, *Biochemistry* **3**:1728–1732 (1964).
126. B. L. Vallee, Active center of carboxypeptidase A′, *Federation Proc.* **23**:8–17 (1964).
127. J. E. Coleman and B. L. Vallee, Metallocarboxypeptidase-inhibitor complexes, *Biochemistry* **3**:1874–1879 (1964).
128. H. Neurath, Procarboxypeptidases: Structure, mechanism of, action and phylogenetics, *Proceedings of the Seventh International Congress of Biochemistry, Tokyo Symposium* 3, p. 151, Science Council, Japan (1967).
129. R. T. Simpson and B. L. Vallee, Iodocarboxypeptidase, *Biochemistry* **5**:1760–1767 (1966).
130. H. I. Lehrer, H. V. Vunakis, and G. D. Fasman, Carboxypeptidase A. Studies with poly α-amino acids, *J. Biol. Chem.* **240**:4585–4590 (1965).
131. W. A. Krivoy and D. Kroeger, The preservation of bradykinin by phenolthiathines *in vitro*, *Brit. J. Pharmacol.* **22**:329–341 (1964).
132. E. G. Erdos and H. Y. T. Yang, *Hypotensive Peptides* (E. G. Erdos, N. Back, and F. Sicuteri, eds.), pp. 235, Springer-Verlag, New York (1966).
133. E. G. Erdos, H. Y. T. Yang, L. L. Taque, and N. Manning, Carboxypeptidase in blood and other fluids-III, *Biochem. Pharmacol.* **16**:1287–1298 (1967).
134. E. G. Erdos, Hypotensive peptides, *Advan. Pharmacol.* **4**:1 (1966).
135. Symposium, Vasoactive peptides, *Federation Proc.* **27**:49–99 (1968).
136. F. LaBella, S. Vivian, and G. Ceveen, Abundance of cystathione in the pineal body. Free amino acids and related compounds of bovine pineal, anterior and posterior pituitary, and brain, *Biochem. Biophys. Acta* **158**:286–288 (1968).
137. C. H. Li, Current concepts on the chemical biology of pituitary hormones, *Perspectives Biol. Med.* **11**:498–521 (1968); also Symposium "growth hormone" (M. Sonnberg, ed.), *Ann. N.Y. Acad. Sci.* **148**:284–571 (1968).
138. M. D. Hollenberg and D. B. Hope, The isolation of the native hormone-binding proteins from bovine pituitary posterior lobes. Crystallization of neurophysin-I and II as complexes with [8-arginine]-vasopressin, *Biochem. J.* **106**:557 (1968).
139. M. Ginsburg and K. Jayasena, The distribution of proteins that bind hypophysial hormones, *J. Physiol.* **197**:65–76 (1968).
140. E. L. Smith, Dipeptidases, *in Methods in Enzymology* (S. P. Colowick and N. O. Kaplan, eds.), Vol. 2, pp. 93–109, Academic Press, New York (1955).
141. J. I. Harris, C. H. Li, P. G. Condliffe, and N. G. Pon, Action of carboxypeptidase on hypophysial growth hormone, *J. Biol. Chem.* **209**:133–143 (1954).
142. M. Sonnenberg, M. Kikutani, C. A. Free, A. C. Nadler, and J. M. Dellacha, Chemical and biological characterization of clinically active tryptic digests of bovine growth hormone, *Ann. N.Y. Acad. Sci.* **148**:532–538 (1968).
143. J. Ramachandran and C. H. Li, in *Advances in Enzymology* (F. F. Nord, ed.), Vol. 29, pp. 391, Wiley–Interscience, New York (1967).
144. P. Jouan and J. C. Rocaboy, Étude de l'activité peptidasique de la glande pinéale du Porc, *Compt. Rend. Soc. Biol.* **160**:859–862 (1966).
145. T. P. J. Vanha-Perttulla and V. K. Hopsu, Esterolytic and proteolytic enzymes of the rat adenohypophysis. I. Studies with homogenate and its fractions, *Histochemistry* **4**:372–378 (1965).
146. T. P. J. Vanha-Perttula and V. K. Hopsu, Esterolytic and proteolytic enzymes of the rat adenohypophysis. II. Chromatographic fractionation and characterization of enzyme activities hydrolysing leucyl-β-naphthylamide, *Ann. Med. Exptl. Fenniae* **43**:32–39 (1965).

147. J. K. McDonald, S. Ellis, and T. J. Reilly, Properties of dipeptidyl arylamidase I of the pituitary. Chloride and sulfhydryl activation of seryltyrosyl-β-naphthylamide, *J. Biol. Chem.* **241**:1494–1501 (1966).
148. J. K. McDonald, T. J. Reilly, B. B. Zeitman, and S. Ellis, Dipeptidyl arylamidase II of the pituitary. Properties of lysylalanyl-β-naphthylamide hydrolysis: inhibition by cations, distribution in tissues, and subcellular localization, *J. Biol. Chem.* **243**: 2028–2037 (1968).
149. S. Ellis and J. M. Nuenke, Dipeptidyl arylamidase III of the pituitary. Purification and characterization, *J. Biol. Chem.* **242**:4623–4629 (1968).
150. G. G. Glenner, L. A. Cohen, and J. E. Folk, The enzymatic hydrolysis of amino acid-β-naphthylamides. I. Preparation of amino acid and dipeptide-β-naphthylamides, *J. Histochem. Cytochem.* **13**:57–64 (1965).
151. S. Kakiuchi and H. H. Tomizawa, Properties of a glucagen degrading enzyme from beef liver, *J. Biol. Chem.* **239**:2160–2164 (1964).
152. R. E. Smith and M. G. Farguhar, Lysosome function in the regulation of the secretory process in cells of the anterior pituitary gland, *J. Cell Biol.* **31**:319–347 (1966).
153. H. Heller and K. Lederis, Characteristics of isolated neurosecretory vesicles from mammalian neural lobes, *in Neurosecretion* (H. Heller and R. B. Clark, eds.), pp. 35–50, Academic Press, New York (1962).
154. H. Sachs, Studies on the intracellular distribution of vasopressin, *J. Neurochem.* **10**:289–297 (1963).
155. C. W. M. Adams, Histochemistry of the cells in the nervous system, *in Neurohistochemistry* (C. W. M. Adams, ed.), pp. 253–331, Elsevier, Amsterdam (1965).
156. R. Ortmann, *in Handbook of Physiology* (J. Field, ed.) Sec. 1, Vol. II, pp. 1034, Williams and Wilkins, Baltimore (1960).
157. H. F. Moyano, Effects of dehydration on rats hypothalamic acid-phosphatase, *Experientia* **23**:529–530 (1967).
158. Y. Arai and T. Kusama, Leucine aminopeptidase in the supraoptic and paraventricular nuclei of the hypothalamus in normal and dehydrated rats, *Proc. Japan Acad.* **41**:734–736 (1965).
159. L. Avvy, Histochemical demonstration of enzymatic activities in neurosecretory centres of some homoiothermic animals, *in Neurosecretion* (H. Heller and R. B. Clark, eds.), pp. 215–225, Academic Press, New York (1962).
160. H. Tuppy, *in Polypeptides which affect smooth muscles and blood vessels* (M. Schachter, ed.), pp. 49–58, Pergamon Press, Oxford (1959).
161. K. C. Hooper, The distribution of hypothalamic peptidases in pregnant and non-pregnant dogs, *Biochem. J.* **90**:584–587 (1964).
162. W. A. Krivoy, The preservation of substance P by lysergic acid diethylamide, *Brit. J. Pharmacol.* **12**:361–364 (1957).
163. K. C. Hooper, The catabolism of some physiologically active polypeptides by homogenate of dog hypothalamus, *Biochem. J.* **83**:511–517 (1962).
164. K. C. Hooper, The enzymatic inactivation of some physiologically active polypeptides by different parts of the nervous system, *Biochem. J.* **88**:398–404 (1963).
165. A. V. Schally, E. E. Muller, A. Arimura, T. Saito, S. Sarvano, C. Y. Bowers, and S. L. Steelman Growth hormone-releasing factor (GRF): Physiological and biochemical studies with GRF preparations of bovine and porcine origin, *Ann. N.Y. Acad. Sci.* **148**:372–373 (1968).
166. S. Watanabe and S. M. McCann, Localization and mechanism of action of follicle stimulating hormone-releasing factor (FSH-RF) as determined *in vitro* assay, *Federation Proc.* **26**:365 (1967).

167. R. Deuben and J. Meites, Stimulation of pituitary growth hormone release by a hypothalamic extract *in vitro*, *Endocrinology* **74**:415 (1964).
168. A. V. Schally, E. E. Muller, S. Sarwano, T. Saito, and T. W. Redding, *In vitro* studies with hypothalamic releasing factors, *Federation Proc.* **26**:365 (1967).
169. L. Thieblot, P. Bastide, S. Blaise, J. Boyer, and G. Dastugue, Enzymic activities of the pineal gland, *Ann. Endocrinol.* **26**:313–314 (1965).
170. W. W. Tourtellotte, A selected review of reactions of the cerebrospinal fluid to disease, *in Neurological Diagnostic Techniques* (W. S. Field, ed.), pp. 1–25, Charles C. Thomas, Springfield, Illinois (1968).
171. W. Heyde, Zur Kenntis der proteasen der cerebrospinal flussig keit, *Z. Neurol.* **138**:536–543 (1932).
172. R. Abderhalden, Vorkommen von peptidases in c.s.f., *Fermentforschung* **17**:173 (1943).
173. K. Stern, A. M. Cullen, V. T. Barber, and R. Richer, Peptidases in the cerebrospinal fluid, *Can. Med. Assoc. J.* **63**:473–476 (1950).
174. J. B. Green and M. Perry, Leucine aminopeptidase activity in c.s.f., *Neurology* **13**: 924–926 (1963).
175. L. F. Chapman and H. G. Wolff, Studies of proteolytic enzymes in cerebrospinal fluid patients with chronic schizophrenic reactions. A preliminary report, *Biological Psychiatry*, pp. 130–141, Grune & Stratton, New York (1959).
176. P. Wiechert, Vorkommen und Aktivitat von peptidasen in liquor cerebrospinalis, *Acta Biol. Med. Ger.* **16**:11–14 (1966).
177. P. J. Riekkinen and U. K. Rinne, Fractionation of peptidase and esterase activities of human cerebrospinal fluid, *Brain Res.* **9**:136–144 (1968).
178. A. Nakao and E. R. Einstein, Chemical and immunological studies with a dialyzable encephalitogenic compound from bovine spinal cord, *Ann. N.Y. Acad. Sci.* **122**: 171–181 (1965).
179. E. R. Einstein, J. Csejtey, and N. Marks, Degradation of the encephalitogen by purified brain acid proteinase, *FEBS Letters* **1**:191–195 (1968).
180. G. Porcellati and R. H. S. Thompson, The effect of nerve section on the free amino acids of nervous system, *J. Neurochem.* **1**:340–347 (1957).
181. R. E. McCamen and E. Robins, Quantitative biochemical studies of Wallerian degeneration in the peripheral and central nervous system-I. Chemical constituents. II. Twelve enzymes, *J. Neurochem.* **5**:18–41 (1959).
182. T. Vanha-Perttula, V. K. Hopsu, and G. G. Glenner, Enzymes in hog kidney hydrolyzing amino acid naphthylamides, *J. Histochem. Cytochem.* **14**:314–325 (1966).
183. V. K. Hopsu-Havu, K. K. Makinen, and G. G. Glenner, Formation of bradykinin from kalliden-10 by aminopeptidase B, *Nature* **212**:1271–1272 (1966).
184. J. W. Ryan and J. K. McKenzie, Properties of renin substrate in rabbit plasma with a note on its assay, *Biochem. J.* **108**:687–692 (1968).
185. L. T. Skeggs, K. E. Lentz, J. R. Kalm, and H. Hochstrasser, Kinetics of the reaction of renin with nine synthetic peptide substrates, *J. Exptl. Med.* **128**:13–34 (1968).
186. Hopsu-Harvu, P. Rintola, and G. G. Glenner, A hog-kidney aminopeptidase liberating N-terminal dipeptides, Partial purification and characterization, *Acta Chem. Scand.* **22**:299–308 (1968).
187. V. S. Mathur and J. M. Walker, Oxytocinase in plasma and placenta in normal and prolonged labour, *Brit. Med. J.* **3**:96–97 (1968).
188. S. P. Bessman and R. Baldwin, Imidazole aminoaciduria in cerebromacular degeneration, *Science* **135**:789–790 (1962).
189. A. Meister, *in Biochemistry of the Amino Acids* (A. Meister, ed.), 2nd ed., Academic Press, New York (1965).

190. A. Lajtha, Alteration and pathology of cerebral protein metabolism, *Intern. Rev. Neurobiol.* **7**:1–40 (1964).
191. L. Heppel, Selective release of enzymes from bacteria, *Science* **156**:1451–1455 (1967).
192. A. B. Pardee, L. S. Prestidge, M. B. Whipple, and J. Dreyfuss, A binding site for sulfate and its relation to sulfate transport into Salmonella typhimuriam, *J. Biol. Chem.* **241**:3962–3969 (1966).
193. J. R. Piperno and D. L. Oxender, Animo acid-binding protein released from *Escherichia coli* by osmotic shock, *J. Biol. Chem.* **241**:5732–5733 (1966).
194. H. R. Kaback and E. R. Stadtman, Glycine uptake in *Escherichia coli*-II. Glycine uptake and metabolism by an isolated membrane preparation, *J. Biol. Chem.* **243**: 1390–1400 (1968).
195. Y. Anraku, The reaction and restoration of galactose transport in osmotically shocked cells of *Escherichia coli*, *J. Biol. Chem.* **242**:793–780 (1967).
196. W. Kundig, S. Ghosh, and S. Roseman, Phosphate bound to histidine in a protein as an intermediate in a novel phospho-transferase system, *Proc. Natl. Acad. Sci.* **52**:1067–1074 (1964).
197. E. Kennedy, Recent progress in the biochemistry of membranes, *Proceedings of the Seventh International Congress of Biochemistry of the Plenary Sessions I.U.B.*, Vol. 36, pp. 51–62, Science Council Japan (1967).
198. H. N. Christensen and M. L. Rafn, Uptake of peptides by a free-cell neoplasm, *Cancer Res.* **12**:495–497 (1952).
199. F. R. Leach and E. F. Snell, The absorption of glycine and alanine and their peptides by Lactobacillus casei, *J. Biol. Chem.* **235**:3523–3531 (1960).
200. D. Kessel and M. Lubin, On the distinction between peptidase activity and peptide transport, *Biochim. Biophys. Acta* **71**:656–663 (1963).
201. T. D. Brock and S. O. Wooley, Glycylglycine uptake in streptococci and a possible role of peptides in amino acid transport, *Arch. Biochem. Biophys.* **105**:51–57 (1964).
202. J. W. Payne and C. Gilvary, Oligopeptide transport in *Escherichia coli*, *J. Biol. Chem.* **243**:3395–3403 (1968).
203. R. K. Crane, Structure and functional organization of an epithelial cell brush border, *in Intracellular Transport* (K. B. Warren, ed.), pp. 71–102, Academic Press, New York (1966).
204. A. Eicholz, Studies on the organization of the brush border in intestinal epithelial cells-V. Subfractionation of enzymatic activities of the microvillus membrane, *Biochim. Biophys. Acta* **163**:101–107 (1968).
205. P. Emmelot, and A. Visser, and E. L. Benedetti, Studies on plasma membranes-VII. A leucyl-β-naphthylamidase containing repeating units of the surface of isolated liver and hepatoma plasma membranes, *Biochim. Biophys. Acta* **150**:364–375 (1968).
206. J. J. Pisano, J. D. Wilson, L. Cohen, D. Abraham, and S. Udenfriend, Isolation of γ-aminobutyrylhistidine (Homocarnosine) from brain, *J. Biol. Chem.* **236**:499–502 (1961).
207. D. Abraham, J. J. Pisano, and S. Udenfriend, The distribution of homocarnosine in mammals, *Arch. Biochem. Biophys.* **99**:210–213 (1962).
208. N. Marks and A. Lajtha, Protein breakdown, *in Handbook of Neurochemistry* (A. Lajtha, ed.), Vol. V, Plenum Press, New York (in preparation).
209. N. Cuzin, N. Kretchmer, R. E. Greenberg, R. Hurwitz, and F. Chapeville, Enzymatic hydrolysis of N-substitutal aminoacyl-*t*RNA, *Proc. Natl. Acad. Sci.* **58**:2079–2083 (1967).
210. H. Kossel and U. L. RajBhandarz, Studies on polynucleotides. Enzymatic hydrolysis of N-acylaminoacyl-transfer-RNA, *J. Mol. Biol.* **35**:539 (1968).

211. J. C. Brown and P. Doty, Protein factor requirement for binding of messenger RNA to ribosomes, *Biochem. Biophys. Res. Commun.* **30**:284–291 (1968).
212. J. Lucas-Lenand and A. L. Haenni, Requirement of guanosine 5′-tri-phosphate for ribosomal binding of aminoacyl-L-*s*RNA, *Proc. Natl. Acad. Sci.* **59**:554–559 (1968).
213. R. K. Datta, N. Marks, and A. Lajtha, Protein breakdown in peripheral nerve and lobster ventral cord (unpublished results), see Axoplasmic transport, *Neurosci. Res. Progr. Bull.* **5**:(4),341 (1968).
214. F. H. Leibach and F. Brinkley, γ-Glutamyl transferase of swine kidney, *Arch. Biochem. Biophys.* **127**:292–301 (1968).
215. A. Yaron and D. Mylnar, Aminopeptidase P., *Biochem. Biophys. Res. Commun.* **32**:658–663 (1968).

*Chapter 6*

# BIOCHEMISTRY OF SELECTED AMINO ACIDS

**Harold J. Strecker**

*Albert Einstein College of Medicine*
*Yeshiva University, New York, New York*

## I. INTRODUCTION

In the highly informative and thorough *Biochemistry of the Amino Acids* by Meister, a steeply rising curve is presented, plotting the total number of new amino acids reported from about 1810 to the 1960's.[1] The accretion of knowledge of amino acid metabolism probably follows a similar curve and has made it necessary for most writers on the subject, even in the relatively limited area of application to the nervous system, to confine their description and discussion to convenient fragments. For this reason this section will deal only with selected aliphatic amino acids confining attention to alanine, aspartic acid, asparagine, serine, threonine, leucine, isoleucine, valine, proline, hydroxyproline, arginine, and some related derivatives. The other amino acids are dealt with in appropriate sections of this handbook. For guidance and instruction the writer has leaned heavily on certain reviews. Foremost among these is the aforementioned authoritative text by Meister.[1] Other important sources include *The Enzymes*, edited by Boyer, Lardy, and Myrbäck, *Amino Acid Pools*, edited by Holden; *Protein and Amino Acid Nutrition*, edited by A. A. Albanese; *New Methods of Nutritional Biochemistry*, edited by A. A. Albanese; and *Mammalian Protein Metabolism*, edited by H. N. Munro and J. B. Allison.

The citations from these authoritative sources are referred to at the appropriate places in this text. The reliance on these books and reviews has made it unnecessary to list all of the references which might be pertinent to the material that is quoted or described. Instead, in most cases only the review is cited and it is left to the interested reader to find the reference to the relevant publication. In a further effort to save on the accumulation of too exhaustive a bibliography, a selection has been made among the many publications that may relate to any amino acid. The papers so chosen are those which seem to be of unusual interest or because they are the latest in the field and thus cite the earlier work or because of some bias on the part of this writer. Apologies are offered to readers as well as authors if some important findings have been inadvertently omitted from this review because of lack of thoroughness in reading all of the literature.

## II. OCCURRENCE OF ALIPHATIC AMINO ACIDS

### A. Distribution and Content of Aliphatic Amino Acids in the Normally Functioning Nervous System

Ion-exchange chromatography has replaced paper chromatography and microbiological determination as the method most commonly used for identification and separation of amino acids. This technique has made it easier to determine, quantitatively, known amino acids present in the nervous system and has led to the identification of amino acids and derivatives not previously described. A concise description of methodology is found in *Methods of Biochemical Analysis.*[2] The content of amino acids has been determined in brain of cat,[3] fish, amphibia, reptiles, aves,[4] mammals including man,[5-9] nervous systems at different stages of development and maturation,[10-12] and various physiological and pathological states.[5,6,8,13] An account of the details of these investigations is beyond the scope of this article and may be found elsewhere in these volumes. However, the amino acid content of tissues, both free and bound, depends presumably, among other things, on various metabolic processes and for this reason some of the findings are included here. The free amino acids found in highest concentrations in brain of most species are glutamate, glutamine, $\gamma$-aminobutyrate, glycine, serine, alanine, and threonine, and these amino acids may be found chiefly in the high-speed supernatant fraction of brain homogenates.[9] Cystathionine and taurine also are present in high concentrations in the brains of some species.[14] The relative proportions of amino acids found in bound form in brain appears to be about the same as in other tissues,[15] which is not unexpected since these figures are average values for different proteins and cell types.

The concentrations of serine, methionine, and creatinine in the nervous system seem fairly high relative to other tissues and may be indicative of importance, quantitatively, as has been so often discussed for glutamate, $\gamma$-aminobutyrate, and glutamine. However, considerable variations are seen in different physiological or pathological states. On the low side, relative to the amounts in other tissues, are glutathione and cysteine-cystine. No amino acids have been reported to be exclusively present in the nervous system. Conversely, it may be that no amino acid found in other tissues is entirely absent from the nervous system, although the precision of analysis may be insufficient to yield unequivocal data in all cases. The question of biochemical individuality[16] of amino acid patterns of the nervous system apparently has not been studied.

The concentrations of amino acids in the brain may be altered also by the postmortem treatment of the tissue. For example, Mangan and Whittaker[9] compared the recovery of amino acids from brains of guinea pigs which were sacrificed by freezing in liquid nitrogen followed by extraction of the frozen brain with brains obtained from animals sacrificed in an unspecified manner and from which extracts were prepared from sucrose homogenates. The values for glutamate, glutamine, $\gamma$-aminobutyrate, valine, serine, and alanine were appreciably different for the two methods of preparation. For example, the second method had resulted in a complete loss of valine and a 100% increase of alanine. Methods of killing have varied widely and include exsanguination either with or without prior anesthesia, anesthesia alone, decapitation with or without prior anesthesia, and freezing.[1-14] The chief methods used for extraction of free amino acids consist of homogenizing the tissue with picric acid, ethanol, sulfosalicylic acid, ethanol-HCl, acetone-HCl, or perchloric acid.[1-14] The procedures used gave, on the whole, good recovery of added amino acids. Nevertheless, because of the possibility of postmortem metabolism, it may not be correct to conclude that these values are the same as those extant *in vivo*.

## 1. *Methods of Isolation and Determination of Aliphatic Amino Acids*

Older methods for separation and identification for amino acids are still practiced, and some refinements have been made. Sequential one-dimensional chromatography using two different solvent systems applied to bands of amino acids on paper has been used to separate ten amino acids from brain.[17] High-voltage electrophoresis is commonly employed[18] as well as a combination of electrophoresis and solvent chromatography.[19,20] Thin-layer chromatography[19,21] and gas–liquid chromatography[22] also are under development.

After separation, the amino acids are quantitatively determined, usually by reaction with ninhydrin. A number of variations are described in the literature. Ninhydrin reagents with copper or cadmium have been found useful for staining amino acids on paper.[18]

Increased sensitivity for detecting or determining amino acids by chromatographic methods or spectrophotometrically has been gained by forming derivatives which can be determined at low concentrations. 1-Dimethylamino napthalene-5-sulfonyl chloride (or amide) forms fluorescent derivates with amines and amino acids.[23,24] Trinitrobenzene-1-sulfonic acid produces colored products with amines, amino acids, and peptides[25] and modifications of this method are available.[26] The Folin-phenol reagent in the presence of divalent cobalt ions can be used for all amino acids. Although this method is relatively insensitive, $NH_3$ or urea do not interfere.[27] Methods have been developed for separation of 2,4-dinitrophenyl derivatives of amino acids.[28-30]

A variety of methods and reagents are available for determinations of specific amino acids either individually or in groups. Bacterial amino acid decarboxylases have been used for lysine, arginine, histidine, ornithine, tyrosine, glutamate, aspartate, and diaminopimelate.[31] Proline, ornithine, lysine, and hydroxylysine give characteristic colors with ninhydrin in acid,[32] and modifications of this reaction have been applied to the estimation of cyclic imino acids.[33] Isatin, under some conditions, has been employed for determination of proline and hydroxyproline.[34] The colorimetric determination originally described for hydroxyproline has been modified often.[36–38] Apparently, individual tissues contain individual interfering substances. Some extra specificity may be obtained by enzymatic analysis.[39] The β-hydroxy-α-amino acids, serine, threonine, and hydroxylysine usually are measured by reactions with the products obtained by oxidation of the hydroxyamino acid with periodate.[40,41] These methods have been examined and modified many times.[42] The basic amino acids have been estimated after being decarboxylated by specific decarboxylases, and the resulting amines have been treated with dinitrofluorobenzene.[43] Decarboxylation of basic amino acids by specific decarboxylases is also required for a conductometric microdetermination.[44] The estimation of arginine by the Sakugichi color reaction[45] still appears to be useful.[46] Methods for guanidine derivatives have been reviewed.[47] Related to arginine are the carbamyl derivatives citrulline and urea, for which colorimetric micromethods are available.[48] Asparagine and glutamine have been determined as trinitrophenyl derivatives.[49] Specific methods for glycine, alanine, and the branched chain amino acids are not so readily available. A reaction of glycine and other amino acids with *o*-phthalaldehyde has been described.[50] It may be possible to elaborate specific enzymatic determinations of alanine[51] and the branched chain amino acids from the sensitive assays used for transaminases acting on these amino acids.[52–54] Reactions with reagents to yield products used for spectrophotometric analysis are described in a recent review.[2]

The extensive development of analytical methodology, which includes work with amino acids, is continually reviewed in such publications as the annual reviews section of *Analytical Chemistry*, *Advances in Chromatography*, and *Advances in Analytical Chemistry and Instrumentation.*

## 2. *Unusual Amino Acids and Derivatives*

In addition to the 23 amino acids which commonly are present in proteins, a large number of amino acids either free or combined are found in other natural products. The total of these natural amino acids is approaching 200.[1] Most of these were discovered in plants. Some, which are normal constituents of animal tissues, undoubtedly are metabolic products of the more common amino acids. For example, α-aminoadipic acid, pipecolic acid, homoarginine, homocitrulline, and cadaverine may be derived from lysine,[1,55] ornithine from arginine or proline,[1,56] putrescine from ornithine,[1] $\Delta^1$-pyrroline-5-carboxylate from proline or ornithine,[1] sarcosine from choline,[1] etc. Compared to the total number isolated from natural sources, relatively few amino acids and derivatives have been reported thus far in animal nervous tissue. These amino acids and derivatives include β-alanine, aminoethanol, β-aminoisobutyric acid, citrulline, creatinine, felinine, ornithine, α-aminobutyrate, γ-aminobutyrate, α-aminoadipate, β-aminoisobutyrate, β-hydroxy-γ-aminobutyrate, serine phosphate, hydroxylysine phosphate, cystathionine, cadaverine, putrescine, taurine, tyramine, γ-guanidinobutyrate, dihydroxyphenylalanine, sarcosine, cysteinesulfinic acid, hypotaurine, cysteic acid amide, methionine sulfoxide, guanidoacetic acid, betaines, and homarine.[7,8,57–60] It seems likely that the development of more sensitive and precise methods of detection and determination will reveal more of these products. The growth of understanding of metabolic pathways probably will serve also to integrate these compounds into the pattern of cellular changes in the nervous system and elsewhere.

## B. Distribution and Content of Aliphatic Amino Acids Under Abnormal or Pathological Conditions

In contrast to the relative difficulty in observing abnormalities of function in most organs, dysfunctions of the nervous system often result in dramatic changes. The administration of drugs or other agents which interfere with metabolic or physiological processes or applications of stressful conditions bring about effects which are readily discernible even by inexperienced and untrained observers. Attempts to relate these physiological and/or psychological responses to biochemical events have occupied many investigators and have been important stimuli to the development of neurochemistry as a branch of biochemistry.

It is well known that nervous system function is dependent on the oxidation of glucose and that this function fails rapidly when carbohydrate metabolism is impaired. This knowledge has directed attention to the pathways of carbohydrate metabolism as possible loci for effects of toxic compounds and of the so-called neurotropic drugs. It seems unlikely that any neurotropic drug acts solely on cells of the nervous system; the name reflects rather the direction of interest of investigators.

Up to 10% of the glucose utilized by the brain may be converted to amino acids as judged by experiments with labeled glucose.[61] As will be discussed later, more recent work has indicated that exchange reactions can account for a considerable incorporation of labeled carbon from glucose into amino acids. Nevertheless, there seems no reason to doubt that the carbon of glutamate and glutamine are obtained from the metabolism of glucose via glycolysis and the citric acid cycle. Glutamic acid is highly concentrated in brain, is a primary product of nitrogen incorporation from ammonia, and is readily formed from glucose by way of α-ketoglutarate.

In spite of the difficulty in sorting out exchange from *de novo* synthesis, it seems highly probable that glutamate is very rapidly formed from glucose. These points have been made many times [compare Strecker[62]]. In turn, glutamate can transfer nitrogen by transamination to a variety of α-keto acids to synthesize nonessential amino acids. Thus, inhibitory or stimulatory effects on amino acid metabolism of drugs, poisons, stresses, and pathological processes may be the secondary consequences of a primary interruption of normal carbohydrate metabolism. As an example, the complicated effects of potassium ion and of amytal on the presumed synthesis of glutamate, glutamine, γ-aminobutyrate, and alanine could be explained as arising from the influence of these two compounds on glycolysis, the citric acid cycle, and oxidative phosphorylation.[63] Other neurotropic agents may affect amino acid metabolism by hindering membrane events or ATP synthesis.[63,64] Amino acid changes which probably are related to alterations in carbohydrate metabolism have been reported to occur with a large assortment of neurotropic drugs classified as stimulants, depressants, convulsants, etc.[65–68] The results reported deal mainly with changes in glutamate, glutamine, alanine, γ-aminobutyrate, and aspartate. The alterations that take place in each of these amino acids vary with the agent administered, and explanations of these effects depend on analysis of carbohydrate metabolism under the conditions employed. Allied to these investigations are experiments dealing with the effects of insulin, electroshock anoxia, hyperoxia, thyroid hormone, and perhaps spreading depression brought about by KCl or electric stimulation.[67–73] In addition to the alterations in the levels of amino acids which are most directly related to glycolysis or the citric acid cycle, modifications in concentrations of other nonessential amino acids such as glycine, proline, or serine were seen also. These are to be expected by virtue of interrelated metabolism with glutamic acid or alanine.

However, shifts in the concentrations of essential amino acids also were observed. Infusion of $NH_4Cl$ increased histidine and decreased methionine and cystathionine. Anoxia increased leucine and tyrosine but decreased methionine and cystathionine. Aminooxyacetic acid, an inhibitor of transaminases, increased lysine.[67] Change in the content of amino acids such as lysine, valine, leucine, tyrosine, and taurine were obtained by administration of methionine sulfoximine, thiosemicarbazide, pentylenetetrazol, picrotoxin, or fluoroacetic acid derivatives.[67] Threonine and cystine levels were affected by hyperoxia.[69] Modifications of the concentrations of both essential amino acids and nonessential amino acids which are not directly formed from glutamate or alanine were seen in spreading depression, in hypothyroidism,[70,71] after methionine sulfoximine administration,[74] during "death" and resuscitation,[75] and after chilling or exercise.[76] These shifts in amino acids may be related to protein degradation or synthesis, transport across cell membranes, or perhaps even to as yet unknown disturbances of enzyme systems metabolizing these amino acids.

# III. ALIPHATIC AMINO ACIDS AND NUTRITION

## A. Amino Acid Requirements of the Nervous System: Biochemical and Physiological Effects of Alteration of Amino Acid Nutrition

Eight amino acids—lysine, threonine, tryptophan, methionine, phenylalanine, leucine, valine, and isoleucine—are required for both growth and nitrogen balance by all species that have been examined. In addition, arginine is essential in all species except in the adult rat, man, dog, and mouse. Histidine is also required by all except adult man; it seems to be necessary for the human infant.[1,77] Glycine and glutamate, though not entirely essential, are required for maximum growth in the chick.[78] Administration of disproportionate amounts of amino acids has produced depression of appetite, retardation of growth, reduction in nitrogen content, and alteration in enzymes. The manifestations of these amino acid imbalances are thus quite similar to those produced by amino acid deficiences.[79,80] The differences in amino acid requirements by species probably also hold for individual organs and tissues within an animal. This is indicated by the observations that the organs and tissues of animals respond in an individual manner to fasting and to varying protein intake.[81-83] In this regard brain seems to be more resistant than other tissues to variations in protein intake. Although recent work with labeled amino acids has disproved the idea that brain proteins were metabolically inert,[84] there is considerable evidence that the total protein content of the nervous system is remarkably constant under nutritional conditions in which the protein of other organs varies widely.[85-88] Deprivation of individual essential amino acids in some cases resulted in characteristic pathological and histological signs which were rather generalized, as might be expected.[1] For example, young rats on a valine-deficient diet showed "a lack of coordinated movement, staggering gait and sensitivity to touch"[89] and men on amino acid-deficient diets showed "nervous irritability."[90] Rats on low-protein diet performed less well on visual discrimination learning tests.[91]

Direct effects involving amino acids in the nervous system seem to be more evident with the hereditary metabolic diseases which arise from a lack of an enzyme in the biosynthetic pathway for a nonessential amino acid or the catabolic pathway for an essential amino acid or a defect in transport of an amino acid.[92] The protein–calorie deficiency diseases kwashiorkor and marasmus result in pathological symptoms referable to the nervous system, although weight losses of brain tissue are minimal.[93] Platt and colleagues have investigated central nervous system pathology induced by experimental protein deficiency. They found that the growth of the central nervous system was less severely affected than other organs by protein–calorie deficiency, but nevertheless the nervous system was malfunctioning as judged by obvious neuropathological signs, by E.E.G. patterns, and by histological changes.[94,95]

With respect to the content of free amino acids in brain, short-term deprivation experiments led to changes in relatively few amino acids. After 7 days of fasting, lower values were found for proline, lysine, histidine, valine, and phenylalanine. Nitrogen deprivation decreased the concentrations of valine, lysine, tyrosine, and threonine.[96] A diet lacking methionine resulted in 2 weeks in a decrease of arginine and tryptophan and an increase of histidine.[97] Nitrogen deprivation for periods up to 6 weeks led to decreases of threonine, aspartate, tyrosine, phenylalanine, and lysine and to increases in histidine, methionine, serine, and arginine.[98] The results of these experiments and others[99] demonstrated that the concentrations of amino acids in the tissues changed appreciably during the experimental period. The finding that an amino acid was present at the same concentration at the beginning and end of an experimental period of, e.g., 6 weeks does not mean that it was the same at the end of 1 week. It seems probable that in many instances the concentration of amino acid at any time may modify, by repression or derepression, enzyme systems which are involved in the synthesis or utilization of this amino acid. The changes in enzyme concentrations in turn may further modify the amino acid level, resulting in cyclic changes. A single "time lapse" view of such interacting systems at work may perhaps be seen in the report that rats on low protein diets had a decreased level in the brain of alanine, aspartic acid, and $\gamma$-amino butyrate together with decreased activity of alanine aminotransferase, aspartic aminotransferase, and glutamic decarboxylase.[91]

Excess protein intake also may have deleterious consequences. Growth depression, organ atrophy, lesions, and other toxic manifestations have been reported to follow ingestion of large amounts of almost any amino acid. These toxicities may be related to those reactions obtained from amino acid imbalances and antagonisms.[80] Physiological effects on the brain also may follow from amino acid imbalances. For example, behavioral changes were obtained from methionine feeding.[100] Injection of phenylalanine into rats resulted in fluctuating levels of a number of amino acids in the brain and in more striking changes in fetal brain after administration to the maternal animal.[101,102] These results of phenylalanine "imbalance" can be contrasted to the effects of phenylalanine deficiency on cerebral levels of amino acids and on leucine turnover in cerebral protein.[103,104]

Many studies have been conducted on the impact of vitamin deficiencies on the central nervous system.[105,106] Some vitamins may be involved in reactions of amino acid metabolism. For example, pyridoxine deficiency which leads to seizures also leads to changes in concentrations of some free amino acids in brain, probably by virtue of its function in transamination reactions.[107,108] Other vitamins also may have direct or indirect effects on amino acid metabolism.[109]

In this age it is trite to state that the well-being of the central nervous system, just as any other organ or tissue, depends on the maintenance of homeostasis. Narrowing our concern to the nutritional requirements of the brain for amino acids, the question of the mechanisms involved should be considered. Studies of recent years have made it clear that in a large measure these mechanisms are based on control of metabolic pathways. This control is exercised chiefly by systems of metabolite inhibition or stimulation of enzymatically catalyzed reactions or by inhibition or stimulation of enzyme synthesis.[110] The research in this area has used mainly tissues with rapid turnover of protein and nucleic acids, i.e., microorganisms, animal cells in culture, and to a lesser extent liver and kidney. There is no reason to believe, however, that the relative resistance of brain to effects of nutritional alterations indicate a lack or an inertia of control devices. On the contrary, it may be suggested that the apparent metabolic inertia together with its demonstrated capacity for turnover,[84] indicate a high degree of metabolic and structural control.

## IV. METABOLISM OF ALIPHATIC AMINO ACIDS

### A. Synthesis of Amino Acids

#### 1. *General Remarks*

The classification by the early nutritionists of amino acids into essential and nonessential[111] led to the understanding that animals lacked the enzymatic equipment to synthesize the so-called essential amino acids. (The aliphatic group within the scope of this chapter includes lysine, threonine, leucine, valine, and isoleucine.) Considerable interest was aroused therefore by the reports that 1-day-old mouse brain promoted the incorporation of carbon from glucose and other precursors into almost all the amino acids, both essential and nonessential, free and protein-bound.[112-114] However, a recent report from part of the same group did not confirm the earlier findings for glucose conversion to essential amino acids.[115] Many investigators have found rapid conversion *in vivo* of glucose carbon to nonessential amino acids in brain of mouse,[116,117] of rat both normal[118-121] and hepatectomized,[121] of cat,[122] and of sheep.[123] Perfusion of cat brain with solutions containing labeled glucose has also been studied.[124] There is no general agreement, perhaps due to different routes of administration and methods of handling the tissue, as to which specific amino acids are most rapidly and most highly labeled. Nevertheless, alanine, glutamate, glutamine, aspartate, and $\gamma$-aminobutyrate have been isolated with quite high specific activities and proline, arginine, serine, glycine, glutathione, and *N*-acetyl aspartic acid with somewhat lower specific activities. In addition, carbon from labeled fatty acids and labeled pyruvate was incorporated into amino acids in sheep brain[123] and rat brain,[125] respectively. Investigations with brain slices or homogenates have provided similar data,[62,126-130] as also have experiments with excised ganglia.[131] The experimental data thus obtained has led most investigators to suggest that glucose in brain is rapidly metabolized to pyruvate and α-ketoglutarate. The carbon of these two α-keto acids is then incorporated into alanine and glutamate, respectively, either by transamination or reductive amination in the case of the latter or by exchange reactions catalyzed by transaminases.

The observed rapidity of the conversion of glucose carbon to dicarboxylic amino acid carbon in brain has prompted some investigators to conclude that glucose is converted to amino acids faster in brain than in other organs.[118,132] The transaminases involved are glutamate-aspartate and glutamate-alanine. The activities of these enzymes in brain tissue have been studied intensively both by direct and indirect methods, and they do not appear to be more active or present in higher concentrations than in other tissues.[133-136] Evidence has been provided that the incorporation of glucose carbon into dicarboxylic acids depends on exchange catalyzed by transaminases. The rates of exchange are 3–8 times as fast as the maximal rate of utilization of the two substrates.[135,137] The observed high rate of $^{14}C$ incorporation into amino acids is presumed to depend on the rapidity of conversion in brain of glucose to acetyl-CoA, the large pool size of glutamate, and the large blood supply of the brain.[136] Nevertheless, there can be no question that synthesis of some nonessential amino acids in brain can proceed from glucose by way of pyruvate and α-ketoglutarate to form alanine and glutamate, respectively. Considerable evidence has been obtained to support the suggestion[62] that glutamate is formed primarily by reductive amination and that transamination from glutamate to keto acids is responsible for the flow of amino groups, resulting in synthesis of certain nonessential amino acids.

These reactions have been demonstrated in brain preparations as well as in other tissues.[134,138] Rat brain mitochondria have been shown to catalyze an almost stoichiometric conversion of glutamate to aspartate.[134] Other amino acids which might be formed by transamination from glutamate include alanine,[61,116-118,121,124-127,129,130,139] serine,[117,124,140,141] glycine,[124] and ornithine.[137] In addition, glutamate probably can form proline[116] and perhaps arginine.[116,142] Other metabolic roles of glutamate are discussed elsewhere in this book. It seems reasonable to conclude that a number of nonessential amino acids—but chiefly glutamate, aspartate, and alanine—can be synthesized in brain through reactions interconnecting with the tricarboxylic acid cycle and glycolysis.[143] Those agents which influence the levels of glycolytic intermediates such as hormones, drugs, physical and psychological stress, nutrition, and coenzyme levels can cause immediate and sometimes drastic changes in the concentrations of the brain's amino acids, all of which are no doubt involved in metabolic flux. These changes may be seen not only with the amino acids which are present in high concentrations[120,121,144-146] but even, for example, with proline which decreased markedly in the brain of mice subjected to inhalation of the convulsive agent Indaklon.[147] The rapid fixation of $^{14}CO_2$ into the dicarboxylic amino acids in nervous tissues[148,149] is a further example of the close interrelationship of amino acid and citric acid cycle metabolism.

The remainder of this section will deal with the synthesis of aspartate, asparagine, serine, alanine, proline, hydroxyproline, and arginine, with some remarks on a few related compounds.

## 2. *Aspartate*

As indicated previously, aspartate is formed from oxaloacetate by transamination with glutamate. Aspartate-transaminase is present in rat cerebral homogenates at about the same concentration as in liver[91] and is found in both the mitochondrial and soluble fractions.[150,151] The enzyme in mitochondria is latent and can be "activated" by various procedures which disrupt mitochondrial structure.[150–152] Very little work seems to have been done with the brain enzyme(s). The report of multiple forms of the enzyme in brain parallels the findings in other tissues,[1,153,154] and it may be presumed that the requirements for pyridoxal phosphate and other properties of the brain enzyme(s) will be similar to those described for the more highly purified enzymes of other tissues.[154,155] The equilibrium constant for the reaction is 3 to 5 in the direction of aspartate and $\alpha$-ketoglutarate.[156] It is interesting in spite of compartmentalization of enzymes and probably of substrates as well that the concentrations in brain of the four substrates of the reaction are not far from satisfying the equilibrium constant.[91,157]

## 3. *Asparagine*

The mechanism of asparagine biosynthesis in animal tissues is still not fully known. In microorganisms the reaction appears to be similar to that catalyzed by glutamine synthetase and requires ATP, aspartate, and $NH_3$.[1,158] In animals, asparagine may be synthesized from glutamine and aspartate,[159,160] but no information is available for brain.

## 4. *Serine*

Extracts of mouse brain have been used to study the biosynthesis of serine.[140] The pathway was represented as glucose → D-glycerate-3-phosphate → hydroxypyruvate phosphate → serine phosphate → serine. The amino group donor (to hydroxypyruvate phosphate) was shown to be glutamate. A possible alternative scheme, which is found in bacteria, parallels the above pathway but uses the corresponding nonphosphorylated intermediates and uses alanine as the amino donor. However, the transamination reaction between alanine and hydroxypyruvate to form serine does not occur to any appreciable extent in dog or beef brain.[141]

## 5. *Alanine*

A discussion of the formation of alanine can more or less parallel that for aspartate. Carbon from labeled glucose is rapidly incorporated into alanine in brain tissue, as noted before, and transamination of pyruvate with glutamate is catalyzed by brain preparations.[133] Mitochondrial bound and soluble isozymes of alanine aminotransferases have been described for rat liver[161] and pig heart.[162] A similar distribution is indicated for brain.[143,151] The enzyme(s) seems to be less active in brain than in liver, in contrast to aspartate aminotransferase.[136,143] However, the mitochondrial alanine aminotransferase of rat liver is quite labile,[161] and the corresponding brain enzyme might be even more so.[143] The equilibrium constant for the reaction is 1.6 in the direction of alanine + α-ketoglutarate formation.[163] As in the case of aspartate aminotransferase, the concentrations of the substrates in brain are not far from satisfying the equilibrium constant.[91,157]

### 6. *Proline*

Proline can be formed from either glutamate or arginine. The first steps in the pathway from glutamate, which probably involves reduction to glutamic-γ-semialdehyde, has been demonstrated only in bacteria.[164,165] However, nutritional and isotopic data from animal experiments are in accord with this pathway in animal tissues as well.[166,167] Glutamic-γ-semialdehyde is spontaneously cyclized to $\Delta^1$-pyrroline-5-carboxylic acid, which is then reduced by a DPNH (TPNH) requiring enzyme ($\Delta^1$-pyrroline-5-carboxylate reductase) to proline.[168] An alternative pathway from arginine involves arginase, ornithine-δ-transaminase, and $\Delta^1$-pyrroline-5-carboxylate reductase.[169] Indirect evidence for synthesis of proline from glutamate in brain is provided by the finding that carbon from labeled glucose is incorporated into proline.[118] With reference to the second pathway, arginase and ornithine-δ-transaminase are both found in brain.[170–172] The final step in this pathway, although not yet directly demonstrated in brain, is common to the first pathway as well.

### 7. *Hydroxyproline*

At least two isomers of hydroxyproline are present in animal tissues, 4-hydroxy-L-proline and 3-hydroxy-L-proline.[1,173] These hydroxyprolines are found mainly in connective tissue and may not be a component of any characteristic brain protein. Nevertheless, this amino acid cannot be dismissed as being unimportant to the nervous system. Hydroxyprolinemia associated with mental retardation has been reported.[174] Hydroxyproline is synthesized from proline in an unique manner, in that the hydroxylation step appears to take place after the proline is in peptide linkage in a so-called protocollagen.[175,176] The possibility exists also of synthesis of free hydroxyproline in animal tissues. An enzyme has been reported that condenses pyruvate and glyoxalate to form γ-hydroxy-α-ketoglutarate.[177] This product can be transaminated to γ-hydroxyglutamate,[178] which may be reduced to hydroxyproline by systems analogous to those described for reduction of glutamate to proline.[179]

### 8. *Arginine*

This amino acid may be synthesized from either glutamate or proline. The common intermediates derived from either of these two amino acids are glutamic-$\gamma$-semialdehyde or $\Delta^1$-pyrroline-5-carboxylic acid, which reach equilibrium with each other rapidly and spontaneously.[180] This intermediate(s) may be derived from proline by a reaction catalyzed by the proline oxidase system of mitochondria.[181] This enzyme has been reported to be in brain.[182] Glutamic-$\gamma$-semialdehyde (-$\Delta^1$-pyrroline-5-carboxylate) can be transaminated by glutamate to form ornithine, as catalyzed by ornithine-$\delta$-transaminase, an enzyme found in brain as well as in other tissues.[172] Evidence for the conversion of ornithine to arginine in brain is somewhat equivocal. Labeled proline injected intracisternally has been shown to be converted to labeled arginine in rat brain, and these results have been cited as support for the existence of the urea cycle in brain.[142] Three enzymes of the urea cycle indeed have been directly demonstrated in brain preparations. These are arginase, argininosuccinate synthetase, and argininosuccinase.[170,171] However, the enzymes synthesizing citrulline from ornithine could not be found in rat brain.[171] The urea cycle in the nervous system is discussed elsewhere in this book. Figure 1 shows major pathways of synthesis in animals of selected aliphatic amino acids.

### 9. *Other Amino Acids*

Amino acids and derivatives which could be related to the aliphatic amino acids discussed within the scope of this chapter and which are found in the nervous system include $\beta$-alanine, aminoethanol,* $\beta$-aminoisobutyrate, citrulline, ornithine, $\alpha$-aminobutyrate,* $\alpha$-aminoadipate,* $\beta$-hydroxy-$\gamma$-aminobutyrate,* serine phosphate, hydroxylysine phosphate, cadaverine, putrescine, $\gamma$-guanidinobutyrate,* sarcosine,* guanidoacetate,* creatine* and betaines.*[7,8,57-60] The compounds marked with an asterisk arise from metabolic routes involving aliphatic amino acids which will be discussed below. Ornithine and citrulline formation already have been discussed. Putrescine can be formed by decarboxylation of ornithine or by hydrolysis of agmatine.[183] Although these reactions have not been described as yet in animal tissues, the repeated finding of putrescine in animal organs including brain, and the need for this compound for synthesizing spermine and spermidine[184] make it likely that an enzymatic system for its production will be found. $\beta$-Alanine and $\beta$-aminoisobutyrate are derived from degradation of pyrimidines, the former from dihydrouracil by way of $\beta$-ureidopropionic acid and the latter from thymine by way of $\beta$-ureidoisobutyric acid.[1] Serine phosphate, as described above, is an intermediate in the biosynthesis of serine.

## B. Degradative Reactions of Aliphatic Amino Acids

### 1. *General Remarks*

General metabolic reactions involving aliphatic amino acids include transamination, decarboxylation, oxidative and nonoxidative deamination,

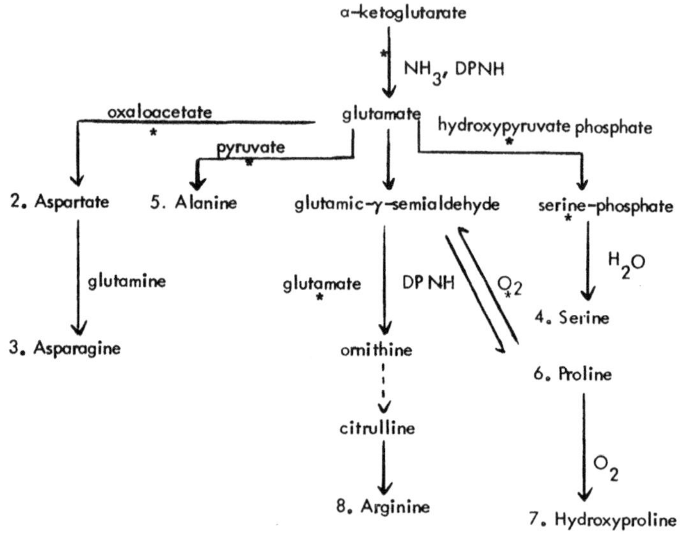


Fig. 1. Major pathways of synthesis in animals of selected aliphatic amino acids. The numbers relate to the descriptions in the text. An asterisk denotes that the enzyme system is known to be present in brain.

oxygenation, racemization, and epimerization.(1) The question of a general L-amino acid oxidase with broad specificity has concerned many investigators.(184–186) Both L- and D-amino acid oxidases have been isolated and purified from animal tissues.(187–189) The crystalline L-amino acid oxidase from rat kidney oxidizes many monoamino monocarboxylic acids but is inactive toward L-serine, L-threonine, L-aspartic acid, L-glutamic acid, and L-lysine.(190) This enzyme has very low activity, and most authors have dismissed it as physiologically unimportant.(187,188) Instead it is considered that oxidations of L-amino acids take place via transamination.(191) For example, liver and kidney mitochondria have been shown not to oxidize L-amino acids unless an α-keto acid such as α-ketoglutarate, pyruvate, or oxaloacetate is present.(192,193) Transamination followed by oxidation of the ketoacid formed can account for disposal of the carbon chains of amino acids but leaves the nitrogen still in organic linkage. From the previous discussion on the interaction of amino acid metabolism and the tricarboxylic acid cycle, it appears most likely that α-ketoglutarate, oxaloacetate, or pyruvate can function as primary amino group acceptors.(136,143) Transamination has been reported to occur between a large number of amino acids and α-ketoglutarate or pyruvate as acceptors.(1,155) Unfortunately, relatively few of these reactions have been isolated and studied, and it is possible that some of the reactions described represent the sums of coupled reactions. However, in many cases it seems that glutamate or alanine are formed as primary products of amino group transfer from other amino acids. The further metabolism of glutamate as described earlier appears to lead to aspartate.[134,138] Thus the fate of amino groups obtained from L-amino acids may be ultimately decided by the fate of aspartate and alanine, which are discussed below.

With respect to brain, although the early work of Weil-Malherbe on the uniqueness of glutamate oxidation in brain[194] has been repeatedly cited, recent work with more sensitive assays have demonstrated oxidation of other L-amino acids although the mechanisms are unknown. Oxidative deamination was obtained with rat brain "breis" and the following aliphatic L-amino acids: alanine, aspartate, leucine, valine, arginine, lysine, and ornithine.[195,196] Glycine and L-alanine labeled in the carboxyl group yielded labeled $CO_2$ with rat brain homogenate.[197,198] Labeled $CO_2$ was recovered also from incubation with homogenates of brain of 1-day-old mice of the L forms of glutamate, alanine, leucine, valine, isoleucine, serine, threonine, glycine, proline, arginine, and lysine, although the last two were oxidized very poorly.[199] A number of reports have appeared on the oxidation of aspartate, alanine, and glycine in brain.[200-203] Brain tissue of Elasmobranch and Teleost fish formed $NH_3$ from glycine, aspartate, isoleucine, alanine, glutamate, ornithine, lysine, arginine, serine, and cysteine.[204]

D-Amino acid oxidase has been reported in the brains of cattle, sheep, mouse, rat, human, and dog.[205-207] The demonstration of this enzyme in the nervous system explains the oxidation of D-amino acids by brain preparations,[196,197] but the physiological function still is unknown.

The remainder of this section will deal with individual metabolic reactions in animal tissues of the following amino acids: aspartic, asparagine, alanine, serine, threonine, leucine, isoleucine, valine, proline, hydroxyproline, arginine, and lysine.

## 2. *Aspartic Acid*

The irreversible removal of the nitrogen from aspartate would seem to be most effectively accomplished by the condensation of this amino acid with citrulline and the sequence of reactions leading to urea.[208] As noted previously the enzymes catalyzing the urea cycle from citrulline to urea are found in brain.[170,171] If an alternative pathway existed for obtaining citrulline, an explanation could be provided for the function of this group of enzymes in the absence of the complete ornithine cycle. Aspartate can transfer its amino group to inosinic acid in a sequence of reactions leading to adenylate.[208] This pathway, which is widely distributed, probably occurs also in brain.[209] The amino group derived from aspartate finally can yield $NH_3$ by the action of adenylic deaminase, which is active in brain.[210,211] Aspartate is also the source for nitrogen atom 1 of the purine ring and furnishes 3 carbon atoms and 1 nitrogen atom to the pyrimidine ring.[1,208] It seems likely that these synthetic reactions also can take place in brain cells.[212-214] Aspartate transcarbamylase, which catalyzes the first step in the synthesis of pyrimidines from aspartate and carbamyl phosphate, is found in rat brain.[215] However, carbamyl phosphate synthetase is not found in brain.[171] Although is it possible that this relatively unstable compound is synthesized *de novo* in liver and then carried to brain cells through the blood–brain barrier, alternate precursors could be considered.[1,215]

It has been proposed that β-alanine in brain is produced from aspartate by decarboxylation.[209] However, there is no evidence for this reaction in animal tissues and, as noted before, β-alanine could arise from the degradation of uracil.[1] Other reactions involving aspartate, such as reduction to aspartic-semialdehyde and deamination to fumarate, have not been described in animals.[1] Utilization of aspartate in the brain for peptides such as *N*-acetyl aspartate and aspartylglutamate will be discussed elsewhere.

### 3. *Asparagine*

Although this was the first amino acid to be isolated from natural sources, it is perhaps the most neglected. Asparagine is deaminated to aspartic acid by a rather widely distributed asparaginase, which has been reported to be present also in brain tissue where it increases in activity during experimental allergic encephalomyelitis.[216] Asparaginase activity in brain also was decreased by barbiturates administered *in vivo*. The depth of sleep corresponded to the decrease of activity.[217] This enzyme is arousing some interest because of its antitumor activity.[218] Asparagine can transfer its α-amino group to various keto acid acceptors to form α-ketosuccinamic acid, which can then be hydrolyzed to oxaloacetic acid and ammonia. This enzyme is not very specific for either the amino donor or acceptor.[1,219]

### 4. *Alanine*

If transamination between glutamate and pyruvate is postulated as a mechanism exclusively for synthesis of alanine, then there seems to be no pathways for degradation comparable to those available to aspartate. The data on L-alanine dissimilation are in accord with the intermediate formation of pyruvate, presumably by transamination.[194,197,201,203] Thus it must be concluded on the basis of the available information that the direction of function of alanine aminotransferase must depend on other mechanisms which act to drive the reaction either way.

### 5. *Serine*

Serine is interconvertible with glycine in a reaction requiring tetrahydrofolic acid. The responsible enzyme, serine transhydroxymethylase, contains pyridoxal phosphate as prosthetic group.[220] Indirect evidence for this enzyme in brain was obtained from experiments with labeled precursors that formed serine and glycine, which were almost equally labeled.[142,221,222] Serine dehydrase (or dehydratase), a pyridoxal phosphate enzyme, deaminates serine to pyruvate and $NH_3$.[223] A highly purified preparation from rat liver also catalyzed synthesis of cystathionine from homocysteine and L-serine, and deaminated L-threonine,[224] but experiments have been reported which indicate that cystathionine synthetase and serine dehydrase are separate enzymes.[225] A sheep liver enzyme is known which deaminates serine but not threonine[226] and a human liver preparation which deaminates threonine but does not synthesize cystathionine.[227] In relation to these findings homogenates of rat brain deaminated serine at a higher rate than any other amino acid.[228] However, the responsible enzyme systems have not been isolated. Serine is used also for the synthesis of cysteine, cystathionine, and intermediates of lipid biosynthesis that in turn lead to choline, betaines, and sarcosine.[1] These reactions will be considered elsewhere in this book.

## 6. *Threonine*

Threonine dehydrase (or dehydratase) deaminates threonine to α-ketobutyrate and $NH_3$. Serine dehydrase activity appears to be always associated with the threonine enzyme,[223,224,226,229] but a serine dehydrase has been reported which does not act on threonine.[226] Transamination of the product, α-ketobutyrate, would form α-aminobutyrate, which is found in brain.[57] Cleavage of threonine to glycine and acetaldehyde has been studied with liver and kidney preparations.[230,231] The enzyme in rat liver was shown recently to act only on allothreonine, and the physiological function is unknown.[232] Preparations of liver from a number of animals oxidize threonine to amino acetone,[233,234] and this reaction occurs also in rat brain.[235] It was noted that rat brain homogenates do deaminate threonine.[228]

## 7. *Proline*

As described previously, proline can be used for synthesis of both arginine and hydroxyproline. A further cellular option is oxidation to glutamate. The first step in this pathway is catalyzed by proline oxidase, which is common to the pathway to arginine and results in $\Delta^1$-pyrroline-5-carboxylate.[181] This product, or its straight-chain form, glutamic-γ- semialdehyde, can then be oxidized by a DPN (TPN)-dependent dehydrogenase to yield glutamate.[236] This enzyme has not yet been reported in brain tissue, but indirect evidence is provided by the high yield of labeled glutamate obtained from labeled proline.[142]

## 8. *Hydroxyproline*

This amino acid is readily oxidized by kidney or liver preparations to $\Delta^1$-pyrroline-3-hydroxy-5-carboxylate, and it has been suggested that this reaction is catalyzed by proline oxidase.[237,238] However, brain preparations have been reported to oxidize only proline and not hydroxyproline.[184] $\Delta^1$-Pyrroline-3-hydroxy-5-carboxylate is oxidized by liver preparations to hydroxyglutamate.[239] Hydroxyglutamate is decarboxylated by brain preparations[240] to 2-hydroxy-4-aminobutyrate, which in turn is converted to malic acid.[241] The intact rat, or preparations from rat liver but not brain, convert hydroxyglutamate to glyoxalate plus alanine.[242,243]

### 9. *Arginine*

This unusually interesting compound appears to have diverse and important functions in different life forms. Arginine phosphate probably plays the same role in invertebrate muscle as creatine phosphate in vertebrate.[244] In vertebrates it is required for the synthesis of creatine, a reaction sequence in which an amidine group is transferred to glycine to form guanidinoacetic acid, which in turn is methylated to yield creatine.[245] Transamidination can occur also with a number of other acceptors to form a variety of guanidino compounds.[245] Arginine functions in the transport and excretion of nitrogen in ureotelic organisms, as a key member of the urea cycle, the hydrolysis by arginase yielding urea and ornithine.[208] Other reactions of arginine in animal tissues include reductive condensation with pyruvate to form octopine[246] and formation of α-keto-δ-guanidinovalerate by either transamination[1] or oxidative deamination.[190] In brain tissue there is evidence thus far only for arginase[170,171] and for a transamination to γ-aminobutyrate to yield γ-guanidinobutyrate.[247] The carbon skeleton (ornithine) derived from the arginase reaction can then be utilized via ornithine-δ transaminase and either $\Delta^1$-pyrroline-5-carboxylate dehydrogenase or $\Delta^1$-pyrroline-5-carboxylate reductase to form glutamate and proline respectively.[166,236]

### 10. *Leucine, Isoleucine, and Valine*

The metabolism of these three amino acids is quite similar, especially in the early stages. The first step is in all likelihood a transamination with α-ketoglutarate, which, at least in pig heart, is catalyzed by an enzyme which can react almost equally well with all 3 branched chain amino acids.[248,249] A separate transaminase for leucine is found also in rat liver.[250] The α-keto acids formed from these three amino acids are degraded by somewhat analogous pathways. Leucine finally yields acetoacetic acid and acetyl CoA, valine forms succinyl CoA and $CO_2$, and isoleucine forms acetyl CoA, propionyl CoA, and $CO_2$.[1] Other products that can arise from these amino acids are isoprenoid compounds from leucine, β-alanine from isoleucine, and β-aminoisobutyrate from valine.[1] In brain slices from rabbit all three amino acids, uniformly labeled with $^{14}C$ yielded $^{14}CO_2$, although no increase of oxygen uptake was seen.[251] Labeled leucine was incorporated into brain cholesterol *in vitro*,[252] and the rate of incorporation into lipid and of $^{14}CO_2$ formation was decreased in brains of rats on a phenylalanine deficient diet.[253]

### 11. *Lysine*

A series of papers beginning in 1948 provided evidence for a pathway of lysine degradation in the animal to pipecolic acid, α-aminoadipic acid, and glutaric acid. These results were obtained by using labeled compounds for metabolic studies with intact rats and with preparations of liver from guinea pig and rat.[254,255] The proposed pathway, in part, consisted of the following steps:[1]

1. Lysine ⇌ α-keto-ε-aminocaproic acid (+$NH_3$)
2. α-Keto-ε-aminocaproic acid ⇌ $\Delta^1$-piperideine-2-carboxylic acid
3. $\Delta^1$-Piperideine-2-carboxylic acid (+DPNH) ⇌ pipecolic acid
4. Pipecolic acid (+O) ⇌ $\Delta^1$-piperideine-6-carboxylic acid
5. $\Delta^1$-Piperideine-6-carboxylic acid ⇌ α-aminoadipic-δ-semialdehyde
6. α-Aminoadipic-δ-semialdehyde (+DPN) ⇌ α-aminoadipic acid
7. α-Aminoadipic acid ⇌ α-ketoadipate (+$NH_3$)
8. α-Ketoadipate ⇌ glutaryl CoA (+$CO_2$)

Thus far only step 3 and the products of some of the reactions from steps 4–8 have been demonstrated *in vitro* in animal tissues.[1,254,255] Recently the isolation of the compound saccharopine, formed by a condensation of lysine and α-ketoglutarate catalyzed by liver mitochondria, has led to the proposal of a more direct pathway from lysine to glutaryl CoA.[256] The hydrolysis of saccharopine, which is also an intermediate in lysine biosynthesis,[257] would yield α-aminoadipic-δ-semialdehyde directly and bypass the steps involving pipecolic acid.

Another pathway for lysine catabolism is indicated by the observation that homocitrulline and homoarginine were excreted after feeding lysine to humans and that labeled homoarginine and homocitrulline were found in the liver and kidneys of rats after administration of labeled lysine.[255] These compounds could be formed by transamination of lysine.[242] Very little has been done with lysine metabolism in the nervous system, although α-aminoadipic acid has been identified in rabbit, rat, and guinea pig brain.[60] Figures 2 and 3 show major pathways of degradation of selected aliphatic amino acids in animals.

## V. AMINO ACID METABOLISM AND DISEASES OF THE NERVOUS SYSTEM

Many excellent publications reviewing the area of amino acidopathies and cerebral dysfunction are available. Two recent ones which this writer has found useful are by Berlet[89] and by Efron.[258] Relevant to the aliphatic amino acids discussed in the present article, the following amino acidopathies are accompanied by symptoms of cerebral dysfunction: β-alaninemia, argininosuccinic aciduria, citrullinemia, maple syrup disease, prolinemia, hydroxyprolinemia, hypervalinemia, and hyperlysinemia. All of these diseases appear to be due to an enzyme deficiency that is inherited. All except one are characterized by an accumulation of certain amino acids in the blood. The exception is argininosuccinic aciduria, in which disease argininosuccinate apparently is not reabsorbed by the kidneys and is excreted in abnormally large amounts. In most of these conditions an aminoaciduria is also seen. In some cases amino acids, in addition to the one associated with the disease, are also excreted at an elevated level. This excessive excretion is blamed on interference by one amino acid with renal transport and perhaps metabolism of others.

Hyper-β-alaninemia was found in a patient with somnolence and seizures. β-Alinane and γ-aminobutyrate were elevated in the plasma. It was proposed that the defective enzyme is β-alanine transaminase.[259]

Hyperlysinemia associated with mental retardation has been reported in several patients.[260,261] The responsible enzyme defect was not determined, and it was suggested that decreased utilization of lysine for protein synthesis was at fault. A patient with congenital lysine intolerance and ammonia intoxication was shown to have a deficiency of lysine dehydrogenase in her liver. This enzyme, which catalyzes an NAD-dependent oxidation of L-lysine to α-ketocaproic acid, had not been reported before.[262] Hypervalinemia associated with drowsiness and abnormal EEG patterns has been reported. It was proposed that the defective step was in deamination of valine, since the keto acid derived from valine was not found when valine loads were given.[263] It is interesting that diets rich in branched-chain amino acids given to infantile monkeys caused mental retardation, convulsions, and bizarre behavior.[264] A leucine-induced hypoglycemia, which sometimes is associated with mental retardation, has been found in a number of children.[265] Leucine and other amino acids stimulate insulin secretion even in normal subjects, and there may be no relationship between the mental retardation and the hypoglycemia.[266] Two types of hyperprolinemia have been reported. In one there is a deficiency of proline oxidase and in the other a deficiency of $\Delta^1$-pyrroline-5-carboxylate dehydrogenase.[182] Hydroxyprolinemia was ascribed to a lack of hydroxyproline oxidase; this enzyme is claimed to be different than proline oxidase.[174] The disorders of the urea cycle, citrullinemia, and argininosuccinicaciduria as well as the classic maple syrup disease have been discussed very thoroughly in many places (compare, e.g., Duncan [267]).

Although these various acidopathies are relatively uncommon, they are also of interest because of the contribution to tracing of recessive and sex-linked genetic characteristics and to elucidation of biochemical pathways in the human. An example is the conclusion that proline oxidase and hydroxyproline oxidase are separate enzymes.[182] As more diseases are discovered it may be expected that other pathways of amino acid metabolism will become known and new relationships between amino acid, lipid, and carbohydrate metabolism will be uncovered. Whether any of these diseases involve the brain as a primary target organ is uncertain. In all of them very general symptomology has been described. Perhaps stress

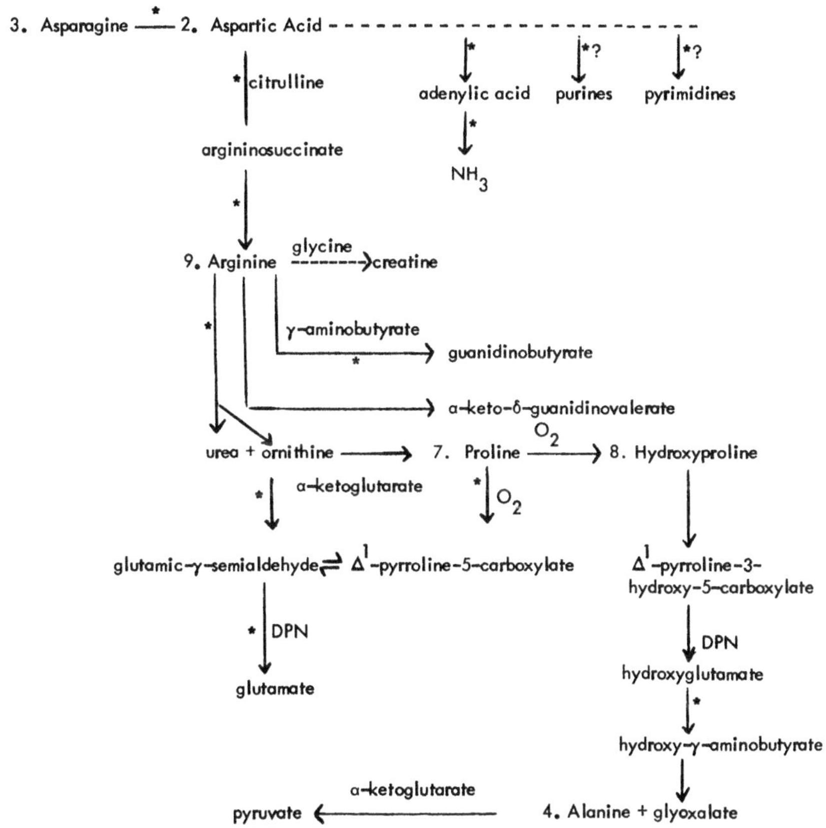


Fig. 2. Major pathways of degradation in animals of selected alipathic amino acids. The numbers relate to the descriptions in the text. An asterisk denotes that the enzyme system is known to be present in brain. A question mark denotes that indirect evidence suggests that the pathway or enzyme is present in brain.

has been put on cerebral dysfunction because of the dramatic nature of the clinical signs.

## VI. EPILOGUE

Relevant to the subject matter of this chapter, it seems that the period of accumulation of data of concentrations of amino acids or of metabolic pathways for amino acids or of the enzymes involved in these metabolic pathways is ending. The exquisite refinements of technique that have enabled determination of amino acids and enzymes even in individual cells of the nervous system may yet provide more precise values or may indicate the presence of metabolites or enzymes heretofore thought to be absent.

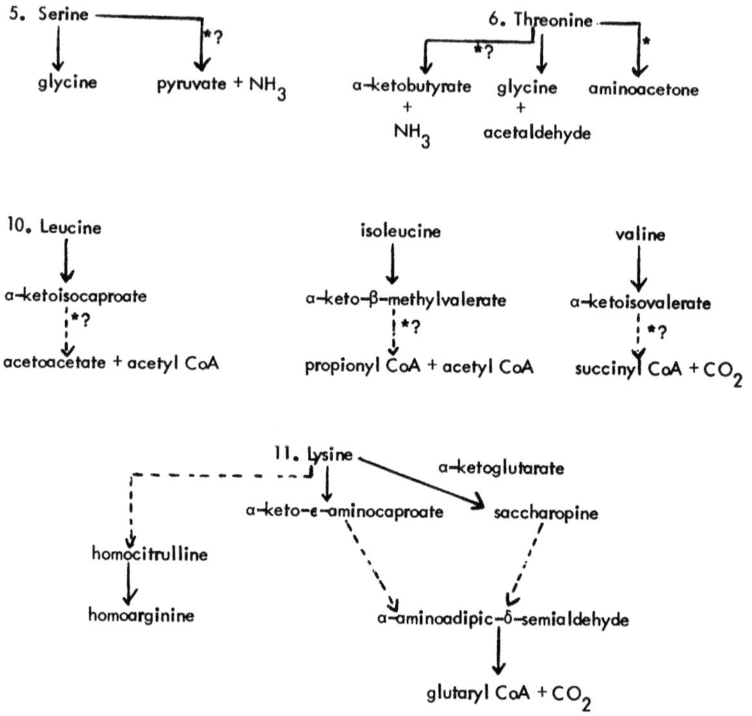


Fig. 3. Major pathways of degradation in animals of selected alipathic amino acids. The numbers relate to the descriptions in the text.

Nevertheless, it seems unlikely that further determinations of single-point temporal and spatial values will further increase our knowledge. Cellular components in rapidly metabolizing tissues are known to be affected greatly in concentration by nutrition, diurnal rhythms, season, temperature, physiology, psychology, age, genetic strain, etc. Although the nervous system seems to provide a more constant environment than other tissues, there is no reason to doubt that flux also exists in the former tissue.

It seems to this writer that researchers in neurochemistry will increasingly concern themselves with attempts to interrelate the various conditions which bring about this flux in order to gain understanding of the mechanism of control involved in the regulation of synthesis and degradation of cellular components. For example, now that it is known that two enzymes of the urea cycle are very low in brain whereas they are very high in liver, the cause of this will be sought. At what stage of development of the nervous system did this relative repression of these two enzymes take place? Are there nutritional and hormonal or only genetic determinants involved? If the glutamate metabolic family (i.e., glutamate, glutamine, $\gamma$-aminobutyrate, proline, and ornithine) really is quantitatively more important in brain, does this not depend on the relationship between the enzymes concerned in the metabolism of this group of amino acids? What are the factors controlling these relationships? To answer questions of this type it seems likely that the contributions of the nutritionists and biologists working in differentiation and development will be used increasingly by the biochemist and molecular biologist and applied to the nervous system as they have been already applied to studies with microorganisms and, to a lesser extent, to other tissues of animals. It seems a pity that amino acid research in the nervous system has always lagged behind that in the liver, for example, and in many cases been merely repetitive of the approaches used for other organs. Important reasons for this lag centred perhaps on the apparent relative metabolic inertia of the nervous system and the difficulty in separation of the various cells and cellular structures. The latter problem has been mainly overcome. The former problem should now begin to prove a challenge to uncover the meaning of this apparent inertia.

## VII. REFERENCES


1. A. Meister, *Biochemistry of the Amino Acids*, Academic Press, New York (1965).
2. S. Jacobs, *in Methods of Biochemical Analysis* (D. Glick, ed.), Vol. 14, pp. 177–202, Interscience, New York (1966).
3. H. H. Tallan, S. Moore, and W. H. Stein, Studies on the free amino acids and related compounds in the tissues of the cat, *J. Biol. Chem.* **211**:927–939 (1954).
4. N. Okumura, S. Otsuki, and T. Aoyama, Studies on the free amino acids and related compounds in the brains of fish, amphibia, reptile, aves and mammal by ion-exchange chromatography, *J. Biochem.* (*Tokyo*) **46**:207–212 (1959).
5. N. Okumura, S. Otsuki, and A. Kameyama, Studies on free amino acids in human brain, *J. Biochem.* (*Tokyo*) **47**:315–320 (1960).
6. Y. Yamamoto, A. Mori, and D. Jinnai, Alterations of amino acids in rabbit brain caused by repetitive convulsive fits and comparsion of amino acid contents in epileptic and nonepileptic human brain, *J. Biochem.* (*Tokyo*) **49**:368–372 (1961).
7. N. Robinson and C. B. Williams, Amino acids in human brain, *Clin. Chim. Acta* **12**:311–317 (1965).
8. J. C. Dickinson and P. B. Hamilton, The free amino acids of human spinal fluid determined by ion-exchange chromatography, *J. Neurochem.* **13**:1179–1187 (1966).
9. J. L. Mangan and V. P. Whittaker, The distribution of free amino-acids in subcellular fractions of guinea-pig brain, *Biochem. J.* **98**: 128–137 (1966).
10. A. R. Dravid and W. A. Himwich *in Progress in Brain Research* (W. A. Himwich and H. E. Himwich, eds.), Vol. 9, pp. 170–173, American Elsevier, New York 1964.
11. M. Wender and Z. Waligóra, The content of amino acids in the proteins of the developing nervous system of the guinea pig—*I. J. Neurochem.* **7**:259–263 (1961).
12. M. Wender and Z. Waligóra, The content of amino acids in the proteins of the developing nervous system of the guinea pig—II, *J. Neurochem.* **9**: 115–118 (1962).
13. P. Mandel, Y. Godin, J. Mark, and C. Kayser, The distribution of free amino acids in the central nervous system of garden dormice during hibernation, *J. Neurochem.* **13**:533–536 (1966).

14. H. H. Tallan, S. Moore, and W. H. Stein, L-Cystathionine in human brain, *J. Biol. Chem.* **230**:707–716 (1958).
15. H. G. Knauff, G. Mayer and D. Marx, Über die aminosäurezusammensetzung der gehirnproteine, Hoppe-Seylers, *Z. Physiol. Chem.* **326**:78–88 (1961).
16. R. J. Williams, *Biochemical Individuality*, Wiley, New York (1956).
17. R. S. DeRopp and E. H. Snedeker, Sequential one-dimensional chromatography: analysis of free amino acids in the brain, *Anal. Biochem.* **1**:424–432 (1960).
18. S. Blackburn *in Methods of Biochemical Analysis* (D. Glick, ed.), Vol. 13, pp. 1–45, Interscience, New York (1965).
19. R. L. Munier and G. Sarrazin, Chromato-electrophorese des aminoacides en couche mince de poudre de cellulose, I, II, III, *Bull. Soc. Chim. France* 1363–1369 (1966).
20. S. Samuels and S. S. Ward, Aminoaciduria screening by thin-layer high voltage electrophoresis and chromatography on microplates, *J. Lab. Clin. Med.* **67**:669–677 (1966).
21. M. Brenner, A. Wiederwieser, and G. Pataki *in Thin-layer Chromatography* (E. Stahl, ed.), pp. 391–440, Springer-Verlag, New York (1965).
22. B. Weinstein *in Methods of Biochemical Analysis* (D. Glick, ed.), Vol. 14, pp. 203–323, Interscience, New York (1966).
23. W. R. Gray and B. S. Hartley, A fluorescent end-group reagent for proteins and peptides, *Biochem. J.* **89**:59P (1963).
24. N. Seiler and M. Wiechmann, Quantitative bestimmung von aminen und von aminosäuren als 1-dimethylan ino-naphthalin-5-sulfon-saureamide auf Dünnschichtchromatogrammen, *Z. Anal. Chem.* **220**:109–127 (1966).
25. K. Satake, T. Okuyama, M. Ohashi, and T. Shinoda, The spectrophotometric determination of amine, amino acid and peptide with 2,4,6,-trinitrobenzene-1-sulfonic acid, *J. Biochem.* (*Tokyo*) **47**:654–660 (1960).
26. L. C. Mokrasch, Use of 2,4,6-trinitrobenzenesulfonic acid for the coestimation of amines, amino acids and proteins in mixtures, *Anal. Biochem.* **18**:64–71 (1967).
27. S. Matsushita, N. Iwami, and Y. Nitta, Colorimetric estimation of amino acids and peptides with the Folin phenol reagent, *Anal. Biochem.* **16**:365–371 (1966).
28. N. A. Matheson, An improved method of separating amino acids as *N*-2,4-dinitrophenyl derivatives, *Biochem. J.* **88**:146-155 (1963).
29. N. A. Matheson, The isolation of amino acids from mixtures as their *N*-2,4-dinitrophenyl derivatives, *Biochem. J.* **94**:513–517 (1965).
30. N. A. Matheson and M. Sheltawy, Determination of amino acids as 2,4-dinitrophenyl derivatives, *Biochem. J.* **98**:297–302 (1966).
31. E. F. Gale *in Methods of Biochemical Anaylsis* (D. Glick, ed.), pp. 285–306, Interscience, New York (1957).
32. F. P. Chinard, Photometric estimation of proline and ornithine, *J. Biol. Chem.* **199**:91–95 (1952).
33. K. A. Piez, F. Irreverre, and H. L. Wolff, The separation and determination of cyclic imino acids, *J. Biol. Chem.* **223**:687–697 (1956).
34. D. Kruze and P. Wierzchowski, Determination of traces of proline in biological fluids, *Anal. Biochem.* **19**:226–233 (1967).
35. R. E. Neuman and M. A. Logan, The determination of hydroxyproline *J. Biol. Chem.* **184**:299–306 (1950).
36. D. J. Prockop and S. Udenfriend, A specific method for the analysis of hydroxyproline in tissues and urine, *Anal. Biochem.* **1**:228–239 (1960).
37. E. C. LeRoy, E. D. Harris, Jr., and A. Sjoerdsma, A modified procedure for radioactive hydroxyproline assay in urine and tissues after labeled proline administration, *Anal. Biochem.* **17**:377–382 (1966).

38. H. E. Firschein and J. P. Shill, The determination of total hydroxyproline in urine and bone extracts, *Anal. Biochem.* **14**:296–304 (1966).
39. C. L. Rosano, Enzymic method for determination of hydroxyproline, *Anal. Biochem.* **15**:341–345 (1966).
40. B. H. Nicolet and L. A. Shinn, The determination of serine by use of periodate, *J. Biol. Chem.* **139**:687–692 (1941).
41. W. R. Frisell and C. C. Mackenzie *in Methods of Biochemical Analysis* (D. Glick, ed.), Vol. 6, pp. 63–77, Interscience, New York (1958).
42. D. R. Keeney and J. M. Bremner, A simple steam distillation method of estimating β-hydroxy-α-amino acids, *Anal. Biochem.* **18**:274–285 (1967).
43. J. Hutzler, M. Odievre, and J. Dancis, Analysis for lysine, arginine, histidine and tyrosine in biological fluids, *Anal. Biochem.* **19**:529–541 (1967).
44. K. Beyermann and E. Knoll, Enzymatische mikrobestimmung basischer aminosäuren, *Z. Anal. Chem.* **219**:13–22 (1966).
45. S. Sakaguchi, Über eine neue farbenreaktion von protein und arginin, *J. Biochem. (Tokyo)* **5**:25–31 (1925).
46. K. R. Bhattacharya, J. Datta, and D. K. Roy, Application of Sakaguchi reaction to the quantitative estimation of arginine: a method involving paper chromatography, *Arch. Biochem. Biophys.* **84**:377–392 (1959).
47. J. F. Van Pilsum *in Methods of Biochemical Analysis* (D. Glick, ed.), Vol. 7, pp. 193–215, Interscience, New York (1959).
48. D. Hunninghake and S. Grisolia, A sensitive and convenient micromethod for estimation of urea, citrulline and carbamyl derivatives, *Anal. Biochem.* **16**:200–205 (1966).
49. J. D. Broome, A method for estimating free asparagine and glutamine in biological fluids as trinitrophenyl derivatives, *Nature* **211**:602–604 (1966).
50. A. R. Patton and E. M. Foreman, Glycine reagent for paper chromatograms, *Science* **109**:339 (1949).
51. R. L. Young and O. H. Lowry, Quantitative methods for measuring the histochemical distribution of alanine, glutamate and glutamine in brain, *J. Neurochem.* **13**:785–794 (1966).
52. R. T. Taylor and W. T. Jenkins, Leucine aminotransferase I. Colorimetric assays, *J. Biol. Chem.* **241**:4391-4395 (1966).
53. L. T. Graham, Jr., and M. H. Aprison, Fluorometric Determination of aspartate, glutamate and γ-aminobutyrate in nerve tissue using enzymic methods, *Anal. Biochem.* **15**:487–497 (1966).
54. T. Matsuzawa and N. Katunuma, Colorimetric assays for serum alanine transaminase and lactic dehydrogenase using diazonium zinc salt, *Anal. Biochem.* **17**:143–153 (1966).
55. W. L. Ryan and I. C. Wells, Homocitrulline and homoarginine from lysine, *Science* **144**:1122-1123 (1964).
56. A. D. Smith, M. Benziman, and H. J. Strecker, The formation of ornithine from proline in animal tissues, *Biochem. J.* **104**:557–564 (1967).
57. H. H. Tallan *in Amino Acid Pools* (J. T. Holden, ed.), pp. 471–485, Elsevier, Amsterdam (1962).
58. F. Irrevere, R. L. Evans, A. R. Hayden, and R. Silber, Occurrence of gamma-guanidinobutyric acid, *Nature* **180**:704–705 (1957).
59. E. Roberts, S. Frankel and P. J. Harman, Amino Acids in nervous tissue, *Proc. Soc. Exptl. Biol. Med.* **74**:383–387 (1950).
60. T. Takao, Isolation of L-α-aminoadipic acid from hog liver, *Biochim. Biophys. Acta* **117**:490–492 (1966).

61. M. M. Kini and J. H. Quastel, Carbohydrate-amino acid inter-relations in brain cortex *in vitro*, *Nature* **184**:252–256 (1959).
62. H. J. Strecker *in Metabolism of the Nervous System* (D. Richter ed.), pp. 459–473, Pergamon Press, New York (1957).
63. J. H. Quastel *in Ultrastructure and Metabolism in the Nervous System* (S. R. Korey, A. Pope, and E. Robins, eds.), pp. 57–66, Williams and Wilkins, Baltimore (1962).
64. J. H. Quastel *in Metabolic Inhibitors* (R. M. Hochster and J. H. Quastel, eds.), Vol. 2, pp. 517–538, Academic Press, New York (1963).
65. R. S. DeRopp and E. H. Snedeker, Effects of drugs on amino acid levels in brain: Excitants and depressants, *Proc. Soc. Exptl. Biol. Med.* **106**:696–700 (1961).
66. S. I. Singh and C. L. Malhotra, Amino acid content of monkey brain-IV, Effects of chlorpromazine on some amino acids of certain regions of monkey brain, *J. Neurochem.* **14**:135–140 (1967).
67. J. K. Tews and W. E. Stone *in Progress in Brain Research* (W. A. Himwich and J. P. Schade, eds.), Vol. 16, pp. 135–163, American Elsevier, New York (1965).
68. N. Okumura, S. Otsuki, and H. Nasu. The influence of insulin hypoglycemic coma, repeated electroshocks and chlorpromazine or $\beta$-phenylisopropylmethylamine administration on the free amino acids in the brain, *J. Biochem.* (*Tokyo*) **46**:247-252 (1959).
69. T. N. Pogorelova, Amino acid content in parts of the brain in hyperoxia, *Dokl. Akad. Nauk SSR* **167**:1421–2 (1966).
70. A. E. R. De Guglielmone and C. J. Gomez, Influence of neonatal hypothyroidism on amino acids in developing rat brain, *J. Neurochem*, **13**:1017–1025 (1966).
71. A. Van Harreveld and M. Kooiman, Amino acid release from the cerebral cortex during spreading depression and asphyxiation, *J. Neurochem.* **12**:431–439 (1965).
72. R. M. C. Dawson, Cerebral amino acids in fluoroacetate-poisoned, anaesthetised and hypoglycaemic rats, *Biochim. Biophys. Acta* **11**:548–552 (1963).
73. H. S. Bachelard, M. K. Gaitonde, and R. Vrba, The effect of psychotropic drugs on the utilization of glucose carbon atoms in the brain, heart and liver of the rat, *Biochem. Pharmacol.* **15**:1039–1043 (1966).
74. J. K. Tews and W. E. Stone, Effects of methionine sulfoximine on levels of free amino acids and related substances in brain, *Biochem. Pharmacol.* **13**:543–545 (1964).
75. M. S. Gaevskaya, *Biochemistry of the Brain During the Process of Dying and Resuscitation*, Monograph, Consultants Bureau Translation, New York (1963).
76. J. N. Williams, Jr., P. E. Schurr, and C. A. Elvehjem, The influence of chilling and exercise on free amino acids in cat tissues, *J. Biol. Chem.* **182**:55–59 (1950).
77. L. E. Holt, Jr., P. Gyorgi, E. L. Pratt, S. E. Snyderman, and W. M. Wallace, *Protein and Amino Acid Requirements in Early Life* New York University Press, New York (1960).
78. H. J. Almquist, *in Protein and Amino Acid Nutrition* (A. A. Albanese, ed.), pp. 349–380, Academic Press, New York (1959).
79. H. E. Clark, *in New Methods of Nutritional Biochemistry* (A. A. Albanese, ed.), Vol. 2, pp. 123–159, Academic Press, New York (1965).
80. A. E. Harper, *in Mammalian Protein Metabolism* (H. N. Munro and J. B. Allison, eds.), Vol. 2, pp. 87–134, Academic Press, New York (1964).
81. T. Addis, L. J. Poo, and W. Lew, The quantities of protein lost by the various organs and tissues of the body during a fast, *J. Biol. Chem.* **115**:111–116 (1936).
82. T. Addis, L. J. Poo, and W. Lew, The rate of protein formation in the organs and tissues of the body I. After casein refeeding, *J. Biol. Chem.* **116**:343–352 (1936).
83. T. Addis, D. D. Lee, W. Lew, and L. J. Poo, The protein content of the organs and tissues at different levels of protein consumption, *J. Nutr.* **19**:199–205 (1940).

84. H. Waelsch and A. Lajtha, Protein metabolism in the nervous system, *Physiol. Rev.* **41**:709–736 (1961).
85. A. Neuberger and F. F. Richards in *Mammalian Protein Metabolism* (H. N. Munro and J. B. Allison, eds.), Vol. I, pp. 243–296, Academic Press, New York (1964).
86. N. Barzoni, M. Cafiero, S. Di Bella, E. De Mori, and M. A. Grillo, Effect of a protein deficient diet upon some enzymatic activities of the brain, lung and kidney of rats, *Experientia* **8**:306–307 (1952).
87. W. W. Wainio, J. B. Allison, L. T. Kremzner, E. Bernstein, and M. Aronoff, Enzymes in protein depletion, III—Enzymes of brain, kidney, skeletal muscle and spleen, *J. Nutr.* **67**:197–204 (1959).
88. P. Lehr and J. Gayet, Response of the cerebral cortex of the rat to prolonged protein depletion I—Tissue weight, nitrogen, deoxyribonucleic acid and proteins, *J. Neurochem.* **10**:169–176 (1963).
89. W. C. Rose and S. H. Eppstein, The dietary indispensability of valine, *J. Biol. Chem.* **127**:677–684 (1939).
90. W. C. Rose, R. L. Wixom, H. B. Lockhart, and G. F. Lambert, The amino acid requirements of man. XV. The valine requirement; summary and final observations, *J. Biol. Chem.* **217**:987–995 (1955).
91. R. Rajalakshmi, K. R. Govindarajan, and C. V. Ramakrishnan, Effect of dietary protein content on visual discrimination, learning and brain biochemistry in the albino rat, *J. Neurochem.* **12**:261–271 (1965).
92. H. H. Berlet, *in Progress in Brain Research* (W. A. Himwich and J. P. Schade, eds.), Vol. 16, pp. 184–215, American Elsevier, New York (1965).
93. F. Viteri, M. Behar, and G. Arrozone, *in Mammalian Protein Metabolism* (H. N. Munro and J. B. Allison, eds.), Vol. II, pp. 523–568, Academic Press, New York (1964).
94. B. S. Platt, C. R. C. Heard, and R. J. C. Stewart, *in Mammalian Protein Metabolism* (H. N. Munro and J. B. Allison, eds.), Vol. II, pp. 445–521, Academic Press, New York (1964).
95. B. S. Platt, *in Chemical Pathology of the Nervous System* (J. Folch-Pi, ed.), pp. 114–118, Pergamon Press, New York (1961).
96. H. T. Thompson, P. E. Schurr, L. M. Henderson, and C. A. Elvehjem, The influence of fasting and nitrogen deprivation on the concentration of free amino acids in rat tissues, *J. Biol. Chem.* **182**:47–53 (1950).
97. A. E. Denton, J. N. Williams, Jr., and C. A. Elvehjem, The influence of methionine deficiency on amino acid metabolism in the rat, *J. Biol. Chem.* **186**:377–385 (1950).
98. P. Mandel and J. Mark, The influence of nitrogen deprivation on free amino acids in rat brain, *J. Neurochem.* **12**:987–992 (1965).
99. P. Lehr and J. Gayet, Response of the cerebral cortex of the rat to prolonged protein depletion II. Free aspartic, glutamic and $\gamma$-aminobutyric acids, *J. Neurochem.* **13**:805–810 (1966).
100. S. S. Kety, *in Ultrastructure and Metabolism of the Nervous System* (S. R. Korey, A. Pope, and E. Robins, eds.), pp. 311–324, Williams & Wilkins, Baltimore (1962).
101. M. J. Carver, Influence of phenylalanine administration on the free amino acids of brain and liver in the rat, *J. Neurochem.* **12**:45–50 (1965).
102. M. J. Carver, J. H. Copenhaver, and R. A. Serpan, Free amino acids in foetal rat brain, Influence of L-phenylalanine, *J. Neurochem.* **12**:857–861 (1965).
103. S. Roberts, Regulation of cerebral metabolism of amino acids–II. Influence of phenylalanine deficiency on free and protein-bound amino acids in rat cerebral cortex: Relationship to plasma levels, *J. Neurochem.* **10**:931–940 (1963).

104. S. Roberts and B. S. Morelos, Regulation of cerebral metabolism of amino acids-IV. Influence of amino acid levels on leucine uptake, utilization and incorporation into protein *in vivo*, *J. Neurochem.* **12**:373–387 (1965).
105. H. M. Sinclair, Vitamins and the nervous system, *Brit. Med. Bull.* **12**:18–23 (1956).
106. H. M. Sinclair, *in Chemical Pathology of the Nervous System* (J. Folch-Pi, ed.), pp. 98–113, Pergamon Press, New York (1961).
107. J. K. Tews and R. A. Lovell, The effect of a nutritional pyridoxine deficiency on free amino acids and related substances in mouse brain, *J. Neurochem.* **14**:1–7 (1967).
108. D. B. Hope, Cystathionine in brain and liver from pyridoxine-deficient rats, *J. Physiol.* **141**:31P–32P (1958).
109. A. A. Albanese, *in Protein and Amino Acid Nutrition* (A. A. Albanese, ed.), pp. 1–9, Academic Press, New York (1959).
110. Cold Spring Harbor Symposium of Quantitative Biology, *Cellular Regulatory Mechanisms*, Vol. 26, The Biological Laboratory, Cold Spring Harbor, Long Island, New York (1961).
111. W. C. Rose, The nutritive significance of the amino acids, *Physiol. Rev.* **18**:109–136 (1938).
112. M. E. Rafelson, Jr., R. J. Winzler, and H. E. Pearson, A virus effect on the uptake of $C^{14}$ from glucose *in vitro* by amino acids in mouse brain, *J. Biol. Chem.* **193**:205–217 (1951).
113. R. J. Winzler, K. Moldave, M. E. Rafelson, Jr., and H. E. Pearson, Conversion of glucose to amino acids by brain and liver of the new-born mouse, *J. Biol. Chem.* **199**:485–492 (1952).
114. K. Moldave, M. E. Rafelson, Jr., D. Lagerborg, H. E. Pearson, and R. J. Winzler, *In vitro* conversion of radioglucose to free and protein-bound amino acids by virus-infected mouse brain, *Arch. Biochem. Biophys.* **50**:383–391 (1954).
115. H. H. Sky-Peck, C. Rosenbloom, and R. J. Winzler, Incorporation of glucose into the protein-bound amino acids of one-day-old mouse brain *in vitro*, *J. Neurochem.* **13**:223–228 (1966).
116. L. B. Flexner, J. B. Flexner, and R. B. Roberts, Biochemical and physiological differentiation during morphogenesis-XXII, Observations on amino acid and protein synthesis in the cerebral cortex and liver of the newborn mouse, *J. Cellular Comp. Physiol.* **51**:385–403 (1958).
117. R. B. Roberts, J. B. Flexner, and L. B. Flexner, Biochemical and physiological differentiation during morphogenesis-XXIII, Further observations relating to the synthesis of amino acids and proteins by the cerebral cortex and liver of mouse, *J. Neurochem.* **4**:78–90 (1959).
118. J. Lindsay and H. S. Bachelard, Incorporation of $^{14}C$ from glucose into α-keto acids and amino acids in rat brain and liver *in vivo*, *Biochem. Pharmacol.* **15**:1045–1052 (1966).
119. R. Vrba, M. K. Gaitonde, and D. Richter, The conversion of glucose carbon into protein in the brain and other organs of the rat, *J. Neurochem.* **9**:465–475 (1962).
120. R. M. O'Neal and R. E. Koeppe, Precursors *in vivo* of glutamate, aspartate and their derivatives of rat brain, *J. Neurochem.* **13**:835–847 (1966).
121. E. V. Flock, G. M. Tyce, and C. A. Owens, Jr., Utilizations of [$U^{14}C$] glucose in brain after total hepatectomy in the rat, *J. Neurochem.* **13**:1389–1406 (1966).
122. M. K. Gaitonde, S. A. Marchi, and D. Richter, The utilization of glucose in the brain and other organs of the cat, *Proc. Roy. Soc.* (*London*) *Ser. B* **160**:124–136 (1964).
123. R. M. O'Neal, R. E. Koeppe, and E. I. Williams, Utilization *in vivo* of glucose and volatile fatty acids by sheep brain for the synthesis of acidic amino acids, *Biochem. J.* **101**:591–597 (1966).

124. S. S. Barkulis, A. Geiger, Y. Kawakita, and V. Aguilar, A study on the incorporation of $^{14}C$ derived from glucose into the free amino acids of the brain cortex, *J. Neurochem.* **5**:339–348 (1960).
125. H. Busch, Studies on the metabolism of pyruvate-2-$C^{14}$ in tumor-bearing rats, *Cancer Res.* **15**:365–374 (1955).
126. A. Beloff-Chain, R. Catanzaro, E. B. Chain, I. Masi, and F. Pocchiari, Fate of uniformly labeled $^{14}C$ glucose in brain slices, *Proc. Roy. Soc. (London) Ser. B* **144**: 22–28 (1955).
127. H. Busch, M. H. Goldberg, and D. C. Anderson, Substrate effects on metabolic patterns of pyruvate 2-$C^{14}$ in tissue slices, *Cancer Res.* **16**:175–181 (1956).
128. Y. Tsukada, Y. Nagata, S. Hirano, and G. Takagaki, Glucose metabolism and amino acid in brain slices, *J. Biochem. (Tokyo)* **45**:979–984 (1958).
129. A. Beloff-Chain, R. Catanzaro, E. B. Chain, L. Longinatti, I. Masi, and F. Pocchiari, The influence of glucose on acetate, alanine and pyruvate metabolism in rat cerebral cortical slices, *Proc. Roy. Soc. (London) Ser. B* **156**:168–171 (1962).
130. O. Gonda and J. H. Quastel, Transport and metabolism of acetate in rat brain cortex *in vitro*, *Biochem. J.* **100**:83–94 (1966).
131. Y. Nagata, Y. Yokoi, and Y. Tsukada, Studies on free amino acid metabolism in excised cervical ganglia from the rat, *J. Neurochem.* **13**:1421–1431 (1966).
132. O. Z. Sellinger, R. Catanzaro, E. B. Chain, and F. Pocchiari, The metabolism of glutamate and aspartate in rat cerebral cortical slices, *Proc. Roy. Soc. (London) Ser. B* **156**:148–162 (1962).
133. P. P. Cohen and G. L. Hekkus, Rate of transamination in normal tissues, *J. Biol. Chem.* **140**:711–724 (1941).
134. H. A. Krebs and D. Bellamy, The interconversion of glutamic acid and aspartic acid in respiring tissues, *Biochem. J.* **75**:523–529 (1960).
135. R. Balázs and R. J. Haslam, Exchange transamination and the metabolism of glutamate in brain, *Biochem. J.* **94**:131–141 (1965).
136. R. J. Haslam and H. A. Krebs, The metabolism of glutamate in homogenates and slices of brain cortex, *Biochem. J.* **88**:566–578 (1963).
137. R. W. Albers and W. B. Jakoby *in Inhibition in the Central Nervous System and γ-Aminobutyric Acid* (E. Roberts, ed.), pp 468–470, Pergamon Press, London (1960).
138. P. Borst and E. C. Slater, The oxidation of glutamate by rat-heart sarcosomes, *Biochim. Biophys. Acta.* **41**:170–171 (1960).
139. M. Rušcak and E. Macejová, Formation of L-α alanine and γ-aminobutyric acid in rat cortical slices in relation to the substrate and pH of the medium, *Physiol. Bohemslov.* **14**:266–275 (1965).
140. W. F. Bridger, The biosynthesis of serine in mouse brain extracts, *J. Biol. Chem.* **240**:4591–4597 (1965).
141. D. A. Walsh and H. J. Sallach, Comparative studies on the pathways for serine biosynthesis in animal tissues, *J. Biol. Chem.* **241**:4068–4076 (1966).
142. M. B. Sporn, W. Dingman, and A. De Falco, A method for studying metabolic pathways in the brain of the intact animal, the conversion of proline to other amino acids, *J. Neurochem.* **4**:141–147 (1959).
143. R. Balázs, Control of glutamate metabolism. The effect of pyruvate, *J. Neurochem.* **12**:63–76 (1965).
144. J. E. Cremer, Amino acid metabolism in rat brain studied with $^{14}C$-labeled glucose, *J. Neurochem.* **11**:165–185 (1964).
145. R. S. De Ropp and E. H. Snedeker, Effect of drugs on amino acid levels in the rat brain: Hypoglycemic agents, *J. Neurochem.* **7**:128–134 (1961).

146. H. S. Bachelard and J. R. Lindsay, Effects of neurotropic drugs on glucose metabolism *in vivo*, *Biochem. Pharmacol.* **15**:1053–1058 (1966).
147. B. Sacktor, J. E. Wilson, and C. G. Tiekert, Regulation of glycolysis in brain, *in situ*, during convulsions, *J. Biol. Chem.* **241**:5071–5075 (1966).
148. S. Berl, G. Takagaki, D. D. Clarke and H. Waelsch, Carbon dioxide fixation in the brain, *J. Biol. Chem.* **237**:2570–2573 (1962).
149. L. J. Côté, Sze-Chuh Cheng, and H. Waelsch, $CO_2$ fixation in the nervous system—II. Environment effects on $CO_2$ fixation in lobster nerve, *J. Neurochem.* **13**:281–287 (1966).
150. O. Z. Sellinger, D. L. Rucker, and F. De Balbian Verster, Cerebral lysosomes—I. A comparative study of lysosomal N-acetyl-β-D-glucosaminidase and mitochondrial aspartic transaminase of rat cerebral cortex, *J. Neurochem.* **11**:271–280 (1964).
151. L. Salganicoff and E. De Robertis, Subcellular distribution of the enzymes of the glutamic acid, glutamine and γ-aminobutyric acid cycles in rat brain, *J. Neurochem.* **12**:287–309 (1965).
152. O. Z. Sellinger and D. L. Rucker, Latency and solubilization of the mitochondrial aspartate transaminase of rat cerebral cortex, *Biochim. Biophys. Acta* **67**:504–507 (1963).
153. H. Wada and Y. Morino, *in Vitamins and Hormones* (R. S. Harris, I. G. Wool, and J. A. Loraine, eds.), Vol. 22, pp. 411–444, Academic Press, New York (1964).
154. O. Bodansky, M. K. Schwartz, and J. S. Nisselbaum, *in Advances in Enzyme Regulation* (G. Weber, ed.), Vol. 4, pp. 229–315, Pergamon Press, New York (1965).
155. A. Meister, *in The Enzymes* (P. D. Boyer, H. Lardy, and K. Myrbäck, eds.), Vol. 6, pp. 193–217, Academic Press, New York (1962).
156. S. F. Velick and J. Vavra, *in The Enzymes* (P. D. Boyer, H. Lardy, and K. Myrbäck, eds.), Vol, 6, pp. 219–247, Academic Press, New York (1962).
157. C. E. Frohman, J. M. Orten, and A. H. Smith, Chromatographic determination of the acids of the citric acid cycle in tissues, *J. Biol. Chem.* **193**:277–283 (1951).
158. J. M. Ravel, S. J. Norton, J. S. Humphreys, and W. Shive, Asparagine biosynthesis in Lactobacillus arabinosus and its control by asparagine through enzyme inhibition and repression, *J. Biol. Chem.* **237**:2845–2849 (1962).
159. L. Levintow, Evidence that glutamine is a precursor of asparagine in a human cell in tissue culture, *Science* **126**:611–612 (1957).
160. S. M. Arfin, Asparagine synthesis in the chick embryo liver, *Biochim. Biophys. Acta* **136**:233–244 (1967).
161. R. W. Swick, P. L. Barnstein, and J. L. Stange, The metabolism of mitochondrial proteins I. Distribution and characterization of the isozymes of alanine aminotransferase in rat liver, *J. Biol. Chem.* **240**:3334–3340 (1965).
162. N. Katunuma, K. Mikumo, M. Matsuda, and M. Okada, Differences between the transaminases in mitochondria and soluble fraction I. Glutamic-pyruvic transaminase, *J. Vitaminol.* **8**:68–73 (1962).
163. H. L. Segal, D. S. Beattie, and S. Hopper, Purification and properties of liver glutamic-alanine transaminase from normal and corticoid-treated rats, *J. Biol. Chem.* **237**:1914–1920 (1962).
164. H. J. Vogel and B. D. Davis, Glutamic-γ-semialdehyde and $\Delta^1$-pyrroline-5-carboxylic acid, intermediates in the biosynthesis of proline, *J. Am. Chem. Soc.* **74**:109–112 (1952).
165. H. J. Strecker, The interconversion of glutamic acid and proline I. The formation of $\Delta^1$-pyrroline-5-carboxylic acid from glutamic acid in *Escherichia coli*, *J. Biol. Chem.* **225**:825–834 (1957).

166. M. R. Stetten, *in Amino Acid Metabolism* (W. D. McElroy and H. B. Glass, eds.), pp. 277–290, The Johns Hopkins Press, Baltimore (1955).
167. H. J. Vogel, *in Amino Acid Metabolism* (W. D. McElroy and H. B. Glass, eds.), pp. 335–346. The Johns Hopkins Press, Baltimore (1955).
168. J. Peisach and H. J. Strecker, The interconversion of glutamic acid and proline V. The reduction of $\Delta^1$-pyrroline-5-carboxylic acid to proline, *J. Biol. Chem.* **237**: 2255–2260 (1962).
169. H. J. Strecker and E. E. Eliasson, Ornithine-δ-Transaminase activity during the growth cycle of Chang's liver cells, *J. Biol. Chem.* **241**:5750–5756 (1966).
170. S. Ratner, H. Morell, and E. Corvalho, Enzymes of arginine metabolism in brain, *Arch. Biochem. Biophys.* **91**:280–289 (1960).
171. H. C. Buniatian and M. A. Davtian, Urea synthesis in brain, *J. Neurochem.* **13**: 743–753 (1966).
172. C. Peraino and H. C. Pitot, Ornithine-δ-transaminase in the rat, I. Assay and some general properties, *Biochim. Biophys. Acta* **73**:222–231 (1963).
173. J. D. Ogle, R. B. Arlinghaus, and M. A. Logan, 3-Hydroxyproline, a new amino acid of collagen, *J. Biol. Chem.* **237**:3667–3673 (1962).
174. M. L. Efron, E. M. Bixby, and C. V. Pryles, Hydroxyprolinemia, *New Engl. J. Med.* **272**:1299–1308 (1965).
175. L. N. Lukens, The size of the polypeptide precursor of collagen hydroxyproline, *Proc. Natl. Acad. Sci. U.S.* **55**:1235–1243 (1966).
176. K. I. Kivirikko and D. J. Prockop, Purification and partial characterization of the enzyme for the hydroxylation of proline in protocollagen, *Arch. Biochem. Biophys.* **118**:611–618 (1967).
177. K. Kuratomi and K. Fukunaga, The metabolism of γ-hydroxyglutamate in rat liver I. Enzymic synthesis of γ-hydroxy-α-ketoglutarate from pyruvate and glyoxalate, *Biochim. Biophys. Acta* **78**:617–628 (1963).
178. K. Kuratomi, K. Fukunaga, and Y. Kobayashi, The metabolism of γ-hydroxyglutamate in rat liver II. A transaminase concerned in γ-hydroxyglutamate metabolism, *Biochim. Biophys. Acta* **78**:629–639 (1963).
179. A. Goldstone and E. Adams, Further metabolic reactions of γ-hydroxyglutamate. Amidation to γ-hydroxyglutamine; possible reduction to hydroxyproline, *Biochem. Biophys. Res. Commun.* **16**:71–76 (1964).
180. H. J. Strecker, The interconversion of glutamic acid and proline II. The preparation and properties of $\Delta^1$-pyrroline-5-carboxylic acid, *J. Biol. Chem.* **235**:2045–2050 (1960).
181. A. B. Johnson and H. J. Strecker, The interconversion of glutamic acid and proline IV. The oxidation of proline by rat liver mitochondria, *J. Biol. Chem.* **237**:1876–1882 (1962).
182. M. L. Efron, Familial hyperprolinemia, *New Engl. J. Med.* **272**: 1243–1254 (1965).
183. H. Tabor and C. W. Tabor, Spermidine, spermine and related amines, *Pharmacol. Rev.* **16**:245–300 (1964).
184. H. A. Krebs, Metabolism of amino acids—III. Deamination of amino acids, *Biochem. J.* **29**:1620–1644 (1935).
185. F. Bernheim and M. L. C. Bernheim, The purification of the enzymes which oxidize certain amino acids, *J. Biol. Chem.* **109**:131–140 (1935).
186. A. E. Braunstein and R. M. Asarkh, The mode of deamination of L-amino acids in surviving tissues, *J. Biol. Chem.* **157**:421–422 (1945).
187. H. A. Krebs, *in The Enzymes* (J. B. Sumner and K. Myrbäck, eds.), Vol. 2, pp. 499–535, Academic Press, New York (1951).
188. A. Meister and D. Wellner, *in The Enzymes* (P. D. Boyer, H. Lardy, and K. Myrbäck eds.), Vol. 7, pp. 609–648, Academic Press, New York (1963).

189. P. Boulanger and R. Osteux, Action de la L-aminoacide-déshydrogénase du foie de dindon (Meleagris gallopavo L.) sur les acides amines basiques, *Biochim. Biophys. Acta* **21**:552–561 (1956).
190. M. Nakano and T. S. Danowski, Crystalline mammalian L-amino acid oxidase from rat kidney mitochondria, *J. Biol. Chem.* **241**:2075–2083 (1966).
191. A. E. Braunstein, *in Advances in Enzymology* (F. F. Nord, ed.), Vol. 19, pp. 335–389, Interscience, New York (1957).
192. K. H. Bässler and C. H. Hammar, Aminosäurestoffwechsel in Zellfraktionen der Rattenleber Transaminieriengen und Oxydation von L-Aminosäuren, *Biochem. Z.* **330**:550–564 (1968).
193. F. J. R. Hird and D. J. Morton, The oxidation of L-amino acids by mitochondria from rat liver, *Biochim. Biophys. Acta* **38**:222–229 (1960).
194. H. Weil-Malherbe, Studies on brain metabolism I. The metabolism of glutamic acid in brain, *Biochem. J.* **30**:665–676 (1936).
195. S. Edlbacher and O. Wiss, Zur Kenntnis des Abbaues der Aminosäuren im tiereschen Organismus, 3. Über den oxydativen Abbau der Aminosäuren im Gehirn, *Helv. Chim. Acta* **27**:1060–1073 (1944).
196. S. Edlbacher and O. Wiss, Zur Kenntnis des Abbaues der Aminosäuren im tierischen Organismus, 4. Über den oxydativen Abbau der Aminosäuren im Gehirn, *Helv. Chim. Acta* **27**:1824–1831 (1944).
197. F. Friedberg, On the dissimilation of DL-alanine-1-$C^{14}$ by rat brain homogenates, *Biochim. Biophys. Acta* **11**:308–309 (1953).
198. L. C. Leeper, V. J. Tulane, and F. Friedberg, Metabolism of glycine-1-$C^{14}$ and glycine-2-$C^{14}$ in rat brain homogenates, *J. Biol. Chem.* **203**:513–517 (1953).
199. B. Shepartz, Oxidation of L-amino acids in homogenates of immature brain, *Biochim. Biophys. Acta* **53**:602–603 (1961).
200. G. G. Shamkulashvili, Oxidative metabolism of aspartic acid in brain slices, *Soobshch. Akad. Nauk. Gruz. SSR* **42**:105–110 (1966).
201. J. N. Potanos, A. D. Friedman, and S. Graff, Some aspects of tricarboxylic acid metabolism in the central nervous system, *Neurology* **10**:213–216 (1960).
202. G. Simon, J. B. Drori, and M. M. Cohen, Mechanism of conversion of aspartate into glutamate in cerebral-cortex slices, *Biochem. J.* **102**:153–162 (1967).
203. K. F. Swaimann and J. M. Milstein, Oxidative decarboxylation of aspartate, alanine and glycine in developing rabbit brain, *Biochim. Biophys. Acta* **93**:64–70 (1964).
204. F. Salvatore, V. Zappia, and C. Costa, Comparative biochemistry of deamination of L-amino acids in elasmobranch and teleost fish, *Comp. Biochem. Physiol.* **16**:303–309 (1965).
205. M. Hori, D-Amino acid oxidase in the brain of the dog, *Seishin Shinkeigaku Zasshi* **67**:548–553 (1965).
206. A. H. Neims, W. D. Zieverink, and J. D. Smilack, Distribution of D-amino acid oxidase in bovine and human nervous tissues, *J. Neurochem.* **13**:163–168 (1966).
207. D. B. Goldstein, D-Amino acid oxidase in brain: Distribution in several species and inhibition by pentobarbitone, *J. Neurochem.* **13**:1011–1016 (1966).
208. S. Ratner, *in The Enzymes* (P. Boyer, H. Lardy, and K. Myrbäck, eds.), Vol. 6, pp. 495–513, Academic Press, New York (1962).
209. P. A. Komitiani, Amino acid metabolism in homogenates of muscular and nervous tissue in connection with adenylic acid reamination, *Biokhimiya* **24**:729–737 (1959).
210. Y. P. Lee, *in The Enzymes* (P. Boyer, H. Lardy, and K. Myrbäck, eds.), Vol. 4, pp. 279–283, Academic Press, New York (1962).

211. M. K. Malysheva, Purification and study of some properties of the adenylic acid deaminase of brain, *Ukr. Biokhim. Zh.* **37**:370–378 (1965).
212. H. McIlwain, *Biochemistry and the Nervous System*, Little Brown, Boston (1964).
213. W. Wells, D. Gaines, and H. Koenig, Studies of pyrimidine nucleotide metabolism in the central nervous system-I. Metabolic effects and metabolism of 6-azauridine, *J. Neurochem.* **10**:709–723 (1963).
214. J. Abelskov, Succinyl adenosine, a new substance in the human cerebrospinal fluid, *Biochim. Biophys. Acta* **32**:566 (1959).
215. J. M. Lowenstein and P. P. Cohen, Studies on the biosynthesis of carbamyl aspartic acid, *J. Biol. Chem.* **220**:57–70 (1956).
216. E. Valovicova, J. Rajcani, T. Tursky, and M. Brozman, Asparaginase in experimental allergic encephalomyelitis, *Z. Ges. Exptl. Med.* **140**:256–267 (1966).
217. L. I. Miloslavskaia, The effect of barbiturates on the activity of cerebral asparaginase and glutaminase, *Biokhimiya* **23**:347–350 (1958).
218. H. M. Suld and P. A. Herbut, Guinea pig serum and liver asparaginases, purification and antitumor activity, *J. Biol. Chem.* **240**:2234–2241 (1965).
219. A. E. Braunstein and H. T. Seng, The scope of amino donor specificity of glutamine transaminase and asparagine transaminase, *Biochim. Biophys. Acta* **44**:187–189 (1960).
220. D. Greenberg, *in Annual Reviews of Biochemistry* (E. E. Snell, J. M. Luck, P. D. Boyer, and G. MacKinney, eds.), Vol. 33, pp. 633–666, Annual Reviews, Palo Alto, California (1964).
221. K. Moldave, R. J. Winzler, and H. E. Pearson, The incorporation *in vitro* of $C^{14}$ into amino acids of control and virus-infected mouse brain, *J. Biol. Chem.* **200**: 357–365 (1953).
222. H. H. Sky-Peck, H. E. Pearson, and D. W. Visser, Incorporation of glucose-U-$C^{14}$, glucose-1-$C^{14}$ and glucose-6-$C^{14}$ *in vitro* into the protein-bound amino acids of one day-old mouse brain, *J. Biol. Chem.* **223**:1033–1041 (1956).
223. A. S. M. Selim and D. M. Greenberg, An enzyme that synthesizes cystathionine and deaminates L-serine, *J. Biol. Chem.* **234**:1474–1480 (1959).
224. A. Nagabhushanam and D. M. Greenberg, Isolation and properties of a homogeneous preparation of cystathionine synthetase-L-serine and L-threonine dehydratase, *J. Biol Chem.* **240**:3002-3008 (1965).
225. M. Suda, *in Advances in Enzyme Regulation* (G. Weber, ed.), Vol. 5, pp. 181–209, Pergamon Press, New York (1967).
226. F. W. Sayre and D. M. Greenberg, Purification and properties of serine and threonine dehydrases, *J. Biol. Chem.* **220**:787–799 (1956).
227. S. H. Mudd, J. D. Finkelstein, F. Irreverre, and L. Laster, Threonine dehydratase activity in humans lacking cystathionine synthetase, *Biochem. Biophys Res. Commun.* **19**:665–670 (1965).
228. F. Salvatore, V. Zappia, and R. Cortese, Studies on the deamination of L-amino acids in mammalian tissues, *Enzymologia* **31**:113–127 (1966).
229. J. S. Nishimura and D. M. Greenberg, Purification and properties of L-threonine dehydrase of sheep liver, *J. Biol. Chem.* **236**:2684–2691 (1961).
230. A. E. Braunstein and G. Ya. Vilenkina, Enzymic formation of glycine from serine, threonine and other hydroxy-amino acids in animal tissue, *Dokl. Akad. Nauk SSSR* **66**:243–246 (1949).
231. S. C. Lin and D. M. Greenberg, Enzymatic breakdown of threonine by threonine aldolase, *J. Gen. Physiol.* **38**:181–196 (1954).
232. T. N. Prutasova, Paths of enzymic breakdown and regulation of metabolism of threonine stereoisomers in rat liver, *Biokhimiya* **30**:836–843 (1965).

233. G. Urata and S. Granick, Biosynthesis of α-aminoketones and the metabolism of aminoacetone, *J. Biol. Chem.* **238**:811–820 (1963).
234. D. Hartshorne and D. M. Greenberg, Studies on liver threonine dehydrogenase, *Arch. Biochem. Biophys.* **105**:173–178 (1964).
235. M. L. Green and W. H. Elliot, The enzymic formation of aminoacetone from threonine and its further metabolism, *Biochem. J.* **92**:537–549 (1964).
236. H. J. Strecker, The interconversion of glutamic acid and proline III. $\Delta^1$-Pyrroline-5-carboxylic acid dehydrogenase, *J. Biol. Chem.* **235**:3218–3223 (1960).
237. E. Adams and A. Goldstone, Hydroxyproline metabolism II. Enzymatic preparation and properties of $\Delta^1$-pyrroline-3-hydroxy-5-carboxylic acid, *J. Biol. Chem.* **236**: 3492–3498 (1960).
238. J. V. Taggart and R. B. Krakaur, Studies on the cyclophorase system V. The oxidation of proline and hydroxyproline, *J. Biol. Chem.* **177**:641–653 (1949).
239. E. Adams and A. Goldstone, Hydroxyproline metabolism IV. Enzymatic synthesis of γ-hydroxyglutamate from $\Delta^1$-pyrroline-3-hydroxy-5-carboxylate, *J. Biol. Chem.* **235**:3504–3512 (1960).
240. L. P. Bouthillier and Y. Binette, Decarboxylation of γ-hydroxyglutamate to α-hydroxy-γ-aminobutyrate in rat brain, *Can. J. Biochem. Physiol.* **39**:1930–1932 (1961).
241. L. P. Bouthillier, J. J. Pushpathadam, and Y. Binette, Study of the metabolism of 2-hydroxy-4-aminobutyric acid, a product of γ-hydroxyglutamic acid decarboxylation, *Can. J. Biochem. Physiol.* **44**:171–177 (1966).
242. E. F. Dekker and U. Maitra, Conversion of γ-hydroxyglutamate to glyoxalate and alanine; Purification and properties of the enzyme system, *J. Biol. Chem.* **237**:2218–2227 (1962).
243. L. P. Bouthillier, Y. Binette, and G. Pouliot, Transformation de l'acide γ-hydroxyglutamique en alanine et en acide glyoxylique, *Can. J. Biochem. Physiol.* **39**:1595–1603 (1961).
244. E. Baldwin, *Dynamic Aspects of Biochemistry*, Cambridge University Press, Cambridge, England (1963).
245. J. B. Walker, *in Comparative Biochemistry of Arginine and Derivatives* (G. E. W. Wolstenholme and M. P. Cameron, eds.), pp. 43–55, Little Brown, Boston (1965).
246. N. Van Thoai, *in Comparative Biochemistry of Arginine and Derivatives* (G. E. W. Wolstenholme and M. P. Cameron, eds.), pp. 3–13, Little Brown, Boston (1965).
247. J. J. Pisano, D. Abraham, and S. Udenfriend, Biosynthesis and disposition of γ-guanidinobutyric acid in mammalian tissues, *Arch. Biochem. Biophys.* **100**:323–329 (1963).
248. A. Ichihara and E. Koyama, Transaminase of branched chain amino acids I. Branched chain amino acids-α-ketoglutarate transaminase, *J. Biochem.* (*Tokyo*) **59**:160–169 (1966).
249. R. T. Taylor and W. T. Jenkins, Leucine aminotransferase II. Purification and characterization, *J. Biol. Chem.* **241**:4396–4405 (1966).
250. A. Ichihara, H. Takahashi, K. Aki, and A. Shirai, Transaminase of branched chain amino acids II. Physiological change in enzyme activity in rat liver and kidney, *Biochem. Biophys. Res. Commun.* **26**:674–678 (1967).
251. K. F. Swaiman and J. M. Milstein, Oxidation of Leucine, isoleucine and related ketoacids in developing rabbit brain, *J. Neurochem.* **12**:981–986 (1965).
252. J. J. Kabara and G. T. Okita, Brain Cholesterol: Biosynthesis with selected precursors *in vivo*, *J. Neurochem.* **7**:298–304 (1961).

253. S. Roberts, K. Seto, and B. H. Hanking, Regulation of cerebral metabolism of amino acids I. Influence of phenylalanine deficiency on oxidative utilization *in vitro*, *J. Neurochem.* **9**:493–501 (1962).
254. H. Borsook, C. L. Deasy, A. J. Haagen-Smit, G. Keighley, and P. H. Lowy, The degradation of L-lysine in guinea pig liver homogenate: formation of a-aminoadipic acid, *J. Biol. Chem.* **176**:1383–1394, 1395–1400 (1948).
255. M. Rothstein, K. E. Cooksey, and D. M. Greenberg, Metabolic conversion of pipecolic acid to α-aminoadipic acid, *J. Biol. Chem.* **237**:2828–2830 (1962).
256. K. Higashino, K. Tsukada, and I. Lieberman, Saccharopine, a product of lysine breakdown in mammalian liver, *Biochem. Biophys. Res. Commun.* **20**:285–290 (1965).
257. H. P. Broquist and J. S. Trupin, *in Annual Reviews of Biochemistry* (J. M. Luck, P. D. Boyer, G. MacKinney, A Meister, and E. E. Snell, eds.), Vol. 35, pp. 231–274, Annual Reviews, Palo Alto, Calfornia (1965).
258. M. L. Efron, Aminoaciduria, *New Engl. J. Med.*, **272**: 1058–1065, 1107–1113 (1965).
259. C. R. Scriver, S. Pruschel, and E. Davies, Hyper β-alaninemia associated with β-aminoaciduria and γ-aminoaciduria, somnolence and seizures, *New Engl. J. Med.* **274**:635–643 (1966).
260 H. Ghadini, V. I. Binnington, and P. Pecora, Hyperlysinemia associated with retardation, *New Engl. J. Med.* **273**:723–729 (1965).
261. N. C. Woody, Hyperlysinemia, *A.M.A. Am. J. Dis. Child.* **108**:543–553 (1965).
262. W. Burgi, R. Richterich, and J. P. Columbo, L-Lysine dehydrogenase deficiency in a patient with congenital lysine intolerance, *Nature* **211**:854–855 (1966).
263. Y. Wada, Idiopathic hypervalinemia: Valine and α-keto acids in blood following an oral dose of valine, *Tokuhu J. Exptl. Med.* **87**:322–331 (1965).
264. H. A. Waisman, T. Gerritsen, D. F. Boggs, V. J. Polidora, and H. F. Harlow, Mental retardation in monkeys: II. Branched chain amino-aciduria and keto-aciduria, *A.M.A. Am. J. Dis. Child.* **104**:488–490 (1962).
265. W. Cochrane, Idiopathic infantile hypoglycemia and leucine sensitivity, *Metabolism* **9**:386–399 (1960).
266. S. S. Fajans, J. C. Floyd, Jr., R. F. Knopf, and J. W. Conn, *in Recent Progress in Hormone Research* (G. Pincus, ed.) Vol. 23, pp. 617–656, Academic Press, New York (1967).
267. G. G. Duncan, *Diseases of Metabolism*, W. B. Saunders, Philadelphia (1964).

# Chapter 7
# METABOLISM OF AROMATIC AMINO ACIDS

**Gordon Guroff and Walter Lovenberg**

*Laboratory of Biochemical Sciences*
*National Institute of Child Health and Human Development*
*and Experimental Therapeutics Branch National Heart Institute*
*National Institutes of Health*
*Bethesda, Maryland*

## I. UPTAKE OF THE AROMATIC AMINO ACIDS BY BRAIN

The central nervous system depends entirely upon the plasma amino acid pool for its supply of aromatic amino acids. These amino acids enter the brain substance readily from the blood. It is clear that this ready uptake is due to the functioning of transport mechanisms specific for amino acids. Related aromatic acids and amines are virtually excluded from entry into the brain.

The characteristics of these active transport mechanisms have been studied *in vivo* and *in vitro*. The uptake of tyrosine into rat brain, for example is rapid, concentrative, and stereospecific both in the whole animal[1,2] and in brain slices.[3] Active uptake of phenylalanine into the brain substance of rats[2,4,5] and dogs[6] and into the cerebrospinal fluid of dogs[7] can also be easily demonstrated. Tryptophan is taken up by brain *in vivo*[2,8,9] and 5-hydroxytryptophan uptake has been shown in intact animals and in brain slices.[10–13] Histidine is known to enter the brain from the plasma more readily than most other amino acids.[14] Mouse brain slices are able to accumulate histidine against a 10-fold concentration gradient.[15–19] Histidine uptake by rat brain slices is energy dependent, $Na^+$-requiring, and inhibited by ammonium chloride.[20] In general, then, aromatic amino acids are accumulated in brain by active transport mechanisms which are stereospecific, energy-requiring, and concentrative.

It seems clear that the rate of amino acid transport into brain plays a crucial role in regulating the levels of free amino acids in the tissue.[21,22] Experiments have been presented which suggest that the uptake of aromatic amino acids into brain is slower than uptake into other tissues.[2,9] Competition studies show that many of the aromatic amino acids share the same transport mechanism[2,11–13,15,17,23–25] and that high levels of one can lower the uptake of others. It is considered likely that the high levels of aromatic amino acids found in blood in conditions such as phenylketonuria cause alterations in the rate of uptake and in the levels of other amino acids in brain due to competition for entry. It may be that these changes are related to the mental deficiency which is characteristic of phenylketonuria.

## II. PHENYLALANINE

### A. Major Metabolic Route in Animals

Phenylalanine is an essential amino acid for animals, and enzymes concerned with its biosynthesis have not been observed in animal tissue. Incubation of minces of the brains of newborn mice with radioactive glucose led to the incorporation of radioactivity into phenylalanine.[26] This result was not obtained with minces of adult brain or with the newborn mice *in vivo*. The investigators concluded that the observed incorporation was due to exchange reactions and not to *de novo* synthesis.

The major route of phenylalanine metabolism in animals is its hydroxylation to tyrosine. The enzyme responsible for this conversion, phenylalanine hydroxylase, is a mixed function oxidase requiring molecular oxygen and a tetrahydropteridine cofactor. The enzyme has been purified from liver and its characteristics described.[27,28] Although certain brain enzymes have weak phenylalanine hydroxylase activity (see below), no enzyme similar to the liver phenylalanine hydroxylase has been observed in brain.

Free phenylalanine is found in brain and plasma of animals. In rats the levels are approximately 0.09 mM in brain and 0.07 mM in plasma.[29,30]

### B. Alterations in Phenylketonuria

Phenylketonuria is a recessive genetic disease in man in which homozygotes lack the enzyme phenylalanine hydroxylase. This results in markedly elevated phenylalanine levels in plasma and tissues. With a block in the major metabolic route, the amounts of normally minor metabolic products in the urine of these patients is greatly increased. These products are probably formed by enzymes normally acting primarily on other substrates. Most either do not occur naturally or occur in amounts much lower than those seen in phenylketonuria. Whether any or all of them are formed exclusively or even partially in brain is not known.

Among the products of phenylalanine metabolism which have been observed are phenylpyruvic acid,[31] phenyllactic acid,[31] phenylacetic acid,[32] and phenethylamine.[33] These certainly could arise through the action of the transaminases, decarboxylases, etc., known to be in brain (see below). Other products known to be formed from phenylalanine in large amounts in phenylketonurics are *o*-hydroxyphenylacetic acid,[34] phenylacetylglutamine,[35] *N*-acetylphenylalanine,[36] and hippuric acid.[37] The derangements in phenylalanine metabolism occurring in phenylketonuria have been thoroughly reviewed,[38] as have the accompanying alterations in tryptophan metabolism.[39]

One of the most unusual metabolites found in phenylketonuric individuals involves the combination of phenethylamine and pyridoxal in a Schiff-base linkage. This metabolite, pyridoxylidene-β-phenethylamine, was initially observed in the urine of phenylketonurics.[40] Most recently, the metabolite has been completely characterized and shown to occur in the brains of rats given high phenylalanine diets.[41] Also, it has been shown that this metabolite inhibits brain pyridoxal kinase.[42] Whether, as has been suggested,[43] a metabolite of this nature plays any role in the mental retardation characteristic of phenylketonuria remains to be seen.

## III. TYROSINE

### A. Major Metabolic Route in Animals

The formation of tyrosine from phenylalanine by phenylalanine hydroxylase is the major source of this amino acid for animals, other than the dietary intake. Tyrosine levels in brain and plasma of rats are approximately 0.12 and 0.10 mM respectively.[29,30] The primary route of metabolism of tyrosine is via transamination to *p*-hydroxyphenylpyruvic acid. The subsequent degradation of this acid to homogentisic acid and then, by ring cleavage, to dicarboxylic acids occurs mainly in the liver.[44,45] The transamination of tyrosine has been shown in brain tissue but the subsequent degradative steps have not been observed.

### B. Metabolism of Tyrosine in Brain and the Biosynthesis of Norepinephrine

Transamination of tyrosine by preparations from dog brain was observed several years ago.[46] It was subsequently shown that brain preparations could carry on the transamination of several aromatic amino acids at varying rates.[47] More recent detailed reports have described the separation of brain transaminase activity into three separate fractions.[48] One of the fractions is specific for 3,4-dihydroxyphenylalanine. It has been purified 38-fold and uses α-ketoglutarate as a preferred acceptor for the amino group.[49] Another of the fractions is most active in the transamination of tyrosine and slightly less active with phenylalanine. This enzyme also shows a preference for α-ketoglutarate as acceptor.[50] This enzyme has been purified approximately 100-fold from rat brain but still contains some of the third fraction, an enzyme rather specific for the transamination of tryptophan and 5-hydroxytryptophan.

The conversion of tyrosine to catecholamines seems to be the major route of tyrosine metabolism in brain and other specialized tissues. Although it is a minor pathway in terms of the metabolism of tyrosine in the whole animal, it serves as the sole metabolic source of these important compounds. That the overall pathway occurs in brain has been generally known for several years from evidence with brain slice preparations[51–53] and brain perfusions.[54,55]

The first step in this pathway, the hydroxylation of tyrosine to 3,4-dihydroxyphenylalanine (DOPA), is catalyzed by the enzyme tyrosine hydroxylase. This enzyme was first obtained in cell-free form in 1964[56] and has since been obtained from brain, heart, and adrenal. The enzyme requires molecular oxygen, ferrous iron, and a reduced pteridine[57] and can be shown to incorporate molecular oxygen into the meta position of tyrosine.[58] The enzyme is clearly different from tyrosinase, the widely distributed, rather nonspecific phenol-oxidizing enzyme.

The enzyme is inhibited by various tyrosine analogs, the most potent being 3-iodotyrosine and α-methyltyrosine.[59] These inhibitors have been shown to be competitive with substrate. Phenylalanine is also competitive with the substrate tyrosine and is itself hydroxylated in the para-position at a slow rate.[60,61] The tyrosine formed is readily converted to DOPA. It is of interest that the enzyme does not form *m*-tyrosine and does not hydroxylate *m*-tyrosine to DOPA.

Certain tryptophan derivatives also inhibit the enzyme but act by another mechanism.[62] Catechol compounds are also potent inhibitors of tyrosine hydroxylase.[59] They appear to compete with the pteridine cofactor for a binding site on the enzyme.

The inhibition by catecholamines, including norepinephrine, has led to the suggestion that the enzyme is subject to feedback control. Several lines of evidence indicate that the hydroxylation of tyrosine is the rate-limiting step in the overall pathway leading to the production of catecholamines[63] and that catecholamines in turn control the rate of their own synthesis.

The decarboxylation of 3,4-dihydroxyphenylalanine (DOPA) to 3,4 dihydroxyphenethylamine (DOPAmine) is carried out by an enzyme first described in 1939.[64] It is now clear the the enzyme is a typical amino acid decarboxylase requiring pyridoxal phosphate and having action against a wide variety of aromatic amino acid substrates. In brain, the enzyme decarboxylates DOPA most effectively but also acts on tryptophan, 5-hydroxytryptophan, and phenylalanine.[65] The subject of amino acid decarboxylases is reviewed in detail in Chapter 8 of Vol. IV of this series.

Dopamine-β-hydroxylase, the enzyme completing the formation of norepinephrine, has been demonstrated in brain.[66] The enzyme has not been further studied in this tissue due to technical difficulties in its assay and purification and due to the presence of potent natural inhibitors.[67] In brain, the enzyme appears to be highest in hypothalamus and caudate nucleus. In adrenal, from which the enzyme has been extensively purified,[68–71] the reaction requires molecular oxygen and ascorbic acid. The enzyme contains copper ions, and its action is markedly enhanced by fumaric acid. The enzyme is relatively nonspecific, catalyzing the side-chain hydroxylation of a number of phenylethylamine derivatives.[72]

## IV. TRYPTOPHAN

### A. Source of Brain Tryptophan

Tryptophan is an essential amino acid in animals and is thought to be derived entirely from the diet. Tryptophan is not believed to be synthesized in animals. Free tryptophan is present in the plasma of animals. The concentrations in man[73] and rat[2] range from 0.05 to 0.10 μmole/ml. The levels in brain are similar to those in plasma.

### B. Metabolism of Tryptophan in Brain

Little is known about the gross metabolism of tryptophan in brain. The amino acid is presumably incorporated into the proteins which are synthesized in brain. The kynurenine pathway is the major degradative route of this amino acid in animals, but earlier reports[74] indicated that tryptophan pyrrolase, the initial enzyme of this pathway, is confined to liver. This enzyme has not been observed in brain, although recently it has been reported in the eye,[75] the intestine,[76] and the kidney.[77] It would appear that further investigation of nervous tissue for this enzyme is justified in view of the observations of Benasi *et al.*,[78] which suggest that schizophrenic patients excrete a greater amount of kynurenine pathway metabolites after a loading dose of tryptophan than do normals. Although these authors concluded that the patients could not utilize the tryptophan metabolites as well as normals, the alternate possibility exists that they have enhanced levels of tryptophan pyrrolase. Also of interest was the recent observation of kynurenine formylase in embryonic chick brain,[79] suggesting that the product of tryptophan pyrrolase can be further metabolized by brain.

Another route of tryptophan degradation is via transamination. It is known that rat brain extracts can utilize both tryptophan and 5-hydroxytryptophan as amino donors in transaminase reactions.[47] The aromatic amino acid transaminases have been purified from rat brain and partially resolved.[48,50] One of these transaminases exhibits a substrate preference for tryptophan and for oxaloacetic acid as the amino group acceptor. It is not known, however, whether transaminases play a significant role in the metabolism of tryptophan in the brain.

Tryptophan can also be decarboxylated in the brain. While the level of tryptamine in brain is ordinarily below detection, administration of tryptophan and a monoamine oxidase inhibitor results in the appearance of tryptamine in the brain of guinea pigs, rabbits, and dogs.[80] It is also known that the aromatic L-amino acid decarboxylase from dog brain utilizes tryptophan as a substrate.[65] Although this presumably represents a minor pathway, it is possible that the tryptamine formed by this reaction is physiologically important. Oates and Sjoerdsma[81] have reported that the administration of tryptophan to individuals receiving monoamine oxidase inhibitors results in unusual neurological symptoms which may reflect tryptamine formation in the central nervous system.

The hydroxyindole pathway which normally constitutes about 1% of the total tryptophan metabolism in man[82] is the most widely studied aspect of tryptophan metabolism in brain and nervous tissue. Interest in this area has resulted from the probable role of serotonin as a possible neurohumoral agent in certain neurons and as a precursor for melatonin in the pineal. The initial and rate-limiting reaction in the hydroxyindole pathway is the hydroxylation of tryptophan. Because the hydroxylation of tryptophan could not be detected in brain tissue in early studies, it was suggested[83] that tryptophan is hydroxylated elsewhere in the body and transported to the brain as either serotonin or 5-HTP. However, the use of intracerebral infusion techniques[84-86] permitted the clear demonstration that tryptophan could be hydroxylated in the brain. At about this time Grahame-Smith reported the conversion of tryptophan-$^{14}C$ to 5-hydroxytryptophan-$^{14}C$ by brain homogenates.[87,88] These findings were confirmed in several other laboratories,[89-91] but in all these reports the rate of tryptophan hydroxylation by brain tissue appeared to be extremely low.

Work reported by Lovenberg *et al.*[92,93] and by Hosada and Glick[94] showed that malignant mouse mast cell tumors contained a soluble tryptophan hydroxylase that exhibited the properties of a typical aromatic ring hydroxylase and was stimulated by 2-mercaptoethanol. Using a sensitive radioassay technique and adding 2-mercaptoethanol and a reduced pteridine to the extracts, it has now been found that brainstem and pineal gland of rats and rabbits have an easily measured tryptophan hydroxylase activity.[95] The rate of hyroxylation observed under these conditions is sufficient to account for the amount of serotonin present in brain.

Attempts to study tryptophan hydroxylase have resulted in conflicting reports of its characteristics and subcellular distribution. It is generally agreed that when the enzyme is assayed in the absence of 2-mercaptoethanol and reduced pteridine the activity is associated with a particulate fraction of the cell. Grahame-Smith[96] has recently found that about 45% of the activity is in the particulate fraction. Upon further fractionation this activity was nearly equally divided between the myelin and the synaptosome fractions. The enzyme activity appears to be largely extracted into the soluble fraction of the homogenates, however, when assayed under optimal conditions.[97] Furthermore, the activity remaining with the particle appears to be easily solubilized when these particles are suspended or dialyzed in a hypo-osmotic medium.[96,98]

Trytophan hydroxylases from brain and pineal have been purified slightly by ammonium sulfate fractionation and a preliminary characterization completed.[98] The enzyme appears to be a typical aromatic ring hydroxylase requiring both molecular oxygen and a reduced pteridine. Tryptophan hydroxylase is inhibited by iron chelators and is stimulated by the inclusion of iron in the reaction mixtures,[98] as are the hydroxylases acting on phenylalanine[27] and tyrosine.[57]

Of interest is the finding that enzyme preparations from beef pineal exhibit approximately equal amounts of tryptophan and phenylalanine hydroxylase activities.[99] The brain preparations, on the contrary, have almost no activity toward phenylalanine (tryptophan/phenylalanine activity $> 100$). It is not known at this stage whether the pineal contains two hydroxylases or if the substrate specificity of the tryptophan hydroxylases from the two sources is different.

Evidence that brain tryptophan hydroxylase is associated with the normal serotonin of brain tissue is accumulating. The work of Grahame-Smith[96] shows that the regional distribution of tryptophan hydroxylase activity in brain coincides with that of 5-hydroxytryptophan decarboxylase activity and with serotonin content. Furthermore, administration of *p*-chlorophenylalanine to rats appears to result in a nearly complete inactivation of brain tryptophan hydroxylase and a simultaneous loss of brain serotonin.[100] The tryptophan hydroxylase activity of rat brain slowly returns to normal over the course of two weeks following a single injection of *p*-chlorophenylalanine. The return of the serotonin level to normal occurs at the same rate as the return of the enzyme activity,[100] thus providing good evidence for the postulate that tryptophan hydroxylase activity controls the production of serotonin in the brain *in vivo*. The mechanism of *p*-chlorophenylalanine inhibition has not yet been established.

The finding that catecholamines, particularly norepinephrine, are strong inhibitors of brainstem and pineal tryptophan hydroxylase *in vitro* suggests that catecholamines may exert some physiologic control over serotonin synthesis in the whole animal.[98]

The second step in the hydroxyindole pathway, the decarboxylation of 5-hydroxytryptophan, is catalyzed by aromatic L-amino acid decarboxylase. This enzyme is present in nervous tissue and has been partially purified from dog brainstem.[65] The enzyme in brain is similar to the decarboxylase derived from guinea pig kidney. The activity of this enzyme in brain is considerably higher than that of tryptophan hydroxylase and it exerts little influence on the overall rate of serotonin synthesis. The enzyme, however, has the same general anatomical distribution in the brain as does serotonin.[101] The metabolism of serotonin via oxidative deamination is discussed in Chapter II of Vol. IV of this series.

In the pineal gland serotonin also serves as a precursor for the pineal hormone melatonin. Serotonin is *N*-acetylated in the gland by an enzyme requiring acetyl coenzyme A as the acetyl donor.[102,103] *N*-acetyl serotonin is next *O*-methylated by hydroxyindole-*O*-methyl tranferase of pineal tissue using *S*-adenosyl methionine as the methyl donor.[104] The latter enzyme catalyzes the rate-limiting step in melatonin synthesis, and it exhibits a diurnal variation of activity in response to light and darkness.[105] This variation in enzyme activity explains the observed diurnal variation in the melatonin content of the pineal.

## V. HISTIDINE

### A. Source of Brain Histidine

Although histidine is considered an essential amino acid in most animals it has been shown that brain tissue from newborn mice has the ability to incorporate $^{14}C$ from glucose into histidine both *in vivo* and *in vitro*.[26] It is generally assumed, however, that histidine is not synthesized in the central nervous system, but is supplied from the plasma.

### B. Metabolism of Histidine in Brain

Histidine in brain is utilized for protein synthesis and some specialized pathways. The major degradative routes, the urocanic acid pathway and the imidazole pyruvic acid pathway,[47] do not appear to exist in brain. Indeed, Neame[16] chose histidine for amino acid transport studies because it was not appreciably metabolized by brain slices. Oxidative deamination of D-histidine but not of L-histidine has been observed in rat brain extracts.[106]

The decarboxylation of histidine has been investigated in brain and nerve because of the possible role of histamine as a neurohumoral agent. Brain and spinal cord contain relatively low levels of histamine (0.1 to 1.0 $\mu g/g$). Many nerves, however, contain considerably more (2 to 30 $\mu g/g$) and postganglionic sympathetic fibers contain up to 60 $\mu g/g$.[107] While it is still not clear whether histamine can or does serve as a neurotransmitter,[108] experiments in von Euler's laboratory[107] suggest that stimulation of the splenic nerve results in the liberation of histamine from the cut end.

Histamine is formed in tissues by a histidine decarboxylase which appears to utilize pyridoxal phosphate as its cofactor.[109] Mammalian tissues are known to have at least two types of histidine decarboxylase. These enzymes exhibit different *p*H optima and substrate affinity and have different activators.[110] It has been shown that one of these histidine decarboxylating activities is due to the nonspecificity of aromatic L-amino acid decarboxylase,[65,111] while the other appears to be a specific histidine decarboxylase. The specific enzyme is typified by the activity found in mast cells and rat stomach. It is not clear which enzyme is responsible for histamine synthesis in nerve tissue. Rat brain histidine decarboxylase has a *p*H optimum similar to that of the specific decarboxylase,[110] while in beef stellate ganglion the activity appears to be due to the nonspecific aromatic L-amino acid decarboxylase.[112] The distribution of histidine decarboxylase and aromatic L-amino acid decarboxylase appears to be very similar in brain and nerve tissue.[109] It is likely that, in fact both types of enzymes are present since mast cells are known to be present in brain[113] and since aromatic L-amino acid decarboxylase is known to be functional in the formation of serotonin and catecholamines in nerve tissue. Considerably more definitive work is needed in this area.

*N*-acetylhistidine was first identified in brain extracts of *Rana esculenta*[(114)] and later was found in the brain and eye of numerous fish, reptiles, and amphibia[(115)] but not in birds or mammals.[(115)] Concentrations were found to be as high as 900 μg/g in brain and up to 2700 μg/g in the lens. Spinal cord and peripheral nerve contain 4 to 20 μg/g. Enzyme systems for the synthesis[(116)] and degradation[(117)] of *N*-acetylhistidine have been studied in brain extracts of the killifish. Acetylcoenzyme A appears to function as the specific acetyl donor in the transacetylation reaction. This reaction has a *p*H optimum of about 7.7 and appears to require at least one of the ions present in Ringers solution.[(116)] The biological role of acetylhistidine is unknown, although Baslow has recently suggested that it may play a role in histidine transfer between the lens and the interocular fluid.[(118)] On the other hand, Hanson indicates that *N*-acetylhistidine may serve as an acetyl donor or acetyl reserve for acetylcholine synthesis in brain.[(119)]

Histidine also participates in the formation of several oligopeptides in brain. This work has been discussed fully elsewhere in this series (see the first volume, Chapter 4) and in a recent review.[(120)] Both homocarnosine (γ-aminobutyrylhistidine) and homoanserine (γ-aminobutyrylmethylhistidine) have been identified as natural constituents of beef brain.[(121,122)] An enzyme has been described which methylates either carnosine or homocarnosine[(123)] to form anserine or homoanserine, respectively. This enzyme is present in the brains and muscles of young rats and has been studied in the pectoral muscle of the chick. The enzyme, carnosine-*N*-methyl transferase, has been purified to some extent and is known to require *S*-adenosyl methionine as its methyl donor.

Both 1-methyl histidine and 3-methyl histidine have also been shown to occur in bovine brain.[(122)] It is not clear, however, whether these amino acids are formed directly by methylation or are degradative products of anserine and homoanserine. In this regard it is significant that L-histidine is not a substrate for purified imidazole-*N*-methyl transferase (histamine methylating enzyme) of brain.[(124)]

## VI. REFERENCES


1. M. A. Chirigos, P. Greengard, and S. Udenfriend, Uptake of tyrosine by rat brain *in vivo*, *J. Biol. Chem.* **235**:2075–2079 (1960).
2. G. Guroff and S. Udenfriend, Studies on aromatic amino acid uptake by rat brain *in vivo*, *J. Biol. Chem.* **237**:803–806 (1962).
3. G. Guroff, W. King, and S. Udenfriend, The uptake of tyrosine by rat brain *in vitro*, *J. Biol. Chem.* **236**:1773–1777 (1961).
4. A. Lajtha and J. Toth, Uptake and transport of amino acids by the brain, *J. Neurochem.* **8**:216–225 (1961).

5. A. Lajtha and J. Toth, The efflux of intracerebrally administered amino acids from the brain, *J. Neurochem.* **9**:199–212 (1962).
6. R. J. Schain, J. H. Copenhaver, and M. J. Carver, Entry of L [$^{14}$C] phenylalanine and α-amino [$^{14}$C] isobutyric acid into the cerebrospinal fluid and brain of dogs, *J. Neurochem.* **14**:195–201 (1967).
7. M. J. Carver, R. J. Schain, and J. H. Copenhaver, Entry of L-phenylalanine-$C^{14}$ into brain and cerebrospinal fluid, *Proc. Soc. Exptl. Biol. Med.* **122**:75–78 (1966).
8. G. W. Ashcroft, D. Eccleston, and T. B. B. Crawford, 5-Hydroxyindole metabolism in rat brain. A study of intermediate metabolism using the technique of tryptophan loading—I, *J. Neurochem.* **12**:483–492 (1965).
9. J. W. Daly, A. B. Mauger, and O. Yonemitsu, The synthesis and metabolism of 2,3-dihydro-L-tryptophan and 2,3-dihydro-5-hydroxy-DL-tryptophan, *Biochemistry* **6**:648–654 (1967).
10. S. Schanberg and N. J. Giarman, Uptake of 5-hydroxytryptophan by rat brain, *Biochim. Biophys. Acta* **41**:556–558 (1960).
11. C. M. McKean, S. M. Schanberg, and N. J. Giarman, A mechanism of the indole defect in experimental phenylketonuria, *Science* **137**:604–605 (1962).
12. S. M. Schanberg, A study of the transport of 5-hydroxytryptophan and 5-hydroxytryptamine (serotonin) into brain, *J. Pharmacol. Exptl. Therap.* **139**:191–200 (1963).
13. S. E. Smith, Uptake of 5-hydroxy [$^{14}$C] tryptophan by rat and dog brain slices, *Brit. J. Pharmacol.* **20**:178–189 (1963).
14. P. Handler and H. Kamin, The metabolism of parentally administered amino acids, *J. Biol. Chem.* **188**:193–205 (1951).
15. K. D. Neame, Phenylalanine as inhibitor of transport of amino acids in brain, *Nature* **192**:173–174 (1961).
16. K. D. Neame, Uptake of amino acids by mouse brain slices, *J. Neurochem.* **6**:358–366 (1961).
17. K. D. Neame, Effect of amino acids on uptake of L-histidine by rat brain slices, *J. Neurochem.* **11**:67–76 (1964).
18. K. D. Neame, Uptake of histidine, histamine and other imidazol derivatives by brain slices, *J. Neurochem.* **11**:655–662 (1964).
19. K. D. Neame, Effect of neutral alpha- and omega-amino acids and basic alpha-amino acids on uptake of L-histidine by intestinal mucosa, testis, spleen, and kidney *in vitro*: A comparison with effect in brain, *J. Physiol. (London)* **185**:627–645 (1966).
20. D. F. deAlmeida, E. B. Chain, and F. Pocchairi, Effect of ammonium and other ions on the glucose-dependent active transport of L-histidine in slices of rat-brain cortex, *Biochem. J.* **95**:793–796 (1965).
21. G. Levi and A. Lajtha, Cerebral amino acid transport *in vitro*—II, *J. Neurochem.* **12**:639–648 (1965).
22. G. Levi, J. Kandera, and A. Lajtha, Control of cerebral metabolite levels. I. Amino acid uptake and levels in various species, *Arch. Biochem. Biophys.* **119**:303–311 (1967).
23. D. Y. Hsia, K. Nishimura, and Y. Brenchley, Mechanism for the decrease of brain serotonin, *Nature* **200**:578 (1963).
24. D. Y. Hsia, P. Justice, J. L. Berman, and Y. Brenchly, Brain serotonin in experimental tyrosinosis, *Nature* **202**:495–496 (1964).
25. R. J. Schain, J. H. Copenhaver, and M. J. Carver, Inhibition by phenylalanine of the entry of 5-hydroxy-tryptophan-1-$C^{14}$ into cerebrospinal fluid, *Proc. Soc. Exptl. Biol. Med.* **118**:184–186 (1965).
26. R. J. Winzler, K. Moldave, M. E. Rafelson, and H. E. Pearson, Conversion of glucose to amino acids by brain and liver of the newborn mouse, *J. Biol. Chem.* **199**:485–492 (1952).

27. S. Kaufman, *in Oxygenases* (O. Hayaishi, ed.), pp. 129–180, Academic Press, New York (1962).
28. S. Kaufman, *in Methods in Enzymology* (S. P. Colowick and N. O. Kaplan, eds.), Vol. 5, pp. 809–816, Academic Press, New York (1962).
29. P. E. Schurr, H. T. Thompson, L. M. Henderson, and C. A. Elvehjem, A method for the determination of free amino acids in rat organs and tissues, *J. Biol. Chem.* **182**: 29–37 (1950).
30. P. E. Schurr, H. T. Thompson, L. M. Henderson, J. N. Williams Jr., and C. A. Elvehjem, The determination of free amino acids in rat tissues, *J. Biol. Chem.* **182**:39–45 (1950).
31. G. A. Jervis, Excretion of phenylalanine and derivatives in phenylpyruvic oligophrenia, *Proc. Soc. Exptl. Biol. Med.* **75**:83–89 (1950).
32. L. I. Woolf, Excretion of conjugated phenylacetic acid in phenylketonuria, *Biochem. J.* **49**:IX (1951).
33. J. A. Oates, P. Z. Nirenberg, J. B. Jepson, A. Sjoerdsma, and S. Udenfriend, Conversion of phenylalanine to phenethylamine in patients with phenylketonuria, *Proc. Soc. Exptl. Biol. Med.* **112**:1078–1081 (1963).
34. M. D. Armstrong, K. N. F. Shaw, and K. S. Robinson, Studies on phenylketonuria. II. The excretion of *o*-hydroxyphenylacetic acid in phenylketonuria, *J. Biol. Chem.* **213**: 797–804 (1955).
35. L. I. Woolf and D. G. Vulliamy, Phenylketonuria with a study of the effect upon it of glutamic acid, *Arch. Disease Childhood* **26**:487–494 (1951).
36. F. B. Goldstein, Studies on phenylketonuria. II. The excretion of *N*-acetyl-L-phenylalanine in phenylketonuria, *Biochim. Biophys. Acta* **71**:204–206 (1963).
37. H. D. Grümer, Formation of hippuric acid from phenylalanine labelled with carbon-14 in phenylketonuric subjects, *Nature* **189**:63–64 (1961).
38. S. P. Bessman, Some biochemical lessons to be learned from phenylketonuria, *J. Pediat.* **64**:828–838 (1964).
39. M. T. Yarbro and J. A. Anderson, L-Tryptophan metabolism in phenylketonuria, *Pediatrics* **68**:895–904 (1966).
40. Y. H. Loo and P. Ritman, New metabolites of phenylalanine, *Nature* **203**:1237–1239 (1964).
41. Y. H. Loo, Characterization of a new phenylalanine metabolite in phenylketonuria, *J. Neurochem.* **14**:813–819 (1967).
42. Y. H. Loo and V. P. Whittaker, Pyridoxal kinase in brain and its inhibition by pyridoxylidene-β-phenylethylamine, *J. Neurochem.* **14**:997–1011 (1967).
43. Y. H. Loo and P. Ritman, Phenylketonuria and vitamin $B_6$ function, *Nature* **213**: 914–916 (1967).
44. W. E. Knox, *in Methods in enzymology* (S. P. Colowick and N. O. Kaplan, eds.), Vol. 2, pp. 287–300, Academic Press, New York (1955).
45. A. Meister, *Biochemistry of the Amino Acids*, 2nd ed., Vol. 2, pp. 894–908, Academic Press, New York (1965).
46. Z. N. Canellakis and P. P. Cohen, Purification studies of tyrosine-α-ketoglutaric acid transaminase, *J. Biol. Chem.* **222**:53-62 (1956).
47. R. Haavaldsen, Transamination of aromatic amino acids in nervous tissue, *Nature* **196**:577–578 (1962).
48. F. Fonnum, R. Haavaldsen, O. Tangen, Transamination of aromatic amino acids in rat brain, *J. Neurochem.* **11**:109–118 (1964).
49. F. Fonnum and K. Larsen, Purification and propeties of dihydroxyphenylalanine transaminase from guinea pig brain, *J. Neurochem.* **12**:589–598 (1965).

50. O. Tangen, F. Fonnum, and R. Haavaldsen, Separation and purification of aromatic amino acid transaminases from rat brain, *Biochim. Biophys. Acta* **96**:82–90 (1965).
51. D. T. Masuoka, W. G. Clark, and H. F. Schott, Biosynthesis of catecholamines by rat brain tissue *in vitro*, *Rev. Can. Biol.* **20**:1–5 (1961).
52. D. T. Masuoka, H. F. Schott, and L. Petriello, Formation of catecholamines by various areas of cat brain, *J. Pharmacol. Exptl. Therap.* **139**: 73–76 (1963).
53. N. T. Iyer, P. L. McGeer, and E. G. McGreer, Conversion of tyrosine to catecholamines by rat brain slices, *Can. J. Biochem. Physiol.* **41**:1565–1570 (1963).
54. E. G. McGeer, G. M. Ling, and P. L. McGeer, Conversion of tyrosine to catecholamines by cat brain *in vivo*, *Biochem. Biophys. Res. Commun.* **13**:291–296 (1963).
55. P. L. McGeer and E. G. McGeer, Formation of adrenaline by brain tissue, *Biochem. Biophys. Res. Commun.* **17**:502–507 (1964).
56. T. Nagatsu, M. Levitt, and S. Udenfriend, Conversion of L-tyrosine to 3,4-dihyroxyphenylalanine by cell-free preparations of brain and sympathetically innervated tissue, *Biochem. Biophys. Res. Commun.* **14**:543–549 (1964).
57. T. Nagatsu, M. Levitt, and S. Udenfriend, Tyrosine Hydroxylase. The initial step in norepinephrine biosynthesis, *J. Biol. Chem.* **239**:2910–2917 (1964).
58. M. Levitt, J. Daly, G. Guroff, and S. Udenfriend, Isotope studies on the mechanism of action of adrenal tyrosine hydroxylase, *Arch. Biochem. Biophys.* **126**:593–598 (1968).
59. S. Udenfriend, P. Zaltzman-Nirenberg, and T. Nagatsu, Inhibitors of purified beef adrenal tyrosine hydroxylase, *Biochem. Pharmacol.* **14**:837–845 (1965).
60. D. Zhelyaskov, M. Levitt, and S. Udenfriend, Tryptophan derivatives as inhibitors of tyrosine hydroxylase *in vivo* and *in vitro*, *Mol. Pharm.* **4**:445–451 (1968).
61. M. Ikeda, M. Levitt, and S. Udenfriend, Hydroxylation of phenylalanine by purified preparations of adrenal on brain tyrosine hydroxylase, *Biochem. Biophys. Res. Commun.* **18**:482–488 (1965).
62. M. Ikeda, M. Levitt, and S. Udenfriend, Phenylalanine as substrate and inhibitor of tyrosine hydroxylase, *Arch. Biochem. Biophys.* **120**:420–427 (1967).
63. S. Udenfriend, Tyrosine hydroxylase, *Pharmacol. Rev.* **18**:43–51 (1966).
64. P. Holtz, Dopadecarboxylase, *Naturwissenschaften* **27**:724–725 (1939).
65. W. Lovenberg, H. Weissbach, and S. Udenfriend, Aromatic L-amino acid decarboxylase, *J. Biol. Chem.* **237**:89–93 (1962).
66. S. Udenfriend and C. R. Creveling, Localization of dopamine-β-oxidase in brain, *J. Neurochem.* **4**:350–352 (1959).
67. T. Nagatsu, H. Kuzuya, and H. Hidaka, Inhibition of dopamine β-hydroxylase by sulfhydryl compounds and the nature of the natural inhibitors, *Biochim. Biophys. Acta* **139**:319–327 (1967).
68. E. Y. Levin, B. Levenberg, and S. Kaufman, The enzymatic conversion of 3,4-dihydroxyphenylethylamine to norepinephrine, *J. Biol. Chem.* **235**:2080–2086 (1960).
69. S. Friedman and S. Kaufman, 3,4-Dihydroxyphenylethylamine β-hydroxylase: A copper protein, *J. Biol. Chem.* **240**:PC 552–554 (1965).
70. S. Friedman and S. Kaufman, 3,4-Dihydroxyphenylethylamine β-hydroxylase, *J. Biol. Chem.* **240**:4763–4773 (1965).
71. M. Goldstein, E. Lauber, and M. R. McKereghan, Studies on the purification and characterization of 3,4-dihydroxyphenylethamine β-hydroxylase, *J. Biol. Chem.* **240**: 2066–2072 (1965).
72. C. R. Creveling, J. W. Daly, B. Witkop, and S. Udenfriend, Substrates and inhibitors of dopamine β-oxidase, *Biochim. Biophys. Acta* **64**:125–134 (1962).
73. W. H. Stein and S. Moore, The free amino acids of human blood plasma, *J. Biol. Chem.* **211**:915–926 (1954).

74. W. E. Knox, *in Methods in Enzymology* (S. P. Colowick and N. O. Kaplan, eds.), Vol. 2, pp. 242–253, Academic Press, New York (1955).
75. W. Ciusa and G. Barbiroli, Metabolismo Del Triptofano Nella Lente Cristalline, *Boll. Soc. Ital. Biol. Sper.* **42**:995–998 (1966).
76. S. Yamamoto and O. Hayaishi, Tryptophan pyrrolase of rabbit intestine, *J. Biol. Chem.* **242**:5260–5266 (1967).
77. I. G. Aslanyan, Tryptophan pyrrolase in kidney, *Dokl. Akad. Nauk Arm. SSR* **43**: 215–219 (1966).
78. C. A. Benasi, P. Benasi, G. Allegri, and P. Ballaun, Tryptophan metabolism in schizophrenic patients, *J. Neurochem.* **7**:264–270 (1961).
79. B. Peterkofsky, unpublished observations.
80. S. M. Hess, B. G. Redfield, and S. Udenfriend, The effect of monoamine oxidase inhibitors and tryptophan on the tryptamine content of animal tissues and urine, *J. Pharmacol. Exptl. Therap.* **127**:178–181 (1959).
81. J. A. Oates and A. Sjoerdsma, Neurologic effects of tryptophan in patients receiving a monoamine oxidase inhibitor, *Neurology* **10**:1076–1078 (1960).
82. A. Sjoerdsma, H. Weissbach, and S. Udenfriend, A clinical, physiologic and biochemical study of patients with malignant carcinoid, *Am. J. Med.* **20**:520-532 (1956).
83. A. Yuwiler, E. Geller, and G. G. Slater, On the mechanism of the brain serotonin depletion in experimental phenylketonuria, *J. Biol. Chem.* **240**:1170–1174 (1965).
84. E. M. Gal, M. Morgan, S. K. Chatterjee, and F. D. Marshall, Jr., Hydroxylation of tryptophan by brain tissue *in vivo* and related aspects of 5-hydroxytryptamine metabolism, *Biochem. Pharmacol.* **13**:1639–1653 (1964).
85. S. Consolo, S. Garattini, R. Ghielmetti, P. Morselli, and L. Valzelli, The hydroxylation of tryptophan *in vivo* of brain, *Life Sci.* **4**:625–630 (1965).
86. L. J. Weber and A. Horita, A study of 5-hydroxytryptamine formation from L-tryptophan in the brain and other tissues, *Biochem. Pharmacol.* **14**:1141–1149 (1965).
87. D. G. Grahame-Smith, The enzymic conversion of tryptophan into 5-hydroxytryptophan by isolated brain tissue, *Biochem. J.* **92**:52P (1964).
88. D. G. Grahame-Smith, Tryptophan hydroxylation in brain, *Biochem. Biophys. Res. Commun.* **16**:586–592 (1964).
89. H. Green and S. L. Sawyer, Demonstration, characterization and assay procedure of tryptophan hydroxylase in rat brain, *Anal. Biochem.* **15**:53–64 (1966).
90. E. M. Gal, J. C. Armstrong and B. Ginsberg, The nature of *in vitro* hydroxylation of L-tryptophan by brain tissue, *J. Neurochem.* **13**:643–654 (1966).
91. A. Nakamura, A. Ichiyama, and O. Hayaishi, Purification and properties of tryptophan hydroxylase in brain, *Federation Proc.* **24**:604 (1965).
92. W. Lovenberg, R. J. Levine, and A. Sjoerdsma, A tryptophan hydroxylase in cell-free extracts of malignant mouse mast cells, *Biochem. Pharmacol.* **14**:887–889 (1965).
93. T. Sato, W. Lovenberg, E. Jequier, and A. Sjoerdsma, Characterization of a tryptophan hydroxylating enzyme from malignant mouse mast cells, *Eur. J. Pharmacol.* **1**: 18–25 (1967).
94. S. Hosada and D. Glick, Studies in histochemistry. LXXIX. Properties of tryptophan hydroxylase from neoplastic murine mast cells, *J. Biol. Chem.* **241**:192–196 (1966).
95. W. Lovenberg, E. Jequier, and A. Sjoerdsma, Tryptophan hydroxylation: Measurement in pineal gland, brainstem, and carcinoid tumor, *Science* **155**:217–219 (1967).
96. D. G. Grahame-Smith, The biosynthesis of 5-hydroxytryptamine in brain, *Biochem. J.* **105**:351–360 (1967).

97. D. S. Robinson, W. Lovenberg, and A. Sjoerdsma, Subcellular distribution and properties of rat brain stem tryptophan hydroxylase, *Arch. Biochem. Biophys.* **123**: 419–421 (1968).
98. W. Lovenberg. E. Jequier, and A. Sjoerdsma, Tryptophan hydroxylation in mammalian systems, *Advan. Pharmacol.* **6A**:21–36 (1968).
99. E. Jequier, W. Lovenberg, and A. Sjoerdsma, unpublished data.
100. E. Jequier, W. Lovenberg, and A. Sjoerdsma, Tryptophan hydroxylase inhibition: The mechanism by which *p*-chlorophenylalanine depletes rat brain serotonin, *Mol. Pharmacol.* **3**:274–278 (1967).
101. S. Udenfriend, *in 5-hydroxytryptamine* (G. P. Lewis, ed.), pp. 43–49, Pergamon Press, New York (1958).
102. H. Weissbach, B. G. Redfield, and J. Axelrod, Biosynthesis of melatonin: Enzymic conversion of serotonin to *N*-acetylserotonin, *Biochim. Biophys. Acta.* **43**:352–353 (1960).
103. H. Weissbach, B. G. Redfield, and J. Axelrod, The enzymic acetylation of serotonin and other naturally occurring amines, *Biochim. Biophys. Acta* **54**:190–192 (1961).
104. J. Axelrod and H. Weissbach, Purification and properties of hydroxyindole O-methyl transferase, *J. Biol. Chem.* **236**:211–213 (1961).
105. J. Axelrod, R. J. Wurtman, and S. H. Snyder, Control of hydroxyindole O-methyl transferase activity in the rat pineal gland by environmental lighting, *J. Biol. Chem.* **240**:949–954 (1965).
106. S. Edlbacher and O. Wiss, Zur Kenntnis des Abbaues der Aminosauren im tierischen Organismus; Uber den oxydativen Abbau der Aminosauren im Gehirn *Helv. Chim. Acta* **27**:1824–1831 (1944).
107. U. S. von Euler, Histamine and nerves, *Ciba Found. Symp. Histamine*, pp. 235–241 (1955).
108. J. P. Green, Histamine and the nervous sytem, *Federation Proc.* **23**:1095–1102 (1964).
109. P. Holtz and E. Westerman, Uber Die Dopadecarboxylase and Histidin-Decarboxylase Des Nevengewebes, *Arch. Exptl. Path. Pharmakol.* **227**:538–546 (1958).
110. R. W. Schayer, Histidine decarboxylase of rat stomach and other mammalian tissues, *Am. J. Physiol.* **189**:533–536 (1957).
111. H. Weissbach, W. Lovenberg, and S. Udenfriend, Characteristics of mammalian histidine decarboxylating enzymes, *Biochim. Biophys. Acta* **50**:177–179 (1961).
112. E. Werle, A. Schaur, and H. W. Buhler, Klassifizierung histidindecarboxylierender Gewebs- und Batkerienenzyme, *Arch. Intern. Pharmacodyn.* **145**:198–207 (1963).
113. M. A. Kelsall and P. Lewis, Mast cells in the brain, *Federation Proc.* **23**:1107–1108 (1964).
114. A. Anastasi, P. Correale, and V. Erspamer, Occurrence of *N*-acetylhistidine in the central nervous system and the eye of *Rana esculenta*, *J. Neurochem.* **11**:63–66 (1964).
115. V. Erspamer, M. Roseghini, and A. Anastasi, Occurrence and distribution of *N*-acetylhistidine in brain and extracerebral tissues of poikilothermal vertebrates, *J. Neurochem.* **12**:123–130 (1965).
116. M. H. Baslow, *N*-acetyl-L-histidine synthetase activity from the brain of the killifish, *Brain Res.* **3**:210–213 (1966).
117. M. H. Baslow and J. F. Lenney, α-*N*-acetyl-L-histidine amidohydrolase activity from the brain of the skipjack tuna *Katsuwonus pelamis*, *Can. J. Biochem. Physiol.* **45**:337–340 (1967).
118. M. H. Baslow, *N*-acetyl-L-histidine metabolism in the fish eye, *Exptl. Eye Res.* **6**: 336–342 (1967).

119. A. Hanson, *N*-acetylhistidine as a donor of acetyl in the biosynthesis of acetylcholine in the brain in fish, *Acta Chem. Scand.* **20**:159–164 (1966).
120. S. Tsunoo and K. Horisaka, A review of recent studies on free histidine and its derivatives in the brain, *Showa Igakukai Zasshi* **27**:523–535 (1967).
121. J. J. Pisano, J. D. Wilson, L. Cohen, D. Abraham, and S. Udenfriend, Isolation of α-aminobutyrylhistidine (homocarnosine) from brain, *J. Biol. Chem.* **236**:499–502 (1961).
122. T. Nakajima, F. Wolfram, and W. G. Clark, The isolation of homoanserine from bovine brain, *J. Neurochem.* **14**:1107–1112 (1967).
123. I. R. McManus, Enzymatic synthesis of anserine in skeletal muscle by *N*-methylation of carnosine, *J. Biol. Chem.* **237**:1207–1211 (1962).
124. D. D. Brown, R. Tomchick, and J. Axelrod, The distribution and properties of a histamine-methylating enzyme, *J. Biol. Chem.* **234**:2948–2950 (1959).

*Chapter 8*

# SULFUR AMINO ACIDS


## M. K. Gaitonde

*Medical Research Council*
*Neuropsychiatric Research Unit*
*Carshalton, Surrey, England*


## I. INTRODUCTION

Several sulfur amino acids and sulfur compounds are found in mammalian tissues. While some find their origin in the diet, other sulfur amino acids are formed *in vivo* from methionine in the tissues. Thus it is known that methionine is converted into homocysteine, cystathionine, cysteine, hypotaurine, and taurine. These metabolites are formed in the course of transferring a methyl group to other compounds. The mechanism of demethylation and the subsequent metabolism of the demethylated product, homocysteine, is now well established. The enzyme systems in most cases were first studied in liver preparations. The demonstration that $^{35}$S-methionine is converted into $^{35}$S-cysteine and $^{35}$S-taurine by rat brain *in vitro* and *in vivo* gave evidence that the sulfur amino acids are metabolized also in the mammalian brain. Several subsequent studies have shown similarities between the metabolism of methionine in liver and in brain, but they have also revealed some characteristic differences in the metabolism of sulfur amino acids in the brain: (1) the cystathionine and taurine concentrations are much higher in the brain than in the liver, (2) the enzyme cysteine sulfinic acid decarboxylase is predominantly a particulate deaminated to form isethionic acid by rat brain and heart and not by liver. An interesting feature of sulfur amino acid metabolism is that many of the enzyme systems involved in the conversion of methionine into its several metabolites require pyridoxal phosphate (vitamin $B_6$) as a cofactor. Whereas in liver this cofactor is tightly bound to some of these enzymes, the corresponding enzymes in the brain are bound loosely to this cofactor, and their activity in the brain can be demonstrated *in vitro* only by adding the cofactor. In view of the dependence of sulfur amino acid metabolism on vitamin $B_6$, it is likely that in the brain it will be influenced by the metabolism of other compounds that also require vitamin $B_6$ as a cofactor, e.g., in the production of amines (catecholamines, indolamines, and GABA) from amino acids by decarboxylases and possibly in the subsequent inactivation of amines by the amine oxidase.

The metabolism of methionine is reviewed briefly in this chapter to indicate what is known from the study of the liver enzymes and to compare it with parallel studies, where available, of the enzyme systems of the brain. The reader's attention is drawn to the reviews on metabolism of methionine and its metabolites.[1-9]

## II. METHIONINE

### A. Metabolism

Methionine, $CH_3SCH_2CH_2CH(NH_2)COOH$, is metabolized *in vivo* by three main pathways: It may serve as (1) a precursor in protein synthesis, as (2) a methyl donor in all biological systems and at the same time as a precurser of cysteine for protein synthesis, or as (3) a source of energy by oxidation (to sulfate and carbon dioxide; Fig. 1) of substrates formed during its metabolic conversion to taurine and of substrates formed after its direct deamination or transamination.

#### 1. *Protein Synthesis*

When methionine is utilized for protein synthesis it is activated at the carboxyl group[10,11] and the resulting methionyladenylate is incorporated into the complex of soluble RNA and the transfer enzyme[11,12]

$$CH_3SCH_2CH_2CH(\overset{+}{N}H_3)CO—O—\underset{\underset{O}{\|}}{\overset{\overset{\bar{O}}{|}}{P}}—\text{adenosine}$$

#### 2. *Methylation and Transsulfuration*

Several earlier observations showed that a soluble liver enzyme of rat, pig, guinea pig, and dog catalyzed the transfer of methyl group of L-methionine to guanidoacetate (glycocyamine)[13] and nicotinamide[14] in the presence of ATP, or a system generating ATP, and Mg. Nicotinamide was not methylated by L-methionine in the presence of enzyme preparation of the liver of horse, sheep, ox, rabbit, pigeon, or duck.[14] In subsequent work it was shown that ATP activates the sulfur atom of methionine to form S-adenosylmethionine, which then serves as the immediate methyl donor in all biological systems.[15,16]

$$\text{Methionine} + ATP \xrightarrow[\text{GSH}]{\text{Mg}} \underbrace{\text{adenosine}—\overset{+}{S}CH_3CH_2CH_2CH(NH_2)COOH}_{\text{S-Adenosylmethionine}} + PP + P_i$$

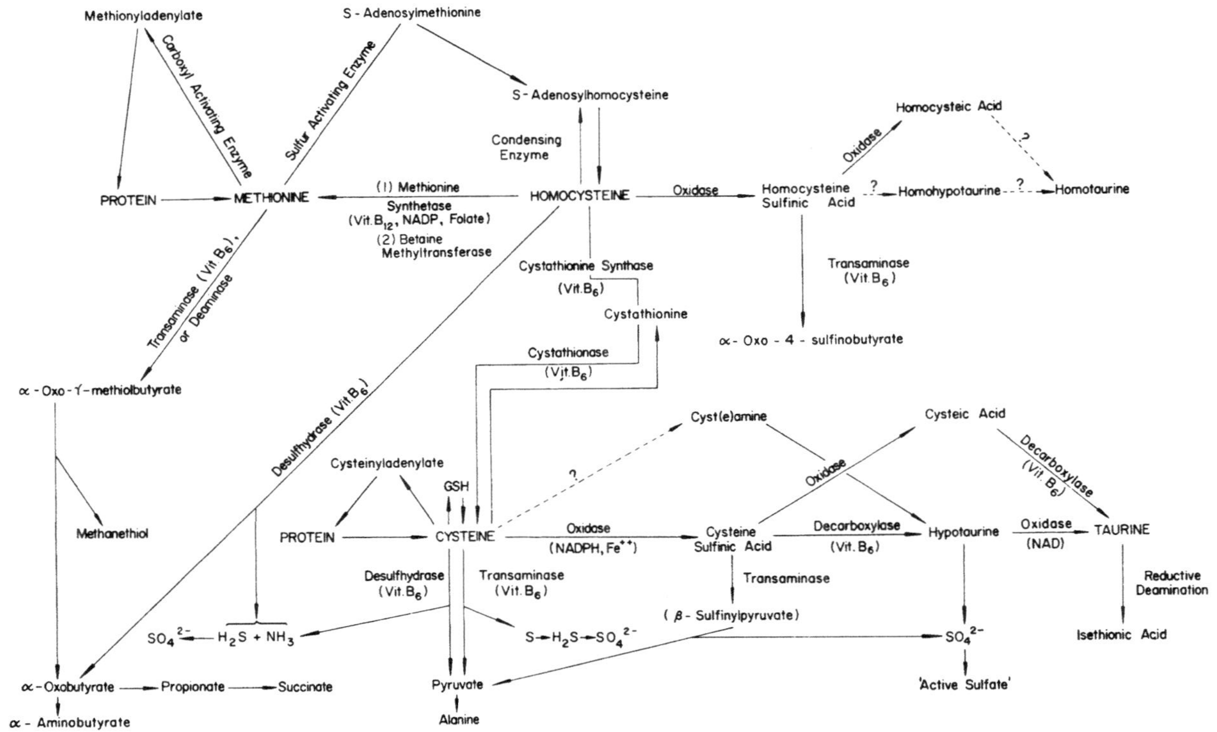


Fig. 1. Metabolic pathways of methionine. Note that, as a result of transsulfuration by the action of cystathionase, the carbon chain of serine, which condenses with homocysteine to form cystathionine, contributes to the carbon chain of cysteine.

The energy in the methylsulfonium bond in S-adenosylmethionine is approximately equivalent to that in the pyrophosphate bond[17] in ATP. The sulfonium compound rapidly donates its methyl group to acceptor molecules and is thereby converted to S-adenosylhomocysteine.[18] Thus, in the absence of ATP and Mg, the transfer of methyl group of S-adenosylmethionine was catalyzed by the specific methyltransferase to nicotinamide,[19] guanidoacetate,[20] and probably to dimethylaminoethanol.[21] The methionine activating enzyme, S-adenosylmethionine synthase, required thiol groups (GSH, cysteine, and homocysteine) and other reducing agents for the activity and was found in rat, pig, and rabbit liver.[15,17] The development of a microassay procedure using $^{14}CH_3$-methionine showed that this enzyme was present in several organs of rat, ox, lamb, hog, horse, guinea pig, hamster, rabbit, monkey and man.[22] The relative activities (mμmoles S-adenosylmethionine formed per milligram protein per hour) of methionine activating enzyme in brain and liver were rat, 3.3, 34; monkey, 4.0, 37; and man, 1.6, 9.7. Other workers reported somewhat higher activity of the enzyme in the brain (24%) in comparison with that found in liver.[22a] The level of methionine activating enzyme in the livers of females was approximately twice that in the livers of males. The activity of the enzyme in males was increased by castration and brought to normal levels by administering 17-ethyl-19-nortestosterone to the castrated males.[23] S-adenosylmethionine synthase activity of the liver was normal in patients with homocystinuria.[24,25]

### 3. *Deamination and Transamination of Methionine*

Among other pathways of methionine metabolism the following two are of lesser quantitative significance: (a) deamination and (b) transamination of methionine into α-oxo-γ-methiolbutyrate, $CH_3SCH_2CH_2COCOOH$.

*a. Deamination.* Methionine is deaminated to the corresponding α-oxo-acid by amino acid oxidase[26] of rat kidney slices (19.6–30.4%) more efficiently than by liver slices (1.4–10.7%). Sheep kidney preparations also converted D-methionine to the corresponding α-oxo acid.[27] The α-oxo acid was isolated from the urine of rat after feeding methionine.[28] Small amounts of the α-oxo acid are probably present in human urine on administration of methionine.[29] The finding that D-methionine is utilized for animal growth indicates that it is deaminated *in vivo* into the α-oxo acid, which is then reaminated to L-methionine.

The α-oxo analogue of methionine is unstable and gives rise to methanethiol when boiled in the presence of acid or base.[30] The α-oxo acid formed from methionine in the presence of rat liver mitochondrial is degraded to methanethiol, which remains bound to mitochondria proteins by disulfide linkage.[27] In contrast, methanethiol formed by tissue slices or *in vivo* is rapidly oxidized to carbon dioxide.[27,31]

*b. Transamination.* The α-oxo acid of methionine may also be formed by transamination of methionine with α-oxoglutarate in the presence of insoluble particulate preparation of rat, pig, and chicken liver and rat kidney. Methionine does not transaminate with oxaloacetate in the presence of rat liver particulate preparation; however, it transaminates with α-oxobutyrate and not with α-oxoglutarate in the presence of rat liver supernatant fraction. The transamination of methionine with pyruvate was observed in the presence of rat liver particulate preparation and not in the presence of homogenates of pig and chicken liver and kidney.[32]

The direct cleavage of methionine into methanethiol probably does not occur in mammals; the thionase preparations of rat liver show no hydrolytic cleavage of methionine.[27]

In patients with liver damage the normal process of demethylation is impaired; methanethiol and dimethyl disulfide exhaled by these patients are responsible for the characteristic odor (foetor hepaticus) of the breath. Dimethylsulfide is toxic and the feeding of excessive amounts caused considerable mortality in rats. The administration of large doses of methionine raised levels of blood α-oxo acid and of urinary inorganic sulfate (acidosis) in man.[29,33]

## B. Biosynthesis

Biosynthesis of methionine is catalyzed by methyltransferase or methionine synthetase.

### 1. *Methyltransferase*

Methionine may be formed by transmethylation of an intact methyl group from betaine,[34] dimethylthetin,[35,36] and dimethyl-β-propiothetin[36] to homocysteine. The enzyme, methyltransferase, required for this transmethylation was found in the liver and kidney of rat, guinea pig, and hog.[21,36]

Homocysteine + betaine → methionine + dimethylglycine
Homocysteine + dimethylthetin → methionine + methioacetic acid

The activity of the enzyme was not detectable in spleen, muscle and pancreas.[36] The enzyme thetin-homocysteine methyltransferase was more stable than betaine-homocysteine methyltransferase at *p*H 4.0. The methyltransferases do not require aerobic conditions for the activity, and they are not inhibited by cyanide, azide, arsenate, or arsenite. The reactions are not reversible. Thetins were not found in several tissues examined.[36] The enzyme, betaine-homocysteine methyltransferase, was dissociated from its cofactor, the nature of which is not known.[21,37] The methylation of homocysteine by thetins resembles methylation of guanidoacetate or dimethylaminoethanol by S-adenosylmethionine since in both reactions a proton is released under physiological conditions.[21] The activity of the two enzymes was of the same magnitude in the liver of control human subjects and in patients with homocystinuria.[38]

Under aerobic conditions choline, $(CH_3)_3{}^+NCH_2CH_2OH$, is rapidly oxidized to betaine, $(CH_3)_3{}^+NCH_2COO^-$, by choline oxidase, and thus it also can serve as a methylating agent for homocysteine. Choline oxidase has limited distribution in the animal kingdom: It was found in the liver of mouse and rat but not in the liver and kidney of guinea pig, rabbit, and chick.[34,39] The oxidation of choline and arsenocholine by the enzyme is inhibited by cyanide, ammonium, trimethylammonium ions and betaine.[39] The experiments *in vivo* showed that as in rats[40] a methyl group of choline was also transferred to homocysteine in guinea pig.[41] The isolation of dimethylglycine gave direct evidence that choline was oxidized to betaine prior to the transfer of methyl groups to homocysteine.[40]

Sulfocholine, $(CH_3)_2{}^+SCH_2CH_2OH$, is also oxidized to dimethylthetin, $(CH_3)_2{}^+SCH_2COO^-$, by rat liver enzyme preparation, which oxidizes choline to betaine. This compound at high concentration is toxic to the animal, but at low levels it supports growth of the animal.[42]

## 2. *Methionine Synthetase*

The synthesis of methionine may occur by a direct synthesis from homocysteine and formaldehyde, formate, or serine by liver extracts of pigeon,[43] pig,[44,47] sheep,[48] chicken,[49] and mouse.[50] Formate and the $\beta$-carbon of serine fed to rats were incorporated into methyl groups of methionine and choline; the incorporation was increased in rats[51] maintained on a diet supplemented with cobalamin (vitamin $B_{12}$). Subsequent work has shown that the incorporation of formaldehyde and serine into methionine is stimulated in the presence of tetrahydrofolate.[43,49,52] Since the suggestion that tetrahydrofolate plays the main role in the transfer of one carbon unit,[53] hydroxymethyltetrahydrofolate has been shown as a nonenzymatically formed intermediate in the conversion of formaldehyde into serine or of serine into glycine[54–56] and into methyl group of methionine in the biosynthesis from homocysteine.[57]

An enzyme system of the pig liver was shown to form 5-methyltetrahydrofolate (methyl-$FH_4$) from formaldehyde and tetrahydrofolate in the presence of NADP[45,57] or NAD:[46] the substitution of NAD for NADP reduced the conversion of formaldehyde into methionine by 75%.[52,57] A methyl derivative of tetrahydrofolate ($FH_4$) was isolated after incubation of $^5N,^{10}N$-methylene-tetrahydrofolate with *Escherichia coli* enzyme system.[58,59] The direct evidence that methyl-$FH_4$ was the intermediate was provided by isolating methyl-labeled methionine after incubating $^{14}C$-methyl-$FH_4$ with acetylhomocysteine[45] or homocysteine[47] in the presence of the pig liver enzyme:

$$\text{HCHO} + \text{tetrahydrofolate} \rightarrow \text{hydroxymethyltetrahydrofolate} \rightarrow {}^5\text{N-methyltetrahydrofolate}$$

5-Methyl-$FH_4$ was found in horse liver,[46] calf liver,[60] and human serum.[61] The enzyme, methionine synthetase, is now known to catalyze the reaction

Homocysteine + $^5$N-methyltetrahydrofolate → methionine + tetrahydrofolate

in the presence of (1) a reducing system, (2) S-adenosylmethionine, and (3) vitamin $B_{12}$ enzyme.

The reducing system is provided directly by adding $FADH_2$ or by generating it in the presence of FAD, NADH, and FAD reductase of *E. coli*, pig heart,[62,63] or hog liver.[46] The reducing system also may be provided directly by NADPH[47] or by generating NADPH by the action of enzymes of the pentose phosphate shunt[44,50] on glucose-6-phosphate or by the action of hydroxymethyltetrahydrofolate dehydrogenase.[57]

The enzyme system isolated from pig liver was shown to require for its activation catalytic amounts of S-adenosylmethionine (S-AM).[47] This finding was confirmed in bacterial enzyme systems that showed several-fold increase in methionine synthesis on the addition of catalytic amounts of S-AM.[62,64,65] The earlier observations of the requirement of ATP and Mg for the activation of methionine biosynthesis was attributed to the supply of catalytic amounts of S-AM by the enzyme system that also contained S-AM synthase and small amounts of methionine.[62,66] S-Adenosylethionine and methyl iodide can partially replace S-AM, but this depends on the particular reducing system employed during incubation of the enzyme preparation.[67,68] The evidence suggests that S-AM is not acting as a methyl donor of methyl groups to homocysteine[64] or as an intermediate[65] or a cofactor in the biosynthesis of methionine.[62]

Methionine synthetase was shown to require methyl-vitamin $B_{12}$[64,65] and not S-AM[66] as the cofactor. The evidence showed that the methyl group of methyl-$FH_4$ was transferred to vitamin $B_{12}$ to form cobalt methylated vitamin $B_{12}$ (cobalamin), which then transferred its methyl group to homocysteine.[64,65,69] A previous suggestion that methyl group was tranferred to S-adenosylhomocysteine[44] was not supported by experimental evidence.[62,70] Indeed, the presence of S-adenosylhomocysteine during incubation showed a decrease in methionine biosynthesis.[45]

A mechanism in which S-AM first activates the enzyme by methylating its cofactor vitamin $B_{12}$, which then accepts the methyl group of methyl $FH_4$ was suggested for the biosynthesis of methionine.[65] The methylation of vitamin $B_{12}$ by S-adenosylmethionine and by $^5$N-methyltetrahydrofolate probably occurs at two different sites.[71] The catalytic role of S-AM also may be explained by assuming that it serves to generate methyl-vitamin $B_{12}$, which is slowly degraded during its function as a cofactor. Several possible mechanisms have been discussed to explain methionine biosynthesis.[66]

While there is good evidence that methionine biosynthesis occurs in mammalian liver and that it apparently is a soluble enzyme system,[57,66] it is not known if such an enzyme system is also present in the mammalian brain. It is known, however, that folate is present in the cerebrospinal fluid of human subjects and its normal level in the CSF is three times that in the serum.[72,73] The administration of anticonvulsant drugs to epileptic patients has been reported to reduce their serum folate level, and this has been suggested to make these patients more prone to mental illness, including schizophrenia-like psychoses.[73]

## C. Concentration of Methionine in Tissues

### 1. *Animal Tissues*

Free methionine content of the tungstate extracts of several rat tissues was determined by microbiological assay.[74,75] The following values were reported (millimicromoles per gram fresh tissue): brain, 81.8–90.6; liver, 123.5–238.0; muscle, 114.0–145.5; spleen, 267–327; and plasma, 61.0–63.6. Two-dimensional chromatography, however, indicated that free methionine was absent or present in only trace amounts in the rat brain extract;[76,77] it was present in detectable amounts in the cow brain extract.[76] The failure to detect free methionine in the rat brain indicated that free methionine in the tissue was present mainly as S-adenosylmethionine or methionyladenylate which, under microbiological assay technique, would be determined as free methionine. Methionyladenylate is labile in the acid medium employed for the tissue extraction and would give rise to free methionine, but S-adenosylmethionine, although stable in the acid medium would decompose partly or completely into methionine, homoserine, or methylthioadenosine during ion exchange chromatography or in the course of preparation of the sample and under conditions employed for paper chromatography.[77,78]

*a. Rat.* The following values are reported for free methionine (millimicromoles per gram wet tissue or 100 ml plasma or urine except where indicated otherwise): brain, 81.8–90.6,[74,75] 20–60,[79] 90.0,[80] 10.0,[81] and 20.0.[82] Free methionine content of the rat brain showed a decrease during development: foetal brain on nineteenth day of gestation, 200; on twenty-first day of gestation 170;[83] newborn rat, 24; 1 day old,10; 7 days old, 12; 14 days old, 15; 21 days old, 25; 30 days old, 3; 60 days old 8; 90 days old, 19; 120 days old, 13; and adult, 10.[81]

Rats fasted for 12 hr showed a 30% decrease of free methionine in the brain,[75] rats maintained on a protein-deficient diet for 3 weeks showed no changes in free methionine content of the brain but after 6 weeks on such diets a significant increase from 90 to 140 mμmoles/g brain was observed.[80] The administration of phenylalanine to pregnant rats caused a significant decrease of free methionine from 170 to 90 mμmoles in the foetal rat brain on the twenty-first day of gestation; however, there was no effect on the free methionine level (4000 mμmoles) of the maternal plasma.[83]

Methionine content of the rat brain showed no changes on administration of drugs:[79] normal, 20–60; imipramine, 50, 70; chlorpromazine, 30, 40; reserpine, 20, 60; and hydroxylamine, 60.

The presence of S-adenosylmethionine was shown in rat tissues.[22a,77,78] The following values are reported for the rat tissues (millimicromoles per gram fresh weight): brain, 26.5,[22a] 31.4,[77] liver, 72,[22a] 83.6;[77] kidney, 51.0; heart, 64.3; spleen, 60; adrenals, 120; serum, 1.2; and ox pineal, 96.[22a] S-Adenosylmethionine was found in different regions of the rat brain: cerebral cortex, 29; cerebellum, 27; brain stem, 33; and basal ganglia, 35.5. Its concentration in the brain decreased during development of the rat after birth (newborn rat, 52.5, and the young adult, 25); a decrease was also observed in the rat liver during development.[22a]

The presence of the methionine-activating enzyme and the absence of cystathionine synthase and cystathionase in the heart[22] indicate that S-adenosylhomocysteine formed after the transfer of methyl group of S-adenosylmethionine is not metabolized further in this tissue by the trans-sulfuration pathway.

*b. Cat.* Free methionine content of several tissues is reported (mμmoles/g):[84] brain, 100; liver, 60.4; kidney, 73.8; plasma, 2690 mμmoles/100 ml; and urine, 73.8 mμmoles/100 ml. The spinal cord of the cat contained 200 mμmoles methionine/g fresh tissue.[85]

*c. Dog.* Free methionine was found in the dog brain (20 mμmoles/g water) and plasma (30 mμmoles/g water).[86]

*d. Ox.* The frontal cortex of the brain contained 13.4 and the anterior pituitary 20.1 mμmoles methionine/g wet tissue; methionine was not detectable in the posterior pituitary and the pineal.[86a]

*e. Dormouse.* The garden dormouse showed no changes of free methionine in the brain (141 mμmoles/g) during hibernation (135 mμmoles/g).[87]

## 2. *Human Tissues*

Free methionine was found in the foetal liver (100 mμmoles/g), adult liver (50 mμmoles/g),[88] and mixed gray and white matter of the brain of a 4-day-old infant (170 mμmoles/g) and of a 25-day-old infant with maple syrup urine disease (180 mμmoles/g).[89]

The following values are reported for methionine in the plasma (mμmoles/100 ml): newborn, 2950,[92] 9–24-month-old children, 300–2900 (fasting) [90], 4–16-year-old children, 2555;[91] and adult, 2140,[92] 1810,[93] 134–670,[94] 1100–3000 (fasting),[90] 2200 (per 100 g water);[95] in the serum: adult, 200–300;[96] and in the blood cells of the adult (mμmoles/100 g water): erythrocytes, 3000,[95] trace;[90] leukocytes, 39,100,[95] 175,000;[90] and platelets, 38,000.[90]

Methionine and homocystine were found in abnormally high concentration in the plasma of patients with homocystinuria[97,98] and in patients with maple syrup urine disease.[94] The following values are reported for the methionine content of the plasma in various diseases (mμmoles/100 ml): normal, 1810; osteomalacia, 2000–2820; hypophosphatasia, 1140; child Fanconi syndrome, 1410; and Hartnup disease, 1140.[93]

The normal human cerebrospinal fluid contained 270,[91,99] 870[100] mμmoles methionine/100 ml. The following values are reported for methionine content of the cerebrospinal fluid of human subjects under various conditions (mμmoles/100 ml): normal, 270;[91,99] mentally retarded, 290; mentally and physically retarded, 190; seizures, 180; febrile convulsions, 210; convulsive disorder, 160, 220; encephalopathy, 220; psychotic, 210; motor accident, 500;[99] phenylketonuria, 230–350; Tay-Sachs disease, 210–280; Huntington's chorea, 280–290, Hurler's syndrome, 200–440; tuberous sclerosis, 180, 200; Mongolism, 220, 280; Schilder's disease, 300; and cerebro-oculorenal disease, 140.[91]

The daily urinary excretion of methionine in the normal subjects was between 20,200 and 80,500 mμmoles in females and 33,600 and 73,700 mμmoles in males.[94]

## III. HOMOCYSTEINE

### A. Cleavage of S-Adenosylhomocysteine

S-Adenosylhomocysteine (S-AH) formed after the transfer of the methyl group of S-adenosylmethionine (S-AM) to the acceptor compounds is hydrolyzed by the rat liver enzyme into homocysteine, $SHCH_2CH_2CH(NH_2)COOH$, and adenosine.[101,102]

With the purified rat liver enzyme the equilibrium was in the direction of synthesis of S-AH from L-homocysteine and adenosine; D-homocysteine, L-cysteine, glutathione, CoA, homocystine, and L-homocysteine thiolactone could not replace L-homocysteine. The enzyme, therefore, has been classified as a condensing enzyme and does not require pyridoxal phosphate as a cofactor. In this respect it differs from other condensing enzymes which require pyridoxal phosphate as a cofactor and also yield a thioether from reactions between (1) homocysteine + serine → cystathionine, (2) $H_2S$ + serine → cysteine, and (3) methanethiol + serine → S-methylcysteine.

The condensing enzyme hydrolyzed S-AH into adenosine and homocysteine if one or the other of the reaction products were removed enzymatically.[101] Provided that this condensing enzyme is responsible for the cleavage *in vivo* of S-AH into homocysteine and adenosine, the activity of this enzyme will depend on the rate of transsulfuration of homocysteine into cysteine via cystathionine or on its rate of utilization for resynthesis of methionine by transmethylation with betaine or thetin or with $^5$N-methyltetrahydrofolate. The presence of the condensing enzyme also may explain why S-AH and not free homocysteine is detectable in the tissue extracts.[78]

Another enzyme which catalyzes the cleavage of glycosyl bond of S-adenosylhomocysteine into adenine and S-ribosylhomocysteine was found in some bacteria.[103] This enzyme has not been shown in mammalian tissue.

## B. Desulfhydration

L-Homocysteine is degraded into α-oxobutyrate, $H_2S$, and $NH_3$ by the specific desulfhydrase of rat liver, kidney, and pancreas. The enzyme does not act upon cysteine and is more labile than cysteine desulfhydrase[104] and probably requires pyridoxal phosphate as a cofactor.

## C. Oxidation and Transamination

Homocysteine is oxidized by rat liver preparation under conditions in which cysteine was completely oxidized to cysteic acid.[105] The final oxidized product of homocysteine is not known. It is likely, however, that homocysteine is also oxidized in a stepwise manner into homocysteine sulfenic acid, homocysteine sulfinic acid, and finally into homocysteic acid. Homocysteine sulfinic acid is not decarboxylated by rat liver homogenate; however, it transaminates with α-oxoglutarate, probably giving rise to α-oxo-4-sulfino-butyrate.[106] L-Homocysteine sulfinic acid is reversibly deaminated by rat liver glutamate dehydrogenase into $NH_3$ and the corresponding α-oxo acid.[107] The activity of the enzyme on L-homocysteine sulfinic acid was about 65% of that on L-glutamate; its affinity for homocysteine sulfinic acid was ten times smaller than for glutamate. The enzyme was inhibited by homocysteic acid.[107]

In view of the observation that homocysteine is oxidized by liver preparations it would be of interest to find out if it is also oxidized by brain preparations and if its pathway of oxidation is similar to that of cysteine. The existence of this path would indicate the possible occurrence of homohypotaurine, $SO_2HCH_2CH_2CH_2NH_2$, and homotaurine, $SO_3HCH_2CH_2CH_2NH_2$, in the brain. Homotaurine has been shown to depress strongly mammalian cortical responses.[108,109]

# IV. CYSTATHIONINE

Cystathionine is synthesized from L-homocysteine and L-serine by the condensing enzyme, cystathionine synthase, and it is cleaved into cysteine and probably homoserine by the cleavage enzyme cystathionase, both found in several mammalian tissues. Of the four possible isomers of cystathionine only L- and L-allo-cystathionine are metabolically active;

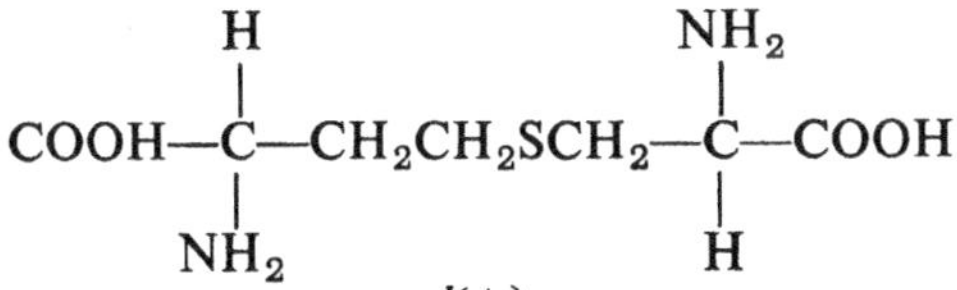

*l*(+)

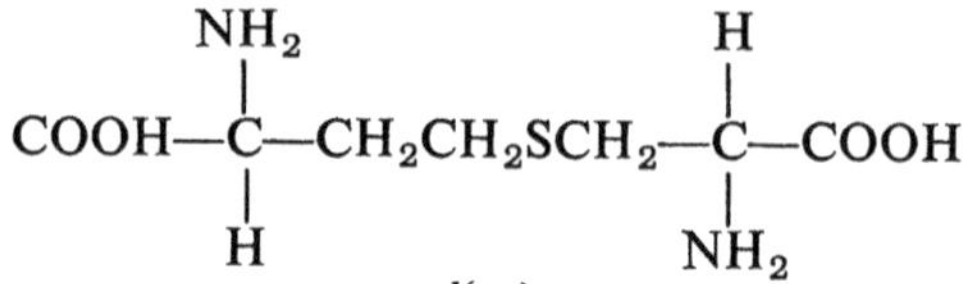

*d*(−)

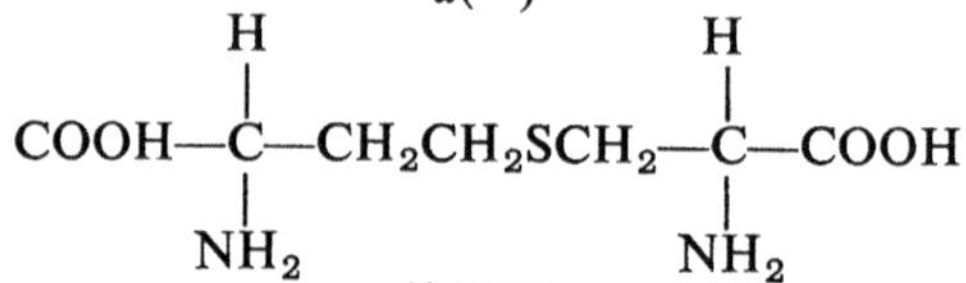

*d*(+)Allo

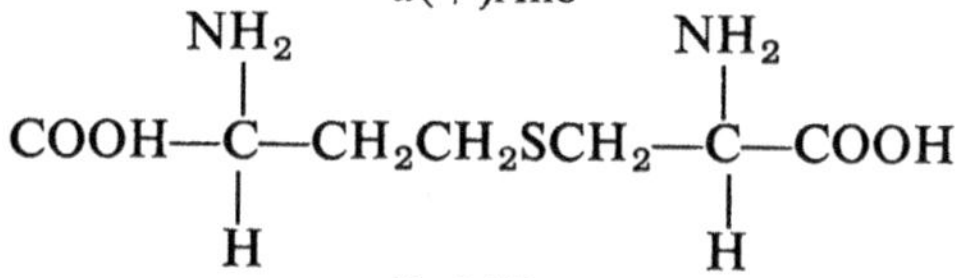

*l*(−)Allo

both supported growth of animals on a cystine-free diet. Moreover, L-allo-cystathionine when fed together with choline also supported growth of the young rat on a methionine-deficient diet.[110] These findings gave evidence that L-cystathionine was converted into L-cysteine and L-allo-cystathionine into D-homocysteine, which also supports growth of the animal.

## A. Cystathionine Synthase

Binkley[111] showed that a rat liver enzyme catalyzed the synthesis of L-cystathionine from homocysteine and serine:

$$\mathrm{COOHCH(NH_2)CH_2CH_2SH + CH_2OHCH(NH_2)COOH}$$
$$\rightarrow \mathrm{COOHCH(NH_2)CH_2CH_2SCH_2CH(NH_2)COOH}$$

The liver enzyme was isolated in a purified form after heating at 50°C for 10 min a liver suspension in 0.2 M KCl or 0.01 M sodium citrate. The final enzyme preparation was free from the cystathionase (cystathionine cleavage enzyme), serine dehydratase and homoserine deaminase. Cystathionine synthase was shown to synthesize cystathionine from homocysteine + serine; it did not catalyze the condensation of methionine + serine, methionine + glycine, or homocysteine + glycine. The enzyme was stable to prolonged dialysis against KCl-citrate solution.

Cystathionine synthase prepared by Selim and Greenberg[112] catalyzed the deamination of serine and threonine as well as the synthesis of cystathionine from serine and homocysteine; it was free from cystathionase, cysteine, and homocysteine desulfhydrase activities. The enzyme showed increased activity in the presence of dimercaptopropanol or mercaptoethanol and was inhibited by cysteine and cyanide. Glutathione, $Mg^{2+}$, and adenylate did not affect the synthase activity. On the other hand, serine dehydratase (serine deaminase) activity residing in the synthase preparation was somewhat activated by GSH, $Mg^{2+}$, $Mn^{2+}$, and adenylate and inhibited by both cysteine and homocysteine. These workers measured cystathionine synthase activity from the disappearance of homocysteine and serine but gave no evidence if cystathionine was actually synthesized and if it was equivalent to the amounts of the substrates that disappeared.

The synthase required pyridoxal phosphate as a cofactor for its activity,[113] and the loss in its activity was restored by adding the cofactor.[112] The cofactor is bound more tightly to the synthase than to the cystathionase. This explains why in pyridoxine deficiency the cystathionase activity of rat liver is reduced more than cystathionine synthase activity.[113]

The suggestion that cystathionine synthase, serine dehydratase, and threonine dehydratase (threonine deaminase) are enzymic activities of one and the same protein[112] is untenable in view of the resolution of cystathionine synthase and serine dehydratase activities in two separate fractions after chromatography on hydroxylapatite columns.[114] Moreover, Binkley had previously shown that his synthase preparation was free from serine dehydratase.[111] That threonine dehydratase is also a separate enzyme is suggested from the observation that a patient deficient in cystathionine synthase possessed normal levels of threonine dehydratase activity.[115] Furthermore, whereas the serine dehydratase and threonine dehydratase preparation of rat liver showed cystathionine synthase activity, the threonine dehydratase preparation of sheep liver possessed no cystathionine synthase activity.[116,117]

The direct evidence for the presence of cystathionine synthase in rat brain was obtained by Hope,[118] who isolated L-cystathionine after incubating L-homocysteine and serine in the presence of the tissue suspension. A microassay procedure[22] based on the conversion of $^{14}C$-serine into $^{14}C$-cystathionine gave the following values (millimicromoles cystathionine formed per milligram protein per 135 min) for mammalian brain and liver, respectively: rat, 84, 415; monkey, 252, 323; and man, 22, 252. The enzyme is not present in rat heart and skeletal muscle[22] and in freshly prepared hemolysates of normal human subjects.[38] Tissue samples stored in the deep freeze were devoid of synthase activity.[38] In homocystinuria, a condition of inborn error of sulfur amino acid metabolism, the defect is attributed to a deficiency of cystathionine synthase in the brain; the liver biopsy samples of these patients showed no synthase activity.[24]

## B. Cystathionase

### 1. *Cystathionase and "Thionase"*

This enzyme first isolated from rat liver was used for studying the cleavage of L-cystathionine and other isomers.[119,120] Cysteine was isolated as one of the products of cleavage of L-cystathionine by several workers: $COOHCH(NH_2)CH_2SCH_2CH_2CH(NH_2)COOH \rightarrow$ cysteine + homoserine (?). The nature of the second primary product of cleavage is not yet established but other products of the cleavage are α-oxobutyrate, $H_2S$, and

$NH_3$. An enzyme preparation, "thionase," was shown to catalyze the cleavage of L-cystathionine and several other S-alkylcysteines (see Table I); such a preparation showed no cleavage of methionine and homocysteine.(27,121) The name thionase was proposed to indicate the enzymic activity of cysteine desulfurase, cysteine desulfhydrase, and cystathionase.(121) A summary of the products of cleavage of several sulfur amino acids by cystathionase is given in Table II.

Cystathionase required pyridoxal phosphate as a cofactor for its activity.(113) It was significantly reduced (60%) in the liver of pyridoxine-deficient rats; the activity was enhanced by small amounts of pyridoxal phosphate, but pyridoxamine phosphate was ineffective or inhibited its activity.(113,129) The purified enzyme preparation was unstable in aqueous solution, inactivated at temperatures below 0°C, destroyed by freezing, and absent in frozen tissue.(121) The enzyme did not lose its activity if heated to 60°C for 5–10 min.(121,130) In the presence of glutathione the activity of aqueous solution remained unchanged for several weeks at 2°C. In the preparation from tissues other than liver, the enzyme activity was not

TABLE I

**The Relative Activities of Rat Liver Cystathionase Preparations**

| Substrate | Reference numbers | | |
|---|---|---|---|
| | 121[a] | 123,124[b] | 126[c] |
| L-Cystathionine | 100 | 100 | 100[d] |
| L-Allocystathionine | 50 | | |
| L-Lanthionine | 100 | 21 | 86 |
| Djenkolic acid | 100 | 38 | |
| L-Cysteine | 50 | 7 | |
| L-Cystine, methylester | 100 | | |
| L-Cystine | | | 76 |
| L-Homoserine | | 126 | |
| S-Methyl-L-cysteine | 50 | 1 | |
| L-Serine | | 3 | |

[a] The activity measured as sulfur released in the presence of glutathione.

[b] The activity measured as α-oxobutyrate formed except in the case of cysteine, which gave pyruvate and $H_2S$. Homoserine was also deaminated. The activity (as keto acid formed) towards L-methionine, DL-ethionine, DL-homocysteine, S-methylcysteine, S-ethylcysteine, and threonine ranged between 0.2 and 0.8%; the enzyme showed no activity toward penicillamine.

[c] The activity measured as keto acid formed.

[d] DL-Allo-cystathionine was used as the substrate in these studies.

## TABLE II

**The Cleavage Products of Several Substrates by Cystathionase Preparations**

| Substrate | Products formed: L-Cysteine (a) | α-Oxo-butyrate (b) | $NH_3$ (c) | $H_2S$ (d) | Pyruvate (e) | Other products (f) | Reference numbers: (a) | (b) | (c) | (d) | (e) | (f) |
|---|---|---|---|---|---|---|---|---|---|---|---|---|
| L-Cystathionine | + | + | + | + | | | 199,120, 122,121, 123 | 122,123, 124 | 122,123 | 121 | | |
| | + | None | None | | None | | 125 | 125 | 125 | | 125 | |
| L-Allo-cystathionine | | | | | | D-Homocysteine | | | | | | 120,121 |
| DL-Allo-cystathionine | | + | | None | + | D-Cysteine | | 126 | | 120 | 126 | 120 |
| L-Djenkolic acid | + | | + | + | + | S-Thiolmethyl cysteine and HCHO | 121,127 | | 127 | 121 | 127 | 127 |
| L-Cysteine | | None | None | + | None | | | 125 | 125 | 125 | 125 | |
| | | | | + | + | | | | | 124 | 124 | |
| | | | | + | | Serine? | | | | 121 | | 121 |
| L-Cystine | | | + | | + | Thiocysteine? | | | 126 | | 126 | 126 |
| Homocysteine | | None | | None | | | | 122, 124 | | | | |
| L-(and meso)-Lathionine | | + | | +(?) | + | | | 124 | | 121 | 126 | |
| Homoserine | | + | + | | | | | 122,123, 124 | 122,128, 123,124 | | | |
| | | | None | | | | | | 125 | | | |
| L-Serine | | | None | | | | | | 125 | | | |

demonstrable in the absence of glutathione.[121] The acetone powders of the tissue homogenates retained the activity for several weeks.[125] The relative activities of this preparation on some substrates are given in Table I.

## 2. *Cystathionase and Homoserine Deaminase*

A crystalline preparation of cystathionase was found to catalyze the cleavage of (1) cystathionine into cysteine, α-oxobutyrate, and $NH_3$ and of (2) homoserine into α-oxobutyrate and $NH_3$. Indeed, the crystalline preparation from rat liver was more active in forming α-oxobutyrate from L-homoserine than from L-cystathionine.[124] The same preparation also showed deaminase activity toward L-homoserine; the relative deaminase activities (in percent) of liver extracts of different species were as follows: rat, 100; ox, 12; sheep, 9; horse, 9; and pig, 7; rat kidney (8 %) and rat heart (1 %) were poor sources of the enzyme.[123] The relative activities of the crystalline cystathionase on some substrates is given in Table I. No data is available on the homoserine deaminase activity of this enzyme in the brain. The crystalline enzyme of the liver also showed cystathionine synthase activity in the presence of cysteine + 2-$^{14}$C-homoserine.[124]

The direct evidence that pyridoxal phosphate is the cofactor for the enzymic activity was demonstrated by resolving the crystalline cystathionase preparation into pyridoxal phosphate and the catalytically inactive apoenzyme. The activity of the enzyme was restored on reconstituting the two components.[124] The crystalline enzyme is inhibited by sulfhydryl reagents, carbonyl reagents, and some chelating agents.[129] Its activity is enhanced in the presence of 2-mercaptoethanol. Since there was no evidence for the formation of homoserine as the second primary product of the cleavage of L-cystathionine, Matsuo and Greenberg proposed the following equation for the cleavage of L-cystathionine by cystathionase:[124]

$$\text{L-cystathionine} \rightarrow \text{cysteine} + \text{CH}{=}\text{CHCH(NH}_2\text{)COOH}$$

The latter intermediate is finally deaminated into α–oxobutyrate and $NH_3$. For the reverse reaction, i.e., for the synthesis of cystathionine, they proposed that homoserine is first dehydrated to this intermediate, which then combines with cysteine in the presence of cystathionine synthase. They further proposed that homoserine required for this synthesis is derived by chemical decomposition[131] or enzymic breakdown of S-adenosylmethionine.[132] Such a mechanism does not take into account the possibility that homoserine, if formed as an intermediate during the cleavage of cystathionine, may be bound to the enzyme where it is deaminated into α-oxobutyrate and $NH_3$. Although there is a general tendency to regard cystathionase, homoserine deaminase, and cysteine desulfhydrase (see below) activities as part of the same enzyme,[129,133–135] it should be pointed out that the first cystathionase preparation described by Binkley and Okeson[125] showed no homoserine deaminase activity. Moreover, they found that cystathionase activity was more stable to heat than homoserine deaminase activity.

### 3. *Cystathionase and Cysteine Desulfhydrase*

Since the observation that the "thionase" preparation of Binkley and Okeson[125] catalyzed the cleavage of cysteine into $H_2S$, cystathionase activity of tissues has been measured by its cysteine desulfhydrase activity. Although first reported as a bacterial enzyme,[136] cysteine desulfhydrase activity was subsequently found in dog liver and several other mammalian species.[137-140] The highest activity was found in rat liver;[140] kidney and pancreas showed some activity,[138] and rat brain showed no activity.[140] The cleavage of L-cysteine by cysteine desulfhydrase gives $H_2S$ and probably amino acrylate $CH_2{=}C(NH_2)COOH$, which is then converted into pyruvate and $NH_3$.[140] The failure to detect the formation of $NH_3$ and pyruvate from L-cysteine in the presence of the desulfhydrase was attributed to the presence of other enzymes utilizing pyruvate and $NH_3$. Cysteine desulfhydrase may be purified from these contaminating enzymes by treatment of the saline extract of the enzyme with chloroform.[140] The cystathionase preparation of Matsuo and Greenberg[124] also catalyzed the cleavage of L-cysteine into $H_2S$, pyruvate, and $NH_3$, thus giving evidence of cysteine desulfhydrase activity in cystathionase preparations. In the absence of pyridoxal phosphate the crystalline cystathionase failed to catalyze the desulfhydration of L-cysteine.[124,129]

Cysteine desulfhydrase was relatively unstable in aqueous solutions.[138,139] Saline extracts may be dialyzed overnight and the activity recovered in a soluble form. The saline extracts stored at 0°C lost most of the activity in 10 days; such extracts were inactivated slowly at 50°C, rapidly at 60°C, and almost completely at 85°C. The enzyme was most active at *p*H 7.4–7.8. Its catalytic activity for the desulfhydration of DL-homocysteine was 10% of that with L-cysteine as the substrate,[140] and it showed no appreciable activity toward D-homocysteine and D-cysteine.[119]

In the presence of small amounts of cyanide (0.001 M) the purified preparations of cystathionase showed an increase in the formation of cysteine from L-cystathionase; there was no effect on desulfhydration of cysteine into $H_2S$.[119,125] At higher concentrations of cyanide (0.007–0.010 M), cysteine desulfhydrase activity on L-cysteine[140] and cystathionase activity on L-allo-cystathionine were inhibited but the activity of cystathionase to form cysteine from L-cystathionine was not affected.[119,125]

Cysteine desulfhydrase activity in the cystathionase preparations of liver is adaptively increased by feeding rats with diets containing methionine [134] and ethionine.[141] The enzymic activity measured *in vitro* is extremely low in livers of rats in pyridoxine deficiency[142] or after thyroxine administration;[134,143,144] the activity was, however, restored on adding pyridoxal phosphate *in vitro*.[143] The enzyme showed an increase in its activity in the livers of rats after thyroidectomy,[134,143,144] probably due to an increased synthesis of the apoenzyme.[145] In pyridoxine-deficient rats there was a decrease in the response of the cysteine desulfhydrase to thyroidectomy, indicating a reduction in the synthesis of the apoenzyme,[146] There is evidence that other apoenzymes, e.g., dihydroxyphenylalanine decarboxylase and cysteic acid decarboxylase, also decrease in livers of pyridoxine-deficient rats.[147]

The activity of cysteine desulfhydrase was high in the liver of the newly hatched chick; it decreased during the first week and thereafter remained constant. The high level of the enzyme activity comparable to that in the liver of the newly hatched chicks was maintained by feeding protein-rich diet to the chicks.[148]

### 4. *Cystathionase and the Cleavage of L-Cystine*

Cavallini showed that cystathionase preparation catalyzed the cleavage of L-cystine into pyruvate, $NH_3$, and thiocysteine, $COOHCH(NH_2)CH_2SSH$, and also suggested that the action of cystathionase on L-cysteine proceeded only after its oxidation into L-cystine.[133,149,150] This claim that cystine is the true substrate for cystathionase has not been confirmed.[151] Liver cystathionase does not catalyze the cleavage of α-substituted cystines.[152] The relative activities of the cystathionase prepared by Cavallini are given in Table I.

An enzyme, L-cystine lyase, from the *Brassica* species has been shown to cleave L-cystine into pyruvate, $NH_3$, and thiocystine; this enzyme preparation was more active in the cleavage of L-cysteine-S-sulfate, $COOHCH(NH_2)CH_2SSO_3H$ (also known as cysteine-S-sulfonate or S-sulfocysteine) and somewhat less active in the cleavage of S-methylcysteine sulfoxide into pyruvate. The enzyme showed no activity to catalyze the cleavage of L-cystathionine, L-allo-cystathionine, L-cysteine, and S-methyl cysteine:[153] This enzyme also required pyridoxal phosphate for its activity and was inhibited by thiol compounds such as cysteine and glutathione.[154] The presence of an enzyme that catalyzes the cleavage of L-cystine into the same products as shown by Cavallini using a rat liver cystathionase preparation raises the possibility of the existence of L-cystine lyase in liver cystathionase preparations. This is also evident from the report that rat liver and kidney preparations, like the L-cystine lyase of the *Brassica* species, catalyze the breakdown of cysteine-S-sulfate into pyruvate, $NH_3$, and thiosulfate.[155]

### 5. *Cystathionase Activity of Tissues*

Using a microassay procedure based on the conversion of 2-$^{14}$C-L-cystathionine into $^{14}$C-α-oxobutyrate, cystathionase activity of tissues of several species has been reported.[22] The activities expressed as mμmoles/mg protein/30 min for brain, liver, and kidney were as follows, respectively: rat, 0.11, 54, 28; monkey, 0.32, 28, 2.60; man (post mortem), 0.15, 16,—; and rabbit,—, 3.4, 1.4.

## C. Concentration of Cystathionine in Tissues

Cystathionine, since long considered as an intermediate in the pathway of conversion of methionine sulfur into cysteine sulfur,[156–159] was isolated as the L-form of cystathionine from human urine in pyridoxine deficiency[160,161] and from the occipital lobe of the human brain.[162] The isolation of the L-isomer from the mammalian tissue and the fact that this isomer is not converted into homocysteine explains why transsulfuration is not reversible in the direction of synthesis of methionine from cystathionine.

Before discussing the data on the cystathionine concentration in the mammalian tissue, it is pertinent to draw attention to the observation that for the complete extraction of cystathionine the tissue should be deproteinized with an acid such as picric acid, perchloric acid, or trichloroacetic acid; the extraction of the tissue with 75% ethanol gave recoveries of less than 10% of the cystathionine added to the tissue.[85]

### 1. *Concentration in Brain and other Tissues*

L-Cystathionine is present in high concentration in human brain and only in small amounts in other human tissues (Table III). Its concentration in human brain is also higher than in the brain of other species[162] (all values given in this section are in μmoles/100 g wet wt.): man (post mortem), 101–255; monkey, 57.6; cow, 5.4; cat, 12.1; rat, 17.5; guinea pig, 13.9; chicken, 2.7; duck, 0.9; and horseshoe crab, 67.5. Lower values (1.8–5.8, μmoles) recently have been reported for rat brain.[164] The true cystathionine concentration is probably a function of the pyridoxine levels in the diet. Rats maintained on a pyridoxine-supplemented diet gave much higher values of cystathionine in the brain (19.3–27.0 μmoles).[165]

**TABLE III**

**Cystathionine Concentration in Human Tissues**

| | μmoles/100 g wet wt. | | | |
|---|---|---|---|---|
| Diagnosis | Brain | Liver | Kidney | Reference numbers |
| Normal (19-year-old boy) | 7.65[a] | | | 166 |
| Normal (control) | 164.0 | 85.5 | | 38 |
| Normal (control) | 25.7–420 | 4.5–12.6 | 0.0 | 38 |
| Homocystinuria | 0.0–1.35 | 4.04–5.4 | 0.0 | 38,163 |
| Cystathioninuria | 278.0[a] | 160.6 | 69.3 | 38 |
| Arteriosclerosis | 115.5 | 3.6 | 3.1 | 162 |

[a] The frontal lobe; otherwise the values for the occipital lobe.

Of the ox tissues examined the pineal showed the highest concentration of cystathionine: pineal, 577; frontal cortex of the brain, 3.15; anterior pituitary, 4.5; and posterior pituitary, 8.55 μmoles/100 g wet tissue.[86a]

Brain biopsy samples obtained from patients with brain tumors contained 126–229 μmoles cystathionine.[164,166] This value may be compared with cystathionine concentration of 15.7–194.3 μmoles for cerebral cortex of brain autopsy samples of patients also with brain tumors.[166] This indicates that high values reported for cystathionine in human brain at autopsy most likely reflect the levels of cystathionine not significantly different from the true levels *in vivo*. Furthermore, no significant post mortem changes in the cystathionine concentration of human brain and rat brain have been observed.[38,164]

## 2. *In Pyridoxine Deficiency*

In pyridoxine deficiency there was a tenfold increase (274–382 μmoles) in the cystathionine concentration of the rat brain;[165] its levels also increased in the rat liver (274 μmoles) and kidney (175 μmoles).[167] These high levels in tissues are probably responsible for the observed increase in the excretion of cystathionine in the urine (cystathioninuria) of pyridoxine-deficient rats.

## 3. *Regional Distribution in the Central Nervous System*

A summary of some of the data available on the regional distribution of cystathionine in human brain and in pyridoxine-deficient rat is given in Table IV. One of the interesting findings of these studies is the observation that cystathionine is present in concentrations 2-to-12-fold higher in white matter than in gray matter of the human brain.[164,168] This finding, while explaining the wide variations in cystathionine concentrations reported by several previous workers, raises the question of the significance of such a high concentration of cystathionine in white matter. The high concentration cannot be attributed to differences in the two enzymic activities, viz., cystathionine synthase and cystathionase, since the two enzymic activities are equally distributed between the white and gray matter.[164] No such difference between the white and gray matter was observed in the cat spinal cord[85] (μmoles/100 g wet wt.): dorsal white, 136; dorsal gray, 162; ventral white, 170; ventral gray, 129; dorsal roots, 3; and ventral roots, 4.

The concentration of cystathionine in the human brain increased during development. Cystathionine was not detectable in the 5-month-old foetus; it was present in significant amounts (63.4 μmoles) only in the pons and medulla of the 8-month-old foetus.[166] In contrast, its concentration in the rat brain decreased during development (μmoles/g wet wt.)[81]: newborn, 0.037; 1 day old, 0.045; 7 day old, 0.048; 21 day old, 0.068; 60 day old, 0.026; and adult; 0.016.

**TABLE IV**

**Regional Distribution of Cystathionine in the Central Nervous System**

| | μmoles/100 g wet wt. | | | |
|---|---|---|---|---|
| | Rat | | Man | |
| Tissue | (Pyridoxine deficient)[165][a] | Whole tissue | Gray matter | White matter |
| Whole brain | 391.4 | | | |
| Cerebrum | 139.5 | | | |
| Cerebellum | 643.3 | 9.45–15.3[38]<br>85.5[168] | 14.4–64.4[164] | 59.0–191.0[164] |
| Frontal lobe | | 7.65[166]<br>51.4–60.4[38] | 6.3–31.1[164]<br>45.0[168] | 96.0–122.5[164]<br>202.4[168] |
| Occipital lobe | | 25.7–420.0[38] | 35.6–90.5[164] | 167–226[164] |
| Caudate nucleus | | 10.7,[166] 157.5[168]<br>38.7–65.7[164] | | |
| Thalamus | | 92.7–104.0[164]<br>180[168] | | |
| Hypothalamus | | 14.9,[166] 45.0[168]<br>15.3–44.1[164] | | |
| Medulla | 301.4 | 23.9, 112.5[168]<br>23.0–45.5[164] | | |
| Pons | | 21.6–35.1[38]<br>12.2–67.0[164]<br>135.0[168] | | |
| Spinal cord | 449.8 | | | |
| Retina | 188.9 | | | |

[a] Superior numbers in parentheses are reference numbers.

### 4. *In Homocystinuria*

The excretion of abnormal amounts of homocystine in the urine of certain mentally retarded patients is recognized as an inborn error of metabolism of the sulfur amino acid. The amount of cystathionine in the brains of these patients was extremely low or it was virtually absent (Table III). This has been attributed to a metabolic defect in the synthesis of cystathionine from homocysteine and serine, as a result of a deficiency of the enzyme cystathionine synthase. The direct evidence for the absence of this enzyme was obtained in a liver biopsy from a patient with homocystinuria.[24] The absence of this enzyme in the autopsy samples of brain, liver, and kidney of both control and patients with homocystinuria has limited the studies of the enzyme. Other enzyme systems such as homocysteine-glutathione transhydrogenase were present in similar amounts both in control and in patients with homocystinuria. Moreover, the fasting methionine levels in these patients were raised; methionine loading showed a delayed disappearance of methionine from the blood of these patients.(38) There was decrease in the excretion of homocystine in their urine following administration of L-serine alone or L-serine together with methionine.(163) The findings indicate that resynthesis of methionine from homocysteine was not affected in the brain and in other organs. Further evidence that transsulfuration pathway was blocked in homocystinuria was obtained by demonstrating that the oxidation of methionine and not of cysteine into inorganic sulfate was impaired in these patients.(169)

A higher homologue of cystathionine recently has been isolated from the urine of seven children with homocystinuria. The compound has been tentatively identified as homolanthionine,

$$COOHCH(NH_2)CH_2CH_2SCH_2CH_2CH(NH_2)COOH,$$

presumed to be formed in the liver by the condensation of homocysteine with homoserine, analogous to the formation of cystathionine by the condensation of homocysteine and serine. Its daily excretion in the patient was 0.07–0.14 μmole/g creatinine. The possibility remains that the isolated compound might be β-methylcystathionine, a product of condensation of homocysteine with threonine.(170)

### 5. *In Cystathioninuria*

In some other mentally retarded patients an increased urinary excretion of cystathionine was reported by Harris, Penrose, and Thomas.(171) In the case reported by Frimpter, Haymovitz, and Horwith,(172) the urinary excretion was 5560 μmoles cystathionine/day; the plasma contained 2.0 μmoles/100 ml, and the CSF, 0.09 μmoles/100 ml. The same patient showed a marked decrease in the excretion of cystathionine on administration of pyridoxine additional to that supplied through the diet.

The analysis of cystathionine in the brain of one of the patients with cystathioninuria gave high values (278 μmoles), but within the range obtained from brains of the control patients (Table III). The cystathionine content of the liver (157 μmoles) of this patient was several fold higher than in the controls (4.5–12.6 μmoles), except in one (apparently normal) case that also gave high values (85.5 μmoles).(38,171) The accumulation of cystathionine in this condition is probably due to a defect at the level of cleavage of cystathionine into cysteine by the enzyme cystathionase.

Conditions of cystathioninuria also have been described in patients with adrenal neuroblastoma and hepatoblastoma.(173) Here again cystathionine levels of the brains of these patients were the same as in the brains of normal patients (31.5 μmoles). In mentally defective patients with cystathioninuria, levels of cystathionine in other tissues, e.g., liver, kidney, and adrenals were not raised.(174)

Cystathioninuria in rats may be produced by feeding a pyridoxine-deficient diet (see above) or by the administration of β-cyanolalanine, $CNCH_2CH(NH_2)COOH$, a neurotoxin of the vetch plants.[175] The neurotoxin was originally thought to produce cystathioninuria by blocking the action of pyridoxal phosphate, but now it has been shown to inhibit the action of cystathionase of the rat liver, probably by competing with the substrate for the enzyme.[176,177]

### 6. *Miscellaneous*

Cystathionine was either absent or present in negligible amounts in the cerebrospinal fluid of the normal human subjects and of patients with phenylketonuria, Tay–Sachs disease, Huntington's chorea, Hurler's syndrome, tuberous sclerosis, Mongolism, Schilder's disease, and cerebro-oculovenal disease.[91] Cystathionine levels of the brain were not affected in rats on administration of imipramine, chlorpromazine, reserpine, or hydroxylamine.[79]

# V. CYSTEINE AND CYSTEINE SULFINIC ACID

## A. Metabolic Pathways of Cysteine

Cysteine, $COOHCH(NH_2)CH_2SH$, may be metabolized by several pathways: (1) It may be utilized directly for the synthesis of tissue proteins or for the synthesis of the tripeptide, glutathione; (2) it may be converted to pyruvate after desulfhydration (see Section IV, B, 3) or transamination[178] and thus serve as a substrate for energy; or (3) it may be oxidized to other sulfur compounds.

Cysteine may be oxidized by the mammalian tissue by two pathways: (1) It may be oxidized to its disulfide form, cystine, which is finally converted to taurine and sulfate, or (2) it may be oxidized stepwise into the corresponding sulfenic, sulfinic, and sulfonic acids. Cysteine sulfinic acid formed then acts as a key intermediate in the formation of taurine.

## B. Oxidation of Cysteine to Cystine and Taurine

Cysteine is oxidized to cystine by cytochrome *c* and cytochrome oxidase.[179] Cystine formed is probably oxidized further by cystine oxidase of the rat liver into cystine disulfoxide, which on decarboxylation and further oxidation is converted into taurine:[105]

Cysteine → cystine → cystine disulfoxide → cystamine disulfoxide
→ taurine

The evidence for this mechanism is based on the observation that liver preparations oxidized cystamine disulfoxide and dithio-dilactic acid but not cystamine and homocystine.

Alternatively, cystine may be slowly hydrolyzed nonenzymatically into cysteine and (cysteine sulfenic acid which is spontaneously converted into) cysteine sulfinic acid. The two products may be oxidized to cysteic acid and/or sulfate.[105] The quantitative significance of this pathway of cysteine oxidation in the mammalian tissue is not known.

## C. Stepwise Oxidation of Cysteine

Pirie postulated that cysteine is oxidized by tissues into sulfate through the intermediate formation of cysteine sulfinic acid; cystine and methionine are also oxidized to sulfate by the same pathway after their conversion into cysteine. The following mechanism was proposed:

$$2RSH \xrightarrow{\text{enzyme}} 2RSOH \xrightarrow[\text{at pH7.4}]{\text{spontaneous}} RSH + RSO_2H \xrightarrow{\text{enzyme}} H_2SO_3 \xrightarrow{\text{enzyme}} H_2SO_4$$

$$\text{Cysteine} \rightarrow \begin{array}{c}\text{cysteine}\\\text{sulfenic}\\\text{acid}\end{array} \longrightarrow \begin{array}{c}\text{cysteine}\\\text{sulfinic}\\\text{acid}\end{array} \longrightarrow \text{sulfite} \longrightarrow \text{sulfate}$$

### 1. *Oxidation to Cysteine Sulfinic Acid*

The direct evidence that cysteine is oxidized to cysteine sulfinic acid was obtained by isolating $^{35}$S-labeled cysteine sulfinic acid from rat liver and kidney following injection of $^{35}$S-labeled cystine.[181] The presence of the enzyme, cysteine oxidase, was shown both in a particulate fraction[182] and in a soluble fraction[183,184] of rat liver. Most of the activity was associated with the soluble fraction; it required oxygen and ferrous ion and reduced NADP or NAD for maximal activity.[183] In the presence of hydroxylamine, there was an accumulation of cysteine sulfinic acid and cysteic acid owing to the inhibition of their further metabolism into taurine[183] and sulfate. The enzyme was highly specific for L-cysteine; D-cysteine, L-cystine, glutathione, and cysteamine formed only small amounts of cysteine sulfinic acid under the same conditions. The activity of the enzyme was blocked by heavy metal reagents (EDTA and *o*-phenanthroline) and sulfhydryl reagents. The oxidation of cysteine to cysteine sulfinic acid was not demonstrated in 0.25 M-sucrose homogenate or its individual subcellular fractions of rat brain, heart, spleen and kidney.[184] The activity of the enzyme, however, was demonstrated if the tissues were homogenized in isotonic KCl solution. The activity in $\mu$moles cysteine sulfinic acid formed /100 mg tissue in the presence of NADP, $Fe^{2+}$, and hydroxylamine was as follows:[185] rat brain, 0.25; liver, 3.70; kidney, 0.30; spleen, 0.23; muscle, 0.10; and blood, 0.14.

### 2. *Oxidation to Cysteic Acid*

The enzyme oxidizing cysteine to cysteic acid was found in the insoluble residue of the saline suspensions of the rat liver. The same preparation also oxidized cystine to cysteic acid, but at a reduced rate—probably through the intermediate formation of cysteine and cysteine sulfenic acid by the non-enzymatic hydrolysis of the disulfide. Under the same conditions, DL-cysteine, homocysteine, and thioacetic acid were also completely oxidized; the end products were not identified. The oxidation of cysteine to cysteic acid presumably occurred through the intermediate formation of cysteine sulfenic and cysteine sulfinic acids. Cysteine sulfinic acid was oxidized to cysteic acid by a very slow reaction in the presence of NAD by a dehydrogenase of rat liver mitochondria.[186,187]

### 3. *Oxidation to Sulfate*

The enzyme oxidizing cysteine to sulfate was demonstrated in rat liver and kidney: The activity in μmoles sulfate formed /100 mg wet wt. was 0.78 in liver and 1.19 in kidney. The enzyme was not detectable in blood, spleen, heart, and lung. Glutathione was hydrolyzed rapidly into the constituent amino acids by extracts of rat kidney and to a lesser degree by extracts of rat liver; cysteine produced was further oxidized into sulfate. The oxidation of cysteine and methionine into sulfate was greater in an atmosphere of 95% oxygen than in air and did not occur if the tissue was finely ground.[180] The enzyme was present mainly in the insoluble fraction of the saline extracts, was highly unstable, lost its activity considerably on washing the insoluble residue with buffered or unbuffered saline, or on dialysis was partially activated in the presence of NAD, and was more active at *p*H 7.3 in the bicarbonate than at *p*H 7.4 in the phosphate medium. Cysteic acid and methionine sulfoxide were not converted into sulfate.[105]

An enzyme system oxidizing cysteine to pyruvate and inorganic sulfate was found localized more in the mitochondrial fraction than in the soluble fraction of the liver;[188] cystine did not act as a substrate unless it was converted into cysteine. The production of sulfate by mitochondrial preparations was not increased on the addition of ATP, NAD, NADP, NADPH, FAD, pyridoxal phosphate, or ascorbic acid nor decreased in pyridoxine-deficient rats. The activity of the enzyme system was considerably reduced on freezing and thawing of the mitochondrial preparation or by adding $Ca^{2+}$, phosphate, or 2,4-dinitrophenol to the mitochondrial preparation. There was a decrease in the formation of sulfate and an increase in cystine. The activity of the enzyme was restored by adding GSH or GSSG.[188] Although several mechanisms (desulfhydration, transamination, or oxidation) are known to convert cysteine into sulfate, the enzyme system responsible for this conversion by the mitochondrial preparation has not been identified.

## D. Metabolism of Cysteine Sulfinic Acid

Cysteine sulfinic acid is metabolized in one of the three pathways: (1) It is oxidized to cysteic acid, (2) it transaminates with α-oxoglutarate or oxaloacetate or is deaminated oxidatively giving rise to β-sulfinylpyruvate, which is then converted into pyruvate and sulfite, and sulfite is further oxidized to sulfate or, (3) it is decarboxylated into hypotaurine, which on further oxidation gives rise to taurine.

The metabolism of cysteine sulfinic acid according to pathway (1) has been referred to above (Sections V,C,1 and 2) and that according to pathway (3) is dealt with in Section VI. The transamination and deamination of cysteine sulfinic acid is considered here:

### 1. *Transamination and Deamination of Cysteine Sulfinic Acid*

Cysteine is oxidized *in vitro* and *in vivo* into sulfate; cysteic acid and taurine were both excluded as possible intermediates in this pathway. This was further confirmed by demonstrating that urinary sulfate isolated from rats given $^{35}$S-taurine orally was essentially free from radioactivity.[189]

The demonstration that cysteine is oxidized to cysteine sulfinic acid and the earlier findings of Medes that cysteine sulfinic acid is converted into sulfate gave evidence for the mechanism suggested by Pirie for the oxidation of cysteine to sulfate. Bergeret and Chatagner confirmed that cysteine sulfinic acid is converted into sulfate and further demonstrated that alanine was also formed as a reaction product. The enzyme system oxidizing cysteine sulfinic acid into sulfate was found in the rat liver and kidney.[2,191–193] The rat liver enzyme is solubilized by grinding the tissue in saline; it is stable and may be salted out with $MgSO_4$ and dialyzed. The enzyme is more active at *p*H 7.3 in the bicarbonate medium than at *p*H 7.4 in the phosphate medium. The enzyme showed an increased conversion of cysteine or cysteine sulfinic acid into sulfate in the presence of NAD.[105,187]

The sulfate formation has been shown to proceed after preliminary transamination of cysteine sulfinic acid with α-oxoglutarate or oxaloacetate.[186,187,194]

α-oxoglutarate ↘ ↙ Cysteine sulfinic acid

Glutamate ↙ ↘ β-sulfinylpyruvate ⇄ pyruvate + $SO_2$ → $SO_3^{2-}$ ⇆ $SO_4^{2-}$

Indeed the classical aspartate-α-oxoglutarate transaminase catalyzed the transamination of cysteine sulfinic acid with α-oxoglutarate faster than the corresponding reaction with aspartate.[186] Because of its efficient transamination with oxaloacetate the intial rate of oxidation of succinate is maintained much longer in the coupled system; also, the rate of oxygen uptake was greater in the presence of cysteine sulfinic acid + succinic acid than with either substrate alone. The mitochondrial preparations of ox heart accumulated pyruvate and sulfite and those of rat heart accumulated pyruvate and sulfate.[186,195] The oxidation of succinate by brain mitochondrial preparations was more than doubled on the addition of cysteine sulfinic acid. This has been attributed to its transamination with oxaloacetate formed by oxidation of succinate.[196]

The activity of the enzyme is not affected in the tissues of the rat in pyridoxine deficiency. The cleavage of β-sulfinylpyruvate into sulfite is catalyzed by manganous ions.[4] The sulfite is oxidized to sulfate by the corresponding oxidase in the liver and kidney.[192,197]

Cysteine sulfinic acid also may be degraded by oxidative deamination into pyruvate, $NH_3$, and sulfate by liver mitochondrial enzyme in the presence of NAD.[4]

## E. Concentration of Cysteine and Cysteine Sulfinic Acid in Tissues

Cysteine sulfinic acid was detected in the rat brain, but its concentration in the tissue is not known.[198] Cysteine and cystine content of several tissues is reported (all values reported below are given in μmoles/100 g wet tissue or 100 ml fluid except where indicated):

### 1. *Animal Tissues*

*a. Rat.* Cystine content of rat brain was 5.0 (control) and 12.44 (insulin coma).[199] Using a specific method for cysteine the following values were reported for rat tissues:[200] cysteine in brain, 5.59, liver, 9.49, and blood, 3.08; cystine in brain, 0.33, liver, 0.62, and blood, 1.88. Cystine and cystathionine content of several areas of the monkey brain, except the midbrain, showed a statistically significant decrease after treatment with reserpine.[201]

*b. Cat.* Several tissues of the cat were analyzed for cystine by ion exchange chromatography. The following values are reported: brain, 4.16; liver, <0.83; kidney, 4.6; plasma, 1.67; and urine, <0.83.[84] It was not present in detectable amounts in the spinal cord.[85]

*c. Dog.* Cystine was found in dog tissues: brain, 10; plasma, 18; and cerebrospinal fluid, 2 (all values in μmoles/100 g water).[86]

*d. Ox.* Significant amounts of (half)-cystine were found in ox tissues:[86a] brain (frontal cortex), 5.0; pineal, 3.75; anterior pituitary, 16.66; and posterior pituitary, 9.17.

### 2. *Human Tissues*

The half-cystine content of the brain was 3.75 in a 4-day-old infant; it was 3.5 in a 25-day-old infant with maple syrup urine disease.[89] Its concentration in the foetal liver was 13 μmoles and in the adult liver, 2 μmoles.[88]

Cysteine was present in measurable amounts (3.3 μmoles) in the plasma of 4- to 16-year-old children.[91] The total content of cysteine plus half-cystine in the fasting children (9–24 months) was 0–4 μmoles and in the fasting adults, 7–10.8 μmoles.[90]

Cystine was found in measurable amounts (1 μmole) in the human cerebrospinal fluid.[100] Other workers found it only in detectable amounts in the cerebrospinal fluid of normal subjects, but its concentration increased in patients with pneumonitis, 0.14; craniosynostosis, 0.08; convulsive disorder, 0.06; and motor accident, 0.03.[99]

### 3. *Cystinuria*

The urinary excretion of cystine is between 161 and 164 μmoles/24 hr (or 10–20 μmoles/g creatinine[96]) in normal human subjects and between 1998 and 5044 μmoles/24 hr in patients with cystinuria.[202–204] An increased urinary excretion of cystine was observed in patients with cystinuria on administration of cysteine but not on administration of cystine or homocystine. The administration of DL-methionine was also without effect on the urinary excretion of cystine by these patients if fed on a high protein diet.[202,205] These patients also excreted abnormal quantities of lysine, arginine and ornithine.[206–208] Moreover, patients with cystinuria (59–880 μmoles/24 hr), Fanconi syndrome, and Wilson's disease recently have been reported to excrete large amounts of a mixed disulfide of cysteine and homocysteine.[204]

No essential difference was found between the normal subjects and patients with cystinuria in their ability to metabolize cysteine and cystine. Dent and his associates suggested that the increase in plasma cystine levels on feeding cysteine and not on feeding cystine was due to the difference in the absorption of these two amino acids by the gut.[202] Cystinuria has been considered as an inherited condition in man.[209]

The increased excretion of cystine and basic amino acids by the patients was probably not due to a mutual inhibition of transport between basic amino acids and cystine or cysteine.[210] The cystine transport *in vitro* in the kidney of the patient was found normal, and it was suggested that the defect was probably in the transport of cysteine.[211] In this connection it is of interest to note that rat kidney slices incorporated more cysteine than cystine and that $^{35}$S-cystine added to the saline medium was incorporated into kidney cortex slices and found predominantly as $^{35}$S-cysteine.[212] High intracellular $^{35}$S-cysteine also was observed on incubating tissue slices of rat brain, liver, diaphragm, and jejunum with $^{35}$S-cystine.[213]

### 4. *Urinary Alkyl Cysteines*

Several S-alkyl cysteines have been isolated:

*a. Isovalthine.* S-(isopropylcarboxymethyl) cysteine

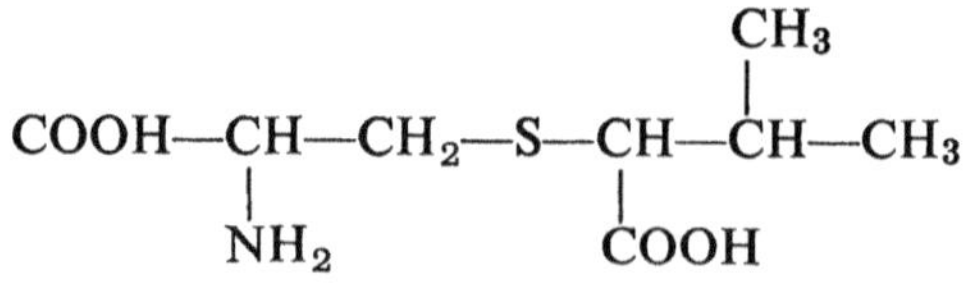


was isolated from the urine of cats and of patients with atherosclerosis, myxoedema, severe diabetes, and hypothyroidism.[214]

*b. Isobuteine.* S-(2-methyl-2-carboxyethyl)-L-cysteine

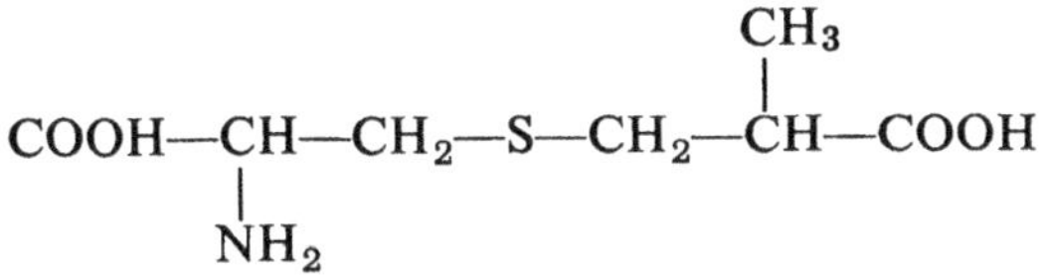

is present in the urine of normal persons, more in children than in adults.[214]

*c. S-(Carboxymethyl)-Cysteine.* This was isolated from radish seedlings,[215] the urine of patients with diabetes and hypertension, and extracts of goat testis and ox liver.[216]

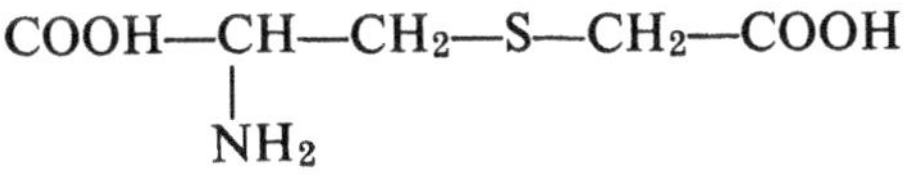

*d. Felinine.* This compound is found in large amounts (482–578 μmoles/24 hr) in cat urine. It is S-hydroxyalkylcysteine.[217]

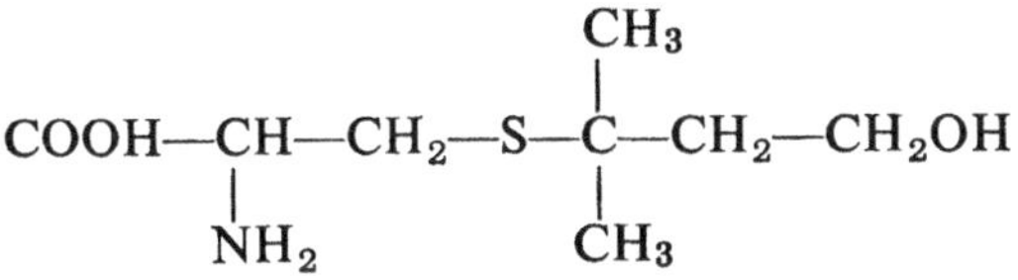

*e. S-Sulfo-L-cysteine.* The enzymic oxidation of sulfite to sulfate is essential in man. The excretion of abnormal quantities of S-sulfocysteine and thiosulfate and the absence of sulfate in the urine of a patient with neurological abnormalities and mental retardation has been attributed to a deficiency of the sulfite oxidase in the tissues, e.g., brain, liver, and kidney. The observed increase in sulfocysteine and thiosulfate was attributed to the binding of large amounts of sulfite to cysteine.[218]

### 5. *Urinary Mixed Disulfides*

A mixed disulfide of cysteine and β-mercaptolactate, 2-hydroxycystine, was isolated from the urine of a mentally defective patient. This compound is formed probably by the disulfide linkage between cysteine and β-mercaptopyruvate followed by the reduction of the keto group or from cystine after partial transamination and reduction of the keto group.[219]

# VI. TAURINE, HYPOTAURINE, CYSTEIC ACID, AND ISETHIONIC ACID

## A. Pathways of Taurine Formation

Taurine may be formed in the mammalian tissues by three pathways: (1) by decarboxylation of cysteine sulfinic acid into hypotaurine and subsequent oxidation of hypotaurine (2-aminoethane sulfinic acid) (2) by further oxidation of cysteine sulfinic acid into cysteic acid and subsequent decarboxylation of cysteic acid into taurine, or (3) by oxidation of cysteine into cystine; cystine formed is probably oxidized to cystine disulfoxide, which on subsequent decarboxylation and oxidation is converted into taurine. To what extent one pathway dominates the other *in vivo* is not known. Taurine is probably formed mainly from hypotaurine.[(220)]

### 1. *Precursors of Taurine*

The experiments on the mechanism of formation of taurocholic acid in the dog and several subsequent nutritional and radioisotopic studies showed that in the mammalian tissue methionine[(221,222)] and cystine[(221,223)] were both readily converted into taurine and sulfate; in contrast, homocystine was not readily converted into taurine.[(221)] Further studies of feeding cysteic acid with cholic acid by mouth to taurine-deficient rats showed an increased excretion of taurocholic acid, suggesting the conversion of cysteic acid into taurine.[(224)] The administration of cysteic acid to rats led to an accumulation of taurine in the liver[(225)] and an increase in the excretion of taurine in the urine.[(226)] Awapara reported an increase in the taurine content of liver on injecting cysteine into rats.[(227)] Another metabolite, 2-aminoethane sulfinic acid normally not detectable in the liver was also found in the liver. These results were confirmed by showing the formation of $^{35}$S-2-aminoethane sulfinic acid and $^{35}$S-taurine in the liver after the administration of $^{35}$S-cysteine to the rat.[(225,228)] The transsulfuration pathway of $^{35}$S-methionine was demonstrated in rat brain *in vitro*[(229)] and *in vivo*.[(230)]

Chatagner and Bergeret found that the aqueous extracts of rabbit liver preparation decarboxylated cysteine sulfinic acid into 2-aminoethane sulfinic acid (cysteamine sulfinic acid), to which they proposed the name hypotaurine.[(231)] Cysteine sulfinic acid when injected into rats was also converted into hyptotaurine.[(231-233)] $^{35}$S-Labeled hypotaurine administered to rats and mice was excreted as $^{35}$S-taurine and $^{35}$S-sulfate in the urine.[(233a)] The presence of both cysteine sulfinic acid and hypotaurine in the rat brain[(198)] gave further evidence of their intermediary role in the formation of taurine previously demonstrated in the dog[(224)] and in the rat.[(226)]

The deficiency or a complete lack of the relevant enzymes (e.g., methionine activating enzyme, cystathionine synthase, cystathionase,[(22)] and cysteine sulfinic acid decarboxylase[(234)]) in the mammalian heart indicates that a large amount of taurine present in this tissue is probably derived from its uptake from the blood. The ventricular tissue of the dog heart *in vitro* has been shown to convert $^{35}$S-L-cystine into $^{35}$S-taurine[(235)]; this suggests that a part of taurine in the heart is synthesized by the tissue by a different pathway.

### 2. *Decarboxylation of Cysteine Sulfinic and Cysteic Acids*

The enzyme catalyzing decarboxylation of L-cysteic acid into taurine was demonstrated in the liver of the dog[236] and the rat.[147] The demonstration of the enzyme decarboxylating cysteine sulfinic acid into hypotaurine[190,231,237] raised the question of whether one and the same enzyme catalyzed decarboxylation of cysteic acid and cysteine sulfinic acid; some observation in support of this was provided by Blaschko:[238,239] The activity of the enzyme to decarboxylate cysteic acid and cysteine sulfinic acid was the highest in dog liver and was found in decreasing amounts in the liver of rat, pig, guinea pig, and mouse; the cat and human liver extract contained little or no activity to decarboxylate both cysteic and cysteine sulfinic acids.[234,238,240] The decarboxylase activity was not detectable in the liver extracts of cod (cysteic acid as the substrate)[239] and of horse (cystine sulfinic acid as the substrate).[241] The activity of the enzyme to decarboxylate the two sulfur amino acids was present in the rat,[234,242] rabbit,[243] and dog[234,242] kidney and in the rat,[234] rabbit,[243] dog,[234] cat,[234] and chicken[234] brain. The cat and human kidney,[234,239] and human brain[234] contained little or no activity. The activity of cysteine sulfinic acid decarboxylase in different parts of the 1-year-old calf brain was (mμmoles cysteine sulfinic acid decarboxylated/mg protein/hr): cerebellar cortex, 221; mesencephalon, 214; diencephalon, 158; corpus striatum, 118; cerebral cortex, 68; and cerebral white matter, 38. The activity of this enzyme in diencephalon, mesencephalon, and cerebellum was 23–58% higher in the 1-year-old calf than in 3- to 7-day-old calf.[243a]

In pyridoxine deficiency the activity of the rat liver enzyme to decarboxylate cysteic acid[147,238] and cysteine sulfinic acid[238,244] was completely lost after 1 week on pyridoxine-deficient diet, suggesting that pyridoxal phosphate is the cofactor of the decarboxylase. This was confirmed in several subsequent studies.[240,243,245] The activity of the enzyme depended on the reduced state of the enzyme; the cofactor, pyridoxal phosphate, has been assumed to be linked to the sulfhydryl groups of the enzyme.[243]

The decarboxylase is relatively stable; no significant loss occurred in the activity of dog livers stored at −20°C for a few days. The activity was partially lost if the tissue was processed at room temperature; acetone powders of the liver lost most of the activity.[241] The aqueous extracts lost the decarboxylase activity if dialyzed against distilled water for 18 hr at 0–2°C;[239] such preparations, however, were reactivated by the addition of pyridoxal phosphate and, when used for studies of the decarboxylation of cysteine sulfinic acid, showed a lag period of 20–60 min on incubation. The lag period was shortened on the addition of $Fe^{2+}$ ion or completely abolished on the addition of boiled liver extract, cysteine, or 2,3-dimercaptopropanol. These results were attributed to the production of SH groups required for the action of the enzyme to decarboxylate cysteine sulfinic acid, cysteic acid, and also glutamate.[243,245] The activity of the enzyme was inhibited by *p*-chloromercuribenzoate, iodoacetate, mercuric chloride, cupric ions,[245] isonicotinyl hydrazide, and reversibly by hydroxylamine[246] and cyanide.[239] The dialyzed preparations of the enzyme do not decarboxylate cysteic acid even in the presence of pyridoxal phosphate; the activity, however, was restored immediately on the addition of a reducing agent (e.g., 2,3-dimercaptopropanol[245]). The purified preparations enriched in their activity toward cysteine sulfinic acid and with a fivefold decrease in their activity toward cysteic acid may be obtained[241]; the purified decarboxylase may be assayed for its maximal activity in the presence of pyridoxal phosphate and GSH.[234]

The decarboxylase activity of the enzyme in any one preparation was always greater with cysteine sulfinic acid as the substrate than with cysteic acid.[220] This has been observed with enzyme preparations of the liver[220,234,240] and brain[220,234,245] of several species. The chicken liver was an exception in that its activity to decarboxylate was more than fourfold with cysteic acid than with cysteine sulfinic acid.[234] The higher activity with cysteine sulfinic acid may be due to a slow reduction of the enzyme by cysteine sulfinic acid,[245] but it is also reported that even in the presence of a reducing agent and pyridoxal phosphate the activity of the enzyme to decarboxylate was considerably lower with cysteic acid than with cysteine sulfinic acid.[243]

The activity of the liver decarboxylase in male rats was twofold of that in the female rat.[238,240,247,248] It showed an increase in the liver of ovariectomized female rats.[248]

Comparative studies of the decarboxylase of brain and liver showed some interesting differences between the two tissues: (1) Whereas the apodecarboxylase of cysteine sulfinic acid of rat liver was saturated with the cofactor, pyridoxal phosphate, the corresponding apodecarboxylase of the rat brain was not, the activity being markedly increased on the addition of pyridoxal phosphate to the brain homogenate; (2) in vitamin $B_6$ deficiency there is a complete disappearance of the apodecarboxylase of the rat liver;[147,220] the apodecarboxylase in the rat brain, however, remains intact;[220] and (3) the activity of the enzyme to decarboxylate cysteine sulfinic acid and cysteic acid is greater in the particulate fraction (45%) than in the supernatant fraction (28%) of the rat brain; in the rat liver the activity of the enzyme is predominantly in the supernatant fraction.[245]

Since the suggestion that cysteine sulfinic and cysteic acids are decarboxylated by the same enzyme, Davison postulated that this enzyme was also responsible for the decarboxylation of glutamate in the brain and that the decarboxylase in the brain was different from the decarboxylase in the liver.[245] Subsequent work indicated that the decarboxylation of cysteine sulfinic and cysteic acids is catalyzed by the same enzyme and that the isoenzyme in the liver is different from that in the brain.[234] The earlier observations[240,245] of the nonadditive nature of the decarboxylation of cysteine sulfinic acid, cysteic acid, and glutamate were extended, and it was shown that glutamate inhibited noncompetitively the decarboxylation of cysteine sulfinic acid in the brain, thereby indicating that the glutamate is not decarboxylated by the same isoenzyme in the brain.[234]

### 3. *Oxidative Decarboxylation of Cystine*

*a. Cystine Disulfoxide.* This pathway was postulated by Medes, who suggested that cystine was converted into taurine after its oxidation into the disulfoxide and subsequent decarboxylation (see Section V,B):

$$\text{Cystine} \xrightarrow[\text{oxidase}]{\text{cystine}} \text{cystine disulfoxide} \xrightarrow[\text{decarboxylase}]{\text{cystine}}$$

$$\text{cystamine-disulfoxide} \xrightarrow[\text{and oxidase}]{\text{hydrolase}} \text{taurine}$$

This mechanism was suggested on the basis of the observation that cystine was specifically oxidized and decarboxylated and also that cystamine disulfoxide but not cystamine $(NH_2CH_2CH_2S)_2$ was oxidized by the liver preparations.[105]

There is no further evidence to show whether cystine disulfoxide is an obligatory intermediate in the conversion of cystine to taurine. The disulfoxide replaced cystine in the diet of cystine-deficient rats suggesting that it is reduced to cytine *in vivo*.[210] Cystine disulfoxide administered with cholic acid orally to dogs showed an increase in the excretion of taurocholic acid and an increase in the urinary sulfate.[224] This latter finding may be explained if a part of cystine disulfoxide was oxidized to taurine according to the scheme proposed by Medes and a part after its reduction into cystine.

*b. Cyst(e)amine.* The earlier work on cystamine as a precursor of taurine gave conflicting results regarding its possible role as an intermediate in the formation of taurine. Cystamine was not converted into taurine in dogs *in vivo*[250] but was converted into taurine *in vitro*.[251]

The studies reported by Eldjarn and by Cavallini and his co-workers showed that cystamine and cysteamine were converted into hypotaurine[252,253] and taurine[252] by the rat liver *in vitro* and *in vivo*. A significant amount of $^{35}S$-cystamine, administered to rat, rabbit, and man, was excreted as $^{35}S$-taurine in the urine.[252,254] These observations may be explained on the assumption that cystamine is oxidatively deaminated by a diamine oxidase into a cyclic aminoaldehyde, cystaldimine (1,2-dehydrodithiamorpholine), which on further degradation gives rise to glyoxal, thiocysteamine, cysteamine, and elemental sulfur:

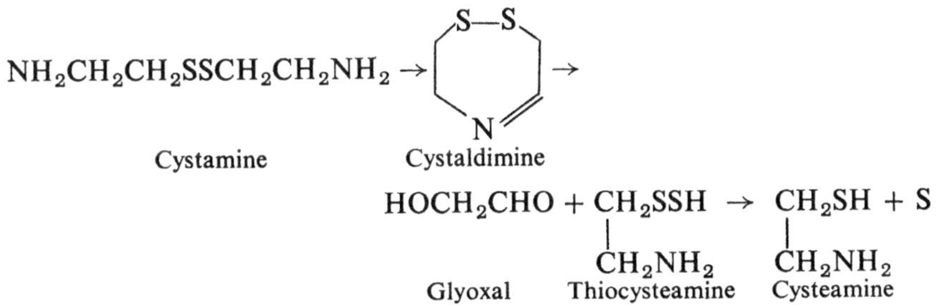

The diamine oxidase was isolated from the hog kidney[255] and was shown to contain pyridoxal phosphate.[256] Thiocysteamine formed may be oxidized into thiotaurine and hypotaurine by a monoamine oxidase found in the tissues of several species[257] or converted directly into cysteamine. Cysteamine formed may be oxidized further by an enzyme to hypotaurine in the presence of a cofactor such as sulfur, sulfide or methylene blue or certain other dyes. The enzyme found in the horse kidney used cysteamine but not cysteine as the substrate. In the absence of the enzyme the same cofactors oxidized cysteamine into cystamine.[258] Further work showed that the oxidation of cysteamine into hypotaurine occurred in the absence of the cofactor provided the substrate concentration was sufficiently low. This finding was attributed to the fact that both cysteamine (the substrate) and cystamine formed by aerobic oxidation were inhibitors of the enzyme.[259] Thus the following mechanism may be proposed for the conversion of cystine into taurine:

$$\begin{array}{lllll} \text{Cystine} & \rightarrow & \text{cystamine} & & \\ \uparrow\downarrow & & \uparrow\downarrow & & \\ \text{Cysteine} & \rightarrow & \text{cysteamine} & \rightarrow \text{hypotaurine} & \rightarrow \text{taurine} \end{array}$$

Cystamine was reduced to cysteamine *in vivo* in dogs and rabbit[260,261]; it was not converted into cystine.[254]

It is evident that cysteamine and cystamine are both converted into taurine. The question therefore arises: Is cysteamine or cystamine formed *in vivo*? No evidence has yet been obtained that they are formed *in vivo*, nor is it demonstrated that cysteine or cystine are decarboxylated by the mammalian tissues *in vitro*. Another aspect of the above studies of cysteamine or cystamine is the observation that both are oxidized into inorganic sulfate in the dog,[250] rat, rabbit, and man.[252,254] The mechanism of this oxidation is not yet known: One possible route is the oxidation of elemental sulfur formed by the degradation of thiocysteamine.

### 4. *Oxidation of Hypotaurine*

Although hypotaurine was shown as a precursor of taurine, the enzyme catalyzing the oxidation was demonstrated recently in the liver homogenates, the activity being dependent on the presence of NAD.[262]

$$NH_2CH_2CH_2SO_2H + NAD^+ + H_2O \rightarrow NH_2CH_2CH_2SO_3H + NADH + H^+$$

There was no oxygen uptake during the reaction; this explains the failure to demonstrate its presence in rat liver, kidney, and muscle by previous workers.[263] The enzyme had an optimum $p$H between 7.5 and 7.8. The purified enzyme was activated in the presence of 2,3-dimercapto-propanol, cysteine, or 2-mercaptoethanolamine and inhibited by the addition of phosphate, potassium cyanide, or ammonium sulfate. NADP was less effective as a cofactor.

## B. Cysteic Acid

### 1. *Metabolism*

Cysteic acid may be metabolized by two pathways: (1) the decarboxylation into taurine or (2) transamination with α-oxo acids. Of the two pathways the first was demonstrated in tissues of several species (see Section VI,A,2). The transamination of cysteic acid with α-oxoglutarate and oxaloacetate was shown to give rise to β-sulfonylpyruvate and glutamate or aspartate. The transaminase was demonstrated using enzyme preparations of pigeon liver, kidney, and breast

$$COOHCH(NH_2)CH_2SO_3H + COOHCOCH_2CH_2COOH \rightleftarrows$$
$$\underset{\beta\text{-sulfonylpyruvate}}{COOHCOCH_2SO_3H} + \underset{\text{glutamate}}{COOHCH(NH_2)CH_2CH_2COOH}$$

and pig heart.[264] Cysteic acid fed orally to dogs was not oxidized to sulfate, but it was excreted in its deaminated form.[265]

The metabolic fate of β-sulfonylpyruvate formed is not known. The mitochondrial preparations of the rat liver did not convert it into pyruvate.[186] One possible pathway of its further metabolism is its reduction into β-sulfonyllactate and the oxidation of β-sulfonyllactate into sulfate by the enzyme that converts β-hydroxysulfonic acids into sulfate.[192]

### 2. *Tissue Concentration*

Cysteic acid was found in the picric acid extract (<0.35 μmoles/g)[266] and the percholic acid extract (0.11 μmoles/g)[80] of the rat brain; it increased on hydrolysis of the picric acid extract (0.94 μmoles/g).[266] Its concentration in the brain was not affected in rats maintained on protein-free diets.[80]

Free cysteic acid was absent in squid blood and axoplasm, but its amide, $CONH_2CH(NH_2)CH_2SO_3H$, was found in squid blood (0.12 μmoles/ml) and in relatively large amounts in squid axoplasm (0.42–4.86 μmoles/g).[267] Rats maintained on a pantothenic acid-deficient diet plus ω-methylpantothenate excreted cysteic acid, taurine, aspartate, and β-alanine in the urine.[268]

## C. Metabolic Fate of Taurine

### 1. *Uptake and Turnover of ³⁵S-Taurine*

$^{35}$S-labeled taurine administered to the chicken[269] and rat[270–273] was found incorporated in all the organs. In 24 hr after the administration of $^{35}$S-labeled taurine to the chicken, about 16% of the injected taurine was excreted in the feces. The uptake of $^{35}S$-taurine by the tissues was as follows (percent per gram tissue): heart, 1.37; duodenum, 0.91; spleen, 0.81; liver, 0.47; kidney, 0.18; blood, 0.13; brain, 0.07; and breast muscle, 0.04.(269) The uptake of $^{35}S$-taurine was 111m $\mu$moles in the heart and 21m $\mu$moles/g in the brain of the rat at 24 hr after the intraperitoneal injection of $^{35}S$-taurine.(273) The turnover of $^{35}S$-taurine varied in different tissues of the rat.(271) The studies *in vitro* demonstrated that taurine was accumulated in the ascites carcinoma cells by an active process of transport; $\beta$-alanine inhibited the process.(274) The preparations of the mouse brain *in vitro* showed a rapid accumulation of taurine; the uptake was greater by the slices of the newborn than of the adult mouse brain. The relative figures for the accumulation of taurine in slice of the adult brain of various species were mouse, 14.3; rat, 13.6; guinea pig, 6.17; hen, 7.28; and frog, 19.0.(275)

The finding that the daily excretion of taurine in the urine (800–2400 $\mu$moles) and in the bile (800–1600 $\mu$moles) of a healthy person suggests that in man about 1600–4000 $\mu$moles of taurine required(252) are either provided by synthesis or through diets. The provision of this amount equivalent to an endogenous synthesis of 23–57 $\mu$moles taurine/24 hr/kg body wt. would amount to 13–31% of the daily requirement(276) of the total sulfur amino acids as cystine (1670 $\mu$moles) and methionine (9400 $\mu$moles). By injecting $^{35}S$-taurine the turnover of taurine in the rat was found to correspond to a half-life of 12–13 days and to an amount of 35 $\mu$moles of taurine synthesized endogenously in 24 hr/100 g body wt. of the rat.(272) This value is in good agreement with the value 22–52 $\mu$moles taurine/24 hr/100 g body wt. of the rat, calculated for the endogenous synthesis of taurine based on the assumption that 13–31% of the daily requirement(252) of the total sulfur amino acids as cystine (42 $\mu$moles) and methionine (335 $\mu$moles) per day for a 250 g rat are converted into taurine.

## 2. *Isethionic Acid*

Taurine, once considered an end metabolic product with no further metabolism, has been shown to give rise to isethionic acid (2-hydroxyethane sulfonic acid) in the dog heart *in vitro*(235) and in the rat heart and brain *in vitro* and *in vivo*.(273)

$$\underset{\text{Taurine}}{SO_3HCH_2CH_2NH_2} \rightarrow \underset{\text{Isethionic acid}}{SO_3HCH_2CH_2OH}$$

Taurine was not converted into isethionic acid in the presence of other tissue preparations of the rat.(273) The administration of isethionic acid orally to dogs did not lead to an excretion of sulfate in the urine.(265)

Isethionic acid is present in the squid blood (0.47 $\mu$moles/ml)(267) and in large amounts in the squid axoplasm (222, 165 $\mu$moles/g)(267,277) dog heart (1.02 $\mu$moles/g),(278) and rat heart (3.38 $\mu$moles/g).(278) It was detected in human urine specimens: Its concentration was 108.6 $\mu$moles/24 hr in patients with overt muscle wasting and 55.5 μmoles/24 hr in patients with muscular atrophy.[279]

## D. Bound and Other Forms of Taurine

Besides its conjugation with cholic acid (taurocholic acid), taurine has been reported in several other bound forms. A peptide fraction isolated from the plasma contained taurine, glycine, alanine, valine, leucine, aspartate, glutamate, and a diamine.[280] It was also found in a conjugated form with ornithine and other amino acids in the chicken liver, kidney, breast muscle, and gizzard muscle. About 7% of the orally administered $^{35}$S-taurine to the chicken was found in this conjugated form.[269] An interesting compound containing mainly phosphate, γ-aminobutyramide, and taurine or taurine containing amide was found in the viscous spiral of the garden spider, *Araneus diademetus.*[281] The presence of a bound form of taurine was indicated in the cat kidney and urine.[84]

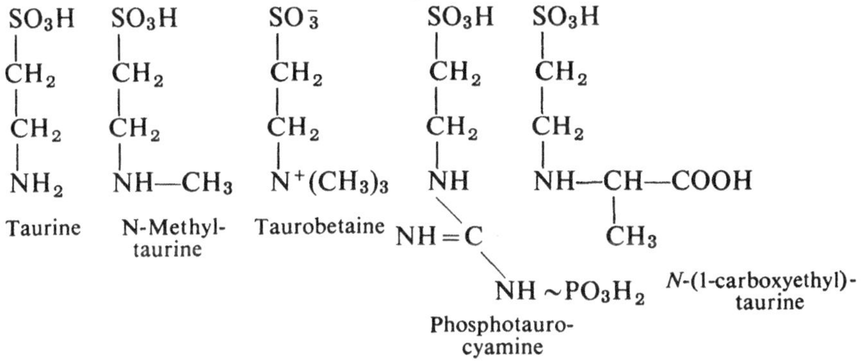

Taurine; N-Methyltaurine; Taurobetaine; Phosphotaurocyamine; *N*-(1-carboxyethyl)-taurine

In the muscle of the marine worms (*Arenicola marina*) taurine is present as taurocyamine and phosphotaurocyamine,[282,283] as carbamyltaurine in the urine of the rat[284] and man,[285] as taurocyamine in the brain (0.03 μmoles/g)[286] and urine[284] of the rat, as taurobetaine in *Briareum asbestinum* (*Pallas*),[287] and as *N*-methyltaurine and di-*N*-methyltaurine[288] and as *N*-(1-carboxyethyl) taurine[289] in red algae.

## E. Concentration of Taurine in Tissues

### 1. *Whole Tissue Distribution*

Large amounts of taurine were isolated from a marine source, abalone, and also from ox bile[290] and several other marine invertebrates[291] and squid axoplasm[277] either as free taurine or in several bound forms. Taurine together with alanine accounted for more than 10% of the total solutes in the axoplasm of *Carcinus* nerve. The following values in μmoles/g wet wt. have been reported: *Carcinus* (crab)[291] whole leg nerve, 65; *Homarus* (lobster)[291] whole leg nerve, 12; *Sepia* (cuttlefish)[291] single axon, 103; and squid axoplasm,[267,277] 76, 107.

Since the demonstration of its presence in the mouse(292) it has been reported in several other species, e.g., in rat,(76,293) cat,(294) rabbit,(295) and cow.(76,296) Taurine was detected in the serum of the adult cow(296) and was found in large amounts in the blood cells of the domestic fowl;(297) in human blood: erythrocytes (<7.0,(95) 3.6(90) μmoles/100 g water), leukocytes (2868,(95) 2603(90) μmoles/100 g water), and platelets (2100 μmoles/100 g water(90); in adult human plasma (6.32–10.4,(93) 1.6–6.4(94) μmoles/100 ml); in the plasma of fasting adults (3.2–13.8 μmoles/100 ml) and 9 to 24-month-old children (1.9–9.1 μmoles/100 ml);(90) in the adult human serum (18.0–21.3 μmoles/100 ml, of the lactating mother);(96) in the human cerebrospinal fluid (0.54,(99) 2.56,(100) 0.58(91) μmoles/100 ml); in rat blood (see below in this section, 3d); in rat plasma on the nineteenth day of gestation (20 μmoles/100 ml);(83) in human milk (19.5–32.5 μmoles/100 ml); and in cow's milk (2.64–28.0 μmoles/100 ml).(96) A brief summary of the distribution of taurine in several organs of the mammalian species is given in Table V. Of the tissues analyzed, heart showed the highest concentration of taurine.

Taurine content (all values reported below are given in μmoles/g fresh tissue) of the brain of several other species has also been reported: hen, 0.97,(303) 1.48,(304) 3.23,(305) 3.82;(275) frog, 0.05,(305) 2.72;(305) catfish,

**Table V**

**Concentration of Taurine in Mammalian Tissues[a]**

| Subject | | (μmoles/g wet wt.) | | | | | | |
|---|---|---|---|---|---|---|---|---|
| | | Brain | Liver | Kidney | Heart | Muscle | Spleen | Testis/uterus |
| Rat: | male | 5.5[b] | 1.6[c] | 7.4[d] | 28.4 | 9.4 | 11.0[e] | 2.2[f] |
| | female | 5.5 | 5.3 | 8.1 | 26.7 | 10.7 | 11.6 | 7.2 |
| Rabbit: | male | 1.1 | 0.0 | 1.9 | 14.6 | 1.3 | 5.7 | |
| | female | 1.8 | 0.0 | 2.4 | 15.8 | 3.7 | 5.6 | 1.6 |
| Guinea pig: | male | 1.0[g] | 0.0 | 1.8 | 9.3 | 9.1 | 5.6 | |
| | female | 0.6 | 0.0 | 1.0 | 12.3 | 9.2 | 5.2 | 3.5 |
| Pig | | | 1.1[h] | 7.4 | 33.7 | 10.4 | 5.1 | |
| Beef | | | 5.3 | | 3.5 | 8.7 | 3.4 | |
| Sheep | | | 3.7 | 4.1 | 7.2 | | 8.8 | |
| Man | | 3.0 | 1.8 | | | | | |

[a] The values given above are taken from Awapara(298) except those for man.(299) Other workers have reported the values given in the following footnotes.

[b] See text Section VI,E,1.

[c] 4.8,(300) 5.5,(299) 4.9(270) for males and 5.6–8.8 for females.(300)

[d] 6.7(270) and 10.1.(301)

[e] 14.5(270) and 17.1.(301)

[f] 3.3(270) and 2.0(301) for testis and 7.6(270) for uterus.

[g] 2.37.(275)

[h] 8.0.(302)

3.4;[305] dog, 1.25,[306] 1.9;[299] rat, 5.5[298] mouse, 8.31,[275] 10.74;[307] cat, 1.9;[84] and garden dormouse, 3.61 (awake) and 3.58 (in hibernation).[87] The dog liver contained 7.2 μmoles/g wet tissue.[299] Different values of taurine have been reported for the rat brain: 2.37,[293] 5.35,[308] 2.57,[303] 3.92,[305] 5.3 (male), 4.3 (female),[270] 5.32,[301] 2.71,[199] 3.36,[309] 3.00,[309a] 2.9–4.8,[272] 4.9,[310] 3.4,[80] 6.63.[275] The combined use of ion-exchange chromatography and a specific spectrophotometric method for taurine gave values of 6.75 μmoles/g of rat brain.[311] Taurine was found in the spinal cord (0.84, 0.49 μmoles) and sciatic nerve (0.52 μmoles) of the hen[303,304] and in the spinal cord (0.05 μmoles)[85] of the cat. The distribution of taurine in other tissues of the male rat except where indicated otherwise was thyroid, 5.0;[270] hypophysis,[270] male, 13.6, and female, 19.5, lung, 10.5,[270] 11.2;[301] and pancreas, 4.6,[270] 2.83.[301] Of the ox tissues examined the pineal contained more taurine: pineal, 5.20; brain (frontal cortex), 0.29; anterior pituitary, 2.88; and posterior pituitary, 3.92.[86a]

## 2. *Regional Distribution in the Brain*

The autopsy samples of various parts of the human brain (19-year-old boy) gave a similar distribution of taurine.[166] The following values in μmoles/g wet tissue were found: frontal lobe, 1.45; caudate nucleus, 1.21; globus pallidus, 1.13; thalamus, 0.74; and hypothalamus, 1.19. The regional distribution for the brain of 5-month-old foetus and 8-month-old foetus also was reported. Some differences in the regional distribution of taurine was observed in the rat brain:[309a,312] hemispheres, 5.34; midbrain, 2.88; cerebellum, 3.10, 4.89; and pons and medulla, 1.06, 1.89.

## 3. *Factors Affecting Taurine Concentration in Tissues*

Several factors are known to affect the levels of taurine in tissues and urine, but most of these factors have no effect on the levels of taurine in the brain.

*a. Dietary Intake.* About 20% of the organic sulfur fed as methionine or cystine to dogs is utilized for the synthesis of taurocholic acid, and 70% is oxidized to sulfate, which is excreted in the urine.[221] The endogenously synthesized taurine corresponds to a value of 35 μmoles taurine/24 hr/100 g body wt.[272] A considerable amount of taurine, however, is consumed through foods of both animal and vegetable origin.[313] Taurine administered to rats and man is excreted in the urine.[254] the urinary output being increased on increasing the amount of taurine administered to the rat.[314] This explains why in some experiments about 63% of the taurine administered to man was presumably excreted as taurine in the urine[315] and in other experiments only 25% taurine was recovered in the urine.[252] The amount eliminated in the urine also differed somewhat depending on whether taurine was administered by mouth (59%), subcutaneously (62%), or intravenously (72%). Its excretion in the urine decreased in rats fed diets low in protein content.(314) This reduction in excretion of taurine suggests a greater utilization of methionine and cystine for various anabolic processes in the tissues. It is known that cholate is formed from cholesterol in several species.(316,317) Thus increased dietary intake of cholesterol would divert a greater proportion of taurine for the formation of bile salts and would reduce the urinary excretion of taurine.(189,314)

*b. Fasting.* The urinary excretion of taurine in normal human subjects was between 280 and 2400 μmoles in males, 216 and 1288 μmoles in females,(94) and 18 and 112 μmoles in 9 to 24-month-old children(90) per day. The lactating mother excreted 30 μmoles taurine/g creatinine at 4 weeks and 670–800 μmoles taurine/g creatinine at 3 months after delivery.(96) In adult male subjects the urinary output of taurine decreased from 1250 μmoles to 578 μmoles during the 24 hr period following fast.(95) In rats fasted for 5–9 days the urinary excretion of taurine increased from the control values of 4 μmoles to 29–45 μmoles/24 hr/200 g body wt.(318) Taurine levels of the liver also increased in male rats, but they showed a decrease in female rats on 7–9 days fasting.(298,318) In contrast, the taurine levels in the brain and in other organs remained unchanged on fasting(298) or on feeding rats with protein-deficient diets.(80)

*c. Age.* There appears to be some variation with age in the amount of taurine. Taurine was not found in the 18–20 hr chick embryo(319) but was present in large amounts in the 8 to 9-day-old chick embryo.(320,321) In the whole embryo it increased from 0.25 μmoles on the first day after incubation to 4.92 μmoles on the third day and to 9.72 μmoles/100 mg dry protein on the fourth day after incubation. Marked changes also were observed in the concentration of taurine of the chick embryo heart during incubation (all values given in μmoles/g wet weight): 4.4 μmoles/g fresh tissue on seventh day; 6.2 μmoles on the eleventh day; 6.9 μmoles on thirteenth day; 8.9 μmoles on fourteenth day; 11.2 μmoles on sixteenth day; 19.0 μmoles in the newly hatched; and 18.0 μmoles in the adult. Taurine concentration is somewhat greater in the brain of 12-day chick embryo (6.4 μmoles) than in the brain of the newly hatched chick (5.5 μmoles).(322) Free taurine is also found in the eggs, larvae, pupae and adult mosquito; its concentration increased during development from the larvae to pupae stage.(323)

In mammals taurine was also detected in the very early stages of embryonic development. It was present in the muscle of the calf embryo.(296) The concentration of taurine decreased in the brain and liver but increased in the heart during development (all values in μmoles/g wet weight): newborn mouse brain, 18.5; adult mouse brain, 8.31;(275) and rat brain at 25 days, 6.94, and at 210 days, 5.32.(301) With increasing age after birth there was a rapid decrease in the taurine content of the rat brain; this decrease was significantly retarded in hypothyroid rats from the twentieth to the one hundredth day after birth.[324] The rat heart showed a consistent increase in the taurine concentration with increasing age[301] or body weight:[272] It was 19.9 μmoles/g at 25 days and 38.6 μmoles/g at 210 days after birth.[301] No marked changes were observed in other tissues, except the ovaries.[272]

The human foetal liver contained 2.4 μmoles and adult human liver, 0.77 moles.[88] The whole brain of the human foetus at 5 months (4.6 μmoles) contained less taurine than at 8 months (5.7 μmoles). However, a decrease of taurine was observed in several parts of the human brain during development, e.g., in the caudate nucleus its concentration decreased from 5.0 μmoles in the 5-month-old foetus to 2.9 μmoles in the 8-month-old foetus; in the 19-year-old boy it was 1.2 μmoles.[166] The concentration of taurine in the human plasma was greater in the newborn (14.1 μmoles/100 ml)[92] than in the plasma of 4 to 16-year-old children (4.5 μmoles/100 ml)[91] and of adults 6.65 μmoles/100 ml).[92]

In normal human subjects the urinary excretion of taurine increased with age.[325]

*d. Sex.* Some differences were observed in the taurine content of the rat liver: male, 1.6 μmoles and female, 5.3 μmoles/g liver (also see Table V). $^{35}$S-Cystine injected into rats was converted into taurine more in the liver, kidney, and brain of the female rat than of the male rat.[271] This observation is in contrast to the reported higher activities of the enzyme decarboxylating cysteic acid and cysteine sulfinic acid in the liver of male than of female rats.[238,240,247,248] The concentration of taurine in the blood however, was, higher in the male than in the female rat (μmoles/100 g of blood cells and plasma, respectively):[270] male, 58, 26, and female, 16, 12.

*e. Pyridoxine Deficiency.* Taurine[244,326] and hypotaurine,[244] the two constituents of the normal rat urine, were not found in the urine of vitamin $B_6$-deficient rats.[244] The urinary excretion of taurine decreased from 149 μmoles/24 hr in the normal rat supplemented with pyridoxine to a value of 88μmoles/24 hr.[327] Taurine content of the brain, however, remained unchanged in pyridoxine deficiency;[307,327,328] it was 5.14 μmoles/g in the normal rat brain and 5.2 μmoles/g in the pyridoxine-deficient rat brain.[327] This finding is consistent with the observed disappearance of the apodecarboxylase of cysteine sulfinic acid from rat liver but not from rat brain.[220]

*f. Vitamin E Deficiency.* In the dystrophic condition produced by vitamin E deficiency there was a decrease in the concentration of several amino acids of the rabbit brain and kidney. The concentration of taurine, however, remained unaltered in the brain and other organs, except liver; its concentration in μmoles/100 mg dry wt. in the normal and vitamin-E deficient rabbits was, respectively:[329] brain, 1.08, 1.04; liver, 0.24, 0.63; kidney, 1.51, 1.75; heart 6.72, 6.6; muscle, 0.74, 0.43; and spleen, 2.74, 2.83.

*g. Radiation.* Normal rats exposed to 500–600 r showed excessive urinary excretion of taurine (40–80 μmoles/100 g body wt./72 hr). This effect was further increased in adrenalectomized rats[330] and in rats with vitamin $B_6$ deficiency.[328] Taurine content of the rat brain, however, was not affected.[328]

*h. Diseases and Mental Disorders.* Taurine content of the human plasma was not appreciably disturbed in various pathological conditions (μmoles/100 ml plasma):[93] normal range, 6.3–10.4; osteomalacia, 5.9–6.5; hypophosphatasia, 8.8; hyperparathyroidism, 7.8; adult Fanconi syndrome, 6.2; child Fanconi syndrome, 6.2–18.4; and Hartnup disease, 5.8. Cerebral cortex of the human brain biopsy samples of patients with brain tumor contained the following amounts of taurine (μmoles/g wet weight): [166] 1.34 (3-year-old girl); 0.72 (17-year-old boy); and epilepsy, 1.14 (35-year-old woman).

The following values have been reported for the concentration of taurine in various diseases (μmoles/100 ml CSF): (1) Normal, 0.58; phenylketonuria, 0.59–0.65; Tay–Sachs disease, (0.36–0.45; Huntington's chorea, 0.58–0.70; Hurler's syndrome, 0.30–0.71; tuberous sclerosis, 0.53, 0.68; Schilder's disease, 0.59; and cerebro-oculorenal disease, 0.28.[91] (2) Normal, 0.54; mentally retarded, 0.53; mentally and physically retarded, 0.70; seizures, 0.62; convulsive disorder, 0.66, 0.86; retardation and convulsions, 0.45; pneumonitis, 0.22; anaemia, 0.46; motor accident, 0.53; traumatic brain injury, 0.85; psychotic, 0.94; and encephalopathy, 0.78.[99]

On administration of chlorambucil to patients with chronic lymphatic leukemia the plasma showed little or no taurine and the leukocytes showed a loss of approximately 85% of taurine.[331]

Changes in the sulfur amino acid levels have been reported in the blood plasma of schizophrenic patients: protein-free filtrates of the plasma showed a higher concentration of taurine and a lower concentration of cyst(e)ine.[325]

*i. Taurinuria.* A high frequency of low taurine excretors was found among the mongoloid.[332] The hand abnormality characterized by a permanent flection of the fingers at the proximal interphalangeal joints was shown in 16 out of 17 people to be associated with a specific overexcretion of taurine in the urine. This condition was inherited as an autosomal dominant characteristic.[333] A specific overexcretion of taurine also was demonstrated in cases of pernicious anaemia and in subacute combined degeneration of the cord.[334]

In cystinuria, urinary excretion of taurine was between 64 and 690 μmoles/day as compared with the normal urinary excretion of 800 to 2400 μmoles/day.[207]

*j. Miscellaneous.* Taurine content of dog brain (1.25 μmoles/g) decreased during seizures (0.89 μmoles/g) and increased during depression (1.64 μmoles/g), produced on administration of $NH_4Cl$; no changes were found post mortem (1.18 μmoles) or under anoxia (1.21 μmoles), ammonia poisoning, electroshock, picrotoxin seizures (1.27 μmoles), and pentylenetetrazole seizures (1.22 μmoles).[306] Rats under insulin coma[199,309] or on administration of reserpine,[266] convulsive hydrazides,[335] and ethanol[336] showed no changes in taurine concentration of the brain. The administration of chlorpromazine to rats on 4 successive days showed no changes in the concentration of taurine in the whole brain, but its concentration decreased from 3.10 to 1.71 μmoles/g in the cerebellum and increased from 2.08 to 2.55 μmoles/g in the cerebral cortex.[309a] Taurine concentration of the brain showed an increase from 4.9 to 6.3 μmoles at 30 min after an injection of 100 mg phenylalanine to male rats weighing between 150 and 175 g.[310] It showed a decrease on administration of hydrazine-type monoamine oxidase inhibitors.[266]

The urinary excretion of taurine in the rat has been reported to vary with the animal strain. It showed a progressive increase in rats during gestation: Taurine was absent, or present in negligible amounts, in the postpartum urine samples.[268]

## VII. PHYSIOLOGICAL ACTION AND FUNCTION OF SULFUR AMINO ACIDS

### A. Toxicity

Feeding of excessive amounts of methionine proved fatal in about 65% of the rats and produced liver atrophy in rats that survived.[337] Large amounts of methionine in the diet produced ketosis[338] and acidosis[29] in man; a rise in blood keto acid levels was observed in patients with liver damage[339] and in hepatic coma.[340] Large amounts of taurine given to man showed no toxic symptoms other than symptoms of diarrhea. Taurine, when fed orally, was found to be toxic for rabbits.[285] These toxic effects in man and rabbits were not confirmed.[341] Taurine added to diets deficient in cystine had no effect on the growth of rats.[342–344]. Rats given excessive amounts of taurine showed no signs of toxicity in contrast to feeding of excessive amounts of cysteic acid[337] and cystine,[345] which produced liver necrosis and cirrhosis. In patients suffering from psoriasis, a daily oral dose of 2 g taurine produced a predictable itching effect in about 1 hr and persisted usually for 24–48 hr.[313]

### B. Intracellular Ions

Since the suggestion that taurine may function in the maintenance of osmotic pressure in the marine invertebrate[346] it also has been considered necessary to provide the anions required to balance the intracellular cations.[291] In the squid axon, taurine (77 μmoles/g) and isethionic acid (222 μmoles/g) account for more than 50% of the total anions (509 μmoles/g).[277] The high concentration of the two sulfonic acids raised the question of whether they were involved in the electrical phenomenon of the nerve.[277] The significance of the occurrence of yet another sulfonic acid, cysteic acid amide,[267] in the squid axon is not known.

## C. Retention of Intracellular Potassium

L-Cysteine inhibits the retention of potassium by brain slices and probably interferes with the intracellular processes involved in the transport of potassium.[347,348] It shows an apparent inhibition of the synthesis *in vitro* of acetylcholine by the enzyme acetyl transferase in the brain by spontaneous transacetylation with acetyl CoA, thus depriving the substrate.[349]

Taurine administered to dogs by intravenous injection depresses the arrhythmia of the heart induced by adrenaline and digoxin; it does not alter the heart rate or blood pressure and shows no toxicity to the heart. Taurine, a zwitterion, is believed to enter the cell as a cyclic structure which, after reductive deamination, yields isethionic acid. The latter binds with potassium, producing a salt and thereby decreasing the cardiac excitability.[350] Sodium salt of ethanesulfonic acid and 3-amino-propane sulfonic acid (homotaurine) were less active than taurine. The infusion of taurine caused an uptake of potassium by the heart.[351] In view of the paucity or virtual absence of the enzymes involved in the transsulfuration of methionine into cysteine and of the enzyme cysteine sulfinic acid decarboxylase, it appears that most of the intracellular taurine of the heart is transported directly from the blood.

In the presence of taurine rat brain slices show a small increase in the uptake of potassium.[347]

## D. Neuronal Excitants or Depressants

In a systematic study of the structure of the amino acids in relation to neuropharmacological action, Curtis and Watkins[108] found that several of the sulfur amino acids on electrophoretic application through micropipettes produced either excitation or inhibition of the cat spinal interneurons. Some of the sulfur amino acids studied by these workers are formed *in vivo* during the metabolism of methionine by transsulfuration pathway and some that are not yet known to be formed *in vivo* may be expected to be formed directly from homocysteine. Thus, on the basis of the general formula of the amino acid

$$X(CH_2)_nCHCOOH \quad (\text{with } NH_2 \text{ on CH})$$

$$\begin{array}{c} X(CH_2)_n\underset{\displaystyle NH_2}{\underset{|}{C}}HCOOH \end{array}$$

the following L-sulfur amino acids produced excitation in intensity similar to that observed with L-aspartate and L-glutamate: cysteine sulfinic acid ($X = SO_2H$, $n = 1$); cysteic acid ($X = SO_3H$, $n = 1$); homocysteine sulfinic acid ($X = SO_2H$, $n = 2$); and homocysteic acid ($X = SO_3H$,

$n = 2$). The presence of homocysteine sulfinic acid in the brain is not yet shown, but in view of the known oxidation of homocysteine by rat liver enzymes it will be of interest to find out if this compound occurs in the brain or if the enzyme oxidizing homocysteine into homocysteine sulfinic acid and homocysteic acid is present in the brain. D-Homocysteic acid ($X = SO_3H$, $n = 2$) was more potent as an excitant than D-glutamic acid ($X = COOH$, $n = 2$). The excitant activity of sulfinic or sulfonic acid decreased with further increase in the carbon chain ($n = 3$). L-Cysteic acid and DL-homocysteic acid produced a prompt depolarization from resting membrane potentials of the guinea pig cortex slices.(352)

On decarboxylation the sulfur amino acids, $X(CH_2)_nCH_2NH_2$, lost excitant action and the sulfonic acids then behaved as depressants. Thus hypotaurine ($X = SO_2H$, $n = 1$), taurine ($X = SO_3H$, $n = 1$), and homotaurine ($X = SO_3H$, $n = 2$) showed a depressant action.(108) It also is shown that the depressant action of taurine resembles that of glycine, a probable inhibitory transmitter in the spinal cord, in its susceptibility to strychnine antagonism; that of homotaurine resembles GABA, another probable inhibitory transmitter of the central nervous system, in not being antagonized by strychnine.(109)

Cystathionine, which does not conform with the above general formula, has been reported to inhibit cat spinal interneurons(353) and toad spinal neurons.(354)

## E. Mental Disorders

The administration of large amounts of methionine (20 g/70 kg/day) to schizophrenic patients maintained on amine oxidase inhibitor drugs has been reported to cause and exacerbation of the symptoms of schizophrenia. These effects have been attributed to a disturbance in the normal methylation processes, especially where biogenic amines act as acceptors of methyl groups.(355,356) Some therapeutic effect of methionine was observed in epileptic children.(357,358) It is known that the administration of large amounts of methionine to normal human subjects produced ketosis(338) and a net increase in the urinary acids (inorganic sulfate).(29) Further studies would reveal if the overall oxidation of methionine to sulfate also is affected in these patients.

Methionine-sulfoximine,

$$COOHCH(NH_2)CH_2CH_2\overset{\overset{\displaystyle O}{\|}}{\underset{\underset{\displaystyle NH}{\|}}{S}}CH_3$$

which produced generalized convulsive seizures in animals(359) and was shown to be an antagonist of methionine, was reported to produce a toxic psychosis in man.(360) It has been claimed to have a therapeutic effect in schizophrenic patients.(361)

## F. Miscellaneous

Both taurine and cystamine reduced blood sugar in the bitch. Guanidotaurine [taurocyamine, $NH_2C(NH)NHCH_2CH_2SO_3H$], produced an increase in blood sugar and then a decrease; however, diguanidocystamine, $[NH_2C(NH)NHCH_2CH_2S]_2$, did not cause a similar initial rise in blood sugar. In contrast, tetramethyldiguanidocystamine caused a significant rise in the blood sugar. Little or no effect on blood pressure was observed on the injection of taurine, guanidotaurine, or methylguanidotaurine, in contrast to a decrease in blood pressure observed on the administration of either cystamine, guanidocystamine, or methylguanidocystamine. Some side effects of toxicity were produced on the injection of methylguanidocystamine.[(362)]

Taurine potentiates the action of insulin in rabbits.[(363)] It acts as a growth factor in the chick during the first few weeks of life by donating its sulfur as a source of sulfate in the synthesis of chondroitin sulfate[(364)] and stimulates the growth of chick embryo tissue cultured *in vitro*.[(365)]

# VIII. CONCLUSION

$^{35}$S-Labeled methionine is converted by brain *in vitro* and *in vivo* into $^{35}$S-labeled cyst(e)ine, taurine, and sulfate. A part of the injected $^{35}$S-methionine and $^{35}$S-cysteine formed from methionine is incorporated into brain proteins. These results show that metabolic pathways of methionine (Fig. 1) in brain are similar to those in liver. Evidence has been obtained for the presence of the following enzymes in the brain: (1) methionine-activating enzyme (methionine $\rightarrow$ S-adenosylmethionine), (2) cystathionine synthase, (3) cystathionase, (4) cysteine oxidase (cysteine $\rightarrow$ cysteinsulfinic acid), (5) cysteinesulfinic acid–oxaloacetate transaminase, and (6) cysteinesulfinic acid decarboxylase. Whether biosynthesis of methionine, catalyzed by methyltransferase or methionine synthetase, occurs in brain is not known. However, it is known that folic acid concentration is significantly greater in the CSF than in the human plasma and, therefore, it will be of interest to find out if methionine synthetase is present in the brain.

A considerable amount of methionine in the brain is present as S-adenosylmethionine; S-adenosylhomocysteine formed after transmethylation is hydrolyzed to homocysteine. The cleavage enzyme catalyzing the hydrolysis has not yet been demonstrated in the brain. Taurine, until recently considered metabolically inert, is converted into isethionic acid in the brain.

The enzyme that oxidizes cysteine to cysteinesulfinic acid is also known to oxidize homocysteine: The question arises as to whether a part of homocysteine is oxidized in the brain to homocysteinesulfinic acid, homohypotaurine, and homotaurine. Since several of the sulfur amino acids show transmitterlike activity either as excitants (cysteinesulfinic acid, homocysteinesulfinic acid, cysteic acid, and homocysteic acid) or as depressants (taurine, hypotaurine, and homotaurine) when applied into the environment of single cells within the spinal cord and brain of the cat, it is not unlikely that a disturbance in the metabolism of methionine may affect the functional activity of the brain.

Several types of sulfur amino acidurias have been reported. Of these, homocystinuria, cystathioninuria, and mixed disulfide (cysteine + $\beta$-mercaptolactate) uria were found in mentally defective patients.

## IX. REFERENCES


1. V. du Vigneaud, *A Trail of Research in Sulfur Chemistry and Metabolism and Related Fields*, Cornell University Press, Ithaca (1952).
2. C. Fromageot, *in The Harvey Lectures*, Vol. 49, pp. 1–36, Academic Press, New York (1955).
3. J. A. Stekol, *in A Symposium on Amino Acid Metabolism* (W. D. McElroy and H. B. Glass, eds.), pp. 509–557, The Johns Hopkins Press, Baltimore (1955).
4. T. P. Singer and E. B. Kearney, *in A Symposium on Amino Acid Metabolism* (W. D. McElroy and H. B. Glass, eds.), pp. 558–590, The Johns Hopkins Press, Baltimore (1955).
5. L. Young and G. A. Maw, *The Metabolism of Sulphur Compounds*, Methuen, London (1958).
6. D. D. Dziewiatkowski, *in Mineral Metabolism* (C. L. Comar and F. Bronner, eds.), Vol. 2, Pt. B, pp. 175–220, Academic Press, New York (1962).
7. A. Meister, *Biochemistry of the Amino Acids*, Vol. 2, pp. 757–818, Academic Press, New York (1965).
8. S. K. Shapiro and F. Schlenk (eds.), *Transmethylation and Methionine Biosynthesis*, University of Chicago Press, Chicago (1965).
9. J. G. Jacobsen and L. H. Smith, Jr., Biochemistry and physiology of taurine and taurine derivatives, *Physiol. Rev.* **48**:424–511 (1968).
10. P. Berg, Participation of adenyl-acetate in the acetate-activating system, *J. Am. Chem. Soc.* **77**:3163–3164 (1955).
11. M. B. Hoagland, E. B. Keller, and P. C. Zamecnick, Enzymatic carboxyl activation of amino acids, *J. Biol. Chem.* **218**:345–358 (1956).
12. M. B. Hoagland, M. L. Stephenson, J. F. Scott, L. I. Hecht, and P. C. Zamecnik, A soluble ribonucleic acid intermediate in protein synthesis, *J. Biol. Chem.* **231**:241–257 (1958).
13. H. Borsook, On the role of the oxidation in the methylation of guanidoacetic acid, *J. Biol. Chem.* **171**:363–375 (1947).
14. G. L. Cantoni, Methylation of nicotinamide with a soluble enzyme system from rat liver, *J. Biol. Chem.* **189**:203–216 (1951).
15. G. L. Cantoni, Activation of methionine for transmethylation, *J. Biol. Chem.* **189**:745–755 (1951).
16. G. L. Cantoni, S-Adenosylmethionine: A new intermediate formed enzymatically from L-methionine and adenosinetriphosphate, *J. Biol. Chem.* **204**:403–416 (1953).
17. G. L. Cantoni and J. Durell, Activation of methionine for transmethylation, *J. Biol. Chem.* **225**:1033–1048 (1957).
18. G. L. Cantoni and E. Scarano, The formation of S-adenosylhomocysteine in enzymatic transmethylation reactions, *J. Am. Chem. Soc.* **76**:4744 (1954).
19. G. L. Cantoni, The nature of the active methyl donor formed enzymatically from L-methionine and adenosinetriphosphate, *J. Am. Chem. Soc.* **74**:2942–2943 (1952).

20. G. L. Cantoni and P. J. Vignos, Jr., Enzymatic mechanism of creatine synthesis, *J. Biol. Chem.* **209**:647–659 (1954).
21. G. A. Maw, Thetin-homocysteine transmethylase, *Biochem. J.* **63**:116–124 (1956).
22. S. H. Mudd, J. D. Finkelstein, F. Irreverre, and L. Laster, Transsulfuration in mammals, *J. Biol. Chem.* **240**:4382–4392 (1965).
22a. R. J. Baldessarini and I. J. Kopin, S-Adenosylmethionine in brain and other tissues, *J. Neurochem.* **13**:769–777 (1966).
23. Y. Natori, Studies on ethionine, *J. Biol. Chem.* **238**:2075–2080 (1963).
24. S. H. Mudd, J. D. Finkelstein, F. Irreverre, and L. Laster, Homocystinuria: An enzymatic defect, *Science* **143**:1443–1445 (1964).
25. J. D. Finkelstein, S. H. Mudd, and L. Laster, Homocystinuria due to cystathionine synthetase deficiency: The mode of inheritance, *Science* **146**:785–787 (1964).
26. E. Borek and H. Waelsch, Metabolism of methionine and its derivatives with tissue slices, *J. Biol. Chem.* **141**:99–103 (1941).
27. E. S. Canellakis and H. Tarver, Studies on protein synthesis *in vitro*. IV. Concerning apparent uptake of methionine by particulate preparation from liver, *Arch. Biochem. Biophys.* **42**:387–398 (1953).
28. H. Waelsch, The excretion of keto acids, *J. Biol. Chem.* **140**:313–314 (1941).
29. J. Lemann, Jr., and A. S. Relman, The relation of sulfur metabolism to acid–base balance and electrolyte excretion: The effects of DL-methionine in normal man, *J. Clin. Invest.* **38**:2215–2223 (1959).
30. H. Waelsch and E. Borek, The stability of the keto acid from methionine, *J. Am. Chem. Soc.* **61**:2252 (1939).
31. E. S. Canellakis and H. Tarver, The metabolism of methyl mercaptan in the intact animal, *Arch. Biochem. Biophys.* **42**:446–455 (1953).
32. E. V. Rowsell, Transamination to pyruvate and some other α-keto-acids, *Nature* **168**:104–106 (1951).
33. F. Challenger and J. M. Walshe, Methyl mercaptan in relation to foetor hepaticus, *Biochem. J.* **59**:372–375 (1955).
34. J. W. Dubnoff, The role of choline oxidase in labilizing choline methyl, *Arch. Biochem.* **24**:251–262 (1949).
35. V. du Vigneaud, A. W. Moyer, and J. P. Chandler, Dimethylthetin as a biological methyl donor, *J. Biol. Chem.* **174**:477–480 (1948).
36. J. W. Dubnoff and H. Borsook, Dimethylthetin and dimethyl-β-propiothetin in methionine synthesis, *J. Biol. Chem.* **176**:789–796 (1948).
37. L. E. Ericson, J. N. Williams, Jr., and C. A. Elvehjem, Studies on partially purified betaine-homocysteine transmethylase of liver, *J. Biol. Chem.* **212**:537–544 (1955).
38. D. P. Brenton, D. C. Cusworth, and G. E. Gaull, Homocystinuria, *Pediatrics* **35**:50–56 (1965).
39. P. J. G. Mann, H. E. Woodward, and J. H. Quastel, Hepatic oxidation of choline and arsenocholine, *Biochem. J.* **32**:1024–1032 (1938).
40. J. A. Muntz, The inability of choline to transfer a methyl group directly to homocysteine for methionine formation, *J. Biol. Chem.* **182**:489–499 (1950).
41. J. W. Dubnoff, Utilization of choline and betaine methyl in the guinea pig, *Arch. Biochem. Biophys.* **22**:474–475 (1949).
42. J. W. Dubnoff, Sulfocholine in methionine synthesis, *Arch. Biochem. Biophys.* **22**:478–480 (1949).
43. P. Berg, A study of formate utilization in pigeon liver extract, *J. Biol. Chem.* **205**:145–162 (1953).
44. W. Wilmanns, B. Rücker, and L. Jeanicke, Zur Biogenese von Methionin, *Hoppe-Seylers Z. Physiol. Chem.* **322**:283–287 (1960).

45. W. Sakami and I. Ukstins, Enzymatic methylation of homocysteine by a synthetic tetrahydrofolate derivative, *J. Biol. Chem.* **236**:PC 50 (1961).
46. K. O. Donaldson and J. C. Keresztesy, Further evidence on the nature of prefolic A, *Biochem. Biophys. Res. Commun.* **5**:289–292 (1961).
47. J. H. Mangum and K. G. Scrimgeour, Cofactor requirements and intermediates in methionine biosynthesis, *Federation Proc.* **21**:242 (1962).
48. A. Nakao and D. M. Greenberg, Co-factor requirements for the incorporation of $H_2{}^{14}CO$ and serine-3-$^{14}C$ into methionine, *J. Am. Chem. Soc.* **77**:6715–6716 (1955).
49. V. M. Doctor, T. L. Patton, and J. Awapara, Incorporation of serine-3-$^{14}C$ and formaldehyde-$^{14}C$ into methionine *in vitro*. I. Role of folic acid, *Arch. Biochem. Biophys.* **67**:404–409 (1957).
50. K. Přistoupilová, Biosynthesis of methionine in the liver of leukaemic mice, *Neoplasma* **14**:353–358 (1967).
51. H. R. V. Arnstein and A. Neuberger, The effect of cobalamin on the quantitative utilization of serine, glycine and formate for the synthesis of choline and methyl groups of methionine, *Biochem. J.* **55**:259–271 (1953).
52. A. Nakao and D. M. Greenberg, Studies on the incorporation of isotope from formaldehyde-$^{14}C$ and serine-3-$^{14}C$ into the methyl group of methionine, *J. Biol. Chem.* **230**:603–620 (1958).
53. R. L. Blakley, The interconversion of serine and glycine: Role of pteroylglutamic acid and other cofactors, *Biochem. J.* **58**:448–462 (1954).
54. L. Jaenicke, Conversion of $\beta$-carbon of serine to $N^{10}$-formyltetrahydrofolic acid, *Federation Proc.* **15**:281 (1956).
55. R. L. Kisliuk, Mechanism of formaldehyde incorporation into serine *Federation Proc.* **15**:289 (1956).
56. R. L. Blakley, The reactive intermediate formed by formaldehyde and tetrahydropteroyl glutamate, *Biochem. Biophys. Acta* **23**:654–655 (1957).
57. A. Stevens and W. Sakami, Biosynthesis of methionine in liver, *J. Biol. Chem.* **234**: 2063–2072 (1959).
58. A. R. Larrabee and J. M. Buchanan, A new intermediate of methionine biosynthesis, *Federation Proc.* **20**:9 (1961).
59. A. R. Larrabee, S. Rosenthal, R. E. Cathou, and J. M. Buchanan, A methylated derivative of tetrahydrofolate as an intermediate of methionine biosynthesis, *J. Am. Chem. Soc.* **83**:4094–4095 (1961).
60. K. Lindstrand, Isolation of methylcobalamin from liver, *Acta Chem. Scand.* **19**:1785–1787 (1965).
61. V. Herbert, A. R. Larrabee, and J. M. Buchanan, Studies on the identification of a folate compound of human serum, *J. Clin. Invest.* **41**:1134–1138 (1962).
62. S. Rosenthal and J. M. Buchanan, Enzymatic Synthesis of the methyl group of methionine, *Acta Chem. Scand.* **17**:S-288–S-294 (1963).
63. H. M. Katzen and J. M. Buchanan, Enzymatic synthesis of the methyl group of methionine, *J. Biol. Chem.* **240**:825–835 (1965).
64. H. Weissbach, A. Peterkofsky, B. G. Redfield, and H. Dickerman, Studies on the terminal reaction in the biosynthesis of methionine, *J. Biol. Chem.* **238**:3318–3324 (1963).
65. M. A. Foster, M. J. Dilworth, and D. D. Woods, Cobalamin and the synthesis of Methionine by *Escherichia Coli*, *Nature* **201**:39–42 (1964).
66. S. S. Kerwar, J. H. Mangum, K. T. Scrimgeour, J. D. Brodie, and F. M. Huennekens, Interrelationship of adenosyl methionine and methyl-$B_{12}$ in the biosynthesis of methionine, *Arch. Biochem. Biophys.* **116**:305–318 (1966).

67. R. T. Taylor and H. Weissbach, Role of S-adenosylmethionine in vitamin $B_{12}$-dependent methionine synthesis, *J. Biol. Chem.* **241**:3641–3642 (1966).
68. R. T. Taylor and H. Weissbach, $N^5$-methyltetrahydrofolate-homocysteine transmethylase, *J. Biol. Chem.* **242**:1517–1521 (1967).
69. R. T. Taylor and H. Weissbach, *Escherichia Coli* B $N^5$-methyltetrahydrofolate-homocysteine vitamin-$B_{12}$ transmethylase: Formation and photolability of a methylcobalamin enzyme, *Arch. Biochem. Biophys.* **123**:109–126 (1968).
70. S. Rosenthal, L. C. Smith, and J. M. Buchanan, Enzymatic synthesis of the methyl group of methionine, *J. Biol. Chem.* **240**:836–843 (1965).
71. R. T. Taylor and H. Weissbach, $N^5$-Methyltetrahydrofolate-homocysteine transmethylase, *J. Biol. Chem.* **242**:1502–1508 (1967).
72. V. Herbert and R. Zalusky, Selective concentration of folic acid activity in cerebrospinal fluid, *Federation Proc.* **20**:453 (1961).
73. E. H. Reynolds, Epilepsy and schizophrenia, *Lancet* **1**:398–401 (1968).
74. P. E. Schurr, H. T. Thompson, L. M. Henderson, and C. A. Elvehjem, A method for the determination of free amino acids in rat organs and tissues, *J. Biol. Chem.* **182**:29–37 (1950).
75. P. E. Schurr, H. T. Thompson, L. M. Henderson, J. N. Williams, Jr., and C. A. Elvehjem, The determination of free amino acids in rat tissues, *J. Biol. Chem.* **182**:39–45 (1950).
76. D. M. Walker, The free amino-acids occurring in the body tissues and blood of the rat and of the cow, *Biochem. J.* **52**:679–683 (1952).
77. M. K. Gaitonde and G. E. Gaull, A procedure for the quantitative analysis of the sulfur amino acids of rat tissues, *Biochem. J.* **102**:959–975 (1967).
78. G. E. Gaull and M. K. Gaitonde, A study of the sulfur amino acids of rat tissues, *Biochem. J.* **102**:294–303 (1967).
79. H. H. Tallan, *in Amino Acid Pools* (J. T. Holden, ed.), pp. 465–470, Elsevier, Amsterdam (1962).
80. P. Mandel and J. Mark, The influence of nitrogen deprivation on free amino acids in rat brain, *J. Neurochem.* **12**:987–992 (1965).
81. H. C. Agrawal, J. M. Davies, and W. A. Himwich, Postnatal changes in free amino acid pool of rat brain, *J. Neurochem.* **13**:607–615 (1966).
82. E. V. Flock, G. M. Tyce, and C. A. Owen, Jr., Utilization of glucose-U-$^{14}C$ in brain after total hepatectomy in the rat, *J. Neurochem.* **13**:1389–1406 (1966).
83. M. J. Carver, J. H. Copenhaver, and R. A. Serpan, Free amino acids in foetal rat brain, *J. Neurochem.* **12**:857–861 (1965).
84. H. H. Tallan, S. Moore, and W. H. Stein, Studies on the free amino acids and related compounds in the tissues of the cat, *J. Biol. Chem.* **211**:927–939 (1954).
85. G. A. R. Johnston, The intraspinal distribution of some depressant amino acids, *J. Neurochem.* **15**:1013–1017 (1968).
86. L. Bito, H. Davson, E. Levin, M. Murray, and N. Snider, The concentrations of free amino acids and other electrolytes in cerebrospinal fluid, *in vivo* dialysate of brain, and blood plasma of the dog, *J. Neurochem.* **13**:1057–1067 (1966).
86a. F. LaBella, S. Vivian, and G. Queen, Abundance of cystathionine in the pineal body. Free amino acids and related compounds of bovine pineal, anterior and posterior pituitary, and brain, *Biochim. Biophys. Acta* **158**:286–288 (1968).
87. P. Mandel, Y. Godin, J. Mark, and C. Kayser, The distribution of free amino acids in the central nervous system of garden dormice during hibernation, *J. Neurochem.* **13**:533–536 (1966).
88. W. L. Ryan and M. J. Carver, Free amino acids of human foetal and adult liver, *Nature* **212**:292–293 (1966).

89. A. L. Prensky and H. W. Moser, Brain lipids, proteolipids, and free amino acids in maple syrup urine disease, *J. Neurochem.* **13**:863–874 (1966).
90. P. Soupart, *in Amino Acid Pools* (J. T. Holden, ed.), p. 220–262, Elsevier. Amsterdam (1962).
91. T. L. Perry and R. T. Jones, The amino acid content of human cerebrospinal fluid in normal individuals and in mental defectives, *J. Clin. Invest.* **40**:1363–1372 (1961).
92. J. C. Dickinson, H. Rosenblum, and P. B. Hamilton, Ion exchange chromatography of the free amino acids in the plasma of the newborn infant, *Pediatrics* **36**:2–13 (1965).
93. D. C. Cusworth and C. E. Dent, Renal clearances of amino acids in normal adults and in patients with aminoaciduria, *Biochem. J.* **74**:550–561 (1960).
94. R. G. Westall, *in Amino Acid Pools* (J. T. Holden, ed.), pp. 195–219, Elsevier, Amsterdam (1962).
95. R. H. McMenamy, C. C. Lund, G. J. Neville, and D. F. H. Wallach, Studies of unbound amino acid distributions in plasma, erythrocytes, leukocytes and urine of normal human subjects, *J. Clin. Invest.* **39**:1675–1687 (1960).
96. M. D. Armstrong and K. N. Yates, Free amino acids in milk, *Proc. Soc. Exptl. Biol. Med.* **113**:680–683 (1963).
97. T. Gerritsen and H. A. Waisman, Homocystinuria, an error in the metabolism of methionine, *Pediatrics* **33**:413–420 (1964).
98. N. A. J. Carson, D. C. Cusworth, C. E. Dent, C. M. B. Field, D. W. Neill, and R. G. Westall, Homocystinuria: A new inborn error of metabolism associated with mental deficiency, *Arch. Disease Childhood* **38**:425–436 (1963).
99. J. C. Dickinson and P. B. Hamilton, The free amino acids of human spinal fluid determined by ion exchange chromatography, *J. Neurochem.* **13**:1179–1187 (1966).
100. D. Mütung and K. N. Shivaram, Quantitative Papierchromatographische Bestimmung der freien Aminosäuren im Liquor cerebrospinalis gesunder Menschen. *Hoppe-Seylers Z. Physiol. Chem.* **317**:34–38 (1959).
101. G. De La Haba and G. L. Cantoni, The enzymatic synthesis of S-adenosyl-L-homocysteine from adenosine and homocysteine, *J. Biol. Chem.* **234**:603–608 (1959).
102. L. E. Ericson, J. N. Williams, Jr., and C. A. Elvehjem, Enzymatic cleavage of S-adenosylhomocysteine and the transfer of labile methyl groups, *Acta Chem. Scand.* **9**:859–860 (1955).
103. J. A. Duerre, A hydrolytic nucleosidase acting on S-adenosylhomocysteine and on 5′-methylthioadenosine, *J. Biol. Chem.* **237**:3737–3741 (1962).
104. C. Fromageot and P. Desnuelle, Actions comparées de differénts systèmes biologiques sur l'homocystéine et la cystéine en anaérobiose: Homocystéine-désulfurase et cystéinedésulfurase, *Bull. Soc. Chim. Biol.* **24**:1269–1273 (1942).
105. G. Medes and N. Floyd, Metabolism of sulfur, *Biochem. J.* **36**:259–270 (1942).
106. B. Jollès-Bergeret and M. Marty-Lopez, Métabolisme de l'acide L-homocysteine sulfinique chez les animaux supérieurs, *Compt. Rend. Ser. D*, **262**:930–932 (1966).
107. B. Jollès-Bergeret, Désamination oxydative de l'aide l-homocystéine-sulfinique par la L-glutamodéshydrogénase de foie de boeuf. Inhibition de l'enzyme par l'acide L-homocystéique, *Biochim. Biophys. Acta* **146**:45–53 (1967).
108. D. R. Curtis and J. C. Watkins, The pharmacology of amino acids related to Gamma-aminobutyric acid, *Pharmacol. Rev.* **17**:347–391 (1965).
109. D. R. Curtis, L. Hosli, and G. A. R. Johnston, Inhibition of spinal neurones by glycine, *Nature* **215**:1502–1503 (1967).
110. W. P. Anslow, Jr., S. Simmonds, and V. du Vigneaud, The synthesis of the isomers of cystathionine and a study of their availability in sulfur metabolism, *J. Biol. Chem.* **166**:35–45 (1946).

111. F. Binkley, Synthesis of cystathionine by preparations from rat liver, *J. Biol. Chem.* **191**:531–534 (1951).
112. A. S. S. M. Selim and D. M. Greenberg, Further studies on cystathionine synthetase-serine deaminase of rat liver, *Biochim. Biophys. Acta* **42**:211–217 (1960).
113. F. Binkley, G. M. Christensen, and W. N. Jensen, Pyridoxine and the transfer of sulfur, *J. Biol. Chem.* **194**:109–113 (1952).
114. C. Brown and S. Mallady, Separation of rat liver serine dehydrase from cystathionine synthetase, *Federation Proc.* **25**:341 (1966).
115. J. D. Finkelstein and S. H. Mudd, Transsulfuration in mammals. The methionine-sparing effect of cystine, *J. Biol. Chem.* **242**:873–880 (1967).
116. J. S. Nishimura and D. M. Greenberg, Purification and properties of L-threonine dehydrase of sheep liver, *J. Biol. Chem.* **236**:2684–2691 (1961).
117. A. Nagabhushanam and D. M. Greenberg. Isolation and properties of a homogeneous preparation of cystathionine synthetase-L-serine and L-threonine dehydratase, *J. Biol. Chem.* **240**:3002–3008 (1965).
118. D. B. Hope, Distribution of cystathionine and cystathionine-synthetase in rat brain, *Federation Proc.* **18**:249 (1959).
119. F. Binkley, W. P. Anslow, Jr., and V. du Vigneaud, The formation of cysteine from 11-S-(β-amino-β-carboxyethyl) homocysteine by liver tissue, *J. Biol. Chem.* **143**:559–560 (1942).
120. W. P. Anslow, Jr., and V. du Vigneaud, The cleavage of the stereoisomers of cystathionine by liver extract, *J. Biol. Chem.* **170**:245–250 (1947).
121. F. Binkley, Enzymatic cleavage of thioethers, *J. Biol. Chem.* **186**:287–296 (1950).
122. W. R. Carroll, G. W. Stacy, and V. du Vigneaud, α-Ketobutyric acid as a product in the enzymatic cleavage of cystathionine, *J. Biol. Chem.* **180**:375–382 (1949).
123. Y. Matsuo and D. M. Greenberg, A crystalline enzyme that cleaves homoserine and cystathionine, I. Isolation procedure and some physicochemical properties, *J. Biol Chem.* **230**:545–560 (1958).
124. Y. Matsuo and D. M. Greenberg, A crystalline enzyme that cleaves homoserine and cystathionine, IV. Mechanism of action, reversibility, and substrate specificity, *J. Biol. Chem.* **234**:516–519 (1959).
125. F. Binkley and D. Okeson, Purification of the enzyme responsible for the cleavage of cystathionine, *J. Biol. Chem.* **182**:273–277 (1950).
126. D. Cavallini, C. De Marco, B. Mondovi, and B. G. Mori, The cleavage of cystine by cystathionase and the transulfuration of hypotaurine, *Enzymologia* **22**:161–173 (1960–61).
127. D. M. Greenberg, P. Mastalerz, and A. Nagabhushanam, Mechanism of djenkolic acid decomposition by cystathionase, *Biochim. Biophys. Acta* **81**:158–164 (1964).
128. F. Binkley and C. K. Olson, Deamination of homoserine, *J. Biol. Chem.* **185**:881–885 (1950).
129. Y.Matsuo and D. M. Greenberg, A crystalline enzyme that cleaves homoserine and cystathionine, III. Coenzyme resolution, activators and inhibitors, *J. Biol. Chem.* **234**:507–515 (1959).
130. Y. Matsuo, M. Rothstein, and D. M. Greenberg, Metabolic pathways of homoserine in the mammal, *J. Biol. Chem.* **221**:679–687 (1956).
131. L. W. Parks and F. Schlenk, The stability and hydrolysis of S-adenosylmethionine; Isolation of S-ribosylmethionine, *J. Biol. Chem.* **230**:295–305 (1958).
132. S. H. Mudd, and G. L. Cantoni, Enzymatic hydrolysis of S-adenosylmethionine, *Federation Proc.* **17**:279 (1958).
133. B. Mondovi and C. De Marco, The mechanism of cysteine-hypotaurine transsulfuration in the presence of cystathionase, *Enzymilogia* **23**:156–162 (1961).

134. F. Chatagner and O. Trautmann, 'Soluble' cysteine desulphurase of rat liver as an adaptive enzyme, *Nature* **194**:1281–1282 (1962).
135. A. Kato, M. Ogura, H. Kimura, T. Kawai, and M. Suda, Control mechanism in the rat liver enzyme system converting L-methionine to L-cystine, *J. Biochem.* (*Tokyo*) **59**:34–39 (1966).
136. H. L. A. Tarr, The enzymic formation of hydrogen sulphide by certain heterotrophic bacteria, *Biochem. J.* **27**:1869–1874 (1933).
137. C. Fromageot, E. Wookey, and P. Chaix, Sur la dégradation anaérobie de la l-cystéine par la desulfurase, nouveau ferment contenu dans le Foie, *Compt. Rend.* **209**:1019–1021 (1939).
138. M. Laskowski and C. Fromageot, Some properties of desulfurase, *J. Biol. Chem.* **140**:663–669 (1941).
139. F. Binkley, On the nature of serine dehydrase and cysteine desulfurase, *J. Biol. Chem.* **150**:261–262 (1943).
140. C. V. Smythe, The utilization of cysteine and cystine by rat liver with the production of hydrogen sulfide, *J. Biol. Chem.* **142**:387–400 (1942).
141. F. Chatagner and O. Trautmann, Effect of DL-ethionine on the level of cystathionase in rat liver, *Nature* **200**:75 (1963).
142. A. Meister, H. P. Morris, and S. V. Tice, Effect of vitamin $B_6$ deficiency on hepatic transaminase and cysteine desulfhydrase systems, *Proc. Soc. Exptl. Biol. Med.* **82**:301–304 (1953).
143. B. Jollès-Bergeret, J. Labouesse, and F. Chatagner, Influence des hormones thyroidiennes sur la désulfuration de la cysteine par la foie de rat. Comparaison du comportement de la désulfuration avec celui d'autres réactions enzymatiques nécessitant le phosphate de pyridoxal, *Bull. Soc. Chim. Biol.* **42**:51–63(1960).
144. F. Chatagner, B. Jollès-Bergeret, and O. Trautmann, Hormones thyroidiennes et enzymes de désulfuration de la L-cysteine du foie du rat, *Biochim. Biophys. Acta* **59**:744–746 (1962).
145. F. Chatagner, O. Durieu-Trautmann, and M-C. Rain, Effects of puromycin and actinomycin D on the increase of cystathionase and cysteine sulphinic acid decarboxylase activities in the liver of thyroidectomized rat, *Nature* **214**:88–90 (1967).
146. F. Chatagner and O. Durieu-Trautmann, Effects of pyridoxine deficiency on the adaptation of rat liver cystathionase to DL-ethionine and to thyroidectomy, *Nature* **207**:1390–1391 (1965).
147. H. Blaschko, C. W. Carter, J. R. P. O'Brien, and G. H. Sloane-Stanley, Pyridoxine in relation to amine formation in the mammalian liver, *J. Physiol.* (*London*) **107**: 18P–19P (1948).
148. M. N. D. Goswami, A. R. Robblee, and L. W. McElroy, Cysteine desulfydrase activity in chick liver, *Arch. Biochem. Biophys.* **70**:80–86 (1957).
149. D. Cavallini, B. Mondovi, C. De Marco, and A. Scioscia-Santoro, Inhibitory effect of mercaptoethanol and hypotaurine on the desulfhydration of cysteine by cystathionase, *Arch. Biochem. Biophys.* **96**:456–457 (1962).
150. D. Cavallini, B. Mondovi, C. De Marco, and A. Scioscia-Santoro, The Mechanism of desulphhydration of cysteine, *Enzymologia* **24**:253–266 (1962).
151. B. Jollès-Bergeret and F. Chatagner, Effect of 2-mercaptoethanol on the deamination of L-cysteine by cystathionase of rat liver, *Arch. Biochem. Biophys.* **105**:640–641 (1964).
152. R. J. Thibert, J. F. G. Diederich, and G. W. Kosicki, Behavior of α-substituted cystines in a cystathionase system and in a pyridoxal phosphate model system, *Can. J. Biochem. Physiol.* **45**:1595–1617 (1967).
153. M. Tishel and M. Mazelis, Enzymatic degradation of L-cystine by cytoplasmic particles from cabbage leaves, *Nature* **211**:745–746 (1966).

154. M. Mazelis, N. Beimer, and R. K. Creveling, Cleavage of L-cystine by soluble enzyme preparations from *Brassica* species, *Arch. Biochim. Biophys.* **120**:371–378 (1967).
155. B. Sörbo, On the metabolism of thiosulfate esters, *Acta Chem. Scand.* **12**:1990–1996 (1958).
156. E. Brand, R. J. Block, B. Kassell, and G. F. Cahill, Carboxymethylcysteine metabolism; Its implications on therapy in cystinuria and on the methionine–cysteine relationship, *Proc. Soc. Exptl. Biol. Med.* **35**:501–503 (1936).
157. V. du Vigneaud, G. B. Brown, and J. P. Chandler, The synthesis of 11-S-(β-amino-β-carboxyethyl) homocysteine and the replacement by it of cystine in the diet, *J. Biol. Chem.* **143**:59–64 (1942).
158. F. Binkley and V. du Vigneaud, The formation of cysteine from homocysteine and serine by liver tissue of rats, *J. Biol. Chem.* **144**:507–511 (1942).
159. F. Binkley, Cleavage of cystathionine by an enzyme system from rat liver, *J. Biol. Chem.* **155**:39–43 (1944).
160. H. Blaschko and D. B. Hope, Excretion of cystathionine in pyridoxine deficiency, *Biochem. J.* **63**:7P (1956).
161. D. B. Hope, L-Cystathionine in the urine of pyridoxine-deficient rats, *Biochem. J.* **66**:486–489 (1957).
162. H. H. Tallan, S. Moore, and W. H. Stein, L-Cystathionine in human brain, *J. Biol. Chem.* **230**:707–716 (1958).
163. T. Gerritsen and H. A. Waisman, Homocystinuria: Absence of cystathionine in the brain, *Science* **145**:588 (1964).
164. H. Shimizu, Y. Kakimoto, and I. Sano, A method of determination of cystathionine and its distribution in human brain, *J. Neurochem.* **13**:65–73 (1966).
165. D. B. Hope, Cystathionine accumulation in the brains of pyridoxine-deficient rats, *J. Neurochem.* **11**:327–337 (1964).
166. N. Okumura, S. Otsuki, and A. Kameyama, Studies on free amino acids in human brain, *J. Biochem.* (*Tokyo*) **47**:315–320 (1960).
167. M. E. Swendseid, J. Villalobos, and B. Friedrich, Free amino acids in plasma and tissues of rats fed a vitamin $B_6$-deficient diet, *J. Nutr.* **82**:206–208 (1964).
168. L. R. Gjessing and A. Torvik, Distribution of cystathionine in human brain, *Scand. J. Clin. Lab. Invest.* **18**:565 (1966).
169. L. Laster, S. H. Mudd, J. D. Finkelstein, and F. Irreverre, Homocystinuria due to cystathionine synthase deficiency: The metabolism of L-methionine, *J. Clin. Invest.* **44**:1708–1719 (1965).
170. T. L. Perry, S. Hansen, and L. MacDougall, Homolanthionine excretion in homocystinuria, *Science* **152**:1750–1751 (1966).
171. H. Harris, L. S. Penrose, and D. H. H. Thomas, Cystathioninuria, *Ann. Human Genetics* **23(4)**:442–453 (1958–1959).
172. G. W. Frimpter, A. Haymovitz, and M. Horwith, Cystathioninuria, *New Engl. J. Med.* **268**:333–339 (1963).
173. L. R. Gjessing, Studies of functional neural tumors, *Scand. J. Clin. Lab. Invest.* **15**:474–478 and 479–482 (1963).
174. L. R. Gjessing, Cystathioninuria in neuroblastoma, *Biochem. J.* **89**:42P (1963).
175. C. Ressler, J. Nelson, and M. Pfeffer, A pyridoxal-β-cyanoalanine relation in the rat, *Nature* **203**:1286–1287 (1964).
176. C. Ressler, J. Nelson, and M. Pfeffer, Metabolism of β-cyanoalanine, *Biochem. Pharmacol.* **16**:2309–2319 (1967).
177. M. Pfeffer and C. Ressler, β-Cyanoalanine, an inhibitor of rat liver cystathionase, *Biochem. Pharmacol.* **16**:2299–2308 (1967).

178. A. Meister, P. E. Fraser, and S. V. Tice, Enzymatic desulfuration of β-mercaptopyruvate to pyruvate, *J. Biol. Chem.* **206**:561–575 (1954).
179. D. Keilin, Cytochrome and intracellular oxidase, *Proc. Roy. Soc. (London) Ser. B* **106**:418–444 (1930).
180. N. W. Pirie, The formation of sulphate from cysteine and methionine by tissues *in vitro*, *Biochem. J.* **28**:305–312 (1934).
181. F. Chapeville and P. Fromageot, La formation de l'acide cystéinesulfinique à partir de la cystine chez le rat, *Biochim. Biophys. Acta* **17**:275–276 (1955).
182. J. Awapara and V. M. Doctor, Enzymatic oxidation of cysteine-$S^{35}$ to cysteine sulfinic acid, *Arch. Biochem. Biophys.* **58**:506–507 (1955).
183. B. Sörbo and L. Ewetz, The enzymatic oxidation of cysteine to cysteine sulfinate in rat liver, *Biochem. Biophys. Res. Commun.* **18**:359–363 (1965).
184. A. Wainer, The production of cysteine sulfinic acid from cysteine *in vitro*, *Biochim. Biophys. Acta* **104**:405–412 (1965).
185. L. Ewetz and B. Sörbo, Characteristics of the cysteinesulfinate-forming enzyme system in rat liver, *Biochim. Biophys. Acta* **128**:296–305 (1966).
186. T. P. Singer and E. B. Kearney, Pathways of L-cysteinesulfinate metabolism in animal tissues, *Biochim. Biophys. Acta* **14**:570–571 (1954).
187. T. P. Singer and E. B. Kearney, Intermediary metabolism of L-cysteinesulfinic acid in animal tissues, *Arch. Biochem. Biophys.* **61**:397–409 (1956).
188. A. Wainer, Mitochondrial oxidation of cysteine, *Biochim. Biophys. Acta* **141**:466–472 (1967).
189. O. W. Portman and G. V. Mann, The disposition of taurine-$^{35}S$ and taurocholate-$^{35}S$ in the rat: Dietary influences, *J. Biol. Chem.* **213**:733–743 (1955).
190. B. Bergeret and F. Chatagner, Désulfinication et décarboxylation enzymatiques de l'acide L-cysténe-sulfinique: Sa transformation quantitative en alanine et en hypotaurine, *Biochim. Biophys. Acta* **9**:141–146 (1952).
191. G. Medes, Metabolism of sulphur, *Biochem. J.* **33**:1559–1569 (1939).
192. M. Heimberg, I. Fridovich, and P. Handler, The enzymatic oxidation of sulfite, *J. Biol. Chem.* **204**:913–926 (1953).
193. P. Fromageot and F. Chapeveille, Oxydation du sulfite en sulfate en présence de protohématine, *Biochim. Biophys. Acta* **50**:325–333 and 334–347 (1961).
194. F. Chatagner, B. Bergeret, T. Séjourné, and C. Fromageot, Transamination et desulfination de l'acide L-cysteine-sulfinique, *Biochim. Biophys. Acta* **9**:340–341 (1952).
195. E. B. Kearney and T. P. Singer, The coupled oxidation of succinate and L-cysteinesulfinate by soluble enzymes, *Biochim. Biophys. Acta* **14**:572–573 (1954).
196. R. Balázs, D. Biesold, and K. Magyar, Some properties of rat brain mitochondrial preparations: Respiratory control, *J. Neurochem.* **10**:685–708 (1963).
197. R. M. MacLeod, W. Farkas, I. Fridovich, and P. Handler, Purification and properties of hepatic sulfite oxidase, *J. Biol. Chem.* **236**:1841–1849 (1961).
198. B. Bergeret and F. Chatagner, Sur la présence d'acide cysteinesulfinique dans le cerveau du rat normal, *Biochim. Biophys. Acta* **14**:297 (1954).
199. H. G. Knauff and F. Böck, Über die freien gehirnaminosäuren und das Äthanolamin der normalen Ratte, Sowie über das Verhalten Dieser stoffe nach experimenteller insulinhypoglykämie, *J. Neurochem.* **6**:171–182 (1961).
200. M. K. Gaitonde, A spectrophotometric method for the direct determination of cysteine in the presence of other naturally occurring amino acids, *Biochem. J.* **104**:627–633 (1967).
201. S. I. Singh and C. L. Malhotra, Amino acid content of monkey brain, *J. Neurochem.* **11**:865–872 (1964).

202. C. E. Dent, B. Senior, and J. M. Walshe, The pathogenesis of cystinuria: II. Polarographic studies of the metabolism of sulphur-containing amino acids, *J. Clin. Invest.* **33**:1216–1226 (1954).
203. G. W. Frimpter, The disulfide of L-cysteine and L-homocysteine in urine of patients with cystinuria, *J. Biol. Chem.* **236**:PC 51–53 (1961).
204. S. W. Frimpter, Cystinuria: Metabolism of the disulfide of cysteine and homocysteine, *J. Clin. Invest.* **42**:1956–1964 (1963).
205. S. A. Lough, W. L. Perilstein, H. J. Heinen, and L. W. Carter, Cystinuria, *J. Biol. Chem.* **139**:487–498 (1941).
206. C. E. Dent and G. A. Rose, Amino acid metabolism in cystinuria, *Quart. J. Med.* **20**:205–219 (1951).
207. W. H. Stein, Excretion of amino acids in cystinuria, *Proc. Soc. Exptl. Biol. Med.* **78**:705–708 (1951).
208. H. L. Yeh, W. Frankl, M. S. Dunn, P. Parker, B. Hughes, and P. Gyorgy, The urinary excretion of amino acids by a cystinuric subject, *Am.J. Med. Sci.* **214**:507–512 (1947).
209. W. E. Knox, *in The Metabolic Basis of Inherited Disease* (J. B. Stanbury, J. B. Wyngaarden, and D. S. Fredrickson, eds.), pp. 1262–1282, McGraw-Hill, New York (1966).
210. L. E. Rosenberg, S. T. Downing, and S. Segal, Competitive inhibition of dibasic amino acid transport in rat kidney, *J. Biol. Chem.* **237**:2265–2270 (1962).
211. M. Fox, S. Thier, L. Rosenberg, W. Kiser, and S. Segal, Evidence against a single renal transport defect in cystinuria, *New Engl. J. Med.* **270**:556–561 (1964).
212. J. C. Crawhall and S. Segal, Dithiothreitol in the study of cysteine transport, *Biochim. Biophys. Acta* **121**:215–217 (1966).
213. J. C. Crawhall and S. Segal, The intracellular cysteine/cystine ratio in kidney cortex, *Biochem. J.* **99**:19C–20C (1966).
214. S. Oomori and S. Mizuhara, Structure of a new sulfur-containing amino acid, *Arch. Biochem. Biophys.* **96**:179–185 (1962).
215. C. Buziassy and M. Mazelis, The formatiom of S-carboxymethyl-L-cysteine by radish (*Raphanus sativus*) seedling homogenates, *Biochim. Biophys. Acta* **86**:185–186 (1964).
216. T. Ubuka, H. Kodama, and S. Mizuhara, Isolation of S-(carboxymethyl) cysteine from urine, *Biochim. Biophys. Acta* **141**:266–269 (1967).
217. R. G. Westall, The amino acids and other ampholytes of urine, *Biochem. J.* **55**:244–248 (1953).
218. S. H. Mudd, F. Irreverre, and L. Laster, Sulfite oxidase deficiency in Man: Demonstration of the enzymatic defect, *Science* **156**:1599–1602 (1967).
219. M. Ampola, E. M. Bixby, J. C. Crawhall, M. L. Efron, R. Parker, W. Sneddon, and E. P. Young, Isolation of a new sulphur-containing amino acid, *Biochem. J.* **107**:16P (1968).
220. B. Bergeret, F. Chatagner, and C. Fromageot, Quelques relations entre le phosphate de pyridoxal et la décarboxylation de l'acide cystéinesulfinique par divers organes du rat normal ou du rat carencé en vitamine $B_6$, *Biochim. Biophys. Acta* **17**:128–135 (1955).
221. R. W. Virtue and M. E. Doster-Virtue, Studies on the production of taurocholic acid in the dog, *J. Biol. Chem.* **119**:697–705 (1937).
222. H. Tarver and C. L. A. Schmidt, Radioactive sulfur studies, *J. Biol. Chem.* **146**:69–84 (1942).
223. M. G. Foster, C. W. Hooper, and G. H. Whipple, The metabolism of bile acids, *J. Biol. Chem.* **38**:379–392 and 421–433 (1919).

224. R. W. Virtue and M. E. Doster-Virtue, Studies on the production of taurocholic acid in the dog, *J. Biol. Chem.* **127**:431–437 (1939).
225. J. Awapara and W. J. Wingo, On the mechanism of taurine formation from cysteine in the rat, *J. Biol. Chem.* **203**:189–194 (1953).
226. D. Cavallini, C. De Marco, and B. Mondovi, Experiments with D-cysteine in the rat, *J. Biol. Chem.* **230**:25–30 (1958).
227. J. Awapara, Alanine and taurine formation from injected cysteine in the rat, *Nature* **165**:76–77 (1950).
228. J. Awapara, 2-Aminoethanesulfinic acid: An intermediate in the oxidation of cysteine *in vivo*, *J. Biol. Chem.* **203**:183–188 (1953).
229. M. K. Gaitonde and D. Richter, *in Metabolism of the Nervous System* (D. Richter, ed.), pp. 449–455, Pergamon Press, London (1957).
230. M. K. Gaitonde and D. Richter, The metabolic activity of the proteins of the brain, *Proc. Roy. Soc.* (*London*) *Ser. B* **145**:83–99 (1956).
231. F. Chatagner and B. Bergeret, Décarboxylation enzymatique, *in vitro* et *in vivo*, de l'acide l'cystéinesulfinique dans le foie des animaux supérieurs, *Compt. Rend.* **232**:448–450 (1951).
232. B. Bergeret, F. Chatagner and C. Fromageot, Desulfinication et décarboxylation de l'acide L-cystéinesulfinique chez l'animal vivant, *Biochim. Biophys. Acta* **9**:147–149 (1952).
233. D. Cavallini, B. Mondovi, and C. De Marco, Il disolfossido délla cistamina nelle urine e nel fegato di ratto in seguito a somministrazione di derivati della cistina, *Giorn. Biochim.* **2**:13–26 (1953).
233a. L. Eldjarn, A. Pihl, and A. Sverdrup, The synthesis of $^{35}$S-labeled hypotaurine and its metabolism in rats and mice, *J. Biol. Chem.* **223**:353–358 (1956).
234. J. G. Jacobsen, L. L. Thomas, and L. H. Smith, Jr., Properties and distribution of mammalian L-cysteine sulfinate carboxy-lyases, *Biochim. Biophys. Acta* **85**:103–116 (1964).
235. W. O. Read and J. D. Welty, Synthesis of taurine and isethionic acid by dog heart slices, *J. Biol. Chem.* **237**:1521–1522 (1962).
236. H. Blaschko, *l*(-)-Cysteic acid decarboxylase of dog's liver, *J. Physiol.* (*London*) **101**:6P–7P (1942).
237. B. Bergeret and F. Chatagner, Utilisation d'echangeurs d'ions pour la séparation de quelques acides aminés formés au cours de la dégradation enzymatique de l'acide cystéinesulfinique. Application à l'isolement de l'hypotaurine (acide-2-aminoéthane-sulfinique), *Biochim. Biophys. Acta* **14**:543–550 (1954).
238. H. Blaschko and D. B. Hope, Enzymic decarboxylation of cysteic and cysteine sulphinic acids, *J. Physiol.* (*London*) **126**:52P–53P (1954).
239. H. Blaschko, *l*(-)-Cysteic acid decarboxylase, *Biochem. J.* **36**:571–574 (1942).
240. D. B. Hope, Pyridoxal phosphate as the coenzyme of the mammalian decarboxylase for L-cysteine sulphinic and L-cysteic acids, *Biochem. J.* **59**:497–500 (1955).
241. B. Sörbo and T. Heyman, On the purification of cysteinesulfinic acid decarboxylase and its substrate specificity, *Biochim. Biophys. Acta* **23**:624–627 (1957).
242. G. Medes and N. Floyd, Metabolism of sulphur. Cysteic acid, *Biochem. J.* **36**:836–844 (1942).
243. B. Bergeret, F. Chatagner, and C. Fromageot, Êtude des decarboxylations de l'acide L-cystéinesulfinique de l'acide L-cystéique et de l'acide L-glutamique par divers organes du lapin. Influence du phosphate de pyridoxal et des groupements thiols, *Biochim. Biophys. Acta* **22**:329–336 (1956).
243a. R. S. Piha and H. Saukkonen, The L-cysteine sulphinic acid decarboxylase activity in various areas of the brain, *Suomen Kemistilehti* **B39**:112–114 (1966).

244. F. Chatagner, H. Tabechian, and B. Bergeret, Répercussion d'une carence en vitamine $B_6$ sur le metabolisme de l'acide-L-cystéinesulfinique, *in vitro* et *in vivo*, chez le rat, *Biochim. Biophys. Acta* **13**:313–318 (1954).
245. A. N. Davison, Amino acid decarboxylases in rat brain and liver, *Biochim. Biophys. Acta* **19**:66–73 (1956).
246. A. N. Davison, The mechanism of the inhibition of decarboxylases by isonicotinyl hydrazide, *Biochim. Biophys. Acta* **19**:131–140 (1956).
247. G. H. Sloane-Stanley, Amino-acid decarboxylases of rat liver, *Biochem. J.* **45**:556–559 (1949).
248. F. Chatagner and B. Bergeret, Décarboxylation de l'acide cysteinesulfinique par le foie et la cerveau du rat male, du rat femelle et du rat femelle ovariectomisé, *Bull. Soc. Chim. Biol.* **38**:1159–1163 (1956).
249. M. A. Bennett, Metabolism of sulphur, *Biochem. J.* **33**:885–892 (1939).
250. R. W. Virtue and M. E. Doster-Virtue, Studies on the production of taurocholic acid in the dog, *J. Biol. Chem.* **126**:141–146 (1938).
251. A. Schöberl, Die Oxydation von Disulfiden zu Sulfonsäuren mit Wasserstoff-superoxyd, *Hoppe-Seylers Z. Physiol. Chem.* **216**:193–201 (1933).
252. L. Eldjarn, The metabolism of cystamine and cysteamine, *Scand. J. Clin. Lab. Invest.* **6**, Suppl. 13 (1954).
253. D. Cavallini, B. Mondovi, and C. De Marco, Ossidazione cistamina a ipotaurina nel ratto, *Ric. Sci.* **24**:2649–2651 (1954).
254. L. Eldjarn, The conversion of cystinamine to taurine in rat, rabbit and man, *J. Biol. Chem.* **206**:483–490 (1954).
255. C. De Marco, G. Bombardieri, F. Riva, S. Dupré, and D. Cavallini, Degradation of cystaldimine, the product of oxidative deamination of cystamine, *Biochim. Biophys. Acta* **100**:89–97 (1965).
256. B. Mondovi, M. T. Costa, A. Finazzi Agro, and G. Rotilio, Pyridoxal phosphate as a prosthetic group of pig kidney diamine oxidase, *Arch. Biochem. Biophys.* **119**:373–381 (1967).
257. C. De Marco, B. Mondovi, R. Scandurra, and D. Cavallini, Further investigation of the thiocysteamine oxidising enzyme, *Enzymologia* **25**:94–104 (1962–1963).
258. D. Cavallini, R. Scandurra, and C. De Marco, The role of sulphur, sulphide and reducible dyes in the enzymic oxidation of cysteamine to hypotaurine, *Biochem. J.* **96**:781–786 (1965).
259. J. L. Wood and D. Cavallini, Enzymic oxidation of cysteamine to hypotaurine in the absence of a cofactor, *Arch. Biochem. Biophys.* **119**:368–372 (1967).
260. Z. M. Bacq, G. Dechamps, P. Fischer, A Herve, H. Le Bihan, J. Lecomte, M. Pirotte, and P. Rayet, Protection against X-rays and therapy of radiation sickness with $\beta$-mercaptoethylamine, *Science* **117**:633–636 (1953).
261. P. Fischer and M. Goutier-Pirotte, Métabolisme de la cystéamine et de la cystamine chez le lapin et la chien, *Arch. Intern. Physiol.* **62**:76–100 (1954).
262. K. Sumizu, Oxidation of hypotaurine in rat liver, *Biochim. Biophys. Acta* **63**:210–212 (1962).
263. D. Cavallini, C. De Marco, B. Mondovi, and F. Stirpe, The biological oxidation of hypotaurine, *Biochim. Biophys. Acta* **15**:301–303 (1954).
264. P. P. Cohen, Transamination with purified enzyme preparations (transaminase), *J. Biol. Chem.* **136**:565–584 (1940).
265. C. L. A. Schmidt and G. W. Clark, The fate of certain sulfur compounds when fed to the dog, *J. Biol. Chem.* **53**:193–207 (1922).
266. E. Mussini and F. Marcucci, *in Amino Acid Pools* (J. T. Holden, ed.), pp. 486–492, Elsevier, Amsterdam (1962).

267. G. G. J. Deffner, The dialyzable free organic constituents of squid blood, a comparison with nerve axoplasm, *Biochim. Biophys. Acta* **47**:378–388 (1961).
268. J. D. Marks and H. K. Berry, *in Amino Acid Pools* (J. T. Holden, ed.), pp. 461–464, Elsevier, Amsterdam (1962).
269. L. J. Machlin and P. B. Pearson, Metabolism of taurine in the growing chicken, *Arch. Biochem. Biophys.* **70**:35–42 (1957).
270. Huynh Vinh An and P. Fromageot, Dosage de la taurine échangeable des organes du rat, *Bull. Soc. Chim. Biol.* **42**:221–226 (1960).
271. J. Awapara, Absorption of injected taurine-$S^{35}$ by rat organs, *J. Biol. Chem.* **225**: 877–882 (1957).
272. P. L. Bouquet and P. Fromageot, Renouvellement de la taurine tissulaire chez le rat, *Biochim. Biophys. Acta* **97**:222–232 (1965).
273. E. J. Peck, Jr., and J. Awapara, Formation of taurine and isethionic acid in rat brain, *Biochim. Biophys. Acta* **141**:499–506 (1967).
274. H. N. Christensen, B. Hess, and T. R. Riggs, Concentration of taurine, β-alanine and triiodothyronine by ascites carcinoma cells, *Cancer Res.* **14**:124–127 (1954).
275. G. Levi, J. Kandera, and A. Lajtha, Control of cerebral metabolite levels, *Arch. Biochem. Biophys.* **119**:303–311 (1967).
276. A. A. Albanese, The amino acid requirements of man, *Advan. Protein Chem.* **3**:227–268 (1947).
277. B. A. Koechlin, On the chemical composition of the axoplasm of squid giant nerve fibers with particular reference of its ion patterns, *J. Biophys. Biochem. Cytol.* **1**:511–529 (1955).
278. J. D. Welty, W. O. Read, and E. H. Shaw, Jr., Isolation of 2-hydroxyethane-sulfonic acid (isethionic acid) from dog heart, *J. Biol. Chem.* **237**:1160–1161 (1962).
279. J. G. Jacobsen, L. L. Collins, amd L. H. Smith, Jr., Urinary excretion of isethionic acid in man, *Nature* **214**:1247–1248 (1967).
280. P. Boulanger and G. Biserte, Chromatographie sur papier des acides aminés et polypeptides des liquides biologiques, *Bull. Soc. Chim. Biol.* **33**:1930–1939 (1951).
281. F. G. Fischer and J. Brander, Eine Analyse der Gespinste der Kreuzspinne, *Hoppe-Seylers Z. Physiol. Chem.* **320**:92–102 (1960).
282. N. van Thoai and J. Roche, Phosphagens of marine animals, *Ann. N.Y. Acad. Sci.* **90**:923–928 (1960).
283. G. E. Hobson, Adenosinetriphosphate-guanidyl phosphokinases in annelid worms, *Biochem. J.* **60**:viii–ix (1955).
284. N. van Thoai, J. Roche, and A. Olumucki, Sur la présence de la taurocyamine (guanidotaurine) dans l'urine de rat et sa signification biochimique dans l'excrétion azotée, *Biochim. Biophys. Acta* **14**:448 (1954).
285. E. Salkowski, Über die Enstehung der Schwefelsäure und das Verhalten des Taurins in thierischen Organismus, *Virchows Arch. Pathol. Anat. Physiol. Klin. Med.* **58**: 460–508 (1873).
286. J. P. Blass, The simple monosubstituted guanidines of mammalian brain, *Biochem. J.* **77**:484–489 (1960).
287. L. S. Ciereszko, P. H. Odense and R. W. Schmidt, Chemistry of coelenterates, *Ann. N.Y. Acad. Sci.* **90**:920–922 (1960).
288. B. Lindberg, Methylated taurines and choline sulphate in red algae, *Acta Chem. Scand.* **9**:1323–1326 (1955).
289. M. Kuriyama, New ninhydrin-reactive substance from red algae, *Nature* **192**:969 (1961).
290. C. L. A. Schmidt and T. Watson, A method for the preparation of taurine in large quantities, *J. Biol. Chem.* **33**:499–500 (1918).

291. P. R. Lewis, The free amino-acids of invertebrate nerve, *Biochem. J.* **52**:330–338 (1952).
292. E. Roberts and S. Frankel, Free amino acids in normal and neoplastic tissues of mice as studied by paper chromatography, *Cancer Res.* **9**:645–648 (1949).
293. J. Awapara, A. J. Landua, and R. Fuerst, Distribution of free amino acids and related substances in organs of the rat, *Biochim. Biophys. Acta* **5**:457–462 (1950).
294. H. H. Tallan, Chromatographic investigation of the free and bound amino acids of animal tissues, *Federation Proc.* **13**:309 (1954).
295. R. Dubreuil and P. S. Timiras, Effect of cortisone on free amino acids in the serum and organs of the rabbit, *Am. J. Physiol.* **174**:20–26 (1953).
296. A. H. Gordon, An investigation of the intracellular fluid of calf embryo muscle, *Biochem. J.* **45**:99–105 (1949).
297. D. J. Bell, W. M. McIndoe, and D. Gross, Tissue components of the domestic fowl, *Biochem. J.* **71**:355–364 (1959).
298. J. Awapara, The taurine concentration of organs from fed and fasted rats, *J. Biol. Chem.* **218**:571–576 (1956).
299. J. C. Jacobsen and L. H. Smith, Jr., Comparison of decarboxylation of cysteine sulphinic acid-1-$^{14}$C and cysteic acid-1-$^{14}$C by human, dog and rat liver and brain, *Nature* **200**:575–577 (1963).
300. J. Bremer, A method for the estimation of the taurine content and its conjugation with cholic acid in rat liver. Bile acids and steroids 22, *Acta Chem. Scand.* **9**:683–688 (1955).
301. J. E. Garvin, A new method for the determination of taurine in tissues, *Arch. Biochem. Biophys.* **91**:219–225 (1960).
302. P. N. Campbell and T. S. Work, Fractionation of the nitrogenous water-soluble constituents of liver, *Biochem. J.* **50**:449–454 (1952).
303. G. Porcellati and R. H. S. Thompson, The effect of nerve section on the free amino acids of nervous tissue, *J. Neurochem.* **1**:340–347 (1957).
304. R. W. R. Baker and G. Porcellati, The separation of nitrogen-containing phosphate esters from brain and spinal cord by ion-exchange chromatography, *Biochem. J.* **73**:561–566 (1959).
305. N. Okumura, S. Otsuki, and T. Aoyama, Studies on the free amino acids and related compounds in the brains of fish, amphibia, reptile, aves and mammal by ion exchange chromatography, *J. Biochem.* (*Tokyo*) **46**:207–212 (1959).
306. J. K. Tews, S. H. Carter, P. D. Roa, and W. E. Stone, Free amino acids and related compounds in dog brain: Post mortem and anoxic changes, effects of ammonium chloride infusion and levels during seizures induced by picrotoxin and by pentylenetetrazol, *J. Neurochem.* **10**:641–653 (1963).
307. J. K. Tews and R. A. Lovell, The effect of a nutritional pyridoxine deficiency on free amino acids and related substances in mouse brain, *J. Neurochem.* **14**:1–7 (1967).
308. G. B. Ansell and D. Richter, A note on the free amino acid content of rat brain, *Biochem. J.* **57**:70–73 (1954).
309. R. S. De Ropp and E. H. Snedeker, Effect of drugs on amino acid levels in the rat brain: Hypoglycemic agents, *J. Neurochem.* **7**:128–134 (1961).
309a. R. S. Piha, S. S. Oja, and A. J. Uusitalo, The effect of chlorpromazine on free amino acids in the rat brain, *Ann Med. Exptl. Biol. Fenniae* (*Helsinki*) **40**:1–28 Suppl. 5 (1962).
310. M. J. Carver, Influence of phenylalanine administration on the free amino acids of brain and liver in the rat, *J. Neurochem.* **12**:45–50 (1965).
311. M. K. Gaitonde, unpublished data.

312. R. K. Shaw and J. D. Heine, Ninhydrin positive substances in different areas of normal rat brain, *J. Neurochem.* **12**:151–155 (1965).
313. D. A. Roe and M. O. Weston, Potential significance of free taurine in the diet, *Nature* **205**:287–288 (1965).
314. O. W. Portmann and G. V. Mann, Further studies of the metabolism of taurine-$S^{35}$ by the rat, *J. Biol. Chem.* **220**:105–112 (1956).
315. C. L. A. Schmidt and E. G. Allen, Further studies on the elimination of taurine administered to man, *J. Biol. Chem.* **42**:55–58 (1920).
316. K. Bloch, B. N. Berg, and D. Rittenberg, The biological conversion of cholesterol to cholic acid, *J. Biol. Chem.* **149**:511–517 (1943).
317. I. Zabin and W. F. Barker, The conversion of cholesterol and acetate to cholic acid, *J. Biol. Chem.* **205**:633–636 (1953).
318. C. Wu, Metabolism of free amino acids in fasted and zein-fed rats, *J. Biol. Chem.* **207**:775–786 (1954).
319. J. Holtfreter, T. R. Koszalka, and L. L. Miller, Chromatographic studies of amino acids in the eggs and embryos of various species, *Exptl. Cell Res.* **1**:453–459 (1950).
320. S. Rosenberg and P. L. Kirk, Isolation of dialyzable growth factor for chick tissue culture. Identification of the ninhydrin reactive band, *Arch. Biochem. Biophys.* **44**:226–229 (1953).
321. W. Hull and P. L. Kirk, Tissue culture studies, *J. Gen. Physiol.* **33**:335–342 (1949–1950).
322. F. Gonzales and J. Awapara, The taurine concentration of developing chick embryo, *Exptl. Cell Res.* **9**:353–355 (1955).
323. D. W. Micks and J. P. Ellis, Free amino acids in the developmental stages of the Mosquito, *Proc. Soc. Exptl. Biol. Med.* **79**:191–193 (1952).
324. A. E. Ramírez de Guglielmone and C. J. Gómez, Influence of neonatal hypothyroidism on amino acids in developing rat brain, *J. Neurochem.* **13**:1017–1025 (1966).
325. A. Agur and A. Pinsky, Comparative study of taurine, sulfur containing amino acids and phospho-lipids in schizophrenic and normal blood and urine, *Israel J. Chem.* **2**:318 (1964)
326. H. Blaschko, S. P. Datta, and H. Harris, Pyridoxin deficiency in the rat: Liver L-cysteic acid decarboxylase activity and urinary amino-acids, *Brit. J. Nutr.* 364–371 (1953).
327. D. B. Hope, The persistence of taurine in the brains of pyridoxine-deficient rats, *J. Neurochem.* **1**:364–369 (1957).
328. E. Nyffenegger, K. Lauber, and H. Aebi, Die Taurinausscheidung normaler und $B_6$-avitaminotischer Ratten nach Ganzkörperbestrahlung, *Biochem. Z.* **333**:226–235 (1960).
329. L. C. Smith and S. R. Nelson, Effect of vit. E-deficiency on free amino acids of various rabbit tissues, *Proc. Soc. Exptl. Biol. Med.* **94**:644–646 (1957).
330. E. I. Pentz, Factors influencing the excretion of taurine in irradiated rats with particular reference to the adrenal glands, *J. Biol. Chem.* **231**:165–174 (1958).
331. G. Rouser, A. J. Samuels, K. Kinugasa, B. Jelinek, and D. Heller, *in Amino Acid Pools* (J. T. Holden, ed.), pp. 396–412, Elsevier, Amsterdam (1962).
332. H. O. Goodman, J. S. King, and J. J. Thomas, Urinary excretion of beta-amino-isobutyric acid and taurine in mongolism, *Nature* **204**:650–651 (1964).
333. N. C. Nevin, L. J. Hurwitz, and D. W. Neill, Familial camptodactyly with taurinuria, *J. Med. Genet.* **3**:265–268 (1966).
334. D. W. Neill and J. A. Weaver, Amino acid and protein metabolism in pernicious anaemia, *Brit. J. Haematol.* **4**:447–455 (1958).

335. K. F. Killam and J. A. Bain, Convulsant hydrazides, *J. Pharmacol. Exptl. Therap.* **119**:255–262 (1957).
336. H. M. Hakkinen and E. Kulonen, The effect of ethanol on the amino acids of the rat brain with a reference to the administration of glutamine, *Biochem. J.* **78**:588–593 (1961).
337. D. P. Earle, Jr., K. Smull, and J. Victor, Effects of excess dietary cysteic acid, *dl*-methionine and taurine on the rat liver, *J. Exptl. Med.* **76**:317–324 (1942).
338. J. A. Barclay, R. G. Kenney, and W. T. Cooke, Treatment of infective hepatitis, *Brit. Med. J.* **2**:298 (1945).
339. D. Seligson, G. J. McCormick, and V. Sborov, Blood ketoglutarate and pyruvate in liver disease, *J. Clin. Invest.* **31**:661 (1952).
340. D. S. Amatuzio and S. Nesbitt, A study of pyruvic acid in the blood and spinal fluid of patients with liver disease with and without hepatic coma, *J. Clin. Invest.* **29**:1486–1490 (1950).
341. C. L. A. Schmidt, E. von Adelung, and T. Watson, On the elimination of taurine administered to man, *J. Biol. Chem.* **33**:501–503 (1918).
342. G. T. Lewis and H. B. Lewis, The metabolism of sulfur, *J. Biol. Chem.* **69**:589–598 (1962).
343. W. C. Rose and B. T. Huddlestun, The availability of taurine as a supplementing agent in diets deficient in cystine, *J. Biol. Chem.* **69**:599–605 (1926).
344. H. H. Mitchell, Cysteine and taurine as substituents for cystine in nutrition, *J. Nutr.* **4**:95–104 (1931).
345. D. P. Earle, Jr. and J. Victor, The effect of various diets on the liver damage caused by excess cystine, *J. Exptl. Med.* **75**:179–189 (1942).
346. A. C. Kurtz and J. M. Luck, Studies on Annelid muscle, *J. Biol. Chem.* **111**:577–584 (1935).
347. K. D. Neame, L-Cysteine as inhibitor of potassium retention by brain slices, *Nature* **203**:1067 (1964).
348. J. Dawson and A. Bone, The effect of thiols and ganglioside on the alterations in water, sodium and potassium distribution produced in brain slices by vasopressin protamine, *J. Neurochem.* **13**:257–268 (1966).
349. J. C. Smith and R. B. Weiskopf, Effect of cysteine on acetylcholine synthesis *Nature* **215**: 1379–1380 (1967).
350. W. O. Read and J. D. Welty, Effect of taurine on epinephrine and digoxin induced irregularities of the dog heart, *J. Pharmacol. Exptl. Therap.* **139**:283–289 (1963).
351. K. H. Byington, Aminosulfonic acids and cardiac arrhythmias, *Dissertation Abstr.* **25**:3075 (1964).
352. H. F. Bradford and H. McIlwain, Ionic basis for the depolarization of cerebral tissues by excitatory acidic amino acids, *J. Neurochem.* **13**:1163–1177 (1966).
353. R. Werman, R. A. Davidoff, and M. H. Aprison, The inhibitory action of cystathionine, *Life Sci.* **5**:1431–1440 (1966).
354. D. E. Curtis, J. W. Phillis, and J. C. Watkins, unpublished observation.
355. W. Pollin, P. V. Cardon, Jr., and S. S. Kety, Effects of amino acid feedings in schizophrenic patients treated with iproniazid, *Science* **133**:104–105 (1961).
356. L. C. Park, R. J. Baldessarini, and S. S. Kety, Methionine effects on chronic schizophrenics, *Arch. Gen. Psychiat.* **12**:346–351 (1965).
357. J. Kaloušek, J. Kohout, Z. Pinta, and F. Boška, The antiepileptic effect of methionine, *Cesk. Neurol.* **26**:21–26 (1963).
358. E. C. Donaghue and D. Richter, unpublished observation.
359. J. A. Wada, H. Ikeda, and K. Berry, Reversible behavioural and electrographic manifestations induced by methionine sulfoximine, *Neurology* **17**:854–868 (1967).

360. I. H. Krakoff, Effect of methionine sulfoximine in man, *Clin. Pharmacol. Therap.* **2**:599–604 (1961).
361. R. G. Heath, W. Nesselhof, Jr., and E. Timmons, D,L-Methionine-*d*,*l*-sulfoximine effects in schizophrenic patients, *Arch. Gen. Psychiat.* **14**:213–217 (1966).
362. D. Ackermann and H. A. Heinsen, Über die physiologische Wirkung des Asterubins und anderer, zum Teil neu dargestellter, schwefelhaltiger Guanidinderivate, *Hoppe-Seylers Z. Physiol. Chem.* **235**:115–121 (1935).
363. A. B. Macallum and C. Sivertz, The potentiation of insulin by sulfones, *Can. Chem. Process.* **26**:569 (1942).
364. W. G. Martin and H. Patrick, The effect of taurine on the sulfate-$^{35}S$ retention by chicks, *Poultry Sci.* **40**:267–268 (1961).
365. M. Kieny, Les besoins nutritifs spécifiques de tibias d'embryon de poulet en culture *in vitro* sur milieux synthétiques, *Compt. Rend. Soc. Biol.* **149**:418–421 (1955).

*Chapter 9*

# THE NATURE OF γ-AMINOBUTYRIC ACID

**Claude F. Baxter**

*Neurochemistry Laboratories*
*Veterans Administration Hospital*
*Sepulveda, California*
*and*
*Department of Physiology*
*UCLA School of Medicine*
*Los Angeles, California*
*and*
*Division of Neurosciences*
*City of Hope Medical Center*
*Duarte, California*

## I. INTRODUCTION

The first reference to γ-aminobutyric acid (GABA) appeared in 1883 when the synthesis of a compound named "Piperidinsaure" was reported.[1] Subsequently, the metabolism of GABA was studied in fungi, bacteria, and plants. Its presence in neurological tissue was not recognized until 1950.[2-4]

High levels of GABA are found in the peripheral nerves of crustaceans and in the central nervous system (CNS)—brain, spinal cord, and retina—of vertebrates. Lower levels have been reported in the lens of the eye and in the "postmortem" primate kidney. With the advent of specific micro assay techniques, low levels of GABA also are being reported in other tissues. It is not known whether GABA is intrinsic to these tissues or originates from bacterial or plant sources in the diet.

Research on GABA received its greatest impetus in 1957 when an inhibitory factor (Factor I) was isolated from brain and identified as consisting primarily of GABA.* In the past ten years thousands of papers have dealt with biochemical, physiological, pharmacological, and behavioral aspects of GABA and closely related compounds. This chapter emphasizes the most recent developments and includes some areas that have received only limited coverage elsewhere. For a fuller account of those aspects of GABA metabolism, physiology, and pharmacology, which are not presented in detail here, the reader should consult some of the excellent reviews that have appeared since 1960.[6-22]

* For reviews of work prior to 1960 see Florey,[15] Baxter and Roberts,[16] and Elliott and Jasper.[7]

## II. SIGNIFICANCE OF GABA

The functions of GABA in the nervous system may be manifold.

### A. Alternate Metabolic Pathway

The significance of GABA in metabolism was the first to be recognized. The carbon chain of α-ketoglutarate (αKG) can be converted via glutamic acid (GA) and GABA to succinate and $CO_2$. Three enzymes are involved:

Glutamic acid decarboxylase (GAD)
[L-glutamic 1-carboxyl-lyase, EC 4.1.1.15]

$$\mathrm{HOOC{-}CH_2{-}CH_2{-}CH(NH_2){-}COOH} \xrightarrow[\mathrm{PyP}]{\mathrm{GAD}} \mathrm{HOOC{-}CH_2{-}CH_2{-}CH_2NH_2} + \mathrm{CO_2}$$

γ-Aminobutyric acid-α-ketoglutaric acid transaminase (GABA-T)
[4-Aminobutyrate: 2-oxoglutarate aminotransferase, EC 2.6.1.19]

$$\begin{matrix} \mathrm{HOOC{-}CH_2{-}CH_2{-}CH_2NH_2} & & \mathrm{HOOC{-}CH_2{-}CH_2{-}CHO} \\ + & \underset{\mathrm{PyP}}{\overset{\mathrm{GABA\text{-}T}}{\longleftrightarrow}} & + \\ \mathrm{HOOC{-}CH_2{-}CH_2{-}C(=O){-}COOH} & & \mathrm{HOOC{-}CH_2{-}CH_2{-}CH(NH_2){-}COOH} \end{matrix}$$

Succinic semialdehyde dehydrogenase (SSA-D)
[Succinate semialdehyde NAD$^+$ oxireductase, EC 1.2.1.16]

$$\begin{matrix} \mathrm{HOOC{-}CH_2{-}CH_2{-}CHO} & & \mathrm{HOOC{-}CH_2{-}CH_2{-}COOH} \\ + & \overset{\mathrm{SSA\text{-}D}}{\longleftrightarrow} & + \\ \mathrm{NAD^+ + H_2O} & & \mathrm{NADH + H^+} \end{matrix}$$

The sum of the three reactions is

$$\begin{matrix} \mathrm{HOOC{-}CH_2{-}CH_2{-}C(=O){-}COOH} & & \mathrm{HOOC{-}CH_2CH_2{-}COOH} \\ & \longrightarrow & \\ \mathrm{+\ NAD^+ + H_2O} & & \mathrm{+\ NADH + H^+ + CO_2} \end{matrix}$$

Thus, in the CNS of mammals as well as the peripheral nerves of arthropods, the "GABA shunt" represents an alternate pathway to the portion of the tricarboxylic acid cycle that leads from α-ketoglutarate to succinate (see Section III, D). The two parallel pathways are shown in Fig. 1 (reactions 1 and 2–4). The individual reaction steps are discussed in Sections IV and V.

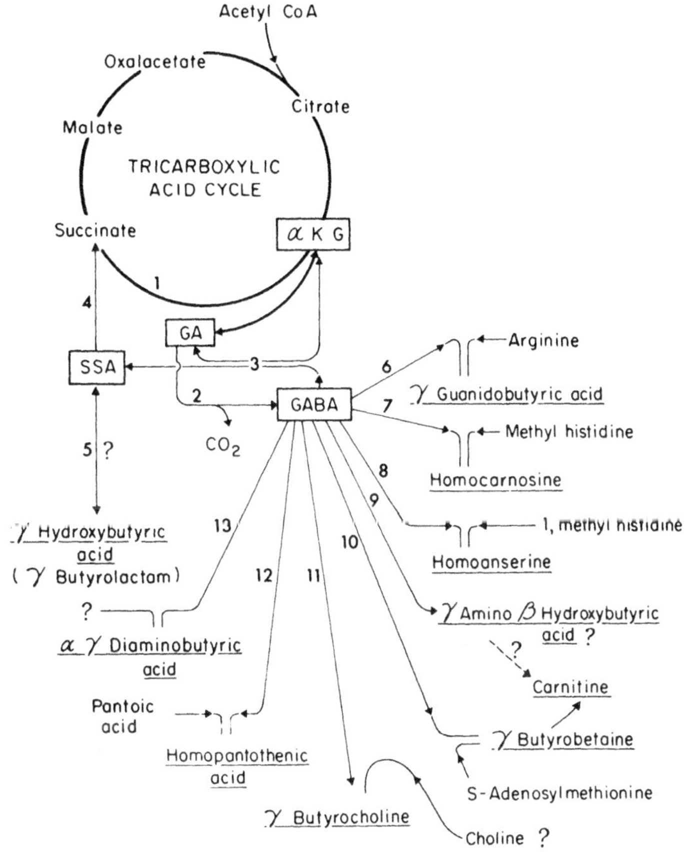


Fig. 1. Proven and suggested pathways of GABA metabolism. Pathways 2–4 represent the "GABA shunt" around pathway 1 of the tricarboxylic acid cycle. Pathways 6 to 13 represent alternate routes of GABA utilization. Quantitatively, these appear relatively unimportant. Pathway 5 is included because of the present pharmacological interest in γ-hydroxybutyric acid.

## B. Specific Inhibitory Transmitter

### 1. *Invertebrates*

In the crustacean nervous system there are distinctly separate excitatory and inhibitory nerve fibers, and many findings support the hypothesis that GABA is a specific inhibitory transmitter at the neuromuscular junction.[8,22–29] In the lobster nervous system, intrinsic GABA is localized almost exclusively within inhibitory axons and nerve cell bodies.[29] A comparison of inhibitory and excitatory neurons revealed that actual GABA levels are 100 times greater, and glutamic acid decarboxylase (GAD) is 10 times more active in the inhibitory axons and their corresponding nerve cell bodies.[24] When inhibitory nerves to lobster muscle are stimulated electrically, GABA is released near the inhibitory synapse; this effect cannot be duplicated by stimulation of the lobster excitatory nerves.[26] Similarly, in the crayfish, foci of localized inhibitory action by GABA coincide exactly with the inhibitory neuromuscular junction.[23] Enzymes for the degradation of GABA also have been found in the tissues of the crustacean nervous system.[28] All these experimental findings satisfy criteria required to designate GABA as an inhibitory transmitter substance.

However, the exact sites of GABA action remain uncertain. Recent evidence suggests that in addition to increasing postsynaptic membrane conductance GABA also may reduce the release of transmitter from the excitatory presynaptic nerve terminals and increase the permeability to chloride ion.[25,27]

### 2. *Vertebrates*

Less complete and more circumstantial evidence links GABA with inhibitory transmitter function in the mammalian nervous system. Several reports indicate that GABA, applied directly to the brain of mammals, has a nonspecific depressant action.[30-33]*

The ionophoretic application of GABA to cortical neurons,[34] Deiter's neurons,[35] and the cuneate nucleus of the brainstem[36] produces a hyperpolarizing action and increased membrane conductance in a way that imitates the potential change seen during synaptic inhibition. Furthermore, both the inhibitory postsynaptic potential (IPSP) and the potential observed by GABA application to a neuron are reversed by the diffusion of chloride ions into the neuron, suggesting that the ionic mechanisms producing both of these potentials might be identical. This fact is further underscored by the observation that the inhibiting action of GABA on evoked potentials in the superior colliculus *in vitro* is abolished by placing the tissue in a chloride-free medium.[37]

Some preliminary studies, attempting to localize GABA at the cellular level in vertebrates, show that the regional distribution is compatible with an inhibitory transmitter function. In the cerebellum of the rabbit, levels of GABA are highest in the layer of cells containing the inhibitory Purkinje cells[22,38] and, in the retina of the frog[39] and rabbit,[22] levels of GABA are highest in the cell layer containing the amacrine cells,[39] believed to be responsible for lateral inhibition in this structure.[40,41]

Even in the spinal cord of the cat, GABA shows highest concentrations in the dorsal gray matter and very low concentrations in the white matter and spinal roots, compatible with the notion that GABA functions as a postsynaptic inhibitory transmitter.[42] However, GABA applied iontophoretically to spinal motoneurons does not produce membrane hyperpolarization (IPSP),[30,32] and the depression in the firing of motoneurons after

* See these earlier papers also for earlier work.

the application of GABA could not be blocked by electrophoretically administered strychnine. Although these observations are unique to the spinal cord neurons, they were originally interpreted as evidence against the function of GABA as an inhibitory transmitter substance throughout the mammalian CNS. More recent data negates this generalization and suggests instead that spinal motoneurons are a special case in which glycine rather than GABA acts as the naturally occurring inhibitory transmitter.[42-45]

## C. Other Possible Physiological Functions

The large quantities of GABA are in sharp contrast to the low levels of acetyl choline found in the CNS of vertebrates. For this reason it is held unlikely that all GABA is associated with chemical transmitter mechanisms in the synaptic regions. The distribution of GABA in the CNS is not only uneven and regional but also distinguishable kinetically as distributed in different metabolic pools.

### 1. *General Inhibition*

It has been suggested that high levels of GABA in the neuronal environment may exert inhibitory effects that are unrelated to GABA's probable role as a specific inhibitory transmitter substance.[31,33]

### 2. *Homeostatic Mechanism*

GABA also has been depicted in the role of a stabilizing agent. For example, a homeostatic function for GABA under hypoxic conditions has been proposed.[46] Exposure of rats to an atmosphere containing only 8% oxygen for 10 min elevates GABA concentrations in brain significantly. The behavioral and functional signs of hypoxia, such as loss of light perception, depression, and visual motor impairment, are strikingly similar to those observed when cerebral GABA levels are elevated artificially (see Sections VI and VII). By contrast, when rats are exposed to conditions of hyperbaric oxygen, levels of GABA are depressed and seizures are induced. A comparison of mice, hamsters, rabbits, rats, and guinea pigs shows that the susceptibility to such seizures (which are highest in the mouse and lowest in the guinea pig) varies inversely with the stability of brain GABA levels.[47] This correlation does not hold for the absolute levels of GABA.

### 3. *Osmotic Regulation*

Osmotic or ionic changes in the external environment will produce corresponding changes in the internal environment of some animals. In the toad (*Bufo boreas*) GABA appears to participate in the osmotic regulation of brain tissue,[48] and the occurrence of substantial quantities of GABA in the lens of the vertebrate eye[49,50] suggests a similar function in this nerveless structure.

### 4. *Protein Metabolism*

GABA is not known to be incorporated into the proteins of nervous tissue. Nevertheless, studies in two laboratories have shown that *in vitro* GABA exerts a stimulatory effect on protein synthesis or turnover in ribosomes and mitochondria of immature rat brain.[51,52] This stimulation appears to occur during the earlier steps of protein synthesis,[53] but the mechanism of action and the specificity of stimulation for particular proteins remains to be determined.

### 5. *Metabolic Products of GABA*

A number of claims have been made suggesting that derivatives of GABA are biologically active in the nervous system and that they might be important in neuronal physiology. These derivatives are discussed in Section V, I.

## III. GABA

## A. Occurrence and Distribution

### 1. *Regional Distribution*

Levels of GABA have been measured in the CNS of many animal species.[54] They range from $<1$ to $>8$ $\mu$moles/g tissue (wet wt.). Individual brain areas may be considerably higher or lower. The occurrence of GABA in peripheral nerves of invertebrates was mentioned in Section II, B, 1. Small amounts of GABA have been found in tissues outside the nervous system. Kidney tissue from infants and children, sampled at the time of autopsy, contained levels of GABA up to 0.4 mM, or about one-fifth the average concentration in brain tissue.[45] No GABA-peptides have been detected in kidney tissue and it is likely that some of the GABA in this organ is exogenous in origin.

Levels of GABA in different areas of brain have been determined for man, monkey, cow, dog, cat, rabbit, rat, and other species.[9,56–63] Although the studies of *gross* regional distribution have shown sizable differences in levels of GABA from one brain area to another, they have yielded little insight into the possible functions of GABA.

In more refined studies sections of brain areas were isolated, either by microdissection or by histological sectioning.[64] It was in this way that the absence of GABA in the optic nerve was first discovered[65] and the preferential localization of GABA in specific cell layers of the cerebellum, spinal cord, and retina was demonstrated (see Section II, B, 2).

### 2. *Cellular and Subcellular Distribution*

The type of cell in which GABA occurs has been investigated with equivocal results. On the basis of studies in which intact neurons are allowed to degenerate in the cat's medial geniculate body, GABA appears to be distributed equally between neurons and glia.[66] By contrast, a comparative study between normal tissue and glial tumors indicated quite clearly that the level of GABA in the tumor was significantly lower, thus inviting the speculation that GABA was preferentially located in neuronal cells.[67]

Studies of the subcellular distribution have been hampered by the possible diffusion of GABA during the fractionation procedure.[68-72] This, as well as the various factors which influence the binding of GABA to particulates (see Section III, B), may account for the lack of definitive localization of GABA in a particulate fraction. Although a fairly high level of GABA is found in nerve ending fractions from guinea pig brain, this was attributed to the occlusion of cytoplasm within the nerve endings and the organized cell fragments.[71] However, the enrichment of GABA, GA, and GAD in mouse brain synaptosomes is not paralleled by an enrichment of lactic dehydrogenase—a cytoplasmic marker—in this fraction.[73] Additional experimental evidence is required to relate GABA *specifically* to the nerve ending particles fraction. Even so, this fraction contains, after separation, sufficient GABA to produce effects on single cortical neurons of the guinea pig,[74] and it has been calculated that the GABA released per second from micropipettes by electroendosmosis in these experiments corresponds to the GABA contained in 350 synaptosomes. This is considered within physiological limits.

## B. "Free" and "Bound" GABA Pools

A "free" amino acid is an operational designation that indicates that the amino acid has been extracted into a particle-free fraction of an aqueous tissue suspension. Similarly, a "bound" or "occluded" amino acid is one that has a fixed close relationship to some particulate structure and is found in the particulate fraction of a tissue suspension. Therefore, the degree to which an amino acid appears "free" or "bound" in a tissue homogenate depends upon the suspending medium and the conditions of the extraction procedure. There is no assurance that the extractability of an amino acid is related in a clearly defined way to its state in living tissue.

Some years ago GABA (measured as Factor I) was shown to exist in a "bound" form and to be released when brain tissues were suspended in a salt-free sucrose solution.[75] When the brains of some drug-treated rats were homogenized in a modified crayfish saline, the ratio of "bound" to "free" GABA appeared to have been changed. Insulin and thiosemicarbazide preferentially depleted the "bound" form, whereas hypoxia, hydroxylamine, and iproniazid preferentially increased the "free" form.[75]

Other studies, in which [$^{14}$C] GABA was used as a marker, have defined more clearly the characteristics of the binding phenomenon. The affinity of GABA for the sedimentable fraction is organ specific and presumably the result of selective binding sites on membranes or selective occlusion by particles of brain tissue. The process is nonenzymatic and requires no energy source. It has a $pH$ optimum between $pH$ 7.3 and 7.5[76] which includes the isoelectric point of GABA ($pH$ 7.38). The binding of endogenous GABA to mitochondria, microsomes, and nerve ending fragments is $Na^+$ dependent.[73,77] The binding site is quite specific as judged from inhibitor studies with sulfhydril reagents, surface active agents, and GABA analogues.[76] At $1 \times 10^{-3}$ M concentration, chlorpromazine $\alpha$-$\gamma$-diaminobutyric acid, and imipramine inhibit the binding of GABA to synaptic vessicles from mouse brain by more than 75%.[22]

## C. GABA Transport

### 1. *Movements from Blood to Brain and Within Brain*

All available evidence shows that GABA injected into the bloodstream does not elevate levels of GABA in brain tissue.[13] It has been assumed therefore, that blood-borne GABA cannot enter brain tissue. Studies of regional uptake in rats suggests that this is not strictly true for all areas of the brain and that in the olfactory lobe and an area close to the midbrain tegmentum some slight elevation does occur when 1 g GABA/kg is injected intraperitoneally.[78] Isotope studies indicate that even if no actual elevation of GABA in brain tissue occurs there is some penetration of [$^{14}$C]-GABA from blood into brain. This penetration is greatly increased when the integrity of the blood–brain barrier is altered, either by local freezing,[79] by a lipid solvent such as chloroform/methanol[80] or ethylchloride,[81] or by pretreating an animal with such agents as methoxypyridoxine.[82] It is also possible that during pathological conditions and during hypoxia[79] the blood–brain barrier to GABA may be altered.

When isotopically labeled GABA is injected stereotaxically into the brain ventricles[83] of rats or mice *in vivo* or administered by ventriculocisternal perfusion to cats[84] some of the radioactivity penetrates slowly into the adjacent brain tissue.

It is estimated by electrophysiological criteria that GABA, applied to the subdural surface in a 2.5% solution, penetrates the rat brain to a depth of 800 $\mu$ in 1 hr.[31] Similar evidence is obtained when rat or mouse brains are examined, radioautographically, after the injection of tracer doses of 2[$^{14}$C]-GABA into the lateral ventricle (Fig. 2).[78]

Some species differences exist. After 2 hr of ventriculocisternal perfusion of cat brains with 5 mM [$^{14}$C]-GABA, the isotope had not penetrated into either the white subcortical or the white internal layers.[84] Under these conditions it was supposed that the rate of exchange from the ventricular space into the ependyma of the caudate nucleus (periventricular section of the brain) was 1.2 $\mu$moles of GABA every 2 hr. This rate is much lower than that found for brain tissue *in vitro* (see Section V, F, 1).

### 2. *GABA Movement Out of the CNS*

There is evidence that endogenous GABA in cat brain can be released from the perforated pial surface of the cerebral cortex[85] and that

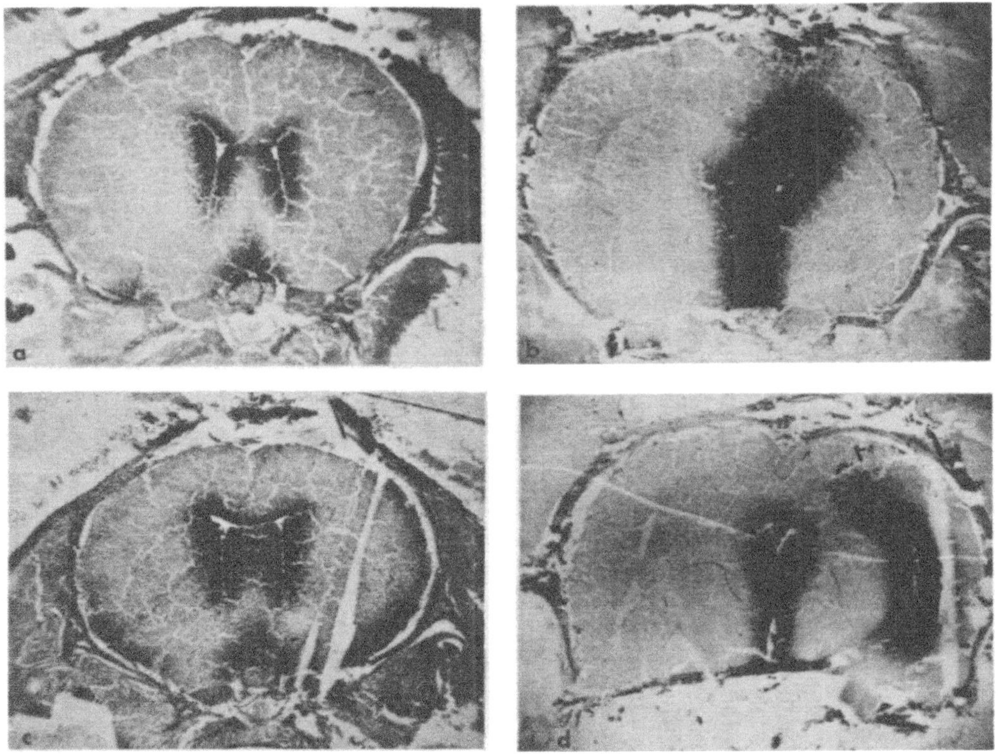

Fig. 2. Radioautographic distribution pattern of 2-[$^{14}$C] GABA in lateral sections of brain after intraventricular injection. Injections of 0.5 $\mu$C 2-[$^{14}$C] GABA in 10λ made stereotaxically into rat and by modified freehand technique[379] into lateral ventricle of mouse. Note slow penetration rate of isotope into tissues of both mouse and rat brain and lack of bilateral distribution in the rat. (a) Mouse, 3 min after injection; (b) rat, 3 min after injection; (c) mouse, 15 min after injection; (d) rat, 15 min after injection.

isotopically labeled GABA, injected into the brain ventricle of a mouse, will be excreted in part into the urine(78); a little unchanged labeled GABA even accumulates in liver and kidney tissues. In rats and mice an appreciable amount of intravenously injected GABA accumulates in these tissues,(86,87) as well as in the adrenal glands. Although release of endogenous GABA from brain tissue into the bloodstream of intact animals has *not* been demonstrated, the above results suggest the possibility that kidney and other organs might acquire some blood-borne GABA originating from dietary sources.

### 3. *GABA Movement into Brain Tissues in Vitro*

It is paradoxical that brain tissue, which *in vivo* takes up only extremely limited amounts of exogenous amino acids, should be a tissue that *in vitro* is one of the most active in absorbing amino acids from the medium against a concentration gradient.(88–90) GABA accumulates to a greater extent than any other amino acid, so that under aerobic conditions and in the presence of glucose, brain slices accumulate up to 40 times the GABA concentration in the medium(90) provided the concentration in the medium is low. Most *in vitro* studies have measured the uptake of GABA by slices of cerebral cortex. In the guinea pig the dorsal part of both the diencephalon and the mesencephalon have a greater capacity to accumulate GABA than does the cortex.(91) Environmental conditions influence uptake. The absence or presence of high levels of potassium ions in the medium inhibit uptake of GABA. Detailed studies of GABA uptake by mouse brain slices indicate considerable specificity for this mechanism. No neutral $\alpha$-amino acids inhibit its uptake competitively and only histidine, $\beta$-alanine, and $\alpha$-$\gamma$-diaminobutyric acid have a slight inhibitory effect.(92)

### 4. *Model for GABA Transport Into Brain Particles*

The transfer of GABA *in vitro* from an external medium (free GABA pool) into membrane-bound particles (bound GABA pools) has been analyzed kinetically.(77,93,94) Two or more metabolic GABA pools can be distinguished within these particles. In a nonmetabolizing system at 4°C, the particles contain one metabolic GABA pool that equilibrates rapidly with the GABA of the external medium. The size of this pool has been calculated to constitute approximately 25% of the total GABA in the particles. The remaining metabolic pools within the particles equilibrate more slowly with the GABA of the external medium.

A model of a carrier-mediated diffusion process has been proposed to account for the accumulation of GABA within the brain particles and the behavior of the GABA pools.(93) The model has the following features:

1. At an extracellular site of a membrane where the environmental level of $Na^+$ ions is high, extraneuronal GABA attaches itself to a binding site on a mobile protein carrier within the membrane.

2. The carrier-bound GABA molecules at the outer side of the membrane are in rapid equilibrium with GABA in the extracellular medium and kinetically constitute the "rapidly equilibrating GABA pool" of the particles.

3. The GABA that is bound to the carrier near the inner surface of the membrane is in rapid equilibrium with the GABA inside the particle, which represents the slowly equilibrating GABA pools. The carrier movement across the inside of the membrane is restricted by a barrier.

4. In the absence of metabolic activity, the rate of GABA transfer within the membrane via the carrier is much slower than the rate of equilibration between bound GABA on the inner and outer sides of the membrane with the internal and external environment, respectively.

5. If GABA is metabolized inside a particle, such as a synaptosome the low concentration of $Na^+$ inside the particle would tend to favor dissociation from the carrier, and the GABA released could be metabolized. In all of this process the $Na^+$ gradient would support a mechanism through which extraneuronal GABA could be removed rapidly to an intraparticulate environment for metabolism. The metabolism of GABA within the particles could generate energy for ATP synthesis, which in turn is required to maintain the $Na^+$ gradient.

This model is supported by experimental results (presented in Section III, B) as well as by the following additional data from *in vitro* experiments[22,73,93,95]:

1. There is no GABA metabolism in particulate fractions of mouse brain in the absence of $Na^+$. Presumably, this reflects the inability of GABA to enter the particles under these conditions.[94]

2. Blockage of GABA breakdown with aminooxyacetic acid results in a net accumulation of GABA within the particles.

3. The $Na^+$ gradient from outside the particle to its inside environment is steeply downhill and would support a mechanism by which extraneuronal GABA could be removed rapidly to an intraparticulate environment for metabolism.

4. Particles suspended in an $Na^+$-free medium lose intrinsic GABA to the medium more slowly than if $Na^+$ is included in the surrounding fluid.[73,95] This has been interpreted to indicate that in the absence of adequate levels of Na, less intrinsic GABA is bound to the carrier near the inside membrane surface of the particle, thereby preventing its transport through the membrane to the surrounding medium.[73]

Within the limitations of the artificial *in vitro* environmental conditions (which do not necessarily reflect conditions *in vivo*), there are no experimental results that conflict with the basic concepts of the model.[22,73,93,95]

### D. The "GABA Shunt Pathway"

In other sections of this chapter the individual enzymatic steps that constitute the "GABA shunt" are described in some detail (see Sections IV and V). As an overall reaction mechanism, the shunt pathway is pulled in the direction of succinate through the activity of succinic semialdehyde dehydrogenase. There is no accumulation of any succinic semialdehyde. Since glutamate is continually regenerated in the transamination reaction (Fig. 1), the net effect of the "GABA shunt" is to convert α-ketoglutarate to succinate with the evolution of $CO_2$ and the reduction of NAD. Carbon atoms of α-ketoglutarate are thus shunted around a portion of the Krebs cycle sequence involving the enzymes α-ketoglutarate dehydrogenase (two enzymes) and succinyl thiokinase. The last-mentioned enzyme catalyzes the reaction

$$\text{Succinyl CoA} + \text{GDP} + \text{Pi} \rightarrow \text{succinate} + \text{GTP} + \text{CoA}$$

Thus, energetically, the Krebs cycle pathway is the more efficient, producing four ATP equivalents instead of the three ATP equivalents generated by the GABA pathway.

Conflicting experimental evidence leaves unresolved whether GABA, as sole substrate *in vitro*, can or cannot support[(98–100)] the respiration of brain-tissue slices.[(96,97)] Similarly, some studies with brain mitochondria indicate that GABA will support oxidative phosphorylation;[(96,101)] other equally plausible studies have provided evidence to the contrary.[(102–104)] The contradictory results may be a function of the methods employed in preparing tissue slices and mitochondria and the composition of the incubation medium (see Sections IV, B and V, B).

There is no doubt, however, that glucose and pyruvate carbon atoms are converted rapidly in brain tissue *in vivo* to the amino acids that are closely related to the tricarboxylic acid cycle: alanine, aspartate, glutamate, glutamine, and GABA.[(105–108)] In the opinion of some, these amino acids *may* represent "obligatory intermediates" for the metabolism of some of the carbohydrate in brain to $CO_2$,[(106)] but this point of view is disputed by others who feel that most of the amino acid labeling is the result of exchange reactions.[(109,110)] Within 2.5 min after the intravenous injection U[$^{14}$C]-glucose into rats, 24–36% of the total radioactivity in brain is in the five amino acids listed above. After 30 min, 36% of the radioactivity in acid soluble metabolites is contained in glutamic acid and 4% in GABA.[(108)]

An assessment of the metabolic importance of the GABA shunt pathway in brain tissues has been attempted both *in vitro* and *in vivo*, and estimates of the extent to which glucose carbon atoms are metabolized via the GABA shunt in brain tissue range from 4% to more than 44%.[(110–113)] This great discrepancy in results appears to be due to a number of factors, the most important of which is cellular heterogeneity of brain tissue. Levels of enzymes, cofactors, and substrates vary widely throughout the brain, and it seems reasonable to assume that the same variability applies to the relative importance of the GABA shunt, not only in different brain areas but also in different cell types or symbiotic cell clusters. One measure of the heterogeneity is the apparent intracellular compartmentalization of metabolites and enzyme systems.

A number of investigators have interpreted both *in vitro* and *in vivo* results to show that GA is compartmentalized in brain tissue.[48,56,107,108,114–116] Some reports show that, after the injection of [$^{14}$C]-glucose into rats, the specific activity of GABA in brain is higher than GA (its only precursor).[108] In other reports the reverse appears to be the case.[117] In order to explain a higher specific activity of a product compared to its precursor, it has been assumed that the product is formed primarily from an isolated compartment within the total precursor pool and that this isolated compartment is more highly labeled with isotope than the total pool. In analog studies of cerebral glutamate metabolism with a digital computer, it was necessary to assume the existence of two *or more* kinetically distinct GA pools in order to simulate the experimental results in the literature.[118]

Rats injected with [$^{14}$C]-glucose and subsequently killed by immersion in liquid nitrogen have in brain tissues a ratio of relative specific activities for GABA/GA that is less than one; yet, when identical animals are decapitated while warm, the ratio of relative specific activities for GABA/GA in their brain tissues is *greater* than one. It is possible, therefore, that in the case of the GA-GABA interrelationship the apparent compartmentation of the GA pool is revealed primarily by the metabolic changes that take place immediately after decapitation.[117] It also has been suggested that these results reveal compartmentation of metabolites to be nothing but post-mortem artefacts. Indeed, some investigators believe that *in vivo* GABA need not originate from specific GA pool compartments.[119,120] It will be apparent from the data in Chapter 19 of Volume II that the evidence against GA compartmentalization involves some debatable interpretation of data, while the case for compartmentalization *in vivo*, if not conclusive, is much stronger. The possibility that GABA is formed equally well from all of the GA pools is unlikely on the basis of experimental evidence and calculations of flux through the GABA shunt, which assume that GABA originates from a homogeneous glutamate pool, are subject to error.

Another major cause for uncertainty in evaluating the importance of the GABA pathway is the inability of many experiments to distinguish between transamination and "exchange" reactions. *In vitro* studies may well give erroneous results because of artificial substrate conditions. Changes in $K^+$ concentration may double the rate of the GABA shunt.[112] Levels of pyruvate and TCA intermediates,[121] as well as the functional state of the tissue, may influence the relative importance of the GABA pathway. The relative levels of pyridine nucleotides, especially NAD and NADH, may regulate the extent to which carbon atoms flow through the GABA shunt and the regular Krebs cycle.[122] Thus, the quantitative importance of the GABA shunt pathway in the overall metabolism of nervous tissue *in vivo* remains an enigma.

## E. Methods for Measuring GABA

### 1. *Chromatographic and Electrophoretic Techniques*

Separation procedures and assay techniques for amino acids have been adapted for use with brain tissue extracts. In addition to two-dimensional paper chromatography and high-voltage paper electrophoresis,[123] special techniques for brain extracts using sequential one-dimensional chromatography,[123a] thin-layer chromatography,[124,125] and combinations of the above methods[126] have been employed. All these methods separate many amino acids, of which GABA is just one. Separations are not always complete, and quantitation by determining ninhydrin color is reliable only if suitable standards are run together with every sample and many environmental variables are rigorously controlled.[78,95,127]

Automated amino acid analyzers can detect and measure GABA, but the identification is made by $R_f$, and sensitivity *usually* permits detection of no less than 0.01 $\mu$ moles. GABA also has been isolated by means of a series of short resin columns through which the sample is washed. Eventually, GABA is eluted, selectively, from the final column.[128] A somewhat simpler two-column system, specifically designed for GABA, also has been worked out.[129] All the techniques mentioned are more or less cumbersome and are used primarily with samples for which it is necessary to obtain information concerning several ninhydrin-positive components besides GABA. They also are used when GABA and other amino acids have to be recovered from extracts to determine their isotopic content. If such procedures are carried out with [$^{14}$C]-labeled amino acids using paper chromatography or electrophoresis for isolation, it is preferable to employ trinitrobenzene-1-sulfonic acid as the detector-spray reagent in order to avoid removal of any [$^{14}$C]-labeled carboxyl groups by the interaction with ninhydrin reagent. Only the carboxyl group of GABA (unlike those of $\alpha$-amino acids) is *not* removed by ninhydrin.[130]

### 2. *Biological Assay*

When Factor I and GABA were first investigated in animal systems, it was popular to assay these substances with biological detectors. The crayfish stretch receptor used initially[131] was replaced by the hindgut of the crayfish (*Procambarus clarkii*).[132] More recently the feasibility for the use of roundworm (*Ascaris lumbricoides*) muscle has been established.[133] The main shortcoming of biological assay techniques for GABA is their lack of specificity. Various compounds that are structurally related to GABA have similar if somewhat lesser effects upon the detector system.[132]

### 3. *Enzymatic Assay*

If only GABA is to be determined quantitatively, the enzymatic assay is the method of choice.* The bacterium *Pseudomonas fluorescens* (EC), if

* For a comparison of different methods see Häkkinen and Kulonen.[134]

grown on a medium containing GABA or pyrrolidine, contains an enzyme system that will convert GABA to succinate via a transamination reaction with α-ketoglutarate and an oxidative step that is coupled to the reduction of NADP.[135] Both the initial rate of NADP reduction and the maximum amount of NADPH formed are proportional to the amount of GABA present, provided there is an excess of α-ketoglutarate, NADP, and enzyme system.[136] A micro assay modification permits the detection of as little as $2 \times 10^{-11}$ moles of GABA in an alcoholic extract of neuronal tissue.[137] The high sensitivity of the micro assay is obtained through converting the NADPH by conventional methods to a highly fluorescent product,[138] which is proportional to the GABA in the sample.

Both macro and micro enzymatic assay procedures have been used extensively, and a lyophilized form of the enzyme system is available commercially. Unfortunately, lyophilization reduces the enzyme activity considerably, and homemade aqueous preparations that are stored in the frozen state contain at least 100% more activity.[78]

## 4. *Tissue Collection for GABA Assay*

The accuracy of determining GABA levels is dependent to no small extent upon the sampling procedure and the way biological tissue samples are treated immediately following their removal from the living animal. When freshly excised brain tissues or decapitated heads are left at room temperature for any length of time, the level of GABA increases while GA decreases a corresponding amount.[108,117,139–143] This effect can cause serious errors in the estimation of *in vivo* GABA levels, especially where detailed dissections of brains at room temperature are involved. In beef brain at 24°C, levels of GABA are doubled within 3 hr.[139] Two independent studies on rat brain[139,140] showed that when animals were frozen in liquid nitrogen or liquid air, brain levels of GABA were lowest (around 170 μg/g tissue wet wt.). When animals were decapitated and the heads dropped directly into liquid nitrogen, levels of GABA were around 185 μg/g tissue wet wt. When brain extracts were prepared 2 min after decapitation at room temperature, GABA levels reached 265 μg/g tissue wet wt. A further increase to 355 μg GABA/g tissue wet wt. was observed when rat brains were kept at room temperature for 1 hr after decapitation. In rats treated with thiosemicarbazide (which inhibits GAD *in vivo*, see Section IV, C), this systematic postmortem increase in GABA concentrations was almost completely prevented.[139]

The increase in GABA after decapitation is primarily an increase in the "free" pool of GABA. Rat cerebral hemispheres, after homogenization in a cold saline solution, contained a GABA pool of which 35–40% appeared to be in the "bound" form.[141] When hemispheres were left standing at room temperature for 2 hr, the overall increase in GABA was 73%. The size of the "free" GABA pool had increased by 95%, while the "bound" pool was enlarged by only 37% from its original size.* The experimental evidence leaves little doubt that the true *in vivo* levels of GABA in nervous tissue are lower than the values actually reported in most of the literature. By the same token, the true *in vivo* levels of GA are higher than the values quoted.

The rapid changes in pool sizes following decapitation also may explain some of the inconsistencies that have been reported from investigators who have studied *in vivo* metabolism of [$^{14}$C]-glucose in the CNS.[108,117,142]

## IV. GABA SYNTHESIS

### A. Glutamic Acid Decarboxylase (GAD)

In the nervous system, glutamic acid (GA) is decarboxylated irreversibly by the enzyme GAD to GABA and carbon dioxide (Fig. 1). The enzyme that is found in animals almost exclusively in the nervous system requires pyridoxal phosphate (PyP) as a coenzyme. Very minute amounts of GAD apoenzyme have been detected in the cerebrospinal fluid. Regional levels of GAD apoenzyme reflect, quite accurately, the regional levels of GABA. This is shown for the guinea pig brain (Fig. 3)[78] and is equally true for other species and interspecies comparisons.[144] The close correlation between the level of apoenzyme and product, in almost all of the brain areas tested, suggests that GAD is the only significant enzymatic pathway leading to GABA in nerve tissues.

### B. General Properties of GAD

Mammalian GAD in crude brain extracts is quite unstable, and purification requires protection of the labile reactive sites on the apoenzyme. This can be accomplished by working with buffer solutions containing PyP at $1 \times 10^{-4}$M, and amino-ethylisothiuronium bromide (AET) at $1 \times 10^{-3}$M (a reagent protecting sulfhydryl and possibly other reactive groups) and by protecting the enzyme from exposure to light.[145] A 150–200 fold purification of the protected GAD apoenzyme can be achieved when a *crude* mitochondrial fraction from mouse brain is extracted and fractionated successively on calcium phosphate gel, DEAE cellulose, and Sephadex columns.[145]

The *p*H optimum varies from 6.4 for crude GAD[146] to 7.2 for the most highly purified enzyme with $K_m$ values changing from $3 \times 10^{-3}$ to $8 \times 10^{-3}$M. The shifts in *p*H and $K_m$ suggest that there are configurational changes in the enzyme protein during purification. Below 40°C the purified GAD is reasonably stable. Above 40°C and below 60°C the enzyme is stable only when attached to calcium phosphate gel.

* Calculated from Elliott *et al.*[141]

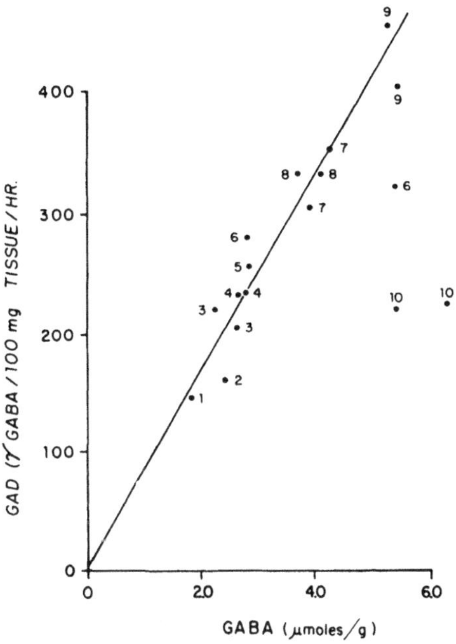


Fig. 3. Correlation of GABA levels and GAD apoenzyme in mature guinea pig brain areas. Levels of GAD were measured in fortified brain homogenates under conditions of optimal PyP and GA concentrations. Enzyme rats were linear during incubation at 37°C. Samples for GABA analysis were dissected and deproteinized within 3–4 min after decapitation. Symbols used for brain areas are 1, medulla; 2, pons; 3, cerebellum; 4, cortex; 5, caudate nucleus; 6, midbrain; 7, diencephalon; 8, inferior colliculi; 9, superior colliculi; and 10, olfactory lobes.

## C. Inhibition of GAD

### 1. *General Inhibitory Agents*

Mammalian GAD is inhibited competitively by halogen anions at 0.05 M concentration. The inhibition by $Cl^-$ decreases with elevated *p*H levels. Since GA is a potential excitatory agent and GABA is inhibitory, it has been suggested that the regulation of GAD by $Cl^-$ concentrations might have physiological significance by controlling the ratio of GA to GABA at specific nerve endings.[(22)] *In vitro*, mammalian GAD is inactivated by a variety of sulfhydryl reagents and by oxidation.[(146)] Both these effects can be partially prevented by adding glutathione or other SH group protective agents. The enzyme is inhibited also by estrogens, salicylate, and biologically important phenolic acids.[(147)]

### 2. *Inhibition by PyP Analogues and Carbonyl Reagents*

The pyridoxal phosphate cofactor of GAD is only loosely linked to the apoenzyme. Dietary $B_6$ deficiency *in vivo* as well as a large number of carbonyl trapping agents inhibit the enzyme.*

On the basis of data available for GAD and studies made on the interaction of amino acid end groups with pyridoxal phosphate in other Vitamin $B_6$ dependent enzyme systems, it has beeen suggested that the

* For review and details see Roberts *et al.*[(15)]

aldehyde group of the coenzyme can be linked to the protein of the GAD apoenzyme, either through an aldimine linkage to two ε-amino groups of lysine or through a thiazolidino linkage to one cysteine sulfhydryl and one ε-amino group of lysine.[15] Compatible with such an hypothesis are spectral data and *in vitro* studies with vitamin $B_6$ analogues and sulfhydryl and carbonyl trapping agents. The latter inhibitors interact with the aldehyde group of the PyP coenzyme and presumably with some amino acid end groups of the apoenzyme.* The quantitative values for some inhibitors of the mouse brain enzyme are shown in Fig. 4.

Although carbonyl trapping agents, such as hydroxylamine and *O*-substituted derivatives (such as aminooxyacetic acid) inhibit GAD in an *in vitro* system, they appear to have little if any effect upon the enzyme *in vivo*.[148] Hydrazines may inhibit GAD directly *in vitro*, but it is possible that *in vivo* their effect on GAD is indirect through a limitation in the availability of coenzyme. The synthesis of pyridoxal phosphate from pyridoxal and ATP is mediated by the enzyme pyridoxal kinase, and it has been found that hydrazine, hydroxylamine, and isonicotinic acid hydrazide are 100–1000 times more potent in inhibiting pyridoxal kinase than GAD *in vitro*.[149] Pyridoxal potentiates the inhibition of GAD by a number of hydrazines,[150] and there is evidence to indicate that this effect *in vivo* might be due to the formation of the pyridoxal hydrazone [151,152,153] (presumably through the interaction of the amine group of the hydrazine with the aldehyde group of pyridoxal). Pyridoxal semicarbazone is 30 times more potent in inhibiting pyridoxal kinase than is semicarbazide.[154]

These results led to the hypothesis that *in vivo* the carbonyl trapping agents or their pyridoxal derivatives might inhibit pyridoxal kinase preferentially, which then would lead to a deficiency of PyP. This in turn,

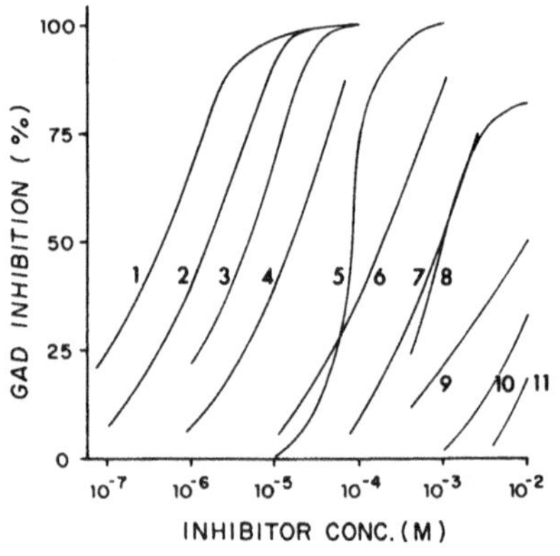


Fig. 4. Inhibition of mouse brain GAD by various molar concentrations of inhibitory agents *in vitro*. The inhibitors shown are: 1, aminooxyproprionic acid; 2, aminooxyacetic acid; 3, (+) 2-aminooxypropionic acid; 4, (−) 2-aminooxypropionic acid; 5, hydroxylamine; 6, nitroso *R* salt; 7, 1,2-napthoquinone-4-sulfonic acid; 8, D penicillamine; 9, L penicillamine; 10. D cysteine; and 11, L cysteine. Compiled and redrawn from Roberts and Simonsen.[146]

* For review and details see Roberts *et al.*[15]

would inhibit selectively enzyme systems, such as GAD, which have a weak affinity for the PyP cofactor. The observation that L-glutamic acid hydrazone which *in vivo* depresses GABA levels dramatically, fails to inhibit pyridoxal kinase *in vitro*(155) places some doubt upon the general application of this widely quoted hypothesis. To complicate matters even more, other hydrazones such as cyclic phospho-5-pyridoxal hydrazone,(156) $\gamma$-glutamyl hydrazone,(151) pyridoxal-1,1-dimethyl hydrazone,(152) and pyridoxal-cyanacetyl hydrazone(157) can take the place of pyridoxal phosphate *in vitro* and activate GAD. This activation does not appear to be the result of pyridoxal phosphate liberation.(158)

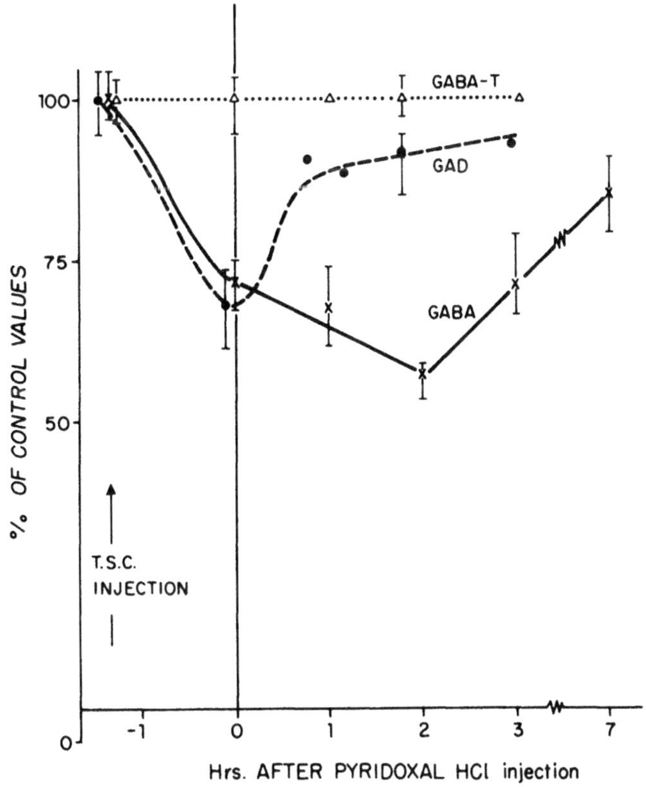


Fig. 5. Effect of pyridoxal injection upon rats pretreated with thiosemicarbazide. Fed, female Sprague-Dawley rats (200 ± 15 g) were injected ip with thiosemicarbazide (TSC) (20 mg/kg, *p*H 8.0). Just before the time that convulsant activity was expected, the rats were injected with pyridoxal HCl (50 mg/kg, *p*H 4.5). No rats convulsed at any time during experiment. Rats were decapitated at times indicated, and GAD, GABA, and GABA-T were measured. Individual points represent average of from three to eight determinations with different animals.

### 3. *Methodological Limitations for in Vivo Studies of GAD*

Not only carbonyl trapping agents such as those described above and semicarbazide, thiosemicarbazide, isoniazid, and isopropylhydrazine[159] will inhibit GAD *in vivo*, but also sulfhydryl reagents such as penicillamine[160] and vitamin $B_6$ antagonists such as ω-methyl pyridoxine[159] and toxopyrimidine.[161] These "*in vivo*" results are most frequently arrived at by measuring GABA and GAD activity levels in brain homogenates or brain extracts of drug-treated animals. Thus, the *actual* enzymatic measurement is made *in vitro*, which may result in erroneous results.

For example, the level of GABA and the activity of GAD decline in the brain of rats injected with thiosemicarbazide (TSC).[162] Subsequently, if the rats are injected with pyridoxal (to prevent convulsions), GABA levels continue to decline,[163] although the activity of GAD appears to return toward normal levels (Fig. 5). This effect cannot be attributed to GABA diffusion out of the brain or to an increased rate of GABA degradation[78] and probably represents a methodological artefact in the measurement of GAD or GABA.

*In vivo*, GAD would not be reactivated if pyridoxal or pyridoxal phosphate fails to reach the GAD apoenzyme site because of some membrane barrier. When GAD levels are measured *in vitro*, however, homogenization has destroyed the membrane barrier and the cofactor and GAD apoenzyme can combine to form the reactivated holoenzyme. Similarly, the lower levels of GABA in brains of TSC treated rats may be the result of postmortem changes in the brains of the normal controls. This possibility is based upon the findings that GABA levels are elevated within minutes after decapitation in the brains of normal controls but not in the brains of TSC treated rats.[16] Any attempts to correlate the action of pharmacologically active compounds *in vivo* with their effect upon GABA metabolism (*in vivo* or *in vitro*) must take cognizance of discrepancies, such as those illustrated above, and it is not surprising that studies of mechanisms by which carbonyl reagents interact with vitamin $B_6$-dependent enzymes *in vivo* seem to be beset by paradoxes (also see Sections VII, A and B).

## D. Comparative Aspects of GAD

The foregoing discussion was directed toward mammalian GAD. Studies also have been performed with GAD from crustacean nervous tissue that show there are many similarities, but also some fundamental differences, between the two enzymes.[164]

While the purified mouse brain enzyme is not affected by high levels of GABA,[145] a 39-fold purified GAD from lobster nerve is distinctly inhibited by its product. The Ki was calculated as $1.25 \times 10^{-3}$ M GABA.[146] It is suggested that product inhibition of GAD may function as a regulatory mechanism for the selective accumulation of GABA in inhibitory neurons of the lobster. This would be similar to the role that has been suggested for $Cl^-$ in the mammalian system[22,165] (see Section IV, C). The lobster GAD is unaffected by anions but has an absolute requirement for $K^+$.[164] No such need has been demonstrated for mammalian GAD. Finally, the pyridoxal phosphate cofactor appears to be bound somewhat more tightly to the lobster apoenzyme than to the mammalian enzymes that have been studied.[164] These comparisons may apply only to mature mammalian brain. The newly discovered GAD in kidney tissues and embryonic chick brain[165a] is activated by $Cl^-$ and has other properties which are different from the enzymes described above.

GAD has been found in the nervous system of all vertebrate and invertebrate species that have been tested. The level of the apoenzyme appears to be higher (per unit wet weight of brain tissue) in the small mammals as compared to the large mammals. GAD activity and GABA levels appear to decrease in some of the lower invertebrates.[166]

## E. Developmental Aspects of GAD and GABA

Studies on the development of GABA levels with age have been made for the brains of mice, rats, rabbits, chickens, cats, dogs, and bullfrogs.[144,167–181] However, corresponding studies for GAD apoenzyme activity have been conducted only with rodents and in the chick.[22,144,167–173] In all systems and brain areas studied there is a progressive increase of both enzyme and substrate until mature levels are reached. In the cortex of the rabbit the greatest increase in GAD activity coincides most closely with the fastest growth of the surface area of the dendrites. Changes in GABA levels are correlated most closely with the proportional volume of the dendrites. Adult EEG patterns and maturation of the cell body nucleus coincide with the attainment of adult GABA levels.[167] Similarly, the maximal increase in GABA occurs in the neo-cortex of the kitten during a period of maximal elaboration of the dentritic system of pyramidal neurons.[174] In the chick, optic lobe levels of GAD and GABA can first be detected in the 7-day-old embryo, and the increase of GAD activity is paralleled by a proportional elevation of GABA until approximately 5–10 days after hatching when adult levels are attained. Similar patterns of development were observed for other brain areas, although adult levels were not all attained at the same time.[144] In the cerebellum, GAD activity increases most rapidly just before hatching, a period corresponding also to the greatest acceleration in dendritic arborizations.[22] In the rat, the time period of greatest change in GAD levels is during the second and third week after birth.[169,170]

It is of interest to note that of all amino acid systems studied in the brain, a reasonable correlation between an enzyme and its product during development has been found only for GAD and GABA.

## F. Localization of GAD

The CNS of several adult mammalian species, including man, has been studied for the regional distribution of GAD.[182,183] In general, areas with high levels of gray matter appear to have high levels of GAD (also see Fig. 2). In the cerebellum of the rabbit, GAD is associated most closely with the Purkinje cell layer, which is believed to be a region of numerous inhibitory synapses.[38] In the spinal cord of the cat, dorsal gray contains twice as much GAD as ventral gray matter.[184]

Evidence regarding the subcellular distribution of GAD is somewhat misleading. When a crude brain homogenate is subjected to differential centrifugation, GAD migrates in the gravitational field with the crude mitochondrial fraction.[68] However, further analyses of this fraction, mainly on discontinuous sucrose gradients, left no doubt that GAD is a cytoplasmic enzyme contained preferentially in the presynaptic nerve ending fraction.[69,70,165,185,186] Hypo-osmotic shock treatment releases the GAD from this particulate fraction into the supernatant.[69]

The distribution of GAD on a sucrose gradient appears to be biphasic,[187] with about 40% of the total GAD in a homogenate sedimenting with the nerve ending fraction.[188,189] In rat or guinea pig brain, the actual amount of GAD that is contained in the particulate fraction, or released by osmotic shock, appears to be $Ca^{2+}$ dependent.[69,190] A 4 mM concentration of $CaCl_2$ in a water or sucrose medium prevents any substantial quantity of GAD from being released from the particulate fraction into the supernatant, unless chelating agents that complex with $Ca^{2+}$ or detergents that disperse membrane structures are used. The detergent method is the more effective of the two in releasing the particulate-bound enzyme into the supernatant.[190] The calcium effect may not be valid for all species; it could not be confirmed using a synaptic vessicle preparation from mouse brain.[73]

GAD has been reported to be associated also with other particulate fractions of the cell.[189] It seems probable that the enzyme is absorbed to the surface of such particles. The site of GAD apoenzyme biosynthesis has not been determined.

## G. GAD Levels and Physiological Parameters

Aside from changes in GAD activity during development and dietary vitamin $B_6$ deficiency, levels of the enzyme have been reported to be elevated during hibernation[191,192] and decreased after adrenalectomy[193] and in animals on a low protein diet.[194] GAD has been measured also after x-irradiation[195] as a function of environmental complexity,[196] after maximal electroshock treatment,[197] during insulin shock hypoglycemia, etc. (Also see Sections VI, B and C.)

## H. Assay Methods for GAD

The level of GAD apoenzyme in neuronal tissues has been determined by measuring GABA formation fluorometrically, by manometric measurements, or by isotope techniques. In all these methods, homogenates or fractions derived from homogenates of mammalian neuronal tissues are incubated under anaerobic conditions at 37°C in a phosphate buffer at a *p*H between 6.0 and 6.5. Levels of glutamate substrate vary from 0.03 to 0.1 M, and an excess of pyridoxal phosphate cofactor (final concentration approximately $1 \times 10^{-4}$ M) is also present. In most assays, mercaptoethanol or some other sulfhydryl reagent is added to stabilize the enzyme. Under optimal conditions GAD activity is linear for more than 1 hr. The lobster GAD is measured under similar conditions except that the incubation temperature is usually 22°C and the buffer *p*H is 7.2.

When GAD activity is measured by product formation, the level of GABA in a system must be determined before and after incubation. Assay techniques for GABA are discussed elsewhere (see Section III, E). A refinement that permits GAD activity determinations in as little as 3 μg dry wt. of brain tissue is based upon the conversion of GABA under alkaline conditions to a highly fluorescent GABA-ninhydrin complex. The method is limited for use under rather specific conditions.[198]

The manometric determination of GAD activity in nervous tissue, based upon the $CO_2$ evolved by the decarboxylase reaction, was first described in 1951.[199] It has been widely used. By using 1[$^{14}C$]-labeled glutamic acid substrate and measuring the $^{14}CO_2$ evolved, the sensitivity of this method is greatly enhanced and offers the possibility to do analyses with very small quantities of tissue. The vessels for this reaction, the $^{14}CO_2$ absorbent used in conjunction with scintillation counting, as well as other details of the method, have been modified repeatedly.[146,164,168,183]

## V. GABA DEGRADATION

### A. γ-Aminobutyric-α-Ketoglutaric Transaminase (GABA-T)

The major pathway of GABA degradation is via transamination with α-ketoglutarate (αKG) to form glutamate and succinic semialdehyde (Pathway 3, Fig. 1). The GABA-T apoenzyme requires PyP as a cofactor.[200,201] The GABA-T reaction is reversible *in vitro*,[202] and an interchange of $C^{14}$ atoms between GABA and succinic semialdehyde, as well as between L-glutamate and αKG,[109,203] has been demonstrated. However, there is no evidence that the reaction goes toward GABA when it is coupled *in vitro* or *in vivo* to succinic semialdehyde dehydrogenase (see Section V, H). Nonenzymatic transamination of GABA with pyridoxal or PyP can be induced *in vitro*, and the imine intermediate has been isolated;[204] but the rate of the reaction at physiological temperatures and substrate concentrations is so slow that its contribution to the biological transamination would go undetected by present assay techniques.

### B. General Properties of GABA-T

GABA-T appears to be a particulate enzyme to which most of the PyP cofactor is tightly bound. It is found not only in brain tissue but also in other tissues of the body.[205,206,207] When [$^{14}C$]-labeled GABA is injected intraperitoneally into mature rats, it is metabolized in body tissues other than brain, as evidenced by the $^{14}C$ label appearing rapidly in the respired $CO_2$, in tricarboxylic acid cycle intermediates, in closely related amino acids, and in the glycogen of the liver.[208,209]

### 1. pH Optima and Isoenzymes

Fairly crude brain extracts have been used to determine the properties of GABA-T. *p*H optima have been determined for beef brain,[201] monkey brain[210] rat brain,[211] and mouse brain transaminase[212] and are 8.2, 8.2, 8.6, and 8.0, respectively. In the lobster nerve the *p*H optimum of GABA-T is substrate dependent. In the presence of 2 mM αKG it is 8.5, while at 20 mM αKG it is 9.0.[28] This shift has been attributed to the large difference in the $K_i$ of the transaminase under the two *p*H conditions.

In a more highly purified mammalian GABA-T system, at least four isoenzymes can be identified as anionic or cationic components. Some of these components appear to form aggregates. There are differences in the profile of GABA-T isoenzymes isolated from intra- and extramitochondrial fractions. The anionic components, which show optimal activity at a lower *p*H than the cationic components, are found distributed to a greater extent in the extramitochondrial fractions.[212a]

Several studies on the activity of GABA-T have been made in media containing sucrose. These investigations may contain methodological artefacts caused by the effect of sucrose upon the enzyme.[207] At a concentration of 0.32 M sucrose, almost 50% of the GABA-T activity *in vitro* is suppressed. However, the addition of 0.5% Triton X (a detergent) restores complete activity.

### 2. Ionic Requirements

$Na^+$ is needed for the metabolism of GABA by brain particles,[94] presumably to facilitate movement of GABA to intraparticulate enzyme sites (see Section III, C). The addition of $K^+$ ions to the incubation medium of mammalian brain slices stimulates, briefly, metabolism through the "GABA shunt" pathway[112] and promotes release of GABA from the tissue into the medium. The effect is antagonized by $Ca^{2+}$.[213] The susceptibility of GABA-T to ionic influences opens the possibility that the activity of the enzyme *in vivo* may be modified by neuronal activity.

### 3. Enzyme Specificity and Mechanisms

GABA-T acts upon a number of amino acid substrates. The crude enzyme transaminates α-γ-diaminoglutaric acid faster than GABA and β-alanine almost as rapidly as GABA. Other ω-amino acids and GABA choline are transaminated at a lower rate.[201,205] By contrast, there appears to be an almost absolute specificity for α-ketoglutarate as the amino acceptor, both in the mammalian and crustacean enzymes.[28,201]

High $K_m$ values indicate a rather low affinity of the mammalian enzyme for both GABA and αKG.[201,207,212] The activity of GABA-T from lobster nerve is inhibited at a substrate concentration of $2 \times 10^{-3}$ M αKG (the $K_m$ reported for GABA-T from mouse brain). The inhibition is competitive with GABA, which suggests that both substrates compete for the same or closely adjacent enzyme sites. On the basis of a detailed kinetic analysis, a catalytic mechanism has been proposed by which each substrate reacts *alternately* with the enzyme, thus eliminating the necessity for binding both substrates to the enzyme simultaneously.(28) Such mechanisms have been suggested also for other transaminase reactions.(135,214) Alternately, a ternary complex between the enzyme and two substrates has been proposed for the GABA-T from mouse brain, also on the basis of kinetic data.(212)

## C. GABA-T Inhibition

It was discovered some years ago that GABA biosynthesis was impaired by some convulsant states,(162,215) and some fortuitous experiments correlated the overall elevation of GABA with anticonvulsant activity.(216) These observations stimulated interest in finding ways to elevate GABA levels within the CNS. Since in mature animals the blood–brain barrier prevents the transfer of appreciable quantities of GABA from the blood into most areas, attempts were made to elevate intrinsic levels of GABA *in vivo* by inhibiting GABA-T selectively.

*In vitro* there are many agents that will inhibit GABA-T from both mammalian and crustacean nervous tissues. They include substrate analogues, carbonyl trapping agents,(17) substances that will react with protein sulfhydryl groups,(28,201) and many compounds that interact with the protein apoenzyme.

In the living animal the effects of any inhibitor or inhibitor-$B_6$ complex will be modified by physiological *p*H values, the rate of detoxication, the rate of inhibitor penetration through the membrane barriers that compartmentalize specific enzyme systems, the rates of competitive reactions for the inhibitor, and many other factors. Even if a reagent reaches the correct metabolic compartment, there is no assurance that it will interact with GABA-T selectively.

### 1. *Substrate Analogues*

In theory, GABA analogues offer the best possibility for *specific* GABA-T inhibition, and *in vitro* they do inhibit the enzyme. However, GABA analogues—in which the hydrogen atom on the γ-carbon is replaced by a methyl, ethyl, or phenyl group, their corresponding lactams,(217) GABA esters, and N-substituted GABA derivatives(148)—are all equally ineffective *in vivo* when administered by intraperitoneal injections.

### 2. *Effect of Carbonyl Reagents on GABA-T in Vivo*

Hydroxylamine ($NH_2OH$),(148,216) aminooxyacetic acid (AOAA),(148,218,219) hydrazines,(220) γ-glutamyl hydrazide,(221) and cycloserine,(222) when administered *in vivo*, inhibit GABA-T and elevate levels of GABA in brain tissue. GAD remains completely unaffected or in some cases is slightly depressed. A number of similar compounds have been reported to elevate intrinsic levels of GABA *in vivo* (see Sections VII, A and B), but they have not been tested specifically for inhibition of GABA-T. As in the case of GAD (see Section IV, C), their effect may be indirect. Some of the carbonyl trapping agents and $B_6$ antagonists, which inhibit GABA-T, also may be limiting the amount of PyP available for combination with the GABA-T apoenzyme. For example, aminooxyacetic acid, besides inhibiting GABA-T directly, also forms an oxime with pyridoxal, causing a vitamin $B_6$ deficiency.[223] In addition, it acts as a powerful inhibitor of pyridoxal kinase.[224]

### 3. *Preferential GABA-T Inhibition*

Certain drugs that inhibit both GAD and GABA-T *in vitro* interact preferentially with one of the two enzymes *in vivo*. As mentioned above, this selectivity of a drug may be related in part to the differing subcellular distribution of the two enzymes. It also has been suggested that an apoenzyme to which a cofactor (or cofactor–inhibitor complex) is bound firmly would be more difficult to reactivate with an added cofactor than an apoenzyme to which a cofactor (or cofactor–inhibitor complex) is loosely attached. GABA-T binds PyP firmly, whereas GAD binds PyP loosely. The preferential inhibition of GABA-T by some carbonyl reagents therefore may be related to the ability of uncomplexed PyP to displace the PyP-inhibitor complex from the enzyme surface of GAD but not from GABA-T. There is experimental evidence to support this hypothesis. For example, the *in vitro* inhibition of GAD by hydroxylamine ($NH_2OH$) can be reversed by the addition of PyP. Under similar conditions, GABA-T inhibition by $NH_2OH$ is irreversible.[148] It is not known, however, whether elevation of cerebral GABA levels, by the administration of $NH_2OH$ or related compounds *in vivo*, involves a brief inhibition of both GAD and GABA-T, with rapid reactivation of GAD by intrinsic PyP. Several observations *in vivo* cannot be explained readily on the basis of the hypothesis discussed above. We have no adequate answers yet to questions such as "Why does hydroxylamine inhibit GABA-T and elevate levels of GABA in the brain of rats, cats, monkeys, etc. but not in mice?"[148,225] or "Why is the inhibition of GABA-T by hydrazines potentiated by the simultaneous administration of pyridoxal?"[220]

## D. Comparative Aspects of GABA Degradation

See Sections V, B, E, and H for pertinent material.

## E. GABA-T During Development

The activity of GABA-T enzyme in brain has been measured systematically in a few studies during embryogenesis and development. In the chick embryo the enzyme has been detected on the tenth and eleventh day.

Activity in the optic lobe reaches adult levels 5–10 days after hatching.(144) In whole chick brain, maximum levels have been reported just before hatching.(226) In the rat cerebral cortex, levels of GABA-T are low at birth but increase about tenfold during development. The greatest increase in apoenzyme occurs between the fifth and thirtieth day after birth. Mature levels are reached somewhere between 1 and 3 months.(207,227) Both in chick and rat changes in GABA-T follow the same pattern as GAD during development. Studies on five brain areas of the nervous system in rabbits have been made, and a similar pattern is apparent in all areas, excepting the spinal cord, where mature levels of GABA-T are reached 15 days after birth.(207)

## F. Localization of GABA-T

### 1. *Regional Distribution*

There are large regional variations in the levels of GABA-T apoenzyme in the brain and spinal cord of mammals.(63,184,210,207,228–231) Levels of the enzyme in white matter are extremely low(210)—below the level of detection by histochemical procedures. However, after blocking GABA-T with aminooxyacetic acid, GABA levels are elevated in some regions of white matter that are known to contain only low levels of GAD and GABA under normal conditions. This suggests that the entire GABA system, including GABA-T, might be present in most of the white matter.(38) In the gray matter of the monkey brain, levels of GABA-T vary over a 35-fold range,(63,210) being lowest in layers 5 and 6 of the occipital cortex (9 mmoles/kg/hr) and highest in the dorsal horn of the lumbar spinal cord (315 mmoles/kg/hr).(210)

In the brainstem of the mouse, GABA metabolism is particularly high in the olives, hypoglossal nucleus, motor nuclei of the vagus, nuclei of the trigeminal nerve, dorsal tegmental nuclei, and trapezoid nuclei. In the rabbit brainstem GABA-T has a similar distribution. Within the brain and brainstem areas, cells in contact with the cerebrospinal fluid or blood appear to contain high levels of GABA-T.(231) These observations have given rise to the suggestion that this strategically localized enzyme acts as the blood–brain barrier to GABA. The hypothesis has some support from the observations that [$^{14}$C] GABA injected into rat ventricle penetrates extremely slow into brain tissue(76,138) (Fig. 2) but lacks support from experiments in which GABA-T inhibitors, such as AOAA, failed to modify the blood–brain barrier to GABA.(232)

In the rabbit cerebellum, which has been studied in detail, GABA metabolism appears highest in Purkinje cell cytoplasm, Golgi cell bodies, and mossy fiber endings. All these structures are related to inhibitory function. The granular layer, containing supposedly excitatory cells, shows very little GABA-T activity.(38) It is not known whether GABA-T is preferentially localized in pre- or postsynaptic regions.

## 2. *Subcellular Distribution*

In contrast to GAD, GABA-T is a mitochondrial enzyme.[69,185] Mitochondria released by osmotic shock from presynaptic nerve endings contain about 20% of the GABA-T activity, while the pooled neuronal, dendritic, glial, and endothelial mitochondria from all other brain cells and cell structures contain approximately 80% of the activity.[233] Enzymes in the two mitochondrial populations differ in their susceptibility to metabolic inhibitors and activators. Glutamic and aspartic acid in the incubation medium inhibit GABA-T in synaptosomal mitochondria but stimulate GABA-T in mitochondria of the neuronal perikaria and glial cells.[185] (Also see Section V, H.) In studies where GABA-T isoenzymes were isolated, differential *p*H effects were noted in testing the enzyme fractions derived from mitochondrial and extramitochondrial compartments.[212] These properties are an indication of potential regulatory mechanisms for GABA-T and, by implication, regulation of GABA levels in different parts of the subcellular environment of brain tissues.

## G. Assay Methods for GABA-T

There are five methods that have been used to determine GABA-T in neuronal tissue *in vitro*:

1. The earliest is based upon the chelating properties of α- and γ-amino acids. α-Amino acids chelate copper strongly and quantitatively, whereas γ-amino acids do not. Thus, in the reaction

$$\text{GABA} + \alpha\text{-ketoglutarate} \xleftrightarrow{\text{GABA-T}} \text{L-glutamate} + \text{succinic semialdehyde}$$

only L-glutamate chelates copper. The soluble copper glutamate chelate is measured by further complexing the copper with a phenanthroline reagent (neocuproin), and the intense yellow color produced is measured spectrophotometrically.[201]

2. An alternate method measures the formation of succinic semialdehyde—the other product of the reaction. The aldehyde is condensed with 3,5-diaminobenzoic acid to form a quinaldine derivative that can be assayed spectrofluorometrically.[210]

3. More recently, succinic semialdehyde has been coupled to an $NAD^+$-linked succinic semialdehyde dehydrogenase derived from guinea pig kidney. The dehydrogenase reaction pulls the GABA-T reaction, and the NADH formed is proportional to the rate of GABA-T. NADH is measured fluorometrically after treatment with alkaline peroxide.[211]

4. In a more histological and semiquantitative approach, GABA-T can be measured by coupling the oxidation of succinic semialdehyde to the reduction of a tetrazolium salt. Brain sections are incubated in contact with an agar-saline-gel *p*H 7.4 containing GABA, α-ketoglutarate, NAD, and 2,2′ di-*p*-nitrophenyl 5,5-diphenyl-3,3 (3,3′ dimethoxy-4,4′ biphenylene) ditetrazolium chloride (abbreviated Nitro BT). The Nitro BT is reduced and

deposited as grains of insoluble formazan in the gel according to the following reactions:

$$\text{Succinic semialdehyde} + NAD^+ + H_2O \rightleftharpoons \text{succinic} + NADH + H^+$$

$$NADH + \text{Nitro BT} + H^+ \rightarrow NAD^+ + \text{Nitro } BTH_2 \text{ (formazan)}$$

$$\text{Sum: Succinic semialdehyde} + \text{Nitro BT} + H_2O \rightarrow \text{Succinate} + \text{Nitro } BTH_2 \text{ (formazan)}$$

After completion of incubation, sections are allowed to dry and then are placed briefly and successively into 100% ethanol and xylene to remove the faint and diffuse background color. The method is fairly specific and proportional to GABA-T activity in the brain sections. Substrates of β-alanine, aspartate, glutamate, and glycine do not stimulate formazan production.[(228)]

5. An isotope method provides the most sensitive means available at the present time to measure quantitatively GABA-T activity *in vitro*. It takes advantage of the "exchange" reaction mentioned earlier[(203)] (see Section V, A).

$$5[^{14}C]\ \alpha\text{-Ketoglutarate} + \text{GABA} \rightleftharpoons 5[^{14}C]\ \text{glutamate} + \text{succinic semialdehyde}$$

Using radioactive α-ketoglutarate, GABA-T activity is proportional to the radioactive glutamate formed. The latter is isolated quantitatively by a simple column technique in which the acidic, deproteinized extract of the reaction mixture is poured onto a column of Dowex 50 × 8 in the $H^+$ form. The unreacted α-ketoglutarate is washed off the column with water and the $[^{14}C]$ glutamic acid is collected in another fraction by elution off the column with 2 N ammonium hydroxide.[(229)]

In one modification of this method, labeled $[^{14}C]$ GABA is used as a substrate and the radioactivity of the products is measured in the initial water washings of the Dowex column.[(28)] This modification of the micro-technique has the disadvantage that the specific activity of the GABA used will be modified by the endogenous GABA in the tissue preparations, a factor for which a correction must be made.

It is customary in all of the above methods to add optimal amounts of substrate in order to maintain valid enzyme kinetics. It is probable, however, that substrate concentrations rather than GABA-T enzyme levels are the rate-limiting factors *in vivo*. For this reason the effects of drugs upon the metabolism of $[^{14}C]$ GABA in brain tissue *in vivo* have been studied. Most measurements obtained in such studies reflect not only the activity of GABA-T but the metabolism of GABA along a pathway of many reaction steps.

### H. Succinic Semialdehyde Dehydrogenase (SSA-D)

GAD, GABA-T, and SSA-D are the three enzymes that comprise the "GABA shunt" pathway around a portion of the tricarboxylic acid cycle.

SSA-D has been studied in detail in lobster nerve[234] and in the brain tissues of rats,[69,235,236] monkeys,[237,238] and humans.[239,240] Overall localization of the enzyme appears similar to that of GABA-T, both on a regional[240] and subcellular distribution[69] basis, and it is as an adjunct to GABA-T that it is considered here.

### 1. *Properties*

SSA-D is a mitochondrial enzyme that can be solubilized and activated by a detergent (Triton X).[236] The enzyme has a *p*H optimum between 8.6 and 9.4. SSA-D from mammalian brain has great substrate specificity both for SSA and $NAD^+$. No other aldehyde tested or $NADP^+$ could be substituted.[237] The crustacean enzyme is less specific; 5 mM concentrations of acetaldehyde are oxidized at one-third to one-half the rate usually observed with 0.064 mM SSA. The enzyme also can operate at one-third the optimal rate with 3 mM $NADP^+$.[234]

The $K_m$ is $1 \times 10^{-5}$M (or less) for SSA and around $10^{-4}$ for $NAD^+$. Absolute values depend upon the assay conditions. These low $K_m$ values may explain why SSA itself is not usually detected in brain tissues.[238] SSA-D is inhibited by its own substrate SSA at concentrations in excess of $10^{-4}$M. The inhibition is believed to be due to a reaction of the enzyme–substrate complex with a second molecule of SSA. Kinetic observations fit this hypothesis.[236]

SSA-D apoenzyme contains an essential SH group. SH inhibitors, such as *p*-chloromercuribenzoate and *N*-ethylmaleamide, inhibit competitively with $NAD^+$. Thus, both thiol reagents (such as $\beta$-mercaptoethanol) and the presence of $NAD^+$ activate the enzyme and protect it from inactivation. Arsenite is inhibitory. Mammalian SSA-D is subject to monovalent cation activation by either $K^+$ or $Na^+$ with maximal activity at 50 mM. It is also activated by temperatures above 37°C. This has led to assay procedures of this enzyme at 44°C.[237] By contrast, the crustacean SSA-D is not stimulated by monovalent cations, and concentrations in excess of 50 mM are inhibitory. Although preincubation with $NAD^+$ activates the enzyme, no actual temperature activation has been demonstrated.

Activity measurements of SSA-D are made by different investigators under different incubation conditions, but the rate has always been followed by the formation of NADH, spectrophotometrically or spectrofluorometrically. In one instance[234] the production of [$^{14}$C] succinate from 1[$^{14}$C] SSA was measured.

### 2. *Purification and Distribution*

SSA-D has been purified 250-fold by conventional means (salt fractionation and column separation). No isoenzymes could be detected. Levels of SSA-D have been reported for various mammalian brains. Under *in vitro* assay conditions, rat brain oxidizes 200 $\mu$moles SSA/g wet wt./hr at 37°C[236] and 2000 $\mu$moles SSA/g dry wt./hr at 44°C.[237] Comparable to the latter rate are (in micromoles SSA per gram dry weight per hour) guinea pig, 1400; mouse, 1100; and rabbit, 750. In the cerebellar layers of the monkey, they range from 1500 in the molecular layer to 300 in the white matter. Values for human brain are comparable.[240]

The developmental aspects of SSA-D may be complex.[235] In the whole rat brain there is a discontinuous increase in levels from birth to 90 days, with the greatest increase occurring from 0 to 30 days. A more detailed study of the cerebellum shows that during the first 90 days after birth SSA-D disappears from the external granular layer, remains virtually unchanged in white matter (on a fat-free basis), and increases 3- to 4-fold in the molecular and internal granular layers.

### 3. *Possible Alternate Pathway for SSA Metabolism*

Recently an alternate pathway for SSA has been proposed in which SSA is reduced to form γ-hydroxybutyric acid by an enzyme that could not be distinguished from lactic dehydrogenase.[241] It is claimed that γ-hydroxybutyric acid is present in mammalian brain at a level of 1 mM,[241] but this claim is contested[242] on the basis of methodological considerations. γ-Hydroxybutyric acid is easily converted to γ-butyrolactone and vice versa.[243] Both these compounds have strong depressant, hypnotic, and sleep-inducing properties, and the dispute as to whether the acid or lactone is the biologically or pharmacologically active form in the brain has not been resolved.[244,245]

The pharmacological properties and interactions of γ-hydroxybutyrate and γ-hydroxybutyrolactone are described in an ever-growing literature.[246] There is no evidence to suggest that any physiological or pharmacological effects of γ-hydroxybutyrate are related to GABA, and there is no agreement as to whether GABA levels of brain tissue change after the injection of hypnotic doses of γ-hydroxybutyric acid.[247–249] It is conceivable that under specific conditions drugs that effect GABA metabolism may exert their effect via SSA[250] through γ-hydroxybutyric acid or its lactam. For example, with high convulsant doses of aminooxyacetic acid[251] such a mechanism cannot be ruled out.

The metabolism of γ-hydroxybutyric acid in brain tissue remains unknown. Except for some evidence that indicates that γ-hydroxybutyric acid can be converted to GABA without involving glutamic acid,[249] no other pathway of metabolism is established. Isotopically labeled γ-hydroxybutyric acid is not converted to succinate[243,252] but might proceed via β-oxidation to 3.4-dihydroxybutyric acid.

## I. Products of GABA Metabolism via Alternate Pathways

GABA-T is the major pathway by which GABA is metabolized, but it is by no means the only one (Fig. 1). Alternate routes of metabolism *are surmised primarily by the products found* and by the demonstration of necessary enzymatic mechanisms to form these products in nervous tissues.*

Although GABA can be incorporated into the amino acid chains of larger peptides in bacteria (*Escherichia coli*),[254] no evidence for the existence of such molecules in nervous tissue has been forthcoming. However, GABA has been found combined with any one of three basic amino acids, namely, arginine, histidine, and l-methyl-histidine.

### 1. γ-*Guanidobutyric Acid (GABA-Arginine)*

The reaction mechanism

$$\text{GABA} + \text{arginine} \rightarrow \gamma\text{-guanidobutyric acid} + \text{ornithine}$$

occurs in brain as well as in many other mammalian organs.[255] The transamidinase enzyme involved is not substrate specific, and glycine is ten times as active as GABA. Also $\beta$-alanine, $\delta$-aminovaleric acid, and lysine have demonstrable activity. $\gamma$-Guanidobutyric acid

$$HOOC—(CH_2)_3—N—C(=NH)—NH_2$$

is present not only in mammalian tissues[256] but also in fish[257] and crustaceans. In the latter group, in many invertebrates, and in some avian tissues the mode of synthesis does not involve GABA but is the product of the oxidative deamination of arginine through the action of L-amino acid oxidase.[258] There is no evidence of transamidinase activity in the crustacean nervous system.[259] Hydrolysis of the acid by $\gamma$-guanidinobutyrase in liver yields urea and GABA.[255,260] The enzyme involved is distinct from arginase. No data concerning the occurrence of this enzyme in nervous tissue of mammals are in the literature.

The physiological function of $\gamma$-guanidobutyric acid is unknown. It is inhibitory when applied to the peripheral crustacean synapse.[261] It has an excitatory action when applied to the mammalian cerebral cortex by augmenting primarily the secondary negative wave of the direct cortical response.[262,263] Injected intracisternally into rabbits (5 mg/kg) it produces convulsions. Large amounts of $\gamma$-guanidobutyric acid also have been found at the epileptogenic focus of human cerebral cortex.[264]

### 2. *Homocarnosine (GABA-Histidine)*

This GABA peptide has been identified, crystallized,[265] and measured in concentrations from 3 to 50 μmoles/100 g wet wt. in brain tissues from primates, ruminants (bovine), rodents (rat, guinea pig, and rabbit), and some birds (chicken). In other vertebrate species, levels in whole brain were

* For data prior to 1960 see Pisano *et al.*[253]

below 0.5 μmoles/100 g of tissue (cat, dog, hog, duck, mackerel, and crucian carp).[266,267] In human brain, homocarnosine

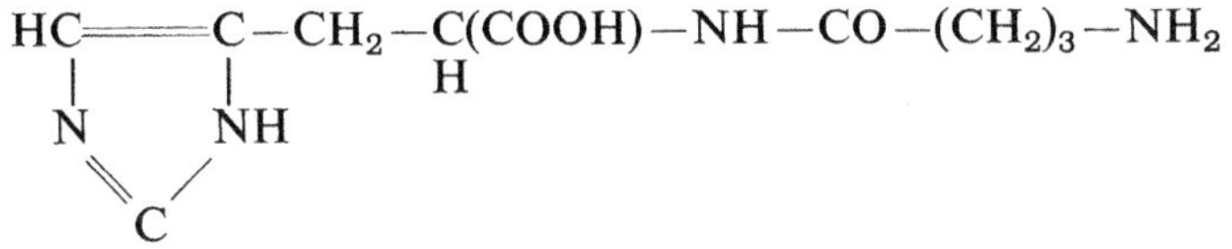

was highest in the thalamus, hypothalamus, and cerebellum (above 40 μmoles/100 g tissue) and lowest in the corpus callosum (less than 25 μmoles/100 g tissue). These values are higher than for any other species tested. The dipeptide is found exclusively in the central nervous system, but the potential for its synthesis also exists in muscle tissue if GABA substrate is provided from a dietary source.[268] Homocarnosine has been separated by ion exchange chromatography and paper electrophoresis. The isolated homocarnosine is diazotized, and as little as 0.02 μmoles is measured colormetrically.[266,267]

Studies of the enzymatic mechanisms involved in the synthesis and degradation of homocarnosine have not yet reached the biochemical literature.

### 3. *Homoanserine (GABA-1-Methyl Histidine)*

This compound was detected in[265] and subsequently isolated, crystallized, and identified[269] from bovine brain tissue. Both GABA and 1-methyl histidine are found in mammalian brain.[270] The enzyme carnosine-*N*-methyl transferase, also found in brain, catalyzes the synthesis of homoanserine

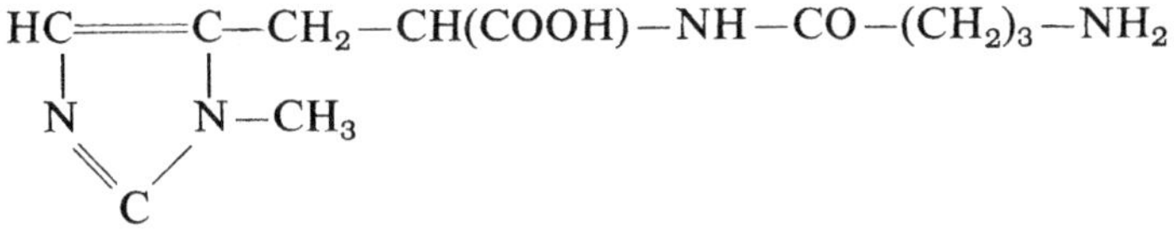

from these two precursors.[269]

### 4. *γ-Amino β-Hydroxybutyric Acid (GABOB)*

Considerable controversy exists regarding GABOB, concerning both its existence and its mechanism of action in the CNS. It was originally detected in the nervous tissue of the rat,[271] and subsequently very high levels were reported in the temporal lobe of cattle brain, from which it was crystallized.[272]

From a chemical point of view, the identification of GABOB

$$HOOC{-}CH_2{-}CHOH{-}CH_2{-}NH_2$$

in the early studies was incomplete. More rigorous investigations failed to substantiate the existence of GABOB.[21,273] If it does exist in rat brain tissue, it is present at extremely low levels. Synthesis of GABOB in brain mitochondria by a hydroxylation mechanism was reported by two groups and then negated by a third,[273] and claims of a methylation reaction that converts GABOB to carnitine also remain unsubstantiated.[253] GABOB has been used clinically in the treatment of epilepsy,[274] and other effects have been attributed to its action. Unfortunately, reports on the studies of GABOB contain so many irreconcilable discrepancies that a true evaluation is not possible at this time.

### 5. γ-*Butyrobetaine and Carnitine*

The presence of γ-butyrobetaine

$$HOOC-(CH_2)_3-N^+-(CH_3)_3$$

and carnitine in brain tissues had been established[275-277] and the occurrence of a number of related betaines reported. These compounds are not unique to brain tissues, and levels of carnitine

$$HOOC-CH_2-CHOH-CH_2-N^+-(CH_3)_3$$

are 3–10 times higher in liver and muscle than in brain.[277] GABA is trimethylated in brain tissue *in vitro* to form γ-butyrobetaine,[278] and the hydroxylation of this product to carnitine has been demonstrated both *in vivo* and *in vitro*.[279-281] However, when [$^{14}$C]-labeled GABA is administered peripherally to rats or mice, no significant amount of isotope is incorporated into the carnitine of brain tissue.[279,280] Such a result may be attributable to permeability barriers and the rapid metabolism of GABA, but it leaves in doubt the *in vivo* conversion of GABA to carnitine.

The neuropharmacological and possible neurophysiological properties of γ-butyrobetaine, carnitine, and their CoA esters, acetyl-L-carnitine and acetyl-L-carnityl CoA, are described as "acetylcholine-like."[282,283] This description fits the acetylcholine-like activity in presynaptic nerve endings from rat brain that has been attributed to the CoA esters of γ-butyrobetaine and L-carnitine.[282] A neurological interrelationship between carnitine and acetylcholine may be indicated by the observation that the administration of L-carnitine depresses levels of acetylcholine in brain and other tissues.[284] This depression is specific for the L-isomer.

### 6. γ-*Aminobutyryl Choline (GABCh)*

A metabolic interrelationship between acetylcholine and GABA is suggested by the occurrence of GABCh in the brain tissues of homeotherms.[285,286] The compound has been crystallized and identified. Levels of 0.15 μmoles/g tissue (wet wt.) have been reported for rat brain. Although hydrolysis of GABCh

$$NH_2-(CH_2)_3-COO-(CH_2)_2-N^+-CH_3)_3$$

by brain tissue could not be demonstrated, the choline esterase of plasma does degrade the compound.[287]

GABCh does not pass the blood–brain barrier.[288] Nevertheless, it has anticonvulsant properties when it is injected peripherally at doses ranging from 2 to 30 mg/kg body weight into a variety of animals.[289] It also lowers blood pressure. Its effect upon cerebral activity when perfused or applied directly to the brain is complex. It has a decided suppressive action against the local electrical response to direct stimulation of the cortex.[290] Other modes of action, which may be cholinergic or anticholinergic in nature, depend upon the brain structure involved.[288,291,292]

### 7. *Homopantothenic Acid*

Evidence for the presence in brain of this reaction product of GABA with pantothenic acid is limited and not based upon chemical isolation and identification.[253] A very small amount of radioactivity from 1-[$^{14}C$] GABA is incorporated *in vitro* by rat brain homogenates into a compound that elutes from an ion-exchange column in a fraction that corresponds to homopantothenic acid.[293] The coupling of pantoic acid with GABA to form homopantothenic acid

$$HOOC-(CH_2)_3-N-CO-CHOH-C-(CH_3)_2-CH_2OH$$

occurs in the presence of an enzyme found in the acetone powder extracts of calves brain.[294]

### 8. *α-γ-Diaminobutyric Acid*

This pharmacologically toxic compound,

$$HOOC-CHNH_2-CH_2-CHNH_2$$

which may compete with GABA in some reaction mechanisms,[295] recently has been reported to be a constituent of bovine brain.[296]

## VI. SOME PROPERTIES OF AND CORRELATES WITH GABA

### A. Effects of GABA on Metabolism

A variety of metabolic reactions in brain tissues may be affected by GABA. The stimulation of protein synthesis by GABA in a ribosomal system from immature rat brain already has been mentioned[51,52] (see Section II, C). No mechanism to explain this effect has been proposed, but it is known that one or more of the steps leading to the incorporation of amino acids into *t*RNA is stimulated by GABA.[53] The incorporation of $^{32}P$ into nucleic acids of rabbit brain was measured *in vivo* and produced no evidence that GABA influenced nucleic acid metabolism.[297] However, an interaction of GABA with brain proteins in which ammonia is evolved has been reported.[298]

Various studies on carbohydrate metabolism indicate that GABA stimulates the uptake of glucose into some tissues.[299] In brain mitochondria, GABA at 3.25 $\mu$moles/ml enhances glucose utilization and lactate formation, while at four times this concentration it is inhibitory. These effects are attributed to the influence of GABA on mitochondrial membrane permeability.[300] GABA also has been reported to stimulate some of the enzymatic steps in glycolysis.[301–303] Injected intraperitoneally, GABA stimulates glycogen degradation in liver and muscle[304] and produces hyperglycemia. In the absence of glucose, GABA in brain cortex slices may stimulate the utilization of GA and glutamine, but in the presence of glucose GABA promotes the accumulation of the two nitrogenous compounds and the detoxication of ammonia.[305]

## B. GABA levels in CNS and Physiological Correlates

In contrast to drugs (see Section VI, A), physiological stress causes changes in overall GABA levels that are compatible with the concept that GABA is correlated with the inhibitory regulation of neuronal activity. Thus, elevated levels of GABA are usually accompanied by sedated or depressed states, while animals with low cerebral GABA levels are often hyperexcitable and prone to convulsions. Some examples are summarized below:

### 1. *$CO_2$ Tension*

In an environment in which the partial pressure of $CO_2$ is 12.5% or more, rats become quite refractory to convulsant stimuli. At the same time, levels of GABA in brain tissues are elevated. When animals conditioned to such an environment are subsequently transferred rapidly to a normal environment low in $CO_2$, spontaneous convulsions are correlated with depressed levels of GABA in brain tissues.[306] It is conceivable that the observed changes in GABA reflect alterations in the intracellular size of the bicarbonate pool which, in turn, may be responsible for alterations in the intracellular *p*H of brain tissues.

### 2. *High Oxygen Pressure (HOP)*

The effect of high oxygen pressure has been of considerable clinical interest.[307] as well as a problem in the practice of deep sea diving and space exploration. Many biochemical effects of HOP have been noted.[308] Rats in an environment of 6 atm oxygen become hypersusceptible to convulsions, and their cerebral levels of GABA are depressed by 19–35% within 33 min.[309,310] As mentioned earlier (Section II, C), this susceptibility is species-dependent and correlated with the stability of brain GABA concentrations during HOP exposure.[47] The injection of GABA intraperitoneally protects animals against HOP-mediated convulsions, but this protective effect is not specific for GABA and can be achieved also with other amino acids and salts.[311,312]

The metabolic site of the HOP effect is not known. *In vitro* HOP inhibits GAD[146,309] as well as enzymes requiring an α-lipoic acid cofactor (i.e., α-ketoglutaric acid dehydrogenase and pyruvate oxidase).[313] Both *in vitro* and *in vivo* anoxic conditions favor GAD. A 1-min exposure to oxygen tensions, equivalent to altitudes of about 15,000 ft, elevates GABA levels in the brain of rats by about 30%.[314] The anticonvulsant protective effect, which GABA provides to animals exposed to HOP, may be related to the GABA shunt. Succinate itself has anticonvulsant properties against HOP seizures.[315] Finally, HOP would tend to move the equilibrium between NAD⇌NADH to the left, while metabolism of GABA would move it back to the right. It is possible that the equilibrium between NAD and NADH in brain tissue is related to convulsant states.[316]

### 3. *Sleep*

Only limited evidence links GABA with sleep. Both the deprivation of sleep[317,318] and the induction of sleep with strong light[319] appear to result in an elevated level of GABA in rat brain. When the brainstem is transected at the level of the upper midbrain (cerveau isole) in cats, the electrocorticogram (ECG) recordings show constant spindle patterns characteristic of sleep. This condition is referred to as "activated sleep" and is accompanied by the release of GABA at a rate of 2 $\mu g/hr/cm^2$ of perforated pial cortical surface. This rate is three times greater than that found in normal, neurotaxically intact, aroused animals or in animals with a brainstem transection in the upper cervical cord (encephale isole) with aroused ECG patterns. The release of GA changes in an opposite direction to that of GABA. During "activated sleep" the rate of release drops from about 9 to 6 $\mu g$ GA/hr/$cm^2$.[85] Possible changes in GAD activity remain to be investigated.

### 4. *Hibernation*

A hibernating mammal is characterized by a drastically reduced body temperature, a profound depression of the metabolic rate, and an extended state of torpor. Under these conditions the brains of ground squirrels,[192] golden hamsters,[191] and garden dormice[310] show a significant elevation in GABA content. This also corresponds to an apparent increase in GAD apoenzyme. However, when body temperature and metabolic rate are depressed for short periods of time, such as occurs during barbiturate anesthesia, GABA levels in several areas of rat brain are decreased.[321]

### 5. *Hormone Mediated Changes*

Adrenalectomy, cold exposure,[322] and insulin treatment[323] lower

GABA levels, while most of the adrenal steroid hormones have little, if any, effect upon GABA levels except in adrenalectomized animals where the hormones tend to return GABA to normal levels.[322]

## C. Correlations of GABA with Behavior

In a more subtle way, the social environment might affect both levels of GABA and the effect of GABA administration upon behavior. Levels of GABA in brain are claimed to be correlated with social deprivation in dogs,[324] learning ability of chicks,[325] and genetically determined emotionality of rats.[327] The effect of intracerebral GABA administration is dependent upon prior conditioning in the cat.[326] The multiple effects in the chick[328,329] are discussed elsewhere (see Section VI, D).

## D. Effects of GABA Administered Peripherally

### 1. *Central Effects*

In mature animals the intravenous or intraperitoneal injection of GABA has no significant effects upon evoked cortical potentials, and intracarotid injections have but a brief, transitory effect.[10] This lack of neuronal response is explained in terms of the blood-brain barrier, which generally prevents the accumulation in brain tissue of injected GABA. Some reports claim that large doses of GABA injected intraperitoneally have anticonvulsant effects. With the exception of the nonspecific protective effect of GABA against HOP convulsions,[311,330] these claims remain unconfirmed.

In young or neonatal mammals and birds, the blood–brain barrier is incompletely developed, and GABA administered peripherally elicits profound central effects.[328,329,331] The injection of 1.5 g GABA/kg body weight in chicks produces ataxia, uncoordination, depression, and a lack of responsiveness to visual stimuli.[332] EEG activity is altered in that GABA appears to block the tectal light-evoked response without affecting tectal spontaneous activity.[327,329] Effects on the ERG are complex, profound, and reversible. Since the effect of GABA on the ERG is retained after the optic nerve is severed,[327] it has been suggested that GABA acts at the retinal level, but the results do not exclude an effect also at the tectal level.[22] The disappearance of the evoked response from the optic tectum coincides closely with the disappearance of the *b* wave of the ERG, which is generated within the nuclear layer[333] of the retina.

### 2. *Peripheral Effects in Vertebrates*

The intravenous administration of GABA to mature mammals produces hypotension and affects circulation and related physiological parameters.[334–341] The hypotension produced has been attributed to the action of GABA on peripheral chemoreceptors, autonomic ganglia, or brainstem neurons.[334–336] Central control of blood pressure is believed to be regulated in the medulla, and application of GABA onto or into this area also produces hypotension.[337] Peripheral GABA receptors may be quite specific, since compounds structurally related to GABA have a lesser effect.[338]

GABA affects respiration by direct action upon the sensitivity of pulmonary stretch receptors,[342] and in smooth muscle it is believed to act upon neural elements within the tissue.[17] GABA has little, if any, effect on peripheral nerve fibers and on most autonomic ganglia except for the mesenteric ganglion of the cat, in which transmission is depressed.

### 3. *Peripheral Effects in Invertebrates*

The peripheral effects of GABA in vertebrates have been studied primarily in arthropods, although other phyla occasionally have been used. There is usually no effect on nerve fibers and chemoreceptors, but GABA exerts a profound effect upon stretch receptors, heart, muscle, and gut, particularly in crustaceans. It is likely that the effect of GABA on the neuromuscular junction of insects[343] is similar to that described for crustaceans.

## E. Medicinal Use of GABA

Despite its rapid metabolism outside the mammalian brain,[208] GABA has been tested clinically as an antihypertensive drug[339,340]—apparently with some success. It also has been claimed that GABA administration has beneficial effects on some patients with psychiatric disorders and that it improves the memory of some geriatric patients.[345,346] The evidence for such claims, subjective in most cases, cannot be interpreted as a direct effect of GABA on cerebral tissue. There are no valid data to suggest that the blood–brain barrier to GABA is altered in patients with psychiatric or geriatric disorders or that GABA, administered to patients peripherally, passes the blood–brain barrier to an extent that is quantitatively significant. In other medically related considerations, GABA mechanisms have been implicated as one mode of action of some antimotion sickness drugs.[347]

# VII. PHARMACOLOGICAL STUDIES AND GABA

## A. GABA Levels

Under the physiological conditions discussed in Section VI, high susceptibility to convulsant stimuli and excitability were equated with low levels of GABA, while refractory states to convulsant stimuli and lethargy were equated with high levels of GABA in the CNS. This relationship does not necessarily apply to the effects of pharmaceutical agents whose mechanisms of action on physiological and behavioral parameters may be totally unrelated to incidental changes in GABA or GABA related enzymes. Changes of GABA levels *in vivo* with some drugs are shown in Table I.

**TABLE I**

**Changes of Cerebral Levels of GABA *in Vivo*[a]**

| Animal | Weight, g | Condition[b] | Drug used | Dose, mg/kg body wt. | Route of injection[c] | Δ time, hr[d] | Δ GABA, % of control | References P | References S |
|---|---|---|---|---|---|---|---|---|---|
| | | | *A. Some Compounds Which Elevate GABA Levels* | | | | | | |
| Rat | 160–190 | Fed | Hydroxylamine HCl | 75 | ip | 1.5 | 160 | 216<br>148 | 348<br>349 |
| Monkey | 7800 | Fed | Hydroxylamine ($NH_2OH$) | 7.4 | iv | 1.5 | 264 | 148 | |
| Mouse | 25 | Starved | Aminooxyacetic acid (AOAA) | 25 | ip | 6.0 | 415 | 350<br>219 | |
| Rat | 175–225 | Fed | Aminooxyacetic acid (AOAA) | 50 | ip | 4.0 | 320 | 148<br>219 | |
| Cat | Adult | | Aminooxyacetic acid (AOAA) | 30 | sc | 6.0 | 530 | 219 | |
| Dog | Adult | | Aminooxyacetic acid (AOAA) | 100 | sc | 4.0 | 294 | 219 | |
| Mouse | 18–23 | | Cycloserine | 300 | ip | 2.0 | 169 | 351 | |
| Rat | Adult | | Cycloserine | 600 | ip | 48.0 | 280 | 222 | |
| Guinea Pig | 350 | | Cycloserine | 700 | ip | 3.0 | 170 | 222 | |
| Rat | 60–100 | Fed | ββ′ Iminodipropionitrile | 20 mg/day | sc | (11 days) | 141 | 352 | |
| Rat | 90–100 | Fed | α-γ-Diaminobutyric acid | 50 mg/day | sc | (4 days) | 129 | 352 | |
| Mouse | Adult | Fed | Hydrazinopropionic acid | 20 | sc | 5.5 | 500 | 353 | |
| Mouse | 22–28 | Fed | Hydrazine 2HCl | 100 | ip | 12.0 | 390 | 349 | 354 |

[a] Details of experiments and physiological correlates of drug treatment can be obtained by consulting the references cited. The numerical values given should serve as guidelines only. Many experimental results differ considerably for various brain areas, and in some cases the results of different studies are not in complete accord. Numerical values quoted in the table were collected primarily from the references in the P column. Referenecs containing related data are listed in the S column.

[b] A blank indicates that the nutritional state of the animals is not known.

[c] Abbreviations used are i.p., intraperitoneal; i.v., intravenous; s.c., subcutaneous.

[d] Time at which animal was killed after injection of drug.

**TABLE I**—*continued*

**Changes of Cerebral Levels of GABA** *in Vivo*[a]

| Animal | Weight, g | Condition[b] | Drug used | Dose, mg/kg body wt. | Route of Injection[c] | Δ time, hr[d] | Δ GABA, % of control | References P | References S |
|---|---|---|---|---|---|---|---|---|---|
| Rat | 180–200 | | Hydrazine | 74 | ip | 1.2 | 131 | 355 | 354 |
| | 175–225 | | Hydrazine | 100 | ip | 12.0 | 273 | 349 | |
| Rat | 350–400 | | Hydrazine | 50 | ip | Up to 1.5 | 130 | 220 | |
| | | | Hydrazine + Pyridoxal | 50 + 60 | ip | Up to 1.5 | 62 | 220 | |
| Mouse | 22–28 | Fed | L-glutamic acid-γ-hydrazide (GAH) | 160 | ip | 6.5 | 320 | 221<br>356 | 153 |
| Mouse | 23–30 | Fed | L-glutamic acid-γ-hydrazide + pyridoxal | 60<br>50 | ip | Approximately 0.5 (convulsions) | 38 | 153 | |
| | 23–30 | Fed | pyridoxal phosphate-γ-glutamyl hydrazone | 110 | ip | Approximately 0.5 (convulsions) | 38 | 153 | |
| Mouse | Adult | | phenylzine (phenyl propyl hydrazine) | 70 | ip | 2.0 | 198 | 357 | |
| | | | | 300 | ip | 2.0 | 258 | 357 | |
| Rat | 100–140 | | phenylzine | 50 | ip | 4.0 | 250+ | 356<br>358 | 357 |
| | | | *B. Some Compounds Which Depress GABA Levels* | | | | | | |
| Rat | Adult | | Semicarbazide HCl | 200 | ip | Convulsions | 68 | 215<br>359 | |
| Mouse | 20–25 | | Semicarbazide HCl | 298 | ip | 1.0 | 55 | 349 | |

**TABLE I—*continued***

**Changes of Cerebral Levels of GABA *in Vivo*[a]**

| Animal | Weight, g | Condition[b] | Drug used | Dose, mg/kg body wt. | Route of Injection[c] | Δ time, hr[d] | Δ GABA, % of control | References P | References S |
|---|---|---|---|---|---|---|---|---|---|
| Mouse | 20–25 | Fed | Thiosemicarbazide (TSC) | 24 | ip | 1.0 | 75 | 349<br>356 | |
| Rat | 190–210 | Fed | Thiosemicarbazide (TSC) | 20 | ip | 1.8 | 59–89 | 360<br>61 | |
| Dog | Adult | Starved | Thiosemicarbazide (TSC) | 20 | iv | Approxi-mately 1.0 | 66 | 361 | |
| Mouse | 16–21 | | Thiocarbohydrazide | 4.0 | ip | 1.0 | 47 | 361<br>349 | |
| Rat | 350–400 | Fed | Monomethyl hydrazine | 39 | ip | Up to 1.5 | 60 | 220 | |
| Mouse | 20–25 | | Acetone semicarbazone | 306 | ip | Approxi-mately 0.5 | 51 | 349 | |
| Mouse | 16–21 | | Isonicotinic acid and hydrazide | 200 | ip | Approxi-mately 1.0 | 51–61 | 360<br>362 | 363 |
| Mouse | 20–25 | | 3 Deoxypyridoxine $PO_4$ | 500 | ip | 1.0–1.5 | 60 | 349 | 361<br>322 |
| Mouse | 20 | | 4 Methoxy methyl pyridoxine | 30 | ip | 0.5 | 50 | 82 | 364 |
| Rat | 60–70 | | Toxopyrimidine | 125 | ip | 1.5–2 | 77 | 365 | |
| Mouse | 18–20 | | Reserpine | 5 | ip | 3.0–18.0 | 70 | 366 | |
| Rat | 350–400 | | Reserpine | 5 | ip | 18.0 | 63 | 354 | 367 |
| Monkey | | | Picritoxin | 5.5 | sc | 0.5 | 50–100 | 368 | 369 |
| Mouse<br>Rat | 20<br>200 | | Potassium cyanide | 40.0 | ip | 1 to 3 min Convulsions | 67–75 | 200 | 370 |

Convulsant drugs (such as semicarbazide) lower the level of GABA, while others (such as hydrazine) elevate GABA levels in brain tissues. Anticonvulsant agents [such as hydroxylamine ($NH_2OH$)] and glutamic acid hydrazide (GAH), which elevate GABA, do not protect against the convulsant effect of thiosemicarbazide at times when levels of GABA in the CNS are distinctly elevated. Although GAH, $NH_2OH$, and aminooxyacetic acid (AOAA) all elevate GABA levels, GAH does not protect against metrazol convulsions, but $NH_2OH$ and AOAA do. In some cases, an elevation of GABA and anticonvulsant activity are correlated, but time studies show that often the elevated GABA levels either precede or follow the anticonvulsant effect! It follows that the convulsant or anticonvulsant properties of most drugs cannot be equated with "overall" GABA levels in whole brain.[15,150,220,349,355,360]

It is still possible, however, that some physiological or behavioral effects of drugs are related to a change in a small fraction of the total GABA pool, a fraction that may be bound to specific membrane sites or concentrated at specific nerve terminals in specific regions of the nervous system. Another equally unproven hypothesis is that the balance between a variety of excitatory and inhibitory neurohumors (of which GABA would be just one) determines physiological and behavioral states of an animal. Drugs may exert their effect by altering the ratio of these neurohumors at key loci in the nervous system.

## B. Correlation of GABA Levels with GAD and GABA-T *in Vivo*

In drug studies GABA levels do not necessarily reflect changes that appear to have occurred in GAD or GABA-T activity.[217,220,371] For example, when GABA levels are depressed by a hydrazine and pyridoxal combination,[220] GABA-T appears to be depressed to a greater extent than GAD. Similarly, phenylzine causes GABA levels to be elevated in areas of brain where GABA-T appears to be barely inhibited.[358] Levels of substrates, competition for cofactors, ionic concentrations, *p*H, and other parameters used for the *in vitro* assay of GAD and GABA-T (from tissues of drug-treated animals) may have no relationship to the conditions in the living, intact animal. Consequently, the apparent discrepancies of results may only illustrate limitations in attempting to deduct relative enzyme activities *in vivo* from measurements of optimal enzyme (or apoenzyme) activity in an *in vitro* environment (see Section IV, C, 3).

## C. Flux Through GABA Shunt

It has been theorized that physiological and behavioral parameters are related to the drug-induced changes in the metabolic flux of carbon atoms through the GABA pathway. There is evidence for such a concept.[153] *In vivo* inhibition of GABA metabolism by some anticonvulsant drugs is suggested by changes in the rate of $^{14}CO_2$ evolution from intraventricularly

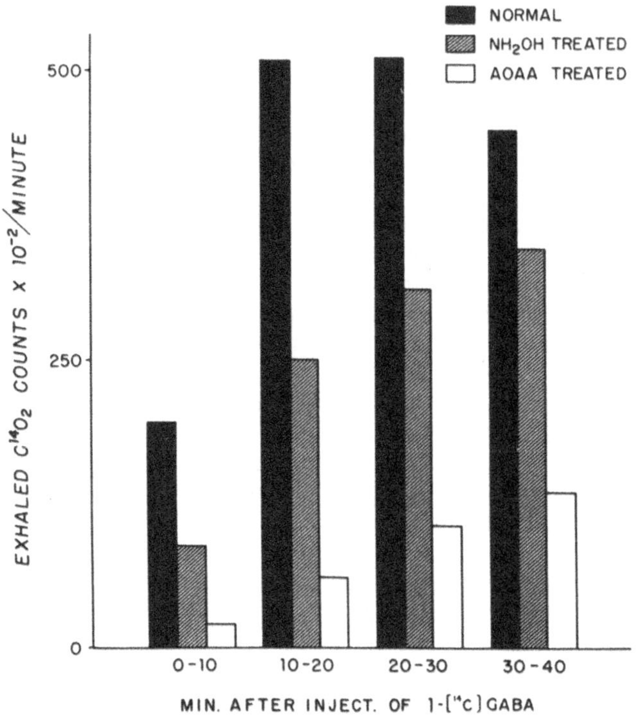


Fig. 6. Metabolism of 1-[$^{14}C$] GABA injected into the lateral brain ventricle of the rat. Although the rates of $^{14}CO_2$ exhaled do not correspond to the total amount of 1-[$^{14}C$] GABA that has been degraded, the difference in the rates is an indication of the size of the GABA pool and the inhibition of GABA metabolism *in vivo* by aminooxyacetic acid(219) and hydroxylamine.(216)

injected [$^{14}C$]-labeled GABA (Fig. 6). Alterations in both the rate of flux via the GABA shunt and the size of the metabolic pool into which [$^{14}C$] GABA was injected must be considered in evaluating these results.

The distribution patterns of [$^{14}C$] label in brain amino acids, after injections of [$^{14}C$] glucose, indicate that the effects of carbonyl trapping agents are not confined to altering the flow of carbon atoms through the GABA shunt but also affect the metabolism of unrelated amino acids.(344) Some experiments do not support the idea that the metabolic flux through the GABA shunt correlates with the pharmacological effect of drugs.(159)

It is likely that real progress will be made in relating conclusively the mechanism of action of any drug to GABA *in vivo*, only when measurements of excitability, convulsant activity, and similar behavioral criteria are refined to the same degree as the measurements of biochemical substrates.

## D. Biochemical Mechanisms of Drug Action *in Vivo*

The mechanisms by which the drugs listed in Table I alter levels of GABA *in vivo* are not fully understood. As an example, AOAA and phenylzine both elevate GABA levels and inhibit GABA-T, but the effect of phenylzine is reversible by some other monamine oxidase inhibitors, whereas the inhibition by AOAA is irreversible.[372] This indicates that on the molecular level the inhibition of GABA-T by these two inhibitors is not alike.

In summary, the following biochemical mechanisms have been suggested to explain the *in vivo* changes in GABA levels resulting from drug administration under various physiological conditions:*

1. The compound involved may be a substrate analogue inhibiting the attachment of substrate to enzyme; both α-γ-diaminobutyric acid and hydrazinoproprionic acid could be classified as GABA analogues, and both inhibit GABA-T.

2. Compounds that interfere with, or occlude, the vitamin $B_6$ coenzyme may inhibit GAD, GABA-T, or pyridoxal kinase. This would include all of the compounds in Table I that are either carbonyl trapping agents or vitamin $B_6$ analogues.

3. Since SSA-D "pulls" GABA-T in a coupled reaction, reagents that alter the normal NAD $\rightleftharpoons$ NADH ratio of the tissue might be expected to affect GABA levels.

4. GABA metabolism is intimately related to the substrates of the tricarboxylic acid cycle. Also, ATP is required in the pyridoxal kinase reaction and NAD in the SSA-D reaction mechanism. Consequently, any drug that affects carbohydrate or energy metabolism may exert an indirect effect upon GABA. For example, insulin hypoglycemia is accompanied by a decrease in cerebral GABA.[323,373] Since blood sugar levels drop in vitamin $B_6$-depleted animals,[323] it is even possible that the effects upon GABA levels by some $B_6$ antagonists are related, in part, to changes in carbohydrate metabolism.

It is often overlooked that the nutritional state of an animal can profoundly alter the action of drugs on GABA levels and metabolism. Sixty to ninety minutes after feeding starved rats with ethanol (500 mg/100 g body weight by stomach tube), a significant but transitory increase of GABA levels in the CNS is observed.[322,374] Brain homogenates from these rats incubated with GA accumulate GABA more rapidly than controls.[375] These and other effects of ethanol on GABA levels[376,377] are not observed when experiments are conducted with fed rats.[322,359,377]

Similarly, the effect of pentobarbital upon GABA levels in *fed* rats is minimal, but the anesthetic produces a significant decrease of GABA, both in the cortex and caudate nucleus of the *postabsorptive* rat.[310] In many

* The relationship of inhibitor structure to its effect upon GAD, GABA-T, and GABA levels in brain tissue is reviewed in detail by others.[15,146,217,378]

pharmacological studies on GABA, the nutritional state of the animals used is not reported (Table I).

5. Many pharmaceutical agents may alter levels of GABA in the CNS by altering temperature, *p*H, $O_2$, or $CO_2$ tension in the internal tissue environment. This would include hormones, carbonic anhydrase inhibitors, anesthetics, and similar compounds.

6. Finally, it is possible that some drugs alter membrane barriers (including the blood–brain barrier) to substrates or to GABA, thereby changing the availability of a limiting substrate or cofactor or permitting the leakage of GABA from its normal sites. A change in the membrane binding site for GABA might act in a similar manner.

## VIII. SUMMARY

GABA has attained, in the past 18 years, an important place in neurochemistry and neurophysiology. The primary mechanisms of its synthesis and degradation and its gross distribution in nervous tissues are established. Its physiological importance as an inhibitory transmitter substance in the peripheral nervous system of crustaceans is generally accepted, and many pieces of evidence suggest a similar function in some parts of the central nervous system of mammals. GABA also may have other regulatory functions. GABA metabolism may be important because it funnels carbon atoms around a portion of the tricarboxylic acid cycle. The possibility that the physiological effects of some drugs might relate to alterations in GABA metabolism remains a possibility. Technological limitations have prevented definitive investigations in this area.

The views expressed in this chapter are based upon evidence available to the author early in 1968. The foundation upon which present knowledge of GABA rests is supported by investigations in many fields. The subject matter of this chapter reflects this interdisciplinary approach.

## IX. ACKNOWLEDGMENTS

I would like to thank my colleague, Dr. R. N. Lolley, for many helpful suggestions as well as Miss L. Eaton and Mrs. R. Bertrand for unstinting help in preparing this manuscript. Bibliographic aid has been received from the UCLA Brain Information Service, which is a part of the Neurological Information Network of the NINDB and is supported under Contract PH-43-66-59. The original work reported and the production of this chapter were supported by NIH Grant # NB-03743.

## X. REFERENCES*

1. K. Schotten, Über die Oxydation des Piperidins, *Berlin Chem. Ges. Ber.* **16**:643–649 (1883); *Moniteur Sci.* **25**:833 (1883).
2. E. Roberts and S. Frankel, γ-Aminobutyric acid in brain. Its formation from glutamic acid, *J. Biol. Chem.* **187**:55–63 (1950).
3. J. Awapara, A. J. Landau, R. Fuerst, and B. Seale, Free γ-aminobutyric acid in brain, *J. Biol. Chem.* **187**:35–39 (1950).
4. S. Udenfriend, Identification of γ-aminobutyric acid in brain by the isotope derivative method, *J. Biol. Chem.* **187**:65–69 (1950).
5. E. Florey, *in Inhibition in the Nervous System and Gamma-Aminobutyric Acid* (E. Roberts, C. F. Baxter, A. Van Harreveld, C. A. G. Wiersman, W. R. Adey, and K. F. Killam, eds.), pp. 72–84, Pergamon Press, New York (1960).
6. C. F. Baxter and E. Roberts, *in The Neurochemistry of Nucleotides and Amino Acids* (R. O. Brady and D. B. Towers, eds.), pp. 127–145, Wiley, New York (1960).
7. K. A. C. Elliott and H. H. Jasper, Gamma-aminobutyric acid, *Physiol. Rev.* **39**: 383–406 (1959).
8. S. W. Kuffler and C. Edwards, Mechanism of γ-aminobutyric acid action and its relation to synaptic inhibition, *J. Neurophysiol.* **21**:589–610 (1958).
9. E. Roberts and E. Eidelberg, Metabolic and neurophysiological roles of γ-aminobutyric acid, *Intern. Rev. Neurobiol.* **2**:279–332 (1960).
10. E. Roberts, C. F. Baxter, A. Van Harreveld, C. A. G. Wiersman, W. R. Adey, and K. F. Killam (eds.), *Inhibition in the Nervous System and Gamma-Aminobutyric Acid*, Pergamon Press, New York (1960).
11. E. Roberts, C. F. Baxter, and E. Eidelberg, *in Structures and Function of the Cerebral Cortex* (D. B. Tower and J. P. Schade, eds.), pp. 392–403, Elsevier, Amsterdam (1960).
12. U. Schwabe, Vorkommen und Bedeutung der γ-Aminobuttersäure im Zentralnervensystem, *Deut. Med. Wochschr.* **86**:2235–2240 (1961).
13. E. Roberts, *in Neurochemistry* (K. A. C. Elliott, I. H. Page, and J. H. Quastel, eds.), 2nd ed., pp. 636–656, Charles C. Thomas, Springfield, Illinois (1962).
14. N. N. Jakovlev (ed.), *The Role of GABA in the Nervous System Activity*, Council of Sechenov, Leningrad Society of Physiologists, Biochemists and Pharmacologists, Leningrad (1964).
15. E. Roberts, J. Wein, and D. G. Simonsen, γ-Aminobutyric acid (γABA), vitamin $B_6$ and neuronal function—A speculative synthesis, *Vitamins Hormones* **22**:503–559 (1964).
16. K. A. C. Elliott, γ-Aminobutyric acid and other inhibitory substances, *Brit. Med. Bull.* **21**:70–75 (1965).
17. D. R. Curtis and J. C. Watkins, The pharmacology of amino acids related to gamma-aminobutyric acid, *Pharmacol. Rev.* **17**:347–391 (1965).
18. H. C. Buniatian (ed.), *Problems in Brain Biochemistry*, Vols. 1 and 2, Academy of Science, Armenian SSR (1964) and (1966).
19. A. A. Galoyan (ed.), *Problems in Brain Biochemistry*, Vol. 3, Academy of Science, Armenian SSR (1967).
20. V. P. Georgiev, Role and Importance of γ-aminobutyric acid for the functions of the nervous system, *Suvremenna Med.* **18**:318–328 (1967).
21. K. A. C. Elliott, Current studies on GABA and other amino acids, *Japan. J. Brain Physiol.* **84**:3116–3126 (1967).

* Compiled in early 1968, with the exception of refs. 52, 53, 184, 353, 379, and 380.

22. E. Roberts and K. Kuriyama, Biochemical–physiological correlations in studies of the $\gamma$-aminobutyric acid system, *Japan. Life Sci. Tokyo* (*Seitai-no-Kagaku*) **15**: 2–27 (1967); *Brain Res.* **8**:1–35 (1968).
23. A. Takeuchi and N. Takeuchi, Localized action of gamma-aminobutyric acid on the crayfish muscle, *J. Physiol.* (*London*) **177**:225–238 (1965).
24. E. A. Kravitz, P. B. Molinoff, and Z. W. Hall, A comparison of the enzymes and substrates of gamma-aminobutyric acid metabolism in lobster excitatory and inhibitory axons, *Proc. Natl. Acad. Sci. U.S.* **54**:778–782 (1965).
25. J. Dudel, The action of inhibitory drugs on nerve terminals in crayfish muscle, *Pfluegers Arch. Ges. Physiol.* **284**:81–94 (1965).
26. M. Otsuka, L. L. Iversen, Z. W. Hall, and E. A. Kravitz, Release of gamma-aminobutyric acid from inhibitory nerves of lobster, *Proc. Natl. Acad. Sci. U.S.* **56**: 1110–1115 (1966).
27. A. Takeuchi and N. Takeuchi, On the permeability of the presynaptic terminal of the crayfish neuromuscular junction during synaptic inhibition and the action of $\gamma$-aminobutyric acid, *J. Physiol.* (*London*) **183**:433–449 (1966).
28. Z. W. Hall and E. A. Kravitz, The metabolism of $\gamma$-aminobutyric acid (GABA) in the lobster nervous system. I. GABA-Glutamate Transaminase, *J. Neurochem.* **14**:45–54 (1967).
29. M. Otsuka, E. A. Kravitz, and D. D. Potter, Physiological and chemical architecture of a lobster ganglion with particular reference to gamma-aminobutyrate and glutamate, *J. Neurophysiol.* **30**:725–752 (1967).
30. D. R. Curtis and J. C. Watkins, *in Inhibition in the Nervous System and Gamma-Aminobutyric Acid* (E. Roberts, C. F. Baxter, A. Van Harreveld, C. A. G. Wiersman, W. R. Adey, and K. F. Killam, eds.), pp. 424–444, Pergamon Press, New York (1960).
31. L. J. Bindman, O. C. J. Lippold, and J. W. T. Redfearn, The non-selective blocking action of $\gamma$-aminobutyric acid on the sensory cerebral cortex of the rat, *J. Physiol.* (*London*) **162**:105–120 (1962).
32. J. M. Crawford and D. R. Curtis, The excitation and depression of mammalian cortical neurons by amino acids, *Brit. J. Pharmacol.* **23**:313–329 (1964).
33. K. P. Bhargava and R. K. Srivastava, Non-specific depressant action of $\gamma$-aminobutyric acid on somatic reflexes, *Brit. J. Pharmacol.* **23**:391–398 (1964).
34. K. Krnjevic and S. Schwartz, The action of $\gamma$-aminobutyric acid on cortical neurons, *Exptl. Brain Res.* **3**:320–336 (1967).
35. K. Obata, M. Ito, R. Ochi, and N. Sato, Pharmacological properties of the postsynaptic inhibition by Purkinje cell axons and the action of $\gamma$-aminobutyric acid on Deiters neurones, *Exptl. Brain Res.* **4**:43–57 (1967).
36. A. Galindo, K. Krnjevic, and S. Schwartz, Micro-ionophoretic studies on neurons in the cuneate nucleus, *J. Physiol.* (*London*) **192**:359–377 (1967).
37. N. Kawai and C. Yamamoto, Effects of $\gamma$-aminobutyric acid on the potentials evoked *in vitro* in the superior colliculus, *Experientia* **23**:822–823 (1967).
38. K. Kuriyama, B. Haber, B. Sisken, and E. Roberts, The $\gamma$-aminobutyric acid system in rabbit cerebellum, *Proc. Natl. Acad. Sci. U.S.* **55**:846–852 (1966).
39. L. T. Graham, Jr., R. N. Lolley, and C. F. Baxter, Effect of illumination upon levels of $\gamma$-aminobutyric acid ($\gamma$ABA) and glutamic acid (GA) in frog retina *in vivo*, *Federation Proc.* **27**:463 (1968).
40. A. L. Byzov, Horizontal Cells of the Retina as Regulators of Synaptic Transmission, *Neurosci. Trans.* **3**:268–276 (1968), trans. from *Fiziol. Zh. SSSR* **53**:1115–1124 (1967).
41. J. E. Dowling and B. B. Boycott, Organization of the Primate retina: Electron. microscopy, *Proc. Roy. Soc.* (*London*) *Ser. B.* **166**:80–111 (1966).

42. L. T. Graham, Jr., R. P. Shank, R. Werman, and M. H. Aprison, Distribution of some synaptic transmitter suspects in cat spinal cord, *J. Neurochem.* **14**:465–472 (1967).
43. R. A. Davidoff, L. T. Graham, Jr., R. P. Shank, R. Werman, and M. H. Aprison, Changes in amino acid concentrations associated with loss of spinal interneurons, *J. Neurochem.* **14**:1025–1031 (1967).
44. R. Werman, R. A. Davidoff, and M. H. Aprison, Inhibition of motoneurones by iontophoresis of glycine, *Nature* **214**:681–683 (1967).
45. D. R. Curtis, L. Hosli, G. A. R. Johnston, and I. H. Johnston, Glycine and spinal inhibition, *Brain Res.* **5**:112–114 (1967).
46. J. D. Wood, A possible role of gamma-aminobutyric acid in the homoestatic control of brain metabolism under conditions of hypoxia, *Exptl. Brain Res.* **4**: 81–84 (1967).
47. J. D. Wood, W. J. Watson, and A. J. Ducker, Oxygen poisoning in various mammalian species and the possible role of gamma-aminobutyric acid metabolism, *J. Neurochem.* **14**:1067–1074 (1967).
48. C. F. Baxter *in Progress in Brain Research* (A. Lajtha and D. H. Ford, eds.), Vol. 29, pp. 429–450, Elsevier, Amsterdam (1968).
49. D. V. N. Reddy, Distribution of free amino acids and related compounds in ocular fluids, lens and plasma of various species, *Invest. Ophthalmol.* **6**:478–483 (1967).
50. J. C. Dickinson, D. G. Durham, and P. B. Hamilton, Ion exchange chromatography of free amino acids in aqueous fluid and lens of the human eye, *Invest. Ophthalmol.* **7**:551–563 (1968).
51. M. K. Campbell, H. R. Mahler, W. J. Moore, and S. Tewari, Protein synthesis systems from rat brain, *Biochemistry* **5**:1174–1184 (1966).
52. S. Tewari and C. F. Baxter, Stimulatory effect of γ-aminobutyric acid upon amino acid incorporation into protein by a ribosomal system from immature rat brain, *J. Neurochem.* **16**:171–180 (1969).
53. C. F. Baxter and S. Tewari, *in Protein Metabolism in the Nervous System* (A. Lajtha, ed.), Plenum Press, New York, in press (1969).
54. H. H. Tallan, *in Amino Acid Pools* (J. T. Holden, ed.) pp. 471–485, Elsevier, Amsterdam (1962).
55. M. Zachmann, P. Tocci, and W. L. Nyhan, The occurrence of γ-aminobutyric acid in human tissues other than brain, *J. Biol. Chem.* **241**:1355–1358 (1966).
56. S. Berl and D. P. Purpura, Regional development of glutamic acid compartmentation in immature brain, *J. Neurochem.* **13**:293–304 (1966).
57. Lj. Kržalić, V. Mandic, and Lj. Mihailovic, On the glutamine and γ-aminobutyric acid contents of various regions of the cat brain, *Experientia* **18**:368 (1962).
58. N. Popov, W. Pohle, V. Rösler, and H. Matthies, Regionale Verteilung von γ-Aminobuttersäure, Glutaminsäure, Asparaginsäure, Dopamin, Noradrenalin und Serotonin im Rattenhirn, *Acta Biol. Med. Ger.* **18**:695–702 (1967).
59. S. I. Singh and C. L. Malhotra, Amino acid content of monkey brain. I. General pattern and quantitative value of glutamic acid/glutamine, gamma-aminobutyric acid and aspartic acid, *J. Neurochem.* **9**:37–42 (1962).
60. Y. Yamamoto, A. Mori, and D. Jinnai, Amino acids in the brain, *J. Biochem.* **49**:368–372 (1961).
61. S. Saito, Y. Tokunaga, and K. Kojima, Effects of several excitants and depressants on glutamic acid and its metabolites concentration in rabbit and rat brain, *Keio J. Med.* **13**:211–235 (1964).

62. Y. Yamamoto, T. Iwado, M. Kitamura, K. Uno, and R. Sugiu, Amino acids in the brain analysis of amino acids in various areas of the cerebral cortex, *Folia Psychiat. Neurol. Japon.* **17**:299–305 (1963).
63. S. Fahn and L. J. Côté, Regional distribution of γ-aminobutyric acid (GABA) in brain of the Rhesus monkey, *J. Neurochem.* **15**:209–213 (1968).
64. H. E. Hirsch and E. Robins, Distribution of γ-aminobutyric acid in the layers of the cerebral and cerebellar cortex. Implications for its physiological role, *J. Neurochem.* **9**:63–70 (1962).
65. E. Florey, *in Inhibition in the Nervous System and Gamma-Aminobutyric Acid* (E. Roberts, C. F. Baxter, A. Van Harreveld, C. A. G. Wiersman, W. R. Adey, and K. F. Killam, eds.), pp. 202–206, Pergamon Press, New York (1960).
66. J. D. Utley, Gamma aminobutyric acid and 5-hydroxytryptamine concentrations in neurons and glial cells in the medial geniculate body of the cat, *Biochem. Pharmacol.* **12**:1228–1230 (1963).
67. M. Wollemann and T. Devenyi, The γ-aminobutyric acid content and glutamate decarboxylase activity of brain tumors, *J. Neurochem.* **10**:83–88 (1963).
68. N. F. Shatunova and I. A. Sytinsky, On the intracellular localization of glutamate decarboxylase and γ-aminobutyric acid in mammalian brain, *J. Neurochem.* **11**: 701–708 (1964).
69. L. Salganicoff and E. De Robertis, Subcellular distribution of the enzymes of the glutamic acid glutamine and γ-aminobutyric acid cycles in rat brain, *J. Neurochem.* **12**:287–309 (1965).
70. H. Weinstein, E. Roberts, and T. Kakefuda, Studies of subcellular distribution of γ-aminobutyric acid and glutamic decarboxylase in mouse brain, *Biochem. Pharmacol.* **12**:503–509 (1963).
71. J. L. Mangan and V. P. Whittaker, The distribution of free amino acids in subcellular fractions of guinea pig brain, *Biochem. J.* **98**:128–137 (1966).
72. R. W. Ryall, The subcellular distribution of acetylcholine substance P, 5-hydroxytryptamine, γ-aminobutyric acid and glutamic acid in brain homogenates, *J. Neurochem.* **11**:131–145 (1964).
73. K. Kuriyama, E. Roberts, and T. Kakefuda, Association of the γ-aminobutyric acid system with a synaptic vessicle fraction from mouse brain, *Brain Res.* **8**: 132–152 (1968).
74. K. Krnjevic and V. P. Whittaker, Excitation and depression of cortical neurones by brain fractions released from micropipettes, *J. Physiol. (London)* **179**:298–322 (1965).
75. K. A. C. Elliott and N. M. Van Gelder, The state of Factor I in rat brain: The effect of metabolic conditions and drugs, *J. Physiol. (London)* **153**:423–432 (1960).
76. K. Sano and E. Roberts, Binding of γ-aminobutyric acid by mouse brain preparations, *Biochem. Pharmacol.* **12**:489–502 (1963).
77. S. Varon, H. Weinstein, T. Kakefuda, and E. Roberts, Sodium-dependent binding of γ-aminobutyric acid by morphologically characterized subcellular brain particles, *Biochem. Pharmacol.* **14**:1213–1224 (1965).
78. C. F. Baxter, unpublished data.
79. P. Strasberg, K. Krnjevic, S. Schwartz, and K. A. C. Elliott, Penetration of blood–brain barrier by γ-aminobutyric acid at sites of freezing, *J. Neurochem.* **14**:755–760 (1967).
80. S. Berl, G. Takagaki, and D. P. Purpura, Metabolic and pharmacological effects of injected amino acids and ammonia on cortical epileptogenic lesions, *J. Neurochem.* **7**:198–209 (1961).
81. D. P. Purpura, M. Girado, T. G. Smith, and J. A. Gomez, Synaptic effects of

systemic γ-aminobutyric acid in cortical regions of increased vascular permeability, *Proc. Soc. Exptl. Biol. Med.* **97**:348–353 (1958).
82. R. P. Kamrin and A. A. Kamrin, The effects of pyridoxine antagonists and other convulsive agents on amino acid concentrations of the mouse brain, *J. Neurochem.* **6**:219–225 (1961).
83. C. F. Baxter, Cerebral metabolism of some amino acids *in vivo*, *Federation Proc.* **22**:301 (1963).
84. E. Levin, C. A. Garcia Argiz, and G. J. Nogueira, Ventriculocisternal perfusion of amino acids in cat brain. II. Incorporation of glutamic acid, glutamine and GABA into the brain parenchyma, *J. Neurochem.* **13**:979–988 (1966).
85. H. H. Jasper, R. T. Khan, and K. A. C. Elliott, Amino acids released from the cerebral cortex in relation to its state of activation, *Science* **147**:1448–1449 (1965).
86. N. M. Van Gelder and K. A. C. Elliott, Disposition of γ-aminobutyric acid administered to mammals, *J. Neurochem.* **3**:139–143 (1958).
87. A. Mori and M. Kosaka, Incorporation of γ-aminobutyric acid (GABA) in the brain and internal organs of mice, *Folia Psychiat. Neurol. Japon.* **15**:92–97 (1961).
88. K. D. Neame, Uptake of L-histidine, L-proline, L-tyrosine and L-ornithine by brain, intestinal mucosa, testis, kidney, spleen, liver, heart muscle, skeletal muscle and erythrocytes of the rat *in vitro*, *J. Physiol.* (*London*) **162**:1–12 (1962).
89. Y. Tsukada, Y. Nagata, S. Hirano, and T. Matsutani, Active transport of amino acid into cerebral cortex slices, *J. Neurochem.* **10**:241–256 (1963).
90. K. A. C. Elliott and N. M. Van Gelder, Occlusion and metabolism of γ-aminobutyric acid by brain tissue, *J. Neurochem.* **3**:28–40 (1958).
91. R. Nakamura and M. Nagayama, Amino acid transport by slices from various regions of the brain, *J. Neurochem.* **13**:305–313 (1966).
92. R. Blasberg and A. Lajtha, Heterogeneity of the mediated transport systems of amino acid uptake in brain, *Brain Res.* **1**:86–104 (1966).
93. H. Weinstein, S. Varon, D. R. Muhleman, and E. Roberts, A carrier-mediated transfer model for the accumulation of $^{14}$-C γ-aminobutyric acid by subcellular brain particles, *Biochem. Pharmacol.* **14**:273–288 (1965).
94. S. Varon, H. Weinstein, C. F. Baxter, and E. Roberts, Uptake and metabolism of exogenous γ-aminobutyric acid by subcellular particles in a sodium-containing medium, *Biochem. Pharmacol.* **14**:1755–1764 (1965).
95. P. Strasberg and K. A. C. Elliott, Further studies on the binding of γ-aminobutyric acid by brain, *Can. J. Biochem.* **45**:1795–1807 (1967).
96. G. M. McKhann and D. B. Tower, Gamma-aminobutyric acid: A substrate for oxidative metabolism of cerebral cortex, *Am. J. Physiol.* **196**:36–38 (1959).
97. Y. Tsukada, Y. Nagata, and G. Takagaki, Metabolism of γ-aminobutyric acid in brain slices, *Proc. Japan Acad.* **33**:510–514 (1957).
98. M. Kurokawa, Metabolic consequences of localized application of electrical pulses to sections of cerebral white matter, *J. Neurochem.* **5**:283–292 (1960).
99. K. A. C. Elliott and F. Bilodeau, The influence of potassium on respiration and glycolysis by brain slices, *Biochem. J.* **84**:421–428 (1962).
100. H. C. Buniatian, V. B. Yeghian, and G. A. Turshian, *in Problems in Brain Biochemistry* (H. C. Buniatian, ed.), Vol. 1, pp. 27–38, Academy of Science, Armenian SSR (1964).
101. B. Sacktor and L. Packer, Reactions of the respiratory chain in brain mitochondrial preparations, *J. Neurochem.* **9**:371–382 (1962).
102. R. Balázs, D. Biesold, and K. Magyar, Some properties of rat brain mitochondrial preparations: Respiratory control, *J. Neurochem.* **10**:685–708 (1963).

103. M. Bacila, A. P. Campello, C. H. M. Vianna, and D. O. Voss, The respiratory chain of rat cerebrum and cerebellum mitochondria: Respiration and oxidative phosphorylation, *J. Neurochem.* **11**:231–242 (1964).
104. H. C. Buniatian, C. G. Movcessian, and M. G. Urgandji, *in Problems in Brain Biochemistry* (H. C. Buniatian, ed.), Vol. 1, pp. 15–26, Academy of Science, Armenian SSR (1964).
105. R. Vrba, Glucose metabolism in rat brain *in vivo*, *Nature* **195**:663–665 (1962).
106. A. Geiger, N. Horvath, and Y. Kawakita, The incorporation of $^{14}C$ derived from glucose into the proteins of the brain cortex at rest and during activity, *J. Neurochem.* **5**:311–322 (1960).
107. J. E. Cremer, Amino acid metabolism in rat brain studied with $^{14}C$-labelled glucose, *J. Neurochem.* **11**:165–185 (1964).
108. M. K. Gaitonde, D. R. Dahl, and K. A. C. Elliott, Entry of glucose carbon into amino acids of rat brain and liver *in vivo* after injection of uniformly $^{14}C$-labelled glucose, *Biochem. J.* **94**:345–352 (1965).
109. R. Balázs and R. J. Haslam, Exchange transamination and the metabolism of glutamate in brain, *Biochem. J.* **94**:131–141 (1965).
110. R. J. Haslam and H. A. Krebs, The metabolism of glutamate in homogenates and slices of brain cortex, *Biochem. J.* **88**:566 (1963).
111. G. M. McKhann, R. W. Albers, L. Sokoloff, O. Mickelsen, and D. B. Tower, *in Inhibition in the Nervous System and Gamma-Aminobutyric Acid* (E. Roberts, C. F. Baxter, A. Van Harreveld, C. A. G. Wiersman, W. R. Adey, and K. F. Killam, eds.), pp. 169–181, Pergamon Press, New York (1960).
112. Y. Machiyama, R. Balázs, and D. Richter, Effect of $K^+$-stimulation on GABA metabolism in brain slices *in vitro*, *J. Neurochem.* **14**:591–594 (1967).
113. Y. Machiyama, R. Balázs, and T. Julian, Oxidation of glucose through the $\gamma$-aminobutyrate pathway in brain, *Biochem. J.* **96**:68P–69P (1965).
114. S. Berl, W. J. Nichlas, and D. D. Clarke, Compartmentation of glutamic acid metabolism in brain slices, *J. Neurochem.* **15**:131–140 (1968).
115. R. W. Albers, G. Koval, G. McKhann, and D. Ricks, *in Regional Neurochemistry* (S. S. Kety and J. Elkes, eds.), pp. 340–347, Pergamon Press, Oxford (1961).
116. S. Berl, A. Lajtha, and H. Waelsch, Amino acid and protein metabolism. VI. Cerebral compartments of glutamic acid metabolism, *J. Neurochem.* **7**:186–197 (1961).
117. F. N. Minard and I. K. Mushahwar, Synthesis of $\gamma$-aminobutyric acid from a pool of glutamic acid in brain after decapitation, *Life Sci.* **5**:1409–1413 (1966).
118. D. Garfinkel, A simulation study of the metabolism and compartmentation in brain of glutamate, aspartate, the Krebs cycle and related metabolites, *J. Biol. Chem.* **241**:3918–3929 (1966).
119. N. Seiler, H. Moller, and G. Werner, Der Einbau von Glucose-Kohlenstoff in die freien Aminosäuren des Mausegehirns unter dem Einfluss von Pyritinol, *Hoppe-Seylers Z. Physiol. Chem.* **348**:675–579 (1967).
120. J. R. Lindsay and H. S. Bachelard, Incorporation of $^{14}C$ from glucose into $\alpha$-keto acids and amino acids in rat brain and liver *in vivo*, *Biochem. Pharmacol.* **15**: 1045–1052 (1966).
121. R. Balázs, Control of glutamate metabolism. The effect of pyruvate, *J. Neurochem.* **12**:63–76 (1965).
122. G. M. McKhann and D. B. Tower, The regulation of $\gamma$-aminobutyric acid metabolism in cerebral cortex mitochondria, *J. Neurochem.* **7**:26–32 (1961).
123. I. Smith, *Chromatographic and Electrophoretic Techniques*, Vols. 1 and 2, Interscience, New York (1960).

123a. R. S. De Ropp and E. H. Snedeker, Sequential one-dimensional chromatography: Analysis of free amino acids in the brain, *Anal. Biochem.* **1**:424–432 (1960).
124. K. Crowshaw, S. J. Jessup, and P. W. Ramwell, Thin-layer chromatography of 1-dimethylaminonaphthalene 5-sulphonyl derivatives of amino acids present in superfusates of cat cerebral cortex, *Biochem. J.* **103**:79–85 (1967).
125. S. Voigt, M. Solle, and K. Konitzer, Dünnschichtchromatographische Abtrennung von γ-Aminobuttersäure aus Hirnextrakten, *J. Chromatog.* **17**:180–182 (1965).
126. E. Gründig, Trennungsgang zur quantitativen Bestimmung von Keto-und Aminosäuren in kleinen Mengen des Liquor cerebrospinalis, *Clin. Chim. Acta* **7**:489–505 (1962).
127. J. F. Thompson and C. J. Morris, Determination of amino acids from plants by paper chromatography, *Anal. Chem.* **31**:1031–1037 (1959).
128. S. Berl and H. Waelsch, Determination of glutamic acid, glutamine, glutathione and γ-aminobutyric acid and their distribution in brain tissue, *J. Neurochem.* **3**:161–169 (1958).
129. R. P. Sandman, The determination of gamma-aminobutyric acid in brain, *Anal. Biochem.* **3**:158–163 (1962).
130. C. F. Baxter and I. Senoner, Liquid scintillation counting of $^{14}C$ labeled amino acids on paper, using trinitrobenzene-1-sulfonic acid and a modified combustion apparatus, *Anal. Biochem.* **7**:55–61 (1964).
131. E. Florey and K. A. C. Elliott, *in Methods in Medical Research* (J. H. Quastel, ed.), Vol. 9, pp. 196–202, Year Book Medical Publishers, Chicago (1961).
132. E. Florey, A new test preparation for bioassay of Factor I and gamma-aminobutyric acid, *J. Physiol.* **156**:1–7 (1961).
133. A. S. F. Ash and J. F. Tucker, Inhibition of *Ascaris* muscle by γ-aminobutyric acid: A possible new assay method, *Nature* **209**:306–307 (1966).
134. H. M. Häkkinen and E. Kulonen, Comparison of various methods for the determination of γ-aminobutyric acid and other amino acids in rat brain with reference to ethanol intoxication, *J. Neurochem.* **10**:489–494 (1963).
135. E. M. Scott and W. B. Jakoby, Soluble γ-aminobutyricglutamic transaminase from Pseudomonas fluorescens, *J. Biol. Chem.* **234**:932–936 (1959).
136. C. F. Baxter, *in Methods in Medical Research* (J. H. Quastel, ed.), Vol. 9, pp. 192–195, Year Book Medical Publishers, Chicago (1961).
137. L. T. Graham and M. H. Aprison, Fluorometric determination of aspartate, glutamate and γ-aminobutyrate in nerve tissue using enzymatic methods, *Anal. Biochem.* **15**:487–497 (1966).
138. O. H. Lowry, N. R. Roberts, and J. I. Kapphahn, The fluorometric measurement of pyridine nucleotides, *J. Biol. Chem.* **224**:1047–1064 (1957).
139. R. A. Lovell, S. J. Elliott, and K. A. C. Elliott, The γ-aminobutyric acid and Factor I content of brain, *J. Neurochem.* **10**:479–488 (1963).
140. V. Chmelar, I. M. Hais, and M. Hodanova, γ-aminobutyric acid content in rat brain processed under different temperature conditions, *Acta Biochim. Polon.* **11**: 327–335 (1964).
141. K. A. C. Elliott, R. T. Khan, F. Bilodeau, and R. A. Lovell, Bound γ aminobutyric and other amino acids in brain, *Can. J. Biochem.* **43**:407–416 (1965).
142. F. N. Minard and I. K. Muchahwar, The effect of periodic convulsions induced by 1,1-dimethylhydrazine on the synthesis of rat brain metabolites from [$2^{14}C$] glucose, *J. Neurochem.* **13**:1–11 (1966).
143. J. K. Tews, S. H. Carter, P. Dante Roa, and W. E. Stone, Free amino acids and related compounds in dog brain: Post-mortem and anoxic changes, effects of ammonium chloride infusion and levels during seizures induced by picrotoxin and by pentylenetetrazol, *J. Neurochem.* **10**;641–653 (1963).

144. B. Sisken, E. Roberts, and C. F. Baxter, *in Inhibition in the Nervous System and Gamma-Aminobutyric Acid* (E. Roberts, C. F. Baxter, A. Van Harreveld, C. A. G. Wiersman, W. R. Adey, and K. F. Killam, eds.), pp. 219–225, Pergamon Press, New York (1960).
145. J. P. Susz, B. Haber, and E. Roberts, Purification and some properties of mouse brain L-glutamic decarboxylase, *Biochemistry* **5**:2870–2876 (1966).
146. E. Roberts and D. G. Simonsen, Some properties of L-glutamic decarboxylase in mouse brain, *Biochem. Pharmacol.* **12**:113–134 (1963).
147. R. E. Tashian, Inhibition of brain glutamic acid decarboxylase by phenylalanine, valine and leucine derivatives: A suggestion concerning etiology of the neurological defect in phenylketonuria and branched chain ketonuria, *Metabolism* **10**:393–402 (1961).
148. C. F. Baxter and E. Roberts, Elevation of $\gamma$-aminobutyric acid in brain: Selective inhibition of $\gamma$-aminobutyric-$\alpha$-ketoglutaric acid transaminase, *J. Biol. Chem.* **236**:3287–3294 (1961).
149. D. B. McCormick, B. M. Guirard, and E. E. Snell, Comparative inhibition of pyridoxal kinase and glutamic acid decarboxylase by carbonyl reagents, *Proc. Soc. Exptl. Biol. Med.* **104**:554–557 (1960).
150. P. Holtz and D. Palm, Pharmacological aspects of Vitamin $B_6$, *Pharmacol. Rev.* **16**:113–178 (1964).
151. R. Tapia, M. Perez de la Mora, and G. H. Massieu, Modifications of brain glutamate decarboxylase activity by pyridoxal phosphate-$\gamma$-glutamyl hydrazone, *Biochem. Pharmacol.* **16**:1211–1218 (1967).
152. M. A. Medina, H. D. Braymer, and J. L. Reeves, *In vitro* reversal of glutamate decarboxylase inhibition induced by 1,1-dimethylhydrazine, *J. Neurochem.* **9**: 307–312 (1962).
153. R. Tapia and J. Awapara, Formation of $\gamma$-aminobutyric acid (GABA) in brain of mice treated with L-glutamic acid, $\gamma$-hydrazide and pyridoxal phosphate-$\gamma$-glutamyl hydrazone, *Proc. Soc. Exptl. Biol. Med.* **128**:218–221 (1967).
154. D. B. McCormick and E. E. Snell, Pyridoxal kinase of human brain and its inhibition by hydrazine derivatives, *Proc. Natl. Acad. Sci. U.S.* **45**:1371–1379 (1959).
155. D. B. McCormick and E. E. Snell, Pyridoxal Phosphokinase. II. Effect of inhibition, *J. Biol. Chem.* **236**:2085–2088 (1961).
156. P. Gonnard and S. Fenard, Décarboxylase glutamique et hydrazones de phospho-5-pyridoxal, *J. Neurochem.* **9**:135–142 (1962).
157. P. Gonnard and J. Duhalt, Action de lay cyanacetylhydrazone de phospho-5-pyridoxal sur la glutamate decarboxylase cerebrale. Modification du trace electrocorticographique du rat, *J. Neurochem.* **13**:407–412 (1966).
158. K. Makino, Y. Ooi, M. Matsuda, and T. Kuroda, *in Chemical and Biological Aspects of Pyridoxal Catalysis* (E. E. Snell, P. M. Fasella, A. Braunstein, and A. Fanelli-Rossi, eds.), pp. 291–304, Pergamon Press, New York (1963).
159. F. Rosen, R. J. Milholland, and C. A. Nichol, *in Inhibition in the Nervous System and Gamma-Aminobutyric Acid* (E. Roberts, C. F. Baxter, A. Van Harreveld, C. A. G. Wiersman, W. R. Adey, and K. F. Killam, eds.), pp. 338–343, Pergamon Press, New York (1960).
160. M. Matsuda and K. Makino, Action of penicillamine on mice with special reference to its effect on glutamic acid decarboxylase in brain, *Biochim. Biophys. Acta* **48**:192–193 (1961).
161. G. Rindi, V. Perri, and U. Ventura, The action of toxopyrimidine on glutamic decarboxylase and on some transaminase of rat tissue, *Ital. J. Biochem.* **8**:149–163 (1959).

162. K. F. Killam, Convulsant hydrazides. II. Comparison of electrical changes and enzyme inhibition induced by the administration of thiosemicarbazide, *J. Pharmacol. Exptl. Therap.* **119**:263–271 (1957).
163. C. F. Baxter and E. Roberts, *in Amino Acid Pools* (J. T. Holden, ed.), pp. 499–508 Elsevier, Amsterdam (1962).
164. P. B. Molinoff and E. A. Kravitz, The metabolism of γ-aminobutyric acid (GABA) in the lobster nervous system—glutamic decarboxylase, *J. Neurochem.* **15**:391–409 (1968).
165. E. Jenny, Subcellulare Lokalisation und partielle Reinigung von Glutaminsäure-decarboxylase (GAD E.C. 4.1.1.15) aus Kalbshirnrinde, *Helv. Physiol. Pharmacol. Acta* **24**:96–97 (1966).
165a. B. Haber, K. Kuriyama, and E. Roberts, An anion-stimulated L-glutamic decarboxylase in non-neuronal tissue: Occurrence and subcellular localization in mouse kidney and developing chick embryo brain, *Biochem. Pharmacol.*, in press (1969).
166. E. Roberts, *in Comparative Neurochemistry* (D. Richter, ed.), pp. 167–178, Macmillan, New York (1964).
167. J. P. Schade and C. F. Baxter, Changes during growth in the volume and surface area of cortical neurons in the rabbit, *Exptl. Neurol.* **2**:158–178 (1960).
168. B. Sisken, K. Sano, and E. Roberts, γ-aminobutyric acid content and glutamic decarboxylase and γ-aminobutyrate transaminase activities in the optic lobe of the developing chick, *J. Biol. Chem.* **236**:503–507 (1961).
169. S. M. Bayer and W. C. McMurray, The metabolism of amino acids in developing rat brain, *J. Neurochem.* **14**:695–706 (1967).
170. C. J. Van den Berg, G. M. J. Van Kempen, J. P. Schade, and H. Veldstra, Levels and intracellular localization of glutamate decarboxylase and γ-aminobutyrate transaminase and other enzymes during the development of the brain, *J. Neurochem.* **12**:863–869 (1965).
171. E. Roberts, P. J. Hartman, and S. Frankel, Gamma-aminobutyric acid content and glutamic decarboxylase activity in developing mouse brain, *Proc. Soc. Exptl. Biol. Med.* **78**:799–803 (1951).
172. E. L. Avenirova, E. V. Bogadnova, I. A. Sytinsky, and G. M. Taokacheva, Activity of the enzymes of GABA metabolism in developing rat brain, *Zh. Evolyutsionni Biokhim. Fiziol.* **2**:493–495 (1966).
173. W. A. Himwich, Biochemical and neurophysiological development of the brain in the neonatal period, *Intern. Rev. Neurobiol.* **4**:117–158 (1962).
174. S. Berl and D. P. Purpura, Postnatal changes in amino acid content of kitten cerebral cortex, *J. Neurochem.* **10**:237–240 (1963).
175. H. C. Agrawal, J. M. Davis, and W. A. Himwich, Postnatal changes in free amino acid pool of rat brain, *J. Neurochem.* **13**:607–615 (1966).
176. A. R. Dravid, W. A. Himwich, and J. M. Davis, Some free amino acids in dog brain during development, *J. Neurochem.* **12**:901–906 (1965).
177. S. S. Oja, Postnatal changes in the concentration of nucleic acids, nucleotides and amino acids in the rat brain, *Ann. Acad. Sci. Fennicae: Ser. A. V.* **125**:1–69 (1966).
178. A. E. Ramires de Guglielmone and C. J. Gomez, Influence of neonatal hypothyroidism on amino acids in developing at brain, *J. Neurochem.* **13**:1017–1025 (1966).
179. H. C. Agrawal, J. M. Davis, and W. A. Himwich, Postnatal changes in free amino acid pool of rabbit brain, *Brain Res.* **3**:374–380 (1967).
180. L. Mihailović and L. Kržalić, Changes in glutamic acid glutamine and gamma-aminobutyric acid concentrations during postnatal maturation of the cat brain, *Acta Med. Jugoslav* **18**:150–156 (1964).

181. E. Roberts, I. P. Lowe, L. Guth, and B. Jelinek, Distribution of $\gamma$-aminobutyric acid and other amino acids in nervous tissue of various species, *J. Exptl. Zool.* **138**:313–328 (1958).
182. P. B. Müller and H. Langemann, Distribution of glutamic acid decarboxylase activity in human brain, *J. Neurochem.* **9**:399–401 (1962).
183. R. W. Albers and R. O. Brady, The distribution of glutamic decarboxylase in the nervous system of the rhesus monkey, *J. Biol. Chem.* **234**:926–928 (1959).
184. L. T. Graham and M. H. Aprison, Distribution of some enzymes associated with the metabolism of glutamate, aspartate, $\gamma$-aminobutyrate and glutamine in cat spinal cord, *J. Neurochem.* **16**:559–566 (1969).
185. G. M. J. Van Kempen, C. J. Van den Berg, H. J. Van der Helm, and H. Veldstra, Intracellular localization of glutamate decarboxylase, $\gamma$-aminobutyric transaminase and some other enzymes in brain tissue, *J. Neurochem.* **12**:581–588 (1965).
186. V. P. Whittaker, The application of subcellular fractionation techniques to the study of brain function, *Progr. Biophys. Biophys. Chem.* **15**:39–96 (1965).
187. R. Balázs, D. Dahl, and J. R. Harwood, Subcellular distribution of enzymes of glutamate metabolism in rat brain, *J. Neurochem.* **13**:897–905 (1966).
188. S. Løvtrup, The subcellular localization of glutamic decarboxylase in rat brain, *J. Neurochem.* **8**:243–245 (1961).
189. P. Gonnard and L. E. A. Rodrigues, Localisation intracellulaire de la glutamate décarboxylase cérébrale, *Bull. Soc. Chim. Biol.* **49**:815–823 (1967).
190. F. Fonnum, The distribution of glutamate decarboxylase and aspartate transminase in subcellular fractions of rat and guinea pig brain, *Biochem. J.* **106**:401–412 (1968).
191. J. D. Robinson and R. M. Bradley, Cholinesterase and glutamic decarboxylase levels in the brain of the hibernating hamster, *Nature* **197**:389–390 (1963).
192. Lj. T. Mihailovic, Lj. Kržalić, and D. Cupic, Changes of glutamine, glutamic acid and GABA in cortical and subcortical brain structures of hibernating and fully aroused ground squirrels (*Citellus citellus*), *Experientia* **21**:709–710 (1965).
193. L. Pandolfo and S. Macaione, Influence della surrenectoma sulfa attivita della GABA transaminasi e della glutamico decarbossilasi di corteccia cerebrale di ratto, *Giorn. Biochim.* **13**:256–261 (1964).
194. R. Rajalakshmi, K. R. Govindarajan, and C. V. Ramakrishnan, Effect of dietary protein content on visual discrimination learning and brain biochemistry in the albino rat, *J. Neurochem.* **12**:261–271 (1965).
195. I. A. Sytinskii and K. C. Shang, Activity of glutamate decarboxylase and the level of $\gamma$-aminobutyric acid in the brain of rats after whole body x-irradition, *Nervnaya Sistema, Leningr. Gos. Univ., Fiziol. Inst.* **7**:47–50 (1966).
196. E. Geller, A. Yuwiler, and J. F. Zolman, Effects of environmental complexity on constituents of brain and liver, *J. Neurochem.* **12**:949–955 (1965).
197. A. K. Pfeifer, E. Sátory, and E. S. Vizi, Studies on the glutamic acid decarboxylase activity of rat brain in the state of inhibition following electroshock, *Arch. Itern. Pharmacodyn.* **138**:230–238 (1962).
198. I. P. Lowe, E. Robins, and G. S. Eyerman, The fluorimetric measurement of glutamic decarboxylase and its distribution in brain, *J. Neurochem.* **3**:8–18 (1958).
199. E. Roberts and S. Frankel, Glutamic acid decarboxylase in brain, *J. Biol. Chem.* **188**:789–795 (1951).
200. Y. Nishizawa, T. Kodama, and S. Konishi, Brain $\gamma$-aminobutyric-$\alpha$-ketoglutaric transaminase, *J. Vitaminol.* **5**:117–128 (1959).
201. C. F. Baxter and E. Roberts, The $\gamma$-aminobutyric acid-$\alpha$-ketoglutaric acid transaminase of beef brain, *J. Biol. Chem.* **233**:1135–1139 (1958).

202. S. P. Bessman, J. Rosesn, and E. C. Layne, γ-aminobutyric acid—glutamic acid transamination in brain, *J. Biol. Chem.* **201**:385–391 (1953).
203. R. W. Albers and W. B. Jakoby, Exchange reactions catalyzed by γ-aminobutyric-glutamic transaminase, *Biochim. Biophys. Acta* **40**:457–461 (1960).
204. F. Olivo, C. S. Rossi, and N. Siliprandi, *in Chemical and Biological Aspects of Pyridoxal Catalysis* (E. E. Snell, P. M. Fasella, A. Braunstein, and A. Fanelli-Rossi, eds.), pp. 91–101, Pergamon Press, New York (1963).
205. K. Nakamura and F. Bernheim, Transaminase activity of naturally occuring inhibitory substances and effects of some drugs, *Japan. J. Pharmacol.* **11**:141–150 (1962).
206. F. Caciappo, L. Pandolfo, and G. Di Chiara, Transamination reaction between 4-aminobutyric acid and α-ketoglutaric acid in certain rat tissues, *Boll. Soc. Ital. Biol. Sper.* **36**:465–467 (1959).
207. G. M. J. Van Kempen, γ-Aminobutyraat transaminase, Een onderzoek over Eigenschappen en localisatie in hersenweefsal, Thesis, University of Leiden, Holland (1964).
208. W. E. Wilson, R. J. Hill, and R. E. Koeppe, The metabolism of γ-aminobutyric acid $4C^{14}$ by intact rats, *J. Biol. Chem.* **234**:347–349 (1959).
209. E Roberts, M. Rothstein, and C. F. Baxter, Some metabolic studies of γ-aminobutyric acid, *Proc. Soc. Exptl. Biol. Med.* **97**:796–802 (1958).
210. R. A. Salvador and R. W. Albers, The distribution of glutamic-γ-aminobutyric transaminase in the nervous system of the Rhesus monkey, *J. Biol. Chem.* **234**:922–925 (1959)
211. F. N. Pitts, Jr., C. Quick, and E. Robins, The enzymic measurement of γ-aminobutyric-α-oxoglutaric transaminase, *J. Neurochem.* **12**:93–101 (1965).
212. A. Waksman and E. Roberts, Purification and some properties of mouse brain γ-aminobutyric-α-ketoglutamic acid transaminase, *Biochemistry* **4**:2132–2139 (1965).
212a. A. Waksman and M. Bloch, Identification of multiple forms of aminobutyrate transaminase in mouse and rat brain: Subcellular localization, *J. Neurochem.* **15**:99–105 (1968).
213. R. Rybova, Effect of cations on γ-aminobutyric acid level in slices of brain cortex, *Nature* **185**:542–543 (1960).
214. B. Bulos and P. Handler, Kinetics of beef heart glutamic-alanine transminase, *J. Biol. Chem.* **240**:3283–3294 (1965).
215. K. F. Killam and J. A. Bain, Convulsant hydrazides. I. *In vitro* and *in vivo* inhibition vitamin $B_6$ enzymes by convulsant hydrazides, *J. Pharmacol. Exptl. Therap.* **119**: 255–262 (1957).
216. C. F. Baxter and E. Roberts, Elevation of γ-aminobutyric acid in rat brain with hydroxylamine, *Proc. Soc. Exptl. Biol. Med.* **101**:811–815 (1959).
217. G. Carvajal, M. Russek, R. Tapia, and G. Massieu, Anticonvulsive action of substances designed as inhibitors of γ-aminobutyric acid-α-ketoglutaric acid transaminase, *Biochem. Pharmacol.* **13**:1059–1069 (1964).
218. N. M. Van Gelder, The effect of aminooxyacetic acid on the metabolism of γ-aminobutyric acid in brain, *Biochem. Pharmacol.* **15**:533–539 (1966).
219. D. P. Wallach, Studies on the GABA pathway. I. The inhibition of γ-aminobutyric acid-α-ketoglutaric acid transaminase *in vitro* and *in vivo* by U7524 (Amino-oxyacetic acid), *Biochem. Pharmacol.* **5**:323–331 (1961).
220. M. A. Medina, The *in vivo* effects of hydrazines and vitamin $B_6$ on the metabolism of gamma-aminobutyric acid, *J. Pharmacol. Exptl. Therap.* **140**:133–137 (1963).
221. G. H. Massieu, R. I. Tapia, H. O. Pasantes, and B. G. Ortega, Convulsant effect of L-glutamic acid-γ-hydrazide by simultaneous treatment with pyridoxal phosphate, *Biochem. Pharmacol.* **13**:118–120 (1964).

222. O. T. Dann and C. E. Carter, Cycloserine inhibition of gamma-aminobutyric-alpha-keto-glutaric transaminase, *Biochem. Pharmacol.* **13**:677–684 (1964).
223. J. P. DaVanzo, R. J. Matthews, G. A. Young, and F. Wingerson, Studies on the mechanism of action of aminooxyacetic acid. II. Possible pyridoxine deficiency as a mechanism of action of aminooxyacetic acid toxicity, *Toxicol. Appl. Pharmacol.* **6**:396–401 (1964).
224. J. P. DaVanzo, L. Kang, R. Ruckart, and M. Daugherty, Inhibition of pyridoxal phosphokinase by aminooxyacetic acid, *Biochem. Pharmacol.* **15**:124–126 (1966).
225. J. D. Gabourel, Anticonvulsant properties of diacetylmonoxime (DAM), *Biochem. Pharmacol.* **5**:283–286 (1960).
226. H. C. Buniatian and R. R. Nercessian, *in Problems in Brain Biochemistry* (H. C. Buniatian, ed.), Vol. 1, pp. 5–13, Academy of Science, Armenian SSR (1964).
227. C. J. Van den Berg and G. M. J. Van Kempen, Glutamate decarboxylase and γ-aminobutyric transaminase in developing rat brain, *Experientia* **20**:375–376 (1964).
228. N. M. Van Gelder, The histochemical demonstration of γ-aminobutyric acid metabolism by reduction of a tetrazolium salt, *J. Neurochem.* **12**:231–237 (1965).
229. J. J. Sheridan, K. L. Sims, and F. N. Pitts, Jr., Brain γ-aminobutyrate-α-oxoglutarate transaminase. II. Activities in twenty-four regions of human brain, *J. Neurochem.* **14**:571–578 (1967).
230. A. Waksman and C. Faienza, Identification de la transaminase glutamique-γ-aminobutyrique dans le cerveau humain, *Clin. Chim. Acta* **5**:450–452 (1960).
231. N. M. Van Gelder, A comparison of γ-aminobutyric acid metabolism in rabbit and mouse nervous tissue, *J. Neurochem.* **12**:239–244 (1965).
232. M. A. Fisher, D. Q. Hagen, and R. B. Colvin, Aminooxyacetic acid: Interactions with gamma-aminobutyric acid and the blood-brain barrier, *Science* **153**:1668–1670 (1966).
233. L. Salganicoff and E. De Robertis, Subcellular distribution of glutamic decarboxylase and gamma-aminobutyric alpha-ketoglutaric transaminase in rat brain, *Life Sci.* **1**:85–91 (1963).
234. Z. W. Hall and E. A. Kravitz, The metabolism of γ-aminobutyric acid (GABA) in the lobster nervous system. II. Succinic semialdehyde dehydrogenase, *J. Neurochem.* **14**:55–61 (1967).
235. F. N. Pitts, Jr., and C. Quick, Brain succinate semialdehyde dehydrogenase. II. Changes in the developing rat brain, *J. Neurochem.* **14**:561–570 (1967).
236. C. Kammeratt and H. Veldstra, Characterization of succinate semialdehyde dehydrogenase from rat brain, *Biochim. Biophys. Acta* **151**:1–10 (1968).
237. F. N. Pitts, Jr., and C. Quick, Brain succinate semialdehyde dehydrogenase. I. Assay and distribution, *J. Neurochem.* **12**:893–900 (1965).
238. R. W. Albers and G. J. Koval, Succinic semialdehyde dehydrogenase: Purification and properties of the enzyme from monkey brain, *Biochim. Biophys. Acta* **52**:29–35 (1961).
239. L. J. Embree and R. W. Albers, Succinic semialdehyde dehydrogenase from human brain, *Biochem. Pharmacol.* **13**:1209–1217 (1964).
240. A. L. Miller and F. N. Pitts, Jr., Brain succinate semialdehyde dehydrogenase. III. Activities in twenty-four regions of human brain, *J. Neurochem.* **14**:579–584 (1967).
241. W. N. Fishbein and S. P. Bessman, γ-hydroxybutyrate in mammalian brain. Reversible oxidation by lactic dehydrogenase, *J. Biol. Chem.* **239**:357–361 (1964).
242. N. J. Giarman and R. H. Roth, Differential estimation of gamma-butyrolactone and gamma-hydroxybutric acid in rat blood and brain, *Science* **145**:583–584 (1964).

243. R. H. Roth and N. J. Giarman, Preliminary report on the metabolism of γ-butyrolactone and γ-hydroxybutric acid *Biochem. Pharmacol.* **14**:177–178 (1965).
244. S. P. Bessman and S. J. Skolnik, Gamma-hydroxybutyrate and gamma-butyrolactone: Concentration in rat tissue during anesthesia, *Science* **143**:1047–1054 (1964).
245. R. H. Roth, J. M. R. Delgado, and N. J. Giarman, γ-Butyrolactone and γ-hydroxybutric acid. I. The pharmacologically active form, *Intern. J. Neuropharmacol.* **5**:421–428 (1966).
246. H. Laborit, Sodium 4–hydroxybutyrate, *Intern. J. Neuropharmacol.* **3**:433–451 (1964).
247. N. J. Giarman and K. T. Schmidt, Some neurochemical aspects of the depressant action of γ-butyrolactone on the central nervous system, *Brit. J. Pharmacol.* **20**: 563–568 (1963).
248. G. D. Pietra, G. Illiano, V. Capano, and R. Rava, *In vivo* conversion of γ-hydroxybutyrate into γ-aminobutyrate, *Nature* **210**:733–734 (1966).
249. C. Mitoma and S. E. Neubauer, Gamma-hydroxybutric acid and sleep, *Experientia* **24**:13–23 (1968).
250. D. P. Wallach, Studies on the GABA pathway. II. The lack of effect of pyridoxal phosphate on GABA-KGA transaminase inhibition induced by aminooxyacetic acid, *Biochem. Pharmacol.* **8**:328–331 (1961).
251. J. P. DaVanzo, R. J. Matthews, and J. E. Stafford, Studies on the mechanism of action of aminooxyacetic acid. I. Reversal of aminooxyacetic acid-induced convulsions by various agents, *Toxical. Appl. Pharmacol.* **6**:388–395 (1964).
252. S. S. Walkenstein, R. Wiser, C. Gudmundsen, and H. Kimmel, Metabolism of γ-hydroxybutric acid, *Biochim. Biophys. Acta* **86**:640–642 (1964).
253. J. J. Pisano, J. D. Wilson, and S. Udenfriend, *in Inhibition in the Nervous System and Gamma-Aminobutyric Acid* (E. Roberts, C. F. Baxter, A. Van Harreveld, C. A. G. Wiersman, W. R. Adey, and K. F. Killam, eds.), pp. 226–235, Pergamon Press, New York (1960).
254. D. Grünberger, J. Černá, and F. Šorm, Intracellular peptides of *Escherichia coli*, *Experientia* **16**:54 (1960).
255. J. J. Pisano, D. Abraham, and S. Udenfriend, Biosynthesis and disposition of γ-guanidinobutyric acid in mammalian tissues, *Arch. Biochem. Biophys.* **100**:323–329 (1963).
256. F. Irrevere, R. I. Evans, A. R. Hayden, and R. Silber, Occurrence of gamma-guanidinobutyric acid, *Nature* **180**:704–705 (1957).
257. R. Baret, M. Mourgue, and A. Broc, Sur la répartition de la γ-guanidobutyrase et de l'arginase chez les poissons, *Compt. Rend. Soc. Biol.* **156**:1117–1119 (1962).
258. N. Thoai, J. Roche, and Y. Robin, Metabolisme des derives guanidyles. I. Degradation de l'argenine chéz les invertebres marins, *Biochim. Biophys. Acta* **11**:403–411 (1953).
259. Z. W. Hall, The metabolism of gamma-aminobutyric acid in the lobster nervous system, PhD Thesis P51, Harvard University, Cambridge, (1966).
260. R. Baret, M. Mourgue, and A. Broc, Metabolism of γ-guanidobutyric acid in certain vertebrates and invertebrates, *Compt. Rend. Soc. Biol.* **159**:703–705 (1965).
261. C. Edwards and S. W. Kuffler, The blocking effect of γ-aminobutyric acid (GABA) and the action of related compounds on single nerve cells, *J. Neurochem.* **4**:19–30 (1959).
262. H. Takahashi, B. Arai, and C. Koshino, Effects of guanidinoacetic acid, γ-guanidinobutyric acid and γ-guanidinobutyryl methylester on the mammalian cerebral cortex, *Japan. J. Physiol.* **11**:403–409 (1961).

263. D. P. Purpura, M. Girado, and H. Grundfest, Central synaptic effects of $\omega$-guanidino acids and amino acid derivatives, *Science* **127**:1179–1181 (1958).
264. D. Jinnai, A. Sawai, and A. Mori, $\gamma$-guanidinobutyric acid as a convulsive substance, *Nature* **216**:617 (1966).
265. A. Kanazawa, Y. Kakimoto, E. Miyamoto and I. Sano, Isolation and identification of homocarnosine from bovine brain, *J. Neurochem.* **12**:957–958 (1965).
266. D. Abraham, J. J. Pisano, and S. Udenfriend, The distribution of homocarnosine in mammals, *Arch. Biochem. Biophys.* **99**:210–213 (1962).
267. A. Kanazawa and I. Sano, A method of determination of homocarnosine and its distribution in mammalian tissues, *J. Neurochem.* **14**:211–214 (1967).
268. D. Abraham, J. J. Pisano, and S. Udenfriend, Synthesis of homocarnosine in muscle *in vivo*, *Biochim. Biophys. Acta* **50**:570–572 (1961).
269. T. Nakajima, F. Wolfgram, and W. G. Clark, The isolation of homoanserine from bovine brain, *J. Neurochem.* **14**:1107–1112 (1967).
270. H. H. Tallan, S. Moore, and W. H. Stein, Studies on the free amino acids and related compounds in the tissues of the cat, *J. Biol. Chem.* **211**:927–939 (1954).
271. T. Hayashi, The inhibitory action of $\beta$-hydroxy-$\gamma$-aminobutyric acid upon the seizure following stimulation of the motor cortex of the dog, *J. Physiol.* **145**:570–578 (1959).
272. K. Ohara, I. Sano, H. Koizumi, and K. Nishinuma, Free $\beta$-hydroxy-$\gamma$-aminobutyric acid in brain, *Science* **129**:1225–1226 (1959).
273. C. Mitoma, *in Inhibition in the Nervous System and Gamma-Aminobutyric Acid* (E. Roberts, C. F. Baxter, A. Van Harreveld, C. A. G. Wiersman, W. R. Adey, and K. F. Killam, eds.), pp. 236–237, Pergamon Press, New York (1960).
274. A. Nishimato, A. Mori, and H. Takashita, Treatment of epilepsy by intraspinal injection of gamma-amino-beta-hydroxybutyric acid (GABOB), *Folia Psychiat. Neurol. Japon.* **17**:351–357 (1964).
275. E. A. Hosein, The isolation of $\gamma$-butyrobetaine, crotonbetaine and carnitine from brains of animals killed during induced convulsions, *Arch, Biochem. Biophys.* **100**: 32–35 (1963).
276. E. A. Hosein and P. Proulx, Isolation and probable functions of betaine esters in brain metabolism, *Nature* **187**: 321–322 (1960).
277. M. A. Mehlman and G. Wolf, Studies on the distribution of free carnitine and the occurrence and nature of bound carnitine, *Arch. Biochem. Biophys.* **98**:146–153 (1962).
278. E. A. Hosein, M. Smart, K. Hawkins, S. Rochon, and Z. Strasberg, The enzymatic synthesis of $\gamma$-butyrobetaine and its CoA ester derivative, *Arch. Biochem. Biophys.* **96**:246–251 (1962).
279. G. Lindstedt and S. Lindstedt, Studies on the biosynthesis of carnitine, *J. Biol. Chem.* **240**:316–321 (1965).
280. J. Bremer, Carnitine precursors in the rat, *Biochim. Biophys. Acta* **57**:327–335 (1962).
281. G. Lindstedt and S. Lindstedt, On the hydroxylation of $\gamma$-butyrobetaine to carnitine *in vitro*, *Biochem. Biophys. Res. Commun.* **7**:394–397 (1962).
282. E. A. Hosein and P. Proulx, Acetylcholine-like activity in subcellular particles isolated from rat brain, *Arch. Biochem. Biophys.* **106**:267–274 (1964).
283. E. A. Hosein and A. Orzeck, Acetylcholine-like activity of acetyl-1-carnityl-CoA in the brains of narcotized rats, *Nature* **210**:731–732 (1966).
284. W. D. Thomitzek, H. Winter, and E. Strack, Effect of carnitine on the acetylcholine content in the brain and heart muscle *in vivo* and on the acetylation of sulfanilamide in the liver, *Acta Biol. Med. Ger.* **16**:342–350 (1966).

285. K. Kuriaki, T. Yakushiji, T. Moro, T. Shimizu, and Sh. Saji, Gamma-aminobutyryl choline, *Nature* **181**:1336–1337 (1958).
286. H. Kewitz, Detection of (gamma) γ-amino-η-butyrylcholine in the brain of warm-blooded animals, *Arch. Exptl. Pathol. Pharmakol.* **237**:308–318 (1959).
287. H. Kewitz, Bestimmung und Hydrolyse von 4-Aminobutyrylcholin *Arch. Exptl. Pathol. Pharmakol.* **240**:16–17 (1960).
288. R. Gryglewski, T. Marczyhski, and J. Trabka, The action of γ-aminobutyryl choline on the central nervous system, *Dissertationes Pharm.* **17**:135–144 (1965).
289. H. Ashida, N. Takeuchi, A. Mori, and D. Jinnai, Anti-convulsive action of gamma-aminobutyryl choline, *Nature* **206**:514–515 (1965).
290. H. Takahashi, A. Nagashima, and C. Koshino, Effect of γ-aminobutyryl-choline upon the electrical activity of the cerebral cortex, *Nature* **182**:1443–1444 (1958).
291. B. Holmstedt and F. Sjoqvist, Pharmacological properties of γ-aminobutyryl-choline, a supposed inhibitory neurotransmitter, *Biochem. Pharmacol.* **3**:297–304 (1960).
292. A. J. Hance, W. D. Winters, P. Bach-y-Rita, and K. F. Killam, A neuropharmacological study of gamma-aminobutyrylcholine, acid, physostigmine and atropine, *J. Pharmacol. Exptl. Therap.* **140**:385–395 (1963).
293. P. Boulanger, G. Biserte, and M. Davril, Combinasion de l'acide γ-aminobutyrique radioactif dans le tissue cérébral du rat *in vitro*, *Compt. Rend.* **260**:5918–5919 (1965).
294. C. McF. Desha and R. Fuerst, Chemical and enzymatic synthesis of γ-pantothenate, *Biochim. Biophys. Acta* **86**:33–38 (1964).
295. D. H. Kessel, Effect of 2, 4-diaminobutyrate in cerebral metabolism of gamma-aminobutyric acid, *Dissertation Abstr.* **21**:291–292 (1960).
296. T. Nakajima, F. Wolfgram, and W. G. Clark, Identification of 1,4-methylhistamine, 1,3-diaminopropane and 2,4-diaminobutyric acid in bovine brain, *J. Neurochem.* **14**:1113–1118 (1967).
297. Z. S. Gershenovich, A. A. Krichevskaya, and V. I. Shunskaya, Specificity of relations between γ-aminobutyric acid and brain proteins, *Dokl. Akad. Nauk SSSR* **162**:1415–1417 (1965).
298. Zh. A. Chalabyan, Effect of γ-aminobutyric acid on nucleic acid metabolism in the rabbit brain, *Ukr. Biokhim. Zh.* **36**:367–372 (1964).
299. H. C. Buniatian, Studies of the role of γ-aminobutyric acid in carbohydrate metabolism, Report presented to the fifth International Congress of Biochemistry, Moscow (August 10–16, 1961).
300. S. G. Movcessian and M. G. Urgandjian, *in Problems in Brain Biochemistry* (H. C. Buniatian, ed.), Vol. 1, pp. 87–96, Academy of Science, Armenian SSR (1964).
301. M. G. Urgandjian and S. G. Movcessian, *in Problems in Brain Biochemistry* (H. C. Buniatian, ed.), Vol. 2, pp. 63–72, Academy of Science, Armenian SSR (1966).
302. A. Mori, Influence of γ-aminobutyric acid and substances possessing similar chemical structure on hexokinase of the brain and heart muscle, *J. Biochem. Tokyo* **45**: 985 (1958).
303. M. Ruščak, E. Macejová, and D. Ruščaková, Effect of L-glutamic acid and γ-aminobutyric acid on glycolysis in slices and mitochondria of the rat brain central nervous system, *Physiol. Bohemoslov.* **13**:156–160 (1964).
304. B. A. Kazarian and E. A. Gulian, *in Problems in Brain Biochemistry* (H. C. Buniatian, ed.), Vol. 1, pp. 73–77, Academy of Science, Armenian SSR (1064).
305. H. C. Buniatian, The role of γ-aminobutyric acid in the metabolism of glutamic acid and glutamine in brain, *J. Neurochem.* **10**:461–469 (1963).

306. D. M. Woodbury and R. Karler, The role of carbon dioxide in the nervous system, *Anesthesiology* **21**:686–703 (1960).
307. R. E. Whalen, H. Heyman, and H. Saltzman, The protective effect of hyperbaric oxygenation in cerebral anoxia, *Arch. Neurol.* **14**:15–20 (1966).
308. F. Dickens, *in Neurochemistry* (K. A. C. Elliott, I. H. Page, and J. H. Quastel, eds.) 2nd Ed., pp. 851–869, Charles C. Thomas, Springfield, Illinois (1962).
309. G. V. Shcherbakova, Activity of glutamic decarboxylase and the content of $\gamma$-aminobutyric acid in rat brain at various levels of functional state caused by increased oxygen pressure, *Dokl. Akad. Nauk SSSR* **146**:1213–1215 (1962).
310. J. D. Wood and W. J. Watson, Gamma-aminobutyric acid levels in the brain of rats exposed to oxygen at high pressures, *Can. J. Biochem. Physiol.* **41**:1907–1913 (1963).
311. F. V. DeFeudis and K. A. C. Elliott, Delay or inhibition of convulsions by intraperitoneal injections of diverse substances, *Can. J. Physiol. Pharmacol.* **45**:857–865 (1967).
312. J. D. Wood and W. J. Watson, Molecular structure-activity relationships of compounds protecting rats against oxygen poisoning, *Can. J. Physiol. Pharmacol.* **42**:641–646 (1964).
313. J. J. Thomas, E. M. Neptune, and H. C. Sudduth, Toxic effects of oxygen at high pressure on the metabolism of D-glucose by dispersion of rat brain, *Biochem. J.* **88**:31–45 (1963).
314. E. D. Avenirova, B. M. Savin, and I. A. Sytinskii, Effect of anoxia and acceleration on content of glutamic and gamma-aminobutyric acids in brain tissue, *Federation Proc. Trans. Suppl.* **24**:T809–T811 (1964).
315. A. P. Sanders, I. H. Hall, and B. Woodhall, Succinate: Protective agent against hyperbaric oxygen toxicity, *Science* **150**:1830–1831 (1965).
316. B. Chance, D. Jamieson, and H. Coles, Energy-linked pyridine nucleotide reduction: Inhibitory effects of hyperbaric oxygen *in vitro* and *in vivo*, *Nature* **206**: 257–263 (1965).
317. D. Mićić, V. Karadžić, and Lj. M. Rakić, Changes of gamma-aminobutyric acid, glutamic acid and aspartic acid in various brain structures of cats deprived of paradoxical sleep, *Nature* **215**:169–170 (1967).
318. P. Mandel and Y. Godin, Sur l'intervention possible de l'acide $\gamma$-aminobutyrique dans le phénomène de sommeil, *Compt. Rend. Soc. Biol.* **158**:2475–2476 (1964).
319. Y. Godin and P. Mandel, Distribution des acides aminés libres dans de système nerveux central du rat au cours du sommeil et de l'état de veille prolongée, *J. Neurochem.* **12**:455–460 (1965).
320. P. Mandel, Y. Godin, J. Mark, and C. Kayser, The distribution of free amino acids in the central nervous system of garden dormice during hibernation, *J. Neurochem.* **13**:533–537 (1966).
321. C. F. Baxter, Regional changes in amino acid composition during anesthesia in the central nervous system of rats, *Proc. West. Pharmacol. Soc.* **9**:52–53 (1966).
322. G. Rindi and U. Ventura, Influence of adrenolectomy of adrenal cortex hormones and of cold on the $\gamma$-aminobutyric acid and glutamic acid content of rat brain, *Ital. J. Biochem.* **10**:135–146 (1961).
323. G. H. Massieu, B. G. Ortega, A. Syrguin, and M. Tuena, Free amino acids in brain and liver of deoxypyridoxine-treated mice subject to insulin shock, *J. Neurochem.* **9**:143–151 (1962).
324. H. C. Agrawal, M. W. Fox, and W. A. Himwich, Neurochemical and behavioral effects of isolation-rearing in the dog, *Life Sci.* **6**:71–78 (1967).
325. N. W. Scholes and L. G. Wheaton, Critical period for detour learning in developing chicks, *Life Sci.* **5**:1859–1865 (1966).

326. E. R. John, K. F. Killam, B. M. Wenzel, and R. B. Tschurgi, *in Inhibition in the Nervous System and Gamma-Aminobutyric Acid* (E. Roberts, C. F. Baxter, A. Van Harreveld, C. A. G. Wiersman, W. R. Adey, and K. F. Killam, eds.), pp. 554–561, Pergamon Press, New York (1960).
327. J. T. Rick, A. K. Huggins, and G. A. Kerkut, The comparative production of γ-aminobutyric acid in the Maudsley reactive and non-reactive strains of rat, *Comp. Biochem. Physiol.* **20**: 1009–1012 (1967).
328. N. W. Scholes and E. Roberts, Pharmacological studies of the optic system of the chick: Effect of γ-aminobutyric acid and pentobarbital, *Biochem. Pharmacol.* **13**: 1319–1329 (1964).
329. S. Z. Kramer, P. A. Sharman, and J. Seifter, Effect of gamma-aminobutyric acid (GABA) and sodium-L-glutamate (Glutamate) on the visual system and EEG of chicks, *Intern. J. Neuropharmacol.* **6**:463–472 (1964).
330. J. D. Wood, W. J. Watson, and N. E. Stacey, A comparative study of hyperbaric oxygen-induced and drug-induced convulsions with particular reference to γ-aminobutyric acid metabolism, *J. Neurochem.* **13**:361–370 (1966).
331. D. P. Purpura and M. W. Carmichael, Characteristics of blood–brain barrier to gamma-aminobutyric acid in neonatal cat, *Science* **131**:410–412 (1960).
332. N. W. Scholes, Effect of parenterally administered gamma-aminobutyric acid on the general behavior of the young chick, *Life Sci.* **4**:1945–1949 (1965).
333. K. T. Brown and T. N. Wiesel, Localization of origins of electroretinograms components by intraretinal recording in the intact cat eye, *J. Physiol.* **158**:257–280 (1961).
334. K. A. C. Elliott and F. Hobbiger, Gamma-aminobutyric acid: Circulatory and respiratory effects in different species: Reinvestigation of the anti-strychorine effect in mice, *J. Physiol.* **146**:70–84 (1959).
335. H. C. Stanton, Mode of action of gamma amino butyric acid on the cardiovascular system, *Arch. Intern. Pharmacodyn.* **143**:195–204 (1963).
336. H. Takahashi, C. Koshino, and O. Ikeda, Relationship between the hypotensive activity and chemical structure of γ-aminobutyric acid in the rabbit, *Japan J. Physiol.* **12**:97–105 (1962).
337. H. Takahashi, M. Tiba, T. Yamazaki, and F. Noguchi, On the site of action of γ-aminobutyric acid on blood pressure, *Japan. J. Physiol.* **8**:378–390 (1958).
338. H. C. Stanton and F. H. Woodhouse, The effect of *gamma* amino-η-butyric acid and some related compounds on the cardiovascular system of anesthetized dogs, *J. Pharmacol. Exptl. Therap.* **128**:233–242 (1960).
339. G. Baldrighi and L. Tronconi, Ricerche sull'effectto antii pertensivo indotto dall' associazione mebutamato-GABA (acido gamma-amino-butirrico) nell'ipertensione essenziale, *Clin. Terap.* **37**:207–234 (1966).
340. T. Hayashi, γ-Aminobutyric acid and its derivatives in mental health, *Trans. Symp. Carl Neuberg Soc. Intern. Sci. Relat. Progr. Biochem. Ther.* **3**: 160–170 (1966).
341. S. A. Mirzoyan and V. P. Akopyan, The effect produced by gamma-aminobutyric acid on the cerebral circulation and oxygen tension in the brain, *Farmak. Toksikol.* **30**:572–574 (1967).
342. A. B. Drakontides, Effect of gamma-aminobutyric acid on pulmonary stretch receptors in the cat, *Am. J. Physiol.* **199**:749–752 (1960).
343. G. A. Kerkut and R. J. Walker, The effect of iontophoretic injection of L-glutamic acid and amino-η-butyric acid on the miniature end-plate potentials and contractures of the coxal muscle of the cockroach *Periplaneta americana* L, *Comp. Biochem. Physiol.* **20**:999–1003 (1967).
344. R. Tapia, H. Pasantes, B. G. Ortega, and G. H. Massieu, Effects *in vivo* and *in vitro*

of L-glutamic acid-γ-hydrazide on metabolism of some free amino acids in brain and liver, *Biochem. Pharmacol.* **15**:1831–1845 (1966).
345. N. Shibata, S. M. Shimizu, M. Kubo, H. Takahashi, Y. Yamaguchi, T. Ezoe, and Y. Tsukada, *in Inhibition in the Nervous System and Gamma-Aminobutyric Acid* (E. Roberts, C. F. Baxter, A. Van Harreveld, C. A. G. Wiersman, W. R. Adey, and K. F. Killam, eds.), pp. 579–581, Pergamon Press, New York (1960).
346. J. Feitelevich, γ-Aminobutyric acid-pyridoxine in the treatment of states of irritability and connected nervous disorders, *Semana Med.* (*Buenos Aires*) **120**:685–699 (1962).
347. C. E. Giurgea, F. E. Moeyersoons, and A. C. Evraerd, A GABA related hypothesis on the mechanism of action of the antimotion sickness drugs, *Arch. Intern. Pharmacodyn.* **166**:238–251 (1967).
348. R. A. Ferrari and A. Arnold, The effect of central nervous system agents on rat brain γ-aminobutyric acid levels, *Biochim. Biophys. Acta* **52**:361–367 (1961).
349. E. W. Maynert and H. K. Kaji, On relationship of brain γ-aminobutyric acid to convulsions, *J. Pharmacol. Exptl. Therap.* **137**:114–121 (1962).
350. K. Kuriyama, E. Roberts, and M. K. Rubinstein, Elevation of γ-aminobutyric acid in brain with amino-oxyacetic acid and susceptibility to convulsive seizures in mice: A quantitative reevaluation, *Biochem. Pharmacol.* **15**:221–226 (1966).
351. P. Scotto, P. Monaco, V. Scardi, and V. Bonavita, Neurochemical studies with L-cycloserine, a central depressant agent, *J. Neurochem.* **10**:831–839 (1963).
352. F. Vivanco, F. Ramos, and C. Jimenez-Diaz, Determination of γ-aminobutyric acid and other free amino acids in whole brains, of rats poisoned with β,β′-iminodipropionitrile and α,γ-diaminobutyric acid with or without administration of thyroxine, *J. Neurochem.* **13**:1461–1467 (1966).
353. N. Van Gelder, Hydrazinoproprionic acid: A new inhibitor of aminobutyrate transaminase and glutamate decarboxylase, *J. Neurochem.* **15**:747–757 (1968).
354. F. D. Marshall, Jr., and W. C. Yockey, The effect of various agents on the levels of homocarnosine in rat brain, *Biochem. Pharmacol.* **17**:640–642 (1968).
355. T. Uchida and R. D. O'Brien, The effects of hydrazines on rat brain 5-hydroxytryptamine, norepinephrine and gamma-aminobutyric acid *Biochem. Pharmacol.* **13**:725–730 (1964).
356. R. Tapia, H. Pasantes, M. Perez de la Mora, B. G. Ortega, and G. H. Massieu, Free amino acids and glutamate decarboxylase activity in brain of mice during drug-induced convulsions, *Biochem. Pharmacol.* **16**:483–496 (1967).
357. P. Oehme and W. Kalusa, Zur Veränderung des Gamma-Aminobuttersäure-Spiegels durch Monaminoxidasehemmer, *Acta Biol. Med. Ger.* **17**:K37–K38 (1966).
358. N. Popov, W. Pohle, and H. Matthies, Einfluss von Phenelzin und Aminooxyessigsäure auf die γ-Aminobuttersäure, Glutaminsäure-Decarboxylase und γ-Aminobuttersäure-α-Ketoglutersäure Transaminase in verschiedenen Regionen des Rattenhirns, *Acta Biol. Med. Ger.* **20**:509–516 (1968).
359. E. Mussini and F. Marcucci, *in Amino Acid Pools* (J. T. Holden, ed.), pp. 486–492, Elsevier, Amsterdam (1962).
360. C. F. Baxter and E. Roberts, Demonstration of thiosemicarbazide induced convulsions in rats with elevated brain levels of γ-aminobutyric acid, *Proc. Soc. Exptl. Biol. Med.* **104**:426–427 (1960).
361. H. Balzer, P. Holtz, and D. Palm, Untersuchungen über die biochemische Grundlagen der konvulsiven Wirking von Hydraziden, *Arch. Exptl. Pathol. Pharmakol.* **239**:520–552 (1960).
362. T. Hado, Effect of isonicotinic acid hydrazide on the decarboxylation of glutamic acid in brain, *Nagoya Med. J.* **5**:45–55 (1959).

363. Y. Nishizawa, T. Kodama, and S. Sugahara, Brain γ-aminobutyric acid during and on suppression of the running fit, *J. Vitaminol.* **6**:236–239 (1960).
364. N. Popov, W. Pohle, V. Rösler, and H. Matthies, Wirkung von Phenelzin auf den Gehalt am γ-Aminobuttersäure in 11 Regionen des Rattenhirns, *Acta Biol. Med. Ger.* **20**:365–370 (1968).
365. G. Rindi and G. Ferrari, The γ-aminobutyric acid and glutamic acid content of brains of rats treated with toxopyrimidine, *Nature* **183**:608–609 (1959).
366. H. Balzer, P. Holtz, and D. Palm, Reserpin und γ-aminobuttersuäre-Gehalt des Gehirns, *Experientia* **17**:38–39 (1961).
367. S. I. Singh and C. L. Malhotra, Amino acid content of monkey brain. III. Effect of reserpine on some amino acids of certain regions of monkey brain, *J. Neurochem.* **11**:865–872 (1964).
368. I. A. Sytinsky and N. T. Thinh, The distribution of γ-aminobutyric acid in the monkey brain during picritoxin-induced seizures, *J. Neurochem.* **11**:551–556 (1964).
369. S. Saito and Y. Tokunaga, Some correlations between picritoxin-induced seizures and γ-aminobutyric acid in animal brain, *J. Pharmacol. Exptl. Therap.* **157**:546–554 (1967).
370. T. Turský and V. Šajter, The influence of potassium on the γ-aminobutyric acid level in rat brain, *J. Neurochem.* **9**:519–523 (1962).
371. V. Bonavita, R. Guarneri, and P. Monaco, Neurophysiological and neurochemical studies with the isonicotinoly-hydrazone of pyridoxal-5-phosphate, *J. Neurochem.* **11**:787–792 (1964).
372. H. Matthies and N. Popov, Die Beeinflussung der Wirkung von Phenelzin, Phenylpropylhydrazine und Aminooxyessigsäure des Rattenhirns, durch Monamine Oxydase-Hemmstoffe, *Acta Biol. Med. Ger.* **20**:371–378 (1968).
373. H. G. Knauff and F. Bock, Über die freien Gehirnaminosäuren und das Athanolamine der normalen Ratte sowie über das Verhalten dieser Stoffe nach experimenteller Insulinhypoglykamie, *J. Neurochem.* **6**:171–182 (1961).
374. E. S. Higgins, The effect of ethanol on GABA content of rat brain, *Biochem. Pharmacol.* **11**:394–395 (1962).
375. H. M. Häkkinen and E. Kulonen, Amino acid metabolism in various fractions of rat brain homogenates with special reference to the effect of ethanol, *Biochem. J.* **105**:261–269 (1967).
376. H. M. Häkkinen and E. Kulonen, Increase in the γ-aminobutyric acid content of rat brain after ingestion of ethanol, *Nature* **184**:726–727 (1959).
377. E. R. Gordon, The effect of ethanol on the concentration of γ-aminobutyric acid in the rat brain, *Can. J. Physiol. Pharmacol.* **45**:915–918 (1967).
378. H. Matthies and N. Popov, Die Bedeutung der chemischen Struktur von Monoaminoxydase-Hemmstoffen dur ihre Wirkung auf den γ-Aminobuttersäure-Gehalt des Rattenhirns, *Acta Biol. Med. Ger.* **18**:617–624 (1957).
379. W. G. Clark, C. A. Vivonia, and C. F. Baxter, Accurate freehand injection into the lateral brain ventricle of the conscious mouse, *J. Appl. Physiol.* **25**:319–321 (1968).

# Chapter 10
# GLUTAMATE AND GLUTAMINE

C. J. van den Berg

*University of Groningen*
*Groningen, The Netherlands*

## I. INTRODUCTION

Glutamate and glutamine are present in nervous tissue in large amounts. Both amino acids play a role in a large number of reactions in brain. The incorporation of carbon or nitrogen from a variety of precursors in these amino acids is fast and extensive, indicating a rapid turnover of these compounds. The conversion of glutamate to glutamine is thought to be important for the detoxification of ammonia in the brain.

Glutamate, applied in small amounts at the surface of neurons, induces an increase in the frequency of action potentials. In high amounts a block results. When applied at the surface of the cortex in large amounts, glutamate causes spreading depression. Glutamine is less active or even inactive. Both compounds therefore might have important biochemical and physiological roles.

This chapter surveys the metabolism of these amino acids, with emphasis on the relation between metabolism and function. Valuable reviews are written by Waelsch[1], Weil-Malherbe[2,3], Strecker[4], and Curtis and Watkins.[5] The last work deals exclusively with pharmacological and physiological aspects.

## II. GLUTAMATE AND GLUTAMINE IN NERVOUS STRUCTURES

### A. Concentration

The concentration of glutamate in the brain of mammals is about 10 mM, while the glutamine concentration is about 4 mM.[6] In lower animals—toad,[7] tortoise,[8] and frog[8]—the glutamate concentration is less, but in some fishes[9] glutamate occurs in a concentration of about 18 mM. The glutamine concentration is in most cases 30–50% of the glutamate concentration.

Extensive regional studies have been carried out by Berl[10-12] in the rat and cat. Regional variations were found, but the relative amounts of glutamate and glutamine were about the same in the regions studied. Glutamate occurs in about the same amounts in the cerebellum, hippocampus, mesodiencephalon, and cerebral cortex of the cat; the concentration in the brainstem is much lower.[11,12] Young and Lowry[13] and Graham *et al.*[14] analyzed the glutamate and glutamine concentrations in small regions of Ammon's horn of the rabbit and the spinal cord of the cat. At this microscale level differences between glutamate and glutamine were observed. The difference between white and gray structures for glutamate was larger than for glutamine. These data show that both compounds are not uniformly distributed.

Glutamate and glutamine are present in the cervical sympathetic ganglion, vagal nerve, and sciatic nerve of the rabbit in smaller amounts than in the brain.[15] In the cervical sympathetic ganglion glutamine has the same level as glutamate.[15] Peripheral nerve of the lobster,[16] the sciatic nerve of the bullfrog,[17] ganglia and nerves from Helix aspersa,[18] ganglia from Mytilus edulis, Periplaneta americana, and Carcinus maenas[18] contain both amino acids. Cell bodies from neurons in lobster ganglia with an inhibitory role have a high $\gamma$-aminobutyrate content, while excitatory cells have a low $\gamma$-aminobutyrate content. The glutamate concentration in the two types of cells is about the same.[19] Glutamine occurs in about the same concentration in fibers with large and small amounts of $\gamma$-aminobutyrate; glutamate is somewhat higher in fibers with a low level of $\gamma$-aminobutyrate.[16]

It appears that both amino acids do occur in measurable amounts in all nervous structures investigated. However, variations in the relative amount of both amino acids do exist.

## B. Developmental Studies

During the rapid growth phase of the brain, glutamate increases two- to threefold (Berl and Purpura[20] for cat, Bayer and McMurray[21] for rat, and Dravid and Jilek[22] for dog). Glutamine does not change very much in this period.[20,21] In extensive regional studies in the cat, Berl[12] found an increase in glutamate in all regions studied. Glutamine did not change much in the hippocampus; it increased and thereafter decreased in the mesodiencephalic parts of the brain. The glutamine concentration in the cerebellum was at birth even higher than the glutamate concentration; it rose to a peak at about 3 weeks and then decreased rapidly. In the brainstem glutamine was high in the early neonatal period, remained unchanged for about 3 weeks, and decreased thereafter steadily. The changes found for aspartate resemble those found for glutamate.

## C. Intracellular Localization

The relative amount of glutamate and glutamine in a mitochondrial fraction of brain is about the same as that found for potassium and lactate dehydrogenase.[23] Potassium and lactate dehydrogenase are markers for cytoplasm.[23] Both amino acids therefore are mainly present in the cytoplasm of the cell. One cannot be sure, however, that one or both of these amino acids have not left their normal intracellular site during the preparation of the fractions. The equal incorporation of $^{14}C$ from glucose-U-$^{14}C$ in amino acids from the mitochondrial fraction and from the supernatant fraction[24] does suggest that there is extensive relocation of these amino acids during the preparation of the fractions; otherwise, one has to assume that *in vivo* glutamate leaves the mitochondria very quickly. There is no doubt that most of the glutamate is labeled from mitochondrial α-oxoglutarate.

## III. UPTAKE AND RELEASE FROM THE BRAIN

### A. Brain–Blood

Schwerin *et al.*[25] reported no increase of glutamate in the brain after the injection of large amounts of glutamate. These findings have been verified many times. Only in immature animals has an increase of glutamate in the brain after the injection of glutamate been found.[26] Five minutes after the intravenous injection of labeled glutamate in the rat, labeled glutamate and glutamine were found in the brain.[27] Injection of high amounts of unlabeled glutamine did not affect the $^{14}C$ uptake of the brain very much, indicating that glutamate as such enters the brain. In mice, Roberts *et al.*[28] estimated the contribution of plasma glutamate to the glutamate pool of the brain to be around 10%. O'Neal *et al.*[29] found the same in the rat.

Glutamine–$^{14}C$ penetrates the brain.[27,29] O'Neal *et al.*[29] found that glutamine and glutamate were about equally taken up or exchanged. Arterial-venous differences for glutamate and glutamine were measured in humans by Adams *et al.*[30] and Knauff *et al.*[31] and in rabbits by Rodnight *et al.*[32] Adams *et al.* reported glutamate uptake and glutamine release; Knauff *et al.* found uptake of glutamate and glutamine. Rodnight *et al.* found no arterial–venous differences, but they used paper chromatography, which is not accurate. Sacks[33] found a very low $^{14}CO_2$ production from glutamate-$^{14}C$ by the human brain. The labeled compound was given intravenously and arterial–venous differences of $^{14}CO_2$ were measured.

Berl *et al.*[34] injected glutamate-$^{14}C$ intracisternally into the rat. The specific activity of glutamine in the brain, 1–2 min after the injection was higher than the specific activity of glutamate. The total amount of $^{14}C$ in glutamate and glutamine remains constant up to 5 min after the injection. The specific activity (counts per minute per micromole) of blood glutamine is always lower than the specific activity of brain glutamine. The specific activity of glutamine increases fastest in the brain. The fall in specific activity of blood glutamine is much faster than that of brain glutamine. These results suggest that only a small fraction of the glutamine leaves the brain under normal conditions. The kinetics of the glutamine and glutamate labeling in brain also suggest that glutamine is a precursor of glutamate. Intracerebrally injected glutamine-$^{14}C$ in fact is rapidly converted in the brain to glutamate-$^{14}C$ and other amino acids.(34) Intravenously injected glutamine-$^{14}C$ is also converted in the brain to glutamate-$^{14}C$ and other amino acids.(29)

$^{15}N$ of ammonium salts introduced in the cat appears to a larger extent in the α-amino group of glutamine than in the α-amino group of glutamate and aspartate.(35) The labeling pattern of the blood amino acids differs from that of the brain amino acids. This suggests again that brain glutamine is not a major source of the glutamine in the blood.

### B. Brain–CSF

The concentration of glutamine in human CSF is 0.5 mM; of glutamate, 0.007 mM. The plasma/CSF ratio for glutamine is one; for glutamate, eight.(36) The concentration of most amino acids in the CSF is much lower than in the blood.(36)

The disappearance of glutamate and glutamine from the ventricles of the cat perfused with these amino acids was measured by Levin *et al.*(37,38) The rate of disappearance was in the first hour of perfusion higher than for inulin; later it was the same. There is some uptake by the brain parenchyma, but it is slow and only 3–4 mm deep. The authors conclude that there is some exchange of amino acids between the CSF and the periventricular parenchyma without a net increase of the amino acid concentration. The recovery of label from the perfusate was somewhat lower, but not significantly, than the recovery of the amino acids chemically determined, suggesting some, but only a small, contribution of unlabeled amino acids from the brain or from the blood.

### C. Release from Cortex

Jasper *et al.*(39) found a release to glutamine and glutamate from the cortex. Van Harreveld and Kooyman(40) also observed some release of amino acids from the cortex, but suggest that the blood is the source of the amino acids. This is also suggested by Crowshaw.(41)

## IV. ENZYMES INVOLVED IN THE METABOLISM OF GLUTAMATE AND GLUTAMINE

### A. Properties of Some Individual Enzymes

#### 1. *Transaminating Enzymes*

Glutamate, aspartate, γ-aminobutyrate, alanine, valine, leucine, isoleucine, tryptophan, 5-hydroxytryptophan, phenylalanine, tyrosine, cystine, cysteic acid, methionine, and ornithine are nitrogen donors for α-oxoglutarate in transamination reactions in brain.[42] Glutamate and aspartate are the most active nitrogen donors, while α-oxoglutarate is very likely the most important amino acceptor in brain. Oxaloacetate is more active as amino acceptor from glutamate than pyruvate.[43] Aspartate transaminase,[42] alanine transaminase,[42] and γ-aminobutyrate-α-oxoglutarate transaminase[42] are probably separate enzymes. A number of separate enzymes also exist for the transamination of aromatic amino acids.[44] Both the mitochondrial and supernatant type of aspartate transaminase occur in brain.[45,46]

## 2. *Glutamate Dehydrogenase*

The isolated liver enzyme operates with NAD and NADP as coenzyme.[47] The activity of glutamate dehydrogenase in a variety of organs, brain included, is proportional to the level of NADP and NADP-isocitrate dehydrogenase.[48] This led Klingenberg *et al.*[48] to suggest that *in vivo* glutamate dehydrogenase uses NADP as a coenzyme. The activity of glutamate dehydrogenase is affected by a number of naturally occuring substances.[47] ADP, e.g., activates the enzyme; this ADP activation also occurs with the brain enzyme.[49] The Km for ammonia, 8 mM,[49] is much higher than the normal ammonia concentration, 0.36 mM[9]. The activity of the enzyme thus will increase or decrease steeply with small changes in the ammonia concentration. In the liver, glutamate increases and α-oxoglutarate decreases when ammonia levels are increased.[50] The situation in the brain is less clear (see later).

The NAD/NADH ratio, calculated with the equilibrium constant given by Williamson *et al.*,[50] with the concentration of α-oxoglutarate 0.127 mM[51], ammonia 0.36 mM,[9] and glutamate 10–12 mM, is around 1. This is much lower than the value of 6.2 observed *in vivo*.[49] The NADP/NADPH ratio calculated in this way is the same as that observed *in vivo*.[49] These calculations are valid only when the reaction partners are in equilibrium and homogeneously distributed in the brain. This last assumption is probably not valid.

## 3. *Glutamine Synthetase*

The existence of this enzyme in brain was shown for the first time by Krebs.[52] Krebs[52] and Wu[53] found the enzyme in the brain of a large number of animals. In most animals the enzyme also is present in the liver. It is possible that in some animals the brain is the source of glutamine for protein synthesis or other purposes. This, if true, is a unique situation.

The enzyme from *E. Coli* is inhibited by a number of compounds, in such a way that the inhibition by one compound is independent from and additive with the inhibition by another compound.[54] The *Coli* enzyme contains at least six such independent sites. The situation with the brain enzyme is not clear. The brain enzyme is inhibited by some di- and tri-nucleotides and by glycine and alanine.[55] These same compounds also inhibit the enzyme from *E. Coli.*[54]

For the brain enzyme at *p*H 6.0 the Km for ammonia is 0.39 mM, the Km for ATP is 1.67 mM, and the Km for glutamate is 5.4 mM. The Km for glutamate at *p*H 7.4 is 1.5 mM.[55] Glutamine synthetase is saturated with ammonia at lower ammonia levels than glutamate dehydrogenase. Methionine sulfoximine is an inhibitor of the brain enzyme.[55]

### 4. *Glutaminase*

The first complete description of this enzyme in brain was published by Krebs.[52] He found that it was strongly inhibited by D- and L-glutamate. Phosphate activates the enzyme four to fivefold.[56] The kidney enzyme is activated by citrate, α-oxoglutarate, and some other citric acid cycle intermediates and by amino acids.[58] This activation depends upon the phosphate concentration. The enzyme from brain and kidney has allosteric properties.[57,58] This implies that glutaminase might have a regulatory role *in vivo*.

## B. Intracellular Localization

About 40% of the aspartate transaminase is present in a crude mitochondrial fraction.[59,60] Part of this activity is in the nerve endings, and this is the supernatant type of the enzyme. The particulate type is present in the mitochondria.[46] Most of the alanine transaminase also is in the supernatant.[59,60]

Glutamate dehydrogenase is only present in the crude mitochondrial fraction. In sucrose gradients of this fraction the enzyme is localized in a higher density region than succinate dehydrogenase.[59,61] Glutamate dehydrogenase and γ-aminobutyrate-α-oxoglutarate transaminase have the same localization.[59] Recent experiments[61] have shown that the acetyl-CoA synthetase and glutamate dehydrogenase also have the same localization.

About 25% of the glutamine synthetase is present in the crude mitochondrial fraction; the largest amount of the enzyme is located in the microsomal and soluble fraction.[59,61] A part of the enzyme present in the crude mitochondrial fraction is probably located in membranes of synaptic complexes.[62] Half or more of the glutamate decarboxylase sediments in the crude mitochondrial fraction.[59,60] In sucrose gradients glutamate decarboxylase is located between acetylcholinesterase and succinate dehydrogenase.[59,60] This is generally interpreted as a localization of the enzyme in a special type of nerve endings.[59] More definite proof is needed. Glutaminase has the same distribution as succinate dehydrogenase.[59,61]

In summary it can be said that glutaminase and the mitochondrial type of aspartate transaminase have exactly the same intracellular distribution, which may be different from the distribution of other enzymes involved in glutamate metabolism.

## C. Biochemical Topography

Lowry *et al.*[63] measured aspartate transaminase and glutamate dehydrogenase in dorsal root ganglion cells and their capsules. The transaminase to dehydrogenase ratio was 10–14 for the ganglion cells and about 2 for the capsules cells. The distribution of glutamate dehydrogenase and fumarate hydratase in layers of the visual and motor cortex, are not the same.[64] In tracts of the white matter McDougal *et al.*[65] found a correlation between the distribution of aspartate transaminase and some mitochondrial enzymes, such as fumarate hydratase and malate dehydrogenase. The changes in activity of glutamate dehydrogenase in layers of the cortex during development are not the same as found for malate dehydrogenase.[66] The distribution of aspartate transaminase in layers of Ammon's horn is about the same as the distribution of malate dehydrogenase,[67] but it is different from the distribution of glutamate dehydrogenase.[13]

It is clear from this discussion that aspartate transaminase has the same distribution as most mitochondrial enzymes but that glutamate dehydrogenase differs from this pattern. This is consistent with the differences observed in the intracellular localization described earlier. Glutamate decarboxylase and $\gamma$-aminobutyrate transaminase both occur mostly in gray matter, but their distribution is not the same.[68,69]

Glutamine synthetase in the cat is high in the cortex and lower in other structures. Pons, medulla, and corpus callosum contain about half the activity of the cortex.[70] This enzyme has not been measured at the microchemical level. Glutaminase is about equally active in white and gray structures.[71] It is evident that considerable differences exist in the anatomical distribution of the enzymes involved in the metabolism of glutamate and glutamine.

## D. Developmental Changes

Mitochondrial and cytoplasmic aspartate transaminase,[45,72] glutamate dehydrogenase,[66] glutamine synthetase,[73] glutamate decarboxylase,[74] and $\gamma$-aminobutyrate transaminase[74] increase in activity during the development of the brain of the rat in about the same fashion. The activity of these enzymes in the brain of adult animals is about four times higher than in the brain of newborn animals. The rate of increase of glutamate decarboxylase is not exactly the same as the rate of increase of $\gamma$-aminobutyrate transaminase.[74] The glutamate decarboxylase to $\gamma$-aminobutyrate transaminase ratio changes during development, especially in the rabbit and chicken.[74] These small differences might be important.

## E. Activity of These Enzymes in Brain

The activity of aspartate transaminase is high in brain, about 8000 $\mu$mol/g wet wt./hr,[75] which is about 60–80% of the activity found in the liver.[43,76] Glutamate dehydrogenase in the brain of the rat has an activity of about 900 μmol/g wet wt./hr; this is 14% of the activity found in the liver.[50] The activity of glutamine synthetase in the cat brain is about 200 μmol/g wet wt./hr.[70] In most species the brain contains the highest amount of this enzyme.[53] Tursky and Valovičová[71] found in the brain of guinea pig a glutaminase activity of about 100 μmol/g wet wt./hr. The activity of glutamate decarboxylase in the rat brain is about 50 μmol/g wet wt./hr, while the γ-aminobutyrate transaminase activity is about 150 μmol/g wet wt./hr.[60]

These data are of course only approximate, since optimal conditions were not always used, but they give some indication of the maximal capacities of the various pathways.

## V. METABOLIC ASPECTS *IN VITRO*

### A. Isolated Preparations

The rate of glutamate disappearance in isolated mitochondria is almost identical to the rate of aspartate formation.[77,78] There is almost no formation of ammonia, even in the presence of malonate.[78] In liver the addition of malonate increases the formation of ammonia.[79] The backward reaction of the tricarboxylic acid cycle does not operate in isolated mitochondria.[77] In these isolated preparations there is a rapid exchange transamination between glutamate and α-oxoglutarate.[76] These results suggest that glutamate oxidation by isolated brain mitochondria proceeds almost completely by the transamination pathway. The relatively low glutamate dehydrogenase activity could explain the fact that even when the aspartate transaminase is inhibited, glutamate oxidation does not proceed by the dehydrogenase pathway. Another, more likely, explanation is that the two enzymes do not share the same glutamate and α-oxoglutarate pools. Aspartate transaminase and glutamate dehydrogenase do not have the same intracellular distribution. A third possiblity is that glutamate dehydrogenase *in situ* is specific for NADP and that NADPH is not effectively oxidized by brain mitochondria.

Lofrumento *et al.*[78] and Sactor and Packer[80] found glutamine oxidation by isolated mitochondria. Ammonia was formed from the glutamine.[78]

### B. Metabolism of Glutamate and Glutamine in Slices

Glutamate, and less so aspartate, increase the oxygen consumption of brain slices in the presence and absence of glucose.[81-83] Other amino acids are inactive. Brain slices produce ammonia in the absence ofglucose.[84] The addition of glucose[84] or glutamate[52] prevents this increase. Krebs[52] found that glutamine is formed. The sum of amide-N (glutamine) plus ammonia remains approximately constant. 2 mM D-glutamate completely blocks the ammonia formation from glutamine added to the medium. It does not affect the endogenous formation of ammonia.[85] 10 mM D-glutamate decreases the ammonia accumulation in slices incubated in glucose-free media.[86] L-glutamate reduces the ammonia accumulation somewhat more. There is a large increase in glutamine. This glutamine increase does not occur in the slices incubated with D-glutamate. D-glutamate inhibits glutaminase. It is very likely that high amounts of D-glutamate reduce the hydrolysis of glutamine.

Labeled glutamate added to slices is converted to labeled asparate, glutamine, and other products;[87] labeled aspartate is converted to glutamate;[87] and labeled glucose is converted to glutamate, aspartate, and glutamine.[88] $^{15}N$ from ammonium nitrate added to slices is incorporated into glutamate, aspartate, the amino and amide group of glutamine, and some other amino acids.[89] $\gamma$-Aminobutyrate also was found to be labeled in all these experiments. Reichelt *et al.*[90] found incorporation of $^{14}C$ from glutamine-U-$^{14}C$ in glutamate, aspartate, and other compounds in brain slices. The specific activity of glutamate and glutamine, after an incubation period of 1 hr, were almost the same. The incorporation of $^{14}C$ from glutamine into glutamate was decreased by the addition of glutamate. These results strongly suggest that glutaminase is active under the conditions of these experiments.

Simon *et al.*[91] incubated brain slices with aspartate labeled in different positions and studied the incorporation of $^{14}C$ into the five carbon atoms of glutamate. About 10% of the label appeared in the glutamate molecule in unusual positions, suggesting the presence of some, still unknown, metabolic pathways.

The incorporation of $^{14}C$ from glutamate into glutamine is very much dependent on the condition of the slice.[92] These experiments show that it is very difficult at the present time to give an exact interpretation of the results obtained with slices. Therefore this chapter will not review all the work done on the effects of inhibition on the metabolism of glutamate and glutamine in slices. From the literature it seems that the conversion of glutamate to glutamine, or better the incorporation of $^{14}C$ from glutamate or other precursors into glutamine, is very sensitive to many agents.

## VI. METABOLIC ASPECTS *IN VIVO*

### A. Intact Animals

About 70% of the $^{14}C$ present in the acid-soluble fraction of rat brain 20 min after the injection of glucose-U-$^{14}C$ is in the amino acid fraction.[93] This is much higher than in other tissues. Not only is the total amount of $^{14}C$ in the amino acid fraction of brain higher than in the liver, but even the specific activities of glutamate, glutamine, and aspartate are much higher in the brain than in the liver.[24]

Glucose-1-$^{14}C$ is also incorporated into brain glutamate to a larger extent than into glutamate in other tissues of the rat.[94] $^{14}C$ from glucose-1-$^{14}C$ accumulates more in glutamate, aspartate, and glutamine in the brains of mice than from glucose-2-$^{14}C$.[95] This difference arises as a result of the turning of the tricarboxylic acid cycle. This fact, in combination with the high specific activity observed for these amino acids, strongly suggests that a large proportion of glutamate and aspartate is more or less in equilibrium with tricarboxylic acid cycle intermediates.[96]

There are many precursors of glutamate. These precursors can be divided into two groups: One group leads to a specific activity of glutamate higher than the specific activity of glutamine; the other group leads to a specific activity of glutamate lower than the specific activity of glutamine. Glucose-U-$^{14}C$,[24] glucose-2-$^{14}C$,[29] and lactate,[29] are all better precursors for glutamate than for glutamine. A specific activity of glutamine higher than the specific activity of glutamate is observed after bicarbonate-$^{14}C$,[97] citrate-2,5-$^{14}C$,[29] butyrate-1- and -3-$^{14}C$,[29] acetate-1-$^{14}C$,[29,95] acetate-2-$^{14}C$,[95] glutamate-U-$^{14}C$,[34,100] aspartate-U-$^{14}C$,[11] $\gamma$-aminobutyrate-$^{14}C$,[98] and leucine-U- $^{14}C$.[99] These results can be interpreted in terms of compartmentation of the metabolism of glutamate: There are at least two glutamate pools differing with respect to precursors and subsequent fate. This subject is covered in detail by Berl and Clarke in this Handbook ("Compartmentation of Amino Acid Metabolism," Chapter 19, Volume II, pp. 447–472).

It is important to note that the large pool of glutamate is probably in close contact with $\alpha$-oxoglutarate from the tricarboxylic acid cycle, while the small pool is synthesized from $\alpha$-oxoglutarate and subsequently converted to glutamine. $^{15}N$ from ammonium salt is preferentially incorporated into the small pool.[35] This suggests that glutamate dehydrogenase has a function in the synthesis of the small pool, while asparate transaminase has a function in the labeling (exchange transamination) of the large glutamate pool. Acetate-$^{14}C$, $\gamma$-aminobutyrate-$^{14}C$, and $^{15}N$ from ammonia are specific precursors of the small glutamate pool. Acetyl-CoA synthetase, $\gamma$-aminobutyrate transaminase, and glutamate dehydrogenase have the same intracellular localization (see earlier). The time course of labeling of glutamate and aspartate after the administration of acetate suggests that glutamine is in these experiments the precursor of glutamate and aspartate.[95,96]

One might suggest that carbon from a variety of sources ends up in glutamine and that glutamine transports this carbon to the tricarboxylic acid cycle or out of the brain. In the latter case two amino groups are removed at the same time.

## B. Perfused Brain

$^{14}C$ from glucose-U-$^{14}C$ is incorporated into the free amino acids in the perfused cat brain.[101] The specific activities of glutamate and aspartate are almost equal. $^{14}C$ from bicarbonate-$^{14}C$ is also incorporated in this system.[102] The specific activity of glutamine is never higher than the specific activity of glutamate, contrary to the observations made *in vivo*.

It seems that the perfused cat brain is not entirely normal. The glutamate–glutamine system, or the small glutamate pool, is very sensitive to still ill-defined influences. A similar phenomenon was observed by Berl *et al.*,[92] *in vitro*, as described earlier.

## VII. PHARMACOLOGICAL ASPECTS

A great many reports are published on the effect of pharmacological agents on the concentrations of amino acids in brain. Most of these reports are contradictory, and will not be reviewed here in detail but only in regard to a few aspects.

### A. Methionine Sulfoximine

Methionine sulfoximine induces convulsive seizures, episodic behavioral change, and other symptoms in a number of animals.[103] The substance inhibits glutamine synthetase *in vitro*[55,104] and *in vivo*.[105] *In vitro* it inhibits the enzyme in a competitive way with respect to glutamate, but incubation of the enzyme in the presence of Mg and ATP results in inactivation of the enzyme.[106] *In vivo* the interaction of methionine sulfoximine with the enzyme leads to a irreversible inactivation. $^{14}C$-methionine sulphoximine injected into an animal is to some extent concentrated in the nerve-ending fraction[62] and the microsomal fraction,[105] the sites where glutamine synthetase is present.

Animals killed during convulsions have a decreased level of glutamine in the brain.[107,108] In dogs[107] there is an increase in the concentration of ammonia, but this increase is much smaller than the decrease of glutamine. Glutamate is somewhat decreased, and aspartate is about 50% reduced. There is a substantial (34%) decrease in the total amount of amino and amide nitrogen in the brain.[107]

The simultanous injection of methionine and methionine sulfoximine prevents the onset of convulsions; nevertheless, the glutamine level in these animals is decreased to about the same level as in animals not treated with methionine.[108] The level of glutamine also was decreased by subconvulsive doses of methionine sulphoximine. With convulsive doses the glutamine level in the brain was still decreased hours after the end of the convulsions.[108] Glutamate applied in fairly high amounts to the cortex of a 15-day-old chick embryo causes siezures. The same amount of glutamate causes seizures in a 20-day-old embryo only when the animal is pretreated with methionine sulphoximine.[109] Glutamate applied to the cortex of the rabbit and cat mixes with the small glutamate pool and is rapidly converted to glutamine.[11,110] The potentiation of the glutamate effect by methionine sulfoximine, just described, and the rapid conversion of glutamate to the inactive glutamine suggest that glutamine synthetase inactivates glutamate.

Folbergova's[108] results are not compatible with this hypothesis, unless one assumes that there are two glutamine pools. A dual origin of glutamine is possible and would make the interpretation of the fast incorporation of $^{14}C$ from glucose into glutamine easier, as already indicated.

## B. Ammonium Salts

The level of glutamine is increased in dogs[111,112] and cats[35] treated with ammonium salts. There is a small or no decrease of glutamate. Aspartate decreases about 50%. Pyruvate is about doubled, while the α-oxoglutarate level is reduced to about 50%. The level of ATP is not changed.[115]

Arginine affords protection against the lethal effect of ammonium salts. In protected animals there is still an increase in the level of glutamine, but the ammonia level does not increase as much as in the unprotected animals.[116] The toxicity of ammonium salts is decreased by the previous injection of methionine sulphoximine.[117] The increase in glutamine is less in these pretreated animals than in the untreated controls.[117] The ammonia level in the methionine sulphoximine-treated animals is as high as in the animals treated with a $LD_{50}$ doses of ammonium salts, but the methionine sulfoximine-treated animals survive the treatment. The level of ammonia in the brain, therefore, is not directly related to the toxic effects of ammonia salts.

Bessman[114,118] tried to explain the pathology of hepatic coma by assuming that the increased uptake of ammonia by the brain led to a overloading of the capacity of the glutamate-to-glutamine conversion system. This, then would lead to a decreased concentration of tricarboxylic acid cycle intermediates and a decrease in the synthesis of ATP in the brain. Fazekas *et al.*[119] were unable to confirm the elevated uptake of ammonia by the brain in patients with hepatic coma.[114] α-Oxoglutarate is not decreased in the brain of rat after brief ammonia intoxication.[113] Decreases in α-oxoglutarate were observed only after perfusion with ammonia[111] or after multiple injections of ammonia salts.[114]

The brain has a capacity to fix carbon dioxide. Some evidence for an increase in the synthesis of tricarboxylic acid cycle intermediates has been obtained.[97] However, no changes in the labeling pattern of glutamate after the injection of pyruvate-2-$^{14}C$ or glucose-2-$^{14}C$ consistent with an increased synthesis of tricarboxylic acid cycle intermediates was found by McMillan and Mortensen[120] and O'Neal and Koeppe.[29] Carbon dioxide fixation at the pyruvate level results in an incorporation of $^{14}C$ from these precursors in the two and three position of glutamate. The labeling pattern of glutamine has not been determined.

The total amount of $^{14}C$ from glucose-U-$^{14}C$ incorporated into glutamine in the brain of hepatectomized rats is only slightly lower than in the controls, while the incorporation of $^{14}C$ in glutamate and aspartate is 30–60% reduced.[121] The specific activity of glutamine in the hepatectomized animals is about half that of the controls. These data suggest a small increase in the rate of glutamine formation, but also that glutamine accumulates in the brain. No differences in the specific activities of glutamate and glutamine after the injection of glucose-2-$^{14}C$ were observed in rats treated with a lethal dose of ammonium acetate.[29] The increase of glutamine in these experiments, however, was small. It is possible that the tricarboxylic acid cycle in brain is not directly affected by ammonia, but the small changes in the glutamate pool or changes in the rate of synthesis of this pool or changes in the rate conversion of this pool to glutamine are related to the functional changes in brain induced by ammonia salts or hepatic coma. The protection afforded by methionine sulphoximine against the toxicity of ammonia salts strongly suggests one of these last possibilities. The small glutamate pool is involved in a complex metabolic network of so-called minor pathways, but it is possible that small changes in this network might influence a number of biochemical or physiological processes.

## C. Insulin

The concentration of glucose and tricarboxylic acid cycle intermediates in the brain of the mouse are lowered tremendously during insulin coma.[51] There is a gradual decrease of glutamate and glutamine, while aspartate increases.[122] According to Dawson,[123] glutamine does not decrease, the glutamate level decreases 30–40%, and the ammonia level remains as in the control animals. The sum of glutamate and aspartate remains approximately constant. There seems to be a shift from glutamate to aspartate and almost no deamination of glutamate. The difference observed between the intracellular localization of the aspartate transminase and glutamate dehydrogenase (see earlier) explains this situation very well; there is little or no glutamate dehydrogenase in the bulk of the mitochondria from brain.

The decrease of α-oxoglutarate is about 40%, while the oxaloacetate levels are not changed.[51] The relative decrease of α-oxoglutarate and increase of aspartate suggest that glutamate, aspartate, α-oxoglutarate, and oxaloacetate are kept in equilibrium by the aspartate transaminase.[51] It was suggested earlier in this chapter that a large fraction of the glutamate and aspartate pools are in rapid equilibrium with each other.

## D. Convulsive Agents

The existence of a relationship between the metabolism of amino acids and the occurrence of convulsions has been suggested many times (see Tews and Stone[124] for a review). Picrotoxin has no effect on the level of glutamate and glutamine; pentylenetetrazol causes a very slight or no decrease of glutamate[112,125]; thiosemicarbazide has no effect on the glutamate and glutamine levels in the dog,[126] and during acute Telodrin intoxication, glutamate is decreased and glutamine is increased.[127]

If one assumes that convulsions are the result of energy changes, the relatively small changes in glutamate indicate that glutamate is not a reservoir of energy in the brain; otherwise one would expect a much larger decrease. The brain apparently is able to utilize only a small amount of the energy available in glutamate, but oxidation does not proceed further than aspartate. During fluoroacetate convulsion there is a decrease of aspartate, a small decrease of glutamate, and no change in the glutamine levels in the dog. Ammonia increases severalfold. In mice a small increase in glutamine was observed in addition to the other changes,[(124)] In the rat Dawson[(128)] found a decrease of glutamate and no change of glutamine.

During convulsions caused by 1,1-dimethylhydrazine in the rat there is a large reduction of the incorporation of $^{14}C$ from glucose-2-$^{14}C$ into brain amino acids. The amount of $^{14}C$ in glutamine was reduced slightly more than that in glutamate.[(129)] Pentylenetetrazol has no striking effect on the specific activities of glutamate and aspartate in the cat brain perfused with glucose-U-$^{14}C$.[(101)]

Sacktor *et al.*[(130)] studied the changes of a number of constituents in brain in the first minutes after the injection of indoklon, a powerful convulsant. There is a rapid increase in the glycolysis rate. The concentrations of glutamate and aspartate do not change at all. The levels of oxaloacetate and α-ketoglutarate were not determined in this study. A small decrease in proline was observed. Carbon from proline is incorporated in glutamate in the brain.[(130a)] It has been suggested that during convulsions "cold" carbon from endogenous sources is converted to carbon dioxide.[(101)] No definite pattern arises from these findings. The brain does not react biochemically in the same manner to all convulsive agents.

### E. Anesthetic Agents

Dawson[(131)] reported a reduction of the glutamate level and an increase of the aspartate level in the brain of rats treated with thiopentone. The level of glutamine was not changed. The decrease in glutamate was much less when the body temperature was maintained.[(131)] The specific activity of glutamate and glutamine 20 min after the injection of glucose-U-$^{14}C$ is much lower in pentobarbitone-treated rats than in controls.[(132)] This author observed a similar effect in mice treated with thiopental (unpublished).

## VIII. EFFECTS OF GLUTAMATE AND GLUTAMINE *IN VITRO* AND *IN VIVO*

### A. Effects *in Vitro*

#### 1. *Biochemical Effects*

Glutamate is accumulated against a concentration gradient by brain slices.[(133)] Potassium is transported at the same time.[(134,135)] Considerable swelling occurs when the slices are incubated with glutamate.[133,135] When corrections were made for this swelling it was found that the intracellular concentration of potassium was not changed.[135] The slices were incubated in these experiments 15 min or longer.

The swelling caused by glutamate is much larger than the swelling caused by a number of other amino acids or glycolytic and tricarboxylic cycle intermediates.[136] Although glutamate is oxidized by slices, it does not support its own accumulation.[133] There is a rapid decrease of phosphocreatine and acid labile phosphates (mainly ATP) in slices treated with glutamate.[83,137] The incorporation of radioactive phosphate is also much reduced.[138] Glutamine does not increase in the first minute after the addition of glutamate, but the rate of decrease of P-creatine is highest in this time interval.[83] In the second minute glutamine starts to increase. One minute after the addition of glutamate (5 mM) there is already a measurable increase of sodium in the slice[137]; the increase thereafter is much slower. The first effect of glutamate on the level of ions in the retina is an increase of sodium and a decrease of potassium; potassium starts to increase later.[139] In a subsequent study,[140] the same author, Ames, found that glutamate increased the efflux of potassium and the influx of sodium, chloride, and calcium. His interpretation is that glutamate increases the permeability of cell membranes in the retina. He suggests also a specific reaction between calcium, glutamate, and some component of the membrane.

Orrego and Lippman[141] recently found a strong inhibition of protein synthesis in cortex slices after the addition of glutamate and a number of other acidic amino acids.

It is clear from this discussion that a large number of reactions in the slices are affected by the addition of glutamate. It should be noted that some of the biochemical changes are extremely rapid.

## 2. *Physiological Effects*

Depolarization of membrane potentials was found after the addition of L-glutamate to slices by Hillman and McIlwain.[142] No changes in membrane potentials were observed after the addition of D-glutamate. Gibson and McIlwain[143] made a more detailed study. They observed membrane potentials of about −50 to −60 mV, with a duration of in general not more than 4 min. This membrane potential dropped 20–30 mV upon the addition, with micropipettes, of a small amount of glutamate. The change in membrane potential with time varied in the same way as the concentration of glutamate, calculated with a simple diffusion equation. The concentration of glutamate needed to give a just observable depolarization of 3 mV was calculated to be 0.2 mM. This figure compares closely with the threshold value for glutamate for stimulation of cortical neurons *in vivo*—0.14 mM.[144] The normal membrane potential returned after a few minutes, and washout with medium increased this return to normal values. Bradford and McIlwain[137] compared the effects of a number of amino acids, known to excite neurons *in vivo*, on membrane potentials *in vitro* and some biochemical parameters. Aspartate, cysteine, glutamate, DL-homocysteine and $\alpha$-aminoadipate caused membrane depolarizations. All these amino acids decreased the amount of phosphocreatine and increased the amount of inorganic phosphate. They also increased the sodium concentration of the slice.

Application of the Nernst equation to the data obtained led to the conclusion that glutamate increased fivefold the tissue's permeability to sodium relative to that of potassium. The sodium entering the tissue as a result could accelerate its Na-K dependent adenosine triphosphatase activity, thus explaining the loss of energy-rich phosphates.[137]

Glutamine is inactive physiologically.[137] *In vivo* the conversion to glutamine of glutamate applied to the cortex is extremely rapid, as has been noted already. *In vitro* there is no increase in tissue glutamine in the first minute after the addition of glutamate.[83] The fact that the change in membrane potential with time caused by the addition of glutamate can account for the time course does not suggest that the action of glutamate is terminated by an inactivation to glutamine. Addition of ammonia to the slice has no effect on the membrane depolarization caused by glutamate.[142]

A striking observation in these experiments is that, although the concentration of glutamate in the brain is around 10 mM, the addition of glutamate to a tissue concentration of 0.1–0.2 mM causes a rapid depolarization. Glutamate must be kept well within the cells.

## B. Effects *in Vivo*

Glutamate applied to neurons excites them, and applied to the cortex it causes spreading depression.[5] The threshold concentration for the excitation of neurons is about 0.1 mM, as already stated. The increase in the frequency of firing of neurons is dependent upon the dose; in high concentration it may produce a complete depolarization block.[145] A very large percentage of the neurons investigated reacts to glutamate.[145] Pretreatment of rat with thiosemicarbazide—an inhibitor of glutamate decarboxylase—enhances the effect of glutamate. This enhancement is abolished by the concomitant microelectrophoresis of pyridoxal-5-phosphate.[145] These results could be explained by assuming that the active glutamate is inactivated by decarboxylation, but one can also assume that the firing frequency of a neuron is dependent upon a balance between glutamate and $\gamma$-aminobutyrate, this latter amino acid exerting a inhibitory action.

D-glutamate is more potent as an inducer of spreading depression in the cerebral cortex of the rabbit than L-glutamate.[146] The initial changes in electrolyte concentration in retinas incubated with D-glutamate were similar to those with L-glutamate. Significant differences between the two optical isomers were found with incubation times of 10 min and longer.[139]

D-glutamate does not change membrane potentials *in vitro*.[142] When iontophoretically applied to cortical neurons, D-glutamate causes only a slight excitation of the cells.[144]

## IX. MISCELLANEOUS REMARKS

Activation of glutamate and glutamine to amino acyl RNA compounds should occur in brain. Acetylation of glutamate might occur, since N-acetyl glutamate has been isolated from brain.[90] A large number of glutamate-containing peptides also have been isolated from brain. $^{14}C$-glutamate is incorporated in some of them.[90] Glutamine is a source of nitrogen for purines, pyrimidines, NAD, and glucosamine phosphate. Synthesis of glucosamine phosphate does occur in brain.[147] Some of these reactions might be more important than generally assumed.

## X. CONCLUDING REMARKS

A large fraction of the total amount of glutamate and aspartate present in the brain is in close contact with tricarboxylic acid cycle intermediates. The main considerations are the rapid incorporation of $^{14}C$ from glucose-$^{14}C$, the time course of $^{14}C$ incorporation from glucose-1-$^{14}C$ and -2-$^{14}C$, and the fact that glutamate, $\alpha$-oxoglutarate, aspartate, and oxaloacetate occur in concentrations required by the aspartate transaminase equilibrium. Large changes in the glutamate and aspartate concentration do not seem to occur in brain. The total amount of $\alpha$-amino nitrogen in these amino acids is fairly constant. This fact, in combination with the differences observed in the intracellular localization of glutamate dehydrogenase and aspartate transaminase, suggests that both amino acids are not used to a great extent for the production of useful energy in the brain. Glutamate and aspartate might provide some stability to the tricarboxylic acid cycle, but this kind of hypothesis is vague and difficult to prove.

The small glutamate pool seems to be situated in the center of an extensive network of so-called minor pathways. This small glutamate pool is preferentially used for the synthesis of glutamine. The enzymes involved—glutamate dehydrogenase and glutamine synthetase—probably have characteristics peculiar for enzymes with a regulatory role. Part of the glutamine might leave the brain under certain conditions, but it seems that glutamine is also deamidated to glutamic acid. The activity of the enzyme involved—glutaminase—is probably controlled by a number of metabolites. Glutaminase has the same intracellular localization as succinate dehydrogenase, but not the same as glutamate dehydrogenase. If this glutamate—glutamine cycle is really operating in brain, then one has to assume that these amino acids are transported from one cellular site to another, as all the enzymes involved have a different intracellular localization.

Interference with the glutamate-to-glutamine conversion might result in a number of functional changes in the brain as a whole, or functional changes in the brain might induce changes in the glutamate-to-glutamine conversion. It is not even certain that changes in glutamate or glutamine or the glutamate-to-glutamine conversion are directly coupled to these functional changes. Changes in the large network of minor pathways leading to glutamate and glutamine as a whole might be more important. A large number of different states of this complex network are possible, and a closer look at these metabolic pathways might provide more insight into the relation between function and metabolism of the nervous system.

A link between metabolism and function might be the control of the level of the physiologically active glutamate. It can be argued that the small glutamate pool is the physiologically active pool. It has not been shown that *in vivo* glutamate has a role in the regulation of the frequency of firing or any other physiological properties of neurons.

## XI. REFERENCES


1. H. Waelsch, Glutamic acid and cerebral function, *Advan. Protein Chem.* **6**:299–341 (1951).
2. H. Weil-Malherbe, Significance of glutamic acid for the metabolism of nervous tissue, *Physiol. Rev.* **30**:549–568 (1950).
3. H. Weil-Malherbe, *in Neurochemistry* (K. A. C. Elliot, I. H. Page, and J. H. Quastel, eds.), pp. 321–330, Charles V. Thomas, Springfield, Illinois (1962).
4. H. J. Strecker, *in Metabolism of the Nervous System* (D. Richter, ed.), pp. 459–474, Pergamon Press, New York (1957).
5. D. R. Curtis and J. C. Watkins, The pharmacology of amino acids related to gamma-aminobutyric acid, *Pharmacol. Rev.* **17**:347–391 (1965).
6. H. H. Tallan, *in Amino Acid Pools* (J. T. Holden, ed.), pp. 471–485, Elsevier, Amsterdam (1962).
7. C. F. Baxter and C. L. Ortiz, Amino acids and the maintenance of osmotic equilibrium in brain tissue, *Life Sci.* **5**:2321–2329 (1966).
8. N. Okumura, S. Otsuki, and T. Aoyama, Studies on the free amino acids and related compounds in the brains of fish, amphibia, reptile, aves and mammal by ion exchange chromatography, *J. Biochem. (Tokyo)* **46**:207–212 (1959).
9. Y. Tsukada, *in Progress in Brain Research* (T. Tokizane and J. P. Schadé, eds.), Vol. 21A, pp. 268–291, Elsevier, Amsterdam (1966).
10. S. Berl and H. Waelsch, Determination of glutamic acid, glutamine, glutathione and γ-aminobutyric acid and their distribution in brain tissue, *J. Neurochem.* **3**:161–169 (1958).
11. S. Berl, Compartmentation of glutamic acid metabolism in developing cerebral cortex, *J. Biol. Chem.* **240**:2047–2054 (1965).
12. S. Berl and D. P. Purpura, Regional development of glutamic acid compartmentation in immature brain, *J. Neurochem.* **13**: 293–304 (1966).
13. R. L. Young and O. H. Lowry, Quantitative methods for measuring the histochemical distribution of alanine, glutamate and glutamine in brain, *J. Neurochem.* **13**:785–793 (1966).

14. L. T. Graham, R. P. Shank R. Werman, and M. H. Aprison, Distribution of some synaptic transmitter suspects in cat spinal cord, *J. Neurochem.* **14**:465–472 (1967).
15. Y. Nagata, Y. Yokoi, and Y. Tsukada, Studies on free amino acid metabolism in excised cervical sympathetic ganglia from the rat, *J. Neurochem.* **13**:1421–1431 (1966); Y. Tsukada, personal communication.
16. E. A. Kravitz, S. W. Kuffler, and D. D. Potter, Gamma-aminobutyric acid and other blocking compounds in crustacea, III, *J. Neurophysical.* **26**:739–751 (1963).
17. L. G. Abood, I. Koyama, and V. Thomas, Relationship of depolarization to phosphorus metabolism and transport in excitable tissues, *Am. J. Physiol.* **207**:1435–1440 (1964).
18. A. K. Huggins, J. T. Rick, and G. A. Kerkut, A comparative study of the intermediary metabolism of L-glutamate in muscle and nervous tissue, *Comp. Biochem. Physiol.* **21**:23–30 (1967).
19. M. Otsuka, E. A. Kravitz, and D. D. Potter, Physiological and chemical architecture of a lobster ganglion with particular reference to gamma-aminobutyrate and glutamate, *J. Neurophysiol.* **30**:725–752 (1967).
20. S. Berl and D. P. Purpura, Postnatal changes in amino acid content of kitten cerebral cortex, *J. Neurochem.* **10**: 237–240 (1963).
21. S. M. Bayer and W. C. McMurray, The metabolism of amino acids in developing rat brain, *J. Neurochem.* **14**:695–706 (1967).
22. A. R. Dravid and L. Jílek, Influence of stagnant hypoxia (oligaemia) on some free amino acids in rat brain during ontogeny, *J. Neurochem.* **12**:837–843 (1965).
23. J. L. Mangan and V. P. Whittaker, The distribution of free amino acids in subcellular fractions of guinea-pig brain *Biochem. J.* **98**:128–137 (1966).
24. M. K. Gaitonde, D. R. Dahl, and K. A. C. Elliott, Entry of glucose carbon into amino acids of rat brain and liver *in vivo* after injection of uniformly $^{14}C$-labelled glucose, *Biochem. J.* **94**:345–352 (1965).
25. P. Schwerin, S. P. Bessman, and H. Waelsch, The uptake of glutamic acid and glutamine by brain and other tissues of the rat and mouse, *J. Biol. Chem.* **184**:37–44 (1950).
26. W. A. Himwich, J. C. Peterson, and M. L. Allen, Hematoencephalic exchange as a function of age, *Neurology* **7**:705–710 (1957).
27. A. Lajtha, S. Berl, and H. Waelsch, Amino acid and protein metabolism of the brain—IV, *J. Neurochem.* **3**:322–332 (1959).
28. R. B. Roberts, J. B. Flexner, and L. B. Flexner, Biochemical and physiological differentiation during morphogenesis—XXIII, *J. Neurochem.* **4**:78–90 (1959).
29. R. M. O'Neal and R. E. Koeppe, Precursors *in vivo* of glutamate, aspartate and their derivatives of rat brain, *J. Neurochem.* **13**:835–847 (1966).
30. J. E. Adams, H. A. Harper, G. S. Gordan, M. Hutchin, and R. C. Bentinck, Cerebral metabolism of glutamic acid in multiple sclerosis, *Neurology* **5**:100–107 (1955).
31. H. G. Knauff, U. Gottstein, and B. Miller, Untersuchungen über den Austausch von freien Aminosäuren und Harnstoff zwischen Blut und Zentralnervensystem, *Klin. Wochschr.* **42**:27–39 (1964).
32. R. Rodnight, H. McIlwain, and M. A. Tresize, Analysis of arterial and cerebral venous blood from the rabbit, *J. Neurochem.* **3**:209–218 (1959).
33. W. Sacks, Cerebral metabolism of isotopic lipid and protein derivatives in normal human subjects, *J. Appl. Physiol.* **12**:311–318 (1958).
34. S. Berl, A. Lajtha, and H. Waelsch, Amino acid and protein metabolism—VI. Cerebral compartments of glutamic acid metabolism, *J. Neurochem.* **7**:186–197 (1961).

35. S. Berl, G. Takagaki, D. D. Clarke, and H. Waelsch, Metabolic compartments *in vivo*, *J. Biol. Chem.* **237**:2562–2569 (1962).
36. J. C. Dickinson and P. B. Hamilton, The free amino acids of human spinal fluid determined by ion exchange chromatography, *J. Neurochem.* **13**:1179–1187 (1966).
37. E. Levin, G. J. Nogueira, and C. A. Garcia Argiz, Ventriculo-cisternal perfusion of amino acids in cat brain—I, *J. Neurochem.* **13**:761–767 (1966).
38. E. Levin, C. A. Garcia Argiz and G. J. Nogueira, Ventriculo cisternal perfusion of amino acids in cat brain—II. *J. Neurochem*, **13** 979–988 (1966).
39. H. H. Jasper, R. T. Khan, and K. A. C. Elliott, Amino acids released from the cerebral cortex in relation to its state of activation, *Science* **147**: 1448–1449 (1965).
40. A. Van Harreveld and M. Kooiman, Amino acid release from the cerebral cortex during spreading depression and asphyxiation, *J. Neurochem.* **12**:431–439 (1965).
41. K. Crowshaw, S. J. Jessup, and P. W. Ramwell, Thin-layer chromatography of 1-dimethylaminonaphthalene-5-sulphonyl derivatives of amino acids present in superfusates of cat cerebral cortex, *Biochem. J.* **103**:79–85 (1967).
42. R. W. Albers, G. J. Koval, and W. B. Jakoby, Transamination reactions of rat brain *Exptl. Neurol.* **6**:85–89 (1962).
43. J. Awapara and B. Seale, Distribution of transaminase in rat organs, *J. Biol. Chem.* **194**:497–502 (1952).
44. O. Tangen, F. Fonnum, and R. Haavaldsen, Separation and purification of aromatic amino acid transaminases from rat brain, *Biochim. Biophys. Acta* **96**:82–90 (1965).
45. G. Amore and V. Bonavita, Aspartate aminotransferase in the brain of the developing rat, *Life Sci.* **4**:2417–2424 (1965).
46. F. Fonnum, The distribution of glutamate decarboxylase and aspartate transaminase in subcellular fractions of rat and guinea-pig brain, *Biochem. J.* **106**:401–412 (1968).
47. C. Frieden, *in The Enzymes*, 2nd ed. (P. D. Boyer, H. Lardy, and K. Myrbäck, eds.), Vol. 7, pp. 3–24, Academic Press, New York (1963).
48. M. Klingenberg and D. Pette, Proportions of mitochondrial enzymes and pyridine nucleotides, *Biochem. Biophys. Res. Commun.* **7**:430–432 (1962).
49. L. Garcia-Bunuel, D. B. McDougal, H. B. Burch, E. M. Jones, and E. Touhill, *J. Neurochem.* **9**:589–594 (1962).
50. D. H. Williamson, P. Lund, and H. A. Krebs, The redox state of free nicotinamide-adenine dinucleotide in the cytoplasm and mitochondria of rat liver, *Biochem. J.* **103**:514–527 (1967).
51. N. D. Goldberg, J. V. Passonneau, and O. H. Lowry, Effects of changes in brain metabolism on the levels of citric acid cycle intermediates, *J. Biol. Chem.* **241**: 3997–4003 (1966).
52. H. A. Krebs, Metabolism of amino acids, *Biochem. J.* **29**: 1951–1969 (1935).
53. C. Wu, Glutamine Synthetase—I. A comparative study of its distribution in animals and its inhibition by DL-allo-δ-hydroxylysine, *Comp. Biochem. Physiol.* **8**:335–351 (1963).
54. C. A. Woolfolk, and E. R. Stadtman, Regulation of glutamine synthetase, *Arch. Biochem. Biophys.* **118**:736–755 (1967).
55. K. Schnackerz, and L. Jaenicke, Reinigung und Eigenschaften der Glutamin-Synthetase aus Schweinehirn, *Z. Physiol. Chem.* **347**:127–144 (1966).
56. J. P. Greenstein and F. M. Leuthardt, Effect of phosphate and other anions on the enzymatic desamidation of various amides, *Arch. Biochem. Biophys.* **17**:105–114 (1948).

57. E. Kvamme, G. Svenneby, and B. Tveit, *in Molecular Basis of Some Aspects of Mental Activity* (O. Walaas, ed.), Vol. 1, pp. 211–219, Academic Press, London (1966).
58. E. Kvamme, B. Tveit, and G. Svenneby, Glutaminase from pig kidney, an allosteric protein, *Biochem. Biophys. Res. Commun.* **20**:566–572 (1965).
59. L. Salganicoff and E. De Robertis, Subcellular distribution of the enzymes of the glutamic acid, glutamine and γ-aminobutyric acid cycles in rat brain, *J. Neurochem.* **12**:287–309 (1965).
60. G. M. J. van Kempen, C. J. van den Berg, H. J. van der Helm, and H. Veldstra, Intracellular localization of glutamate decarboxylase, γ-aminobutyrate transaminase and some other enzymes in brain tissue, *J. Neurochem.* **12**:581–588 (1965).
61. A. Neidle, C. J. van den Berg, and A. Grynbaum, The heterogeneity of rat brain mitochondria isolated on continuous sucrose gradients, *J. Neurochem.* **16**:225–234 (1969).
62. E. De Robertis, O. Z. Sellinger, G. Rodríguez de Lores Arnaiz, M. Alberici, and L. M. Zieher, Nerve endings in methionine sulphoximine convulsant rats, a neurochemical and ultrastructural study, *J. Neurochem.* **14**:81–89 (1967).
63. O. H. Lowry, N. R. Roberts, and M. W. Chang, The analysis of single cells, *J. Biol. Chem.* **222**:97–107 (1956).
64. E. Robins, D. E. Smith, K. M. Eydt, and R. E. McCaman, The quantitative histochemistry of the cerebral cortex—II, *J. Neurochem.* **1**:68–76 (1956).
65. D. B. McDougal, E. M. Jones, and U. I. Sila, *in Ultrastructure and Metabolism of the Nervous System* (S. R. Corey, A. Pope, and E. Robins, eds.), Vol. 40, pp. 182–188, Williams & Wilkins, Baltimore (1962).
66. R. E. Kuhlman and O. H. Lowry, Quantitative histochemical changes during the development of the rat cerebral cortex, *J. Neurochem.* **1**:173–180 (1956).
67. O. H. Lowry, *in Morphological and Biochemical Correlates of Neural Activity* (M. M. Cohen and R. S. Snider, eds.), pp. 178–191, Harper & Row, New York (1964).
68. R. A. Salvador and R. W. Albers, The distribution of glutamin-γ-aminobutyrate transaminase in the nervous system of the rhesus monkey, *J. Biol. Chem.* **234**: 922–925 (1959).
69. R. W. Albers and R. O. Brady, The distribution of glutamic decarboxylase in the nervous system of the rhesus monkey, *J. Biol. Chem.* **234**:926–928 (1959).
70. S. Berl, Glutamine synthetase. Determination of its distribution in brain during development, *Biochemistry* **5**:916–922 (1966).
71. T. Turský and E. Valovičová, Asparaginase in the brain of guinea pig in experimental allergic encephalomyelitis, *J. Neurochem.* **11**:99–108 (1964).
72. J. M. Pasquini, B. Kaplún, C. A. Garcia Argiz, and C. J. Gomez, Hormonal regulation of brain development, *Brain Res.* **6**: 621–634 (1967).
73. C. Wu, Glutamine synthetase, *Arch. Biochem. Biophys.* **106**:394–401 (1964).
74. C. J. van den Berg, G. M. J. van Kempen, J. P. Schadée, and H. Veldstra, Levels and intracellular localization of glutamate decarboxylase and γ-aminobutyrate transaminase and other enzymes during the development of the brain, *J. Neurochem.* **12**:863–869 (1965).
75. O. H. Lowry, N. R. Roberts, and C. Lewis, The quantitative histochemistry of the retina, *J. Biol. Chem.* **220**:879–892 (1956).
76. R. Balázs and R. J. Haslam, Exchange transamination and the metabolism of glutamate in brain, *Biochem. J.* **94**:131–141 (1965).
77. R. Balázs, Control of glutamate oxidation in brain and liver mitochondrial systems, *Biochem. J.* **95**:497–508 (1965).
78. N. E. Lofrumento, G. De Gregorio, S. Papa, C. Serra, and E. Quagliariello,

Metabolismo dell'acido glutammico nei mitochondri de cervello, *Boll. Soc. Ital. Biol. Sper.* **40**:1452–1455 (1964).
79. E. J. De Haan, J. M. Tager, and E. C. Slater, Factors affecting the pathway of glutamate oxidation in rat-liver mitochondria, *Biochim. Biophys. Acta* **131**:1–13 (1967).
80. B. Sacktor and L. Packer, Reactions of the respiratory chain in brain mitochondrial preparations, *J. Neurochem.* **9**:371–382 (1962).
81. H. A. Krebs, Metabolism of amino acids, *Biochem. J.* **29**:1620–1644 (1935).
82. H. Weil-Malherbe, Studies on brain metabolism, *Biochem. J.* **30**:665–676 (1936).
83. R. J. Woodman and H. McIlwain, Glutamic acid, other amino acids and related compounds as substrates for cerebral tissues: Their effects on tissue phosphates, *Biochem. J.* **81**:83–93 (1961).
84. F. Dickens and G. D. Greville, Metabolism of normal and tumour tissue, *Biochem. J.* **27**:1123–1133 (1933).
85. H. Weil-Malherbe and R. H. Green, Ammonia formation in brain, *Biochem. J.* **61**:210–218 (1955).
86. G. Takagaki, S. Hirano, and Y. Nagata, Some observations on the effect of D-glutamate on the glucose metabolism and the accumulation of potassium ions in brain cortex slices, *J. Neurochem.* **4**:124–134 (1959).
87. O. Z. Sellinger, R. Catanzaro, E. B. Chain and F. Pocchiari, The metabolism of glutamate and aspartate in rat cerebral cortical slices, *Proc. Roy. Soc. (London) Ser. B* **156**:148–162 (1962).
88. A. Beloff-Chain, R. Catanzaro, E. B. Chain, I. Masi, and F. Pocchiari, Fate of uniformly labelled $^{14}C$ glucose in brain slices, *Proc. Roy. Soc. (London) Ser. B* **144**: 22–28 (1955).
89. E. B. Chain, M. Chiozzotto, F. Pocchiari, C. Rossi, and R. Sandman, Participation of the ammonium ion in the transformation of glucose to amino acids in brain tissue, *Proc. Roy. Soc. (London) Ser. B* **152**:290–297 (1960).
90. K. L. Reichelt and E. Kvamme, Acetylated and peptide bound glutamate and aspartate in brain, *J. Neurochem.* **14**:987–996 (1967).
91. G. Simon, J. B. Drori, and M. M. Cohen, Mechanism of conversion of aspartate into glutamate in cerebral-cortex slices, *Biochem. J.* **102**:153–162 (1967).
92. S. Berl, W. J. Nicklas, and D. D. Clarke, Compartmentation of glutamic acid metabolism in brain slices, *J. Neurochem.* **15**:131–140 (1968).
93. R. Vrba, M. K. Gaitonde, and D. Richter, The conversion of glucose carbon into protein in the brain and other organs of the rat, *J. Neurochem.* **9**:465–475 (1962).
94. H. Busch, E. Fujiwara, and L. M. Keer, Metabolic patterns for glucose-1-$^{14}C$ in tissues of tumor-bearing rats, *Cancer Res.* **20**:50–57 (1960).
95. C. J. van den Berg, P. Mela, and H. Waelsch, On the contribution of the tricarboxylic acid cycle to the synthesis of glutamate, glutamine and aspartate in brain, *Biochem. Biophys. Res. Commun.* **23**:479–484 (1966).
96. C. J. van den Berg, Lj. Kržalić, P. Mela, and H. Waelsch, Compartmentation of glutamate metabolism in brain. Evidence for the existence of two different tricarboxylic acid cycles in brain, *Biochem. J.* **113**:281–290 (1969).
97. H. Waelsch, S. Berl, C. A. Rossi, D. D. Clarke, and D. P. Purpura, Quantitative aspects of $CO_2$ fixation in mammalian brain *in vivo*. *J. Neurochem.* **11**:717–728 (1964).
98. C. F. Baxter, Intrinsic amino acid levels and the blood-brain barrier, *Progr. Brain Res.* **29**:429–444 (1968).
99. S. Roberts and B. S. Morelos, Regulation of cerebral metabolism of amino acids—IV, *J. Neurochem.* **12**:373–387 (1965).

100. E. Roberts, M. Rothstein, and C. F. Baxter, Some metabolic studies of $\gamma$-aminobutyric acid, *Proc. Soc. Exptl. Biol. Med.* **97**:796–802 (1958).
101. S. S. Barkulis, A. Geiger, Y. Kawakita, and V. Aguilar, A study on the incorporation of $^{14}C$ derived from glucose into the free amino acids of the brain cortex, *J. Neurochem.* **5**:339–348 (1960).
102. S. Otsuki, A. Geiger, and G. Gombos, The metabolic pattern of the brain in brain perfusion experiments *in vivo*—I, *J. Neurochem.* **10**:397–404 (1963).
103. M. Proler and P. Kellaway, The methionine sulfoximine syndrome in the cat, *Epilepsia* **3**:117–130 (1962).
104. O. Z. Sellinger and P. Weiler, The nature of the inhibition *in vitro* of cerebral glutamine synthetase by the convulsant, methionine sulfoximine, *Biochem. Pharmacol.* **12**:989–1000 (1963).
105. C. Lamar and O. Z. Sellinger, The inhibition *in vivo* of cerebral glutamine synthetase and glutamine transferase by the convulsant methionine sulfoximine, *Biochem. Pharmacol.* **14**:489–506 (1965).
106. O. Z. Sellinger, Inactivation of cerebral glutamine synthetase by DL-methionine-DL-sulfoximine, *Biochim. Biophys. Acta* **132**:514–516 (1967).
107. J. K. Tews and W. E. Stone, Effects of methionine sulfoximine on levels of free amino acids and related substances in brain, *Biochem. Pharmacol.* **13**:543–544 (1964).
108. J. Folbergrova, Free glutamine level in the rat brain in vivo after methionine sulphoximine administration, *Physiol. Bohemoslov.* **13**:21–27 (1964).
109. A. P. C. Bot and J. P. Schadé, personal communications.
110. R. L. Potter and A. Van Harreveld, The effect of metrazol on the glutamate metabolism of cerebral cortex, *J. Neurochem.* **9**:105–112 (1962).
111. G. M. Clark and B. Eiseman, Studies in ammonia metabolism, *New Engl. J. Med.* **259**:178–180 (1958).
112. J. K. Tews, S. H. Carter, P. D. Roa, and W. E. Stone, Free amino acids and related compounds in dog brain: Post-mortem and anoxic changes, effects of ammonium chloride infusion, and levels during seizures induced by picrotoxin and by pentylenetetrazol, *J. Neurochem.* **10**:641–653 (1963).
113. J. Shorey, D. W. McCandless, and S. Schenker, Cerebral $\alpha$-ketoglutarate in ammonia intoxication, *Gastroenterology*, **53**:706–711 (1967).
114. S. P. Bessman, *in Proceedings of the Fourth International Congress of Biochemistry* (F. Brücke, ed.), Vol. 3, pp. 141–145, Pergamon Press, London (1959).
115. S. Schenker and J. H. Mendelson, Cerebral adenosine triphosphate in rats with ammonia-induced coma, *Am. J. Physiol.* **206**:1173–1176 (1964).
116. J. P. Du Ruisseau, J. P. Greenstein, M. Winitz, and S. M. Birnbaum, Studies on metabolism of amino acids and related compounds *in vivo*, *Arch. Biochem. Biophys.* **68**:161–171 (1957).
117. K. S. Warren and S. Schenker, Effect of an inhibitor of glutamine synthesis (methionine sulfoximine) on ammonia toxicity and metabolism, *J. Lab. Clin. Med.* **64**:442–449 (1964).
118. S. P. Bessman and A. N. Bessman, The cerebral and peripheral uptake of ammonia in liver disease with an hypothesis for the mechanism of hepatic coma, *J. Clin. Invest.* **34**:622–628 (1955).
119. J. F. Fazekas, H. E. Ticktin, W. R. Ehrmantraut, and R. W. Alman, Cerebral metabolism in hepatic insufficiency, *Am. J. Med.* **21**:843–849 (1956).
120. P. J. McMillan and R. A. Mortensen, The metabolism of brain pyruvate and acetate in the tricarboxylic acid cycle, *J. Biol. Chem.* **238**:91–93 (1963).
121. E. V. Flock, G. M. Tyce, and C. A. Owen, Utilization of (U-$^{14}C$) glucose in brain after total hepatectomy in the rat, *J. Neurochem.* **13**:1389–1406 (1966).

122. R. S. De Ropp and E. H. Snedeker, Effect of drugs on amino acid levels in the rat brain: Hypoglycemic agents, *J. Neurochem.* **7**:128–134 (1961).
123. R. M. C. Dawson, Studies on the glutamine and glutamic acid content of the rat brain during insulin hypoglycaemia, *Biochem. J.* **47**:386–391 (1950).
124. J. K. Tews and W. E. Stone, Free amino acids and related compounds in brain and other tissues: effects of convulsant drugs, *Progr. Brain Res.* **16**:135–163 (1965).
125. R. P. Kamrin and A. A. Kamrin, The effects of pyridoxine antagonists and other convulsive agents on amino acid concentrations of the mouse brain, *J. Neurochem.* **6**:219–225 (1961).
126. P. D. Roa, J. K. Tews, and W. E. Stone, A neurochemical study of thiosemicarbazide seizures and their inhibition by amino-oxyacetic acid, *Biochem. Pharmacol.* **13**:477–487 (1964).
127. D. E. Hathway and A. Mallinson, Chemical studies in relation to convulsive conditions, *Biochem. J.* **90**:51–60 (1964).
128. R. M. C. Dawson, Cerebral amino acids in fluoroacetate-poisoned, anaesthetised and hypoglycaemic rats, *Biochim. Biophys. Acta* **11**:548–552 (1953).
129. F. N. Minard and I. K. Mushahwar, The effect of periodic convulsions induced by 1, 1-dimethylhydrazine on the synthesis of rat brain metabolites from (2-$^{14}$C) glucose, *J. Neurochem.* **13**:1–11 (1966).
130. B. Sacktor, J. E. Wilson, and C. G. Tiekert, Regulation of glycolysis in brain, *in situ*, during convulsions, *J. Biol. Chem.* **241**:5071–5075 (1966).
130a. M. B. Sporn, W. Dingman, and A. Defalco, A method for studying metabolic pathways in the brain of the intact animal, *J. Neurochem.* **4**:141–147 (1959).
131. R. M. C. Dawson, The metabolism and glutamic acid content of rat brain in relation to thiopentone anaesthesia, *Biochem. J.* **49**:138–144 (1951).
132. H. S. Bachelard and J. R. Lindsay, Effects of neurotropic drugs on glucose metabolism in rat brain *in vivo*, *Biochem. Pharmacol.* **15**:1053–1058 (1966).
133. J. R. Stern, L. V. Eggleston, R. Hems, and H. A. Krebs, Accumulation of glutamic acid in isolated brain tissue, *Biochem. J.* **44**:410–418 (1949).
134. C. Terner, L. V. Eggleston, and H. A. Krebs, The role of glutamic acid in the transport of potassium in brain and retina, *Biochem. J.* **47**:139–149 (1950).
135. H. M. Pappius and K. A. C. Elliott, Factors affecting the potassium content of incubated brain slices, *Can. J. Biochem. Physiol.* **34**:1053–1064 (1956).
136. P. Joanny, H. Hillman, and J. Corriol, The effect of some amino acids, glycolytic intermediates and citric acid cycle intermediates on the swelling, and the potassium and sodium concentrations, of guinea pig cerebral cortex slices *in vitro*, *J. Neurochem.* **13**:371–374 (1966).
137. H. F. Bradford and H. McIlwain, Ionic basis for the depolarization of cerebral tissue by excitatory acidic amino acids, *J. Neurochem.* **13**:1163–1177 (1966).
138. M. M. Cohen and H. P. Cohen, The effect of glutamic acid on phosphorus metabolism in cerebral tissue preparations, *J. Neurochem.* **13**:811–818 (1966).
139. A. Ames, Studies on water and electrolytes in nervous tissue, *J. Neurophys.* **19**: 213–223 (1955).
140. A. Ames, Y. Tsukada, and F. B. Nesbett, Intracellular $Cl^-$, $Na^+$, $K^+$, $Ca^{2+}$, $Mg^{2+}$, and P on nervous tissue; response to glutamate and to changes in extracellular calcium, *J. Neurochem.* **14**:145–159 (1967).
141. F. Orrego and F. Lippmann, Protein synthesis in brain slices, *J. Biol. Chem.* **242**: 665–671 (1967).
142. H. H. Hillman and H. McIlwain, Membrane potentials in mammalian cerebral tissues *in vitro*: Dependence on ionic environment, *J. Physiol.* **157**:263–278 (1961).
143. I. M. Gibson and H. McIlwain, Continuous recording of changes in membrane

potential in mammalian cerebral tissues in vitro; Recovery after depolarization by added substances, *J. Physiol.* **176**:261–283 (1965).
144. K. Krnjević and J. W. Phillis, Iontophoretic studies of neurones in the mammalian cerebral cortex, *J. Physiol.* **165**:274–304 (1963).
145. F. A. Steiner and K. Ruf, Excitatory effects of L-glutamic acid upon single unit activity in rat brain and their modification by thiosemicarbazide and pyridoxal-5′-phosphate, *Helv. Physiol. Pharmacol. Acta* **24**:181–192 (1966).
146. A. Van Harreveld, Compounds in brain extracts causing spreading depression of cerebral cortical activity and contraction of crustacean muscle, *J. Neurochem.* **3**: 300–315 (1959).
147. N. Canal and L. Frattola, Hexosamine synthesis in nervous tissue, *Med. Exptl.* **8**:129–134 (1963).

*Chapter 11*

# GLYCINE: ITS METABOLIC AND POSSIBLE TRANSMITTER ROLES IN NERVOUS TISSUE*

**M. H. Aprison, Robert A. Davidoff, and Robert Werman**

*The Institute of Psychiatric Research and Departments of Psychiatry and Biochemistry*
*Indiana University Medical Center*
*and Veterans Administration Hospital*
*Indianapolis, Indiana*

## I. INTRODUCTION

Glycine is found in most tissues. Its synthesis has been demonstrated in animals, plants, and microorganisms. It has been shown to be present in large amounts in central nervous tissue.[1-4] Although glycine is the simplest amino acid from a structural point of view, its intermediary metabolism is far from simple and involves a rather complex series of reactions. However, few of these metabolic steps have been elucidated in the nervous system. Recently it has been discovered that glycine is probably a chemical transmitter in the cat spinal cord.[2,3,5-8] This transmitter function was found as a result of a combined neurochemical and neurophysiological approach directed to the problem of the identification of CNS transmitters.[9,10] It was made possible both by methodological advances in allied disciplines and by the convergence of several lines of biochemical and physiological effort—namely, the great impetus given to the study of amino acids in small amounts of nervous tissue by the application of ion-exchange chromatography, and the development of iontophoresis and equilibrium potential measurements as tools to study the effects of biochemical compounds applied onto synaptic membranes.

In this chapter we propose to show how these methodological advances have produced evidence that glycine has a possible physiological role in cat

* The investigations in which the authors participated were supported in part by Research Grant MH-03225-07,08 from the National Institute of Mental Health, Research Grant NB-07307-01, and Career Development Award NB 14-815-03 (R.W.) from the National Institutes of Health, U.S. Public Health Service.

spinal cord. The distribution and metabolism of glycine in the nervous system will be reviewed to this end.

## II. DETERMINATION OF GLYCINE IN NERVOUS TISSUE

Glycine can be quantitatively measured in nervous tissue by a colorimetric method involving diazotization of glycine to glycollic acid and reaction of the product with sulfuric acid to yield formaldehyde. Formaldehyde is then reacted with chromotropic acid in the presence of sulfuric acid to give a purple color. The original method of Giroux and Puech[11] was scaled down and with some modifications has been used to measure glycine in the 1 to 40 $\times$ $10^{-9}$ mole range even in the presence of other free amino acids.[2,3] The techniques of paper,[12] column,[13] and thin-layer[14] chromatography have been used extensively to measure small amounts of glycine when it is present among other amino acids. These latter methods can suffer from losses in recovery during chromatography.

## III. METABOLISM OF GLYCINE IN NERVOUS TISSUE

### A. Biosynthesis and Utilization

In mammals, glycine is a nonessential amino acid and can be synthesized from glucose and other substrates in nervous tissue. In addition, glycine may be transported to brain and spinal cord from blood. It has been known since the late 1950's that systematically administered labeled glycine can enter nerve cells.[15] Whether synthesis of glycine in nervous tissue or transport from blood is the major source of this amino acid is unknown. Furthermore, it is not known whether glycine in nervous tissue exists in separate metabolic and transmitter compartments or pools. In addition, if such functional pools exist the source of glycine in each may or may not be different.

The pathways shown in Fig. 1 give an outline of those reactions which glycine probably undergoes in nervous tissue. Synthesis of glycine from glucose by the pathway via D-3-phosphoglyceric acid and serine appears to be the best documented in the CNS. Rapid labeling of amino acids including glycine in brain has been shown to occur after an intravenous injection of glucose-U-$^{14}$C.[16,17] It has been demonstrated[17] that 1-day-old mouse brain minces could convert glucose–3,4–$^{14}$C into alanine, serine, and glycine (these three were the only labeled amino acids isolated from protein). Two routes from glucose to serine have been described in non-nervous tissue. They are referred to as the "phosphorylated" (utilizing phosphoglyceric acid dehydrogenase) and the "nonphosphorylated" (utilizing D-glycerate-dehydrogenase) pathways.[18] In the former pathway, phosphoglyceric acid dehydrogenase catalyzes the NAD-dependent oxidation of 3-phosphoglyceric acid, yielding 3-phosphohydroxypyruvic acid. Transamination and

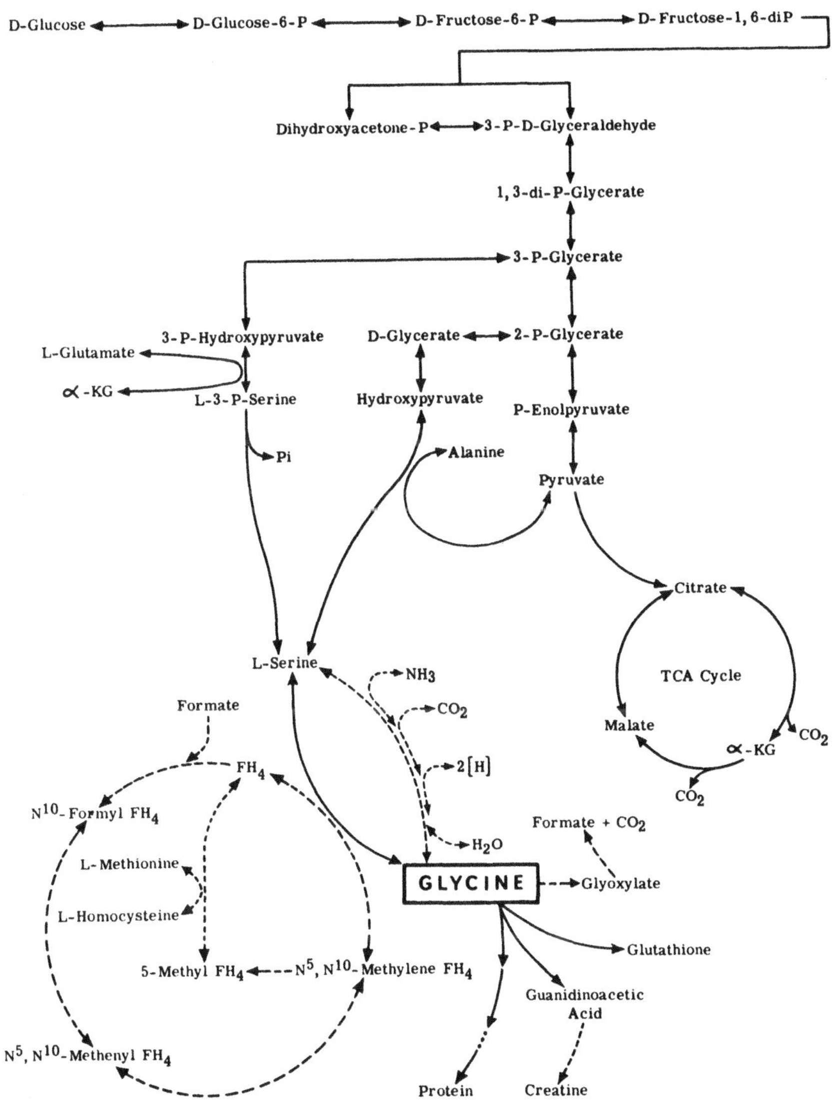


Fig. 1. Pathways illustrating the synthesis and utilization of glycine in nervous tissue. The solid lines delineate reactions which have been claimed to occur. The broken lines indicate probable pathways which have not yet been demonstrated in nervous tissue.

hydrolysis of the phosphate ester complete the synthesis of serine. Recently, Bridgers[18] demonstrated that a cell-free extract of adult mouse brain can synthesize serine from glucose with 3-phosphoglycerate and 3-phosphoserine as intermediates (see Fig. 1). This evidence suggests that in mouse brain the phosphorylated pathway is operative. Whether it is the system utilized in most mammalian nervous systems remains to be shown.

Some data on the synthesis of glycine in the lobster nerve cord[19] have appeared recently which have shown that the label from either glucose-6-$^{14}$C or pyruvate-1-$^{14}$C can be incorporated into glycine (as well as into alanine, glutamic acid, and aspartic acid). These data suggest that the pathway from glucose to serine is via hydroxypyruvate.[19] However, the contribution of this pathway to glycine synthesis is small in this species.

The conversion of serine to glycine in non-nervous tissue has been shown to take place via the following two reactions[20,21]:

$$\text{Serine} + FH_4 \xleftrightarrow{E_1;\ \text{pyridoxal phosphate};\ Mn^{2+}} \text{glycine} + {}^5N,{}^{10}N\text{-methylene } FH_4 \quad (1)$$

$$\text{Serine} + CO_2 + NH_3 + 2[H] \leftrightarrow 2\ \text{glycine} + H_2O \quad (2)$$

where $E_1$ is the enzyme serine transhydroxymethylase, $FH_4$ is tetrahydrofolic acid, and $^5N,^{10}N$-methylene $FH_4$ is $^5N,^{10}N$-methylenetetrahydrofolic acid. Isotopic measurements have shown a small amount of intraconversion *in vitro* of glycine and serine in rat cortical slices.[22] It was postulated that both pathways were operative, but since the first reaction is so ubiquitous, it is probable that it is the main pathway in the CNS.

Glycine is involved in numerous pathways, but information about these pathways in neural tissue is limited. Results from work *in vitro* with guinea pig[23] and mouse[17] cortex, rabbit sciatic nerve,[24] and rat brain *in vivo*[15,25] have given evidence that glycine-$^{14}$C is incorporated into brain proteins. For example, in the latter studies, different rates of uptake were noted in different brain areas. Labeling of protein in spinal cord was higher than in cerebrum. Koenig[26] showed that glycine-2-$^3$H given by intracisternal injection was rapidly incorporated into proteins of nerve cells and glia in cat spinal cord. That neurons effect a more rapid turnover of newly formed protein than glia was suggested by the finding that neurons lost their radioactivity more rapidly than did glia. Glycine can also be incorporated in the tripeptide glutathione in brain.[27,28]

Nakada and Weinhouse[29] reported that $^{14}CO_2$ was liberated at a low rate from rat brain slices incubated with glycine-1-$^{14}$C. Liberation of $^{14}CO_2$ from homogenates did not occur. In kidney and liver, these same authors identified glyoxylate ($\alpha$-carbon) and formate as intermediates in the oxidation of the glycine $\alpha$-carbon; these compounds were not identified in brain. In studies with rat brain homogenates,[30] no label was found in $CO_2$ when glycine-2-$^{14}$C was used as a substrate; the label was incorporated into glutamic acid and proteins. When glycine-1-$^{14}$C was employed, some $^{14}CO_2$ was measured. However, it was shown that the breakdown of glycine by rat brain cortex slices can be substantially increased by adding glucose to the incubation medium.[22] The effect of glucose was not confined to the carboxyl carbon of glycine since stimulation of $^{14}CO_2$ production from either glycine-1-$^{14}C$ or 2-$^{14}C$ occurred. With the latter substrate, both a lag period of 30–60 min and a reduced $^{14}CO_2$ production were reported. Further, the stimulation by glucose was confined to brain tissue; in liver slices under identical conditions the oxidation of glycine was unaffected whereas with kidney cortex slices 20% inhibition occurred.(22) These few studies suggest that glycine is not readily converted into intermediates of the Krebs cycle or other energy-producing pathways.

Although Borsook and Dubnoff(31) were not able to measure any arginine–glycine transaminase activity in rat brain, Pisano *et al.*(32) did demonstrate the presence of small amounts of this enzyme in brain. They measured guanidino–acetic acid, the product of the reaction of glycine and arginine.

## B. Uptake and Regulation

Since glycine has potent effects on synaptic membranes (*vide infra*), its concentration in brain and spinal cord must be closely regulated. Uptake of amino acids from blood is important in regulation of brain levels. The problems of the blood–brain barrier may be circumvented by studying amino acid uptake *in vitro*. Slices of cerebral cortex from a number of species can rapidly accumulate glycine from an incubating medium against a concentration gradient.(22,33–35) The rate of uptake is exceeded only by GABA and glutamate. The uptake of glycine is accelerated by the presence of glucose and is related to the amount of ATP present.(33) The process is very sensitive to various drugs, metabolic inhibitors, other amino acids, and changes in ionic environment. Obviously the process is complex. These studies suggest that glycine released from nerve endings after stimulation may be removed from extracellular fluid in the synaptic cleft by uptake into the pre- or postsynaptic neuron. A similar phenomenon has been shown for norepinephrine and GABA. Lajtha(36) suggested that the mechanism of uptake of amino acids by slices *in vitro* and by brain *in situ* may be similar in spite of the fact that the controlling factors are different or missing in the former situation. Consequently, glycine may normally be taken up by neurons from plasma and extracellular fluid, although in smaller amounts than by slices.

Arnstein and Neuberger(37) suggested that the formation of serine from glycine is probably an important mechanism regulating the level of glycine in tissue. Simkin and White(38) came to similar conclusions. They found that the conversion of glycine to serine was a rapid, first-order reaction dependent on the concentration of glycine but not of serine. The reverse reaction, however, was found to be of zero order, proceeding at a constant rate regardless of the concentration of the reactants. These conclusions were based on data obtained from non-neural tissue. If it were also true for brain and spinal cord, this concept would fit in with the importance of glycine as a physiologically active compound involved in synaptic inhibition.

# IV. TRANSMITTER FUNCTION OF GLYCINE

## A. Early Studies

The neuropharmacological activities of glycine and other amino acids have been investigated since 1959. Purpura *et al.*[39] demonstrated that a change in the surface electrical activity of cat cortex was produced by topically applied glycine and other inhibitory amino acids. They postulated that these amino acids have specific blocking actions on cortical excitatory synapses. Curtis and Watkins[40] and Krnjević and Phillis[41] mentioned that glycine was appreciably active in inhibiting the firing of neurons when applied by iontophoresis in cat spinal cord and cortex. The reflex activity of amphibian spinal cord was decreased by perfusion with glycine *in vitro*[42] and *in situ*,[43] but neither crayfish stretch-receptor[44] nor nerve–muscle preparation[45] was affected by this amino acid. In addition, no effect was demonstrated on the cerebral cortical activity of rabbit[46] or on the monosynaptic reflex of cat.[47] Most of these studies, however, were primarily interested in demonstrating the structure–activity relationships of various amino acids, and all demonstrated that in a variety of unrelated systems GABA was a more potent pharmacological agent than glycine. Furthermore, leading workers in the field[40,48,49] had strongly stressed that a transmitter role was not likely for amino acids and concluded that the action of these compounds was unrelated to synaptic activity.

## B. Combined Neurochemical and Neurophysiological Approach

By 1965, the advanced state of lumbosacral spinal cord physiology and anatomy in the cat made it possible to predict the neurochemical distribution of the transmitter responsible for disynaptic or "direct" inhibition in the spinal cord. Furthermore, the actions on spinal neurons of most amino acids now suspected to be transmitters had been studied.[49] Proof that a suspected compound is a CNS transmitter can be reduced to three pieces of evidence[9,10,50]: the transmitter suspect (1) should be present in the presynaptic neuron of the synapse under consideration, (2) should produce the same changes in membrane ionic processes as does the transmitter released by stimulating the presynaptic cell, and (3) should be in the extracellular fluid following the activation of the synapse. In 1965, after an analysis of the distribution of most amino acids present in cat lumbar spinal cord, it was noted that the glycine distribution was similar to the distribution postulated for the inhibitory transmitter released from interneurons.[2] On the basis of this finding and the physiological evidence that iontophoresed glycine exerted a marked inhibitory effect on firing of single spinal motoneurons, Aprison and Werman[2] speculated that glycine might be a spinal inhibitory transmitter.

## C. Association of Glycine and Interneurons

An interneuron is interpolated in inhibitory pathways from dorsal root fibers to spinal motoneurons. The processes of interneurons are relatively short and mostly confined to gray matter. On this basis, one would expect a transmitter involved in postsynaptic inhibition to be concentrated in spinal gray matter—the location of the soma and presynaptic terminals of the interpolated interneurons. A high concentration of transmitter in other presynaptic cells had been demonstrated previously.(51) Among the amino acids found in spinal gray matter, the concentration of glycine is the highest.(3) Furthermore, microdissection of cat spinal cord and analysis of its various parts demonstrated that the highest concentration of glycine was in ventral gray matter and the lowest in dorsal and ventral roots which do not contain any interneuron processes (Fig. 2). The intermediate concentration in ventral white matter is compatible with the concentration of propriospinal axons in this region which originate from interneurons. In contrast, glutamic acid, which has excitatory properties when applied to spinal neurons, and glutamine, which is without effect on neuron firing, have entirely different patterns of distribution in spinal cord.(3) It was felt, therefore, that the distribution pattern of glycine reflected an association with specific neuronal elements—interneurons. However, another amino acid with inhibitory properties, GABA, had a similar distribution.(3)

The specific association of glycine with interneurons was demonstrated by studies of the biochemical changes produced when loss of interneurons was effected in cat spinal cord by prolonged anoxia of the cord.(52) In this situation, there were several concomitant occurrences: (1) a significant loss of glycine, but not of GABA, in gray matter and ventral white matter; (2) a significant correlation between the remaining interneurons and the glycine concentration, and (3) reduction of inhibitory reflexes.(5,6) The changes in glycine are best explained by assuming that this amino acid is concentrated in the inhibitory interneurons destroyed by the anoxia, the same ones which are responsible for mediating polysynaptic inhibitory reflexes.

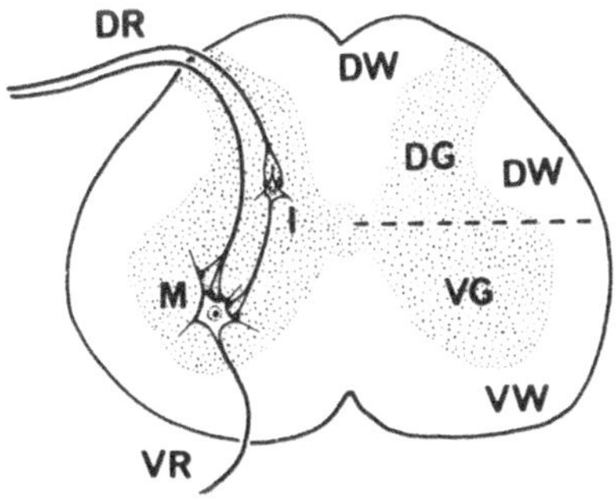


| Tissue Analyzed | Glycine Concentration ($\mu$moles/g) |
|---|---|
| dorsal white (DW) | 3.04 ± 0.26 |
| dorsal grey (DG) | 5.65 ± 0.18 |
| ventral grey (VG) | 7.08 ± 0.31 |
| ventral white (VW) | 4.39 ± 0.26 |
| dorsal root (DR) | 0.64 ± 0.04 |
| ventral root (VR) | 0.64 ± 0.00 |

Fig. 2. Diagrammatic cross section of cat lumbar spinal cord. The synapses between dorsal root fibers and an interneuron ("direct" or disynaptic inhibitory pathway) and a motoneuron (monosynaptic excitatory pathway) are illustrated. Glycine concentrations in roots and various cord areas are given.

## D. Iontophoretic Studies of Glycine

Recently Werman *et al.*[7,8] examined the mechanism of action of glycine on motoneurons of the cat lumbar spinal cord. These studies used an iontophoretic technique to restrict the administration of amino acids to the immediate extracellular environment of an individual motoneuron while recording from a second micropipette located in that particular nerve cell. The intracellular recording allowed precise measurements to be made simultaneously of the changes in membrane properties, especially equilibrium potentials, produced by iontophoresis of glycine and those produced by stimulating inhibitory inputs to the same cell. In this way, the action both of iontophoresed glycine and of the transmitter released from stimulation of the presynaptic cell could be compared.

Iontophoretic release of glycine regularly produced an increase of membrane potential which was related to the amount of iontophoretic current used to release glycine. This hyperpolarization usually reached a maximum value (mean, 3.8 mV) and was accompanied by several phenomena: (1) block of invasion of the soma-dendritic membrane by spikes, (2) decrease in amplitude of both excitatory (EPSP) and inhibitory (IPSP) postsynaptic potentials, and (3) an increased membrane ionic conductance (decreased resistance). Glycine was also shown to saturate receptor sites normally available to inhibitory transmitters. Furthermore, the equilibrium potential of the hyperpolarization produced by glycine was identical with the equilibrium potential of the IPSP. A highly specific measure of the identity of the membrane processes produced by both glycine and the agent released by stimulating inhibitory inputs to a motoneuron can be obtained by equilibrium potential measurements before and after altering the environment of a motoneuron by intracellular injection of bromide and iodide ions.[53] Following this procedure, glycine produced depolarization and the IPSP was in a depolarizing direction as well. Furthermore, both processes again had the same equilibrium potential. Therefore, it was concluded that glycine and the inhibitory transmitter both acted on the synatic membrane in an identical manner. Recently Curtis *et al.*[54,55] provided confirmation for some of these neurophysiological data.

# V. COMPARATIVE DISTRIBUTION OF GLYCINE IN NERVOUS TISSUE

## A. Invertebrates

Values reported for the concentration of glycine in invertebrate nervous tissue are given in Table I. The high value in lobster nerve is of interest, but glycine apparently has little or no effect on the crustacean neuromuscular system.[44,45] Whether glycine has any function in invertebrate nervous tissue other than that of a metabolic intermediate is unknown.

TABLE I

Glycine Concentration in Invertebrates[a]

| Species | Tissue | Concentration | Reference |
|---|---|---|---|
| Squid | Giant axon | 14.0 | 56 |
| | Axoplasm | 11.6 | 57 |
| Cuttlefish | Axon | < 5 | 58 |
| Crab | Leg nerve | < 5 | 58 |
| Lobster | Leg nerve | 35 | 58 |
| Lobster | Ventral nerve cord | 32.4 | 19 |
| Bee | Brain | 5.3 | 59 |

[a] Values in μmoles/g.

## B. Vertebrates

### 1. *Nervous Tissue*

The few studies available which give the distribution of glycine in brain and spinal cord show evidence of marked regional variation in the concentration of this amino acid (Table II). In general, glycine is present in high concentration in the cervical and lumbar enlargements of the spinal cord and in the medulla oblongata. Except for monkey and rabbit, all values are three to five times higher than the values for glycine in supramedullary structures and in cerebellum.

The higher concentrations of glycine in the cervical and lumbar enlargements—compared to the thoracic cord—parallel the greater gray to white matter ratio in these areas.[62] The gray matter of the lumbar spinal cord contains a higher concentration of glycine than does the white matter (Fig. 2). Differences in water and lipid content between white and gray matter do not account for these differences in glycine concentration.[4] The enlargements of the spinal cord are the areas of major motor outflow to the extremities. The marked localization of glycine in these areas correlates with the large numbers of synaptic junctions and interneurons. Compatible with these data were the findings that there were no differences in glycine concentration in various levels of the spinal cord in fish and snake, which do not have limbs.[4]

In view of the probable transmitter function of glycine in lumbar spinal cord and of the recent demonstration that glycine has potent inhibitory effects on neurons in the cuneate nucleus,[63] the high concentration in the medulla oblongata is of interest. The possibility that glycine may play a role in synaptic transmission in the medulla thus exists.

Glycine, as well as other amino acids, apparently is widely distributed in the cytoplasmic fraction of brain homogenates and is not concentrated

TABLE II

**Glycine Concentration in Vertebrate CNS[a]**

| Region | Species and references | | | | | | |
|---|---|---|---|---|---|---|---|
| | Monkey[60] | Cat[4] | Rat[1,4] | Rabbit[61] | Caiman[4] | Pigeon[4] | Bullfrog[4] |
| Cerebrum | 2.68[b] | 1.28 | 0.83, 0.91 | 2.56 | 1.04 | 1.20 | 0.79 |
| Diencephalon | 2.89[c] | 1.60 | —, 0.98 | 2.58 | 1.09 | 1.20 | 0.85 |
| Midbrain | 1.54 | 1.98 | 1.45, 1.70 | 2.30 | 0.98 | 1.40 | 1.38 |
| Medulla | — | 3.46 | 3.63,[d] 3.87 | 3.86 | 3.63 | 4.96 | 3.25 |
| Cerebellum | 2.90 | 0.79 | 1.04, 0.60 | 3.69 | 0.81 | 1.20 | 1.17 |
| Cervical cord | — | 3.68 | —, 3.96 | — | 4.31 | 4.48 | 3.70 |
| Thoracic cord | — | 1.82 | —, 3.42 | — | 3.59 | 3.08 | 3.50 |
| Lumbar cord | — | 4.40 | —, 4.42 | — | 4.94 | 4.40 | 3.96 |

[a] Values in $\mu$moles/g.
[b] Frontal lobe.
[c] Hypothalamus.
[d] Pons plus medulla.

in the synaptic vesicle fraction.[64] However, though only a small proportion of the total amino acid is recovered in the latter fraction, it has been possible to show by means of release from micropipettes that enough potent amino acid is present in extracts of this fraction to produce effects on guinea pig cortical neurons. It is possible that glycine and other active amino acids may adhere to synaptic vesicles. The amount of amino acid associated with the vesicle fraction from spinal cord is unknown.

### 2. *Cerebrospinal Fluid*

The concentration of glycine and other amino acids in CSF has been extensively reviewed.[65,66] Values of glycine concentration vary, but the most recent studies utilizing ion-exchange chromatography reported a glycine concentration of $< 1.0$ $\mu$mole/100 ml human CSF.[67] Aprison and Werman[2] found that cat CSF contained $< 1.5$ $\mu$mole/100 ml; similar values have been measured for dog CSF.[68] The low concentration of glycine in CSF and its very high plasma/CSF ratio of 35[67] is consistent with its potent physiological actions and probable transmitter function. The presence of large amounts of glycine in extracellular spaces might well interfere with normal brain or spinal cord function.

## VI. CHANGES IN PATHOLOGICAL AND EXPERIMENTAL CONDITIONS

### A. Pathological Conditions in Humans

There are at least two disorders of metabolism in which glycine accumulates in appreciable amounts in the blood of children suffering from developmental retardation and various neurological deficits. Idiopathic hyperglycinemia is characterized by recurrent episodes of ketosis, metabolic acidosis, and coma, but it is unlikely that the symptoms of the disease represent simply the effects of excess glycine. The administration of moderate amounts of glycine does not produce any change in the physical state of patients with this condition. There may exist a defect in the conversion of glycine to serine.[69]

A second disorder of metabolism has been recently described in which glycine is the only amino acid elevated in blood. Ketosis and acidosis are not present. On the basis of a decreased urinary oxalic acid excretion, a reduced "glycine oxidase" activity has been postulated.[70]

The concentration of glycine in CSF can be elevated by maintaining high blood levels of this amino acid.[71] In those conditions in which hyperglycinemia occurs, glycine is also present in large amounts in CSF. Thus, in severe liver and kidney disease[72,73] and idiopathic hyperglycinemia[69,70] there are marked elevations of glycine in the CSF. The relationship between the high glycine concentration in the CSF and the clinical state is not clear from these studies.

The concentration of glycine in CSF has been studied in a variety of neurological diseases, but too few cases with any one condition have been reported. However, a significantly increased amount of glycine in the CSF of ten patients with Parkinsonism has been described.[74]

### B. Experimental Studies

Glycine concentration in brain appeared to be unaffected by hypoxia,[75] insulin-induced hypoglycemia,[76,77] deprivation of protein,[78] or infusion of ammonia.[75] There were no early postmortem changes in glycine concentration.[75] A small decrease was produced by a diet deficient in vitamin E,[79] but the changes in a pyridoxine-deficient state were unclear.[80] The changes in glycine reported in convulsive states were variable; for example, the production of epileptogenic lesions in guinea pig cortex failed to alter the concentration of cortical glycine,[81] but a decrease of approximately 50% was reported in the brain of rabbits subjected to repeated electroshock convulsions.[82]

Glycine has potent CNS effects when given systematically. Intravenous administration of large amounts of glycine produces vomiting, diarrhea, and convulsions,[83,84] but in low doses it protects dogs from the effects of several emetic drugs.[84] On this basis, a specific action on the medullary vomiting center has been proposed.[84] Glycine has also been shown to potentiate pentobarbital anethesia[85] and to protect rats against seizures produced by oxygen poisoning and various convulsive drugs.[86]

### C. Postnatal Changes

The brain in adult mice has between 37%[87] and 60%[35] less free glycine than does the brain in newborn mice. However, no significant difference between the concentration of free glycine in adult and immature rat brain was found.[88] Brain slices removed from adult mice incorporate considerably more glycine than slices from newborns do.[35]

## VII. CONCLUDING REMARKS

We have attempted to point out that, although evidence exists for the synthesis of glycine from glucose in mouse brain, very little other information is available for any of its involved reactions or for alternate synthetic pathways in nervous tissue. Furthermore, despite its important role in synthesis of key compounds and its potential physiological importance as an inhibitory transmitter, only scanty information is available concerning its uptake *in vivo* by brain and spinal cord and the regulation of its concentration in the CNS. Nothing is known of its possible existence in pools in nervous tissue although such evidence is available for peripheral tissues.[36]

An exact description of the action of glycine on ionic processes in the membrane of the cat motoneuron has been given. Glycine has been shown to be present in interneurons which are the presynaptic cells involved in spinal inhibition; however, its release from these cells has not been demonstrated as yet. Further, distribution studies suggest that it may have a transmitter function at other sites in the nervous sytem and in other species. There is no detailed information to substantiate these inferences. At present,* the conclusion must be that glycine is an important amino acid in nervous tissue—both from a biochemical and a physiological point of view—and one that warrants further investigations.

## VIII. REFERENCES


1. R. K. Shaw and J. D. Heine, Ninhydrin positive substances present in different areas of normal rat brain, *J. Neurochem.* **12**:151–155 (1965).
2. M. H. Aprison and R. Werman, The distribution of glycine in cat spinal cord and roots, *Life Sci.* **4**:2075–2083 (1965).
3. L. T. Graham, Jr., R. P. Shank, R. Werman, and M. H. Aprison, Distribution of some synaptic transmitter suspects in cat spinal cord: Glutamic acid, aspartic acid, α-aminobutyric acid, glycine, and glutamine, *J. Neurochem.* **14**:465–472 (1967).
4. M. H. Aprison, R. P. Shank, R. A. Davidoff, and R. Werman, The distribution of glycine, a neurotransmitter suspect, in the central nervous system of several vertebrate species, *Life Sci.* **7**:583–590 (1968).
5. R. A. Davidoff, R. P. Shank, L. T. Graham, Jr., M. H. Aprison, and R. Werman, Association of glycine with spinal interneurons, *Nature* **214**:680–681 (1967).
6. R. A. Davidoff, L. T. Graham, Jr., R. P. Shank, R. Werman, and M. H. Aprison, Changes in amino acid concentrations associated with loss of spinal interneurons, *J. Neurochem.* **14**:1025–1031 (1967).
7. R. Werman, R. A. Davidoff, and M. H. Aprison, Inhibition of motoneurons by iontophoresis of glycine, *Nature* **214**:681–683 (1967).
8. R. Werman, R. A. Davidoff, and M. H. Aprison, Inhibitory action of glycine on spinal neurons in the cat, *J. Neurophysiol.* **31**:81–95 (1968).
9. R. Werman and M. H. Aprison, *in Structure and Function of Inhibitory Neuronal Mechanisms* (C. von Euler, S. Skoglund, and U. Soderberg, eds.), pp. 473–486, Pergamon Press, New York (1968).
10. M. H. Aprison and R. Werman, *in Neurosciences Research* (S. Ehrenpreis and O. C. Solnitzky, eds.), Vol. 1, pp. 143–174, Academic Press, New York (1968).
11. J. Giroux and A. Puech, Nouvelle methode de dosage du glycolle, *Ann. Pharm. Franc.* **21**:469–476 (1963).
12. R. J. Block, E. L. Durrum, and G. Zweig, *A Manual of Paper Chromatography and Paper Electrophoresis*, Academic Press, New York (1958).
13. R. B. Hamilton, Ion exchange chromatography of amino acids, *Anal. Chem.* **35**:2055–2064 (1963).
14. K. Randerath, *Thin Layer Chromatography*, Academic Press, New York (1966).
15. G. E. Vladimirov and A. P. Urinson, Glycine metabolism in the cerebral tissue of the rat in normal resting and in amytal-induced sleep, *Biochemistry* (*New York*) **22**:665–670 (1957).
16. H. Busch, E. Fujuwara, and L. M. Keer, Metabolic patterns for glucose-1-$C^{14}$ in tissues of tumor-bearing rats, *Cancer Res.* **20**:50–57 (1960).
17. H. H. Sky-peck, C. Rosenbloom, and R. J. Winzler, Incorporation of glucose into


* This chapter was submitted on February 15, 1968.

the protein-bound amino acids of one-day old mouse brain *in vitro*, *J. Neurochem.* **13**:223–228 (1966).
18. W. F. Bridgers, The biosynthesis of serine in mouse brain extracts, *J. Biol. Chem.* **240**:4591–4597 (1965).
19. R. Gilles and E. Schoffeniels, Action de la veratrine, de la cocaine et de la stimulation electrique sur la synthese et sur le pool des acides amines de la chaine nerveuse ventrale du hombard, *Biochim. Biophys. Acta* **82**:525–537 (1964).
20. R. L. Blakley, The interconversion of serine and glycine: Role of pteroylglutamic acid and other cofactors, *Biochem. J.* **58**:448–462 (1954).
21. H. Kawasaki, T. Sato, and G. Kituchi, A new reaction for glycine biosynthesis, *Biochem. Biophys. Res. Commun.* **23**:227–233 (1966).
22. S. Sved, The metabolism of amino acids in the central nervous system, Ph.D. Thesis, McGill University (1958).
23. K. Mase, U. Takahashi, and K. Ogata, The incorporation of [$^{14}$C] glycine into the protein of guinea pig brain cortex slices, *J. Neurochem.* **9**:281–288 (1962).
24. Y. Takahashi, M. Nomura, and S. Furusawa, *In vitro* incorporation of [$^{14}$C]-amino acids into proteins of peripheral nerve during wallerian degeneration, *J. Neurochem.* **7**:97–102 (1961).
25. N. D. Gracheva, Autoradiographical studies of glycine-$^{14}$C incorporation in some parts of the rat nervous system, *Tsitologiya* **6**:324–330 (1964).
26. H. Koenig, An autoradiographic study of nucleic acid and protein turnover in the mammalian neuraxis, *J. Biophys. Biochem. Cytol.* **4**:785–792 (1958).
27. G. W. Douglas and R. A. Martensen, The rate of metabolism of brain and liver glutathione in the rat studied with $C^{14}$ glycine, *J. Biol. Chem.* **222**:581–585 (1956).
28. Y. Takahashi and Y. Akabane, Incorporation of [$^{14}$C] glycine into glutathione in rat cerebral cortex slices, *J. Neurochem.* **7**:89–96 (1961).
29. H. I. Nakada and S. Weinhouse, Studies of glycine oxidation in rat tissues, *Arch. Biochem. Biophys.* **42**:257–270 (1953).
30. L. C. Leeper, V. J. Tulane, and F. Friedberg, Metabolism of glycine-2-$C^{14}$ in rat brain homogenates, *J. Biol. Chem.* **203**:513–517 (1953).
31. H. Borsook and J. Dubnoff, The formation of glycocyamine in animal tissues, *J. Biol. Chem.* **138**:389–403 (1941).
32. J. J. Pisano, J. D. Wilson, and S. Udenfriend, *in Inhibition in the Nervous System and γ-Aminobutyric Acid* (E. Roberts, ed.), pp. 226–235, Pergamon Press, New York (1960).
33. P. N. Abadom and P. G. Schoefield, Amino acid transport in brain cortex slices. I. The relation between energy production and the glucose-dependent transport of glycine, *Can. J. Biochem. Physiol.* **40**:1575–1590 (1962).
34. R. Blasberg and A. Lajtha, Substrate specificity of steady-state amino acid transport in mouse brain slices, *Arch. Biochem. Biophys.* **112**:361–377 (1965).
35. G. Levi, J. Kandera, and A. Lajtha, Control of cerebral metabolite levels. I. Amino acid uptake and levels in various species, *Arch. Biochem. Biophys.* **119**:303–311 (1967).
36. A. Lajtha, *in International Review of Neurobiology* (C. C. Pfeiffer and J. R. Smythies, eds.), Vol. 6, pp. 1–98, Academic Press, New York (1964).
37. H. R. V. Arnstein and A. Neuberger, The synthesis of glycine and serine by the rat, *Biochem. J.* **55**:271–279 (1953).
38. J. L. Simkin and K. White, The formation of glycine and serine. The influence of the administration of glycine, DL-serine and other compounds on levels of tissue glycine and serine, *Biochem. J.* **67**:287–291 (1957).
39. D. P. Purpura, M. Girado, T. G. Smith, D. A. Callan, and H. Grundfest, Structure–

activity determinants of pharmacological effects of amino acids and related compounds on central synapses, *J. Neurochem.* **3**:238–266 (1959).
40. D. R. Curtis and J. C. Watkins, The excitation and depression of spinal neurones by structurally related amino acids, *J. Neurochem.* **6**:117–141 (1960).
41. K. Krnjević and J. W. Phillis, Iontophoretic studies of neurones in the mammalian cerebral cortex, *J. Physiol.* **165**:274–304 (1963).
42. D. R. Curtis, J. W. Phillis, and J. C. Watkins, Actions of amino-acids on the isolated hemisected spinal cord of the toad, *Brit. J. Pharmacol.* **16**:262–283 (1961).
43. M. Fukuya, Studies on some physiological properties of γ-aminobutyric acid and related compounds, *Japan. J. Physiol.* **11**:126–146 (1961).
44. C. Edwards and S. W. Kuffler, The blocking effect of γ-aminobutyric acid (GABA) and the action of related compounds on single nerve cells, *J. Neurochem.* **4**:19–30 (1959).
45. J. Robbins, The excitation and inhibitions of crustacean muscle by amino acids, *J. Physiol.* (*London*) **148**:39–50 (1959).
46. H. Takahashi, A. Nagashima, C. Koshino, and H. Takahashi, Effects of γ-aminobutyric acid (GABA) γ-aminobutyryl choline (GABA-Ch) and their related substances on the cortical activity, *Japan. J. Physiol.* **9**:257–265 (1959).
47. A. Muneoka, Depression and facilitation of spinal reflexes by systematic omega-amino acids, *Japan. J. Physiol.* **11**:555–563 (1961).
48. D. R. Curtis, J. W. Phillis, and J. C. Watkins, The depression of spinal neurones by γ-amino-n-butyric acid and β-alanine, *J. Physiol.* (*London*) **146**:185–203 (1959).
49. D. R. Curtis and J. C. Watkins, The pharmacology of amino acids related to γ-aminobutyric acid, *Pharmacol. Rev.* **17**:347–391 (1965).
50. R. Werman, A review—Criteria for identification of a central nervous system transmitter, *Comp. Biochem. Physiol.* **18**:745–766 (1966).
51. C. Hebb, *in Handbuch der Experimentellen Pharmakologie* (G. B. Koelle, ed.), Vol. 15, Springer-Verlag, Berlin (1963).
52. S. Gelfan and I. M. Tarlov, Altered neuron population in $L_7$ segment of dogs with experimental hind-limb rigidity, *Am. J. Physiol.* **205**:606–616 (1963).
53. R. Werman, The specificity of molecular processes involved in neural transmission, *J. Theoret. Biol.* **9**:471–477 (1965).
54. D. R. Curtis, L. Hosli, G. A. R. Johnston, and I. H. Johnston, Glycine and spinal inhibition, *Brain Res.* **5**:112–114 (1967).
55. D. R. Curtis, L. Hosli, and G. A. R. Johnston, Inhibition of spinal neurons by glycine, *Nature* **215**:1502–1503 (1967).
56. B. A. Koechlin, On the chemical composition of the axoplasm of squid giant nerve fibers with particular reference to its ion pattern, *J. Biophys. Biochem. Cytol.* **1**:511–529 (1955).
57. G. G. J. Deffner, The dialyzable free organic constituents of squid blood; a comparison with nerve axoplasm, *Biochim. Biophys. Acta* **47**:378–388 (1961).
58. P. R. Lewis, The free amino-acids of invertebrate nerve, *Biochem. J.* **52**:330–338 (1952).
59. N. Frontali, *in Comparative Neurochemistry* (D. Richter, ed.), pp. 185–192, Macmillan Co., New York (1964).
60. S. I. Singh and C. L. Malhotra, Amino acid content of monkey brain. II. Quantitative values of cystine, α-alanine, serine and glycine, *J. Neurochem.* **9**:585–588 (1962).
61. P. Wiechert and P. Schroter, Der Einfluss von γ-aminobuttersäure, L-Glutaminsaure und glycin auf die Blut-Hirn-Schranke und die Enzymaktivitaten des Kaninchengehirnes, *Acta Biol. Med. Ger.* **12**:475–480 (1964).

62. A. M. Lassek, A comparative volumetric study of the gray and white substance of the spinal cord, *J. Comp. Neurol.* **62**:361–376 (1935).
63. A. Galindo, K. Krnjevič, and S. Schwartz, Micro-iontophoretic study on neurones in the cuneate nucleus, *J. Physiol.* **192**:359–377 (1967).
64. J. L. Mangan and V. P. Whittaker, The distribution of free amino acids in subcellular fractions of guinea pig brain, *Biochem. J.* **98**:128–137 (1966).
65. H. G. Knauff, W. Mialkowsky, and H. Zickgraf, Über die freien Aminosäuren des Liquor cerebrospinalis und ihren Hachweis mit kombinierten papier chromatographischen und elektropherographischen Methoden, *Z. Klin. Med.* **155**:483–505 (1959).
66. K. Schreier, *in Amino Acid Pools* (J. T. Holden, ed.), pp. 263–283, Elsevier, Amsterdam (1962).
67. J. C. Dickinson and P. B. Hamilton, The free amino acids of human spinal fluid determined by ion exchange chromatography, *J. Neurochem.* **13**:1179–1187 (1966).
68. J. Logothetis, Free amino acid content of cerebrospinal fluid in humans and dogs, *Neurology* **8**:299–302 (1958).
69. W. L. Nyhan and P. Tocci, Aminoaciduria, *Ann. Rev. Med.* **17**:133–160 (1966).
70. T. Gerritsen, E. Kaveggia, and H. A. Waisman, A new type of idiopathic hyperglycinemia with hypo-oxaluria, *Pediatrics* **36**:882–891 (1965).
71. P. Wiechert, Über die Permeabilität der Blut-Liquao-Schranke für einige Aminosäuren, *Acta Biol. Med. Germ.* **10**:305–310 (1963).
72. J. M. Walshe, Disturbances of amino acid metabolism following liver injury, *Quart. J. Med.* **22**:483–505 (1953).
73. D. Müting and K. N. Shivaram, Quantitative papier chromatographische Bestimmung der freien Aminosäuren in Liquor cerebrospinalis gesunder Menschen, *Hoppe-Seylers Z. Physiol. Chem.* **317**:34–38 (1959).
74. H. Bruck, F. Gerstenbrand, E. Grundig, and R. Teuflmayer, Über Ergebnisse von liquoranalysen beim Parkinson-syndrom, *Acta Neuropathol.* **3**: 638–644 (1964).
75. J. K. Tews, S. H. Carter, P. D. Roa, and W. E. Stone, Free amino acids and related compounds in dog brain: Post-mortem and anoxic changes, effects of ammonium chloride infusion, and levels during seizures induced by picrotoxin and by pentylenetetrazol, *J. Neurochem.* **10**:641–653 (1963).
76. H. G. Knauff and F. Bock, Über die freien Gehirnaminosäuren und das Änthanolamin de normalen Ratte sowie über das Verhalten diesier Stoffe nach experimenteller Insulinhypoglykämie, *J. Neurochem.* **6**:171–182 (1961).
77. R. K. Shaw and J. D. Heine, Effect of insulin on nitrogenous constituents of rat brain, *J. Neurochem.* **12**:527–532 (1965).
78. P. Mandel and J. Mark, The influence of nitrogen deprivation on free amino acids in rat brain, *J. Neurochem.* **12**:987–992 (1965).
79. L. C. Smith and S. R. Nelson, Effect of Vitamin E-deficiency on free amino acids of various rabbit tissues, *Proc. Soc. Exptl. Biol. Med.* **94**:644–646 (1957).
80. J. K. Tews and R. A. Lovell, The effect of a nutritional pyridoxine deficiency on free amino acids and related substances in mouse brain, *J. Neurochem.* **14**:1–7 (1967).
81. Y. Aelony, J. Logothetis, B. Bart, F. Morell, and M. Boris, Free amino acid concentrations in cerebral cortex of guinea pigs with epileptogenic lesions, *Exptl. Neurol.* **5**:525–532 (1962).
82. Y. Yamamoto, A. Mori, and D. Jinnai, Amino acids in the brain. Alteration of amino acids in rabbit brain caused by repetitive convulsive fits and comparison of amino acid contents in epileptic and non-epileptic human brain, *J. Biochem. (Tokyo)* **49**:368–372 (1961).
83. J. H. Lewis, The metabolism of glycine given intravenously at constant rates, *J. Biol. Chem.* **35**:567–576 (1918).

84. R. Koster, Emetic and anti-emetic actions of glycine in dogs, *Arch. Intern. Pharmacodyn.* **150**:384–400 (1964).
85. R. Koster, Effects of glycine on electroshock and barbiturates in mice, *Pharmacologist* **9**:225 (1967).
86. J. D. Wood, W. J. Watson, and N. E. Stacey, A comparative study of hyperbaric oxygen-induced and drug-induced convulsions with particular reference to $\gamma$-aminobutyric acid metabolism, *J. Neurochem.* **13**:361–370 (1966).
87. R. B. Roberts, J. B. Flexner, and L. B. Flexner, Biochemical and physiological differentiation during morphogenesis. XXIII. Further observations relating to the synthesis of amino acids and proteins by the cerebral cortex and liver of the mouse, *J. Neurochem.* **4**:78–90 (1959).
88. H. C. Agrawal, J. M. Davis, and W. A. Himwich, Postnatal changes in free amino acid pool of rat brain, *J. Neurochem.* **13**:607–615 (1966).

Chapter 12

# DEAMINATION OF NUCLEOTIDES AND THE ROLE OF THEIR DEAMINO FORMS IN AMMONIA FORMATION FROM AMINO ACIDS

H. C. Buniatian

*Institute of Biochemistry*
*Academy of Sciences*
*Armenian SSR, Yerevan*

## I. DEAMINATION OF ADENINE NUCLEOTIDES AND NICOTINAMIDE-ADENINE-DINUCLEOTIDES IN BRAIN

Conway and Cooke, studying the deamination of adenosine and of AMP in various tissues, demonstrated that they were also deaminated by brain tissue.[1] In 1953 Muntz[2] showed that in acetone powder extracts of dog brain the level of ammonia (Am) was increased by a number of nucleotides (AMP, ADP, ATP, NAD, and GMP), the most effective being GMP and NAD. ATP enhanced markedly the deamination of only AMP, with a concomitant increase of IMP. Weil-Malherbe[3] and Green purified AMP deaminase (10–20-fold) of ox brain and confirmed the activation of this enzyme by ATP and the formation of IMP from AMP. Other nucleotides ADP, ITP, and IDP—were ineffective. Later on Mendicino and Muntz,[4] using purified preparations of AMP deaminase of dog brain, studied the specific activation of the enzyme by ATP and showed that the ATP remained unchanged at the end of the reaction. The mechanism underlying ATP activation of the enzyme remains to be elucidated.

In the deamination of AMP, conversion of AMP to adenosine and adenosine to Am and inosine is of interest. Brain homogenates besides AMP deaminase contain 5-nucleotidase and adenosine deaminase.[5] The processes of dephosphorylation and Am formation from ATP and ADP are important. In brain homogenates ATP is dephosphorylated rapidly (ATP-ase), with an increase of ADP and AMP (myokinase) in the initial period and IMP and inosine later on. The dephosphorylation of ATP with the formation of ADP, AMP, IMP, and inosine occurs also in mitochondria of rat brain.[5] In brain homogenates a considerable amount of inosine is formed from added AMP; the latter is broken down almost completely within 30 min of incubation. The content of IMP increases for the first 20 min and subsequently decreases gradually during the rest of the incubation period (60 min); in this period considerable amounts of Pi are formed.[5]

Studies of the distribution of enzymes deaminating AMP and adenosine showed AMP deaminase to be associated with the particulate fraction[3] and adenosine deaminase with the soluble fraction.[5] Detailed studies in the laboratory of Palladin[6] indicated that white matter of rabbit brain had higher adenosine deaminase and 5-nucleotidase activities than gray matter. Adenosine deaminase was found mainly in the soluble fraction, its activity being low in nuclei and mitochondria, whereas 5-nucleotidase was localized mainly in mitochondria and microsomes. They demonstrated that ATP-activated AMP-deaminase was distributed in all subcellular particles, especially in mitochondria and in the soluble fraction. In mitochondria, ATP and ADP inhibited 5-nucleotidase activity and AMP was mainly converted to IMP.

The data presented indicate that brain mitochondria are rich in 5-nucleotidase and AMP-deaminase. In this respect the deamination and dephosphorylation of a number of nucleotides in brain mitochondrial preparations is of interest. The investigations of our laboratory (Table I) have shown that in rat brain crude mitochondrial preparations, AMP, ATP, GTP, NAD, NADH, adenosine diphosphate ribose (ADP-R), and FAD produced a considerable amount of Am. The same thing was observed in the presence of added ADP. As seen from Table I, the production of Am from ADP-R and especially AMP coincided with the marked release of Pi, while NAD induced no change in Pi. The formation of Am from AMP in mitochondria should be mainly ascribed to its conversion to adenosine, which is split into inosine and Am[6] by adenosine deaminase. A part of AMP can be directly deaminated to IMP. The latter and IDP were shown in our experiments also to be dephosphorylated intensively in mitochondrial preparations of brain. Obviously ATP and GTP give rise to Am after their preliminary dephosphorylation to AMP and GMP. However, the possibility of the direct deamination of ATP and even more so of ADP cannot be excluded. Thus, for instance, in our experiments on dialyzed extracts of rabbit brain acetone powder,[11] zinc was shown to inhibit the formation of Am from AMP by 30.5% and from ADP by 92.2%, while inhibition of Pi formation from ADP was only by 35%. $Cd^{2+}$ and $Ni^{2+}$ inhibited the formation of Am from AMP by 5.3 and 3.6% and from ADP by 81.4 and 83.1%, respectively, whereas the release of Pi from ADP was inhibited by 18.9 and 20.5%. A similar phenomenon was observed following PCMB. Apparently the deamination of ADP and AMP is catalyzed by different enzymes. Of considerable interest is the formation of Am from NAD.

Of the enzymes splitting NAD, NAD-nucleosidase (NAD-ase) exhibits considerable activity[12-15] in brain tissue. It is localized in microsomes,[16] and it splits NAD to nicotinamide (NA) and ADP-R, its activity being strongly inhibited by NA.[12-16] Brain tissue is devoid of NAD-pyrophosphatase,[14,16] which splits NAD to NMN and AMP. NAD-

TABLE I

**Deamination and Dephosphorylation of Nucleotides in Rat Brain Mitochondrial Preparations[a]**

| Substance | Ammonia ($\mu$g/mg protein) | Pi ($\mu$g/mg protein) |
|---|---|---|
| Incubation mixture (prior incubation) | 0.70 ± 0.6 | 7.40 ± 1.10 |
| Incubation mixture (after incubation) | 2.30 ± 0.7 | 13.20 ± 1.20 |
| AMP | 6.30 ± 0.8 | 42.2 ± 1.40 |
| ATP | 7.08 ± 0.8 | |
| GTP | 6.95 ± 0.23 | |
| NAD | 5.90 ± 0.97 | 13.2 ± 1.20 |
| NAD + nicotinamade | 6.28 ± 0.95 | 13.1 ± 0.7 |
| ADP-R | 5.80 ± 0.65 | 19.7 ± 0.5 |
| NADH | 8.13 ± 0.3 | |
| FAD | 4.47 ± 0.25 | |

[a] The mitochondrial preparations were obtained according to Brody and Bain[(7)] as modified by Palladin and Kirsenko.[(8)] Incubation mixture was composed of 0.1 ml 0.133 *M* potassium phosphate buffer (*p*H 7.4); 0.15 ml 0.2 *M* tris-HCl buffer (*p*H 7.45) 0.1 ml 0.12 *M* $MgSO_4$; and 0.5 ml mitochondrial preparation (3.5–4.5 mg protein). The volume of the reaction mixture was brought to 1.5 ml by the addition of 0.25 *M* sucrose (*p*H 7.4). NAD, NADH, FAD, ADP-R, and AMP (Sigma Co.) were added in 2.86 $\mu$moles; ATP and GTP, in 1 $\mu$mole; and NA, 8.2 $\mu$moles to each test. Incubation 2 hr at 37°. In tests with dephosphorylation, potassium phosphate buffer was omitted. Ammonia was determined by microdiffusion, Pi (according to Lowry and Lopez,[(9)]) and proteins (Lowry *et al.*[(10)]). Results are means plus or minus S.E.M. of five to seven experiments.

pyrophosphorylase, which catalyzes the formation of NAD from ATP and NMN and the cleavage of NAD by PPi to ATP and NMN, is observed exclusively in nuclei.[(17–19)]

Of the products obtained from NAD following the action of these enzymes, ADP-R, AMP, ATP, and NMN can form Am. However, NADase, which produces ADP-R, was inhibited by NA; besides, the formation of Am from ADP-R in the mitochondrial fraction was accompanied by a considerable release of Pi. As to AMP and ATP, they cannot be formed from NAD in brain mitochondria since brain tissue does not possess NAD pyrophosphatase activity and NAD pyrophosphorylase as localized in nuclei. On the other hand, the formation of Am from AMP as well as from ADP and ATP in mitochondria was accompanied by a marked increase of Pi, while no change in Pi was observed following NAD, even in the absence of NA. Thus the formation of Am from NAD, which was studied in our laboratory by Movcessian, cannot be ascribed to the activity of the above-mentioned enzyme systems. It might have been expected that Am could be obtained from NAD through its deamidation; however, as was shown by our experiments, no Am was formed from NA and deamino-NAD (D-NAD) in various brain preparations. The formation of Am following NAD cannot be ascribed to the stimulation of the oxidative deamination of endogenous Glu either, inasmuch as almost an equal amount of Am is formed from NAD in anaerobic conditions. The results obtained leave no doubt that NAD is deaminated in the adenine moiety and converted to D-NAD, in which the adenylic acid residue of NAD is replaced by the inosinic acid residue.

As is shown in Table I, in brain mitochondrial preparations both NAD and NADH were deaminated intensively, NADP being deaminated to a lesser extent. The investigations of Movcessian and Manacian[20] showed that the deamination of NAD occurred also in rat brain homogenates, with the formation of some Pi. The increase of Pi was completely inhibited by NA, and the deamination of NAD proceeded more intensively (Table II).

The results obtained favor the proposition that in brain tissue NAD is split mainly by NADase and that this tissue exhibits no NAD pyrophosphatase activity.

The data presented (Table II) indicate that considerable amounts of Am were formed from AMP with a twofold increase of Pi, NA not affecting the levels of Am and Pi.

**TABLE II**

**Deamination and Dephosphorylation of NAD and AMP by Homogenates of Rat Brain**[a]

| Substance | Ammonia ($\mu$g/mg protein) | Pi ($\mu$g/mg protein) |
|---|---|---|
| Incubation mixture (prior incubation) | 1.06 ± 0.3 | 12.30 ± 2.50 |
| Incubation mixture (after incubation) | 3.03 ± 0.89 | 19.60 ± 3.83 |
| AMP | 10.60 ± 1.50 | 46.70 ± 2.46 |
| NAD | 9.08 ± 0.98 | 27.70 ± 6.23 |
| Nicotinamide | 3.05 ± 0.85 | 18.70 ± 1.25 |
| NAD + nicotinamide | 9.54 ± 1.00 | 17.20 ± 1.00 |

[a] Conditions were the same as in Table I. 0.5 ml of homogenate (150 mg tissue) was added to each test. Results are means plus or minus S.E.M. of four to eight experiments.

The content of NAD in mitochondrial preparations following incubation was estimated by yeast alcoholdehydrogenase reaction (NADH formation) by ultraviolet absorption at 340 mμ. On incubation of mitochondrial fractions with NAD + NA the content of NAD was considerably decreased parallel to the formation of Am. As was mentioned above the production of Am following the addition of NAD may be due to the conversion of NAD to D-NAD only, the activity of the latter toward yeast alcoholdehydrogenase being much lower than NAD.[21,22]

NAD is deaminated to D-NAD by preparations of taka-diastase. Kaplan *et al.*[23] have shown that deaminase obtained from taka-diastase deaminates besides NAD also adenosine, AMP, ADP, ATP, ADP-R, and NADH, i.e., it is a nonspecific enzyme. The results obtained by us indicate that AMP and NAD are deaminated in brain tissue by different deaminases. Thus, for instance, in glycine buffer, brain mitochondrial preparations did not deaminate NAD, while there was an intensive deamination of AMP. Studying the deamination of AMP and NAD in various subcellular fractions, Movcessian demonstrated that the deamination of both nucleotides proceeds almost equally in nuclei and mitochondria. In microsomes AMP was deaminated more intensively (about three times) than NAD, and in the soluble fraction a substantial increase of Am from AMP was observed, while NAD was not deaminated at all. Moreover, ATP increased AMP deaminase activity, but did not affect the NAD deaminase, which is in keeping with the data of Muntz.[2]

## II. DEAMINATION OF AMINO ACIDS IN BRAIN TISSUE

Amino acids may be considered as the eventual main source of Am in the organism. However, the mechanism of Am formation from them, both in brain and in other organs, remains obscure. The brain possesses no L-amino acid oxidase activity. This enzyme exhibits low activity in liver and kidney even at optimal high *p*H values.[24] D-amino acid oxidase, which is considerably more active, was also found in brain tissue.[25–28] However, the role of D-amino acid oxidase in animal tissues remains enigmatic, since neither D-amino acids nor any racemase activity have been found in them. Krebs[24] has shown that Glu is the only amino acid supporting respiration of brain tissue without any increase of free Am, which on the contrary decreases. According to the author this is due to the synthesis of Gln.[29] Weil-Malherbe[30] demonstrated that of the 13 amino acids assayed only Glu underwent oxidative deamination in brain tissue with the formation of α-ketoglutaric acid. In our laboratory the utilization of endogeneous Glu, Asp, and GABA were studied in rat brain homogenates during postnatal development by Aprikian and Paronian.[30a] The results obtained showed that following the incubation of homogenates the disappearance of Glu started in the first week and increased up to the fourth week of postnatal life. Concomitantly an appreciable increase of

Asp was observed, which exceeded the amount of utilized Glu. The utilization of GABA was much less pronounced; it started in the second week and increased up to the fourth week.

The discoveries of Glu dehydrogenase by Euler *et al.*(31) and of the process of transamination by Braunstein and Kritzman(32) in animal tissues were of a great significance in amino acid metabolism. Later on Braunstein *et al.*(33,34) proposed that Am is formed from amino acids through transdeamination, i.e., by virtue of Glu formation from α-ketogluturate and amino acids and deamination of Glu to Am and α-ketogluturate. However, as was shown by Euler *et al.*(31) and later on by others,(35,36) the main role of Glu dehydrogenase is to catalyze the reductive amination of α-ketoglutarate rather than the thermodynamically less favorable oxidative deamination of Glu by NAD. It must also be noted that a number of tissues lack Glu dehydrogenase activity (skeletal, cardiac muscle, and mammillary gland) in spite of the formation of free Am there.

Numerous investigations have shown that in animal tissues Glu is metabolized mainly through Asp. Even in liver, which possesses the highest Glu dehydrogenase activity (mitochondria), Glu was nearly quantitatively converted into Asp.(37,38) In various brain tissue preparations up to 90% of the Glu utilized is converted into Asp.(39–43) The data available indicate that only in aged and disrupted mitochondria, or in the presence of malonate and uncouplers, the transamination pathway of Glu metabolism is suppressed and the deamination pathway is greatly stimulated.(37,38,40,43–45) But in intact mitochondria a considerable part of the amino acid nitrogen can be incorporated through Glu into Asp by the transamination of Glu with oxaloacetic acid, which is again regenerated from the α-ketoglutarate formed. From the evidence presented the conclusion may be drawn that Asp plays an important role in the terminal metabolism of the amino nitrogen of amino acids. Its participation in the synthesis of urea is well known. Its role in the formation of free Am from amino acids is also of interest. In this respect the reamination of IMP by Asp to AMP and the deamination of AMP with the formation of Am and IMP as well as the possibility of the reamination of other deamino-nucleotides by Asp is of interest.

## III. THE ROLE OF INOSINE MONOPHOSPHATE (IMP) AND NICOTINAMIDE-HYPOXANTHINE-DINUCLEOTIDE (DEAMINO-NAD) IN THE FORMATION OF AMMONIA FROM AMINO ACIDS IN BRAIN

The first experimental data concerning the formation of adenylosuccinate (AMP-S) from AMP and fumarate and the cleavage of AMP-S to AMP and fumarate by an enzyme fraction purified from yeast autolysates were obtained by Carter and Cohen.(46–48) Lieberman(49) showed on cell-free extracts of E. coli that IMP is condensed with Asp, forming AMP-S, and that the latter is split into AMP and fumarate by an enzyme from E. coli. According to him no AMP-S formation occurs in the absence of GTP. The amination of IMP by Asp and the formation of AMP-S in muscle preparations has been shown by many authors.(50–54) In the above mentioned investigations of the many amino acids and amino compounds tested only Asp was shown to be effective in the amination of IMP.(49–54)

According to Kometiani *et al.*(55,56) the reamination of IMP by Asp and the deamination of AMP can play an important role in the formation of Am from amino acids. Since NAD is deaminated to D-NAD possessing an IMP residue, it was interesting to study the possibility of its reamination by Asp. The effect of Glu and GABA on this process was also studied. The data obtained(57) (Table III) indicated that in crude rat brain mitochondrial preparations Asp, GABA, D-NAD, D-NAD + NA alone produced an insignificant amount of Am. A significant increase of Am was observed following the addition of D-NAD with GABA and especially with Asp. Having in mind that ATP(50,53) and according to some authors GTP(49,51) participate in the amination of IMP by Asp, in our experiments we studied their effect on the formation of Am in the presence of added Asp and D-NAD. The results presented in Table III indicated that the

**TABLE III**

**The Formation of Ammonia in Rat Brain Mitochondrial Preparations Incubated with Asp, GABA and D-NAD[a]**

| Substance | The difference in ammonia ($\mu$g/mg protein) as compared with incubated controls[b] |
|---|---|
| Asp | 0.38 ± 0.3 |
| GABA | 0.29 ± 0.2 |
| D-NAD | 0.5 ± 0.4 |
| D-NAD + nicotinamide | 0.56 ± 0.3 |
| Asp + D-NAD + nicotinamide | 2.21 ± 0.3 |
| GABA + D-NAD + nicotinamide | 1.39 ± 0.5 |
| ATP | 4.09 ± 1.0 |
| Asp + ATP | 5.19 ± 1.7 |
| GABA + ATP | 4.75 ± 0.35 |
| D-NAD + ATP + nicotinamide | 4.56 ± 0.62 |
| Asp + D-NAD + ATP + nicotinamide | 7.62 ± 0.91 |
| GABA + D-NAD + ATP + nicotinamide | 6.21 ± 0.28 |

[a] Conditions were the same as in Table I. D-NAD was added in 1.4 $\mu$moles; Asp and GABA in 26.0 $\mu$moles. Results are means plus or minus S.E.M. of four to ten experiments.

[b] The level of ammonia in incubated controls was 3.07 ± 0.44. $\mu$g/mg protein.

simultaneous addition of ATP, D-NAD and Asp resulted in a considerable increase of Am. GABA was less effective than Asp. It must be noted that in experiments with ATP a part of the Am formed was confined in Gln; thus the total amount of Am was higher than the values brought in Table III. The studies carried out with Glu[(57)] have shown that it is less effective than Asp in the formation of Am in the presence of D-NAD. Apparently, GABA and Glu participate in the production of Am through the Asp formed from them. In our experiments GTP itself increased the content of Am, but contrary to ATP it had an insignificant effect on the production of Am in experiments with D-NAD and Asp.

In investigations carried out with IMP,[(57)] which was added in amounts twice as high as D-NAD (Table IV), no marked effect on Am formation was observed in the presence of either added Asp or ATP + Asp.

The formation of Am following the addition of D-NAD and Asp to the mitochondrial preparation of the brain considerably increased during a 30 min incubation, when the structural integrity of mitochondria was not impaired. The level of Am remained unchanged on continuing the incubation for 2 hr. In experiments with added D-NAD + Asp + ATP the process of Am formation markedly and gradually increased up to 1 hr but did not change further during the next hour.

The investigations of our laboratory showed that in rat liver mitochondrial preparations D-NAD also stimulates considerably the formation

**TABLE IV**

**The Formation of Ammonia in Rat Brain Mitochondrial Preparations Incubated with Asp and IMP[a]**

| Substance | The difference in ammonia ($\mu$g/mg protein) as compared with incubated controls[b] |
|---|---|
| Asp | 0.31 ± 0.16 |
| IMP | −0.07 ± 0.02 |
| ATP | 3.94 ± 0.35 |
| ATP + IMP | 4.35 ± 0.22 |
| IMP + Asp | 0.88 ± 0.22 |
| ATP + Asp | 5.08 ± 0.33 |
| ATP + IMP + Asp | 5.86 ± 0.22 |

[a] Conditions were the same as in Table I. IMP was added in 2.8 $\mu$moles. Results are means plus or minus S.E.M. of 6 experiments.

[b] The level of Am in incubated controls was 3.08 ± 0.18 $\mu$g/mg protein.

of Am from Asp and Glu, this effect being more pronounced than that obtained in brain mitochondrial preparations. A concomitant and intensive utilization of Asp and Glu took place. In anaerobic conditions the formation of Am in the presence of added D-NAD and Asp proceeded almost to the same degree as in aerobic conditions, while in experiments with D-NAD + Glu it was markedly reduced in anaerobic conditions.[58] This indicates that Glu is oxidized to Asp and the latter together with D-NAD participates in Am formation. It should be noted that when D-NAD was replaced by NAD a much smaller amount of Am was formed from added Glu. This does not favor the oxidative deamination of Glu, since in the activation of Glu dehydrogenase, D-NAD is less effective than NAD.[59]

In liver, IMP was also almost ineffective in the formation of Am from Asp. The ineffectiveness of IMP as compared with D-NAD should be ascribed to the fact observed in our experiments that in mitochondrial preparations IMP and the product of its reamination, AMP, were intensively dephosphorylated (Table I), the dephosphorylation of AMP being more intensive than its deamination. On the other hand, under conditions providing the regeneration of ATP, AMP is almost completely converted to ATP. In the investigations of Newton and Perry with dialyzed extracts of muscle tissue acetone powder in the presence of IMP, Asp, fructose diphosphate, NAD, and $Mg^{2+}$, no AMP was found and the Asp nitrogen was recovered in ATP.[50,52] The same thing was observed in the experiments of Yefimochkina and Braunstein[53] in the presence of phosphoglyceric acid. As was shown above the formation of AMP-S and deamination of AMP are activated by ATP. In mitochondria, during the process of oxidative phosphorylation, AMP is converted to ATP. In ATP deficiency AMP is intensively dephosphorylated and converted to adenosine, the latter being deaminated to inosine,[6] which is not reaminated with Asp.[49] Thus, in mitochondria of brain and liver, processes inducing an exhaustion of the sources of AMP and IMP occur. As to NAD and D-NAD their deamination and reamination are predominant in mitochondria. The reduction of NAD to NADH in mitochondria in all probability does not abolish the stimulatory effect of these nucleotides in the production of Am from Asp, since NADH was deaminated by rat brain mitochondria even to a greater extent than NAD (Table I). In addition, in experiments with NADH + Glu and NADH + GABA the formation of Am was more pronounced than when Glu and GABA were available alone.[57]

The process of the reamination of D-NAD to NAD is most probably similar to the reamination of IMP by Asp, i.e., D-NAD with Asp forms NAD-succinate, which is split to fumarate and NAD, the latter being again deaminated to D-NAD.

To test this point fumarate and NAD were added to rabbit liver mitochondrial fractions, having in mind that fumarate with AMP forms AMP-S.[46,48,49] The formation of Am in the presence of added NAD, AMP, D-NAD, and Asp was studied. The results obtained indicated (Table V) that fumarate strongly inhibited the production of Am from

TABLE V

**The Effect of Fumarate on the Formation of Ammonia from NAD, AMP, and D-NAD + Asp in Liver Mitochondrial Preparations**[a]

| Substance | The difference in ammonia (μg/mg protein) as compared with incubated controls[b] |
|---|---|
| Fumarate | −1.50 ± 0.18 |
| NAD + nicotinamide | 2.80 ± 0.42 |
| NAD + nicotinamide + fumarate | −1.10 ± 0.34 |
| AMP | 4.20 ± 0.60 |
| AMP + fumarate | −1.50 ± 0.16 |
| D-NAD + nicotinamide + Asp + fumarate | −0.80 ± 0.20 |

[a] Conditions were the same as in Table I. Fumarate was added in 26 μmoles. Results are means plus or minus S.E.M. of 6 experiments.
[b] The level of Am in incubated controls was 5.75 ± 0.31 μg/mg protein.

NAD, AMP, and D-NAD + Asp, the addition of which, especially D-NAD + Asp without fumarate, resulted in a substantial increase of Am (Table III). Fumarate decreased the level of Am in every case below control values. The same phenomenon was observed with rat brain mitochondria. The inhibitory effect of fumarate may be explained in terms of the formation of AMP-S and NAD-succinate, which in the presence of excess fumarate were not split to AMP and NAD. In order to confirm the formation of NAD-succinate, Asp was determined in experiments with added NAD + fumarate and AMP + fumarate in deproteinized supernatant fractions following hydrolysis with $Ba(OH)_2$. It is known that the hydrolysis of AMP-S with hydrochloric acid[60,61] or $Ba(OH)_2$[61,62] give rise to Asp. The data presented in Table VI indicated that in experiments with NAD + fumarate and AMP + fumarate the alkaline hydrolysis of the protein-free supernatant fraction brought about the release of considerable amounts of Asp, which might be due to the formation of AMP-S and NAD-succinate from AMP and NAD in the presence of added fumarate. It is worth noting that in experiments in which only NAD was added the level of endogenous Asp was considerably decreased. The results of our other experiments, indicating that during the incubation of the mitochondrial preparations with D-NAD and $^{15}N$-labeled Asp the label was recovered in NAD isolated after the incubation on DEAE-cellulose, and especially in Am formed, also confirm the reamination of D-NAD to NAD by Asp. The demonstration of the label in NAD following the addition of labeled Asp with D-NAD cannot be explained by the incorporation of the

## TABLE VI

**The Release of Asp by Alkali Hydrolysis of Protein-Free Supernatant of Incubated Rabbit Liver Mitochondrial Preparations with NAD + Fumarate and AMP + Fumarate[a]**

| Substance | Asp (μg/mg protein) |
|---|---|
| Incubation mixture alone (prior incubation) | 2.65 ± 0.28 |
| Incubation mixture alone (after incubation) | 3.4 ± 0.08 |
| Incubation mixture + fumarate | 5.1 ± 0.3 |
| Incubation mixture + AMP | 3.4 ± 0.13 |
| Incubation mixture + NAD | 1.0 ± 0.13 |
| Incubation mixture + AMP + fumarate | 9.2 ± 0.5 |
| Incubation mixture + NAD + nicotinamide + fumarate | 11.0 ± 0.4 |

[a] Conditions were the same as in Table I. AMP and NAD were added in 5.86 μmoles, fumarate in 26 μmoles, and NA in 16.4 μmoles. The results are means plus or minus S.E.M. of 5 to 6 experiments.

Asp nitrogen in adenine nucleotides and through them in NAD, because NAD is formed from them only in the nuclei.[17-19]

The results obtained and the data available allow the following mechanism of Am formation from amino acids in brain and most probably in other organs to be proposed:[57,63]

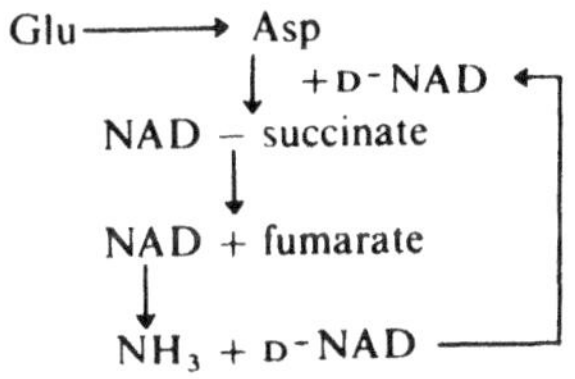


This cyclic process in which D-NAD is involved repeatedly with one Am fabricated for each turn of the cycle provides a mechanism channeling amino groups from amino acids into NAD with subsequent deamination of the latter to Am and D-NAD. Concerning the reamination of D-NAD to NAD in living organisms, the results of Kaplan *et al.*[64] indicating a considerable increase of NAD in mouse liver and kidney and some increase in brain following the injection of D-NAD are of interest. It is worth mentioning that D-NAD was more effective in raising NAD than NAD itself. The relationship between intra- and extramitochondrial pyridine nucleotides is of significance. As was mentioned above, they are synthesized only in nuclei. In the experiments of Purvis and Lowenstein,[65] carried out on liver mitochondria *in vitro*, NAD and NADPH and to a much lesser extent NADH and NADP were transferred across the mitochondrial membrane. According to them, NAD and NADH penetrate into the mitochondria *in vivo* at rates that lead to a 62% replacement of mitochondrial NAD in about 1 hr.

That AMP deamination with IMP reamination has a role in Am formation from amino acids cannot be excluded.[55,56] The rapid incorporation of $^{15}N$ from labeled ammonium citrate in the amino group of skeletal muscle adenine nucleotides,[66] the reamination of IMP by Asp in muscle preparations,[50,51,53,54] and the demonstration of AMP-S in liver[61,62] speak in favor of this mechanism. However, as has been shown in our experiments in brain and liver mitochondria, D-NAD is considerably more effective in the formation of Am in the presence of Asp than IMP. Preliminary experiments carried out by us on brain mitochondria[11] have shown that IDP stimulates more intensively the formation of Am from Asp than IMP. The mechanism underlying the action of IDP remains to be settled.

The formation of D-NAD by mitochondria, the presence of which in considerable amounts was established in our laboratory by Movcessian and Kamalian in liver mitochondria, may be of importance not only in the metabolism of amino acids but in oxido-reduction as well. NAD and D-NAD have a different effect on glycolysis. In crude brain mitochondrial fractions the former enhances lactate formation, while on the addition of D-NAD less lactate is formed but pyruvate and ketoglutarate are considerably increased.[67] The preliminary experiments of Movcessian in our laboratory indicate that D-NAD increases the process of coupled phosphorylation in brain mitochondria.

## IV. REFERENCES

1. E. J. Conway and R. Cooke, The deaminates of adenosine and adenylic acid in blood and tissues, *Biochem. J.* **33**:479–492 (1939).
2. J. A. Muntz, The formation of ammonia in brain extracts, *J. Biol. Chem.* **201**:221–233 (1953).
3. H. Weil-Malherbe and R. H. Green, Ammonia formation in brain. 2. Brain adenylic deaminase, *Biochem. J.* **61**:218–226 (1955).
4. J. Mendicino and J. A. Muntz, The activating effect of adenosine triphosphate on brain adenylic deaminase, *J. Biol. Chem.* **233**:178–183 (1958).
5. R. M. Smillie, The breakdown of adenosine phosphate by brain tissue, *Arch. Biochem. Biophys.* **67**:213–224 (1957).
6. A. V. Palladin, N. M. Polyakova, and M. K. Malysheva, Deaminases of Brain and its Subcellular Particles, in *Chemistry and Function of Nervous System*, pp. 104–109, Proceedings of the International Symposium, September 24–28, 1965, Leningrad, (1967).
7. T. M. Brody and J. A. Bain, A mitochondrial preparation from mammalian brain, *J. Biol. Chem.* **195**:685–696 (1952).

8. A. V. Palladin and O. V. Kirsenko, Adenosine triphosphatase in various cell fractions of the cerebrum, *Biokhimiya* **26**:283–390 (1961).
9. O. H. Lowry and J. Lopez, The determination of inorganic phosphate in the presence of labile phosphate esters, *J. Biol. Chem.* **162**:421–428 (1946).
10. O. H. Lowry, N. J. Rosebrough, A. L. Farr, and R. T. Randal, Protein measurement with the Folin phenol reagent, *J. Biol. Chem.* **193**:265–275 (1951).
11. H. C. Buniatian and A. V. Haroutunian, *Problems of Brain Biochemistry*, Vol. 4, pp. 29–39, Academic Press, Yerevan (1968).
12. P. J. G. Mann and J. H. Quastel, Nicotinamide, cozymase and tissue metabolism, *Biochem. J.* **35**:502–517 (1941).
13. P. Handler and J. R. Klein, The inactivation of pyridine nucleotides by animal tissues *in vitro*, *J. Biol. Chem.* **143**:49–57 (1942).
14. A. Kornberg and O. Lindberg, Diphosphopyridine nucleotide pyrophosphatase, *J. Biol. Chem.* **176**:665–677 (1948).
15. H. McIlwain and R. Rodnight, Breakdown of cozimase by a system from nervous tissue, *Biochem. J.* **44**:470–477 (1949).
16. K. B. Jacobson and N. O. Kaplan, Distribution of enzymes cleaving pyridine nucleotides in animal tissues, *J. Biophys. Biochem. Cytol.* **3**:31–43 (1957).
17. G. H. Hogeboom and W. C. Schneider, The synthesis of diphosphopyridine nucleotides by liver cell nuclei, *J. Biol. Chem.* **197**:611–619 (1952).
18. M. R. Atkinson, J. F. Jackson, and R. K. Morton, Mononucleotide nicotinamide adenylyltransferase of pig liver nuclei. The effect of nicotinamide mononucleotide concentration and pH on dinucleotide synthesis, *Biochem. J.* **80**:318–323 (1961).
19. S. E. Severin, L. A. Zeitlin, and V. J. Telepneva, Enzymatic synthesis of NAD from nicotinamide-mononucleotide and ATP in isolated nuclei of heart, skeletal muscle and liver of rabbit, *Dokl. Akad. Nauk SSSR* **164**:953–955 (1965).
20. S. G. Movcessian and R. F. Manassian, The Deamination and Deamidation of Nicotinamid-Adenine-Dinucleotide and of Its Components in Brain Liver and Renal Tissue Preparations, *Problems of Brain Biochemistry*, Vol. 3, pp. 53–66, Academic Press, Yerevan (1967).
21. F. Schlenk, H. Hellström, and H. V. Euler, Desaminocozymase, *Ber.* **71B**:1471–1479 (1938).
22. M. E. Pullman, S. P. Colowick, and N. O. Kaplan, Comparison of diphosphopyridine nucleotide with its deaminated derivative in various enzyme systems, *J. Biol. Chem.* **194**:593–602 (1952).
23. N. O. Kaplan, S. P. Colowick, and M. M. Ciotti, Enzymatic deamination of adenosine derivatives, *J. Biol. Chem.* **194**:579–591 (1952).
24. H. A. Krebs, Metabolism of amino acids. III. Deamination of amino acids, *Biochem. J.* **29**:1620–1644 (1935).
25. K. Yagi, T. Nagatsu, and T. Ozawa, Inhibitory action of chlorpromazine on the oxidation of D-amino-acid in the diencephalon of the brain, *Nature* **177**:891–892 (1956).
26. J. T. Dunn and G. T. Perkoff, D-amino acid oxidase activity in human tissues, *Biochim. Biophys. Acta* **73**:327–331 (1963).
27. A. H. Neims, W. D. Zievernik, and J. D. Smilack, Distribution of D-amino acid oxidase in bovine and human nervous system, *J. Neurochem.* **13**:163–168 (1966).
28. D. B. Goldstein, D-amino acid oxidase in brain: Distribution in several species and inhibition by pentobarbitone, *J. Neurochem.* **13**:1011–1016 (1966).
29. H. A. Krebs, Metabolism of amino-acids. IV. The synthesis of glutamine from glutamic acid and ammonia, and the enzymic hydrolysis of glutamine in animal tissues, *Biochem. J.* **29**:1951–1969 (1935).

30. H. Weil-Malherbe, The metabolism of the glutamic acid in brain, *Biochem. J.* **30**:665–676 (1936).

30a. G. A. Aprikian and J. A. Paronian, Certain Aspects of Brain Nitrogen Metabolism in Ontogenesis, *Problems of Brain Biochemistry*, Vol. 3, pp. 67–82, Academic Press, Yerevan (1967).

31. H. V. Euler, E. Adler, G. Günter, and B. B. Das, Über den enzymatischen Abbau und Aufbau der Glutaminsäure. II In tierischen Geweben, *Hoppe Seyler's. Z. Physiol. Chem.* **254**:61–103 (1938).

32. A. E. Braunstein and M. G. Kritsman, Formation and breakdown of amino acids by intermolecular transfer of the amino group, *Nature* **140**:503–504 (1937).

33. A. E. Braunstein and R. M. Asarkh. On the mechanism of deamination of L-amino acids in liver and kidney, *Biokhimiya* **9**:337–359 (1944).

34. A. E. Braunstein and R. M. Asarkh, The mode of deamination of L-amino acids in surviving tissues, *J. Biol. Chem.* **157**:421–422 (1944).

35. J. A. Olson and C. B. Anfinsen, The crystallization and characterization of glutamic acid dehydrogenase, *J. Biol. Chem.* **197**:67–79 (1952).

36. H. J. Strecker, Glutamic dehydrogenase, *Arch. Biochem. Biophys.* **46**:128–140 (1953).

37. P. Borst, The pathway of glutamate oxidation by mitochondria isolated from different tissues, *Biochim. Biophys. Acta* **57**:256–269 (1962).

38. N. Katunuma and M. Okada, An alternative coupling reaction of mitochondria-bound transaminase and tricarboxylic acid cycle and its metabolic role, *J. Vitaminol* **8**:309–314 (1962).

39. H. A. Krebs and D. Bellamy, The interconversion of glutamic and aspartic acid in respiring tissues, *Biochem. J.* **75**:523–529 (1960).

40. B. J. Haslam and H. A. Krebs, The metabolism of glutamate in homogenates and slices of brain cortex, *Biochem. J.* **88**:566–578 (1964).

41. H. M. Diomin, S. S. Musaelian, V. C. Carapetian, E. U. Osipova, and G. A. Hagopian, The effect of $\gamma$-aminobutyric acid on the metabolism of glutamic acid aspartic acid, alanine and on the neutralization of ammonia in brain tissue, *Problems of Brain Biochemistry*, Vol. 1, pp. 45–59, Academic Press, Yerevan (1964).

42. R. Baläzs, Control of glutamate metabolism. The effect of pyruvate, *J. Neurochem.* **12**:63–76 (1965).

43. R. Baläzs, Control of glutamate oxidation in brain and liver mitochondrial systems, *Biochem. J.* **95**:497–508 (1965).

44. A. F. Müller and F. Leuthardt, Die umwandlung der Glutaminsüre in Asparaginsäure in den Mitochondria der Leber, *Helv. Chim. Acta* **33**:268–273 (1950).

45. E. J. De Hahn, J. M. Tager, and E. C. Slater, Factors affecting the pathway of glutamate oxidation in rat liver mitochondria, *Biochim. Biophys. Acta* **131**:1–13 (1967).

46. C. E. Carter and L. H. Cohen, Enzymatic synthesis of adenylosuccinic acid, *Federation Proc.* **14**:189–190 (1955).

47. C. E. Carter and L. H. Cohen, Enzymatic synthesis of adenylosuccinic acid, *J. Am. Chem. Soc.* **77**:499–500 (1955).

48. C. E. Carter and L. H. Cohen, The preparation and properties of adenylosuccinase and adenylosuccinic acid, *J. Biol. Chem.* **222**:17–30 (1956).

49. J. Lieberman, Enzymatic synthesis of adenosine-5-phosphate from inosine-5-phosphate, *J. Biol. Chem.* **223**:327–339 (1956).

50. A. A. Newton and S. V. Perry, Incorporation of nitrogen-15 in 6-$NH_2$ group of adenosine triphosphate by muscle extracts, *Nature* **179**:49–50 (1957).

51. A. A. Newton and S. V. Perry, The incorporation of $N^{15}$ into adenine nucleotides and their formation from inosine monophosphate by skeletal muscle preparations, *Biochem. J.* **74**:127–136 (1960).

52. C. L. Davey, Synthesis of adenylosuccinic acid in preparations of mammalian skeletal muscle, *Nature* **183**:995–996 (1959).
53. E. F. Yefimochkina and A. E. Braunstein, The amination of inosinic acid to adenylic acid in muscle extracts, *Arch. Biochem. Biophys.* **83**:350–352 (1959).
54. C. L. Davey, The amination of inosine monophosphate in skeletal muscle, *Arch. Biochem. Biophys.* **95**:296–304 (1961).
55. P. A. Kometiani, Studies of transformations of amino acids in nervous and muscle tissues in connection of reamination of adenylic system, *Biokhimiya* **24**:729–737 (1959).
56. P. A. Kometiani, Relationship of transformations of amino acids with the metabolism of ammonia in brain, Ukrain. *Biokhim. Zh.* **37**:721–733 (1965).
57. H. C. Buniatian and S. G. Movcessian, The deamination and reamination of nicotinamide-adenine-dinucleotides in brain tissue, *Problems of brain biochemistry*, Vol. 2, pp. 5–22, Academic Press, Yerevan (1966).
58. S. G. Movcessian and H. C. Buniatian, *Problems of brain biochemistry*, Vol. 4, Academic Press, Yerevan (1968).
59. S. A. Olson and C. B. Anfinsen, Kinetic and equilibrium studies on crystalline L-glutamic acid dehydrogenase, *J. Biol. Chem.* **202**:841–856 (1953).
60. C. E. Carter, Synthesis of 6-succinoaminopurine, *J. Biol. Chem.* **223**:139–146 (1956).
61. E. Okuhara and R. G. Hansen, Reinvestigation of bovine liver nucleotide-peptide and separation of adenylosuccinic acid 6-succinaminopurine-9-ribosyl-5-phosphate, *J. Biochem. (Japan)* **55**:287–292 (1961).
62. W. K. Yorlik, The occurrence of adenine and adenyl-succinic acid in mammalian liver, *Biochim. Biophys. Acta* **22**:211–212 (1956).
63. H. C. Buniatian and S. G. Movcessian, The formation of free ammonia in brain tissue from amino acids, *Abstracts V*, p. 1022, Seventh International Congress of Biochemistry, Tokyo (1967).
64. N. O. Kaplan, A. Goldin, S. R. Humphreys, M. M. Ciotti, and F. S. Stolzenbach, Pyridine nucleotide synthesis in the mouse, *J. Biol. Chem.* **219**:287–298 (1956).
65. J. L. Purvis and J. M. Lowenstein, The relation between intra and extramitochondrial pyridine nucleotides, *J. Biol. Chem.* **236**:2794–2803 (1961).
66. H. M. Kalckar and D. Rittenberg, Rejuvenation of muscle adenylic acid nitrogen *in vivo* studied with isotopic nitrogen, *J. Biol. Chem.* **170**:455–459 (1947).
67. S. G. Movcessian and M. G. Urgandjian, The Role of $\gamma$-Aminobutyric Acid in the Regulation of Glycolysis in Nervous Tissue, *Problems of Brain Biochemistry*, Vol. 2, pp. 63–73, Academic Press, Yerevan (1966).

*Chapter 13*

# CEREBROSIDES AND SULFATIDES


## Norman S. Radin

*Mental Health Research Institute*
*University of Michigan*
*Ann Arbor, Michigan*


Cerebrosides contain three moieties: a fatty amine, a fatty acid, and a hexose. The fatty amine is usually sphingosine, an 18-carbon straight-chain primary amine characterized by 2 hydroxyls (one of them allylic) and 1 amine group. A small amount of dihydrosphingosine occurs also. The fatty acid is of the straight-chain type and may have a hydroxyl group in the 2-position. If the cerebroside contains a hydroxy fatty acid, it is referred to as a hydroxy cerebroside. The sugar moiety is galactose or glucose. Since the occurrence of the latter in brain has only recently been soundly established, the older literature refers to the galactocerebroside as cerebroside; however, it is becoming increasingly important to mention the hexose in the name, as in glucocerebroside. A general structure is shown here:

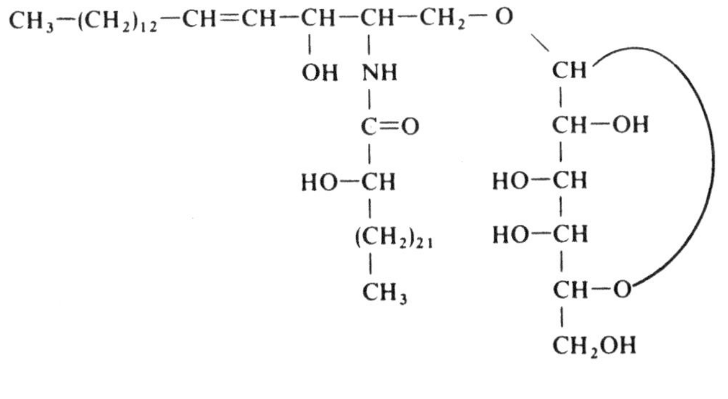


If the hexose is absent from the structure, we have a ceramide (or hydroxy ceramide, if the fatty acid is of the hydroxy type). Thus cerebrosides are sometimes called galactosyl or glucosyl ceramide. Other names are ceramide galactoside or ceramide galactose. If the fatty acid is absent from the structure, we have psychosine, or glycosyl sphingosine.

A name recently proposed for dihydrosphingosine is sphinganine; sphingosine is called 4-sphingenine.[1]

Sulfatides are galactocerebrosides esterified at the 3-position of the hexose with sulfuric acid, as the mono ester. They are also called cerebroside sulfate.

# I. BIOSYNTHESIS

## A. Galactocerebroside

This cerebroside appears to be made by the following reactions:

Sphingosine + UPDGal → psychosine + UDP (1)

Psychosine + fatty acyl-CoA → Cerebroside + CoA (2)

Reaction 1 was demonstrated by Cleland and Kennedy[2] in the rapidly sedimenting microsomes of rat and guinea pig brain. A mixture of galactose-1-phosphate and UDPGlu served as galactose donor just as well as UDPGal, due to the presence of galactose-1-P uridyl transferase. Both manganese and magnesium ions stimulated the synthesis, as did Tween 20. The sphingosine was dispersed in the Tween as a cloudy suspension. Use of additional Tween, which improves the appearance of the emulsion, resulted in inhibition. Tween alone, without sphingosine, also inhibited the endogenous blank considerably.

The psychosine forming system, incubated with galactose-1-P as the source of $^{14}C$, showed good activity over a *p*H range of 6.6 to 9.1, with a modest optimum between 8 and 9. It is interesting that dihydrosphingosine was almost as good an acceptor as sphingosine and that a variety of abnormal analogues also showed some activity. Thus, the *cis* isomer (sphingosine is normally *trans*), the acetylenic isomer, the isomer lacking the C-3 hydroxyl, and the *threo* form (C-3 inverted) were all active, presumably forming the abnormal analogues of psychosine. Phytosphingosine, ordinarily considered to be the plant version of sphingosine, was a fair acceptor. This is of more interest today, since phytosphingosine has been found in animal tissues.

In whole homogenate, *N*-acetyl-DL-*threo*-*trans*-sphingosine was a good acceptor, and it is unfortunate that the product was not identified. Perhaps the acetyl group was removed by hydrolysis and psychosine was formed from the resultant sphingosine; alternatively, the acetyl sphingosine may have condensed directly with the UDPGal to form a cerebroside.

The authors noted that the activity of the ceramide as acceptor disappeared on fractionation of the homogenates and that with the sphingosines increased. Some neutral galactolipid was also formed by the microsomal fraction, starting with sphingosine. Psychosine formation was confirmed by isolating the lipid via a cation exchange column, but the neutral lipid was not identified. In similar experiments in my laboratory, Dr. Pierre Morell has confirmed the formation of psychosine by thin-layer chromatography and identified its carbohydrate moiety as galactose. A number of neutral lipids are also formed in the system.

Reaction 2 has been described only in a preliminary way in a note by R. O. Brady.[3] Rat brain particles covering a wide span were used (6000–100,000 g) together with labeled stearoyl-CoA, ATP, $Mg^{2+}$, and Tween 20. Incubations ran 3 hr under nitrogen. It was found that addition of psychosine yielded cerebroside, which was characterized by elution from Florisil, passage through mixed ion exchange resins, and recrystallization to constant specific activity. Virtually all the radioactivity was present in the fatty acid portion of the cerebroside.

Omission of ATP from the medium reduced the formation of cerebroside, presumably by allowing more endogenous loss of the CoA ester. Sphingosine had some ability to replace psychosine, and more if UDPGlu was also present. However, since the sugar in this batch of cerebroside was not identified, and since the epimerase in this preparation probably could not convert UDPGlu to UDPGal in the absence of DPN, there is the strong possibility that the labeled cerebroside was actually glucocerebroside.

It would be important to confirm the utilization of psychosine in this reaction by starting with labeled psychosine. Similarly, there is a need to confirm the first step with labeled sphingosine.

Eliasson[4] showed that rat sciatic nerve homogenate could also utilize stearoyl-CoA, as well as oleoyl-CoA, to make cerebrosides. Psychosine, Tween 20, and ATP were included in the incubation mixture and oxygen–carbon dioxide was used as the atmosphere, so anaerobic conditions do not seem to be essential. An interesting aspect of this study is that the nerve from diabetic rats barely utilized the saturated CoA derivative while oleoyl-CoA was taken up normally. It would be interesting to know if this difference is reflected in the cerebroside content and fatty acid distribution of human diabetics.

The conclusions of the above studies are rendered more difficult to interpret by very recent work in this laboratory by Dr. Pierre Morell.[35] We have found that brain microsomes incubated with ceramides and labeled galactose-1-P (plus UDPGlu, ATP, and $Mg^{2+}$) formed only hydroxy cerebroside. A mixture of hydroxy and nonhydroxy ceramides or hydroxy ceramide alone was an effective galactose acceptor, while pure nonhydroxy ceramide was ineffective. Addition of Tween 20 greatly inhibited the synthesis, and it was necessary to offer the ceramide to the microsomes in the form of a very thin coating on Celite. The cerebroside formed was shown to be galactocerebroside (not gluco-), and the position of the labeled galactose was shown to be correct by hydrolyzing the cerebroside and characterizing the resultant psychosine.

## B. Sulfatides

This group of lipids appears to be made by direct sulfation of galactocerebrosides by PAPS (3′-phosphoadenosine-5′-phosphosulfate), the universal sulfate donor. PAPS is formed from ATP and sulfate by enzymes in the brain cytosol,[5,6] and the sulfokinase occurs in the microsomal fraction of brain.[5] The kinase could be "solubilized" by sonication of the microsomes with deoxycholate, with a resultant large increase in specific activity and diminution of endogenous activity. Addition of galactocerebroside (not glucocerebroside) emulsified in BRIJ-96 greatly stimulated formation of sulfolipid. The diminution of endogenous activity by extraction with deoxycholate is presumably due to removal of insoluble cerebroside, normally occurring in microsomes.

The incubation system used by McKhann and Ho[7] consisted of the microsomal extract, cerebroside emulsion, PAPS, ATP, sulfate, KCl, and imidazole buffer, *p*H 7.4. Maximal activity was observed in rat brain at 20 days of age, with sharply reduced activity in younger and older animals.

A somewhat different system also was used to demonstrate sulfatide formation.[8] The mixed microsomes and cytosol of sheep brain white matter were fractionated to yield the material precipitating between 30 and 60% of ammonium sulfate saturation. This material, on incubation with Tris buffer (*p*H 7.4), labeled PAPS, EDTA, and cysteine or glutathione, formed labeled sulfatide. Both EDTA and thiol were quite stimulatory, and the preparation apparently contained enough cerebroside as acceptor lipid. To show that cerebroside was indeed the acceptor, Balasubramanian and Bachhawat treated the enzyme preparation with galactose oxidase, which is known to oxidize the 6-position of the galactose moiety.[9] This reduced the amount of sulfatide formed by 90%. Unfortunately, activity could not be restored by addition of a cerebroside emulsion; from this observation, the authors concluded that only protein-bound cerebroside is active as acceptor. The destruction of endogenous cerebroside by the galactose oxidase was not actually demonstrated.

Since McKhann's group was able to get good stimulation of sulfatide formation by addition of a cerebroside emulsion, the failure of the other group to get such an effect may be due either to use of the wrong detergent (Triton X-100 vs BRIJ-96) or to the incorrect amount of detergent or to the use of an enzyme preparation that is very sensitive to detergents. No mention is made of a control run in which Triton was incubated with the normal synthesizing system.

A very recent study of the sulfation reaction by Cumar *et al.*[10] showed that the deoxycholate extract of rat brain microsomes could transfer sulfate even in the presence of a very high detergent concentration: 25 mg of Tween 20 in each milliliter of incubation mixture. The only other adjuvant present was $Mg^{2+}$. Not only did the addition of galactocerebroside yield lipoidal sulfate, but so did psychosine and lactosyl ceramide (galactosylglucosyl ceramide). Some stimulation was shown also by sphingosine, trihexosyl ceramide, and tetrahexosyl ceramide. The product of cerebroside sulfation was identified as sulfatide, while the product of psychosine sulfation was shown to be a more polar lipid, possibly psychosine sulfate. Transferase activity was found in virtually all subcellular fractions.

The effectiveness of the many galactolipids as acceptors suggests that brain may contain small amounts of other sulfolipids. The 3-sulfate of lactosyl ceramide indeed has been found in kidney.[11] The apparent ability of psychosine to form a sulfate indicates, as Cumar *et al.* suggest, that sulfation may precede acylation in the formation of sulfatide. It is unfortunate that none of the studies compared nonhydroxy cerebroside with hydroxy cerebroside as sulfate acceptor. Perhaps only the latter can accept sulfate, and nonhydroxy sulfatide is made by acylation of psychosine sulfate. Some support for this idea comes from a study by Kopaczyk and Radin[12] in which lignoceroyl psychosine (a nonhydroxy cerebroside) was injected into the brains of young rats. While considerable breakdown to ceramide was demonstrated (as well as utilization of the ceramide for sphingomyelin formation), very little conversion to sulfatide was found.

### C. Glucocerebroside

Trace amounts of glucosyl ceramide have been found in bovine brain by Nishimura and Yamakawa.[13] The fatty acid distribution in this material was not typical of sphingolipids, being relatively rich in oleate and linoleate. However, since glucocerebroside can act as an acceptor for galactose,[14] it is possible that part of the normally present glucolipid is a precursor of gangliosides.

The only information available from *in vitro* work comes from an abstract by Basu,[14] describing findings with particles from embryonic chick brain. Labeled UDPGlu was shown to form glucocerebroside, and the endogenous activity was greatly increased by addition of an emulsion of ceramide. Neither $Mg^{2+}$ nor $Mn^{2+}$ stimulated the reaction, and EDTA did not inhibit it. Since Den and Kaufman recently showed that some of the enzymes involved in ganglioside biosynthesis reside in the synaptosomes,[15] it is possible that the glucosyl transferase is also located there.

## II. BIODEGRADATIONS

### A. Galactocerebroside

It was evident from a turnover study with labeled galactose[16] that this moiety of brain cerebroside undergoes turnover in young rats. Part of the turnover is evidently due to sulfatide formation. Another part is probably due to formation of 6-fatty acyl cerebrosides.[17] Perhaps the major conversion route is hydrolysis to ceramide, demonstrated first *in vivo*[12] and then in the cytosol of rat and pig brain.[18] Subsequent work[19,20] has shown that the galactosidase occurs primarily in lysosomes, judging by its subcellular distribution and low optimal *p*H (4.5).

Assay for the galactosidase is carried out in the presence of cholate or taurocholate and citrate buffer, using as substrate cerebroside emulsified in Triton X-100, Tween 20, and G-2159 (a polyoxethylene stearate with a high HLB value). In crude preparations, the addition of free fatty acid stimulates activity appreciably. Both hydroxy and nonhydroxy cerebrosides were found to undergo hydrolysis to the corresponding ceramide.

Because the radioactive substrate is somewhat diluted by endogenous cerebroside in brain preparations, it was necessary in developing an assay procedure to remove most of the lipid. This was done[(21)] by digesting the brain homogenate with a mixture of pancreatic enzymes, taurocholate being added to protect the cerebrosidase. The cerebrosidase was then removed from most of the bile salt and reaction products by precipitation at *p*H 3.

A similar procedure was used to purify the enzyme.[(19)] Crude rat brain lysosomes were extracted with sodium taurocholate, and the extract was treated with pancreatic enzymes in the cold while simultaneously dialyzing the reaction mixture. The digestion–dialysis removed 90% of the protein and roughly half the cholate; it also modified a good deal of the lipid present. After removal of additional cholate with a Sephadex column, the enzyme was purified further by density gradient electrophoresis and DEAE-Sephadex chromatography. The apparent purification was about 300-fold; an exact figure cannot be given because of the presence of cerebroside in the original homogenate and because the cholate extraction and digestion increase the total activity.

The ion exchange purification step rendered the cerebrosidase unstable to storage; addition of glycerol (to yield 50%) made it storable indefinitely at −20°. More recent work[(22)] has shown that taurocholate also is a good stabilizer and can be used during the ion exchange procedure. It is of interest that much of the brain hydrolytic activity toward nitrophenyl-β-D-galactoside was removed during the purification, and it is thereby evident that at least two β-galactosidases exist in brain.

A search for inhibitors showed that mercaptoethanol had no effect, but mercuribenzoate was a good inhibitor. γ-D-Galactonolactone, which appears to be a good inhibitor of animal galactosidases, brought the activity down to 10% at a concentration of 0.8 mM. Sephadex chromatography indicated the molecular weight of the enzyme to be about 50,000.

A striking characteristic of the cerebrosidase was that little activity could be demonstrated in the absence of bile salt, and maximal stimulation by sodium cholate was achieved at a concentration that was far above the solubility of cholic acid, which precipitates at *p*H 4.5. One explanation of the phenomenon is that the extra cholate is stimulatory simply because it adds cations to the medium; the stimulatory effect of high salt concentration may be exerted by a salting-out action on the enzyme so that the enzyme is more readily attached to the insoluble substrate. Another explanation is that the extra cholic acid increases the surface area of the system, with consequent adsorption of the substrate and enzyme onto the surface of the bile acid crystals. Examination of the supernatant in the cholate system showed that the enzyme and substrate were both precipitated together with the cholic acid.

The need by a brain enzyme for a bile acid seems anomalous, since the bile acid concentration in brain must be very low. Perhaps the hydrolase normally functions at suboptimal speeds and its full capacity appears only when the blood–brain barrier is lowered and allows entry of more bile acid from the blood. Of course, it is possible that the true activating material in brain is actually an entirely different substance.

The question of how does a lysosomal enzyme contact its lipoidal substrate was investigated in a study[21] of two mouse mutant strains, "quaking" and "jimpy." The concentration of brain cerebroside was very low in these neurologically affected animals, and the concentration of cerebrosidase was found to be normal in the jimpy mice while only moderately low (17%) in the quaking mice. However, the relative amount of hydrolase that could be brought into a dispersed state (by sonication of the homogenate) was much higher in the former group (44%). This may mean that the cerebrosidase in the jimpy animals is more active than normally because of the greater ease with which their lysosomes disintegrate. It is important that the dispersibility of other acid hydrolases (nitrophenyl galactosidase, nitrophenyl glucosidase, and aryl sulfatase A) was normal in all groups.

## B. Sulfatide

The sulfatase acting on cerebroside sulfate has been highly purified by Mehl and Jatzkewitz,[23] starting with pig kidney. Unfortunately, the brain enzyme appears to occur in much lower concentration and little study has been made with sulfatide as substrate. However, arylsulfatase A appears to be the same enzyme.[24,25] Arylsulfatases A and B are both active toward 2-hydroxy-5-nitrophenyl sulfate but can be differentiated by physical separation and by differential sensitivity to conditions of incubation. The A enzyme is assayed at low substrate concentration, in the presence of dilute pyrophosphate, 0.84 *M* NaCl, and 0.25 *M* acetate buffer, *p*H 5.[26] (The high salt concentration called for is perhaps, as noted above in the case of cerebrosidase, a general aid for lipid hydrolases.)

The identification of arylsulfatase A as sulfatidase was made in two ways. Patients with metachromatic leukodystrophy exhibit accumulation of sulfatide in various organs and spillage into urine; from this it seemed likely that the sulfatide-degrading enzyme might be lacking in such individuals. In a series of pioneering studies Austin[25] showed that these individuals had typical arylsulfatase B activities but little or no arylsulfatase A activity in a number of organs and urine. Mehl and Jatzkewitz[27] found a lack of activity toward sulfatide in human leukodystrophic kidney. Moser *et al.*[28] studied the turnover of sulfatide in human subjects and found that the leukodystrophic subjects exhibited very slow breakdown of sulfatide (isolated from urine). Thus there is strong presumptive evidence that the two enzymes are identical, requiring only the assumption that a genetic defect can affect only one protein.

A more direct comparison has been made with highly purified sulfatase of kidney.[24] Electrophoresis yielded two peaks, one for the A enzyme and one for the B enzyme. Activity toward sulfatide lay within the A peak, but the two curves were by no means coincident. The authors explain this discrepancy by pointing to the presence of a heat-stable material that migrates just ahead of the sulfatidase peak, probably overlapping it. This material greatly stimulates sulfatidase activity, although it has no effect on arylsulfatase activity. While the assays of the former activity were made in the presence of added "heat-stable factor," the role of this material apparently has not been worked out in a quantitative way and it is possible that suboptimal amounts were added. We have noted in our own work with cerebrosidase[20] that an enzyme preparation can respond differently to detergents when tested against the lipoidal substrate and the artificial, less lipoidal substrate.

When Mehl and Jatzkewitz compared their initial preparation against a more highly purified preparation, which showed only one band by gel electrophoresis, there was no change in the ratio of activities toward sulfatide and nitrocatechol sulfate. Bleszynski[29] has purified ox brain arylsulfatase A very highly, and it will be interesting to see if this preparation is also active toward sulfatide. It will also be interesting to see what the "heat-stable factor" turns out to be. Perhaps it is an acidic lipid that modifies the micellar properties of the lipoidal substrate so as to improve the adsorption of the enzyme to the micelle.

## C. Glucocerebroside

A glucosidase active toward glucocerebroside has been partially purified from ox brain by Gatt.[30] Its optimal *p*H was found to be 5.0 in acetate buffer, and δ-gluconolactone was a strong inhibitor. Like the brain galactocerebrosidase and related glycolipid hydrolases, this enzyme was somewhat stimulated by bile salt (taurocholate) when acting on glucocerebroside. It also hydrolyzed the unnatural substrate, nitrophenyl-β-D-glucoside.

The purification of the enzyme[31] involved preparation of a subcellular fraction (which must include most of the lysosomes), sonication of the particles, and extraction of the sedimentable portion of the sonicate with sodium cholate. The extract was dialyzed against Tris buffer, the enzyme was precipitated with ammonium sulfate, and the precipitate was dialyzed to remove the sulfate. At this point a precipitate formed, containing much of the activity toward nitrophenyl glucoside, while the supernatant held much of the activity toward nitrophenyl galactoside. The precipitated glucosidase was solubilized by extraction with cholate, dialyzed somewhat, and stored at −20°. Because of the slowness with which cholate dialyzes out, it is very likely that the final preparation contained some cholate. While the enzyme was purified only 10-fold over the activity in the brain particles, much of the galactosidase activity was removed.

Glucocerebrosidase may function in the normal degradation of gangliosides, acting on the substrate formed from lactosyl ceramide by a galactosidase. The latter enzyme, lactosyl ceramide galactosidase, has been studied by Gatt and Rapport.[32] It is distinct from galactosyl ceramide galactosidase and probably accounts for most of the activity shown by the brain toward nitrophenyl galactoside.[19,33] If gangliosides are indeed hydrolyzed by a series of enzymes to form lactosyl ceramide, glucosyl ceramide, ceramide, and sphingosine plus fatty acid, one would expect brain ceramide to contain the 20-carbon sphingosine characteristic of gangliosides. This indeed has just been reported: Klenk and Huang[34] isolated stearoyl sphingosine from brain and showed that it contained both $C_{18}$ and $C_{20}$ sphingosine. (Stearic acid is by far the major acid in gangliosides.) The other ceramides isolated, with longer fatty acids and hydroxy acids, contained only $C_{18}$ sphingosine and are therefore related metabolically to cerebrosides. However, it is not yet certain whether the naturally occurring ceramides lie on the synthetic or on the degradative pathways.

## III. REFERENCES


1. IUPAC-IUB Commission on Biochemical Nomenclature, *J. Lipid Res.* **8**:523–524 (1967).
2. W. W. Cleland and E. P. Kennedy, The enzymatic synthesis of psychosine, *J. Biol. Chem.* **235**:45–51 (1960).
3. R. O. Brady, Cerebroside synthesis from psychosine and stearyl-1-$C^{14}$ – CoA, *J. Biol. Chem.* **237**:PC2416–PC2417 (1962).
4. S. G. Eliasson, Lipid synthesis in peripheral nerve from alloxan diabetic rats, *Lipids* **1**:237–240 (1966).
5. G. M. McKhann, R. Levy, and W. Ho, Metabolism of sulfatides. I. The effect of galactocerebrosides on the synthesis of sulfatides, *Biochem. Biophys. Res. Commun.* **20**:109–113 (1965).
6. A. S. Balasubramanian and B. K. Bachhawat, Enzymic degradation of active sulphate, *Biochim. Biophys. Acta* **59**:389–397 (1962).
7. G. M. McKhann and W. Ho, The *in vivo* and *in vitro* synthesis of sulphatides during development, *J. Neurochem.* **14**:717–724 (1967).
8. A. S. Balasubramanian and B. K. Bachhawat, Formation of cerebroside sulphate from 3′-phosphoadenosine 5′-phosphosulfate in sheep brain, *Biochim. Biophys. Acta* **106**:218–220 (1965).
9. B. W. Agranoff, N. S. Radin, and W. Suomi, Enzymic oxidation of cerebrosides: studies on Gaucher's disease, *Biochim. Biophys. Acta* **57**:194–196 (1962).
10. F. A. Cumar, H. S. Barra, H. J. Maccioni, and R. Caputto, Sulfation of glycosphingolipids and related carbohydrates by brain preparations from young rats, *J. Biol. Chem.* **243**:3807–3816 (1968).
11. E. Martensson, Sulfatides of human kidney: Isolation, identification, and fatty acid composition, *Biochim, Biophys. Acta* **116**:521–531 (1966).
12. K. C. Kopaczyk and N. S. Radin, In vivo conversions of cerebroside and ceramide in rat brain, *J. Lipid Res.* **6**:140–145 (1965).

13. K. Nishimura and T. Yamakawa, Isolation of cerebroside containing glucose (glucosyl ceramide) and its possible significance in ganglioside synthesis, *Lipids* **3**:262–266 (1968).
14. S. Basu, Synthesis of glucosylceramide by an enzyme from embryonic chicken brain, *Federation Proc.* **27**:346 (1968).
15. H. Den and B. Kaufman, Ganglioside and glycoprotein glycosyltransferases in synaptosomes, *Federation Proc.* **27**:346 (1968).
16. N. S. Radin, F. B. Martin, and J. R. Brown, Galactolipide metabolism, *J. Biol. Chem.* **224**:499–507 (1957).
17. Y. Kishimoto, M. Wajda, and N. S. Radin, 6-Acyl galactosyl ceramides of pig brain: structure and fatty acid composition, *J. Lipid Res.* **9**:27–33 (1968).
18. A. K. Hajra, D. M. Bowen, Y. Kishimoto, and N. S. Radin, Cerebroside galactosidase of brain, *J. Lipid Res.* **7**:379–386 (1966).
19. D. M. Bowen and N. S. Radin, Purification of cerebroside galactosidase from rat brain, *Biochim. Biophys. Acta* **152**:587–598 (1968).
20. D. M. Bowen and N. S. Radin, Properties of cerebroside galactosidase, *Biochim. Biophys. Acta* **152**:599–610 (1968).
21. D. M. Bowen and N. S. Radin, Hydrolase activities in brain of neurological mutants: cerebroside galactosidase, nitrophenyl galactoside hydrolase, nitrophenyl glucoside hydrolase and sulphatase, *J. Neurochem.* **16**:457 (1969).
22. K. Nishimura and N. S. Radin, unpublished work.
23. E. Mehl and H. Jatzkewitz, Eine Cerebrosidsulfatase aus Schweineniere, *Z. Physiol. Chem.* **339**:260–276 (1964).
24. E. Mehl and H. Jatzkewitz, Cerebroside 3-sulfate as a physiological substrate of arylsulfatase A, *Biochim. Biophys. Acta* **151**:619–627 (1968).
25. J. Austin, D. McAfee, and L. Shearer, Metachromatic form of diffuse cerebral sclerosis, *Arch. Neurol.* **12**:447–455 (1965).
26. H. Baum, K. S. Dodgson, and B. Spencer, The assay of arylsulphatases A and B in human urine, *Clin. Chim. Acta* **4**:453–455 (1959).
27. E. Mehl and H. Jatzkewitz, Evidence for the genetic block in metachromatic leucodystrophy (ML), *Biochim, Biophys. Res. Commun.* **19**:407–411 (1965).
28. H. W. Moser, A. B. Moser, and G. M. McKhann, The dynamics of a lipidosis, *Arch. Neurol.* **17**:494–511 (1967).
29. W. Bleszynski, Purification and significance of four soluble arylsulphatases from ox brain, *Enzymologia* **32**:169–181 (1967).
30. S. Gatt, Enzymic hydrolysis of sphingolipids: hydrolysis of ceramide glucoside by an enzyme from ox brain, *Biochem. J.* **101**:687–691 (1966).
31. S. Gatt and M. M. Rapport, Isolation of $\beta$-galactosidase and $\beta$-glucosidase from brain, *Biochim. Biophys. Acta* **113**:567–576 (1966).
32. S. Gatt and M. M. Rapport, Hydrolysis of ceramide lactoside by an enzyme from rat brain, *Biochem. J.* **101**:680–686 (1966).
33. N. S. Radin, L. Hof, R. M. Bradley, and R. O. Brady, Lactosylceramide galactosidase: comparison with other sphingolipid hydrolases in developing rat brain, *Brain Res.* **14**:497 (1969).
34. E. Klenk and R. T. C. Huang, Zur Kenntnis der Gehirnceramide und der darin vorkommenden Sphingosinbasen, *Z. Physiol. Chem.* **349**:451–454 (1968).
35. P. Morell and N. S. Radin, Synthesis of cerebroside by brain from uridine diphosphate galactose and ceramide containing hydroxy fatty acid, *Biochemistry* **8**:506 (1969).

*Chapter 14*

# GANGLIOSIDES


## Lars Svennerholm

*Department of Neurochemistry*
*Psychiatric Research Centre*
*Faculty of Medicine*
*University of Göteborg, Sweden*


## I. TOPOGRAPHIC DISTRIBUTION OF GANGLIOSIDES

### A. Nomenclature

Gangliosides is the generic term for a glycosphingolipid containing sialic acid. The gangliosides of various sources differ in the patterns of fatty acids and sphingosines and in the number of units in the carbohydrate chain, but only the differences in the latter have been used for the characterization of the gangliosides. All the major gangliosides have in common a neutral carbohydrate moiety, galactosyl-β-(1→3)-*N*-acetyl-galactosaminyl-β-(1→4)-galactosyl-β-(1→4)-glucosyl, which is attached to the terminal, primary hydroxyl of sphingosine. The sialic acid of mammalian brain gangliosides is *N*-acetylneuraminic acid (NAN), which is attached by ketosidic bonds to two main positions: C-3 of galactose and C-8 of another sialic acid.

As in the field of carbohydrate chemistry, there has been a great need for simple generic terms for the different gangliosides instead of the very long systematical designations. The terms mono-, di-, and trisialogangliosides[1] seem to have been generally accepted. Although there are accepted terms for designations of the gangliosides from their number of sialic acids no common system of trivial names or symbols for the individual gangliosides has been accepted. Instead most investigators have preferred to introduce their own system. In the beginning most of the symbols were based on the migratory rate of the ganglioside fraction in thin-layer chromatography with a certain solvent mixture, and the assigned ganglioside was often incompletely defined. Kuhn and Wiegandt,[2] who have isolated and proposed structures for all the common brain gangliosides, and Svennerholm[3] developed code systems that have been used by other research groups, Kuhn and Wiegandt based their original code system on the migratory rate, and Svennerholm based it on the composition of the carbohydrate chain. In Svennerholm's system, G stands for ganglioside;

index M, D, and T for mono-, di-, and trisialosyl groups; and index 1 for the major neutral carbohydrate moiety with four monosaccharide units, 2 for carbohydrate chain lacking the terminal galactose and 3 for the chain lacking galactosyl-*N*-acetylgalactosamine. Wiegandt[4,5] has recently suggested a new code system that is also based on the composition of the carbohydrate chain and in which carbohydrate abbreviations are used as index. The new system is convenient when written but will cause much confusion if used for oral communication. The two last-mentioned code systems are shown in Table I together with the structure formulas of the assigned gangliosides and their generic terms assigned according to the rules suggested by the IUPAC–IUB Commission on Biochemical Nomenclature.[6] Code systems used by other investigators have been tabulated in two recent reviews[7,8] on gangliosides, to which the reader is referred. It should also be mentioned that the sequence of the sugars of the tri- and tetraglycosylgangliosides of extraneural organs are different from that of brain gangliosides. Thus, the code systems suggested are valid only for gangliosides of nervous tissue.

## B. Distribution in Different Regions of Nervous System

Klenk and Langerbeins[9] could demonstrate the occurrence of gangliosides only in the gray matter of normal brain. Because of this finding Klenk[10] called the new lipid a ganglioside. Svennerholm[11] developed the first quantitative method for the analysis of gangliosides and showed that the ganglioside concentration (calculated on a dry weight basis) was about 2% of cerebral cortex and 0.4% of cerebral white matter of human adults. This means that the gangliosides comprise about 6% of the lipids in gray matter but only 0.6% of those in the cerebral white matter. These results were obtained before it was known that the gangliosides constituted a large family of related compounds. During the last years several groups have reported on the ganglioside content (expressed as total lipid-bound sialic acid) and the ganglioside pattern in different regions of the nervous system, in various animals, during development and under various pathological conditions. The most extensive and painstaking determinations have been performed by Suzuki,[13,14] who has developed a quantitative method for the determination of individual gangliosides,[12,13] which is now in common use. In Table II the figures for the ganglioside pattern and content in various mammalians and some other vertebrates, and in Fig. 1 the development patterns of rat brain gangliosides are shown.

The gangliosides occur in the nervous system of vertebrates; studies of several different nonvertebrates have shown a complete lack of gangliosides,[22] although minute amounts of gangliosides in crab have been reported. The concentration of gangliosides is highest in mammalians and birds, while reptiles, amphibians, and fishes contain lower amounts. It is not possible from the available figures to draw any conclusions about species differences with regard to the concentration of gangliosides. The

**TABLE I**

**Gangliosides of Mammalian Brain**

| Chemical structure | Generic term | Code system: Ref 3. | Code system: Ref. 4 |
|---|---|---|---|
| NAN(2 → 3)Gal(β,1 → 4)Glu(1 → 1)Cer | Monosialosyl-lactosylceramide | $G_{M3}$ | $G_{Lact}1$ |
| NAN(2 → 8)NAN(2 → 3)Gal(β,1 → 4)Glu(1 → 1)Cer | Disialosyl-lactosylceramide | $G_{D3}$ | $G_{Lact}2$ |
| GalNAc(β,1 → 4)Gal(β,1 → 4)Glu(1 → 4)Cer<br>(3 ↑ 2) [on Gal]<br>NAN | Monosialosyl-*N*-triglycosylceramide | $G_{M2}$ | $G_{NTrII}1$ |
| GalNAc(β,1 → 4)Gal(β,1 → 4)Glu(1 → 1)Cer<br>(3 ↑ 2) [on Gal]<br>NAN(8 ← 2)NAN | Disialosyl-*N*-triglycosylceramide | $G_{D2}$ | $G_{NTrII}2$ |
| Gal(β,1 → 3)GalNAc(β,1 → 4)Gal(β,1 → 4)Glu(1 → 1)Cer<br>(3 ↑ 2) [on Gal(β,1 → 4)]<br>NAN | Monosialosyl-*N*-tetraglycosylceramide | $G_{M1}$ | $G_{NT}1$ |
| Gal(β,1 → 3)GalNAc(β,1 → 4)Gal(β,1 → 4)Glu(1 → 1)Cer<br>(3 ↑ 2) [on Gal(β,1 → 4)]<br>NAN(8 ← 2)NAN | Disialosyl-*N*-tetraglycosylceramide | $G_{D1a}$ | $G_{NT}2a$ |

TABLE I (cont.)

| Chemical structure | Generic term | Code system: Ref. 3 | Code system: Ref. 4 |
|---|---|---|---|
| Gal(β,1 → 3)GalNAc(β,1 → 4)Gal(β,1 → 4)Glu(1 → 1)Cer<br>(3 ↑ 2) (3 ↑ 2)<br>NAN NAN | Disialosyl-*N*-tetraglycosylceramide | $C_{D1b}$ | $G_{NT}2b$ |
| Gal(β,1 → 3)GalNAc(β,1 → 4)Gal(β,1 → 4)Glu(1 → 1)Cer<br>(3 ↑ 2) (3 ↑ 2)<br>NAN NAN(8 ← 2)NAN | Trisialosyl-*N*-tetraglycosylceramide | $G_{T1}$ | $G_{NT}3$ |
| Gal(β,1 → 3)GalNAc(β,1 → 4)Galβ,1 → 4)Glu(1 → 1)Cer<br>(3 ↑ 2)+NAN (3 ↑ 2)<br>NAN NAN(8 ← 2)NAN | Tetrasialosyl-*N*-tetraglycosylceramide | $G_{Q1}$ | $G_{NT}4$ |

## TABLE II

### The Ganglioside Content in Brain Tissue of Vertebrates

| Material | Total NAN μg/g wet weight | Distribution of NAN% $G_{M3}$ | $G_{M2}$ | $G_{D2}$ | $G_{M1}$ | $G_{D1a}$ | $G_{D1b}$ | $G_{T1}$ | $G_{Q1}$ | Ref. |
|---|---|---|---|---|---|---|---|---|---|---|
| Human | | | | | | | | | | |
| Cerebral cortex | | | | | | | | | | |
| Fetus 5 months | 300 | | | | | | | | | 15 |
| Newborn | 400 | 1.0 | 3.6 | 1.1 | 14.6 | 71.6 | 1.8 | 7.3 | | 3, 15 |
| 2.5 months–8 yr (6 cases) | 738 | 1.0 | 2.3 | 4.0 | 14.2 | 42.5 | 12.7 | 15.9 | 3.6 | 16 |
| 3–5 yr (7 cases) | 686 | | | | | | | | | 22 |
| 8 yr | 797 | | 3.6 | | 17.4 | 39.4 | 19.8 | 15.9 | 2.9 | 14 |
| 44 yr | 1.002 | | 1.3 | | 11.3 | 22.4 | 28.3 | 29.9 | 5.9 | 14 |
| 73 yr | 796 | | 1.7 | | 12.8 | 22.8 | 25.5 | 31.2 | 5.1 | 14 |
| 60–90 yr (40 cases) | 846 | | | | | | | | | 11 |
| White matter | | | | | | | | | | |
| Newborn | 400 | 1.0 | 6.9 | 1.4 | 19.1 | 57.8 | 2.1 | 3.4 | | 3, 15 |
| 2.5 months–8 yr (6 cases) | 60 | 1.0 | 1.5 | 1.5 | 16.1 | 38.5 | 11.6 | 18.1 | 8.2 | 16 |
| 3–5 yr (7 cases) | 269 | | | | | | | | | 22 |
| 8 yr | 73 | | 1.7 | | 20.4 | 42.9 | 12.5 | 16.0 | 5.9 | 14 |
| 44 yr | 156 | | 1.0 | | 9.3 | 14.0 | 31.4 | 38.2 | 6.4 | 14 |
| 73 yr | 191 | | 1.9 | | 12.5 | 18.4 | 30.4 | 27.9 | 7.9 | 14 |
| 60–90 yr (16 cases) | 312 | | | | | | | | | 11 |

**TABLE II (cont.)**

| Material | Total NAN μg/g wet weight | Distribution of NAN% $G_{M3}$ | $G_{M2}$ | $G_{D2}$ | $G_{M1}$ | $G_{D1a}$ | $G_{D1b}$ | $G_{T1}$ | $G_{Q1}$ | Ref. |
|---|---|---|---|---|---|---|---|---|---|---|
| Ox | | | | | | | | | | |
| Cerebral cortex | 940 | | | | | | | | | 18 |
| | 497 | | 2.0 | | 15 | 39 | 15 | 22 | 6 | 4 |
| White matter | 210 | | | | | | | | | 18 |
| | 102 | | | | 26 | 35 | 15 | 20 | 4 | 4 |
| Rat (adult) | 1.047 | | 1.2 | | 13.0 | 32.1 | 20.3 | 27.3 | 6.0 | 13 |
| Mice | 640 | | | | | | | | | 19 |
| Guinea pig | 425 | | | | | | | | | 20 |
| Rabbit | | | | | | | | | | |
| Cerebral cortex | 497 | | 3 | | 15 | 38 | 16 | 23 | 5 | 4 |
| Hen | 590 | | | | | | | | | 19 |
| Chicken | 590 | | | | | | | | | 19 |
| Chicken embryos | | | | | | | | | | |
| 5 days | 23 | | | | | | | | | 21 |
| 8 days | 119 | | | | | | | | | 21 |
| 13 days | 233 | | | | | | | | | 21 |
| 18 days | 314 | | | | | | | | | 21 |
| 21 days | 605 | | | | | | | | | 21 |
| Chicken hatched 2 days | 660 | | | | | | | | | 21 |
| Turtle | 220 | | | | | | | | | 19 |
| Snake | 320 | | | | | | | | | 19 |
| Frog | 220 | | | | | | | | | 19 |
| Fish | 370 | | | | | | | | | 19 |

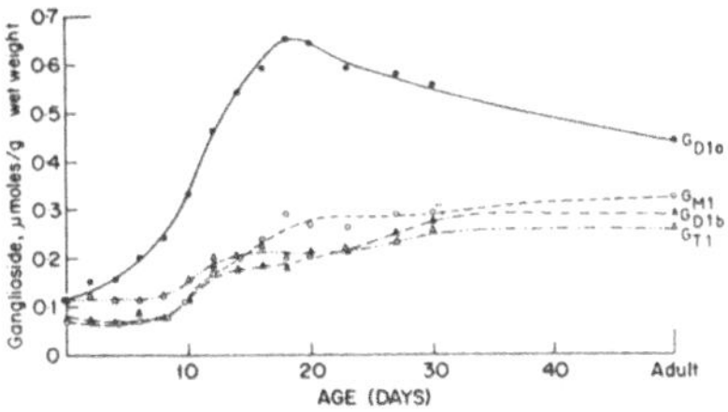


Fig. 1. Developmental patterns of rat brain gangliosides. After Suzuki.[14] The nomenclature is changed in conformation with the manner of this chapter.

varying figures shown in Table II seems instead to demonstrate the variations between different laboratories. The figures for total gangliosides of cerebral cortex found by Suzuki[13,14] agree closely with the data by Svennerholm,[11,15] Tingey,[23] McCluer, Coram and Lee,[24] and Hess and Rolde[18] but are considerably higher than those reported by others.[6,25–29] In the human cerebral white matter Svennerholm[11] found higher figures for total ganglioside sialic acid than Suzuki.[14,16] This difference is most pronounced in immature brain. Some of the discrepancies might be explained by a loss of mainly monosialogangliosides in the purification method used by Suzuki.[12,13] This assumption is supported by our finding of a higher percentage of monosialogangliosides in cerebral white matter.[17]

## C. Developmental Changes of Gangliosides

The ganglioside concentration changes during development. In human brain there is a twofold increase of the ganglioside concentration from the end of the third fetal month to the time of delivery,[16] and Suzuki showed a further twofold increase from birth to the adulthood.[14] In rat brain the time of rapid development begins later. The gangliosides increase slowly after birth for 6–8 days, followed by a period of rapid increase; at the sixteenth day the ganglioside has almost reached its adult level, which is about three times that at birth.[14] Similar studies for rat brain by Pritchard and Cantin[30] and James and Fotherby[27] indicate a more linear increase. Rosenberg and Stern[31] found that the ganglioside content of rat brain increased 50-fold from birth up to the end of the fifth week. A still more rapid increase of the ganglioside content was found in the chicken embryonic brain by Carrigan and Chargaff,[21] a discovery that has been of utmost importance for the elucidation of the biosynthesis of gangliosides.

The ganglioside pattern also changes considerably during development. Svennerholm[3] found that the pattern of human newborn brain was characterized by a very large $G_{D1a}$-fraction while $G_{D1b}$ and $G_{T1}$ were small. In fetal human brain the pattern was more similar to that of adult brain.[15] Suzuki[12] has shown that there is a preponderance for $G_{D1a}$ for almost the entire first decade of life. Later Suzuki showed that the ganglioside pattern of rat brain undergoes a very similar change to that of human brain (Fig. 1).[14] The drastic increase of $G_{D1a}$ occurs during the same period for which Suzuki has shown the largest increase of the ganglioside concentration; he points out that the peak of $G_{D1a}$ corresponds to the height of myelination and speculates that the ganglioside metabolism might be intimately associated with the process of myelination or with its regulation.[14]

The concentration of gangliosides in different areas of the cerebral cortex is very similar, but it is lower in cerebellum.[14] In the basal ganglia the ganglioside content varies inversely to their amounts of myelinated fibers. The ganglioside content of myelin-free ganglia is the same as in cerebral cortex.[14] The ganglioside content is very low in spinal medulla and peripheral nerves—much lower than in the cerebral white matter. The pattern is also different with higher percentages of monosialogangliosides and $G_{D2}$ and $G_{D3}$.[22]

## D. Subcellular Localization

Klenk,[10] in his original paper on gangliosides, stated that the gangliosides were localized mainly in the neuronal cell body. However, Svennerholm,[11] who had found the same concentration of gangliosides in the axonic layer of unmyelinated brain and one-third the concentration in adult white matter as in the cerebral cortex, suggested the gangliosides to be localized also in the axons and the dendrites. The first reports on the distribution of gangliosides in subcellular fractions, which appeared from McIlwain's laboratory,[32,33] suggested that the gangliosides were mainly associated with the microsomal fractions. Wolfe[32] stressed that the gangliosides occurred in pinched off nerve endings that appeared identical with the terminal boutons seen on the dendrites of neurons, and Wherret and McIlwain[33] showed that the gangliosides may be characteristic components of the endoplasmic reticulum and the plasma membrane. Later studies showed a more even distribution of the gangliosides, although an enrichment in the microsomal fraction has been confirmed,[20,34,35] with the highest concentration in a light membrane fraction. A report on the localization of the gangliosides in synaptic vesicles[36] has turned out to be erroneous.[37,38] In a recent short communication Derry and Wolfe[39] reported on the determination of gangliosides in isolated neurons and glial cells. The highest ganglioside concentration was found in the cleaned neuron and adjacent neuropil, while the glial cell contained practically no gangliosides. The recorded ganglioside concentration was astonishingly low, only a little more than 1% of the dry weight, in the neuron and neuropil. This study thus confirms the prevailing view on the localization of the gangliosides in the nervous system.

In general only total gangliosides have been determined in the isolated subcellular fractions, but Eichberg, Whittaker, and Dawson[20] separated the gangliosides by thin-layer chromatography. The distribution of individual ganglioside components did not vary greatly between the various cell fractions, except for a relative increase of monosialogangliosides in the large myelin-particle fraction. This finding has been extended by Suzuki, Poduslo, and Norton,[40] who found almost nothing but ganglioside $G_{M1}$ in purified myelin. Whether this ganglioside was a component of myelin or was axonic material, probably axolemma, attached to the myelin could not be decided. In a recent study, Thompson, Goodwin, and Cumings[41] could not detect gangliosides in normal brain but found rather large amounts of gangliosides in myelin from subjects with leukodystrophy.[42]

## II. BIOSYNTHESIS OF GANGLIOSIDES

### A. Formation of Active Sialic Acids

The biosynthesis of the carbohydrate components of gangliosides results from a sequence of reactions involving the conversion of glucose to the monosaccharides and their derivatives, "activation" of the mono-saccharides, and polymerization of the glycose units. The "activated" monosaccharides are nucleotide derivatives of the sugars. The pathways leading to their formation, except for the sialic acids, were summarized by Roseman several years ago.[44] At that time nothing was known of the intermediary metabolism of sialic acids, and because they play such an important role in the metabolism of gangliosides their biosynthesis will be shortly outlined.

The pathway for the biosynthesis of the sialic acids (Fig. 2) has been outlined by Roseman and co-workers,[45] but important contributions also have been made by Warren and associates.[46,47] Glucose is converted to *N*-acetyl-D-mannosamine in a series of steps. There are two key reactions in the reaction sequence to *N*-acetyl-D-mannosamine. The first is the amination of fructose-6-P to glucosamine-6-P, and the second is the formation of *N*-acetylmannosamine from *N*-acetylglucosamine. The latter reaction can be accomplished in animals by an enzymatic epimerization of *N*-acetylglucosamine in which ATP is a necessary cofactor.

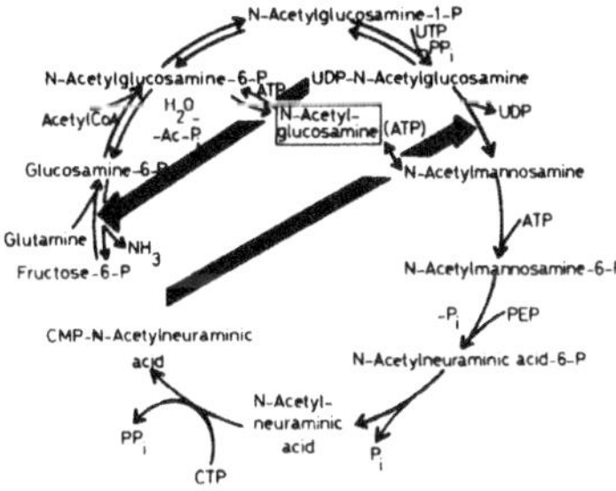


Fig. 2. Pathway of biosynthesis of UDP-*N*-acetylglucosamine and CMP-*N*-acetylneuraminic acid. The long, dark arrows indicate sites of feedback inhibition. Modified after Kornfeld, Kornfeld, Neufeld, and O'Brien.[50]

*N*-acetylmannosamine, catalyzed by a specific kinase, is phosphorylated by ATP to *N*-acetylmannosamine-6-P, which is then condensed with phosphoenolpyruvate, with the loss of its phosphate. *N*-acetylneuraminic acid-9-P is finally dephosphorylated by a specific phosphatase. The formation of sialic acid from *N*-acetylmannosamine has been studied in various animal tissues, and all the enzymes involved were present in different regions of brain.[48] All three reactions are essentially irreversible. Warren and Felsenfeld[49] have shown that in the presence of excess ATP and phosphoenolpyruvate, crude liver extracts quantitatively transformed *N*-acetylmannosamine to *N*-acetylneuraminic acid (NAN).

NAN needs to be activated until it can be transferred to its acceptors. Contrary to the other carbohydrates of ganglioside biosynthesis, which are linked to uridine diphosphate (UDP), NAN is linked to cytosine monophosphate (CMP). The reaction is irreversible because the high-energy phosphoanhydride bond of CTP is broken and a relatively low-energy phosphoester bond of CMP-NAN is formed.

The biosynthetic reactions for NAN are virtually all irreversible and strongly favor the formation of NAN and its incorporation into macromolecules. It is obvious that there is some control mechanism; otherwise the cell could be loaded with bound sialic acid. Control mechanisms of hexosamine and sialic acid metabolism recently have been described for rat liver.[50] Two key steps are inhibited: The formation of glucosamine-6-P by the amination of fructose-6-P is inhibited by UDP-*N*-acetylglucosamine, and the formation of *N*-acetylmannosamine from UDP-*N*-acetylglucosamine, by CMP-NAN. They are both typical examples of feedback inhibition—the end product in a sequence of reactions inhibits the enzyme catalyzing the first step in the reaction, which is unique for the synthesis of the final product. UDP-*N*-acetylglucosamine inhibits new formation of hexosamine and CMP-NAN formation of new sialic acid. Ultimately, CMP-NAN controls the formation of glucosamine-P.

## B. Sialyltransferases

The sialyltransferases are a family of enzymes that catalyze the incorporation of CMP-NAN or CMP-*N*-glycolylneuraminic acid (NGN) into polymers. They can be distinguished from each other on the basis of their specificities toward the acceptor molecules or on the basis of the chemical structure of the products.[51] The acceptor can be an oligosaccharide, a glycoprotein, or a glycolipid.

The first sialyltransferase of mammalian tissue was found in lactating mammary gland,[52] but they since have been detected in several other glands, in kidney, and in brain. The actions of sialyltransferases on ganglioside biosynthesis have been studied systematically only by Roseman and associates.[53,54] Chicken embryonic brain mainly served as the enzymic source. From a brain homogenate in 0.25 *M* sucrose, the fraction sedimenting between 3000 and 39,000 *g* was collected. The enzyme activity was located in a light particulate layer overlaying a heavy residue. After washing, this layer was resuspended in the sucrose and used as enzyme source. Unsuccessful attempts were made to solubilize the transferases. Several glycolipids, oligosaccharides, and enzymatically desialized glycoproteins served as substrates, and CMP-NAN$^{14}$-C was either acetyl- or carboxyl-labeled. Active lipid acceptors were lactosylceramide, *N*-tetraglycosylceramide, and monosialosyl-*N*-tetraglycosylceramide ($G_{M1}$). Partial heat inactivation and substrate competition experiments indicated that there are at least two different enzymes, one acting on lactosylceramide and the other on lactose, glycoproteins, and the other two glycolipids. The sialyltransferase that acted on lactosylceramide had no metal requirements and was comparatively heat stable in contrast to the other, which required $Mg^{2+}$ and was heat-labile. The lowest $K_m$ values were obtained for the glycolipids; *N*-tetraglycosylceramide had the lowest among the glycolipids.

The highest activities of sialyltransferases were detected with embryonic chicken and fetal pig brains. The activities in adult animals were very low and in some animals not detectable. At these experiments a total brain homogenate was used as enzymic source, and in general much higher activities were obtained with *N*-tetraglycosylceramide than with lactosylceramide as acceptor.

The products were characterized after isolation from large-scale incubations. The monosialosyllactosylceramide formed was indistinguishable from authentic ganglioside $G_{M3}$ by thin-layer chromatography and contained the theoretical chemical composition. When *N*-tetraglycosylceramide (CER–TETRA) was used as acceptor, a mono- and disialoganglioside were obtained in a ratio of 9:1. The monosialosyl-*N*-tetraglycosylceramide differed in one very important respect from the normal brain monosialoganglioside ($G_{M1}$): The major part of the sialic acid was easily liberated by neuraminidase. This indicates that the sialic acid was not transferred to C-3 of internal galactose, as suggested by the authors.(53) They also claimed that the monosialosyl-*N*-tetraglycosylceramide had the same $R_F$-value as $G_{M1}$ at thin-layer chromatography, but the actual figure(54) showed a somewhat higher $R_F$-value. There is no evidence that the disialoganglioside formed had the same structure as $G_{D1a}$ or $G_{D1b}$.

The first indication for the incorporation of CMP-sialic acid into a glycolipid was given by Kanfer *et al.*(55) They used rat kidney homogenate as enzyme source and asialoganglioside and aminoglycolipid, probably *N*-tri- and *N*-tetraglycosylceramides obtained by acidic hydrolysis of beef brain gangliosides, as acceptors. They found that more than 20% of the labeled CMP-NAN was incorporated in 3 hr and 80% of the labeling was recovered in monosialosyl-*N*-tetraglycosylceramide ($G_{M1}$). Their results are difficult to interpret and not consistent with those obtained with brain homogenates by Roseman *et al.*(56) The latter found practically no incorporation of CMP-NAN into *N*-triglycosylceramide.

Sialyltransferases of brain have also been studied by Caputto and associates.[57] The transferring activity was found in mitochondria, microsomes, and supernatant fluid. A solubilization of the enzyme was obtained by disruption of microsomes with sodium deoxycholate, and these disrupted microsomes were used to measure the $K_m$ value for the different acceptors. The lowest $K_m$ value was found for lactosylceramide, while the figures were higher for lactosylsphingosine or asialogangliosides (mainly *N*-tetraglycosylceramide) obtained by acidic hydrolysis of a ganglioside mixture. The products of the synthesis were only tentatively identified.

## C. Galactosyltransferases

The particulate fraction from embryonic chicken brain,[53] which contained the sialyltransferases, was shown to also contain galactosyltransferase activity.[56] Incorporation occurred into monosialosyl-*N*-triglycosylceramide ($G_{M2}$), glycoproteins, and *N*-acetylglucosamine, but not into *N*-acetylgalactosamine. Heat inactivation studies and substrate inhibition experiments showed that a galactosyltransferase was active, specifically with $G_{M2}$. It differed from sialyltransferase in the requirement for a cation; $Mn^{2+}$ was most effective, while $Mg^{2+}$ was inactive. Product isolation showed the formed ganglioside to have the same composition and the same chromatographic $R_F$-values as authentic $G_{M1}$.

The glycolipid sialyltransferase activity was fairly constant between the seventh to fifteenth day of embryonic development and then declined. The corresponding activity curve for galactosyltransferase showed a sharp rise in activity between the seventh and eleventh days of development and was then rather constant until hatching. The activity also was high in fetal pig brains, whereas the activity in adult animals was very low.

## D. Other Gangliosidetransferases

Only preliminary results have been reported for the *N*-acetylgalactosaminyltransferase. The same particulate fraction was used as for the other transferases.[53,54] Several different glycolipids were used as acceptors. The highest activity was obtained with monosialosyllactosylceramide ($G_{M3}$), but some activity was also obtained with lactosylceramide and monosialosyl-*N*-triglycosylceramide ($G_{M2}$).[54]

Recent studies by Basu[58] showed that the particulate fraction from embryonic chicken brain also contains a glucosyltransferase that catalyzes the formation of glucosylceramide from UDP-Glu-$^{14}$C and ceramide. This acceptor could not be replaced by sphingosines or galactosylsphingosine; it was not reported whether glucosylsphingosine was tried. When UDP-galactose replaced the other nucleotide sugar no incorporation was detected. The reaction did not require any metal.

A second ganglioside galactosyltransferase also was demonstrated that catalyzed the formation of lactosylceramide from glucosylceramide and

UDP-galactose. A similar transferase that catalyzed the same reaction has been reported in rat spleen by Hauser.[59]

Results from isolation of the particulate fraction of embryonic chicken brain suggested that it may be rich in nerve endings. In a preliminary note Den and Kaufman[60] reported on the determination of glycosyltransferases in synaptosomes. They found that all the glycosyltransferases involved in the synthesis of gangliosides and glycoproteins, except galactosyltransferase, were located primarily in the synaptosomes.

## E. Incorporation of Labeled *in Vivo*

The *in vivo* incorportation of radioactive precursors into brain gangliosides have been studied with glucose (Glu), galactose (Gal), glucosamine, serine, and acetate.[61-65] The first experiments on the incorporation of labeled sugars were performed by Radin, Martin, and Brown.[61] Galactose-1-$^{14}$C was rapidly incorporated into cerebrosides and gangliosides of rat brains, reaching a maximum level about 24 hr after injections. These data were confirmed by Burton and associates,[02] who showed in addition that the incorporation was greatest from 10 to 16 days of age. Maximum incorporation of labeled glucose into brain gangliosides of mice was found to occur between 8 and 30 days of age by Moser and Karnovsky.[63] By the combined use of labeled glucose, galactose, and galactosamine, Burton *et al.*[62] could show that all the sugar moieties of gangliosides—glucose, galactose, galactosamine, and *N*-acetylneuraminic acid—had the same specific activity. The enzymatically stable and labile NAN were labeled to the same extent. This is in contrast to the results by Suzuki and Korey,[64] who found less activity in the enzymatically labile NAN and suggested that the two sialic acids were derived from different pools.

There are also some reports on the incorporation of labeled sugars into individual gangliosides. Burton *et al.*[62] were able to detect incorporation only into $G_{D1a}$ and $G_{D1b}$ in 14-day-old rats, while Suzuki and Korey[64] found essentially the same specific activity in all the four $G_1$-gangliosides (sialosyl-*N*-tetraglycosylceramides) in 7-day-old rats. In a recent study Suzuki[65] found that the relative rate of formation of the $G_1$-gangliosides paralleled the changes in the percentage distribution. It means that the rapid increase in the relative concentration of $G_{D1a}$ during the early postpartum period was reflected by a higher labeling of this ganglioside. The study also showed that the whole pool of any $G_1$-ganglioside could not be precursor to the other. Unfortunately the radioactivity of $G_{M2}$ and $G_{M3}$ was not determined.

## F. The Biosynthesis of Gangliosides

The formation of gangliosides can proceed via two different types of pathways:

1. Sugars are added to a lipid acceptor as oligosaccharide units.

2. Sugars are added to a lipid acceptor as monosaccharide units in a stepwise manner.

There is no evidence that the first route is used, although free nucleotide oligosaccharides have been isolated from mammalian sources.[66] Incorporation studies[64,65] with labeled sugars suggest that it is a possible route, and the extensive studies by Roseman and associates[53,54] have not ruled out the possibility that it may be a route in brains of adult animals.

In the fetal brain tissue, stepwise addition of monosaccharide units to the appropriate glycolipid acceptor seems to be the normal pathway for the biosynthesis of brain gangliosides (Fig. 3). All the steps from ceramide to monosialosyl-*N*-tetraglycosylceramide ($G_{M1}$) have been accounted for, and the enzymes that catalyze the reactions have a high or moderately high specificity. At least their activities are tenfold larger with the appropriate lipid acceptor than with other acceptors.

One crucial problem in the formation of gangliosides has been whether a neutral *N*-tetraglycosylceramide is first formed and the sialic acids then added or the first sialic acid introduced into the lactosylceramide. The studies by Roseman and associates[53,54] have unequivocally shown that the first sialic acid is incorporated at the lactosylceramide stage, under the catalysis by a sialyltransferase with high specificity. The sialyltransferase

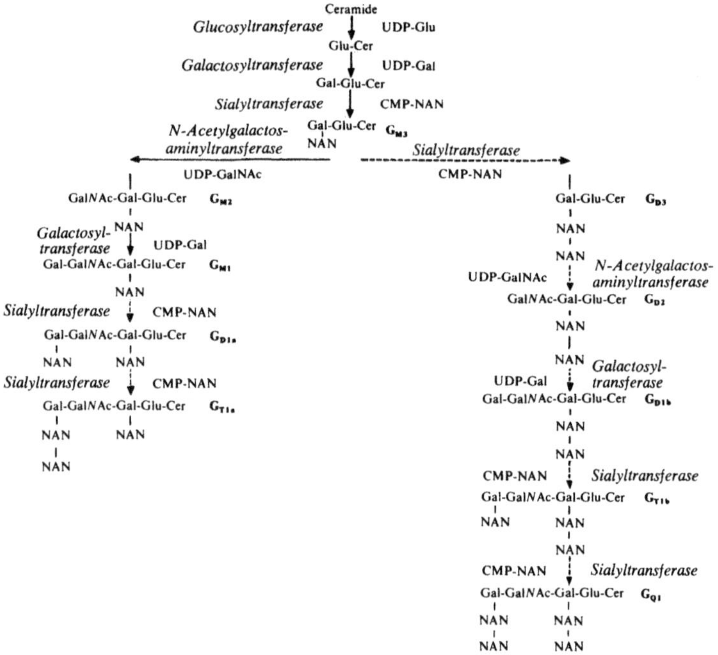


Fig. 3. Pathway for the biosynthesis of gangliosides. Unbroken line: demonstrated step (Kaufman *et al.*,[53] Basu,[54,58]); dashed line: probable step; dotted line: possible step. The structures of gangliosides within brackets have not been verified.

that catalyzes the incorporation of the second sialic acid seems to be rather unspecific and to act on several different acceptors—or else it has not been possible to separate this sialyltransferase from a similar one. This question eventually could be solved if it were possible to elaborate a method for the separation of the different sialyltransferaaes. A prerequisite condition is soluble enzymes. Roseman and associates[53,54] were not able to solubilize the transferases, but it is possible that the procedure used by Caputto and associates[57] could be a useful initial step in a method for separation of sialyltransferases.

The incorporation of the second sialic acid into the gangliosides is the weakest chain in the reaction sequence, and there seem to be no evidence concerning to which sugar of the ganglioside moiety the second sialic acid is linked. The nearly quantitative liberation of the sialic acid from the disialoganglioside formed from *N*-tetraglycosylceramide indicates that the two sialic acids are linked to the terminal galactose. This would suggest that the gangliosides with two sialic acid residues linked to the internal galactose, as $G_{D1b}$ and $G_{T1}$, have been synthesized from disialosyllactosylceramide ($G_{D3}$).

The first two gangliosides in this sequence normally occur in brains. In fetal and newborn brains the concentrations of $G_{D1b}$ and $G_{T1}$ are low, which might be explained by a low activity of the sialyltransferase that catalyzes the incorporation of the second sialic acid into monosialosyllactosylceramide.

The recent observation by Den and Kaufman,[60] that most of the ganglioside transferase activity in chicken brain was recovered from the synaptosome fraction, is extremely interesting. Only a minor amount of the total brain gangliosides can be accounted for in the synaptosome fraction.[20] Their findings suggest that it is possible that the gangliosides are formed in the synaptosomes and then transported centripetally to the neuronal cell body. However, as has been described in Section I,C, the rise in ganglioside concentration is extremely high during a short period of development of the brain. It is possible that ganglioside transferase activities are stimulated by a factor intimately connected with the outgrowth of the neurites and dendrites.

## III. DEGRADATION OF GANGLIOSIDES

### A. Turnover Studies of Gangliosides

The data from the turnover experiments are difficult to interpret, as recently discussed by Burton.[68] When Burton *et al.*[67] used glucosamine and glucose as precursors they found the half-life of gangliosides to be about 24 days, with glucosamine as precursor, and approximately 10 days when the precursor was glucose-1-$^{14}$C. Burton[68] interpreted these data as indicating that the turnover time of the ganglioside carbohydrate moiety is less than 10 days. The glucose and galactose enter large pools, which undergo rapid metabolism since glucose is the primary source of energy for the brain tissue. The reincorporation of labeled glucose and galactose will be very slight. On the other hand there is not much demand for amino sugars, and the pool is small. The amino sugars will be resynthesized into gangliosides and will have considerable activity.

Suzuki[65] confirmed Burton's result for the half-life of gangliosides (25 days) when D-glucosamine-1-$^{14}$C was used, but found 20 days with D-glucose-1-$^{14}$C if rats of the same age were compared.

Suzuki[65] found the turnover of gangliosides to be very low in the first 10 days postpartum. The turnover was fastest between 10 and 20 days postpartum, and then it declined. The turnover rates of the four $G_1$-gangliosides were very similar.

Kishimoto, Davies, and Radin[69] reported a relatively fast turnover also of the ganglioside fatty acids during the same period. The turnover rates were not given but were estimated to be of the same order as for the phosphoglyceride fatty acids. Judging from their curves, the turnover rates then rapidly diminished. From the turnover studies there thus is evidence that the fastest degradation of gangliosides occurs simultaneously with maximal biosynthesis.

## B. Neuraminidases

Neuraminidase (or sialidase, *N*-acetylneuraminylhydrolase, E.C. 3.2.1.18) activity of nervous tissue was first demonstrated by Carubelli, Trucco, and Caputto[70] in rat brain. Morgan and Laurell[71] found that homogenates prepared from human, guinea pig, and bovine brains released sialic acid at incubation. They attributed this finding to the hydrolysis of gangliosides by a mammalian brain neuraminidase, but only total and not ganglioside sialic acid was assayed after enzymic hydrolysis. In other studies gangliosides were added to brain homogenates, and liberated sugars and remaining gangliosides and glycosylceramides were estimated. In this manner Korey and Stein[72] demonstrated the occurrence of a "gangliosidase" in extracts of human brain. In addition to sialic acid these preparations also liberated neutral sugars and sphingosine. Human brain homogenates[73] and pig brain extracts[74] also were shown to degrade $G_{T1}$ and $G_{D1}$-gangliosides to $G_{M1}$-ganglioside.

Carubelli *et al.*[70] used sialosyllactose for the determination of the neuraminidase activity, and in some recent publications the same substrate was used for the determination of neuraminidase activity in mice brain homogenates[75] and rat brain supernatants.[76]

A substantial purification of the enzyme was not reported until recently, when Leibowitz and Gatt[77] described the partial purification of neuraminidase from calf brain. Fresh brains were homogenized with cool acetone; the dried acetone powder was treated with cholate solution twice; and the enzyme was then extracted with Triton X-100. After centrifugation at 100,000 *g* the supernatant was collected. It was still particulate, kept in colloidal suspension by the detergent. The optimal *p*H for hydrolysis of gangliosides was 4.4 using acetate or citrate buffer. The highest activity was obtained with $G_{T1}$ and $G_{D1a}$ as substrates, while somewhat lower figures were obtained with $G_{M3}$ and $G_{D1b}$. The enzyme preparation was inactive against sialosyllactose and sialoglycoproteins. In view of the studies mentioned above it is evident that brain tissue contains at least two different neuraminidases, one possible soluble enzyme which hydrolyze sialic acid-containing saccharides and a second one specific for gangliosides.

Support for the suggestion that there are neuraminidases with low substrate specificity in mammalian tissues is obtained from a recent report by Sandhoff and Jatzkewitz.[78] They purified, from the lysosomal fraction of rat liver, a neuraminidase that liberated sialic acid from sialosyllactose, mucin, and gangliosides. The data are incomplete, but they suggest that mucin was the preferred substrate and that the enzyme preparation also hydrolyzed the other substrates at similar rates.

## C. Glycosidases

$\beta$-Glycosidases are widespread in mammalian tissue. $\beta$-Glucosidase ($\beta$-D-glucoside glucohydrolase, 3.2.1.21) and $\beta$-galactosidase ($\beta$-D-galactoside galactohydrolase, 3.2.1.23) were determined several years ago in many different tissues by Cohen *et al.*[79,80] and Conchie *et al.*[81] They showed the activities in intact tissues or crude enzyme preparations. Several attempts were made later to separate the two enzymes, but without success. In recent studies by Gatt and associates[82] a general purification procedure was elaborated for the hydrolases of gangliosides and neutral sphingoglycolipids in brain tissue. This procedure also was successfully applied for the isolation of the glycosidases.[83] Brain was homogenized in sucrose-EDTA, and the particles that sedimented between 800 *g* and 15,000–18,000 *g* were isolated. The particles were subjected to sonic disintegration and extraction with cholate. These extracts had both $\beta$-glucosidase and $\beta$-galactosidase activities. Acidification of the dialyzed extract to *p*H 5 precipitated the $\beta$-glucosidase activity. The supernatant contained the $\beta$-galactosidase activity, practically free from glucosidase activity. The $\beta$-glucosidase activity was recovered from the precipitate by cholate extraction. The purification of the glycosidases was 10–20-fold. Rat, calf, and beef brains were used for the isolations.

### 1. $\beta$-*Galactosidases*

Lactosylceramide prepared by acidic hydrolysis of brain gangliosides was degraded by the rat $\beta$-galactosidase to glycosylceramide and galactose.[84,85] The reaction was not reversible and required cholate; still higher activities were obtained after the addition of a nonionic detergent to the reaction mixture. The optimum *p*H was 5.0. The reaction was inhibited by galactose, $\gamma$-galactonolactone, ceramide, fatty acids, and sphingosine.

The $\beta$-galactosidase also acted on other glycosphingolipids that had a terminal galactose. Galactose was liberated from the following substrates tested:[86] sialosyl-*N*-tetraglycosylceramide($G_{M1}$), *N*-tetraglycosylceramide (obtained by acidic hydrolysis of $G_{M1}$), triglycosylceramide [galactosyl-β-(1 → 4)-glucosyl-β-(2-*N*-acyl) sphingosine], digalactosylceramide [galactosyl-β-(1 → 4) galactosyl-β-(2-*N*-acyl)-sphingosine]. The enzyme did not remove galactose from galactosylceramide. Thus, it seems to be specific for the hydrolysis of a β-galactosidic linkage in which galactose is bound to another sugar as glucose, galactose, or *N*-acetylgalactosamine (GalNAc). The enzyme cannot be considered absolutely specific for glycosphingolipids, since it also hydrolyzed lactose and *o*- and *p*-nitrophenyl-β-D-galactopyranoside. The $K_m$ value for lactosylceramide was, however, twentyfold less than for *p*-nitrophenylgalactoside, indicating a relatively large specificity.

This β-galactosidase differed from that isolated from the 100,000*g* supernatant by Radin and associates.[87] The latter enzyme degraded galactosylceramide to galactose and ceramide. It had a low specific activity, 100-fold less than the β-galactosidase isolated by Gatt and Rapport.[83] A third type of β-galactosidase has been described by Brady and collaborators.[88] They determined the β-galactosidase content of several tissues in 14-day-old rats. The highest activity was found in small intestine. The enzyme was isolated by a procedure similar to that adopted by Gatt and Rapport.[83] The enzyme solubilized by cholate was precipitated by ammonium sulfate and finally purified with ion exchange cellulose chromatography. It was purified more than 2000 times, but it still had both β-galactosidase and β-glucosidase activity. Since the ratio of activity toward galactosyl- and glycosylceramides remained near unity in the different stages of purification, it was assumed to be a single enzyme. However, it was also active toward some more complex glycosylceramides, including triglycosylceramide and *N*-tetraglycosylceramide. The latter substrate has a terminal β-linked *N*-acetylgalactosamine group, which indicates that the intestinal enzyme also contained β-*N*-acetylgalactosaminidase activity.

In a recent paper, Brady *et al.*[89] purified from the same intestinal extract an enzyme that specifically catalyzed the hydrolysis of the terminal galactose of a triglycosylceramide [galactosyl-β-(1 → 4)-galactosyl-β-(1 → 4)-β-glucosyl-β-(2-*N*-acyl)-sphingosine]. Thus, the enzyme did not catalyze the hydrolysis of two related glycosphingolipids with a terminal galactose: lactosylceramide and galactosylceramide. It is unfortunate that digalactosylceramide was not tested, since the enzyme can be assumed to be specific for the hydrolysis of a β-galactosidic bond in which the galactose is bound to another galactose. The enzyme was reported to occur in several tissues, and next to intestine the highest activity was found in brain. It seems questionable whether the brain enzyme is a specific one or is the same β-galactosidase that Gatt and Rapport[83] prepared from brain and showed to also hydrolyze the triglycosylceramide shown above.

### 2. β-*Glucosidase*

The β-glucosidase isolated by Gatt and Rapport [83] from beef brain hydrolyzed glucosylceramide to ceramide (Cer) and glucose.[90] The hydrolysis was stimulated by a nonionic detergent and cholate. It had optimum *p*H 5.0 with acetate buffer and 5.6 with pyridine buffer. The $K_m$ value was about 0.1 mM. It was inhibited by δ-gluconolactone and sphingosine but not by ceramide. It hydrolyzed nitrophenyl-β-D-glucopyranoside but not the β-glucosidic linkage of several oligosaccharides or methyl-β-glucoside. Thus it was more specific than the brain β-galactosidase.

A glucosylceramide-cleaving enzyme shortly before had been isolated from human spleen tissue by Brady *et al.*[91] This enzyme was isolated from the 100,000 *g* supernatant fraction, and it had a somewhat higher *p*H optimum—6.0. It was inactive toward galactosylceramide, but the specificity toward other β-glucosides was not reported. A decreased activity of this soluble β-glucosidase was subsequently reported by Brady *et al.*[92,93] in Gaucher disease. There did not appear to be any obvious correlation between the diminution in the concentration of enzymatic activity and the onset and progression of the disease.[94] The β-glucosidase diminution in Gaucher disease also has been demonstrated in blood white cells, and a clinical routine procedure has been elaborated for the enzymatic diagnosis of Gaucher disease in blood.[95] There seems to have been no attempt to investigate whether the intestinal β-glucosidase[87] is diminished in Gaucher disease or whether the β-galactosidase activity is diminished. The papers by Brady and associates also have raised the important questions of whether a soluble β-glucosidase occurs in brain that hydrolyzes glucosylceramide or whether all glucosylceramide-cleaving enzyme is particulate.

### 3. β-*N*-*Acetylhexosaminidases*

β-*N*-Acetylhexosaminidases (β-2-acetylamino-2-deoxy-D-glycoside acetylaminodeoxyglycohydrolases, 3.2.1.30) have been demonstrated in many tissues.[81,96–100] The enzyme preparations hydrolyzed the β-glycosidic linkage of both *N*-acetylglucosaminides and *N*-acetylgalactosaminides, and several attempts to separate the two activities failed.[96–100] The presence of β-*N*-acetylhexosaminidase activity in brain tissue was shown by Conchie *et al.*[81] Sellinger *et al.*[101] compared the intracellular distribution in rat cerebral cortex of *N*-acetylglucosaminidase with that of acid phosphatase and found it to coincide with the latter. From these studies they draw the conclusion that *N*-acetylglucosaminidase was a "lysosomal" enzyme. Three β-*N*-acetylhexosaminidases were isolated by Frohwein and Gatt[102] from calf brain. In a fraction sedimenting between 1000 *g* and 25,000 *g* an enzyme was isolated that hydrolyzed the *p*-nitrophenyl derivatives of both β-*N*-acetylglucosamine (*p*H-optimum 4.2 and $K_m$ 0.8 mM) and β-*N*-acetylgalactosamine (*p*H-optimum 3.8 and $K_m$ 0.5 mM). The hydrolysis was inhibited by acetate and free and acetylated glucosamine and galactosamine. Attempts to separate the two enzymatic activities were unsuccessful. The two other enzymes were isolated from the 100,000 *g* supernatant by precipitation with protamine sulfate. The two enzymatic activities could be separated when *N*-acetylglucosaminidase was purified in the presence of *N*-ethylmaleimide.[103]. The two "soluble" hexosaminidases differed from the "particulate" by being less stable and having higher *p*H-optima.

The three hexosaminidases were tested on three glycosphingolipids containing a terminal galactosamine, globoside (*N*-tetraglycosylceramide), ganglioside $G_{M2}$, and the corresponding asialoganglioside, *N*-triglycosylceramide.[104] They were hydrolyzed only by the particulate $\beta$-*N*-acetylhexosaminidase. The two neutral glycolipids were hydrolyzed at similar rates, while the ganglioside was only slowly degraded, after some latency. The ganglioside also inhibited the hydrolysis of the other two aminoglycolipids.

## D. Catabolic Pathway of Brain Gangliosides

As is evident from the preceding sections, brain enzymes for the hydrolysis of gangliosides to ceramide have been purified and partially characterized. Gatt[105] also has isolated a ceramidase from rat brain that catalyzed a reversible reaction in which the amide bond of *N*-acylsphingosine was either hydrolyzed or synthesized. Specificity was shown for the fatty acid component: 16- and 18-carbon fatty acids were easily hydrolyzed. A more complete discussion of this enzyme can be found in the chapters on sphingomyelin and cerebroside metabolism.

Degradation of trisialoganglioside to monosialosyllactosylceramide was recently described, on brain homogenates of human newborns.[73] The gangliosides were prepared from senile human brains in which $C_{20}$-sphingosine comprised 70% of total sphingosine. Because $C_{20}$-sphingosine constituted only 1–2% in the newborn brains used as enzyme source, it was possible to use the $C_{20}$-sphingosine content of the product as a measure of hydrolyzed ganglioside. The proposed schema for the stepwise hydrolysis of gangliosides[73] now has been verified and extended by the isolation and purification of several glycolipid hydrolases by Gatt and associates,[77] and the stepwise degradation of trisialoganglioside to free fatty acid and sphingosine can be outlined (Fig. 4). With the human brain neuraminidase,[106] as with those of bacterial origin,[3] the degradation of trisialoganglioside $G_{T1}$ proceeds mainly over $G_{D1b}$, which means that the sialic acid bound to the terminal galactose is first hydrolyzed. The last sialic acid bound to the internal galactose will not be liberated by neuraminidase of human,[106] rat,[77] or bacterial neuraminidase;[2,3] this was suggested by Kuhn and Wiegandt[2] to depend on steric hindrance from *N*-acetylgalactosamine bound to C4 of internal galactose. A competitive or other form of inhibition by galactosamine and *N*-acetylgalactosamine also has been recently demonstrated.[107] As soon as *N*-acetylgalactosamine has been hydrolyzed, neuraminidase easily liberates the sialic acid. This finding makes it unlikely that the neutral *N*-tri- and *N*-tetraglycosylceramides that occur in brain tissue are formed from gangliosides, as suggested by Jazkewitz *et al.*[108] It is more probable that they have been biosynthesized

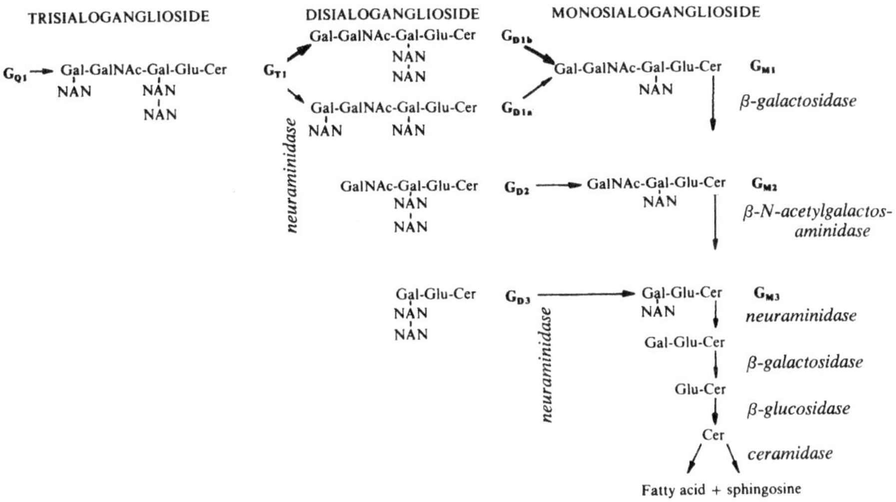


Fig. 4. The biodegradation of brain gangliosides.

by the same enzymes as the gangliosides, except that *N*-acetylgalactosamine is incorporated into lactosylceramide instead for NAN. When the *N*-triglycosylceramide has been formed, the sialyltransferase does not seem to be able to transfer any sialic acid to the internal galactose.[53] Thus, the *N*-acetylgalactosamine linked to C4 of galactose not only prevents the hydrolysis of the sialic acid by neuraminidase, but seems also to be a steric hindrance for the action of sialyltransferase.

## IV. GANGLIOSIDOSES

The research on brain gangliosides was initiated by the important discovery by Klenk[109] that a hitherto unknown glycolipid was accumulated in brain of children with Tay–Sachs disease. It was, however, more than 20 years before it was shown that only a single ganglioside was accumulated in the disorder and that this ganglioside differed from the normal parent ganglioside $G_{M1}$ by lacking the terminal galactose.[110] Since this discovery, two further inherited ganglioside disorders have been detected. Very much as in Tay–Sachs disease, $G_{M1}$-ganglioside is stored instead of $G_{M2}$. This disorder has been described under several different names: biochemically special form of infantile amaurotic idiocy,[111] generalized gangliosidosis,[112] and systematic late infantile lipidoses.[113] The preferred term has been generalized gangliosidosis since it was said to differ from the classical Tay–Sachs disease because of visceral involvement. Since we[114] have been able to demonstrate the accumulation of $G_{M2}$-ganglioside in visceral organs in Tay–Sachs disease, there is no longer any real difference between the two disorders. It would be more convenient, then, to term the disorders after the stored ganglioside—e.g., $G_{M1}$-gangliosidosis and $G_{M2}$-gangliosidosis—as recently suggested by Suzuki and Chen.[115] In the third disorder in which a certain ganglioside was stored, only a single case has been described.[116] Furthermore, the biochemical data are incomplete, because only formalin-preserved brain material could be obtained from this case.[117] One might be entitled to assume that ganglioside $G_{M3}$ was the major stored substance in the brain even though lactosylceramide was found in larger amounts, since formalin preservation leads to liberation of sialic acid. This case differed also from the other two ganglioside disorders by a severe visceral involvement. Because lactosylceramide and sialosyllactosylceramide ($G_{M3}$) are normally the most common glycolipid in visceral organs, a large accumulation of them is to be expected in these organs.

The enzymic lesions in the three gangliosidoses have not yet been demonstrated. An accumulation of a certain ganglioside might be caused by a disturbance of the biosynthesis or of the biodegradation. From a strict biochemical point of view, neither of the two possibilities is more likely than the other. But the clinical features of the disorders give very strong support for a lesion of degradative enzymes. Patients with these disorders appear essentially normal at birth and develop the symptoms of the disease progressively. This is not compatible with a defective formation of gangliosides, in which the symptoms would present themselves at birth.

In the fourth form of inherited ganglioside disorder, infantile Gaucher disease, a strongly diminished activity of β-glucosidase in spleen[93] has been demonstrated. In the brains of three patients the gangliosides and the neutral glycolipids were quantitatively isolated and their composition analyzed.[73] It was shown that except for the occurrence of glucosylceramide which normally can be found only in trace amounts, lactosylceramide and ganglioside $G_{M3}$ were also increased. The ceramide portion of the glucosylceramide was very similar to that of brain gangliosides, but the $C_{20}$-sphingosine content was much lower in glucosylceramide than in the brain gangliosides. These findings were interpreted to suggest glucosyl ceramide of brain to be derived from gangliosides. In Fig. 5 the suggested enzymic lesions and the stored glycolipid in the known form of gangliosidoses are given.

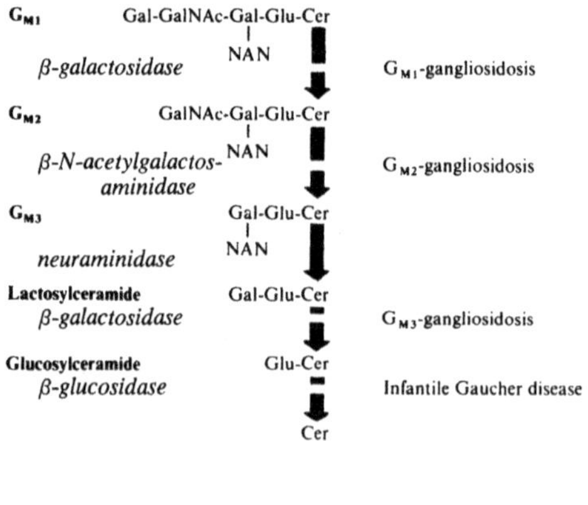


Fig. 5. The enzymic lesions in the gangliosidoses.

## V. REFERENCES


1. L. Svennerholm and A. Raal, Composition of brain gangliosides, *Biochim. Biophys. Acta* **53**:422–424 (1961).
2. R. Kuhn and H. Wiegandt, Die Konstitution der Ganglio-*N*-tetraose und des Ganglioside $G_1$, *Chem. Ber.* **96**:865–880 (1963).
3. L. Svennerholm, Chromatographic separation of human brain gangliosides, *J. Neurochem.* **10**:613–623 (1963).
4. H. Wiegandt, Ganglioside, *Ergeb. Physiol. Biol. Chem. Exptl. Pharmakol.* **57**:190–222 (1966).
5. H. Wiegandt, Struktur and Funktion der Ganglioside, *Angew. Chem.* **80**:89–98 (1968).
6. *The Nomenclature of Lipids* IUPAC–IUB Commission on Biochemical Nomenclature (April, 1967).
7. R. H. McCluer and R. J. Penick, Isolation and Structural Analysis of Brain Gangliosides, *in Inborn Disorders of Sphingolipid Metabolism* (S. M. Aronson and B. W. Volk, eds.), pp. 241–250, Pergamon Press, Oxford (1967).
8. R. Leeden, The chemistry of gangliosides: A review, *J. Am. Oil Chem. Soc.* **43**:57–66 (1966).
9. E. Klenk and H. Langerbeins, Über die Verteilung der Neuraminsäure im Gehirn, *Hoppe-Seyler's Z. Physiol. Chem.* **270**:185–193 (1941).
10. E. Klenk, Über die Ganglioside, eine neue Gruppe von Zuckerhaltigen Gehirnlipoiden, *Hoppe-Seyler's Z. Physiol. Chem.* **273**:76–86 (1942).
11. L. Svennherholm, Quantitative estimation of gangliosides in senile human brains, *Acta Soc. Med. Upsalien.* **62**:1–16 (1957).
12. K. Suzuki, A simple and accurate micromethod for quantitative determination of ganglioside patterns, *Life Sci.* **3**:1227–1233 (1964).
13. K. Suzuki, The pattern of mammalian brain gangliosides—II. Evaluation of the extraction procedures, post mortem changes and the effect of formalin preservation, *J. Neurochem.* **12**:629–638 (1965).
14. K. Suzuki, The pattern of mammalian brain gangliosides—III. Regional and development differences, *J. Neurochem.* **12**:969–979 (1965).
15. L. Svennerholm, The distribution of lipids in the human nervous system—I. Analytical procedure. Lipids of foetal and newborn brain, *J. Neurochem.* **11**:839–853 (1964).
16. K. Suzuki, Ganglioside Patterns of Normal and Pathological Brains, in *Inborn Disorders of Sphingolipid Metabolism* (S. M. Aronson and B. W. Volk, eds.), pp. 215–230, Pergamon Press, Oxford (1967).
17. O. Eeg-Olofsson, K. Kristensson, P. Sourander, and L. Svennerholm, Tay–Sachs disease—A generalized metabolic disorder, *Acta Paediat. Scand.* **55**:546–562 (1968).
18. H. H. Hess and E. Rolde, Fluorometric assay of sialic acid in brain gangliosides, *J. Biol. Chem.* **239**:3215–3220 (1964).
19. N. T. Eldredge, G. Read, and W. Cutting, Sialic acids in the brain and tissues of various animals: analytical and physiological data, *Med. Exptl.* **8**:265–277 (1963).
20. J. Eichberg, Jr. V. P. Whittaker, and R. M. C. Dawson, Distribution of lipids in subcellular particles of guinea-pig brain, *Biochem. J.* **92**:91–100 (1964).
21. O. W. Carrigan and E. Chargaff, Studies on the mucolipids and the cerebrosides of chicken brain during embryonic development, *Biochim. Biophys. Acta* **70**:452–464 (1963).
22. R. Norén and L. Svennerholm, The gangliosides of fish, crab, lobster and octopus, in press.

23. A. Tingey, The results of glycolipid analysis in certain types of lipidosis and leucodystrophy, *J. Neurochem.* **3**:230–237 (1959).
24. R. H. McCluer, E. H. Coram, and H. S. Lee, A silicic acid adsorption method for the determinaton of ganglioside sialic acid, *J. Lipid Res.* **3**:269–274 (1963).
25. J. N. Cumings, H. Goodwin, and G. Curzon, Ganglioside and its relationship to neuraminic acid and hexosamine in the brain, *J. Neurochem.* **4**:234–237 (1959).
26. C. Long and D. A. Staples, Determination of neuraminic acid in crude brain lipids, *Biochem. J.* **73**:385–389 (1959).
27. F. James and K. Fotherby, Distribution in brain of lipid-bound sialic acid and factors affecting its concentration, *J. Neurochem.* **10**:587–592 (1963).
28. J. A. Lowden and L. S. Wolfe, Studies on brain gangliosides—III. Evidence for the location of gangliosides specifically in neurones, *Can. J. Biochem.* **42**:1587–1594 (1964).
29. R. Landholt, H. H. Hess, and C. Thalheimer, Regional distribution of some chemical structural components of the human nervous system—I. RNA and ganglioside sialic acid, *J. Neurochem.* **13**:1441–1452 (1966).
30. E. T. Pritchard and P. L. Cantin, Ganglioside in maturing rat brain, *Nature* **193**: 580–581 (1962).
31. A. Rosenberg and N. Stern, Changes in sphingosine and fatty acid components of the gangliosides in developing rat and human brain, *J. Lipid Res.* **7**:122–131 (1966).
32. L. S. Wolfe, The distribution of gangliosides in subcellular fractions of guinea-pig cerebral cortex, *Biochem. J.* **79**:348–355 (1961).
33. J. R. Wherret and H. McIlwain, Gangliosides, phospholipids, protein and ribonucleic acid on subfractions of cerebral microsomal material, *Biochem. J.* **84**:232–237 (1962).
34. L. M. Seminario, N. Hren, and C. J. Gomez, Lipid distribution in subcellular fractions of the rat brain, *J. Neurochem.* **11**:197–207 (1964).
35. H. Koenig, D. Gaines, T. McDonald, R. Gray, and J. Scott, Studies of brain lysosomes—I. Subcellular distribution of the acid hydrolases, succinate dehydrogenase and gangliosides in rat brain, *J. Neurochem.* **11**:729–743 (1964).
36. R. M. Burton, R. E. Howard, S. Baer, and Y. M. Balfour, Gangliosides and acetylcholine of the central nervous system, *Biochim. Biophys. Acta* **84**:441–447 (1964).
37. V. P. Whittaker, Some properties of synaptic membranes isolated from the central nervous system, *Ann. N.Y. Acad. Sci.* **137**:982–998 (1966).
38. H. Wiegandt, The subcellular localization of gangliosides in the brain, *J. Neurochem.* **14**:671–674 (1967).
39. D. M. Derry and L. S. Wolfe, Gangliosides in isolated neurons and glial cells. *Science* **158**:1450–1452 (1967).
40. K. Suzuki, S. E. Poduslo, and W. T. Norton, Gangliosides in the myelin fraction of developing rats, *Biochim. Biophys. Acta* **144**:375–381 (1967).
41. E. J. Thompson, H. Goodwin, and J. N. Cumings, Caesium chloride in the preparation of membrane fractions from human cerebral tissue, *Nature* **215**:168–169 (1967).
42. J. N. Cumings, E. J. Thompson, and H. Goodwin, Sphingolipids and phospholipids in microsomes and myelin from normal and pathological brains, *J. Neurochem.* **15**: 243–248 (1968).
43. J. N. Cumings, E. J. Thompson, and H. Goodwin, Sphingolipids and phospholipids in microsomes and myelin from normal and pathological brains, *J. Neurochem.* **15**: 243–248 (1968).
44. S. Roseman, Metabolism of connective tissue, *Ann. Rev. Biochem.* **28**:545–578, 1959.
45. G. W. Jourdian and S. Roseman, Intermediary metabolism of sialic acids, *Ann. N.Y. Acad. Sci.* **106**:202–217 (1963).

46. L. Warren, R. S. Blacklow, and C. W. Spearing, Biosynthesis and metabolism of sialic acids, *Ann. N.Y. Acad. Sci.* **106**:191–201 (1963).
47. L. Warren, The Metabolism of Sialic Acids, in *Inborn Disorders of Sphingolipid Metabolism* (S. M. Aronson and B. W. Volk, eds.), pp. 251–259, Pergamon Press, Oxford (1967).
48. R. Joseph and B. K. Bachhawat, Purification and properties of *N*-acetylneuraminic acid-synthesizing enzyme from sheep brain, *J. Neurochem.* **11**:517–526 (1964).
49. L. Warren and H. Felsenfeld, The biosynthesis of sialic acids, *J. Biol. Chem.* **237**: 1421–1431 (1962).
50. S. Kornfeld, R. Kornfeld, E. F. Neufeld, and P. J. O'Brien, The feedback control of sugar nucleotide biosynthesis in liver, *Proc. Nat. Acad. Sci. U.S.* **52**:371–379 (1964).
51. S. Roseman, Studies on the biosynthesis of sialic acid, sialoglycoproteins and gangliosides, *Birth Defects Original Article Ser.* **2**:25–30 (1966).
52. G. W. Jourdian, D. M. Carlsson, and S. Roseman, The enzymatic synthesis of sialyllactose, *Biochem. Biophys. Res. Communs.* **10**:352–358 (1963).
53. B. Kaufman, S. Basu, and S. Roseman, Studies on the biosynthesis of gangliosides, in *Inborn Disorders of Sphingolipid Metabolism* (S. M. Aronson and B. W. Volk, eds.), pp. 193–213, Pergamon Press, Oxford (1967).
54. S. Basu, *Studies on the Biosynthesis of Gangliosides*, Thesis, University of Michigan, Ann Arbor (1966).
55. J. N. Kanfer, R. S. Blacklow, L. Warren, and R. Brady, The enzymatic synthesis of gangliosides I. The incorporation of labelled *N*-acetylneuraminic acid into monosialoganglioside, *Biochem. Biophys. Res. Communs.* **14**:287–291 (1964).
56. S. Basu, B. Kaufman, and S. Roseman, Conversion of Tay–Sachs ganglioside to monosialoganglioside by brain uridine diphosphate-D-galactose glycolipid galactosyltransferase, *J. Biol. Chem.* **240**:PC 4115–4117 (1965).
57. A. Arce, H. F. Maccioni, and R. Caputto, Enzymic binding of sialyl groups to ganglioside derivatives by preparations from the brain of young rat, *Arch. Biochem. Biophys.* **116**:52-58 (1966).
58. S. Basu, Synthesis of glucosylceramide by an enzyme from embryonic chicken brain, *Federation Proc.* **27**:346 (1968).
59. G. Hauser, The enzymatic synthesis of ceramide lactoside from ceramide glucoside and UDP-galactose, *Biochem. Biophys. Res. Commun.* **28**:502–509 (1967).
60. H. Den and B. Kaufman, Ganglioside and glycoprotein glycosyltransferases in synaptosomes, *Federation Proc.* **27**:346 (1968).
61. N. S. Radin, F. B. Martin, and J. R. Brown, Galactolipide metabolism, *J. Biol. Chem.* **224**:499–507 (1957).
62. R. M. Burton, L. Garcia-Bunuel, M. Golden, and Y. Balfour, Incorporation of radioactivity of D-glucosamine-1-$C^{14}$, D-glucose-1-$C^{14}$, D-galactose-1-$C^{14}$, and DL-serine-3-$C^{14}$ into rat brain glycolipids, *Biochemistry* **2**:580–583 (1963).
63. H. W. Moser and M. L. Karnovsky, Studies on the biosynthesis of glycolipids and other lipides of the brain, *J. Biol. Chem.* **234**:1990–1997 (1959).
64. K. Suzuki and S. R. Korey, Study on ganglioside metabolism—I. Incorporation of D-$U^{14}$-glucose into individual gangliosides, *J. Neurochem.* **11**:647–653 (1964).
65. K. Suzuki, Formation and turnover of the major brain gangliosides during development, *J. Neurochem.* **14**:917–925 (1967).
66. G. W. Jourdian, F. Schimizu, and S. Roseman, Isolation of nucleotide-oligosaccharides containing sialic acid, *Federation Proc.* **10**:161 (1961).
67. R. M. Burton, Y. M. Balfour, and J. M. Gibbons, Gangliosides and cerebrosides turnover rates in rat brain, *Federation Proc.* **23**:230 (1964).

68. R. M. Burton, Biochemistry of Sphingosine Containing Lipids in *Lipids and Lipidoses* (E. G. Schetler, ed.), pp. 122–167, Springer Verlag, Berlin, 1967.
69. Y. Kishimoto, W. E. Davies, and N. S. Radin, Turnover of the fatty acids of rat brain gangliosides, glycerophosphatides, cerebrosides and sulfatides as a function of age, *J. Lipid Res.* **6**:525–531 (1965).
70. R. Carubelli, R. E. Trucco, and R. Caputto, Neuraminidase activity in mammalian organs, *Biochim. Biophys. Acta* **60**:196–197 (1962).
71. E. H. Morgan and C.-B. Laurell, Neuraminidase in mammalian brain, *Nature*, **197**: 921–922 (1963).
72. S. R. Korey and A. Stein, Studies in Tay–Sachs diseases—III. Biochemistry. B. Catabolism of gangliosides and related compounds, *J. Neuropathol. Exptl. Neurol.* **22**:67–80 (1963).
73. L. Svennerholm, The Metabolism of Gangliosides in Cerebral Lipidoses, in *Inborn Disorders of Sphingolipid Metabolism* (S. M. Aronson and B. W. Volk, eds.), pp. 169–186, Pergamon Press, Oxford (1967).
74. V. Zambotti, G. Tettamanti, and B. Berra, The evidence of ganglioside-sialidases in the brain of different animals, *Proc. Federation European Biochem. Soc. Vienna*, **A236** (1965).
75. R. T. Kelly and D. Greiff, Neuraminidase and neuraminidase-labile substrates in experimental influenza virus encephalitis, *Biochim. Biophys. Acta* **110**:548–553 (1965).
76. B. H. Taha and R. Carubelli, Mammalian neuraminidase: Intracellular distribution and changes of enzyme activity during lactation, *Arch. Biochem. Biophys.* **119**:55–61 (1967).
77. Z. Leibovitz and S. Gatt, Enzymatic hydrolysis of sphingolipids—VII. Hydrolysis of gangliosides by a neuraminidase from calf brain, *Biochim. Biophys. Acta* **152**: 136–143 (1968).
78. K. Sandhoff and H. Jatzkewitz, A partial particle-bound sialyl lactosideceramide splitting mammalian sialidase, *Biochim. Biophys. Acta* **141**:442–444 (1967).
79. R. B. Cohen, K. Tsou, S. H. Rutenburg, and A. M. Seligman, The colorimetric estimation and histochemical demonstration of β-D-galactosidase, *J. Biol. Chem.* **195**:239–249 (1952).
80. R. B. Cohen, S. H. Rutenburg, K. Tsou, M. A. Woodbury, and A. M. Seligman, The colorimetric estimation of β-D-glucosidase, *J. Biol. Chem.* **195**:607–614 (1952).
81. J. Conchie, J. Findlay, and G. A. Levvy, Mammalian glycosidases. Distribution in the body, *Biochem. J.* **71**:318–325 (1959).
82. S. Gatt, Comparison of Four Enzymes from Brain Which Hydrolyze Sphingolipids, in *Inborn Disorders of Sphingolipid Metabolism* (S. Aronson and B. W. Volk, eds.), pp. 261–266 Pergamon Press, Oxford (1967).
83. S. Gatt and M. M. Rapport, Isolation of β-galactosidase and β-glucosidase from brain, *Biochim. Biophys. Acta* **113**:567–576 (1966).
84. S. Gatt and M. M. Rapport, Enzymic hydrolysis of sphingolipids. Hydrolysis of ceramide lactoside by an enzyme from rat brain, *Biochem. J.* **101**:680–686 (1966).
85. S. Gatt and M. M. Rapport, Hydrolysis of ceramide lactoside and ceramide glucoside by glycosidases from brain, *Israel J. Med. Sci.* **1**:624–627 (1965).
86. S. Gatt, Enzymic hydrolysis of sphingolipids—V. Hydrolysis of monosialoganglioside and hexosylceramides by rat brain β-galactosidase, *Biochim. Biophys. Acta* **137**:192–195 (1967).
87. A. K. Hajra, D. M. Bowen, Y. Kishimoto, and N. S. Radin, Cerebroside galactosidase of brain, *J. Lipid Res.* **7**:379–386 (1966).

88. R. O. Brady, A. E. Gal, J. N. Kanfer, and R. M. Bradley, The metabolism of glucocerebrosides—III. Purification and properties of a glucosyl- and galactosylceramide-cleaving enzyme from rat intestinal tissue, *J. Biol. Chem.* **240**:3766–3770 (1965).
89. R. O. Brady, A. E. Gal, R. M. Bradley, and E. Mårtensson, The metabolism of ceramide trihexosides—I. Purification and properties of an enzyme that cleaves the terminal galactose molecule of galactosyl galactosyl glucosyl ceramide, *J. Biol. Chem.* **242**:1021–1026 (1967).
90. S. Gatt, Enzymic hydrolysis of sphingolipids. Hydrolysis of ceramide glucoside by an enzyme from ox brain, *Biochem. J.* **101**:687–691 (1966).
91. R. O. Brady, J. Kanfer, and D. Shapiro, The metabolism of glucocerebrosides—I. Purification and properties of a glucocerebroside-cleaving enzyme from spleen tissue, *J. Biol. Chem.* **240**:39–43 (1965).
92. R. O. Brady, J. N. Kanfer, and D. Shapiro, Metabolism of glucocerebrosides—II. Evidence of an enzymatic deficiency in Gaucher's disease, *Biochem. Biophys. Res. Commun.* **18**:221–225 (1965).
93. R. O. Brady, J. N. Kanfer, R. M. Bradley, and D. Shapiro, Demonstration of a deficiency of glucocerebroside-cleaving enzyme in Gaucher's disease, *J. Clin. Invest.* **45**:1112–1115 (1966).
94. R. O. Brady, The sphingolipidoses, *New Engl. J. Med.* **275**:312–318 (1966).
95. J. P. Kampine, R. O. Brady, J. N. Kanfer, M. Feld, and D. Shapiro, Diagnosis of Gaucher's disease and Niemann–Pick disease with small samples of venous blood, *Science* **155**:86–88 (1967).
96. J. W. Woollen, R. Heyworth, and P. G. Walker, Studies on glucosaminidase 3. Testicular *N*-acetyl-*β*-glucosaminidase and *N*-acetyl-*β*-galactosaminidase, *Biochem. J.* **78**:111–116 (1961).
97. J. W. Woollen, P. G. Walker, and R. Heyworth, Studies on glucosaminidase 6. *N*-acetyl-*β*-glucosaminidase and *N*-acetyl-*β*-galactosaminidase activities of a variety of enzyme preparations, *Biochem.* **79**:294–298 (1961).
98. P. G. Walker, J. W. Woollen, and R. Heyworth, Studies on glucosaminidase 5. Kidney *N*-acetyl-*β*-glucosaminidase and *N*-acetyl-*β*-galactosaminidase, *Biochem. J.* **79**:288–294 (1961).
99. B. Weissmann, S. Hadjiioannou, and J. Tornheim, Oligosaccharase activity of *β*-*N*-acetyl-D-glucosaminidase of beef liver, *J. Biol. Chem.* **239**:59–63 (1964).
100. E. Buddecke and E. Werries, Reinigung und Eigenschaften einer *β*-*N*-Acetyl-D-hexosaminidase aus Rindermilz, *Z. Naturforsch.* **19b**:798–800 (1964).
101. O. Z. Sellinger, D. L. Rucker, and F. DeBalbian Verster, Cerebral lysomes—I. A comparative study of lysosomal *N*-acetyl-*β*-D-glucosaminidase and mitochondrial aspartic transaminase of rat cerebral cortex, *J. Neurochem.* **11**:271–280 (1964).
102. Y. Z. Frohwein and S. Gatt, Isolation of *β*-*N*-acetylhexosaminidase, *β*-*N*-acetylglucosaminidase, and *β*-*N*-acetylgalactosaminidase from calf brain, *Biochemistry* **6**:2775–2782 (1967).
103. Y. Z. Frohwein and S. Gatt, Enzymatic hydrolysis of sphingolipids—VI. Hydrolysis of ceramide glycosides by calf brain *β*-*N*-acetylhexosaminidase, *Biochemistry* **6**: 2783–2787 (1967).
104. Y. Z. Frohwein and S. Gatt, Separation of *β*-*N*-acetylglucosaminidase and *β*-*N*-acetylgalactosaminidase from calf brain cytoplasm, *Biochim. Biophys. Acta* **128**: 216–218 (1966).
105. S. Gatt, Enzymatic hydrolysis of sphingolipids—I. Hydrolysis and synthesis of ceramides by an enzyme from rat brain, *J. Biol. Chem.* **241**:3724–3750 (1966).

106. R. Öhman, A. Rosenberg, and L. Svennerholm, Human brain sialidase, *J. Biol. Chem.*, submitted.
107. K. Lipovac and A. Rosenberg, The mechanism by which sialic acid in the gangliosides is rendered immune to sialidase, *1st Intern. Meet. Intern. Soc. Neurochem., Strasbourg* **A138** (1967).
108. H. Jatzkewitz, H. Pilz, and K. Sandhoff, The quantitative determination of gangliosides and their derivatives on different forms of amaurotic idiocy, *J. Neurochem.* **12**: 135–144 (1965).
109. E. Klenk, Beiträge zur Chemie der Lipoidosen. I Niemann-Pick'sche Krankheit und amaurotische Idiotie, *Hoppe-Seyler's Z. Physiol. Chem.* **262**:128–143 (1939/1940).
110. L. Svennerholm, The chemical structure of normal human brain and Tay–Sachs gangliosides, *Biochem. Biophys. Res. Commun.* **9**:436–441 (1962).
111. H. Jatzkewitz and K. Sandhoff, On a biochemically special form of infantile amaurotic idiocy, *Biochim. Biophys. Acta* **70**:354–356 (1963).
112. J. S. O'Brien, M. B. Stern, B. H. Landing, J. K. O'Brien, and G. Donnel, Generalized gangliosidosis: Another inborn error of ganglioside metabolism? *Am. J. Dis. Children* **109**: 338–346 (1965).
113. N. K. Gonatas and J. Gonatas, Ultrastructural and biochemical observations on a case of systemic late infantile lipidosis and its relationship to Tay-Sachs disease and gargoylism, *J. Neuropathol. Exptl. Neurol.* **24**:318–340 (1965).
114. O. Eeg-Olofsson, K. Kristensson, P. Sourander, and L. Svennerholm, Tay–Sachs disease, A generalized metabolic disorder, *Acta Paediat. Scand.* **55**:546–562 (1966).
115. K. Suzuki and G. C. Chen, Brain ceramide hexosides in Tay–Sachs disease and generalized gangliosidosis ($G_{M1}$–gangliosidosis), *J. Lipid Res.* **8**;105–113 (1967).
116. L. Jörgenson, T. W. Blackstad, W. Harkmark, and J. A. Steen, Niemann–Pick's disease. Report of a case with chemical evidence of neuronal storage of acid glycolipids. *Acta Neuropathol.* **4**:90–106 (1964).
117. H. Pilz, K. Sandhoff, and H. Jatzkewitz, A disorder of ganglioside metabolism with storage of ceramide lactoside, monosialo ceramide lactoside and Tay–Sachs ganglioside in the brain, *J. Neurochem.* **13**:1273–1282 (1966).

*Chapter 15*

# SPHINGOMYELIN: ENZYMATIC REACTIONS

**Abraham Rosenberg**

*Department of Biological Chemistry*
*Pennsylvania State University*
*Hershey, Pennsylvania*

## I. INTRODUCTION

Total synthesis and total degradation of sphingomyelin can now be accomplished with isolable enzymes. Some of these have been purified and partially characterized. Most preparations are overwhelmingly crude. Nevertheless, substantial progress has been made toward understanding the enzymatic reactions of sphingomyelin. Some highlights of this progress are outlined here.

## II. ENZYMATIC SYNTHESIS OF SPHINGOMYELIN

### A. Synthesis of Sphingomyelin from Ceramide

The first thoroughgoing study on the synthesis of sphingomyelin *in vitro*, at the enzyme level, was carried out relatively recently in the laboratory of E. P. Kennedy.[1] The directive thinking in this study was to attempt a synthesis of sphingomyelin by enzymatic transfer of the phosphorylcholine group of CDPcholine to the primary hydroxyl group of *N*-acyl sphingosine (ceramide). In this way, sphingomyelin would be synthesized in the manner shown in the same laboratory to be effective for the biosynthesis of lecithin,[2] a phosphoglyceride closely related in structure to sphingomyelin.

$$\text{1,2-Diglyceride} + \text{CDPcholine} \rightarrow \text{lecithin} + \text{CMP} \quad (1)$$

$$N\text{-Acyl sphingosine} + \text{CDPcholine} \rightarrow \text{sphingomyelin} + \text{CMP} \quad (2)$$

A necessary precursor for the synthesis of sphingomyelin by reaction (2) is the ceramide molecule. Enzymatic synthesis of ceramide has been achieved in two essentially different ways.

### 1. *Direct Synthesis of Ceramide*

Zabin,[3] in a preliminary fashion, has shown that ceramide is formed when a brain homogenate of 15–20-day-old rats is incubated with palmitoyl CoA in the presence of reactants and cofactors that are necessary for the simultaneous biosynthesis of sphingosine [reaction (3)].

$$\text{Palmitoyl CoA} + \text{serine} + (\text{NADP}) \xrightarrow[\text{nicotinamide}]{Mg^{2+}} \underset{\text{(ceramide)}}{\text{palmitoyl sphingosine}} \quad (3)$$

Palmitoyl CoA acts as the precursor of the hydrocarbon chain of the sphingosyl moiety, and it provides the fatty acyl moiety of the ceramide molecule. Palmitoyl serine is not an intermediate in this reaction. Therefore, it appears that sphingosine is first synthesized *in toto*, and then it reacts with palmitoyl CoA to form ceramide. Sribney[4] also has very briefly described an enzyme system in chicken liver and in rat brain that appears to catalyze the reaction of palmitoyl CoA and free sphingosine to yield ceramide. Gatt[5] has been able to demonstrate the net synthesis of ceramide by reversal of the action of ceramide hydrolase that splits the amide linkage of ceramide to give sphingosine and fatty acid. The enzyme, which will be described further later, was found in a particulate fraction prepared from rat brain. The enzyme required an anionic detergent, cholate, for activation, but apparently it had no metal ion requirement, and the reversible reaction which it catalyzes

$$\text{Fatty acid} + \text{sphingosine} \rightleftharpoons \text{ceramide} + H_2O$$

must depend for its direction upon the availability of water molecules. This, no doubt, is governed by quite complex environmental factors in the cell.[6] Fatty acids of medium chain length (10–14 carbon atoms) gave the highest reaction rate in the direction of ceramide synthesis, although such fatty acids do not generally comprise the major fatty acyl components of the brain sphingolipids. "Specificity" for fatty acid chain lengths may be governed by environmental as well as enzymatic factors, a problem whose solution remains of importance for understanding the modes of biosynthesis and biodegradation of the complex lipids.

### 2. *Indirect Synthesis of Ceramide*

The synthesis of ceramide may be brought about indirectly by the hydrolytic removal of the sugar components of sphingoglycolipid molecules. These molecules are not, as far as is known, directly synthesized from ceramide, but they apparently arise instead by the acylation of glycosylated sphingosine,[7] e.g.,

$$\text{Sphingosine} + \text{UDPgal} \xrightarrow{\text{transferase I}} \text{psychosine} \quad (4)$$

$$\text{Psychosine} + \text{fatty acyl CoA} \xrightarrow{\text{transferase II}} \text{cerebroside} \quad (5)$$

$$\text{Cerebroside} + H_2O \xrightarrow{\text{hydrolase}} \text{galactose} + \text{ceramide} \quad (6)$$

Clear evidence that ceramide, derived from cerebroside, may be utilized for the biosynthesis of sphingomyelin *in vivo* has been obtained by Kopaczyk and Radin.[8] Labeled cerebroside (lignoceryl-1-$^{14}$C psychosine) was emulsified with polyoxyethylene stearate and injected directly into the brain of 15-day-old rats. After 1, 2, and 4 days, the brains were examined for the level of radioactivity incorporated into lipid fractions. Aside from the cerebrosides, which had been injected with label, ceramide had the highest specific activity, with the radioactive label mostly retained in its original position at the first carbon atom of the fatty acyl component. Sphingomyelin also was substantially labeled. In support of *in vitro* findings, an essentially undirectional process appeared to be indicated *in vivo*: Cerebroside was split to ceramide, which was then converted to sphingomyelin, but the reverse process of a splitting of sphingomyelin to ceramide, and the conversion of ceramide to cerebroside, did not appear to take place.

### 3. *Transfer of Phosphorylcholine to Ceramide*

The biosynthesis of the phosphoglyceride, lecithin, is controlled by the action of phosphorylcholine transferase, an enzyme that has been demonstrated in preparations from chicken liver and other sources. In the reaction mediated by this enzyme [reaction (1)] the phosphorylcholine moiety of CDPcholine is transferred to the free primary hydroxyl group of D-1 (or $\alpha$), 2 (or $\beta$)-diglyceride. The structural analogies between lecithin and sphingomyelin render the assumption quite reasonable that sphingomyelin also might be synthesized in the same manner, i.e., by the enzymatic transfer of phosphorylcholine from CDPcholine to the free primary hydroxyl group of ceramide, as outlined in reaction (2). This has been shown to be the case.[1] Chicken liver and other vertebrate organs were used as sources of the enzyme. Enzymatic activity was found in a mixed particulate fraction containing both mitochondria and microsomes. The particles were collected at 18,000 *g* from chicken liver homogenized in 0.25 M sucrose containing 0.001 M ethylenediamine tetraacetate. At *p*H 7.4, in the presence of cysteine, the particulate enzyme preparation catalyzed the following reaction:

$$\text{“Ceramide”} + \text{CDPcholine} \xrightarrow[\text{phosphorycholine : ceramide transferase}]{\text{Mn}^{2+}\ (\text{or Mg}^{2+})} \text{sphingomyelin} + \text{CMP)} \qquad (7)$$

High specificity was shown for CDPcholine as donor and for ceramide as acceptor of the phosphorylcholine group. The nature of the ceramide molecule that acted as the acceptor for the phosphorylcholine group presented an unusual problem. It has been proven by a total synthesis of the molecule[9] that sphingomyelin occurs naturally as the D-enantiomorph, and the asymmetric carbon atoms at positions 2 and 3 in the sphingosine component of the molecule are arranged in the *erythro* configuration.

But Sribney and Kennedy[1] clearly have shown that only the ceramide whose sphingosine moiety has the unnatural, *threo*, configuration can act *in vitro* as an acceptor for phosphorylcholine in the ceramide: CDPcholine phosphorylcholine transferase system that was isolated from chicken liver. Further difficulty was presented by the required type of fatty acyl component in the acceptor ceramide. Ceramides with very short chain acyl components—e.g., *N*-acetyl-DL-*threo* sphingosine—were active as acceptors. With increasing chain length of the fatty acyl component, Tween 20 was needed to disperse the ceramide in order to form an active substrate. With fatty acyl chain lengths greater than octanoyl, however, activity was low even with Tween 20 present. With *N*-palmitoyl-DL-*threo*-sphingosine, the system was completely inactive. Yet palmitic acid often is a major fatty acid component of natural liver sphingomyelin, at least in humans,[10] and also presumably in other vertebrates. Ceramide molecules with *cis*-unsaturated fatty acyl components (*N*-oleyl and *N*-linoleyl) were quite active. It may be that ceramide molecules with *cis*-unsaturated fatty acyl components are more capable of phase mixing with detergent molecules, and therefore they may form mixed micelles in which the substrate ceramide is more readily available for reaction. Sphingomyelin was actively synthesized from ceramide, whose sphingosine component contained a *trans* double bond at position 4, as it occurs naturally in sphingomyelin. The unnatural *cis*-isomeric form of ceramide was inactive. However, there was activity when the *trans* double bond was replaced by a triple bond, presumably because with such a bond the molecule can present a spatial arrangement similar to the arrangement possible with a *trans* double bond. The unexpected reactivity of *threo*- rather than *erythro*-ceramide, and the need for a short chain or unsaturated fatty acyl component rather than a longer chain saturated fatty acyl component, may well be reflections of the better solubility properties *in vitro* of molecules with these unnatural components rather than a case of unexplainable enzyme specificity. It is known that there are striking physical differences between the *threo* and *erythro* forms of sphingomyelin. For example, the diastereoisomer of the former is unresolvable with D- or L-tartrate, while the latter is readily resolvable.[6] Perhaps *threo*-ceramide possesses the physical properties necessary to form an aqueous dispersion in such a way as to favor enzyme action *in vitro*, while in contrast the *erythro* form of the molecule may undergo relatively specific kinds of intermolecular associations that favor reaction within the cell. The available evidence from *in vivo* studies seems to point to the possibility that sphingomyelin actually can be synthesized from either the *threo* or the *erythro* form of ceramide. It has been shown that DL-sphingosine-$^3$H of either the *threo* or the *erythro* form is converted efficiently to ceramide and then to sphingomyelin in the brain of 9-day-old rats.[11] It has not been ascertained whether an inversion of the *threo* to the *erythro* form occurs during the synthesis. However, it is known that no inversion of the *threo* form to the *erythro* form occurs *in vitro* during enzymatic synthesis of sphingomyelin from ceramide.[1] At any rate, there is convincing evidence that conversion of natural ceramide to sphingomyelin occurs quite efficiently in the living animal,[8] and the pathway elucidated by Sribney and Kennedy[1] in all probability represents a major biosynthetic route.

## B. Synthesis of Sphingomyelin from Sphingosylphosphorylcholine

Another pathway for the enzymatic synthesis of sphingomyelin has been demonstrated in the laboratory of Brady.[12] For this pathway, auxiliary confirmatory information from whole animal studies has not yet been sought. The biosynthesis of sphingomyelin was shown to take place by acylation of the amino group of sphingosylphosphorylcholine in a manner analogous to that in which the amino group of galactosyl sphingosine is acylated to yield cerebroside[13] [reactions (8) and (9)].

Stearoyl CoA + sphingosylgalactoside (psychosine) →
*N*-acyl sphingosyl galactoside (cerebroside) + CoASH (8)

Stearoyl CoA + sphingosylphosphorylcholine →
*N*-acyl sphingosylphosphorylcholine (sphingomyelin) + CoASH (9)

The enzyme that catalyzes reaction (9) was found in two particulate fractions prepared from 14-day-old rat brain homogenized in a fivefold volume of 0.25 M sucrose. Activity was found in a fraction sedimenting after 12 min at 8400 *g* and in a second fraction sedimenting after 30 min at 100,000 *g*. It is of interest that the enzyme system that catalyzes the synthesis of sphingomyelin from ceramide and CDPcholine[1] was also found in the same preparations. Thus it appears that alternative systems for the synthesis of sphingomyelin may occur in the same type of cellular structure. The system that synthesizes sphingomyelin by the acylation of sphingosylphosphorylcholine could be solubilized from the 8400 *g* fraction by treatment overnight, at 0°, with a solution of sodium cholate. This soluble preparation, and the 100,000 *g* fraction, both catalyzed the formation of labeled sphingomyelin from synthetic sphingosylphosphorylcholine and stearoyl-1-$^{14}$C-CoA. The radioactive label was retained almost completely in the fatty acyl portion of the newly synthesized sphingomyelin, indicating that the direct transfer of a fatty acyl unit to sphingosylphosphorylcholine most probably occurs, as indicated in reaction (9). The choice of stearoylCoA as the fatty acyl donor for this system is apt. Stearic acid is the major fatty acid fraction in the sphingomyelin of young rat brain.[14] Since the enzyme system of Sribney and Kennedy[1] also was present in the enzyme preparations studied, the possibility exists that the labeled sphingomyelin instead might have been synthesized by another route, namely, the hydrolysis of the sphingosylphosphorylcholine to free sphingosine followed by utilization of the latter to build first ceramide, and then sphingomyelin, by the action of ceramide: CDPcholine cholinephosphotransferase as outlined in reaction (2). However, this possibility is minimized by the finding that ATP failed to enhance the synthesis of sphingomyelin. In the crude system, ATP might have been expected to favor the resynthesis of CDPcholine and, therefore, to help to drive synthesis of sphingomyelin from ceramide. Whether sphingosylphosphorylcholine occurs naturally and how it is synthesized remain problematical. It may be significant in this regard that the 8400 *g* particulate fraction described above was found to be capable of incorporating a fatty acyl group into sphingomyelin without the addition of exogenous sphingosylphosphorylcholine. It would be of more than passing interest to determine whether this finding indicates that endogenous sphingosylphosphorylcholine is present in these particles. In quite recent work by Fujino and Negishi[(15)] it was demonstrated that both *threo*- and *erythro*-sphingosylphosphorylcholine can be enzymatically acylated to yield sphingomyelin. The *erythro* form was somewhat more active than the *threo* form. Compared with palmitoylCoA, oleoylCoA was a poor fatty acyl donor and acetyl CoA was completely inactive.

## III. ENZYMATIC DEGRADATION OF SPHINGOMYELIN

Enzymatic degradation of sphingomyelin appears to commence, as far as is known, with the removal of the hydrophilic portions of the molecule to produce *N*-fatty acyl sphingosine (ceramide). If enzymes exist that initially degrade sphingomyelin by the alternative hydrolytic route, namely, removal of the fatty acyl component to yield sphingosylphosphorylcholine, such enzymes have not yet been demonstrated. It should be borne in mind, however, that the finding of Brady's group,[(12)] mentioned earlier, suggests that fatty acylCoA possibly may react with an endogenous acceptor in brain to form sphingomyelin. Logically, such an acceptor would be sphingosylphosphorylcholine, which could arise by the deacylation of sphingomyelin. As yet, clear-cut evidence along these lines is not available. Generally speaking, bacterial and mammalian enzymes that heretofore have been known to act on sphingomyelin do so by splitting the phosphodiester bridge at the linkage of phosphoric acid to sphingosine, to yield a lipophilic ceramide moiety and a hydrophilic phosphorylcholine moiety. Indications may be found in the older literature[(16)] that enzymes in human brain and liver release phosphoric acid from sphingomyelin, but the evidence is vague and inconclusive and the way in which the release of phosphoric acid takes place is unknown. The action of plant enzymes on sphingomyelin has not received much attention.

### A. Removal of Choline from Sphingomyelin

Recently a relatively specific enzyme has been demonstrated, for the first time, which splits sphingomyelin at the ester linkage between the phosphoryl and choline groups,[(17)] as shown in the following reaction:

$$\text{Sphingomyelin} \xrightarrow{H_2O} N\text{-acylsphingosylphosphate} + \text{choline} \quad (10)$$

Since it splits off choline from a lipid phosphodiester, the new enzyme is considered by its discoverers to belong to the group of enzymes now commonly classified as phospholipases D (phosphatidylcholine phosphatidohydrolases E.C.2.1.4.4). Enzymes so designated generally are found in higher plants.[18] The specific enzyme was obtained, in crude form, from a filtrate of the culture medium of *Corynebacterium ovis*. A partial purification of the enzyme was obtained by fractional precipitation with methanol at low temperature. The action of the enzyme on washed sheep, human, and rabbit erythrocytes in 0.05 M borate buffer at *p*H 7.6 was to substantially reduce their sphingomyelin content without significantly lowering the amounts of other phospholipids (phosphatidyl choline and phosphatidyl ethanolamine). When this enzyme preparation, in the same buffer, was incubated with a sonically produced dispersion of pure sphingomyelin from erythrocytes, the sphingomyelin rapidly was cleaved to yield choline and a new lipid that was identified as *N*-acyl sphingosylphosphate, confirming the process outlined in reaction (10) above. The enzyme was relatively specific for sphingomyelin, and it had no activity toward phosphoglycerides, except for lysophosphatidyl choline. Whether this latter was of the commonly occurring l-alkenyl type, and thus quite akin structurally to sphingomyelin, has not been determined. In view of its inertness toward most phosphatides, including phosphatidyl choline (lecithin), the designation of this enzyme as a phosphatidylcholine phosphatidohydrolase is perhaps somewhat imprecise, and a more apt designation indicating a relative specificity for ceramide-l-phosphorylcholine rather than for 1,2-diacyl or alkenyl, acyl glycerophosphoryl choline molecules may be in order. The routes for further metabolism of the new lipid, *N*-acyl sphingosylphosphate, are not yet known, but in view of the ubiquitous occurrence and often nonspecific nature of cellular phosphomonoesterases, hydrolytic degradation of *N*-acyl sphingosylphosphate to ceramide and inorganic phosphate would not be unexpected.

## B. Splitting of Sphingomyelin to Ceramide and Phosphorylcholine

Aside from the interesting *corynebacterium* enzyme, choline phosphoesterases that display a relative specificity for sphingomyelin have not been described. Generally, the enzymatic degradation of sphingomyelin has been shown to commence with cission of the ester linkage between the sphingosyl and the phosphorylcholine portions of the molecule, as outlined in reaction (11).

$$\text{Sphingomyelin} \xrightarrow[\substack{\text{or specific mammalian}\\ \text{sphingomyelinases}}]{\text{phospholipases C}} \text{ceramide} + \text{phosphorylcholine} \quad (11)$$

Enzymes catalyzing the hydrolysis of this phosphodiester bond have long been known to occur in mammalian organs and in bacteria.[19-22] The bacterial enzymes usually are not specific for sphingomyelin. They will remove phosphorylcholine from phosphatidylcholine (lecithin), and they will attack phosphatidyl ethanolamine, phosphatidyl serine, and phosphatidyl inositol as well, although slowly. It should be pointed out that enzymes of this kind, designated as phospholipases C (phosphatidylcholine choline phosphohydrolases, E.C.3.1.4.3), may actually be a combination of several phospholipases C, each specific for a particular molecular substrate,[23] and there may be a need for further study of the assumed broad specificity of the phospholipases C. The action of phospholipase C from *clostridium* welchii[24] has been studied in some detail.[1] The enzyme is active with sphingomyelin, which contains either *erythro-* or *threo-* sphingosine, and no inversion of the sphingosine from one form to the other occurs as a result of enzyme action. The phospholipases C with broad specificity are generally calcium ion dependent, and their activity is measurably enhanced by the presence of ether.[25] The reader is referred to an excellent descriptive review of enzymes of this type under the older designation of "phosphatidase D".[26] An enzyme having the action of phospholipase C and reported to be specific for sphingomyelin has been demonstrated in the toxin of an $\alpha\beta$ strain of *Staphylococcus pyrogenes*,[27] but evidence pointing to such specificity is lacking, and this enzyme is included for the time being with the other, common, phospholipases C.

## C. Specific Mammalian Sphingomyelinases

Growing out of initial studies that clearly demonstrated the presence of sphingomyelinases in mammalian organs,[28,29] several thoroughgoing investigations of specific mammalian sphingomyelinases have been performed. Two independent reports on rat liver sphingomyelinase have appeared simultaneously. They will be considered together.[30,31] Rat brain[32] and human spleen[33] also have yielded specific sphingomyelinases.

### 1. *Rat Liver Sphingomyelinase*

A sphingomyelinase was isolated from rat liver homogenized with 4 volumes of 0.25 M sucrose. The major part of the enzyme activity sedimented between 600 and 9000 *g*[30] A soluble enzyme was released from this crude particulate fraction by ultrasonic treatment followed by repeated freezing and thawing between the temperatures of −70 and 50°. The enzyme was purified by precipitation at *p*H 5.3 and further purified by fractional precipitation with ammonium sulfate (15–40% saturation) in 0.25 M sucrose solution containing 0.08% Na cholate at *p*H 7.5–7.8. Alternatively, the tissue was blended with 9 volumes of 0.25 M sucrose,[31] and the activity was collected in a crude particulate fraction sedimenting at 9000 *g*. Treatment with 1% Cutscum, a nonionic detergent, solubilized the enzyme so that it could not be sedimented at 100,000 *g* for 1 hr. Purification was achieved by removing insoluble protein that formed after dialysis and again after heating at 45° for 3 min at *p*H 4.5. Further purification was achieved by fractional precipitation with ammonium sulfate between 40 and 50% saturation and by a second dialysis. Measurable activation was obtained by alternate freezing and thawing of the preparation. The enzyme acted optimally between *p*H 5 and 5.5. The reaction proceeded at a linear rate for 2.5–4 hr, depending upon conditions. Although the true concentration of the micellar substrate cannot be readily determined, the variation in the rate of enzyme action with respect to substrate amount per volume permitted calculation of an apparent $K_m$ that fell in the range of 0.2 to 0.9 mM. The enzyme split phosphorylcholine from sphingomyelin but not from its phosphoglyceride analogue, lecithin. In fact, lecithin was found to be a potent competitive inhibitor of the action of the enzyme on sphingomyelin,[31] with a $K_I$ of 0.2 mM. A high specificity was shown for the naturally occurring isomer of sphingomyelin, D-*erythro*-*N*-acyl-l-*O*-phosphorylcholine sphingosine, or for the DL-*erythro* racemate.[30] The presence of the *trans*-4 double bond was unnecessary, the dihydro-*erythro* form being readily attacked. In contrast, DL-*threo*-dihydrosphingomyelin proved to be a poor substrate. Unfortunately, the action of the enzyme on the *threo*-analogue of natural sphingomyelin, D-*threo*-trans-4-*N*-acyl-1-*O*-phosphorylcholine sphingosine, obtainable by the method of Sribney and Kennedy,[1] was not tested, so that the preference of the enzyme for the *erythro* over the *threo* form of sphingomyelin is not yet completely established. The lack of requirement of the hydrolase for a *trans* double bond, and its apparent preference for the *erythro* form of sphingomyelin, contrast sharply with the *in vitro* requirements of the synthetase,[1] which has a requirement for the double bond and appears to act only on the *threo* form. From the physical nature of the particles in which it is found, the way in which it is "solubilized" by detergents, and the persistence of an acid phosphomonesterase in a relatively purified preparation[30] it is reasonable to assume that rat liver sphingomyelinase may be a component of a lysosomal enzyme complex.[34]

## 2. *Rat Brain Sphingomyelinase*

Like the rat liver enzyme, rat brain sphingomyelinase[32] was located in a particulate fraction of intermediate density, sedimenting between 600 and 25,000 *g* in 0.25 M sucrose and $10^{-3}$ M ethylenediamine tetraacetate. Sonic oscillation followed by freezing and thawing between −40 and 37° released the enzyme; so did treatment with 0.5% Na cholate. The enzyme precipitated at *p*H 5. It was active optimally around this *p*H, but only in the presence of a detergent. It cleaved sphingomyelin to ceramide and phosphorylcholine. The apparent $K_m$ for the reaction with rat brain sphingomyelin was 0.13 mM, in the same range as the liver enzyme. Similarly to the latter, the brain enzyme was inhibited by lecithin. Other phosphatides also were inhibitory, as were ceramide, sphingosine, and fatty acid. The mode of action of these lipid inhibitors is uncertain. Among the various possibilities, the simplest is the possibility that the added lipids interfere with the formation of the proper kinds of substrate micelles. Further similarity between the rat brain enzyme and the rat liver enzyme is seen in its lack of requirement for a double bond, since dihydrosphingomyelin is efficiently hydrolyzed by it, and in its preference for the D-*erythro* form of the substrate, the L-*erythro* and DL-*threo* forms being poorly reactive.

### 3. *Human Spleen Sphingomyelinase*

Human spleen also has yielded a specific sphingomyelinase that hydrolyzes sphingomyelin to phosphorylcholine and ceramide.[33] The tissue was blended with 9 volumes of 0.25 M sucrose. The enzyme was found in a particulate fraction sedimenting at 8500 *g* Dialysis of this fraction overnight at 4° against 0.01 M Tris buffer at *p*H 8 released the enzyme in soluble form. Nucleic acids in the extract were degraded with RNAase and DNAase, and the preparation was heated to 45° for 10 min at *p*H 5 in the presence of 0.2 % Triton X-100. Purification was then achieved by fractional precipitation with cold ethanol followed by column chromatography on ECTEOLA-cellulose. The enzyme was eluted at *p*H 7 with 0.1 *M* KCl containing Triton X-100. Like the rat enzymes, the human enzyme was most active near *p*H 5, but it had a considerably lower "$K_m$" of about $3 \times 10^{-5}$ M. The enzyme was inactive with lecithin but was not inhibited by it. Sphingomyelin and dihydrosphingomyelin were hydrolyzed almost equally well. Acetylation of the allylic hydroxyl group, to yield D-*erythro-trans-N*-acyl-1-0-phosphorylcholine-3-0-acetyl sphingosine, inactivated the substrate, as did removal of the fatty acyl component to yield 1-0-phosphorylcholine sphingosine. An enzymatically synthesized *threo* form of sphingomyelin, D-*threo-trans-N*-octanoyl 1-0-phosphorylcholine sphingosine, was a poor substrate. Inorganic phosphate strongly inhibited enzyme action. In this respect, the human spleen enzyme appeared to differ from the rat enzymes. Thus, caution is indicated in making direct interspecies comparisons of the sphingomyelinases that have been demonstrated in various vertebrate organs, e.g., in rat spleen and kidney, human liver and kidney, and chicken liver.

### 4. *Sphingomyelinase in Niemann–Pick Disease*

In certain forms of Niemann–Pick disease, which is characterized by the accumulation of great amounts of sphingomyelin in target organs, a deficiency of sphingomyelinase has been demonstrated in the liver[35] and in the spleen.[33] The implication of these findings is that the massive accumulation of sphingomyelin in certain instances may be traceable to the diminished activity of a necessary catabolic enzyme. Accumulation of sphingomyelin has been found to occur in successive diploid cell cultures from donors with Niemann–Pick disease.[36] It would be of great interest to determine whether a deficiency of sphingomyelinase also occurs and persists in such cultures. Oddly, in the classical and visceral forms of

Niemann–Pick disease, there appears to be an elevated level of a sphingomyelinase of undetermined specificity. This enzyme has a magnesium ion requirement, and it shows optimal activity at *p*H 7.4.[33] Apparently the enzyme does not degrade and remove the excess sphingomyelin that accumulates in the diseased organs.

## IV. THE PROBLEM OF ENZYMATIC SPECIFICITY FOR SPHINGOMYELIN

Although it is most difficult to assess, the problem of specificity of the mammalian enzymes for sphingomyelin merits consideration. Superficially, it would appear that the same phosphodiester bond must be broken in order to release phosphorylcholine from sphingomyelin on the one hand and from lecithin on the other. However, from the point of view of intramolecular associations, sphingomyelin differs from lecithin. In sphingomyelin, the allylic hydroxyl group, at carbon atom 3 of the sphingosine component, can form an ion-dipole association with the ionic oxygen of the phosphate group.[37] Because of this, the surface of a sphingomyelin micelle dispersed in water can bear an induced net positive charge. In contrast, lecithin micelles have a net surface charge of zero. This factor may play a role in the determination of enzyme specificity. From the point of view of the hydrophobic fatty acyl and sphingosyl chains of the sphingomyelin molecule, other factors can come into play. Clearly, the lipophilic associations of sphingomyelin with other membrane lipids must in some way influence its availability for enzymatic reactions. It has been proposed[38] that intermolecular complexes formed by cholesterol, sphingomyelin, and other lipids serves to stabilize them and slow their metabolic, and therefore enzymatic, reactions. Reduction of the double bond in the sphingosine component of sphingomyelin alters neither its mode of packing in a monomolecular film[39] nor its susceptibility to the attack of specific mammalian sphingomyelinases. In contrast, the conformation of the fatty acyl component, as reflected by its degree of *cis* unsaturation, greatly influences the mode of packing of sphingomyelin in a monomolecular film;[39] also, a change from the *threo* to the *erythro* form has a clear influence on physical properties,[9] and in both instances one sees corresponding specificity in the action of sphingomyelinase. Whether these are simply fortuitous relationships must be decided experimentally. It is known that the fatty acid composition of sphingomyelin changes markedly with age[40] in brain but not in other organs.[41] The specificity of individual tissue sphingomyelinases for a given group of fatty acyl components may play a part in this phenomenon.

## V. DEGRADATION OF THE COMPONENTS OF SPHINGOMYELIN

### A. Hydrolysis of Ceramide

A ceramidase has been isolated from rat brain.[5] The enzyme was solubilized by sonication and extraction with cholate. One hundredfold purification of the enzyme was achieved by treatment with Sephadex followed by fractional precipitation with ammonium sulfate. The enzyme reversibly catalyzed reaction (12).

$$\begin{aligned}&N\text{-Acyl sphingosine} + H_2O \rightleftharpoons \text{fatty acid} + \text{sphingosine}\\&N\text{-acyl dihydrosphingosine} + H_2O \rightleftharpoons \text{fatty acid} + \text{dihydrosphingosine}\end{aligned} \quad (12)$$

Other than detergent molecules, cofactors were not required. As with the sphingomyelinases, the reaction catalyzed by ceramidase proceeded optimally near *p*H 5 and had an apparent $K_m$ of 0.3 mM. Specificity was shown for the fatty acyl component of the ceramide, 16- and 18-carbon fatty acids being split off readily, but very short (acetyl) and very long (lignoceryl) fatty acyl components were not hydrolyzed.

### B. Further Degradation of Sphingomyelin Components

A discussion of the oxidative degradation of fatty acids and the many biosynthetic processes in which fatty acids are utilized is beyond the purview of this chapter. For interesting new information and a list of some pertinent references concerning the catabolism of sphingosine, the reader is referred to recent publications.[42-44]

## VI. EPILOGUE

Enzymatic studies with sphingomyelin has been hampered by problems arising from insolubility of both enzyme and substrate. The measurable levels of enzyme activity and the modes of enzyme action sometimes have been found to be subject to bizarre and usually unknown physical influences. In most experiments, artificial detergents were necessary to effect the interaction of enzyme and substrate. It is particularly difficult to extrapolate from the isolated test tube to the level of cellular processes the interpretations of measurements made with such arbitrarily arranged systems. For these reasons, real evaluations of the enzymatic studies described here must await the arrival of considerable new information from turnover experiments in whole animals, and a more intimate knowledge also must be gained of those controlling factors that arise from lipid–lipid, lipid–protein, and other interactions within the cell.

## VII. REFERENCES

1. M. Sribney and E. P. Kennedy, The enzymatic synthesis of sphingomyelin, *J. Biol. Chem.* **233**:1315 (1958).
2. E. P. Kennedy and S. B. Weiss, The function of cytidine coenzymes in the biosynthesis of phospholipides, *J. Biol. Chem.* **222**:193 (1965).

3. I. Zabin, Biosynthesis of ceramide by rat brain homogenates, *J. Am. Chem. Soc.* **79**:5834 (1957).
4. M. Sribney, Enzymatic synthesis of ceramide, *Federation Proc.* **21**:280 (1962).
5. S. Gatt, Enzymatic hydrolysis of sphingolipids. I. Hydrolysis and synthesis of ceramides by an enzyme from rat brain, *J. Biol. Chem.* **241**:3724 (1966).
6. J. Cerbon, NMR studies on the water immobilization by lipid systems *in vitro* and *in vivo*, *Biochim. Biophys. Acta* **144**:1 (1967).
7. W. W. Cleland and E. P. Kennedy, The enzymatic synthesis of psychosine, *J. Biol. Chem.* **235**:45 (1960).
8. K. K. Kopaczyk and N. Radin, *In vivo* conversions of cerebroside and ceramide in rat brain, *J. Lipid Res.* **6**:140 (1965).
9. D. Shapiro and H. M. Flowers, Studies on sphingolipids. VII. Synthesis and configuration of natural sphingomyelins, *J. Am. Chem. Soc.* **84**:1047 (1962).
10. E. Svennerholm, S. Ställberg-Stenhagen, and L. Svennerholm, Fatty acid composition of sphingomyelins in blood, spleen, placenta, liver, lung, and kidney, *Biochim. Biophys. Acta.* **125**:60 (1966).
11. J. M. Kanfer and A. E. Gal, *In vivo* conversion of *erythro* and *threo* DL-sphingosine-$^3$H to ceramide and sphingomyelin, *Biochem. Biophys. Res. Commun.* **22**:442 (1966).
12. R. O. Brady, R. M. Bradley, O. M. Young, and H. Kaller, An alternative pathway for the enzymatic synthesis of sphingomyelin, *J. Biol. Chem.* **240**:PC 3693 (1965).
13. R. O. Brady, Studies on the total enzymatic synthesis of cerebrosides, *J. Biol. Chem.* **237**:PC 2416 (1962).
14. A. Rosenberg and N. Stern, Changes in the sphingosine and fatty acid components of the gangliosides in developing rat brain, *J. Lipid Res.* **7**:122 (1966).
15. Y. Fujino and T. Negishi, Investigation of the enzymatic synthesis of sphingomyelin, *Biochim. Biophys. Acta* **152**:428 (1968).
16. A. Goebel and H. Seckfort, Uber spaltung von sphingomyelinen durch organe, *Biochem. Z.* **319**:203 (1948).
17. A. Soucek, C. Michalec, and A. Souckova, Enzymic hydrolysis of sphingomyelins by a toxin of *Corynebacterium ovis*, *Biochem. Biophys. Acta* **144**:180 (1967).
18. F. M. Davidson and C. Long, The structure of the naturally occurring phosphoglycerides, *Biochem. J.* **69**:458 (1958).
19. S. J. Thannhauser and M. Reichel, Studies on animal lipids. XVI. The occurrence of sphingomyelin as a mixture of sphingomyelin fatty acid ester and free sphingomyelin, demonstrated by enzymatic hydrolysis and mild saponification, *J. Biol. Chem.* **135**:1 (1940).
20. M. G. Macfarlane, The biochemistry of bacterial toxins 2. The enzymic specificity of *clostridium welchii* lecithinase, *Biochem. J.* **42**:581 (1948).
21. Y. Fujino, Studies on the conjugated lipid. IV. On the enzymatic hydrolysis of sphingomyelin, *J. Biochem.* (*Japan*) **39**:55 (1952).
22. K. V. Druzhinina and M. G. Kritzman, Lecithinase in animal fatty tissue, *Biokhimiya* **17**:77 (1952).
23. M. W. Slein and G. F. Logan, Jr., Characterization of the phospholipases of *Bacillus cereus* and their effect on erythrocytes, bone, and kidney cells, *J. Bacteriol.* **90**:69 (1965).
24. M. G. Macfarlane and B. C. J. G. Knight, Biochemistry of bacterial toxins; lecithinase activity of Cl. welchii toxins, *Biochem. J.* **35**:884 (1941).
25. D. J. Hanahan and R. Vercamer, the action of lecithinase D on lecithin. The enzymatic preparation of D-1,2-dipalmitolein and D-1,2-dipalmitin, *J. Am. Chem. Soc.* **76**:1804 (1954).

26. M. Kates, in *Lipide Metabolism* (K. Bloch, ed.), p. 206 *et seq.* Wiley, New York (1963).
27. H. M. Doery, B. J. Magnusson, I. M. Cheyne, and J. Gulasekharan, A phospholipase in staphylococcal toxin which hydrolyzes sphingomyelin, *Nature* **198**:1091 (1963).
28. A Roitman and S. Gatt, Isolation of phospholipase-C from rat brain, *Israel J. Chem.* **1**:190 (1963).
29. M. Heller and B. Shapiro, The hydrolysis of sphingomyelin by rat liver, *Israel J. Chem.* **1**:204 (1963).
30. M. Heller and B. Shapiro, Enzyme hydrolysis of sphingomyelin by rat liver, *Biochem. J.* **98**:763 (1966).
31. J. H. Kanfer, O. M. Young, D. Shapiro, and R. O. Brady, The metabolism of sphingomyelin. I. Purification and properties of a sphingomyelin-cleaving enzyme from rat liver, *J. Biol. Chem.* **241**:1081 (1966).
32. Y. Barenholz, A. Roitman, and S. Gatt, Enzymatic hydrolysis of sphingolipids. II. Hydrolysis of sphingomyelin by an enzyme from rat brain, *J. Biol. Chem.* **241**:3731 (1965).
33. P. B. Schneider and E. P. Kennedy, Sphingomyelinase in normal human spleens and in spleens from subjects with Niemann–Pick's disease, *J. Lipid Res.* **8**:202 (1967).
34. C. de Duve, in *Subcellular Particles* (T. Hayashi, ed.), Ronald, New York (1959).
35. R. O. Brady, J. N. Kanfer, M. B. Mock, and D. Fredrickson, The metabolism of sphingomyelin. II. Evidence of an enzymatic deficiency in Niemann–Pick Disease, *Proc. Natl. Acad. Sci. U.S.* **55**:366 (1966).
36. B. W. Uhlendorf, A. I. Holtz, M. B. Mock, and D. S. Fredricksson, in *Inborn Disorders of Sphingolipid Metabolism* (S. M. Aronson and B. W. Volk, eds.), p. 443, Pergamon, New York (1967).
37. D. O. Shah and J. H. Schulman, The ionic structure of sphingomyelin monlayers, *Biochim. Biophys. Acta* **135**:184 (1967).
38. L. F. Eng and M. E. Smith, The cholesterol complex in the myelin membrane, *Lipids* **1**:296 (1966).
39. J. H. Raper, D. B. Gammack, and G. H. Sloane-Stanley, A study of cerebral sphingomyelins in monomolecular films, *Biochem. J.* **98**:21p (1966).
40. S. Ställberg-Stenhagen and L. Svennerholm, Fatty acid composition of human brain sphingomyelins: Normal variation with age and changes during myelin disorders, *J. Lipid Res.* **6**:146 (1965).
41. E. Svennerholm, S. Ställberg-Stenhagen, and L. Svennerholm, Fatty acid composition of sphingomyelins in blood, spleen, placenta, liver, lung and kidney, *Biochim. Biophys. Acta* **125**:60 (1966).
42. K.-A. Karlsson, B. E. Samuelson, and G. O. Steen, Studies on sphingosines. 15. Degradation of phytosphingosine to hydroxy fatty acid and ethanolamine by the yeast Hansenella ciferrii, *Acta Chem. Scand.* **21**:2566 (1967).
43. K.-A. Karlsson, Studies on sphingosines, the chemical structure of a dienic long chain base of human blood plasma sphingomyelins, *Acta Chem. Scand.* **21**:2577 (1967).
44. Y. Barenholz and S. Gatt, The utilization and degradation of phytosphingosine by rat liver, *Biochem. Biophys. Res. Commun.* **27**:319 (1967).

# Chapter 16
# METABOLISM OF PHOSPHOGLYCERIDES

**R. J. Rossiter and K. P. Strickland**

*Department of Biochemistry*
*University of Western Ontario*
*London, Canada*

## I. INTRODUCTION

Other contributors to this handbook already have clearly outlined the important role of the phosphoglycerides as components of membranes of the cells of the nervous system. This chapter is concerned with the manner in which the phosphoglycerides of the nervous system are formed and broken down. Only a brief discussion will be offered on the turnover and laying down of these lipids in the developing brain, adult brain, and cells such as the Schwann cell or glial cell which are concerned with myelin formation and maintenance.

The phosphoglycerides to be considered may be represented by the general formulas shown in Fig. 1. It can be seen that there are three main types: (a) the diacylglycerophosphatides with fatty acyl groups at the $\alpha'$-(or 1-) and $\beta$-(or 2-) position of glycerol (phosphatidyl group), (b) the plasmalogens, which have an alk-1-enyl group at the $\alpha'$-(or 1-) position of glycerol and a fatty acyl group at the $\beta$-(or 2-) position (phosphatidal group), and (c) the alkyl ether phosphoglycerides, which are similar to the plasmalogens but have the vinyl ether linkage fully saturated. (For references see Chapter 8, Volume I, and Ansell and Hawthorne,[1] Dawson,[2] Strickland,[3] and Rossiter.[4] It should be noted that the nomenclature recently recommended by the IUPAC–IUB Commission on Biochemical Nomenclature* has not been included in Fig. 1, but that this nomenclature is given in parentheses in subsequent figures.

Since the phosphoinositides are considered in detail in the chapter that follows, discussion of their metabolism here is limited only to situations where the metabolism of other phosphoglycerides involves the same precursors or metabolites. In discussion of the formation of the phosphoglycerides, brief consideration is given to early studies with labeled precursors and then reference is made to contributions that are relevant to the establishment of the pathways of phosphoglyceride formation. After this, investigations that contribute to our understanding of these pathways *in situ* are

* In *Biochim. Biophys. Acta.* **152**:1–9 (1968).

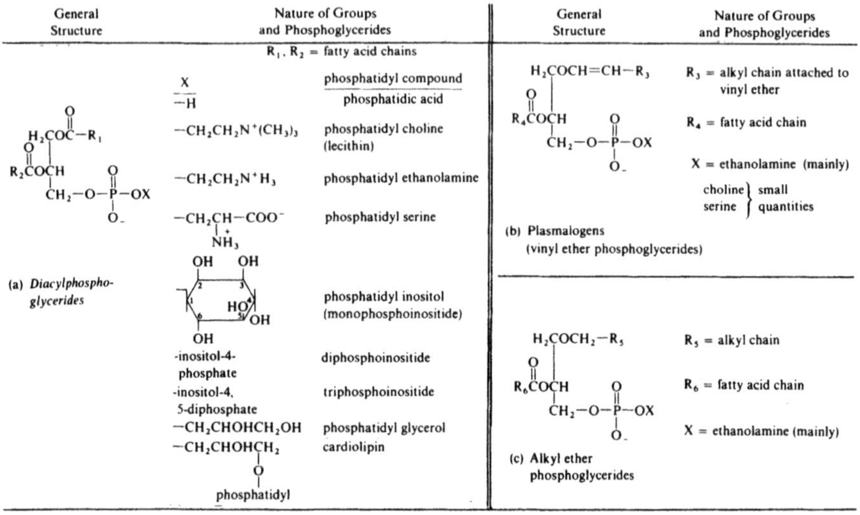


Fig. 1. General formulas for phosphoglycerides of the nervous system. In the figures that follow the nomenclature recommended by the IUPAC–IUB Commission on Biochemical Nomenclature is given in parentheses.

discussed. A somewhat similar but briefer consideration is given to the pathways of breakdown for the phosphoglycerides of the nervous system. The closing sections of this chapter are devoted to the overall dynamics of phosphoglyceride metabolism.

## II. FORMATION

### A. Early Studies

Prior to 1937 it was believed that once phospholipids were formed and laid down during growth they became part of the structural elements of tissues and as such were almost inert metabolically. With the availability of artificially produced radioactive isotopes (in particular $^{32}$Pi), *in vivo* experiments soon demonstrated that phospholipids were readily labeled. Very quickly it was realized that phospholipids were formed at rates in excess of those concerned only with growth. These early experiments proved that phospholipids along with many other cellular constituents were constantly being built up and broken down. Reviews by Ansell and Hawthorne(1) and Rossiter(5,6) describe in detail the contributions that bear directly on the nervous system. Although labeled substances (e.g., $^{32}$Pi or $^{14}$C-fatty acid or glycerol) do not always readily penetrate into brain, it was soon established that once such precursors reached the cells of the nervous system they are incorporated into the appropriate moieties of the phospholipids. This conclusion was reached by experiments carried out both *in vivo* and *in vitro*, with brain slices at first and later with various cell-free preparations. Notable contributions came from the laboratories of Hevesy, Chaikoff, Dawson, Abood, and Rossiter and their associates.[4]

## B. Biosynthesis of Phosphoglycerides

Although the isotope experiments referred to above demonstrated that phospholipids are continually undergoing renewal and breakdown, little information was provided on the metabolic pathways involved. It remained for approaches such as those of Kornberg and Pricer in 1953 and Kennedy and his associates in 1955 to begin to make significant contributions to our understanding of how phosphoglycerides are formed by tissues (described in the reviews of Ansell and Hawthorne,[1] Kennedy,[7,8] Dawson,[2] Rossiter,[5,6,9,10] Rossiter and Strickland,[11] and Strickland[3]). Below is given a brief description of the origin of each of the individual moieties in nervous tissue, followed by a consideration of the manner in which these moieties are utilized to form the various phosphoglycerides. A number of the above reviews relate in particular to the nervous system.[1,5,9]

### 1. *Origin of Individual Moieties*

A variety of studies (*in vivo*, perfusion, and *in vitro*) using isotopically labeled substances indicate that, in general, the nervous system, in common with most tissues, is capable of synthesizing phosphoglycerides from the individual moieties or small molecules that comprise these phospholipids.[1]

*a. Phosphate.* Numerous studies support the view that the phosphate group derives from tissue inorganic phosphate. The latter, despite the blood–brain barrier, presumably largely derives from the inorganic phosphate of the blood plasma.[1] *In vitro* studies from this laboratory[12] and others[1,5] support the view that phosphate ultimately is incorporated by way of adenosine triphosphate.

*b. Glycerol.* Studies with $^{14}C$-labeled glycerol demonstrate that brain preferentially converts this precursor into the nonfatty acid portion of the phosphoglycerides under *in vitro* (slices)[1,5] or *in vivo* (intracisternal injection into cats) conditions.[13] It is possible that a significant part of the glycerol moiety may derive from the triose phosphates of glycolysis. There is good evidence that brain contains a highly active mitochondrial α-glycerophosphate dehydrogenase and a weakly active NAD-dependent dehydrogenase present in the cell sap[14] [reaction (1a), Fig. 2]. All the available evidence indicates that glycerol must be incorporated into the phosphoglycerides by way of L-α-glycerophosphate. The latter may arise directly from glycolysis as indicated or through direct phosphorylation by glycerol kinase [E.C.2.7.1.30; reaction (1b), Fig. 2]. Several reports and recent unpublished work from this laboratory indicate that radioactivity from $^{14}C$-labeled glucose does appear in the nonfatty acid portion of the phosphoglycerides of brain.

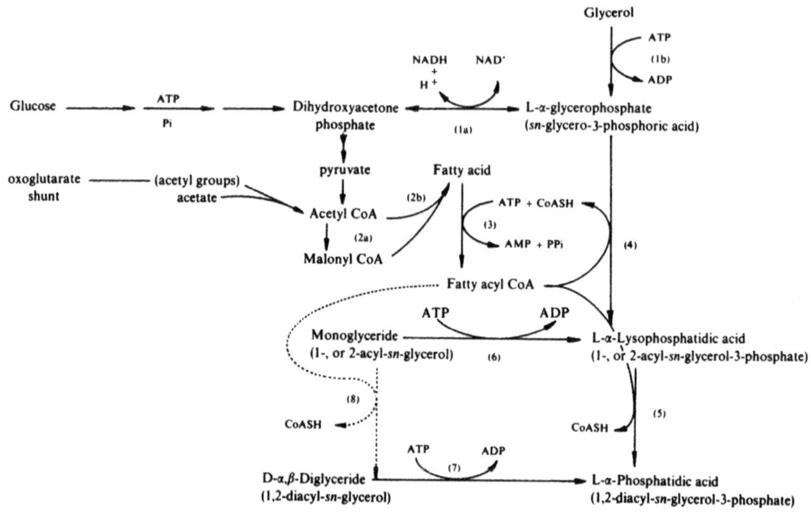


Fig. 2. Biosynthesis of phosphatidic acid.

*c. Nitrogenous Compounds (Choline, Ethanolamine, and Serine).* Experiments carried out by Rossiter and colleagues[5] have shown that incubation of $^{14}C$-labeled bases in brain slice preparations leads to the incorporation of each base into its appropriate phosphoglyceride. However, only limited evidence was obtained for any interconversion among the three nitrogenous components. Choline was incorporated only into the choline-containing phospholipids. Serine was incorporated mainly into the serine phospholipids with a small amount appearing in the ethanolamine phospholipids. Glycine was shown to be incorporated into the serine-containing phospholipids.

These early observations to a large extent have been confirmed by recent reports from the laboratories of Abood[15] and Ansell[16,17] that describe the fate of $^{14}C$-labeled serine, ethanolamine, and choline following intracerebral injection into rats. Abdel-Latif and Abood[15] in studies on developing rat brain showed that radioactivity from L- U-$^{14}C$ serine is incorporated into phosphoglycerides in the following order: phosphatidyl serine > phosphatidyl ethanolamine > phosphatidal serine > sphingomyelin and phosphatidyl choline. Ansell and Spanner[16,17] have assessed the metabolism of both ethanolamine and choline. They have concluded that ethanolamine is incorporated mainly into the ethanolamine phospholipids and choline into the choline phospholipids. There is little or no conversion of ethanolamine to choline through stepwise methylation as occurs in liver.[1,3,9]

*d. Fatty Acid or Alkyl Groups.* Labeled acetate is readily incorporated into the fatty acid portion of the phosphoglycerides of brain slices.[5] *In vivo*, acetate is incorporated into the fatty acids of brain providing it can reach this tissue (see summary by Miyamoto *et al.*[17]). $^{14}C$-labeled linoleate and linolenate are incorporated into the phosphoglycerides of developing chick brain at a rate comparable to that for liver. It is of interest to note that some of the acetyl groups can come from the oxoglutarate shunt.[19] In this shunt the citrate cleavage enzyme is responsible for the formation of acetyl CoA.

There is good evidence that *de novo* synthesis of fatty acid occurs in brain. Brady[20] and others have demonstrated the presence in rat brain of an active fatty acid synthetase that forms long-chain fatty acid ($C_{16}$) from acetyl CoA by the malonyl CoA pathway.

There is reasonable evidence to indicate that the alkenyl and alkyl groups of the ether-containing phosphoglycerides derive from either short-chain fatty acids (acetate)[21] or longer-chain fatty acids (e.g., palmitate).[9,22,23]

## 2. *Biosynthesis of Individual Phosphoglycerides*

*a. Phosphatidic Acid.* L-α-glycerophosphate formed either by way of reaction (1a) or (1b) (Fig. 2) represents the starting point for the formation of phosphatidic acid from small molecules. In 1952 Kornberg and Pricer[24,25] showed that a liver enzyme preparation was capable of forming phosphatidic acid. The synthesis required the formation of fatty acyl CoA [reaction (3), Fig. 2] followed by the stepwise acylation of L-α-glycerophosphate to form L-α-phosphatidic acid [reactions (4) and (5), Fig. 2]. Several studies[12,26,27] with rat brain preparations have shown that phosphatidic acid is formed as described above for liver. The presence of acyl-CoA synthetase activity [E.C.6.2.1.2; reaction (3), Fig. 2] has been demonstrated in rat brain;[28] maximum activity is located in the supernatant obtained on centrifugation of sucrose homogenates. The order of esterification remains in doubt, but the work of Pieringer and Hokin[29] indicates that microsomes of brain can actively acylate lysophosphatidic acid [either α′-(or 1-)-acyl or β-(or 2-)-acyl] as depicted by reaction 5 (Fig. 2). In a similar study on liver, Lands and Hart[30] not only demonstrated that lysophosphatidic acid is readily acylated but that the initial acylation of α-glycerophosphate (Step 4) is slow. This situation also may be true for brain. Recently Martensson and Kanfer[31] succeeded in partially purifying a solubilized enzyme system (L-α-glycerolphosphate acyl transferase; E.C.2.3.1.15) from rat brain that is capable of forming phosphatidic acid from L-α-glycerophosphate and palmitoyl CoA.

Alternate pathways for phosphatidic acid formation have been demonstrated in brain tissue [see reactions (5)–(8), Fig. 2]. In independent studies by Hokin and Hokin[32] and Strickland[3,33] an active diglyceride kinase has been described [reaction (7)] that seems to be stereospecific for D-α, β-diglyceride (1,2-diacyl-*sn*-glycerol). Pieringer and Hokin[34] further established that monoglyceride may be phosphorylated by ATP to form preferentially α′-lysophosphatidic acid [reaction (5)], which can undergo further acylation to form phosphatidic acid. There is little evidence that reaction (8) (Fig. 2) is operative in brain.

*b. Lecithin (Phosphatidyl Choline).* As already noted, labeled choline is incorporated into the lecithin of most tissues including brain. The work of Kornberg and Pricer and others[3] with doubly labeled phosphorylcholine (PC) showed that it was incorporated as a unit into lecithin. Evidence, in general, indicates that PC is formed by action of choline kinase [E.C.2.7.1.32; reaction (9), Fig. 4].[3] Berry, McPherson, and Rossiter[35] and McCaman[36] have studied choline kinase activity in brain and normal and degenerating peripheral nerve. The activity of this enzyme in rabbit brain was relatively much higher than for other tissues.[36] Recently Ansell and Spanner[17] have established that the minimal rate of phosphorylation of choline administered intracerebrally is 0.03 μmoles/g brain/hr.

The mechanism of how PC is incorporated as a unit into lecithin was established as a result of the discovery by Kennedy and Weiss[37] that their ATP sample contained an impurity, CTP, which was essential for the incorporation of PC into lecithin. These workers demonstrated that the intermediate, cytidine diphosphate choline (CMP–PC; Fig. 3a), was formed from CTP and PC by action of the enzyme PC-cytidyl transferase [CTP: choline phosphate cytidylyltransferase, E.C.2.7.7.15; reaction (10), Fig. 4]. This enzyme is present in most mammalian tissues, including brain.[3,37]

CMP–PC has been shown by Kennedy and Weiss[37] to undergo reaction with the "lipid acceptor" D-α,β-diglyceride (1,2-diacyl-*sn*-glycerol) to form lecithin [reaction (11), Fig. 4] in the presence of the surface active agent "Tween 20." Weiss *et al.*[38,39] have partially purified from liver the PC-glyceride transferase (CDP choline: 1,2-diglyceride choline phosphotransferase; E.C.2.7.8.2) enzyme and studied some of its properties. It was shown that the reaction is reversible and that adequately emulsified diglyceride is required. Diglyceride with an unsaturated fatty acid (e.g., oleic) at the 2-position is preferred. The specificity toward cytidine diphosphate compounds has received some study (see Ansell and Hawthorne[1] and Strickland.[3] Intactness of the C-6 amino group seems essential. The

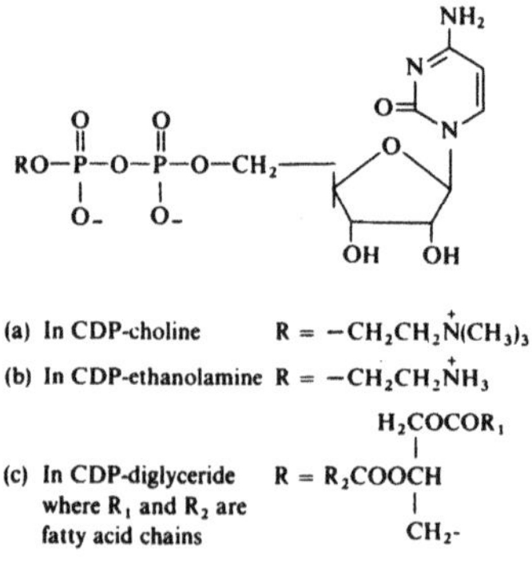


Fig. 3. Structures of cytidine diphosphate (CDP) compounds functioning in the biosynthesis of the phosphoglycerides.

choline portion may be replaced by 2-monomethyl- or 2-dimethyl-aminoethanol. Independent studies[40,41] with nerve and brain preparations provide evidence that lecithin may be formed by the reactions described. McCaman and Cook[42] recently measured the amounts of PC-glyceride transferase activity present in brain in early postnatal development, in discrete morphological subdivisions and in various subcellular fractions. More activity was found in gray matter than white, and most of the activity was confined to the microsomal fraction.

It seems probable that most of the diglyceride for the *de novo* synthesis of lecithin is derived from phosphatidic acid through action of a specific phosphatase [L-α-phosphatidate phosphohydrolase, E.C.3.1.3.4; reaction (12), Fig. 4] discovered by Smith *et al.*[43] in liver tissue. This enzyme has also been demonstrated in homogenates of brain.[44] The enzyme is strongly inhibited by $Mg^{2+}$. It is probable that this inhibition accounts for the accumulation of phosphatidic acid that is observed for many tissue preparations,[3] including brain.[12] Fresh tissues normally have very low amounts of phosphatidic acid.[45] Coleman and Hübscher[46] have studied the distribution of this enzyme and found that next to kidney and intestine brain is most active. Most of the activity was found in the microsomal fraction. More recent evidence with rabbit brain suggests that the enzyme may be associated with the lysosomes.[47] McCaman *et al.*[47] have found that enzyme activity is higher in the early postnatal period (4–20 days of age) and during demyelination and that gray matter areas are two- to four-fold more active than white. The latter authors state that the observed distribution and fluctuations of phosphatidic acid phosphatase might be related to "the regulation of the enzyme activity by a 'feedback' mechanism responsive to the glycerophosphatide content of the tissues involved."

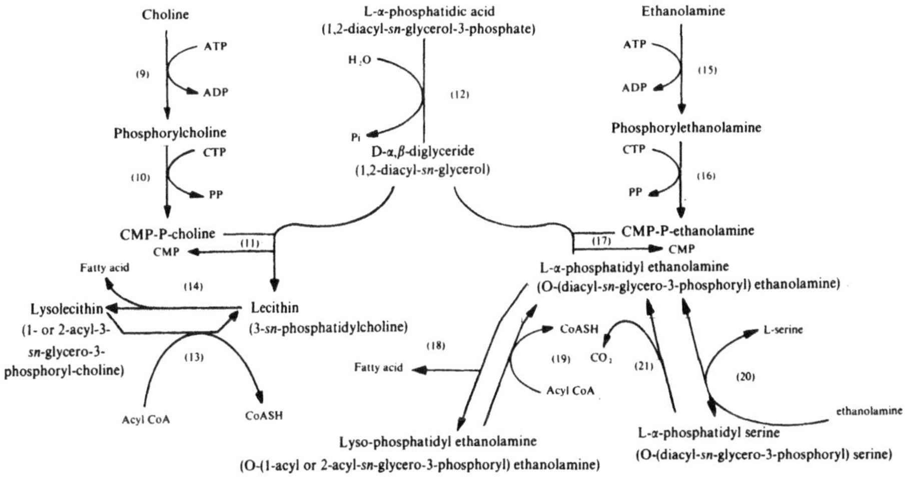


Fig. 4. Biosynthesis of the phosphoglycerides: phosphatidyl choline, phosphatidyl ethanolamine, and phosphatidylserine.

The present evidence favors the view that the *de novo* synthesis of lecithin from small molecules by the nervous system occurs mainly, if not entirely, by the above pathway. Ansell and Spanner[16,17] have not been able to demonstrate that brain can form lecithin by the stepwise methylation of phosphatidylethanolamine (a pathway known to occur in liver). It appears that the first methylation step is not possible.

An alternative pathway that is present in many tissues including brain[10] is that first described by Lands for liver (see Lands and Merkl[48]). This pathway does not lead to *de novo* synthesis from small molecules but results in the formation of lecithin from lysolecithin through acylation by acyl CoA (reaction 13, Fig. 4). Webster and colleagues (see Webster for references[49]) have extensively studied the role of this pathway in brain. They conclude that lysolecithin is a normal minor constituent that possibly arises from phospholipase A-like activity. Brain slices, homogenates, and subcellular particles were shown to be capable of activating fatty acids and incorporating them into lecithin. Although a high degree of specificity is not shown toward the $\alpha'$-(or 1-) or $\beta$-(or 2-) position of the lyso derivative (of four labeled fatty acids, palmitic, stearic, oleic, and linoleic, stearic acid only is preferentially esterified in the 1-position), it seems likely that the combined actions of enzymes catalyzing reactions (13) and (14) in Fig. 4 could be responsible for the redistribution of fatty acids in the lecithins and, as will be described, in other phosphoglycerides.

*c. Phosphatidyl Ethanolamine.* Almost without exception the experiments on lecithin formation have been repeated for phosphatidyl ethanolamine and have provided results which suggest that the latter is formed by pathways similar to those for lecithin[3,10] [see reactions (15)–(19), Fig. 4]. The demonstration that PC was concerned in lecithin biosynthesis led Kennedy and Weiss[37] to investigate whether phosphorylethanolamine (PE) was a precursor of phosphatidyl ethanolamine. Their experiments showed that liver contained enzymes capable of incorporating PE into phosphatidyl ethanolamine and that the mechanism of incorporation involved the formation and participation of cytidine diphosphate ethanolamine (CMP–PE; Fig. 3b). Similar reactions have been shown to occur in brain.[50,51]

Some evidence is available that indicates that alternative pathways exist for the formation of phosphatidyl ethanolamine in the nervous system. Both McMurray[50] and Abdel-Latif and Abood[15] have observed that a significant portion of the label from $^{14}C$-labeled serine is incorporated into phosphatidyl ethanolamine. The latter workers, using $^{14}COOH$-labeled phosphatidyl serine, have demonstrated phosphatidyl serine decarboxylase activity [reaction (20)] in developing brain and shown that the activity is mainly confined to the mitochondria. Support for a second alternative pathway has been provided by Webster.[49] In a recent study on the incorporation of $^{14}C$-labeled fatty acids into the ethanolamine phospholipids of slices of rat cerebrum he obtained results interpretable on the basis that preexisting or newly formed lysophosphatidyl ethanolamine [reaction (18), Fig. 4] may undergo reacylation by specific 1- or 2-acyl transferase enzymes [reaction (19), Fig. 4]. These enzymes, if present, indicate that the alternative pathway described by Merkl and Lands(44) in liver for the formation of phosphatidyl ethanolamine by acylation of the lyso derivative also occurs in brain. This pathway, as for phosphatidylcholine, could provide a means whereby fatty acid distributions could be altered in accordance with the activity and specificity of the acyl transferase and the availability of different fatty acids.

Ansell and colleagues(16,17) in a recent investigation of ethanolamine metabolism in brain *in vivo* concluded from their results that ethanolamine first undergoes phosphorylation to PE [reaction (15), Fig. 4)—possibly catalyzed by an enzyme identical to choline kinase (E.C.2.7.1.32)—and then the PE is incorporated into the ethanolamine lipids by the cytidine pathway [in Fig. 4, reaction (16) is catalyzed by CTP: ethanolamine-phosphate cytidylyltransferase, E.C.2.7.7.14, and reaction (17) is catalyzed by CDP ethanolamine: 1,2-diglyceride ethanolaminephosphotransferase, E.C.2.7.8.1). No evidence was obtained to support the view that in brain ethanolamine can be methylated to form choline, as has been well established for liver.

*d. Phosphatidyl Serine.* All the available evidence supports the view that in mammalian tissues phosphatidyl serine is formed by a pathway differing from that for the corresponding choline and ethanolamine phosphoglycerides (for discussion see Strickland(3)). Attempts to implicate CDP-serine and *O*-phosphoserine have been unsuccessful. The work of Borkenhagen *et al.*(53) and Hübscher(54) indicates that serine is the direct precursor, being incorporated by displacement of ethanolamine from a phosphatidyl group as shown in reaction (20) (Fig. 4). This base exchange reaction is activated by $Ca^{2+}$ ions and does not require ATP or CTP. There is some evidence that CMP stimulates. Borkenhagen *et al.*(53) also have shown that intact phosphatidyl serine can undergo decarboxylation to form phosphatidyl ethanolamine [reaction (21), Fig. 4]. The two reactions (20) and (21) together lead to a phosphoglyceride cycle in which the overall effect is the decarboxylation of free serine to ethanolamine.

Studies on $^{14}C$-serine incorporation into the phosphatidyl serine of rat brain by Pritchard(55) (slices), McMurray(51) (homogenates of developing brain), and Abdel-Latif and Abood(15) (*in vivo* incorporation into various cellular fractions) all have yielded results consistent with the pathway described [i.e., reactions (20) and (21), Fig. 4]. In each of these investigations, in general, it was observed that phosphatidyl serine was most actively labeled but that some of the label ultimately appeared in phosphatidyl ethanolamine. Work carried out in this laboratory (unpublished results of Strickland) also demonstrated that labeled serine, and not phosphoserine, is incorporated into phosphatidyl serine and that CMP is to some extent stimulatory.

*e. Plasmalogens.* As discussed in earlier reviews,[3,10,23] a number of studies with brain preparations and other tissues such as heart have shown that label from fatty acids (e.g., $^{14}C$-acetate, -palmitate) are incorporated into the alkenyl group of ethanolamine and/or choline plasmalogens. However, the nature of the mechanism of incorporation remains obscure. Attempts to demonstrate that the incorporation occurs by way of an aldehyde intermediate have been unsuccessful. Gambal and Monty[22] observed that the appearance of radioactivity from palmitate-1-$^{14}C$ in the alkenyl group of the plasmalogen of a cell-free preparation from rat brain is stimulated by CTP. Evidence for the possible direct involvement of cytosine nucleotides was provided by Kiyasu and Kennedy[56] when they observed that rat liver contains the necessary cholinephosphotransferase and ethanolamine phosphotransferase enzymes to form choline and ethanolamine plasmalogen, respectively, in the presence of added "plasmalogenic" diglyceride (1-alkenyl-2-acylglycerol) [reactions (22) and (23), Fig. 5]. The latter was obtained by action of phospholipase C on choline plasmalogen from beef heart. McMurray[57] has shown that homogenates of developing rat brain are capable of carrying out the reactions described above for the formation of choline and ethanolamine plasmalogens. In a later study Ansell and Metcalfe[58] provided additional evidence that in adult rat brain ethanolamine becomes incorporated into the corresponding plasmalogen by way of CDP-ethanolamine. Although not demonstrated, it is possible that serine plasmalogen is formed by a base exchange reaction [reaction (24), Fig. 5] similar to that for phosphatidyl serine formation.

At present there is no definitive evidence that 1-alkenyl-2-acylglycerol is the normal precursor for the *de novo* synthesis of plasmalogens. Also, regardless of whether such a compound is a precursor, our knowledge of how it may be formed biologically is very scant. Waku and Lands[59] recently demonstrated that deacylated choline plasmalogen (1-alkenyl-glycero-3-phosphorylcholine) can be reacylated in muscle [reaction (25), Fig. 5]. The enzyme preferentially utilized linoleoyl CoA, followed by arachidonoyl CoA and then oleoyl-CoA. These authors summarized the evidence for the possibility that deacylation and reacylation may take place [reactions (25) and (26), Fig. 4], thereby accounting for a possible redistribution of the fatty acids.

There is a distinct possibility that plasmalogens may be formed from other phosphoglycerides as a result of the generation of a vinyl ether bond during a lipid-to-lipid interconversion. Nearly all studies in which comparisons are made of the rate of synthesis of individual plasmalogens to those of the corresponding diacylphosphoglycerides reveal that the rate for plasmalogens is much lower.[16,17,23,52] This makes it unlikely that the diacylglycerophosphatides can derive from plasmalogens, but the reverse may be possible. Recent work by Thompson[60,61] on the slug, *Arion ater*, using doubly-labeled glyceryl ethers has led him to the conclusion that plasmalogens derive from the corresponding alkyl ether phosphoglycerides. This conclusion also has been reached by Horrocks and Ansell[62] for rat brain. The latter authors, on the basis of specific activity–time relationships for incorporation of labeled ethanolamine into diacyl-, alkenyl acyl-, and alkyl acyl-glycero-3-phosphoryl ethanolamine· following intracerebral injection, concluded that the alkenyl, acyl derivative may be formed by desaturation of the alkyl, acyl derivative. Another interesting possibility is that offered by Bickerstaffe and Mead.[63] Following intracerebral injection of palmitaldehyde-1-$^{14}C$ into rats they obtained data that suggested that palmitaldehyde is first oxidized to palmitic acid and then the latter is incorporated into phosphatidylethanolamine. This is followed by reduction at the 1-acyl group to form either an alkenyl group directly or possibly an intermediate alkyl group that is desaturated to an alkenyl group.

It must be concluded that none of the above experiments provide any clear indication of how plasmalogens are formed. A number of possible interconversions are indicated. It seems likely that the phosphoryl-base moiety ultimately is incorporated via a cystosine nucleotide intermediate, but whether this occurs with diacylglycerol, alkenyl, acylglycerol, or alkyl, acylglycerol is not known. It is possible that a glyceryl ether (alkenyl or alkyl) is formed that becomes partially acylated before it is utilized. This is not unlike the proposal of Thompson[60,61] for *Arion ater*. The pathway for glyceryl ether formation is not well understood, but glucose and long-chain fatty acids appear to be precursors (see Fig. 5 and Oswald *et al.*[64] for summary).

*f. Alkyl Ether Phosphoglycerides.* Much of the previous discussion may be applied to the alkyl ether phosphoglycerides. Little is known concerning the *de novo* synthesis of these phosphoglycerides. Studies of

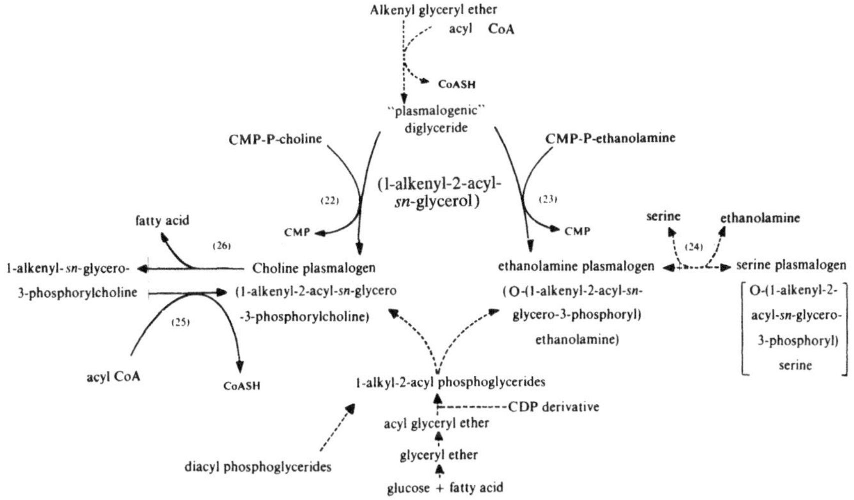


Fig. 5. Biosynthesis of phosphoglycerides: plasmalogens and alkyl ether compounds. Dotted lines show pathways that have not been fully established.

McMurray[57] and Ansell and Metcalfe[58] indicate that labeled ethanolamine from CDP-ethanolamine is incorporated into this fraction of phosphoglyceride. The recent work of Horrocks and Ansell[62] also confirms that ethanolamine (following intracerebral injection of $^{14}C$-ethanolamine into rats) is readily incorporated into the corresponding alkyl ether phosphoglyceride. No precursor–product relationships could firmly be established apart from that already referred to in the preceding section (i.e., alkyl-ether phosphoglyceride → ethanolamine plasmalogen).

*g. Phosphoglycerides Involving the Participation of Cytidine Diphosphate Diglyceride (CDP-Diglyceride).* The phosphoglycerides of the nervous system that involve the participation of CDP-diglyceride in their biosynthesis include monophosphoinositide (phosphatidyl inositol), phosphatidyl glycerophosphate, phosphatidylglycerol, and possibly cardiolipin (diphosphatidyl glycerol).

The observation by McMurray *et al.*[12] in 1957 that the addition of CTP increases the incorporation of $^{32}Pi$ into the monophosphoinositide of cell-free preparations of rat brain suggested the participation of a cytosine nucleotide in the formation of this lipid. More direct evidence for the involvement of a cytosine nucleotide came from the study of Agranoff *et al.*[65] on the incorporation of tritiated inositol into monophosphoinositide by a particulate fraction from guinea pig kidney. Their observations led them to postulate participation of the liponucleotide, CDP-diglyceride. Since added phosphatidic acid and CDP-choline were particularly effective in their system, it was suggested that CDP-diglyceride formed from these two precursors. Somewhat similar observations were obtained for brain by Thompson *et al.*[3,66] but CTP was found to be more effective. Paulus and Kennedy[67] were able to demonstrate conclusively, using liver preparations, that glycerophosphate is incorporated into monophosphoinositide by way of the intermediate, CDP-diglyceride. It was found that the latter is formed from phosphatidic acid and CTP [reaction (27), Fig. 6]. CDP-diglyceride formed either enzymatically or chemically was shown to be converted in the presence of inositol to monophosphoinositide [reaction (28), Fig. 6]. Thompson *et al.*[66] have subsequently shown that for the incorporation of $^{14}C$-labeled phosphatidic acid ($^{14}C$-in glycerol portion) into the monophosphoinositide of rat brain, CTP is specifically required and that for the incorporation of tritiated inositol, CDP-diglyceride (CDP-diplamitin used) is required. This provides evidence that reactions (27) and (28) occur in brain. Additional investigation of the role played by the cytosine nucleotides in monophosphoinositide biosynthesis in brain has been carried out by Possmayer and Strickland.[26] More recently Petzold and Agranoff[68] have investigated the biosynthesis of CDP-diglyceride by a particulate fraction from embryonic chick brain. They conclude that CDP-diglyceride is synthesized by a CTP: 1,2-diacylglycerophosphate-cytidylyl transferase (E.C.2.7.7) reaction. Palmitoyl coenzyme A was found to inhibit this reaction. Further discussion of the phosphoinositides may be found in the following chapter.

In view of the discovery that phosphatidyl glycerophosphate occurs in brain[69] and the facts that small amounts of phosphatidyl glycerol are found in brain mitochondria and that small quantities of cardiolipin may occur, it is of considerable interest to consider briefly the biosynthesis of these phosphatidyl glycerol-containing compounds.

Kiyasu *et al.*[70] were the first to demonstrate that a variety of mammalian tissues (in addition to certain microorganisms) contained phosphatidyl glycerol-synthesizing enzymes. In studies with $^{32}P$- and $^{14}C$-L-α-glycerophosphate, CDP-diglyceride was found to react with L-α-glycerophosphate to form phosphatidyl glycerophosphate [reaction (29), Fig. 6]. The latter was shown to undergo dephosphorylation to form phosphatidyl glycerol [reaction (30), Fig. 6] by an enzyme that is strongly inhibited by $HgCl_2$. These authors also suggested that cardiolipin may be formed by transfer of a phosphatidyl group from CDP-diglyceride to the 3′-position of phosphatidyl glycerol. More recently Lecocq and Ballou[71] suggested that cardiolipin also might be formed from a reaction between phosphatidyl glycerophosphate and the appropriate diglyceride. Work from this laboratory[72] has provided similar evidence that both phosphatidyl glycerophosphate and phosphatidyl glycerol formation can take place in rat brain preparations. The greatest activity is localized in the mitochondrial fraction.

## III. BREAKDOWN

### A. General

In nature it has been found that the enzymatic degradation of the phosphoglycerides can be brought about by the action of four types of hydrolases that can be distinguished according to their site of attack on a lecithin molecule (see reviews by Van Deenen[73,74] and Dawson[2]). It will

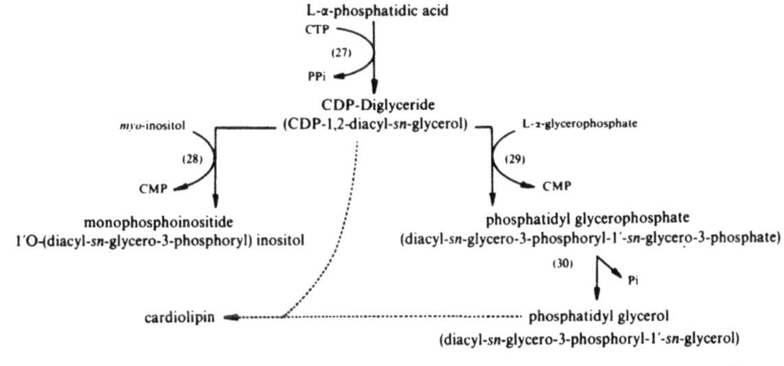


Fig. 6. Biosynthesis of phosphoglycerides: phosphatidyl glycerophosphate, phosphatidyl glycerol, and phosphoinositides. Dotted lines show pathways not fully established.

be seen from Fig. 7 that this molecule can undergo hydrolytic attack at each of the ester bonds. These enzymes are not usually specific for lecithin in that they normally hydrolyze any analogous phosphoglyceride provided the right experimental conditions are used. The cleavage or removal of one acyl group from a phosphoglyceride molecule is catalyzed by phospholipase A (phosphatide acyl hydrolase, 3.1.1.4.). This enzyme, which occurs widely in nature, is most often found to be specific toward the β- (or 2-) ester linkage. More recently reports have appeared that describe a phospholipase A (usually designated $A_1$) specific for the α′- (or 1-) ester linkage. Most tissues are capable of hydrolyzing monoacylphosphoglycerides to a free fatty acid and phosphodiester. The enzyme responsible is phospholipase B (lysolecithin acyl hydrolase, 3.1.1.5). There is some doubt as to whether this enzyme shows an absolute specificity toward α′-acyl-lysophosphoglycerides. Under certain conditions of activation[2,73] what has been termed phospholipase B has been shown to hydrolyze both α′-acyl and β-acyl ester groups of lecithin. Phospholipase C (phosphatidylcholine cholinephosphohydrolase, 3.1.4.3), which occurs in many bacteria and possibly in certain tissues, hydrolyzes lecithin into diglyceride and phosphorylcholine. Phospholipase D (phosphatidylcholine phosphatidohydrolase, 3.1.4.4), which seems to be confined mainly to plants, cleaves the base moiety from lecithin or phosphatidyl ethanolamine. Further details on the specificity of these phospholipases may be found in the reviews by Van Deenen[73,74] or Dawson.[2]

The fatty acids released then enter the normal pathways of metabolism for fatty acids. This involves reactivation by CoA to form acyl CoA derivatives that then can participate in reacylation reactions or undergo oxidation by the now well-established steps of β-oxidation. In animal tissues the phosphodiesters (e.g., glycerylphosphorylcholine and glycerylphosphorylethanolamine) that form are normally acted upon by a specific diesterase (L-3-glycerylphosphorylcholine glycerophosphohydrolase, 3.1.4.2) that removes the base moiety. The glycerophosphate released may be reutilized in a number of the pathways described under biosynthesis, or it may undergo further hydrolysis by widely occurring alkaline and acid phosphomonoesterases (orthophosphoric monoester phosphohydrolase, 3.1.3.1 and 3.1.3.2).

## B. Breakdown of Phosphoglycerides in the Nervous System

Compared to other tissues, relatively little work has been carried out on the breakdown of phosphoglycerides in the nervous system. There is need for the purification and a detailed study of the properties of the hydrolase

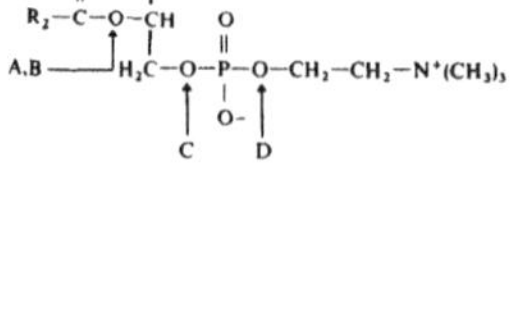


Fig. 7. The hydrolysis of the ester bonds of lecithin by various phospholipases ($A_1$, A, B, C, and D).

enzymes involved. In general the evidence is that the pathway already described for the breakdown of the phosphoglycerides is operative in the nervous system (for reviews see Ansell and Hawthorne,[1] Thompson,[75] and Webster[76]).

### 1. *Lecithin (Phosphatidyl Choline)*

The observation that lysolecithin could be demonstrated in the lipid extracts of brain led Gallai-Hatchard *et al.*[77] to investigate the possible mechanism for its formation. The were able to show that lecithin–deoxycholate emulsions were hydrolyzed to yield lysolecithin (possibly $\alpha'$-acyl) by brain extracts. More recently Gatt and colleagues[78,79] studied a phospholipase A from rat and calf brain, which has been shown to be specific for the $\alpha'$-acyl (or 1-acyl) ester of lecithin. The enzyme, designated phospholipase $A_1$, has been partially purified. The partially purified preparation shows low lipase activity and only slowly hydrolyses either $\alpha'$-acyl or $\beta$-acyl lysolecithins. Phosphatidyl ethanolamine is hydrolyzed at 20–30% of the rate for lecithin. Gatt[79] indicates that he has evidence for the presence of a phospholipase A that is currently under study. This work confirms the earlier suggestion of Robertson and Lands[80] that many tissues including brain may contain both of these enzymes (i.e., A and $A_1$).

The distribution of phospholipase B activity in mammalian tissues has been studied by Marples and Thompson.[81] Brain homogenates, although relatively low in activity, were capable of hydrolyzing $\alpha'$-acyl lysolecithin to glycerylphosphorylcholine. Gray matter was shown to contain considerably more activity than white matter.

Hydrolases capable of further action on glycerylphosphorylcholine have been demonstrated in brain and nervous tissue. Webster *et al.*[82] found that all regions of the nervous system studied possessed a relatively active glycerylphosphorylcholine diesterase that hydrolyzed $\alpha$-glycerylphosphorylcholine to $\alpha$-glycerophosphate and choline. Although the *p*H optimum was at *p*H 9.5, significant activity could be demonstrated at *p*H 7.5. The products of this hydrolysis may either be reutilized in synthesis or alternatively broken down further by acid and/or alkaline phosphomonoesterases that occur widely in all tissues including brain.[83,84] A summary of the steps involved in lecithin breakdown is given in Fig. 8.

Ansell and Spanner[85] in a brief note indicated that brain tissue may possess a weak phospholipase C activity. Under certain defined conditions they observed that the glycerophosphate linkage of phosphatidyl choline was hydrolyzed (1.2 $\mu$moles/g fresh tissue/hr). This enzyme would lead to the release of phosphorylcholine, a normal precursor of lecithin formation.

### 2. *Phosphatidylethanolamine and Phosphatidyl Serine*

Information on the breakdown of these two phosphoglycerides by nervous tissue is much more incomplete than for lecithin. However, the specificities of the phospholipases from animal tissues are generally not so

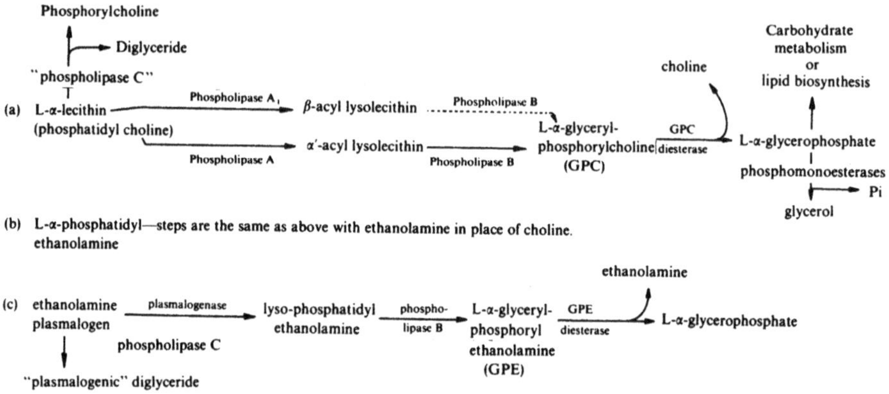


Fig. 8. Scheme showing the breakdown in nervous tissue of (a) lecithin, (b) phosphatidyl ethanolamine, and (c) ethanolamine plasmalogen.

great as to exclude these two phosphoglycerides. A number of the studies on the phospholipase A's of brain (A and $A_1$)[77,79] have provided evidence that phosphatidyl ethanolamine is also hydrolyzed, but possibly at a lower rate. It is generally accepted that the breakdown of phosphatidyl ethanolamine is similar to that for lecithin (see Fig. 8). Whether phosphatidyl serine follows the same pathway is less certain. It has been shown that brain extracts can hydrolyze phosphatidyl serine to a lyso derivative.[77] It is conceivable that the phosphatidyl serine decarboxylase activity (see section on biosynthesis) of nervous tissue is great enough to convert this phosphoglyceride to phosphatidyl ethanolamine. It should be noted that Ansell and Spanner[85] have observed a phospholipase C-like activity that is capable of hydrolyzing the glycerophosphate linkage of phosphatidyl ethanolamine (1.8 $\mu$moles/g fresh tissue hr).

### 3. *Plasmalogen and Alkyl Ether Phosphoglyceride (Ethanolamine Type)*

Although there appears to be no direct evidence that brain phospholipase A of the nervous system will cleave the 2-acyl group from either of these phosphoglycerides, it is reasonable to propose that this may occur in view of the fact that phospholipase A from other sources (snake venoms) will cleave this grouping. Because of the considerable loss of ethanolamine plasmalogen from white matter often observed in demyelinating diseases of the nervous system, Ansell and Spanner[85,86] studied the catabolism of ethanolamine plasmalogen. They reported a phospholipase C-like activity (1.0 $\mu$moles/g fresh tissue/hr)[85] and, what is of considerable interest, a plasmalogenase activity.[86] The latter is a $Mg^{2+}$-dependent enzyme that is specific for the vinyl ether linkage of ethanolamine plasmalogen. The activity was associated with the white matter but was absent from myelin fractions. There was increased activity in certain demyelinating conditions.

Apart from the above reports little is known concerning the catabolism of the alkenyl ether or alkyl ether phosphoglycerides in the nervous system. A summary of the breakdown of ethanolamine plasmalogen is given in Fig. 8.

### 4. *Other Phosphoglycerides*

The breakdown of the phosphoinositides is described in the following chapter. Little work has been done on the breakdown of the phosphatidyl glycerol-containing phosphoglycerides (phosphatidyl glycerol, phosphatidyl glycerophosphate, and cardiolipin) in the nervous system. It seems reasonable to suggest on the basis of studies of the action of phospholipases from other sources on these phosphoglycerides that breakdown may occur through action of appropriate phospholipases (A and B), diesterases, and phosphomonoesterases.

## IV. TURNOVER AND INTERRELATIONSHIPS

Although there are many gaps in our knowledge on the biosynthesis and degradation of the phosphoglycerides of the nervous system, the preceding pages indicate that some information is available concerning certain of the individual steps concerned. It is, however, important to realize that the *in situ* concentrations of the phosphoglycerides reflect a steady-state balance between formation and breakdown. In the limited space available we can only cite a few examples of the approaches now being used to gain further insight into the overall metabolism of the phosphoglycerides.

*In vitro* studies with labeled precursors have proved to be most helpful in the elucidation of the reactions concerned in formation and breakdown. However, they provide limited information in regard to the total metabolism of the phosphoglycerides. For the latter, it is necessary to devise appropriate *in vivo* experiments. A useful approach has been to administer labeled metabolites intracerebrally, thereby permitting materials to reach the nervous system directly. Much useful information also has come from both *in vitro* and *in vivo* studies on young animals (e.g., the rat) over the period when active myelination is in progress. A third helpful approach has been to study the effect of nerve transection both on the formation and the disappearance of phosphoglycerides. A fourth approach is to measure, under a variety of conditions, both the distribution and activity of the individual enzymes involved. Below are cited some examples that illustrate the type of information that can be obtained by these methods.

The many studies with labeled precursors (referred to earlier in this chapter) established not only that the phosphoglycerides were built up from small molecules but that continual turnover of the phosphoglycerides takes place even though at times it may be slow (see Davison[87] or other reviews[1,5]). While interpretation of data from experiments of this kind is difficult, some useful tentative general conclusions may be drawn. For example, experiments such as those described by McMurray[51,57] and Abdel-Latif and Abood[15] on the incorporation of labeled precursors into developing rat brain have given results that are consistent with the view that the synthesis of each of the phosphoglycerides found principally in myelin reaches a maximum when myelination is most active (i.e., at 10–20 days of age). Some increase in synthesis of phosphoglycerides that also are located in nonmyelin structures (phosphatidyl choline and phosphatidyl inositol) also may occur, but the pattern of increase more closely parallels that associated with the development of gray matter components of the nervous system. It is of interest to note that Davison and colleagues[87–89] have carried out extensive studies that indicate that once the myelin lipids, including phosphoglycerides, are laid down there is relatively little turnover (breakdown and renewal) of such lipids. There is also some evidence that this may be true for lipids contributing to the structural elements of brain mitochondria. It should be noted, however, that a more recent assessment by Ansell and Spanner[16] of the relative "inertness" of the ethanolamine-containing phosphoglycerides of myelin in the adult rat brain indicates that there is appreciable turnover at least of a portion of the phosphatidyl ethanolamine and ethanolamine plasmalogen molecules.

Detailed studies such as those undertaken by Ansell and Spanner[16,17] on ethanolamine and choline metabolism in the adult rat brain have provided useful information. Thus, $^{14}C$-ethanolamine when injected intracerebrally followed a pattern of labeling consistent with it undergoing phosphorylation and subsequent incorporation into ethanolamine phosphoglycerides by way of cytosine nucleotides.[16] A brief report on choline metabolism[17] has shown that choline is phosphorylated at a minimal rate of 0.03 $\mu$moles/g brain/hr and that in this form it is readily incorporated into phosphatidyl choline. Lesser amounts of activity appear in sphingomyelin and in choline plasmalogen. The specific activity of the latter, however, was quite high in sharp contrast to that observed for ethanolamine plasmalogen. Extension of this experimental approach with careful interpretation should provide a clearer understanding of the metabolism of these phosphoglycerides *in vivo*.

As a result of studies with homogenates and slices Webster[49] and colleagues provided good evidence for the existence in brain of a monoacyl–diacyl phosphoglyceride cycle. It seems likely that this cycle may be sufficiently active to account for the redistribution of fatty acids in choline and ethanolamine phosphoglycerides. The response would seem to be dependent not only on the activities and specificities of the enzymes involved but also on the acids available in the fatty acid pool.

An approach of considerable interest is that undertaken by McCaman.[36,42,47] He has initiated quantitative study of the activity of individual enzymes concerned in the formation of phosphatidyl choline. Up to the present three enzymes, choline phosphokinase,[36] phosphatidic acid phosphatase,[42] and phosphorylcholine-glyceride transferase[47] have been investigated. These enzymes appear to show certain differences and certain parallels in their distribution. Of a number of tissues studied, the activity of choline phosphokinase in brain was greatest while that for phosphorylcholine-glyceride transferase was intermediate and only about 20% that of liver. Choline phosphokinase is essentially a soluble enzyme, whereas phosphatidic acid phosphatase and the transferase enzyme are particulate. The kinase varies little during maturation compared to the progressive rise of activity that is observed with the other enzymes during the period 4–20 days after birth. All three enzymes show the greatest activity in areas of gray matter where the metabolism of phosphatidyl choline is greatest, and all there enzymes show elevated activities during demyelination.

This approach with carefully and adequately controlled assay systems should provide much information concerning the potential possessed by various areas of the nervous system to provide for the synthesis and degradation of phosphoglycerides. A more complete answer as to the formation and fate of phosphoglycerides *in situ* can come only from a thoughtful fitting together of the various and sometimes conflicting findings from a large number of widely differing experimental approaches.

## V. REFERENCES


1. G. B. Ansell and J. N. Hawthorne, *in Phospholipids, Chemistry, Metabolism and Function, Biochim. Biophys. Acta* Library Vol. 3 Elsevier, Amsterdam (1964).
2. R. M. C. Dawson, *in Essays in Biochemistry* (P. N. Campbell and G. D. Greville, eds.), Vol. 2, pp. 69–115, Academic Press, London and New York (1966).
3. K. P. Strickland, *in Biogenesis of Natural Compounds* (P. Bernfeld, ed.) 2nd ed., pp. 103–205, Pergamon Press, Oxford and New York (1967).
4. R. J. Rossiter, *in Neurochemistry* (K. A. C. Elliott, I. H. Page, and J. H. Quastel, eds.), 2nd ed., pp. 10–54, Charles C. Thomas, Springfield, Illinois (1962).
5. R. J. Rossiter, *in Metabolism of the Nervous System* (D. Richter, ed.), pp. 355–380, Pergamon Press, London and New York (1957).
6. R. J. Rossiter, *in Neurochemistry* (K. A. C. Elliott, I. H. Page, and J. H. Quastel, eds.), 2nd ed., pp. 870–896, Charles C. Thomas, Springfield, Illinois (1962).
7. E. P. Kennedy, Metabolism of lipides, *Ann. Rev. Biochem.* **26**:119–148 (1957).
8. E. P. Kennedy, The metabolism and function of complex lipids, *Harvey Lectures Ser. 57* 143–171 (1961–1962).
9. R. J. Rossiter, *in Metabolism and Physiological Significance of Lipids* (R. M. C. Dawson and D. N. Rhodes, eds.), pp. 511–525, Wiley, London and New York (1964).
10. R. J. Rossiter, *in Metabolic Pathways* (D. M. Greenberg, ed.), 3rd ed. Vol. 2, pp. 70–115, Academic Press, New York (1968).
11. R. J. Rossiter and K. P. Strickland, *in Lipide Metabolism* (K. Bloch, ed.), pp. 69–127, Wiley, London and New York, (1960).
12. W. C. McMurray, K. P. Strickland, J. F. Berry, and R. J. Rossiter, Incorporation of $^{32}$P-labelled intermediates into the phospholipids of cell-free preparations of brain, *Biochem. J.* **66**:634–644 (1957).
13. W. P. Kennedy, *Biosynthesis of Polyphosphoinositides in Brain,* M.Sc. Thesis, University of Western Ontario, London, Canada (1968).

14. G. D. Greville, *in Neurochemistry* (K. A. C. Elliott, I. H. Page, and J. H. Quastel, eds.) 2nd ed., pp. 238–275, Charles C. Thomas, Springfield, Illinois (1962).
15. A. A. Abdel-Latif and L. G. Abood, *In vivo* incorporation of L-[$^{14}$C] serine into phospholipids and proteins of the subcellular fractions of developing rat brain, *J. Neurochem.* **13**:1189–1196 (1966).
16. G. B. Ansell and S. Spanner, The metabolism of labelled ethanolamine in the brain of the rat *in vivo, J. Neurochem.* **13**:873–885 (1967).
17. G. B. Ansell and S. Spanner, The metabolism of $^{14}$C-labelled choline and phosphorylcholine in the brain of the rat *in vivo*, *Biochem. J.* **106**:20P (1968).
18. K. Miyamoto, L. M. Stephanides, and J. Bernshohn, Incorporation of [1-$^{14}$C] linoleate and linolenate into polyunsaturated fatty acids of phospholipids of the embryonic chick brain, *J. Neurochem.* **14**:227–237 (1967).
19. A. F. D'Adamo, Jr., and A. P. D'Adamo, Acetyl transport mechanisms in the nervous system. The oxoglutarate shunt and fatty acid synthesis in the developing rat brain, *J. Neurochem.* **15**:315–323 (1968).
20. R. O. Brady, Biosynthesis of fatty acid II. Studies with enzymes obtained from brain, *J. Biol. Chem.* **235**:3099–3103 (1960).
21. H. Debuch, *in Metabolism and Physiological Significance of Lipids* (R. M. C. Dawson and D. N. Rhodes, eds.), pp. 537–539, Wiley, London and New York (1964).
22. D. Gambal and K. Monty, The biosynthesis of plasmalogens, *Federation Proc.* **18**:232 (1959).
23. E. F. Hartree, *in Metabolism and Physiological Significance of Lipids* (R. M. C. Dawson and D. N. Rhodes, eds.), pp. 207–218, John Wiley, New York (1964).
24. A. Kornberg and W. E. Pricer, Jr., Enzymatic synthesis of phosphorus containing lipides, *J. Am. Chem. Soc.* **74**:1617 (1952).
25. A. Kornberg and W. E. Pricer, Jr., Enzymatic esterification of α-glycerophosphate by long chain fatty acids, *J. Biol. Chem.* **204**:345–357 (1953).
26. F. Possmayer and K. P. Strickland, The incorporation of α-glycerophosphate-$^{32}$P into the lipids of rat brain preparations II. On the biosynthesis of monophosphoinositide. *Can. J. Biochem.* **45**:63–70 (1967).
27. C. M. Redman and L. E. Hokin, Stimulation of the metabolism of phosphatidylinositol and phosphatidic acid in brain cytoplasmic fractions by low concentrations of cholinergic agents, *J. Neurochem.* **11**:155–163 (1964).
28. P. M. Vignais, C. H. Gallagher, and I. Zabin, Activation and oxidation of long chain fatty acids by rat brain, *J. Neurochem.* **2**:283–287 (1958).
29. R. A. Pieringer and L. E. Hokin, Biosynthesis of phosphatidic acid from lysophosphatidic acid and palmitoyl coenzyme A, *J. Biol. Chem.* **237**:659–663 (1962).
30. W. E. M. Lands and P. Hart, Metabolism of glycerolipids: V. Metabolism of phosphatidic acid, *J. Lipid Res.* **5**:81–87 (1964).
31. E. Martensson and J. Kanfer, The conversion of L-glycerol-$^{14}$C 3-phosphate into phosphatidic acid by a solubilized preparation from rat brain, *J. Biol. Chem.* **243**:497–501 (1968).
32. M. R. Hokin and L. E. Hokin, The synthesis of phosphatidic acid from diglyceride and adenosine triphosphate in extracts of brain microsomes, *J. Biol. Chem.* **234**:1381–1386 (1959).
33. K. P. Strickland, Phosphorylation of diglycerides by rat brain, *Can. J. Biochem. Physiol.* **40**:247–259 (1962).
34. R. A. Pieringer and L. E. Hokin, Biosynthesis of lysophosphatidic acid from monoglyceride and adenosine triphosphate, *J. Biol. Chem.* **237**:653–658 (1962).
35. J. F. Berry, C. F. McPherson, and R. J. Rossiter, Chemical studies of peripheral nerve during Wallerian degeneration—IX Choline Kinase, *J. Neurochem.* **3**:65–71 (1958).

36. R. E. McCaman, Intermediary metabolism of phospholipids in brain tissue. Microdetermination of choline phosphokinase, *J. Biol. Chem.* **237**:672–676 (1962).
37. E. P. Kennedy and S. B. Weiss, The function of cytidine coenzymes in the biosynthesis of phospholipides, *J. Biol. Chem.* **222**:193–214 (1956).
38. S. B. Weiss, S. W. Smith, and E. P. Kennedy, The enzymatic synthesis of lecithin from cytidine diphosphate choline and D-1,2-diglyceride, *J. Biol. Chem.* **231**:53–64 (1968).
39. S. B. Weiss, E. P. Kennedy, and J. Y. Kiyasu, The enzymatic synthesis of triglycerides, *J. Biol. Chem.* **235**:40–44 (1960).
40. R. J. Rossiter, I. M. McLeod, and K. P. Strickland, Biosynthesis of lecithin in brain and degenerating nerve. Participation of cytidine diphosphate choline, *Can. J. Biochem. Physiol.* **35**:945–951 (1957).
41. K. P. Strickland, D. Subrahmanyam, E. T. Pritchard, W. Thompson, and R. J. Rossiter, Biosynthesis of lecithin in brain. Participation of cytidine diphosphate choline and phosphatidic acid, *Biochem. J.* **87**:128–136 (1963).
42. R. E. McCaman and K. Cook, Intermediary metabolism of phospholipids in brain tissue. III. Phosphorylcholine—glyceride transferase, *J. Biol. Chem.* **241**:3390–3394 (1966).
43. S. W. Smith, S. B. Weiss, and E. P. Kennedy, The enzymatic dephosphorylation of phosphatidic acids, *J. Biol. Chem.* **228**:915–922 (1958).
44. R. J. Rossiter and K. P. Strickland, Biogenesis of phosphatides and triglycerides, *Ann. N.Y. Acad. Sci.* **72**:790–802 (1959).
45. L. E. Hokin and M. R. Hokin, The presence of phosphatidic acid in animal tissues, *J. Biol. Chem.* **233**:800–804 (1958).
46. R. Coleman and G. Hübscher, Metabolism of phospholipids V. Studies of phosphatidic acid phosphatase, *Biochim. Biophys. Acta.* **56**:479–490 (1962).
47. R. E. McCaman, R. E. Smith, and K. Cook, Intermediary metabolism of phospholipids in brain tissue II. Phosphatidic acid phosphatase, *J. Biol. Chem.* **240**:3513–3517 (1965).
48. W. E. M. Lands and I. Merkl, Metabolism of glycerolipids III. Reactivity of various acyl esters of coenzyme A with $\alpha^1$-acylglycerophosphorylcholine and positional specificities in lecithin synthesis, *J. Biol. Chem.* **238**:898–904 (1963).
49. G. R. Webster, The incorporation of long-chain fatty acids into phospholipids of respiring slices of rat cerebrum, *Biochem. J.* **102**:373–380 (1967).
50. G. B. Ansell and T. Chojnacki, Incorporation of 1-*O*-phosphoryl-2-dimenthylaminoethanol and phosphorylcholine into the phospholipids of brain and liver dispersions, *Nature* **196**:545–547 (1962).
51. W. C. McMurray, Metabolism of phosphatides in developing rat brain—I Incorporation of radioactive precursors. *J. Neurochem.* **11**:287–299 (1964).
52. I. Merkl and W. E. M. Lands, Metabolism of glycerolipids. IV. Synthesis of phosphatidylethanolamine, *J. Biol. Chem.* **238**:905–906 (1963).
53. L. F. Borkenhagen, E. P. Kennedy, and L. Fielding, Enzymatic formation and decarboxylation of phosphatidylserine, *J. Biol. Chem.* **236**:PC28–30 (1961).
54. G. Hübscher, Metabolism of phospholipids VI. The effect of metal ions on the incorporation of L-serine into phosphatidylserine, *Biochim. Biophys. Acta.* **57**:555–561 (1962).
55. E. T. Pritchard, the formation of glycerophosphatides from $C^{14}$-labelled precursors in rat brain slices, *Can. J. Biochem. Physiol.* **36**:1211–1220 (1958).
56. J. Y. Kiyasu and E. P. Kennedy, The enzymatic synthesis of plasmalogens, *J. Biol. Chem.* **235**:2590–2594 (1960).

57. W. C. McMurray, Metabolism of phosphatides in developing rat brain—II. Labelling of plasmalogens and other alkali-stable lipids from radioactive cytosine nucleotides, *J. Neurochem.* **11**:315–326 (1964).
58. G. B. Ansell and R. F. Metcalfe, The labelling of brain phospholipids by cytidine 5′-diphosphate ethanolamine *in vitro*, *Biochem. J.* **98**:22P (1966).
59. K. Waku and W. E. M. Lands, Acyl coenzyme A: 1-Alkenyl-glycero-3-phosphorylcholine acyltransferase action in plasmalogen biosynthesis, *J. Biol. Chem.* **243**:2654–2659 (1968).
60. G. A. Thompson, Jr., The biosynthesis of ether-containing phospholipids in the slug, *Arion ater*. II. The role of glyceryl ether lipids as plasmalogen precursors, *Biochemistry* **5**:1290–1296 (1966).
61. G. A. Thompson, Jr., The biosynthesis of ether-containing phospholipids in the slug, *Arion ater*. III. Origin of the vinylic ether bond of plasmalogens, *Biochim. Biophys. Acta.* **152**:409–411 (1968).
62. L. A. Horrocks and G. B. Ansell, The incorporation of ethanolamine into ether-containing lipids in rat brain, *Lipids* **2**:329–333 (1967).
63. R. Bickerstaffe and J. F. Mead, Metabolism of palmitaldehyde-1-$^{14}$C in the rat brain, *Biochemistry* **6**:655–662 (1967).
64. E. O. Oswald, C. E. Anderson, C. Piantadosi, and J. Lim, Metabolism of alkyl glyceryl ethers in the rat, *Lipids* **3**:51–58 (1968).
65. B. W. Agranoff, R. M. Bradley, and R. O. Brady, The enzymatic synthesis of inositol phosphatide, *J. Biol. Chem.* **233**:1077–1083 (1958).
66. W. Thompson, K. P. Strickland, and R. J. Rossiter, Biosynthesis of phosphatidylinositol in rat brain, *Biochem. J.* **87**:136–142 (1963).
67. H. Paulus and E. P. Kennedy, The enzymatic synthesis of inositol monophosphatide, *J. Biol. Chem.* **235**:1303–1311 (1960).
68. G. L. Petzold and B. W. Agranoff, The biosynthesis of cytidine diphosphate diglyceride by embryonic chick brain, *J. Biol. Chem.* **242**:1187–1191 (1967).
69. M. A. Wells and J. C. Dittmer, The identification of glycerophosphorylglycerol phosphate as the deacylation product of a new brain lipid, *J. Biol. Chem.* **241**:2103–2105 (1966).
70. J. Y. Kiyasu, R. A. Pieringer, H. Paulus, and E. P. Kennedy, The biosynthesis of phosphatidyl glycerol, *J. Biol. Chem.* **238**:2293–2298 (1963).
71. J. Lecocq and C. E. Ballou, On the structure of cardiolipin, *Biochemistry* **3**:979–980 (1964).
72. F. Possmayer, G. Balakrishnan, and K. P. Strickland, The incorporation of labelled α-glycerophosphate into the lipids of rat brain preparations. III. On the biosynthesis of phosphatidyl glycerol, *Biochim. Biophys. Acta*, in press.
73. L. L. M. Van Deenen, *in Metabolism and Physiological Significance of Lipids* (R. M. C. Dawson and D. N. Rhodes, eds.), pp. 155–178, Wiley, London and New York (1964).
74. L. L. M. Van Deenen and G. H. Haas, Phosphoglycerides and phospholipases, *Ann. Rev. Biochem.* **35**:157–194 (1966).
75. R. H. S. Thompson, *in Metabolism and Physiological Significance of Lipids* (R. M. C. Dawson and D. N. Rhodes, eds.), pp. 541–551, Wiley, London and New York (1964).
76. G. R. Webster, Some aspects of lipid metabolism in nervous tissue, *Intern. Rev. Neurobiol.* **3**:293–317 (1961).
77. J. Gallai-Hatchard, W. L. Magee, R. H. S. Thompson, and G. R. Webster, The formation of lysophosphatides from di-acyl phosphatides by brain preparations, *J. Neurochem.* **9**:545–554 (1962).

78. S. Gatt, Y. Barenholz, and A. Roitman, Isolation of rat brain lecithinase-A, specific for the $\alpha^1$-position of lecithin, *Biochem. Biophys. Res. Commun.* **24**:169–172 (1966).
79. S. Gatt, Purification and properties of phospholipase $A_1$ from rat and calf brain, *Biochim. Biophys. Acta* **159**:304–316 (1968).
80. A. F. Robertson and W. E. M. Lands, Positional specificities in phospholipid hydrolyses, *Biochemistry* **1**:804–810 (1962).
81. E. A. Marples and R. H. S. Thompson, The distribution of phospholipase B in mammalian tissues, *Biochem. J.* **74**:123–127 (1960).
82. G. R. Webster, E. A. Marples, and R. H. S. Thompson, Glycerylphosphorylcholine diesterase activity of nervous tissue, *Biochem. J.* **65**:374–377 (1957).
83. K. P. Strickland, R. H. S. Thompson, and G. R. Webster, The Action of the phosphatases of human brain on lipid phosphate esters, *J. Neurol. Neurosurg. Psychiat.* **19**:12–16 (1956).
84. K. P. Strickland, R. H. S. Thompson, and G. R. Webster, Hydrolysis of phosphoryl choline and related esters by the phosphomonoesterases of animal tissues, *Arch. Biochem. Biophys.* **64**:498–505 (1956).
85. G. B. Ansell and S. Spanner, The catabolism of ethanolamine-containing phospholipids in rat-brain preparations, *Biochem. J.* **98**:23P (1966).
86. G. B. Ansell and S. Spanner, Plasmalogenase activity in normal and demyelinating tissue of the central nervous system, *Biochem. J.* **108**:207–209 (1968).
87. A. N. Davison, *in Metabolism and Physiological Significance of Lipids* (R. M. C. Dawson and D. N. Rhodes, eds.), pp. 527–537, Wiley, London and New York (1964).
88. M. L. Cuzner, A. N. Davison, and N. A. Gregson, Turnover of brain mitochondrial lipids, *Biochem. J.* **101**:618–626 (1966).
89. M. L. Cuzner, A. N. Davison, and N. A. Gregson, Chemical and metabolic studies of rat myelin of the central nervous system, *Ann. N.Y. Acad. Sci.* **122**:86–94 (1965).

*Chapter 17*

# METABOLISM OF PHOSPHOINOSITIDES

**J. N. Hawthorne and M. Kai**

*Department of Biochemistry*
*University of Birmingham, England*

## I. INTRODUCTION

Present interest in the inositol lipids of brain began with the isolation of the diphosphoinositide fraction by Folch[1] and the observation by Dawson[2] that $^{32}P$ was rapidly incorporated into these lipids *in vitro*. More recent work has shown that there are three brain phosphoinositides, phosphatidylinositol or monophosphoinositide (I), diphosphoinositide (II), and triphosphoinositide (III). The chemistry and metabolism of these compounds was reviewed in some detail by Hawthorne and Kemp[3] in 1964, so this chapter concentrates on work that has appeared since then, giving brief summaries of earlier studies only where necessary to complete the account.

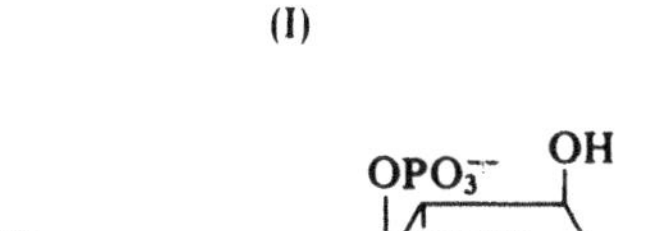

(I)

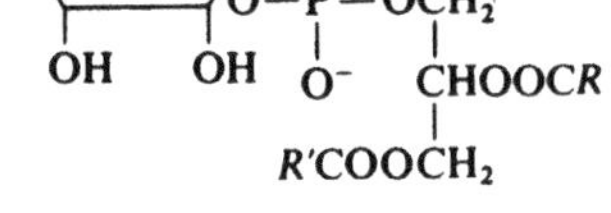

(II)

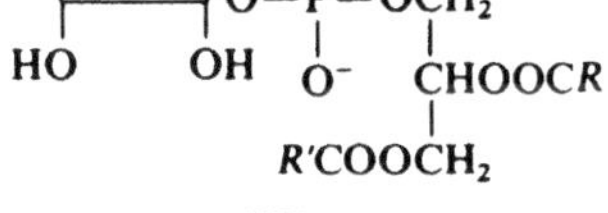

(III)

TABLE I

**Phosphoinositide Levels in Brain**

| | Phosphatidylinositol | Diphosphoinositide | Triphosphoinositide | Reference |
|---|---|---|---|---|
| Micrograms phosphoinositide P per gram fresh brain | | | | |
| Guinea pig[a] | | 11.0 | 54.4 | (57) |
| Rat[a] | | 12.2 | 32.5 | (57) |
| Ox[b] | | 2.8 | 57.2 | (57) |
| Guinea pig | 56.8 | | | (26) |
| Young rat[c] | 55.3 | | | (26) |
| Cat[d] | 36.0 | 3.8 | 25.2 | (65) |
| Rat | 108.0, 93.0 | | | (26) |
| Phosphoinositide P as percent total lipid P | | | | |
| Guinea pig[e] | 3.0 | 0.58 | 2.58 | (57,26) |
| Cat | 1.8 | 0.2 | 1.3 | (65) |

[a] Thiopentone anesthesia, brain fixed *in situ* with liquid $N_2$.
[b] Carried to a laboratory on ice, 45 min *post mortem.*
[c] Three weeks old.
[d] Cerebral hemispheres.
[e] Calculated using a figure of 0.19% fresh weight for total lipid P.

Table I gives concentrations of the three phosphoinositides in brains of different mammals.

## II. BIOSYNTHESIS

### A. Phosphatidylinositol

Myo-Inositol itself can be formed from glucose in rat brain slices,[4] probably by cyclization of glucose 6-phosphate to give D-inositol 1-phosphate as reported for yeast[5] and rat testis.[6] It might be expected that this inositol phosphate would be an intermediate in the biosynthesis of the phosphoinositides, by analogy with phosphoryl choline in the biosynthesis of lecithin. The phosphoinositides, however, have the L-inositol 1-phosphate structure, and the biosynthesis of phosphatidylinositol involves free inositol rather than the phosphate:[7,8]

$$\text{CTP} + \text{phosphatidic acid} \rightarrow \text{CDP-diglyceride} + \text{pyrophosphate} \quad (1)$$

$$\text{CDP-diglyceride} + \text{inositol} \rightarrow \text{phosphatidylinositol} + \text{CMP} \quad (2)$$

The enzymic phosphorylation of inositol has not been convincingly demonstrated, though there is a brief report claiming that it takes place in several tissues.[9] This work does not seem to have been confirmed.

Thompson, Strickland, and Rossiter[10] have shown that the biosynthesis of phosphatidylinositol in rat brain follows reactions (1) and (2). The incorporation of [$^{32}$P]-glycerophosphate into this lipid was stimulated as effectively by CDP-choline and CDP-glycerol as by CTP.[11] The labeling of phosphatidic acid was reduced by CTP in these experiments, the specific activity of this lipid falling to values well below that of phosphatidylinositol. However, the results are no longer considered to indicate an alternative biosynthetic route to phosphatidylinositol,[11] and there is evidence that CDP-glycerol and CDP-inositol are not involved in the synthesis.

The precursor phosphatidic acid can be formed in rat brain by either of the following reactions:[12]

L-1,2-diglyceride + ATP → phosphatidic acid + ADP (3)

L-3-glycerophosphate + 2 acyl CoA → phosphatidic acid + 2 CoA (4)

Experiments of Pieringer and Hokin[13] indicate another synthetic routes but its relative importance is unknown [reactions (5) and (6)].

Monoglyceride + ATP → lysophosphatidic acid + ADP (5)

Lysophosphatidic acid + acyl CoA → phosphatidic acid + CoA (6)

Finally, phosphatidylinositol can be formed by reaction (7) in homogenate of guinea pig brain.[14]

Lysophosphatidylinositol + oleoyl CoA → phosphatidylinositol + CoA (7)

The plasmalogen form of phosphatidylinositol occurs in brain.[15]

## B. Diphosphoinositide

By 1961, work in several laboratories[3] had shown that $^{32}$P was incorporated much more rapidly into the monoesterified phosphates of diphosphoinositide and triphosphoinositide than into the diesterified 1-phosphate. Isotope studies suggested that triphosphoinositide was formed by a two-step phosphorylation of phosphatidylinositol:[16]

Phosphatidylinositol → diphosphoinositide → triphosphoinositide (8)

More direct evidence indicating that the synthesis of diphosphoinositide followed reaction (9) was presented by Michell and Hawthorne[17] for rat liver and by Colodzin and Kennedy[18] and Kai, White, and Hawthorne[19]

Phosphatidylinositol + ATP → diphosphoinositide + ADP (9)

for rat brain. The enzyme concerned, phosphatidylinositol kinase, has a subcellular distribution similar to that of 5′-nucleotidase in liver and brain. The distribution also resembles that of $Na^+/K^+$-stimulated ATPase in brain. The kinase appears to be associated with the plasma membrane of a number of types of cells and with the outer membrane of the synaptosome.[19a] The brain enzyme required $Mg^{2+}$ ions for activity and was inhibited by $Ca^{2+}$ ions in the presence or absence of $Mg^{2+}$. Certain detergents activated phosphatidylinositol kinase. The reaction rate decreased after a short time, probably because a phosphomonoesterase converting diphosphoinositide to phosphatidylinositol was present.

The kinase is widely distributed among mammalian tissues, having been detected in brain, liver, erythrocytes, kidney, lung, heart, and spleen. This raises some interesting questions, since diphosphoinositide is not usually considered to occur in appreciable amounts outside the nervous system. A variety of analytical figures can be found in the literature, probably because of the difficulty in obtaining complete extraction of diphosphoinositide. Seiffert and Agranoff[20] hydrolyzed trichloroacetic acid-treated tissues with 5 N HCl and obtained inositol di- and tri-phosphates in each case. The triphosphate occurred in much higher concentrations in brain than in the other tissues, but the diphosphate figures were as follows (micromoles per gram wet weight of tissue): brain, 1.40; liver, 1.46; kidney, 2.24; heart, 1.63; and lung, 1.45. It is not clear whether all the inositol phosphates found by this method originated from phosphoinositides. The topic cannot be pursued here, but the present discrepancies between different tissue polyphosphoinositide determinations need to be resolved by better analytical methods.

## C. Triphosphoinositide

The kinase responsible for triphosphoinositide synthesis has been studied by Kai, Salway, and Hawthorne.[21] Like phosphatidylinositol kinase, it requires ATP and $Mg^{2+}$ ions [reaction (10)], but in other ways the enzyme is quite different. It is a soluble enzyme, for instance and can be

$$\text{Diphosphoinositide} + \text{ATP} \rightarrow \text{triphosphoinositide} + \text{ADP} \qquad (10)$$

purified by ammonium sulfate fractionation, treatment with ethanol at $-15°$, and chromatography on Sephadex G200.[21] Detergents do not stimulate the kinase, nor do thiol groups appear necessary for activity. Both phosphatidylinositol kinase and diphosphinosoitide kinase are inhibited by $Ca^{2+}$ ions at optimum $Mg^{2+}$ concentration, but only the latter enzyme can be partly activated by $Ca^{2+}$ ions in the absence of $Mg^{2+}$ Neither of the kinases is activated by $Na^+$ or $K^+$ ions.

Triphosphoinositide appears more characteristic of the nervous system than diphosphoinositide. As yet there is little information on the synthesis of triphosphoinositide in other tissues, though erythrocyte ghosts incubated with [$\gamma$-$^{32}$P]ATP incorporated radioactivity into both diphosphoinositide and triphosphoinositide.[22]

Triphosphoinositide is usually considered a myelin component.[23–25] However, incorporation of inorganic $^{32}P$ *in vivo* into triphosphoinositide was not limited to the myelin fraction in young rats.[26] Moreover, studies *in vitro* showed that the kinase responsible for synthesis occurs mainly in the supernatant fraction rather than the myelin.[(21)] The supernatant fraction also contains triphosphoinositide phosphomonoesterase.[(27)] It seems likely, therefore, that the metabolism of triphosphoinositide *in vivo* takes place at the surface of the myelin sheath or in the axolemma.

In this connection, it is interesting that the concentrations in rat brain of both the kinase and the monoesterase concerned with this lipid increase markedly during the period of active myelination (Salway *et al.*[(28)]). Wells and Dittmer[(29)] also showed that polyphosphoinositides accumulate rapidly during the period of myelination.

## III. CATABOLISM

### A. Phosphatidylinositol

The breakdown of phosphatidylinositol takes place largely by the phospholipase C route [reaction (11)] described by Kemp, Hübscher, and Hawthorne[(30)] in rat liver. The enzyme responsible requires $Ca^{2+}$ ions for

$$\text{Phosphatidylinositol} + H_2O \rightarrow \text{L-inositol 1-phosphate} + \text{L-1,2-diglyceride} \quad (11)$$

activity and is specific for inositol lipids, diphosphoinositide and triphosphoinositide also being hydrolyzed with release of diglyceride. Similar enzymes have been reported in guinea pig intestinal mucosa,[(31)] in rat brain,[(32)] and in guinea pig brain.[(33)] This phospholipase C activity partly accounts for the rapid incorporation of phosphate into phosphatidylinositol in several tissues.

### B. Diphosphoinositide and Triphosphoinositide

Thompson and Dawson[(34–36)] have purified extracts of brain acetone powders that catalyze the following reactions:

$$\text{Triphosphoinositide} + H_2O \rightarrow \text{diphosphoinositide} + \text{inorganic phosphate} \quad (12)$$

$$\text{Diphosphoinositide} + H_2O \rightarrow \text{phosphatidylinositol} + \text{inorganic phosphate} \quad (13)$$

$$\text{Triphosphoinositide} + H_2O \rightarrow \text{diglyceride} + \text{inositol triphosphate} \quad (14)$$

$$\text{Diphosphoinositide} + H_2O \rightarrow \text{diglyceride} + \text{inositol diphosphate} \quad (15)$$

The diesterase activity [reactions (14) and (15)] differs from that described by Thompson[(32)] in that there was no absolute requirement for calcium ions.

Salway *et al.*[(27)] recently developed a specific assay method for triphosphoinositide phosphomonoesterase [reaction (12)] and applied it to subcellular fractions from rat brain. The enzyme was found mainly in the supernatant fraction, myelin and synaptosomal fractions being relatively inactive. Unlike the purified acetone powder extract of Dawson and Thompson,[36] the enzyme from the dialyzed supernatant fraction was inhibited by $Na^+$ and $K^+$ ions.

These phosphatases have been studied further by Hauser *et al.*[37] Chang and Ballou[38] recently have shown that the phosphomonoesterase specifically cleaves the phosphate group at the 5-position of the inositol ring in triphosphoinositide, giving diphosphoinositide with the expected 1,4 structure. This has been confirmed by Prottey, Salway, and Hawthorne.[38a]

## C. Other Catabolic Enzymes

The origin of the lysophosphatidylinositol identified in brain[14] is unknown, but it is likely that suitable enzymes of the phospholipase A type could attack phosphatidylinositol, given the right physico-chemical conditions.

Dawson[39] obtained a lysophospholipase preparation from *Penicillium notatum* that removed both fatty acids from phosphatidylinositol. The same type of activity is found in liver.[30]

The manganese ion-activated exchange reaction by which inositol is incorporated into phosphatidylinositol[8] may be due to the reversal of a phospholipase D reaction [reaction (16)] but this is not yet clear. There has

$$\text{Phosphatidylinositol} + H_2O \rightarrow \text{phosphatidic acid} + \text{inositol} \qquad (16)$$

been no convincing demonstration of this type of enzyme. Table II lists activities of some enzymes concerned with phosphoinositide metabolism in rat brain.

**TABLE II**

**Enzymes of Phosphoinositide Metabolism in Rat Brain**

| Enzyme | Rate,[a] millimicromoles substrate converted per hour per milligram protein | Reference |
|---|---|---|
| CDP-diglyceride inositol phosphatidate transferase | 24.9 | (28) |
| Phosphatidylinositol kinase[b] | 3.0, 8.9 | (19, 28) |
| Diphosphoinositide kinase | 16.8, 10.2 | (21, 28) |
| Triphosphoinositide phosphomonoesterase | 2180 | (27) |

[a] All figures are based on homogenates of whole brain.
[b] The higher figure for the activity of this kinase was obtained in the presence of detergent.

# IV. PHOSPHOINOSITIDE METABOLISM AND NERVE FUNCTION

## A. Introduction

The rapidity with which $^{32}P$ is incorporated into the inositol lipids of brain and the fact that acetylcholine can stimulate this incorporation in the case of phosphatidylinositol suggest that the metabolism of these lipids plays an important part in nerve function. Since all are acidic lipids found in the lipoprotein membranes of the cell, one suggestion is that they are able to control the movement of cations. It seems clear that they do not act directly as carriers in the "sodium pump" of any tissues studied so far, but they certainly have some connection with divalent ions such as $Ca^{2+}$.[(3)] The precise function of the phosphoinositides in nerve is still far from clear, but there are a number of interesting experimental observations that may lead to a solution. These are summarized in the following section.

## B. Effects of Acetylcholine on Phospholipid Metabolism

### 1. *Phosphatidylinositol and Phosphatidic Acid*

The pioneer work of the Hokins,[(40)] showing that acetylcholine in the presence of eserine stimulated the incorporation of $^{32}P$ into the phosphatidic acid and phosphatidylinositol of brain slices, has been considered in detail elsewhere.[(3,41)] The effect on phosphatidic acid, but not on phosphatidylinositol, was also obtained in a "microsomal" fraction of brain. It was considered that only synthesis by the diglyceride kinase route was affected[(42)] [Section II,A, reaction (3)] and not the acylation of glycerophosphate [Section II,A, reaction (4)]. The acetylcholine effect was abolished by freezing, which suggested that it depended on intact subcellular structures. Durell and Sodd[(43)] showed that the chief site of the effect was the synaptosomal fraction. Though [$^{32}P$]-ATP generated in the medium could label the phosphatidic acid, this labeling was not increased by acetylcholine. The results suggested that the chief action of acetylcholine was to stimulate the incorporation of [$^{32}P$]-orthophosphate from the medium into the phosphatidic acid of the synaptosomal fraction. A number of mechanisms are possible, assuming that the diglyceride kinase pathway operates. Acetylcholine might increase the permeability of the synaptosomal membrane to phosphate or stimulate the production of ATP by the mitochondria of the nerve endings or directly stimulate diglyceride kinase. These and other possibilities call for further experimental work.

Redman and Hokin[(44)] studied the incorporation of inorganic $^{32}P$ into the phosphatidic acid and phosphatidylinositol of a cell-free homogenate of brain. Labeling of phosphatidylinositol was more readily stimulated by cholinergic agents than labeling of phosphatidic acid. The latter lipid was unaffected by $10^{-5}$ M carbamyl choline, a drug concentration that stimulated phosphatidylinositol labeling by about 40%. Hokin and Hokin,[(45)] working with goose salt gland slices, showed that labeling of both the lipids was stimulated by acetylcholine. A subsequent addition of atropine, the specific acetylcholine antagonist, increased the incorporation of $^{32}P$ into phosphatidylinositol even further, but at the same time the labeling of phosphatidic acid decreased. It was suggested that in the resting state phosphatidylinositol was present at the membrane sites involved and that on stimulation it was converted to diglyceride (phospholipase C), which in turn was phosphorylated to phosphatidic acid (diglyceride kinase). The phosphatidic acid at the active site would be converted to phosphatidylinositol on removal of the stimulus (e.g., by atropine). It is not known whether such a theory could apply to brain or peripheral ganglia,[(46)] but there is no doubt that phosphatidylinositol is involved in the membrane events initiated by acetylcholine.

### 2. *Diphosphoinositide and Triphosphoinositide*

When tissue slices or subcellular fractions are incubated with [$^{32}P$]-orthophosphate the bulk of the radioactivity incorporated into the polyphosphoinositides appears in the monoesterified 4- and 5-phosphates. Palmer and Rossiter,[(47)] using cat brain slices, showed that under condition in which acetylcholine stimulated the incorporation of [$^{32}P$]-orthophosphate into phosphatidylinositol, the labeling of the polyphosphoinositides was unaffected. Similar results were obtained with slices of cat superior cervical ganglia,[(48)] the results being expressed in counts per minute per microgram total lipid P. On the other hand, formation of both diphosphoinositide and triphosphoinositide from labeled ATP in brain homogenates was stimulated by acetylcholine.[(49)] The diphosphoinositide kinase of brain supernatant fraction was not affected.[(21)]

In summary, acetylcholine has a definite effect on the incorporation of orthophosphate into phosphatidylinositol, but no such effect has been seen in the case of the polyphosphoinositides. The observation with labeled ATP[(49)] remains to be explained.

## C. Effects of Other Drugs on Brain Phosphoinositide Metabolism

Ansell[(50)] has reviewed the effects on phospholipid metabolism of various drugs that stimulate or depress nervous activity. Such effects may be produced by a number of mechanisms, and we know little of the exact metabolic changes associated with drug action. Indirect effects on phospholipid metabolism could be produced by changes in brain acetylcholine concentration, for instance. Not all workers seem to realize that the drug concentration used *in vitro* should be similar to that resulting from a therapeutic dose given to a patient. Effects obtained at much higher concentrations are only relevant to a toxicologist.

Chlorpromazine stimulated the uptake of phosphate into the phosphatidylinositol of brain slices and brain mitochondria[(51)] but did not affect the labeling of the polyphosphoinositides.[(47)] The labeling of phosphatidic acid and phosphatidylserine also increased but that of phosphatidylcholine and phosphatidylethanolamine was reduced. Ansell and Dohmen[52] showed that incorporation of $^{32}P$ into a phosphoinositide (probably diphosphoinositide) was decreased by 80% during thiopentenal narcosis. Labeling of other phospholipids also was reduced. Morphine[53] at high concentration ($10^{-2}$ M) increased incorporation of inositol and phosphate into phosphatidylinositol and diphosphoinositide of brain slices, but there was less effect on triphosphoinositide.

## D. Phosphatidylinositol and Sympathetic Ganglia

The elegant studies of Larrabee and his colleagues[54,55] have shown that electrical stimulation of sympathetic ganglia caused an increased incorporation of [$^{32}P$]-orthophosphate into the phosphatidylinositol of the ganglia. Phosphatidic acid was unaffected. Acetylcholine, the transmitter substance in such ganglia, had a similar effect on slices of the tissue,[48] but in this case phosphatidic acid labeling also was increased. Larrabee's work with sympathetic ganglia has the advantages that the ganglia are less complex and anatomically better defined than brain slices and that they can be stimulated under conditions approaching the physiological while action potentials are recorded from the postganglionic nerve.

The phosphatidylinositol effect is postsynaptic, being abolished by suitable concentrations of D-tubocurarine[55] and unaffected by degeneration of presynaptic nerve endings.[48] Autoradiographic studies with [$^3H$]-inositol suggested that the phosphatidylinositol effect was associated with the cytoplasm of the neurons rather than the synaptic membranes.[46] No significant effect of conducted impulses on the labeling of the inositol lipid was found in vagus, sympathetic, or phrenic nerve trunks or in the nonsynaptic ganglion nodosum of the vagus nerve.[55]

Reductions in the calcium ion content of the fluid bathing the ganglia decreased the $^{32}P$ incorporation into phosphatidylcholine, phosphatidylethanolamine, and phosphatidic acid, but not that into phosphatidylinositol.[54] Labeling of the latter compound by $^{14}C$ from [$^{14}C$]-glucose was increased by withdrawal of calcium, though the incorporation of $^{14}C$ into the choline and ethanolamine compounds was not significantly altered. It was suggested that reduction in calcium ion concentration stimulated phosphatidylinositol synthesis, since lack of calcium increases neuronal activity, but that at the same time there was a decreased synthesis of phosphate-containing precursors.

The significance of the phosphatidylinositol effect in ganglionic transmission remains uncertain. It is unlikely to be related to sodium transport. The effect has been compared with that seen on stimulation of secretion in the pancreas. Hokin[46] has suggested that it may be connected with protein transport within the neuron or with an effect of depolarization on intracellular membranes. These rather vague suggestions underline our ignorance of the molecular events associated with ganglionic transmission.

## E. Phosphoinositides and Sodium Transport

### 1. *Introductory*

The most recent evidence indicates that a phosphorylated protein may be the carrier molecule in the "sodium pump." Such a protein is certainly associated with the sodium-dependent ATPase considered part of the transport mechanism. The theory that acidic lipids may be carriers for cations such as $Na^+$ and $K^+$ therefore has been discarded by many workers. It is still possible, however, that phospholipids may be indirectly concerned with the transport of cations. At the very least, they are important structural components of the membranes involved.

The work outlined in this section shows that changes in the concentration of $Na^+$ and $K^+$ ions in brain tissue can affect phosphoinositide metabolism, but none of it suggests a direct connection between these lipids and the transport of the ions.

### 2. *Effect of Electrical Stimulation*

Electrical stimulation is known to cause a loss of potassium ions from brain slices and an influx of sodium ions. After electrically induced convulsions the incorporation of $^{32}P$ *in vivo* into a compound resembling triphosphoinositide was substantially decreased.[(56)] The identification of the inositol lipids in this work was only tentative, however. Brief electrical stimulation under similar conditions did not affect concentrations of polyphosphoinositides in rat brain.[(57)]

### 3. *Effects of $Na^+$ and $K^+$ Ion Concentrations on Phosphoinositides in Brain Slices*

Palmer and Rossiter[(47)] incubated cat brain slices in the presence of [$^{32}P$]-orthophosphate. Increasing the $K^+$ ion concentration of the medium from 5.9 to 105 mM decreased the incorporation of $^{32}P$ into all three phosphoinositides by about 50%. About the same decrease was observed in the phospholipid fraction extracted by neutral chloroform-methanol (i.e., most of the other phospholipids). Hayashi *et al.*[(58)] made similar isotope studies with slices of guinea pig brain. On changing from incubation in choline-Ringer solution, which contained $K^+$ ions but no $Na^+$ ions, to normal Ringer with $Na^+$ ions, the specific activity of the total phospholipid fraction increased threefold. The percentage of the $^{32}P$ found in phosphatidic acid decreased, but that in the polyphosphoinositides increased.

Experiments with subcellular particles or purified enzymes have shown effects of $Na^+$ and $K^+$ ions on the activity of certain enzymes involved in phosphoinositide metabolism. Crude preparations of triphosphoinositide phosphomonoesterase were stimulated by sodium chloride and potassium chloride, though the purified enzyme was not activated in the presence of a *p*H 5.0 supernatant fraction.[(36)] The same enzyme prepared from rat brain without the use of acetone powders was slightly inhibited by KCl after dialysis and considerably inhibited by NaCl.[27] It is impossible to deduce from such results the effect of the monovalent ions *in vivo*, since other proteins and lipids in the cell obviously affect the phosphomonoesterase.

The diphosphoinositide kinase of rat brain homogenate was inhibited by sodium and potassium ions, and the inhibition was partly reversed by adding 0.5mM oubain.[49] The purified kinase was unaffected, however.[21] Phosphatidylinositol kinase was not much affected by KCl.[19]

### 4. *Phosphoinositides and the $Na^+/K^+$-Dependent ATPase*

There is good evidence that the sodium pump in several tissues is associated with ATP hydrolysis, the ATPase system being activated by external $K^+$ and internal $Na^+$ and $Mg^{2+}$. Oubain inhibits this ATPase. Glynn *et al.*[59] prepared the "transport ATPase" from electric eel and torpedo electric organs and exposed the enzyme complex to terminally labeled ATP in the presence of different $K^+$ and $Na^+$ ion concentrations. There was no labeling of phosphoinositides or phosphatidic acid. The authors concluded that phospholipids were not intermediates in the ATPase system. Hayashi *et al.*[60] incubated a guinea pig brain microsomal fraction that contained the transport ATPase, with similarly labeled ATP. In this case labeling of phosphatidic acid and the polyphosphoinositides took place, even in 30 sec at 0°. Sodium ions did not affect the phospholipid labeling, though that of the phosphoinositides was markedly depressed by adding 3 mM $CaCl_2$. The results do not suggest that polyphosphoinositides are intermediates in the transport system. In contrast to this microsomal preparation of Hayashi *et al.*, the diphosphoinositide kinase of rat brain supernatant was not active at 0°.[21] Standefer and Samson,[61] in a brief note, claimed a close relation between polyphospoinositide labeling and the $Na^+/K^+$stimulated ATPase of the brain microsomal fraction.

While it seems unlikely that the phosphoinositides are carriers in the transport system, it is quite possible that these and other phospholipids are necessary components. Work from several laboratories shows that inactivation follows removal of phospholipid from the system by detergents or phospholipases.

## F. Polyphosphoinositides and Divalent Cations

### 1. *Myelin and the Polyphosphoinositides*

There is good evidence that diphosphoinositide and triphosphoinositide are characteristic myelin components. Diphosphoinositide synthesis appears to be associated with plasma membrane in other tissues than nerve, as mentioned in Section II,B. It is interesting, therefore, that myelin itself is considered to be formed from plasma membrane.

Dawson and his colleagues have shown that the polyphosphoinositides are associated with myelin, emphasizing the rapid hydrolysis of these compounds *post mortem*.[57] For accurate analysis, therefore, brains must be frozen in, e.g., liquid nitrogen immediately after death. Much higher levels of the polyphosphoinositides were found in myelinated than in nonmyelinated nerves.[23] Results that indicate that their metabolism takes place in the myelin sheath or a closely related structure were given in Section II,C.

Amaducci *et al.*[62] studied the relation between fiber tract numbers and diameters and proteolipid concentration in different regions of the bovine nervous system. They also studied the phosphatido-peptide fraction that contains polyphosphoinositides and suggested that it might be found in a specific axonal membrane rather than as a universal constituent of the periodically repeating structure of the myelin sheath. This is a difficult area, since the phosphatido-peptide itself may be an artifact of the extraction process.

### 2. *Affinity of Polyphosphoinositides for Divalent Cations*

Early work indicating that divalent cations such as $Ca^{2+}$ and $Mg^{2+}$ are readily bound to the monoesterified phosphate groups of diphosphoinositide and triphosphoinositide were reviewed previously.[3] More recently Dawson[63] has shown that triphosphoinositide has a greater affinity than EDTA for calcium ions. In purified myelin there is a close equivalence between the number of polyphosphosphoinositide monoester phosphate groups and the total $Ca^{2+}$ and $Mg^{2+}$ content.[24]

Addition of these ions to the polyphosphoinositides entirely changes the physical properties of the lipids. Their salts with monovalent ions are relatively hydrophilic molecules readily dispersed in water. The calcium or magnesium salts, on the other hand, are insoluble in water and soluble in organic solvents.

### 3. *Divalent Ions and Polyphosphoinositide Metabolism*

Like most kinases, the enzymes responsible for diphosphoinositide and triphosphoinositide synthesis are dependent on magnesium ions. In the presence of optimum $Mg^{2+}$, calcium ions are inhibitory. The phosphomonoesterase is also activated by magnesium ions; calcium shows no inhibitory effect and, in fact, can activate if insufficient magnesium is present. On the whole, therefore, calcium depresses the synthesis of polyphosphoinositides.

### 4. *Calcium Ions and the Action Potential*

It is generally considered that the presence of calcium ions makes the neuronal membrane less permeable. Several authors have suggested that calcium ions block the "pores" in the axolemma through which sodium ions pass on depolarization. Hodgkin and Keynes[64] showed that calcium ions enter the axoplasm of the squid giant axon during stimulation and suggested that their discharge from the membrane made it more permeable to sodium ions.

## 5. *Reaction Rates*

Table III from the paper of Kai *et al.*[21] compares the rates of synthesis and breakdown of triphosphoinositide with corresponding figures for acetylcholine. It is striking that rates are of the same order in each case and that hydrolysis is more than 100 times as rapid as synthesis. There is no evidence that triphosphoinositide can function as a transmitter like acetylcholine in nerve, but the enormous potential for its destruction is likely to be of physiological significance.

**TABLE III**

**Synthesis and Breakdown of Brain Constituents**

| Enzyme | Rate, millimicromoles converted per gram fresh brain per minute | Reference |
|---|---|---|
| Phosphatidylinositol kinase | 6 | (19) |
| Diphosphoinositide kinase | 26 | (21) |
| Triphosphoinositide phosphomonoesterase | 4000 | (27) |
| Choline acetyltransferase | 50 | Quoted in (21) |
| Acetylcholinesterase | 6000 | Quoted in (21) |

## 6. *Polyphosphoinositides and the Permeability of the Axonal Membrane*

With the facts above in mind, a theory can be presented linking polyphosphoinositide metabolism with the calcium content and permeability of the axonal membrane. A region of the membrane rich in diphosphoinositide (triphosphoinositide would be equally suitable but would give a more confusing diagram) is presented schematically in Fig. 1. Removal of the 4-phosphate groups and their associated calcium ions by the action of the phosphomonoesterase would increase the permeability of this area of the membrane. A tentative scheme linking this permeability change with the action potential and associated events is given in Fig. 2.

At rest the axolemma would contain calcium ions bound to polyphosphoinositide and would be impermeable to sodium ions. Depolarization would allow the triphosphoinositide phosphomonoesterase of the axoplasm to attack membrane polyphosphoinositide, as in Fig. 1. The removal of the charge from the membrane in itself might be sufficient to initiate enzymic hydrolysis. As a result the membrane would become permeable to sodium ions. During the recovery process (Fig. 2, Stage 4) triphosphoinositide would be resynthesized in the membrane, which would now contain little calcium so that the relevant kinases could act at its surface.

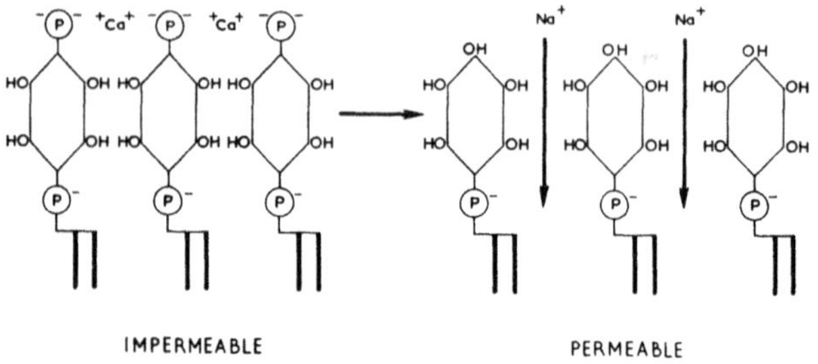


Fig. 1. Possible effect on membrane permeability of the removal of the 4-phosphate of diphosphoinositide by the phosphomonoesterase.

At the same time the localized high sodium content of the axoplasm would inactivate triphosphoinositide phosphomonoesterase.[27] It is of interest that sodium ions promote labeling of polyphosphoinositides by $^{32}P$ in brain slices,[58] as would be expected at Stage 4. The theory also predicts that active extrusion of sodium ions will accompany the resynthesis of triphosphoinositide, though the two processes need not be related. It assumes, of course, that even the so-called nonmyelinated nerves possess a membrane containing polyphosphoinositides. In fact, the bulk of such lipid in a myelinated nerve may be inactive, metabolic activity being restricted to a surface layer or the axolemma alone.

Though the hypothesis of Fig. 2 fits many of the facts, there is as yet no direct evidence for it. As predicted by it, the labelling of triphosphoinositide is more rapid in nerve trunks than in ganglia.[66]

| | | |
|---|---|---|
| 1. REST | CALCIUM TPI: MEMBRANE IMPERMEABLE | — |
| 2. DEPOLARIZATION | TPI → DPI → PI | Monoesterase active; kinase inactive ($Ca^{++}$) |
| 3. CALCIUM LOST SODIUM ENTERS | MEMBRANE PERMEABLE | — |
| 4. RECOVERY: CALCIUM REGAINED | PI → DPI → TPI | Kinases active (low $Ca^{++}$); monoesterase inactive ($Na^{+}$); sodium pump active |
| 1. REST | CALCIUM TPI: MEMBRANE IMPERMEABLE | — |

Fig. 2. Suggested events in the axolemma that accompany the action potential. TPI, triphosphoinositide; DPI, diphosphoinositide; PI, phosphatidylinositol.

## G. Phosphoinositide Metabolism in Hereditary Ataxia

Hereditary ataxia in rabbits progressively involves lateral brainstem and deep cerebellar nuclei. Glycogen deposits have been noted in the areas of the brain involved. More recently[67] it has been claimed that the brainstem of ataxic rabbits contains less of all three phosphoinositides than normal brainstem. Incorporation of labeled inositol into the inositides of brainstem slices was reduced in the affected animals in comparison with control slices from normal rabbits. This interesting work needs to be extended by the direct analysis of the individual inositol lipids and the determination of suitable enzyme activities in the brain areas involved.

## H. Conclusions

The importance of phosphoinositide metabolism in nervous tissue is well established. There are two areas of interest at present. First, phosphatidylinositol is involved in the post synaptic events accompanying transmission of impulses by sympathetic ganglia. Second, the polyphosphoinositides appear important in the axonal membrane, possibly because of their affinity for divalent cations. Our understanding of the molecular events underlying excitation and transmission in nervous tissue will improve as we learn more about these inositol lipids.

# V. REFERENCES


1. J. Folch, Complete fractionation of brain cephalin: Isolation from it of phosphatidyl serine, phosphatidyl ethanolamine and diphosphoinositide, *J. Biol. Chem.* **177**:497–504 (1949).
2. R. M. C. Dawson, The measurement of $^{32}P$ labelling of individual kephalins and lecithin in a small sample of tissue, *Biochim. Biophys. Acta* **14**:374–379 (1954).
3. J. N. Hawthorne and P. Kemp, The brain phosphoinositides, *Advan. Lipid Res.* **2**: 127–166 (1964).
4. G. Hauser and V. N. Finelli, Biosynthesis of free and phosphatide myo-inositol from glucose by mammalian tissue slices, *J. Biol. Chem.* **238**:3224–3228 (1963).
5. J. W. Chen and F. C. Charalampous, D-Inositol 1-phosphate as intermediate in the biosynthesis of inositol from glucose 6-phosphate, *J. Biol. Chem.* **241**:2194–2199 (1966).
6. F. Eisenberg, Jr., and G. H. Bolden, D-myo-Inositol 1-phosphate, an intermediate in the biosynthesis of inositol in the mammal, *Biochem. Biophys. Res. Commun.* **21**: 100–105 (1965).
7. B. W. Agranoff, R. M. Bradley, and R. V. Brady, The enzymatic synthesis of inositol phosphatide, *J. Biol. Chem.* **233**:1077–1083 (1958).
8. H. Paulus and E. P. Kennedy, The enzymatic synthesis of inositol monophosphatide, *J. Biol. Chem.* **235**:1303–1311 (1960).
9. M. Dietz and P. Albersheim, The enzymic phosphorylation of myo-inositol, *Biochem. Biophys. Res. Commun.* **19**:598–603 (1965).
10. W. Thompson, K. P. Strickland, and R. J. Rossiter, Biosynthesis of phosphatidylinositol in rat brain, *Biochem. J.* **87**:136–142 (1963).

11. F. Possmayer and K. P. Strickland, Incorporation of α-glycerophosphate-$^{32}$P into the lipids of rat brain preparations, *Can. J. Biochem.* **45**:53–69 (1967).
12. K. P. Strickland, Phosphorylation of diglycerides by rat brain, *Can. J. Biochem. Physiol.* **40**:247–259 (1962).
13. R. A. Pieringer and L. E. Hokin, Biosynthesis of phosphatidic acid from lysophosphatidic acid and palmityl CoA, *J. Biol. Chem.* **237**:659–663 (1962).
14. R. W. Keenan and L. E. Hokin, The identification of lysophosphatidylinositol and its enzymic conversion to phosphatidylinositol, *Biochim. Biophys. Acta* **60**:428–430 (1962).
15. D. E. Slagel, J. C. Dittmer, and C. B. Wilson, Lipid composition of human glial tumour and adjacent brain, *J. Neurochem.* **14**:789–798 (1967).
16. H. Brockerhoff and C. E. Ballou, Phosphate incorporation in brain phosphoinositides, *J. Biol. Chem.* **237**:49–52 (1962).
17. R. H. Michell and J. N. Hawthorne, The site of diphosphoinositide synthesis in rat liver, *Biochem. Biophys. Res. Commun.* **21**:333–338 (1965).
18. M. Colodzin and E. P. Kennedy, Biosynthesis of diphosphoinositide in brain, *J. Biol. Chem.* **240**:3771–3780 (1965).
19. M. Kai, G. L. White, and J. N. Hawthorne, The phosphatidylinositol kinase of rat brain, *Biochem. J.* **101**:328–337 (1966).
19a. J. L. Harwood and J. N. Hawthorne, Metabolism of phosphoinositides in guinea-pig brain synaptosomes, *J. Neurochem.* **16**:1377–1387 (1969).
20. U. B. Seiffert and B. W. Agranoff, Isolation and separation of inositol phosphates from hydrolysates of rat tissues, *Biochim. Biophys. Acta* **98**:574–581 (1965).
21. M. Kai, J. G. Salway, and J. N. Hawthorne, The diphosphoinositide kinase of rat brain, *Biochem. J.* **106**:791–801 (1968).
22. L. E. Hokin and M. R. Hokin, The incorporation of $^{32}$P from [γ-$^{32}$P]-adenosine triphosphate into polyphosphoinositides and phosphatidic acid in erythrocyte membranes, *Biochim. Biophys. Acta* **84**:563–575 (1964).
23. A. Sheltawy and R. M. C. Dawson, The polyphosphoinositides and other lipids of peripheral nerves, *Biochem. J.* **100**:12–18 (1966).
24. J. Eichberg and R. M. C. Dawson, Polyphosphoinositides in myelin, *Biochem. J.* **96**:644–650 (1965).
25. R. M. C. Dawson, *in Cyclitols and Phosphoinositides* (H. Kindl, ed.), pp. 57–67, Pergamon Press, Oxford (1966).
26. M. Kai and J. N. Hawthorne, Incorporation of injected [$^{32}$P]-phosphate into the phosphoinositides of subcellular fractions from young rat brain, *Biochem. J.* **98**:62–67 (1966).
27. J. G. Salway, M. Kai, and J. N. Hawthorne, Triphosphoinositide phosphomonoesterase activity in nerve cell bodies, neuroglia and subcellular fractions from whole rat brain, *J. Neurochem.* **14**:1013–1024 (1967).
28. J. G. Salway, J. L. Harwood, M. Kai, G. L. White, and J. N. Hawthorne, Enzymes of phosphoinositide metabolism during rat brain development, *J. Neurochem.* **15**:221–226 (1968).
29. M. A. Wells and J. C. Dittmer, A comprehensive study of the postnatal changes in the concentration of the lipids in developing rat brain, *Biochemistry* **6**:3169–3174 (1967).
30. P. Kemp, G. Hübscher, and J. N. Hawthorne, Enzymic hydrolysis of inositol-containing phospholipids, *Biochem. J.* **79**:193–200 (1961).
31. R. S. Atherton, P. Kemp, and J. N. Hawthorne, Phosphoinositide inositolphosphohydrolase in guinea-pig intestinal mucosa, *Biochim. Biophys. Acta* **125**:409–412 (1966).

32. W. Thompson, The hydrolysis of monophosphoinositide by extracts of brain, *Can. J. Biochem.* **45**:853–861 (1967).
33. R. O. Friedel, J. D. Brown, and J. Durell, Monophosphatidylinositol inositol-phosphohydrolase in guinea-pig brain, *Biochim. Biophys. Acta* **144**: 684–686 (1967).
34. W. Thompson and R. M. C. Dawson, The hydrolysis of triphosphoinositide by extracts of ox brain, *Biochem. J.* **91**:233–236 (1964).
35. W. Thompson and R. M. C. Dawson, The triphosphoinositide phosphodiesterase of brain tissue, *Biochem. J.* **91**:237–243 (1964).
36. R. M. C. Dawson and W. Thompson, The triphosphoinositide phosphomonoesterase of brain tissue, *Biochem. J.* **91**:244–250 (1964).
37. G. Hauser, J. Eichberg, S. M. Gompertz, and M. Ross, Studies on Triphosphoinositide Phosphohydrolases of Rat Brain, *Abstracts of the First International Meeting of the International Society for Neurochemistry*, Strasbourg, France (July 1967).
38. M. Chang and C. E. Ballou, Specificity of ox brain triphosphoinositide phosphomonoesterase, *Biochem. Biophys. Res. Commun.* **26**:199–205 (1967).
38a. C. Prottey, J. G. Salway, and J. N. Hawthorne, The structures of enzymically produced diphosphoinositide and triphosphoinositide, *Biochim. Biophys. Acta* **164**: 238–251 (1968).
39. R. M. C. Dawson, Enzymic hydrolysis of monophosphoinositide by phospholipase preparations from *P. notatum* and ox pancreas, *Biochim. Biophys. Acta* **33**:68–77 (1959).
40. L. E. Hokin and M. R. Hokin, Acetylcholine and the exchange of inositol and phosphate in brain phosphoinositide, *J. Biol. Chem.* **233**:818–821 (1958).
41. J. N. Hawthorne, The inositol phospholipids, *J. Lipid Res.* **1**:255–280 (1960).
42. L. E. Hokin and M. R. Hokin, The Mechanism of phosphate exchange in phosphatidic acid in response to acetylcholine, *J. Biol. Chem.* **234**:1387–1390 (1959).
43. J. Durell and M. A. Sodd, Studies on the acetylcholine-stimulated incorporation of radioactive inorganic orthophosphate into the phospholipid of brain particulate preparations, *J. Neurochem.* **13**:487–491 (1966).
44. C. M. Redman and L. E. Hokin, Stimulation of the metabolism of phosphatidylinositol and phosphatidic acid in brain cytoplasmic fractions by low concentrations of cholinergic agents, *J. Neurochem.* **11**:155–163 (1964).
45. M. R. Hokin and L. E. Hokin, *in Metabolism and Physiological Significance of Lipids* (R. M. C. Dawson and D. N. Rhodes, eds.), pp. 423–434, Wiley, New York (1964).
46. L. E. Hokin, Autoradiographic localization of the acetylcholine-stimulated synthesis of phosphatidylinositol in the superior cervical ganglion, *Proc. Natl. Acad. Sci., U.S.* **53**:1369–1376 (1965).
47. F. B. Palmer and R. J. Rossiter, A simple procedure for the study of inositol phosphatides in cat brain slices, *Can. J. Biochem.* **43**:671–683 (1965).
48. L. E. Hokin, Effects of acetylcholine on the incorporation of $^{32}P$ into various phospholipids in slices of normal and denervated superior cervical ganglia of the cat, *J. Neurochem.* **13**: 179–184 (1966).
49. M. Kai, J. G. Salway, R. H. Michell, and J. N. Hawthorne, The biosynthesis of triphosphoinositide by rat brain *in vitro*, *Biochem. Biophys. Res. Commun.* **22**:370–375 (1966).
50. G. B. Ansell, The action of drugs on phospholipid metabolism, *Advan. Lipid Res.* **3**:139–170 (1965).
51. G. B. Ansell and J. N. Hawthorne, *Phospholipids*, p. 356, Elsevier, Amsterdam (1964).
52. G. B. Ansell and H. Dohmen, The metabolism of individual phospholipids in the rat brain during hypoglycemia, anaesthesia and convulsions, *J. Neurochem.* **2**:1–10 (1957).

53. S. J. Mulé, Effect of morphine and nalorphine on the metabolism of phospholipid in guinea pig cerebral cortex slices, *J. Pharmacol. Exptl. Therap.* **154**:370–383 (1966).
54. M. G. Larrabee, J. D. Klingman, and W. S. Leicht, Effects of temperature, calcium and activity on phospholipid metabolism in a sympathetic ganglion, *J. Neurochem.* **10**:549–570 (1963).
55. M. G. Larrabee and W. S. Leicht, Metabolism of phosphatidylinositol and other lipids in active neurones of sympathetic ganglia and other peripheral nervous tissues: the site of the inositide effect, *J. Neurochem.* **12**:1–13 (1965).
56. K. Hayashi, T. Kanoch, S. Schimizer, M. Kai, and S. Yamazoe, The effect of electrically induced convulsions on the incorporation of $^{32}P$ into rabbit brain phospholipids, *J. Biochem., Tokyo* **51**:72–77 (1962).
57. R. M. C. Dawson and J. Eichberg, Diphosphoinositide and triphosphoinositide in animal tissues: extraction, estimation and changes *post-mortem*, *Biochem. J.* **96**: 634–643 (1965).
58. K. Hayashi, Y. Yagihara, I. Nakamura and S. Yamazoe, *Post-mortem* breakdown of phosphoinositides and phosphatidic acid and $^{32}P$ incorporation into phospholipids in various states of brain tissue, *J. Biochem., Tokyo* **60**:42–51 (1966).
59. J. M. Glynn, C. W. Hayman, J. Eichberg, and R. M. C. Dawson, The ATPase system responsible for cation transport in electric organ: exclusion of phospholipids as intermediates, *Biochem. J.* **94**:692–699 (1965).
60. K. Hayashi, Y. Yagihara, I. Nakamura, A. Katagiri, Y. Arakawa, and S. Yamazoe, Incorporation of $^{32}P$ from [$\gamma$-$^{32}P$]-ATP into polyphosphoinositides and phosphatidic acid in subcellular particles of guinea pig brain, *J. Biochem. Tokyo* **62**:15–20 (1967).
61. J. C. Standefer and F. E. Samson, Phosphoinositides related to the ($Na^+$, $K^+$) activated ATPase in rat brain, *Federation Proc.* **26**:765 (1967).
62. L. Amaducci, A. Pazzagli, and G. Pessina, The relation of proteolipids and phosphatidopeptides to tissue elements in the bovine nervous system, *J. Neurochem.* **9**:509–518 (1962).
63. R. M. C. Dawson, "Phosphatido-peptide"-like complexes formed by the interaction of calcium triphosphoinositide with protein, *Biochem. J.* **97**:134–138 (1965).
64. A. L. Hodgkin and R. D. Keynes, Movements of labelled calcium in squid giant axons, *J. Physiol.* **138**:253–281 (1957).
65. R. J. Rossiter and F. B. Palmer, Incorporation of $^{3}H$-myoinositol into the phosphoinositides of cat brain *in vivo*, *Biochem. Z.* **342**:337–344 (1965).
66. Y. Yagihara, J. G. Salway, and J. N. Hawthorne, Incorporation of $^{32}P$ *in vitro* into triphosphoinositide and related lipids of rat superior cervical ganglia and vagus nerves, *J. Neurochem.* **16**:1133–1139 (1969).
67. S. G. Eliasson, J. D. Scarpellini, and R. R. Fox, Inositide metabolism in rabbit hereditary ataxia, *A.M.A. Arch. Neurol.* **17**:661–665 (1967).

*Chapter 18*

# LIPID HAPTENS


## Maurice M. Rapport

*New York State Psychiatric Institute and*
*College of Physicians and Surgeons*
*Columbia University*
*New York, New York*


## I. INTRODUCTION

Although the detection of organ-specific antigens in nervous tissue and their recognition as lipids was a classical problem of immunology, relatively little information is available on this subject in modern immunological texts. In 1926 Brandt, Guth, and Müller[1] reported their observation that antisera prepared by injecting rabbits with a mixture of bovine brain lipid and pig serum reacted with alcoholic extracts of the brains of several animal species (man, beef, pig, and guinea pig). These observations were confirmed and extended by others.[2,3] In 1933 Rudy[4] reported that the immunologically active substance was associated with the cerebroside fraction of brain lipids but that it could not be cerebroside (galactosylceramide) itself. Subsequent events have shown that this conclusion was incorrect,[5] and this error, occurring when it did, probably delayed developments in this area for more than a generation.[6] The main reason for the failure to identify galactocerebroside as the immunologically active lipid may be traced to the auxiliary lipid phenomenon. These first observations, and most of the later ones as well, were based on the method of complement fixation. Although it was not fully appreciated in the early studies, it is now known that detection of the immunological activity of lipids by complement fixation requires the presence of auxiliary lipids such as lecithin and cholesterol and most frequently a combination of both.[7] Moreover, the quantity of auxiliary lipid in the test mixture that gives the most sensitive and intense reactions (optimal reaction) is very much larger than the quantity of lipid antigen, the weight ratio varying from 10- to 150-fold. It therefore can be appreciated that purification efforts based on assays that did not recognize this imbalance would be incorrectly interpreted. For example, fractions that contained barely detectable quantities of cerebroside might be very active, and pure cerebroside was inactive when tested alone.

This would naturally have led to the conclusion that the active antigen or hapten was not cerebroside.

I have used the terms hapten and antigen interchangeably, but it should be noted that Landsteiner's differentiation between these terms is still valid: *Antigen* refers to a complete antigen, namely, a substance that not only reacts with antibody but also provokes antibody formation; hapten, in contrast, is only a partial antigen, a substance that will react with antibody once it has been produced but does not elicit its formation. The immunological function of lipids appears to be limited to that of haptens. One also must distinguish between simple and complex haptens. Simple haptens are substances that combine reversibly with antibody without forming the three-dimensional aggregate of a typical immune complex. Complex haptens, on the other hand, produce immune complexes that are similar to those obtained with complete antigens. Lipid haptens belong to the category of complex haptens.

Although the first lipid hapten to be isolated was cardiolipin,[7,8] the present state of knowledge of lipid haptens originated in the isolation of cytolipin H from human cancer tissue.[9,10] This relatively simple substance, now identified as lactosyl ceramide,[11,12,13] is the component of lipid extracts of human tumor tissues that accounts for most of the antibody

**TABLE I**

**Reactivity of Rabbit Antibovine Brain Serum with Organ Lipids[a]**

| Species | Organ | Units[b] of activity per milligram of total lipid |
|---|---|---|
| Dog | Brain | 530 |
| Goat | Brain | 560 |
| Hamster | Brain | 540 |
| Man | Brain | 600 |
| Monkey | Brain | 650 |
| Ox | Brain | 510 |
| Pig | Brain | 780 |
| Rabbit | Brain | 580 |
| Rat | Brain | 600 |
| Sheep | Brain | 770 |
| Ox | Heart | <10 |
| Ox | Liver | <10 |
| Ox | Lung | <10 |
| Ox | Kidney | <10 |
| Ox | Spleen | 20 |

[a] From Joffe *et al.*[5]

[b] A unit of activity is arbitrarily defined as the quantity of lipid producing an end point of 50% hemolysis in a complement-fixation test using six units of complement[18] and 5 $\mu$l of antiserum.

activity directed against lipid that is found in rabbit antisera to human tumors. Lactosyl ceramide, a glycosphingolipid, is closely related chemically to two other naturally occurring glycosphingolipids of traditional interest to neurochemists, namely, galactosyl ceramide and glucosyl ceramide (glucocerebroside). A direct relationship to the pioneering studies of Goebel and Avery[14] could readily be recognized. Their studies showed that antibodies could be produced against lactose, galactose, glucose, and other saccharides by immunizing rabbits with appropriate saccharide–protein conjugates. When antibody against glucose was prepared in this way, it reacted readily with glucosyl ceramide in the presence of auxiliary lipids.[15] The door was thus opened to the rapid identification of galactosyl ceramide as the brain lipid that accounted for organ specificity of antibrain sera.[5] Independent confirmation was gained through the studies of Niedieck.[16] A most important consequence of this identification was the generalization of the relationship of chemical structure of lipids to their immunological activity: Since the simplest molecules of the glycosphingolipid class could function as haptens, all glycosphingolipids must have immunological activity.

Among the lipids of interest to neurochemists, attention has been focused on the immunological properties only of cerebrosides, gangliosides, and proteolipid. Some information on the immunological reactions of myelin has been acquired through the reactions of its cerebroside determinants. These studies will be described in some detail below.

## II. CEREBROSIDES (GALACTOSYL CERAMIDE; GALACTOCEREBROSIDE)

### A. Demonstration of Haptenic Activity

Identification of cerebroside, a major constituent of mammalian brain, as the lipid substance accounting for organ specificity of antibodies to brain was made through a systematic reinvestigation of the problem.[5] For the test reagent an antiserum to bovine brain was selected that reacted to approximately the same degree with total lipid extracts of brains of ten mammalian species (Table I), i.e., it gave an end-point measurement of 1.28–1.96 $\mu$g of lipid with excess antibody. Since no reactions occurred with lipids from bovine heart, liver, lung, and kidney, the properties of this antiserum represented a clear expression of the problem of brain specificity as it has been experimentally formulated by previous workers. Using the fractionation procedure of Davison and Wajda[17] and testing the individual column fractions both alone and in the presence of tenfold amounts of auxiliary lipid (lecithin plus cholesterol), it was quickly perceived that cerebroside fractions showed a very high degree of reactivity with this antibrain serum. When lipid fractions obtained from brain are tested by the method of complement fixation, one is faced with a special difficulty, namely, the intense anticomplementary activity of fractions containing sulfatide. This interference can be overcome by the addition of very small amounts of lecithin.

Identification of cerebroside as the immunologically active lipid was based on four experimental facts. First, with optimal amounts of auxiliary lipid (100 parts by weight of a mixture of lecithin and cholesterol, 1:1 wt./wt.), 3–5 ng of pure cerebroside were readily measurable with this antiserum. This degree of activity was much greater than that required to account for the activity of alcoholic extracts of brain. Second, the same reactivity was found with pure preparations of cerasine (unsubstituted fatty acid residue), phrenosine (hydroxylated fatty acid residue), and mixed cerebrosides of spinal cord.

Comparisons of the different cerebroside preparations were made using isofixation curves,[18] thus covering the complete immunological range from antibody excess to antigen excess (Fig. 1).

Perhaps it should be mentioned here that in dealing with naturally

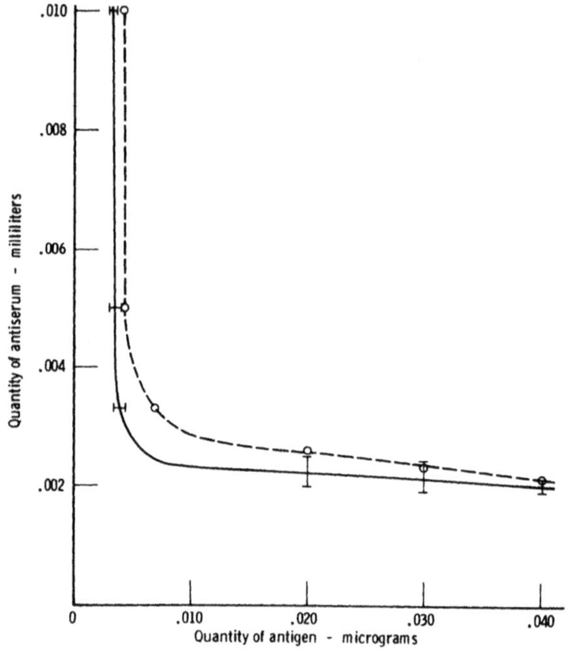


Fig. 1. Isofixation curves[18] of galactocerebrosides with antiserum to bovine brain. The solid line is drawn through points obtained by averaging the three values found with pure phrenosine and pure cerasine from ox brain and highly purified galactocerebroside from ox spinal cord. The range of these values is shown by the short horizontal or vertical lines. The dashed line shows the values found with synthetic phrenosine.

occurring complex lipids, the concept of purity is not the traditional one indicating a homogeneous molecular species. For example, in a "pure" natural glycosphingolipid, considerable variability is found in the fatty-acid chain residue, while a lesser degree of variation is present in the long-chain base residue. The polar moiety, however, is homogeneous. Molecules that are pure in the traditional sense are available only by the preparative methods of organic synthesis.

Third, a sample of phrenosine (lignocerylsphingosyl galactoside) prepared synthetically[19] had approximately the same activity as the natural product (Fig. 1). This eliminated the possibility that some undetected substance that was always associated with the natural products might be the active component. And fourth, since the antiserum gave no reactions with other glycosphingolipids such as glucocerebroside and cytolipin H, the molecular specificity of the system was assured.

Pure galactocerebroside was inactive when tested in the absence of auxiliary lipid; but optimal reactions (combining high intensity and sensitivity) were obtained by adding to the cerebroside a 100-fold quantity of lecithin–cholesterol (1:1 wt./wt.). These proportions of auxiliary lipid were the same as those found best in extensive experiments with the cytolipin H system, and the sensitivities of the individual immunological systems for detection of hapten were also quite similar (3–5 ng with six units of complement).

The accumulated evidence constituted rigorous proof of the indentity of galactocerebroside and the brain hapten and thus provided a very simple solution to a problem that had been posed more than 35 years earlier.

## B. Multiplicity of Antibodies to Cerebroside

Several years later an observation was made that led to an unexpected result.[20] Examination of a number of antibrain and antimyelin sera showed that although their reactivity with total lipid extracts of white matter was high, they gave no reactions with the galactocerebroside test antigen described above (Table II). The most logical explanation for this observation was that a hapten other than cerebroside was present among the lipids in CNS white matter and that some of the antibodies in antisera against brain and myelin were directed against this component. An experimental observation similar to this had led, a number of years earlier, to the detection of a novel hapten, cytolipin G, among lipids of human gastrointestinal tract tissues.[21] The correct explanation in this case was quite different. By separating white matter lipids into a limited number of fractions of defined lipid content and by formulating a number of test antigens from highly purified lipids, it was quickly established that only those containing galactocerebroside were serologically active.[20] The conclusion was reached that some antisera required a different formulation of cerebroside with auxiliary lipid in order to produce an immune complex that would fix complement: Although the most reactive test antigen for some antisera was

TABLE II

**Reactivity of Antibrain Sera with Total White Matter Lipid and with Galactocerebroside–Lecithin–Cholesterol (1:50:50)[a]**

| Antiserum | Antigen titer: With white matter lipid, 1 mg/ml | Antigen titer: With galactocerebroside–lecithin–cholesterol, 1 μg/ml |
|---|---|---|
| Antibovine-766 | 51 (1.96 μg)[b] | 30 (3.33 ng) |
| Antibovine-859 | 64 (1.56 μg) | <1 ( >100 ng) |
| Antihuman-842 | 64 (1.56 μg) | 18 (5.56 ng) |
| Antihuman-1057 | 34 (2.94 μg) | <1 ( > 100 ng) |

[a] From Rapport *et al.*[20] Test with 6 units of complement and 0.01 ml of antiserum.
[b] Numbers in parentheses show quantity of antigen correponding to antigen titer.

cerebroside combined with a 100-fold quantity of auxiliary lipid, other antisera reacted best with a nine- or tenfold quantity.[20]

Since the antibodies in these antisera were clearly of two different kinds, they were distinguished by calling them C type (100-fold auxiliary lipid) and X type (tenfold auxiliary lipid). Three other differences between the two systems were noted (Table III): First, their sensitivities with respect to cerebroside differed almost tenfold; second, the proportions of lecithin and cholesterol in the auxiliary lipid were different; and third, the X system was exquisitely sensitive to inhibition by excess lecithin. Isofixation curves of these two systems are shown in Fig. 2, in relation to the quantity of cerebroside, and in Fig. 3, in relation to the total quantity of lipid (hapten plus auxiliary lipid). Despite the tenfold lower sensitivity of the type-X system for cerebroside, the degree of reactivity found was sufficient to

TABLE III

**Characteristics of Two Types of Antibody–Cerebroside Reactions (Complement-Fixation with Antibrain Sera)**

| Condition | Type C | Type X |
|---|---|---|
| 1. Optimal quantity of auxiliary lipid | 100-fold | 5 to 10-fold |
| 2. Limiting quantity of cerebroside | 5 ng | 50 ng |
| 3. Proportions of lecithin–cholesterol in auxiliary lipid | 1:1 | 1:3 |
| 4. Inhibition by lecithin | not seen | marked |

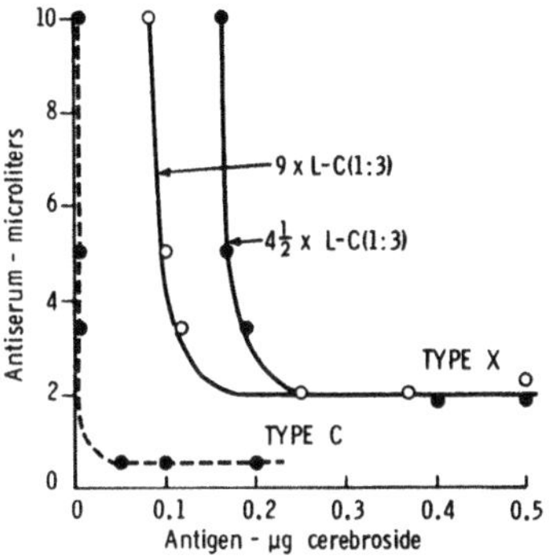


Fig. 2. Isofixation curves of antihuman brain serum (1057) and antigalactocerebroside serum (942) at a sensitivity level of six units of complement. The antihuman brain serum (solid line) illustrates type-X reactivity (test antigen: cerebroside-lecithin-cholesterol, 4:9:27), whereas the antigalactocerebroside serum (broken line) illustrates type-C reactivity (test antigen: cerebroside-lecithin-cholesterol, 1:50:50). Also shown is the appreciably lower reactivity of antigen in the type-X system when the ratio of lecithin-cholesterol to cerebroside is decreased from 9 to 4.5.

account for the reactions between antisera and total white matter lipids. With respect to *total weight of test antigen* (hapten plus auxiliary lipid), the difference in sensitivity between the type-C system and the type-X system is not very large (Fig. 3). Tests with a fairly large number of antisera to either brain particulates, brain lipids, pure myelin, or proteolipid indicate that type-X antibody occurs much more frequently than type C. It should be mentioned that although type-X antibody will not react with the type-C antigen, the reverse is not true; all type-C antisera give substantial reactions with type-X antigen although these are weaker than with type-C antigen.

## C. Impact of the Auxiliary Lipid Phenomenon

The dependence of these immune reactions on the quantity and composition of the auxiliary lipid indicates that no simple mechanism based either on solubilization of the lipid hapten or alteration of particle size will account for the effect. One may suggest rather that auxiliary lipids play a role through the formation of a surface aggregate that permits adequate penetration of antibody molecules into the lipid particle. Whatever the mechanism, the phenomenon itself makes it very difficult to design definitive experiments that will attain either of the two objectives that are of major concern, namely, (1) to establish or rule out the presence of anticerebroside

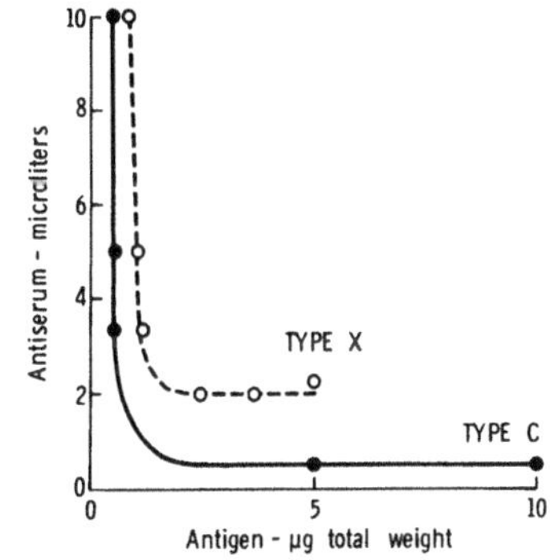


Fig. 3. Isofixation curves of antihuman brain serum (1057) and antigalactocerebroside serum (942) at a sensitivity level of six units of complement. The quantity of antigen is expressed as total weight of lipid (galactocerebroside plus auxiliary lipid).

antibody in certain pathological processes involving nervous tissue and (2) to define rigorously all lipid components of nervous tissue against which antibodies in a particular antiserum may be directed. It would, however, seem that this recent information might well explain observations reported by Schwab[22] that seemed to indicate the presence of more than one kind of organ-specific lipid hapten in brain tissue.

### D. Studies Based on Precipitation Methods

Shortly after the evidence was presented proving that cerebroside was the brain specific hapten, a paper by Niedieck and Kuwert[23] reported that sera of rabbits inoculated with spinal cord and showing typical signs of allergic encephalitis gave precipitation lines in agar gels (double diffusion technique of Ouchterlony) with a lipid mixture containing cerebroside. The testing of this mixture, formulated with cerebroside-lecithin-cholesterol (2.2:1:2.1 wt./wt./wt.), followed previous studies carried out with ethanolic extracts of nervous tissue.[24] A recent paper[16] summarizes in English earlier contributions published in German and also describes elicitation of anticerebroside antibody by intradermal inoculation with a mixture of cerebroside, bovine serum albumin, and complete adjuvant. Since agar diffusion may be much less sensitive than complement fixation,[25] since this method does not lend itself readily to objective quantification, and since precipitation reactions between lipid mixtures and serum proteins may be nonspecific, it is important to confirm results obtained by this method. From a purely physical standpoint, much of the lipid placed in an agar well may not diffuse into the gel unless previously subjected to special treatment. Niedieck produced remarkably transparent and stable aqueous solutions of cerebroside–lecithin–cholesterol aggregates by starting with very concentrated solutions in ethanol, making intermediate dilutions in ethanol, and having the final, aqueous dilution contain 20% ethanol. What factors affect the formation of a reactive, diffusible aggregate is not known, and the interpretation of inhibition experiments, particularly those involving hydrophobic substances, is thus a matter for conjecture.[16]

### E. Nature of Antibodies Reacting with Cerebroside

Both Ig-G (7S) and Ig-M (19S) antibodies were found to react with cerebroside-containing mixtures by both precipitation and complement-fixation techniques,[16] and the two kinds of antibody molecules were found in both anticerebroside sera and EAE sera of rabbits. Differentials in titers suggested that the Ig-G fraction was more effective in precipitation whereas the Ig-M fraction was better for complement-fixation.

### F. Myelin

Although galactocerebroside is present in high concentration in myelin, the chemical architecture of the biomembrane is not well established. The composite bilayer model currently in vogue[26] suggests that cerebroside molecules are not exposed on the surface and leads to the prediction that myelin ought not to react with anticerebroside antibody. This is not the case. The technical problem of working with myelin, which is completely insoluble in aqueous media, is readily surmounted by using the technique of complement fixation. It was found that suspensions of pure myelin reacted with both antibrain serum and antigalactocerebroside serum[27] and that 7 $\mu$g of myelin could be readily measured (Fig. 4). Since myelin is not anticomplementary (no nonspecific interference), the test used for these studies could be made very sensitive (three units instead of six units of complement). This direct reaction with the membrane has considerable importance for the problem of immunologically active lipids. Since auxiliary lipids were not required, it is clear that the mixed lipid array (containing lecithin, cholesterol, and other lipids) as constituted in the intact membrane structure permits an adequate degree of lipid–protein penetration. The need for auxiliary lipids in studies with *pure haptens* and the complications deriving from this phenomenon have had to be introduced because chemical structure and immunological specificity can be evaluated only by starting with pure substances. The fact that quantitative measurements on isolated systems cannot be related directly to the properties of elements of biological ultrastructure only makes the problem of applying our new knowledge more difficult.

By means of absorption studies it was established that antigalactocerebroside serum reacted with myelin solely through galactocerebroside

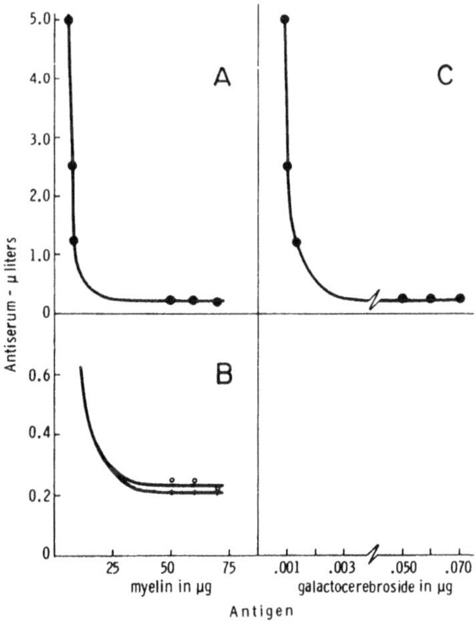


Fig. 4. Isofixation curves of antigalactocerebroside serum with myelin (panels A and B) and galactocerebroside (panel C) at a sensitivity level of three units of complement. In panel A the points represent the average of the two values obtained with two preparations of myelin. In panel B, the ordinate is expanded fivefold to show small differences in the region of antigen excess obtained with the different preparations: (+) represents M-1; (○) represents M-2. The antigen in panel C was galactocerebroside plus 100 parts by weight of auxiliary lipid (lecithin–cholesterol, 1:1 wt./wt.).

determinants, whereas with antibrain serum the evidence suggested that antibodies against other determinants were present as well.[27] The technique used was to remove antibody partially by absorption with varying amounts of myelin or cerebroside, followed by comparison of the efficiency of this process for the two antigens (Fig. 5). These observations were made before it had been established that two kinds of anticerebroside antibodies could be present in such antisera, and the properties of myelin should be re-examined in the light of the newer information. The reaction of myelin with anticerebroside antibodies indicates that the galactose residues are exposed on the membrane surface, and this observation is therefore more in accord with the recent Benson model of membrane architecture[28] than with the classical Davson–Danielli model. Since antigen–antibody reactions can provide information on three-dimensional complementary configuration, the reactions of myelin with antisera of defined chemical valency should be useful in acquiring information on chemical architecture. A recent preliminary report[29] indicates that rat and beef myelins react to a very similar degree with antigalactocerebroside serum and therefore should have very similar molecular arrangements at the surface.

## G. Proteolipid

Proteolipid is a molecule (or preferably a lipid–protein aggregate) that contains much galactocerebroside; its structure should be less complicated than that of myelin since some of the phosphatide and protein and all of the cholesterol have been removed. Its immunological importance stems from its apparent activity in provoking allergic encephalomyelitis in some animal species.[30,31] The immunological properties of proteolipid, a water-insoluble material, have not yet been studied extensively *in vitro*. A preliminary report[32] indicates that proteolipid preparations are anticomple-

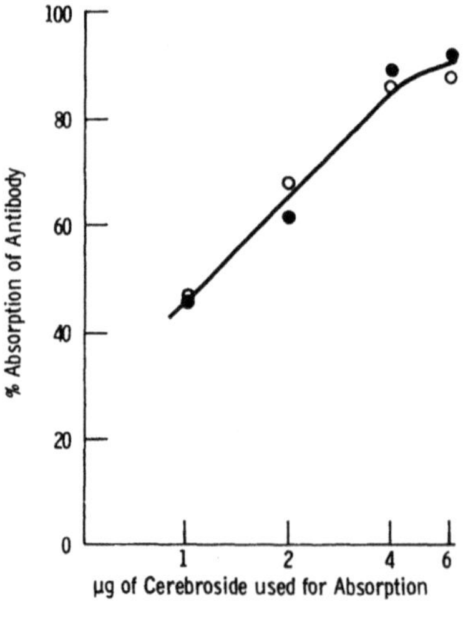


Fig. 5. Graded absorption of anticerebroside serum with increasing amounts of galactocerebroside plus auxiliary lipid (lecithin). Samples of absorbed serum were tested for removal of both anticerebroside antibody (●) and antimyelin antibody (○).

mentary but that this interference can be overcome by adding auxiliary lipids. Complement fixation required the addition of only a small amount of auxiliary lipid (1.5-fold quantity of lecithin–cholesterol, 1:2 wt./wt.). It was indicated that at least some of the reactivity of proteolipid occurred through galactocerebroside determinants.

## III. GANGLIOSIDES

Elicitation of antiganglioside antibodies has been reported by a number of investigators.[33–36] Demonstration of the immunological reaction has been based on a variety of methods including passive hemagglutination (agglutination of ganglioside-coated red cells), precipitation, and complement fixation. Comparisons cannot be readily made among the different reports, and the results obtained so far suggest that gangliosides are relatively weak antigens and that the elicitation of an antibody response to them is attended by unrecognized subtleties. Yokoyama, Trams, and Brady[34] immunized rabbits with red cells coated with ganglioside, and the antisera so obtained were able to agglutinate ganglioside-coated red cells. Although titers were low, Somers, Kanfer, and Brady[37] attempted to use these antisera to acquire some information on ganglioside structure using inhibition methods. The results were not decisive and contributed little to an understanding of the specificity of the reaction. Antisera prepared similarly against the molecule from which the sialic acid residues had been cleaved by acid (asialoganglioside) were inhibited most effectively by lactose, a result that is exceedingly difficult to explain. Sherwin, Lowden, and Wolfe[35] produced antiganglioside serum of much higher titer by multiple injections into the footpads of rabbits of ganglioside in Freund's complete adjuvant with subsequent booster injections given intramuscularly. The specificity of these antisera was not examined. Antisera reacting with gangliosides also were obtained by immunization with temporal lobe or caudate nucleus but not with optic or sciatic nerves. The most substantial report concerning antiganglioside production by hyperimmunization techniques is that of Pascal, Saifer, and Gitlin.[36] They produced precipitating antibodies by intramuscular injection of ganglioside with Freund's adjuvants followed by sustained intravenous injections. Antibodies were produced against ganglioside preparations from either normal human brain, Tay–Sachs' brain, or beef brain in 15 of 16 rabbits. Limited studies of specificity carried out by the Ouchterlony double diffusion technique showed that antibodies against several antigenic components (distinct precipitation lines) were present when the antisera were studied with ganglioside preparations from either normal human or Niemann–Pick brain. Antiserum prepared with Tay–Sachs' ganglioside reacted only with this ganglioside preparation, whereas antiserum prepared with beef brain ganglioside reacted equally well with ganglioside preparations from beef brain and normal human brain but not with ganglioside preparations from Tay–Sachs' brain. Since the term ganglioside is a generic term referring to about eight or nine components that are sialyl derivatives of a number of glycosphingolipid molecules containing two to four monosaccharide residues,[38,39] it can be appreciated that much work remains to be done to define the specificity of antiganglioside systems. The precipitation method is a rather insensitive tool for this purpose since tests are carried out with ganglioside preparations at a concentration of 5 mg/ml.[36] Also one does not know the extent to which different molecular species remain associated with one another during diffusion through the aqueous gel. A preliminary report[40] in which antiserum to bovine brain ganglioside was studied with eight highly purified ganglioside components (obtained from the brain of a patient with subacute sclerosing leukoencephalitis) indicated that only three gave positive reactions. These were $G_2$, $G_{3A}$, and $G_4$: $G_2$ is the ganglioside with two sialic acid residues attached to the internal galactose residue; $G_{3A}$ has two sialic acid residues attached to lactosyl ceramide; and $G_4$, the major monosialoganglioside of normal brain, has the single sialic acid residue also attached to the internal galactose residue. When these molecules were combined with a 100-fold quantity of lecithin-cholesterol (1:3 wt./wt.), 5, 12.5, and 4 ng of the respective components could be detected by complement fixation. Two factors that affected activity were indicated: the initial concentration at which the ganglioside test antigen was prepared and the composition of the auxiliary lipid.

## IV. MISCELLANY

### A. Other Lipids of the Nervous System with Immunological Activity

Three lipids of potential interest because of their occurrence in nervous tissue are "asialoganglioside," cardiolipin, and sulfatide.

Studies of the asialoganglioside of normal brain ganglioside were reported by Somers, Kanfer, and Brady.[37] This compound may be readily obtained from ganglioside preparations through mild acid hydrolysis. The structure is galactose (1→3)-*N*-acetyl galactosamine (1→4)-galactose (1→4)-glucose(1→1′)-ceramide.[38] Antibodies elicited by extended immunization of rabbits with antigen-coated red cells[34] were detected by agglutination reactions with these red cells. The agglutination was inhibited by a number of water-soluble compounds, the most effective being lactose, lactobionic acid, 6-*O*-β-D-galactosyl-D-galactose, and 4-*O*-β-D-galactosyl-*N*-acetyl glucosamine, in that order. (The latter compound appears to have been inadvertently referred to as the galactosamine derivative in a short review.[41]) These results do not lead to any simple conclusions, perhaps because of the inherent complexity of the test method.

Cardiolipin is present in small amounts in brain as a normal component of mitochondria. Although its immunologic properties are well recognized, antisera reacting with cardiolipin are encountered only infrequently as a result of immunization with either subcellular fractions of brain or brain lipids (Graf and Rapport, unpublished).

Sulfatide, which is present in brain in rather high concentrations, has intense anticomplementary activity. Although this compound is a glycosphingolipid, specific antibodies reacting with it were not detected (complement fixation and precipitation) following immunization with carrier protein and Freund's adjuvants.[42] At low salt concentrations, nonspecific precipitation of sulfatide with serum proteins was seen in agar gels.

### B. Lipid Haptens in Pathological Processes

Little information is available to suggest that antibodies directed against lipids play a role in pathological processes in the nervous system. This judgment may be premature since it is apparent that sufficient information is not yet available to permit the formulation of test antigens (by combining haptens with correct amounts of auxiliary lipids) for all types of antibody. Both precipitating and complement-fixing anticerebroside antibodies have been detected in the sera of rabbits afflicted with experimental allergic encephalomyelitis.[16,23,43] However, at the present time antibodies to a basic protein antigen are held responsible for this disease.

A most provocative finding of a preliminary nature is the occurrence, with low incidence, of small amounts of antiganglioside antibodies (detected by passive hemagglutination) in multiple sclerosis and other neurological disorders.[44] In this same publication antibody activity toward asialoganglioside was reported in a number of viral diseases, in Tay–Sachs' disease, and in schizophrenia.

### C. Methodology

Specialized techniques that are useful in studying the immunological properties of lipids recently have been reviewed.[45] Complement fixation is clearly the method of choice for the study of hydrophobic substances since it offers both the required sensitivity and quantification in the regions of antibody excess and antigen excess. Isofixation curves are particularly valuable for judging the properties of an antigen–antibody system and especially for comparing different antigens.[18,46]

## V. EPILOGUE

Although the immunological activity of lipids has been recognized for many years and one of them, cardiolipin, has proven to be exceedingly useful as a diagnostic tool for more than a quarter of a century, the experimental base in this area is very circumscribed. The research effort is still limited to a small number of investigators, perhaps because the problems imposed by working with hydrophobic materials in an aqueous medium severely hamper progress. This latter aspect, which is clearly delineated by the auxiliary lipid phenomenon, was not generally appreciated until several pure lipid haptens became available for study. This occurred when immunological activity was firmly related to lipid structure through the glycosphingolipids.

Galactocerebroside is the simplest of these compounds, and it is also the most easily obtained since it is a major constituent of nervous tissues. This provides an interesting opportunity for the interplay of experimental findings: Advances in our knowledge of *in vitro* behavior of lipid haptens, if obtained with cerebroside, will have immediate implications for the detection of antibody in immunopathological processes involving the nervous system. Chemical advances concerning the molecular arrangement of cerebroside molecules on membrane surfaces will have implications for immunological activity of antibody directed against these cerebroside determinants, and the converse is true.

Although the assignment of immunological specificity to the carbohydrate residue of the cerebroside molecule is a very plausible hypothesis, it has not yet received much experimental challenge. The experiments may appear to be simple, but direct analogy with immunological systems involving water-soluble antigens is not warranted, since consideration must be given to alterations of aggregates of molecules, both in the degree of aggregation and in its character.

However difficult it may be to extrapolate the knowledge acquired with artificial systems to those that occur naturally, the revival of interest in lipid haptens and renewed exploration of the immunological properties of cerebroside offer an unusual potential for acquiring new information about the nervous system.

## VI. ACKNOWLEDGMENTS

Studies described in this review were supported by the National Multiple Sclerosis Society and the National Science Foundation. I am indebted to Dr. Liselotte Graf for assistance in the preparation of this paper.

## VII. REFERENCES

1. R. Brandt, H. Guth, and R. Müller, Zur Frage der Organspezifität von Lipoidantikörpern, *Klin. Wochschr.* **5**:655 (1926).
2. E. Witebsky and J. Steinfeld, Untersuchungen über spezifische Antigenfunctionen von Organen, *Z. Immunitaetsforsch.* **58**:271–296 (1928).
3. F. F. Schwentker and T. M. Rivers, The antibody response of rabbits to injections of emulsions and extracts of homologous brain, *J. Exptl. Med.* **60**:559–574 (1934).
4. H. Rudy, Über die chemische Natur des Hirn Antigens II. Weiteres über die Reindarstellung des Haptens, *Biochem. Z.* **267**:77–88 (1933).
5. S. Joffe, M. M. Rapport, and L. Graf, Immunochemical studies of organ and tumor lipids XII. Identification of an organ specific lipid hapten in brain, *Nature* **197**:60–62 (1963).

6. M. M. Rapport, Ebb-tide antigens in the brain. *Biol. Bull.* **125**:356 (1963).
7. M. M. Rapport, Structure and specificity of the lipid haptens of animal cells, *J. Lipid Res.* **2**:25–36 (1961).
8. M. C. Pangborn, Isolation and purification of a serologically active phospholipid from beef heart, *J. Biol. Chem.* **143**:247–256 (1942).
9. M. M. Rapport, L. Graf, V. P. Skipski, and N. F. Alonzo, Cytolipin H, a pure lipid hapten isolated from human carcinoma, *Nature* **181**:1803–1804 (1958).
10. M. M. Rapport, L. Graf, V. P. Skipski, and N. F. Alonzo, Immunochemical studies of organ and tumor lipids VI. Isolation and properties of cytolipin H, *Cancer* **12**:438–445 (1959).
11. M. M. Rapport, L. Graf, and J. Yariv, Immunochemical studies of organ and tumor lipids IX. Configuration of the carbohydrate residues in cytolipin H, *Arch. Biochem. Biophys.* **92**:438–440 (1961).
12. D. Shapiro and E. S. Rachaman, Total synthesis of cytolipin H, *Nature* **201**:878–879 (1964).
13. M. M. Rapport and L. Graf, Serological activity of cytolipin H (lactocytoside), *Nature* **201**:879–880 (1964).
14. W. F. Goebel, O. T. Avery, and F. H. Babers, Chemo-immunological studies on conjugated carbohydrate-proteins IX. The specificity of antigens prepared by combining p-aminophenol glycosides of disaccharides with protein, *J. Exptl. Med.* **60**:599–617 (1934).
15. M. M. Rapport, The antigenic properties of sphingolipids, *in Brain Lipids and Lipoproteins, and the Leucodystrophies* (J. Folch and H. Bauer, eds.), pp. 83–89, Elsevier, Amsterdam (1963).
16. B. Niedieck, E. Kuwert, O. Palacios, and O. Drees, Immunochemical and serological studies on the lipid hapten of myelin with relationship to experimental allergic encephalomyelitis, *Ann. N.Y. Acad. Sci.* **122**:266–276 (1965).
17. A. N. Davison and M. Wajda, Metabolism of myelin lipids: Estimation and separation of brain lipids in the developing rabbit, *J. Neurochem.* **4**:353–359 (1959).
18. M. M. Rapport and L. Graf, Immunochemical analysis based on complement fixation, *Ann. N.Y. Acad. Sci.* **69**:608–632 (1957).
19. D. Shapiro and H. M. Flowers, Synthetic studies on sphingolipids VI. The total syntheses of cerasine and phrenosine, *J. Am. Chem. Soc.* **83**:3327–3332 (1961).
20. M. M. Rapport, R. Cavanna, and L. Graf, Immunochemical studies of organ and tumor lipids XVII. The existence of two complement-fixing systems involving cerebroside, *J. Neurochem.* **14**:9–18 (1967).
21. L. Graf, M. M. Rapport, and R. Brandt, Immunochemical studies of organ and tumor lipids X. The presence of a novel lipid hapten, cytolipin G, in tumors and tissues of the gastrointestinal tract, *Cancer Res.* **21**:1532–1536 (1961).
22. E. Schwab, Ueber ein neues in der Protagonfraction des Hirnes enthaltenes Hapten, *Z. Immunitaetsforsch.* **87**:426–444 (1936).
23. B. Niedieck and E. Kuwert, Zur Serologie der experimentellen allergischen Encephalomyelitis I. Vergleich der Reactivität von Lipidgemischen und Myelinextrakten mit den entsprechenden Antiseren im Präzipitationstest und in der Komplementbindungsreaktion, *Z. Immunitaetsforsch.* **125**:470–492 (1963).
24. B. Niedieck and K. Lanken, Zur Serologie der experimentellen allergischen Neuritis. Ein Agar Präcipitationstest mit alkoholischen Nervextrakten, *Klin. Wochschr.* **39**: 1164–1169 (1961).
25. L. Graf and M. M. Rapport, Immunochemical studies of organ and tumor lipids XVI. Gel diffusion analysis of the cytolipin K system, *Intern. Arch. Allergy* **28**:171–177 (1965).

26. J. B. Finean, Molecular parameters in the nerve myelin sheath, *Ann. N.Y. Acad. Sci.* **122**:51–56 (1965).
27. M. M. Rapport, L. Graf, L. A. Autilio, and W. T. Norton, Immunochemical studies of organ and tumor lipids XIV. Galactocerebroside determinants in the myelin sheath of the central nervous system, *J. Neurochem.* **11**:855–864 (1964).
28. A. A. Benson, On the orientation of lipids in chloroplast and cell membranes, *J. Am. Oil Chem. Soc.* **43**:265–270 (1966).
29. L. Graf, M. M. Rapport, and R. Cavanna, Serological comparison of rat and beef myelins, *Federation Proc.* **27**:620 (1968).
30. P. K. Olitsky and C. Tal, Acute disseminated encephalomyelitis produced in mice by brain proteolipid, *Proc. Soc. Exptl. Biol. Med.* **79**:50–53 (1952).
31. B. H. Waksman, H. Porter, M. D. Lees, R. D. Adams, and J. Folch, A study of the chemical nature of components of bovine white matter effective in producing allergic encephalomyelitis in the rabbit, *J. Exptl. Med.* **100**:451–471 (1954).
32. R. Cavanna, M. M. Rapport, and L. Graf, Immunological properties of brain proteolipid, *Federation Proc.* **25**:732 (1966).
33. S. Bogoch, Demonstration of serum precipitin to brain ganglioside, *Nature* **185**:393–394 (1960).
34. M. Yokoyama, E. G. Trams, and R. O. Brady, Immunochemical studies with gangliosides, *J. Immunol.* **90**:372–380 (1963).
35. A. L. Sherwin, J. A. Lowden, and L. S. Wolfe, The production of antisera to gangliosides from human nervous tissue, *Can. J. Biochem.* **42**:1640–1642 (1964).
36. T. A. Pascal, A. Saifer, and J. Gitlin, Immunochemical studies of normal and Tay–Sachs' brain gangliosides, *Proc. Soc. Exptl. Biol. Med.* **121**:739–743 (1966).
37. J. E. Somers, J. N. Kanfer, and R. O. Brady, Immunochemical studies with gangliosides II. Investigations of the structure of gangliosides by the hapten inhibition technique. *Biochemistry* **3**:251–254 (1964).
38. R. Ledeen, The chemistry of gangliosides: A review, *J. Am. Oil Chem. Soc.* **43**:57–66 (1966).
39. H. Wiegandt, Ganglioside, *Rev. Physiol. Biochem. Exptl. Pharmacol.* **57**:189–222 (1966).
40. M. M. Rapport, L. Graf, and R. Ledeen, Specificity of antiserum to brain ganglioside, *Federation Proc.* **27**:463 (1968).
41. R. O. Brady, Immunochemical properties of lipids, *J. Am. Oil Chem. Soc.* **43**:67–69 (1966).
42. B. Niedieck, Immunochemical studies on spinal cord sulfatides, *Z. Immunitaetsforsch.* **132**:139–146 (1967).
43. C. G. Honegger, G. Ritzel, and H. P. Rieder, Versuche zur Charakterisierung von Lipid-Haptenen bei EAE, *Z. Immunitaetsforsch.* **126**:49–52 (1963).
44. M. Yokoyama, E. G. Trams, and R. O. Brady, Sphingolipid antibodies in sera of animals and patients with central nervous system lesions, *Proc. Soc. Exptl. Biol. Med.* **111**:350–352 (1962).
45. M. M. Rapport and L. Graf, Preparation and Testing of Lipids for Immunological Study, *in Methods of Immunology and Immunochemistry* (C. A. Williams and M. W. Chase, eds.), Vol. 1, pp. 187–196, Academic Press, New York (1968).
46. M. M. Rapport and L. Graf, Cancer antigens: How specific should they be? *Cancer Res.* **21**:1225–1237 (1961).

*Chapter 19*

# FATTY ACIDS


## Amedeo F. D'Adamo, Jr.

*Saul R. Korey Department of Neurology and Department of Biochemistry*
*Albert Einstein College of Medicine of Yeshiva University*
*New York, New York*


## I. INTRODUCTION

The biochemical maturation of the central nervous system proceeds with profound changes in the metabolism of the fatty acids and associated derivatives. These changes reflect, in part, differences in permeability of the nervous system to fatty acids and to precursors and the need for specific complex fatty acid derivatives for structural elements as development proceeds. The timing of these events differs in the peripheral nervous system.

## II. SOURCES OF FATTY ACIDS IN THE BRAIN

The fatty acids of the nervous system have been shown to arise both by transport from the surrounding fluids and by synthesis *in situ*.

### A. Transport into the Brain

#### 1. *Sources of Artifact in Transport Experiments*

Incorporation of deuterium-labeled fatty acids into rat brain lipids by addition of these labeled lipids to the diet was at the barely detectable level in the early experiments of Sperry, Waelsch, and Stoyanoff.[1] This result has been interpreted as indicating a restricted but nevertheless existent transport of fatty acids into the brain.

Interpretation of transport experiments should take cognizance of the following sources of artifact: (1) Although the deuterium in the experiments cited above is initially on the fatty acid molecule, the lipid can be metabolized in the liver to labeled acetate, which may be transported to the brain and utilized as such. Alternatively, this labeled acetate can be converted by the liver to glucose labeled in positions that also will provide labeled acetyl groups for lipid synthesis when catabolized by the brain.[2] (2) Radioactive bicarbonate will label the brain lipids by pathways not presently known, giving an activity of up to 1% that of the $^{14}CO_2$ pool.[3] (3) The mode of administration may not simulate the true *in vivo* situation. It has been shown that the rate of uptake of palmitic acid complexed with albumin can be increased 2.5-fold, depending on the procedure used for complex formation.[4] (4) Differential absorption of the administered fatty acid may occur. Thus, the uptake of [$^{14}$C] stearic acid from the digestive system proceeds more slowly than labeled oleic acid,[5] making comparisons of the transport of different fatty acids more difficult.

### 2. *Transport of Dietary Fatty Acids*

Alterations in the composition of dietary fatty acids, either by manipulation of the maternal diet prior to gestation or during suckling[6] or by dietary alterations of weaned progeny[7–10] have been shown to lead to significant changes in the fatty acid composition of brain lipids.

A maternal diet containing 10% corn oil, which gives a high linoleic acid ($\omega$-6 series)*-linolenic acid ($\omega$-3 series) ratio, resulted in higher levels of 22 : 5$\omega$6 and lower levels of 22 : 6$\omega$3 in the brain of the newborn rat than a diet (grain based) with a low ratio.[6] That the maternal diet in the immediate postnatal period as well as events *in utero* can modify the brain lipid pattern was shown by transfer of litters at birth to foster mothers on different diets. After 3 weeks of suckling these animals had a brain lipid pattern that reflected the dietary lipids of the foster mother.

Animals that have low 22 : 6$\omega$3 at birth exhibit a marked tendency to accumulate this fatty acid during early development even when the level of linolenic acid in the milk was quite low. This suggests a preferential uptake of this fatty acid in the immediate postnatal period in the rat.

Transport of the polyunsaturated fatty acid also occurs postmyelinization. Mohrhauer and Holman[7] found that rats fed linoleic or arachadonic acid for 100 days show an increase of polyunsaturated fatty acid of the $\omega$-6 series in brain lipids while rats fed linolenic acid show an increase of $\omega$-3 members. Biran and co-workers[8] found a marked decrease in the saturated $C_{16}$ and $C_{18}$ of the neutral lipids of brain after a diet deficient in essential fatty acids or one that contains primarily linolenic acid. Similar changes were not found in the liver or kidney lipids.

The effects of dietary fats on the fatty acid composition of the mitochondria from chicken brain[9] and rat brain and liver[10] and of myelin[11] have been reported. Not only can the fatty acid composition of myelin[11] be altered by the dietary lipids but the thickness of the myelin sheath as well.[12] The relationship between the proportion of linoleic acid in myelin to that in the blood serum is linear after feeding experiments of 34 weeks duration. There are no apparent neurological differences in these animals.[11]

* In the abbreviation for polyunsaturated fatty acids, X : Y$\omega$Z, X is the number of carbon atoms, Y is the number of double bonds, and $\omega$Z is the position of first double bond from the methyl end.

High intake of polyunsaturated fatty acid without adequate protection against autoxidation, however, does facilitate the appearance of encephalomalacia.[13,14]

## 3. *Studies with Labeled Fatty Acids*

Uptake *in vivo* of radioactivity from labeled palmitic, nervonic, and erucic acid, which is not normally found in mammalian tissues, has been studied by Carroll[15] in adult rat. He found similar amounts of radioactivity in a crude lipid extract of the brain from all three fatty acids. Unfortunately, no attempt was made to prove that the radioactivity actually resided either in the specific fatty acid or in the fatty acid fraction.

Laurell[16] specifically isolated the fatty acid fraction after base hydrolysis of rat brain tissue and recovered approximately 0.1% of the label from [1-$^{14}$C] palmitate injected into a neck vein. Allweis and co-workers[17] studied the oxidation of albumin-bound [U-$^{14}$C] palmitic acid by the perfused cat brain preparation. After apparent isotopic equilibration, this fatty acid was the origin of about 2.8% of the $CO_2$ of the perfused brain and about 0.3% of that produced by the cortex, although no uptake from the perfusate could be detected by A-V analysis. The results indicate either that a small uptake or exchange of fatty acid, similar to that observed with glutamate, occurs.

Fatty acid transport into the brain is most certainly age and species dependent. A significantly greater radioactivity from labeled fatty acids was found in brain extracts of 12-day-old rats compared with adult rats.[15] Studies with newborn rats injected with precursor intraperitoneally show a small incorporation of palmitate and linolenate into the ester-linked fatty acids of the brain; a fivefold greater incorporation of linoleate is observed.[18] With all three precursors, all of the radioactivity is in the acyl portion of the molecule. Prior degradation to acetate in other organs was not responsible for incorporation, since with this precursor under the same conditions only 30% of phospholipid radioactivity is in the ester-linked fatty acids.

With the embryonic chick brain, on the other hand, palmitate, stearate, linoleate, and linolenate, injected into the yolk sack, are incorporated to a similar extent in the glycerophosphate lipids. In this species the fatty acid fragment contain 60–70% of the total radioactivity of the compound.[19]

The degree of transport of fatty acid into the brain from the blood, in addition to being dependent on the degree of unsaturation and on development, may be dependent on the chain length. Thus Gatt[20] has found that even after multiple intravenous injections of [1-$^{14}$C] lignoceric acid in adult rats, a chloroform–methanol extract of the brain had almost no radioactivity. The production of $^{14}CO_2$ from this precursor, however, was similar to that found in the previously mentioned study with palmitate,[15] indicating a comparable uptake of these fatty acids in nonnervous tissues.

Definitive and quantitative studies on the transport of the fatty acids into the brain *in vivo* remain to be done. Specifically, tritium-labeled fatty acids, which will avoid the isotope recyclization problem, and regional perfusion studies or brain perfusion techniques in which all extracerebral tissues are removed[21] may be useful in this regard.

### 4. *Other Transport Studies*

A number of studies, some of which will be discussed in connection with metabolism, show that fatty acids can be taken up by the brain when injected intracerebrally. Uptake of fatty acids from the medium has been found with slices from immature[22] and adult brain [23] and of albumin-bound fatty acids by the isolated rat retina[24] and rabbit cornea.[25] In contrast to these observations on uptake, net fatty acid output has been reported to occur in slices;[26] this output is increased in slices exposed to EAE serum.

Radio-iodinated fatty acids, used in the detection of brain tumors, indicate that fatty acids are transported into tumors more rapidly than into the normal brain.[27]

## B. Fatty Acid Biosynthesis

### 1. *Acetyl Transfer Mechanisms*

The biosynthesis of fatty acids requires acetyl groups both for *de novo* synthesis and for the variety of chain elongation steps that lead to the extra long-chain fatty acids. In the nervous system these acetyl groups are primarily derived from glucose via pyruvate oxidation in the mitochondria. It is quite likely, however, that pyruvate oxidation does not provide all the required two-carbon units during all stages of development. For example, acetate,[28] citrate,[29,30] glutamine,[29] or leucine and other compounds[31] taken up from the blood may provide acetyl CoA in the brain.

It is clear that the extramitochondrial acetyl CoA pool for fatty acid biosynthesis is a different two-carbon pool than the mitochondrial pool from which it is derived. Recent experiments indicate that the acetyl CoA required for the functioning of the tricarboxylic acid cycle[32] and for chain elongation also exists in discrete pools within the mitochondria,[33] although all may be derived from the common two-carbon pool provided by pyruvate. The transfer of acetyl CoA from this common pool must be effected by specific carrier systems, as shown in Fig. 1. These can transport the required two-carbon units intact in the form of acetate,[34] acetyl carnitine,[35] acetyl amino acids,[36,37] etc., or as citrate, which can be cleaved in the extramitochondrial compartment to acetyl CoA.[38] Cytoplasmic citrate also may arise by reductive carboxylation of α-ketoglutarate.[29]

### 2. De Novo *Fatty Acid Biosynthesis*

*a. Acetate and Acetyl CoA Metabolism.* Levels of acetyl CoA[39] and acetyl carnitine[40] have been reported in the brain. The subcellular

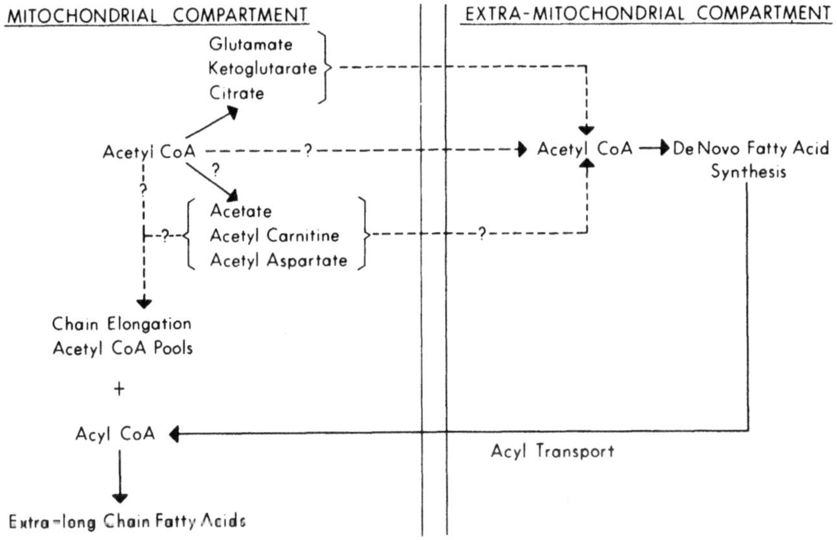


Fig. 1. Acetyl transport mechanisms.

distribution of acetyl CoA synthetase has been studied by several groups. Unpublished experiments quoted by Whittaker indicate that the enzyme is extramitochondrial.(41)

On the other hand, Schubert(42) and Tucek(44) both found the enzyme to be primarily associated with the mitochondrial fraction with minor amounts in the soluble fraction. Unpublished experiments by Whittaker and associates(41) suggest that there is a factor present in particulate fractions of the brain that inhibits the synthesis of acetyl CoA by acetyl CoA synthetase. Most of the activity was found in the fraction rich in nerve endings and was thought to be identical with a mixture of phosphatases. Tucek(43) suggests that this inhibition may be due to the low concentrations of ATP used and the stimulation of an ATPase by the presence of a high concentration of sodium ion. It is of interest that the acetyl CoA synthetase of heart is also inhibited by sodium ions.(44)

Pyruvate decarboxylation to acetyl CoA by brain mitochondria(45) and by normal and thiamine deficient brain *in vivo*(46) has been studied. In light of the ability of glutamate to provide acetyl groups in the cytoplasm,(29) it is of interest that no other tissue given [2-$^{14}$C] pyruvate or its metabolic precursors synthesizes glutamate with as high a percentage of radioactivity in carbon-5 as does the brain.(47,48) The factors that stimulate or inhibit the release of acetyl CoA from isolated brain mitochondria have been examined.(49)

Carnitine acetyl transferase, like acetyl CoA synthetase and pyruvate oxidase, is also primarily in the mitochondrial fraction of brain homogenates.(50) The activity of the enzyme of the rabbit brain cortex at 4 days after birth is only 10% that in adult animals. The bulk of the increase in enzymic activity does not occur in the early postnatal period, when fatty acid biosynthesis is probably greatest, but rather after 32 days of age. Regional distribution studies showed little difference between gray and white matter. In contrast, the activity of the enzyme in the sciatic nerve is less than 8% that of brain.

ATP citrate lyase has been found to be primarily localized in the extramitochondrial fraction in the guinea pig[43] and in the rat brain.[29] In addition, a significant proportion of the enzymic activity was found in the microsomal fraction from sheep caudate nuclei;[43] whether this is an adsorption artifact was not examined. The activity of ATP citrate lyase in the high-speed supernatant fraction of developing rat brain was found to decrease with aging.[29] The highest activities, which are found in the newborn rat, are only one-third of the values observed in rat liver.[51]

*b. Enzyme Systems.* Brady[52] has studied the *de novo* biosynthesis of fatty acids in a high-speed supernatant from immature rat brain. The system from the brain utilizes malonyl CoA but differs markedly from that found in liver since it has the ability to catalyze the reduction and condensation of β-hydroxybutyryl CoA, crotonyl CoA, and octanoyl CoA. Later work by Robinson, Bradley, and Brady[53] showed that 4-carbon acyl CoA derivatives such as acetoacetyl CoA also can be converted to long-chain fatty acids by this preparation.[53]

Conversion of acetate to fatty acids by the high-speed supernatant fraction from brain of adult rats also has been reported;[54] significant incorporation also was found with the mitochondrial fraction. Fatty acid synthesis in fractions of peripheral nerve has been studied by Hughes and Eliasson.[55] Both the homogenate fraction (an 800-g supernatant) and the sheath fraction that contained the Schwann cell nuclei were able to convert labeled acetate into fatty acids.

Tetrolyl CoA[56,57] and propiolyl CoA have been shown to be strong noncompetitive inhibitors of fatty acid synthesis; these compounds appear to inhibit the condensation of acetyl CoA with malonyl CoA binding with the required enzyme sulfhydryl groups.[57]

*c.* In Vitro *Whole Cell Systems.* Considerable evidence exists that the loss of electrical activity in *in vitro* systems has a profound effect on the metabolism of lipid precursors such as acetate and glucose. A marked decrease in [1-$^{14}$C] acetate oxidation was found by electrical stimulation of cerebral cortex tissue.[58] With peripheral nerve, a small increase in acetate oxidation and larger increases in pyruvate and glucose oxidation were observed.[59]

The conversion of acetate to lipids has been shown to be increased in sciatic nerve stimulated at 60 pulses/sec compared to unstimulated controls.[60] Conversion of acetate to fatty acids in brain minces supplemented with various cofactors is only 5% that found by direct intracerebral injection of the precursor.[61] A similar decrease, relative to the *in vivo* results, in the incorporation into fatty acid from acetyl labeled *N*-acetyl aspartic acid has been observed with brain slices. This effect is not limited to fatty acid biosynthesis. Although some *in vivo* synthesis is found in the adult rat brain,[61,62] slices do not convert radioactive acetate into cholesterol.[63]

Finally, Gatt[20] has shown, by using tritium oxide as a brain lipid precursor, that the greatest incorporation into the total lipid extract by the intact rat occurs in animals 15–20 days old. Individual fatty acid-containing lipids such as lecithin, sphingomyelin, and cerebrosides also show maximal incorporation by this age group. Slice experiments with animals of various ages with acetate,[29,64] citrate,[29] and glutamate[29] do not show this pattern. Instead, the highest incorporation is found with the newborn brain, and the conversion drops steadily with aging, again illustrating the difficulty of comparing *in vivo* lipogenesis with *in vitro* experiments.

The conversion of precursors to fatty acid is also dependent on the slice orientation; the conversion of acetate to lipids by the first cortex slice has been found to be twice that of the second cortex slice.[64] Respiration of the two slices is identical.

The *in vitro* conversion of acetate to fatty acids has been studied using single, whole mouse brains.[65] Although this technique gives appreciable incorporation of radioactivity, its usefulness is open to question until additional information on the respiration, glucose utilization, and other biochemical parameters of this preparation are obtained.

## 3. *Chain Elongation and Desaturation*

Kishimoto and Radin[66] have discussed the criteria to be established for determining whether a particular fatty acid is synthesized *de novo* or by chain elongation from radioactive acetate. For C16 : 0, which is made *de novo*, the relative specific activity of the carboxyl group to the total fatty acid chain is always 1 : 8, regardless of how soon after injection the animal is killed. With chain elongation, the specific activity of the carboxyl relative to that of the chain will depend on the time elapsed between injection and sampling.

From these considerations they have suggested that the stearate used for cerebroside synthesis is synthesized *de novo*, while that for ganglioside synthesis is made by chain elongation. They suggest that the liver microsomal system that converts malonyl CoA to stearate also functions in the brain to synthesize extralong-chain fatty acids *de novo*.[67–69] Chain elongation by a mitochondrial enzyme system also has been described in other tissues,[70,71] but this system has not yet been clearly defined in the brain.

It recently has been suggested that the elongation of the $\omega 3$ and $\omega 6$ series of unsaturated fatty acids in the brain may differ both in mechanism and in control factors.[33] Differences in the labeling of arachidonic and docosohexaenoic acid from radioactive acetate have been interpreted as indicating that separate compartments exist for the elongation of these two series of fatty acids.

This data also implies that the sources of the two-carbon units required for elongation of each series is not the same. A similar conclusion on the compartmentalization of carbon pools involved in synthesis of the polyenic acids can be drawn from the data of Klenk.[72] He examined the labeling in the dicarboxylic acids obtained from degradation of the unsaturated fatty acid synthesized from radioactive glucose or acetate by brain slices. If the pools of acetyl units from each precursor were identical, then the ratio of the activity of equivalent fragments in each experiment should be the same. However, although the $C_3$ dicarboxylic acid from acetate is approximately 3 times the activity of that from glucose, the $C_5$ fragment in the first experiment is more than 13 times the activity found in the latter. A similar disparity of labeling exists in the other fragments.

It should be pointed out that exogenous fatty acids may not be transported into the cell compartments where chain elongation takes place. Thus Fulco and Mead[73] have suggested that oleic acid is the precursor of nervonic acid in the brain since, with [1-$^{14}$C] acetate the carboxyl group of nervonic acid is highly labeled. Kishimoto and Radin[74] also find that the carboxyl group of nervonic acid from the brain sphingolipids is highly labeled from carboxyl-labeled oleate. The $\omega$-end has virtually no activity. Thus it appears that exogenous oleate is not used for chain elongation of the sphingolipid fatty acids but is degraded to acetate, which elongates endogenous fatty acids.

The polyunsaturated fatty acids in the developing brain have been determined.[75] The hexaenes comprise the greatest proportion of the polyunsaturated fatty acids and, in contrast to other tissues, the brain contains only traces of dienes but relatively large quantities of tetraenes. The regional distribution of the polyenes is such that the hexaenes predominate in gray matter and the tetraenes in white matter.[76]

The synthesis of the polyunsaturated fatty acids in the brain has not been examined in the same detail as in other tissues. All of the intermediates to the longer-chain polyunsaturated fatty acids have been found in the chick embryo brain by injection of radio-labeled C18 : 2 and C18 : 3 into the yolk sack,[19] suggesting that the pathways in brain are the same as in other tissues. The effect of feeding these polyunsaturated fatty acids in the diet has been previously detailed.

The effects on the desaturation of fatty acids by other fatty acids by microsomal enzymes from liver have been studied by Brenner and Peluffo;[77,78] such studies using systems obtained from brain remain to be done.

## 4. *α-Hydroxy and Odd-Numbered Fatty Acids*

The α-hydroxy acids and the odd-numbered fatty acids are related by a precursor–product sequence.

The odd-numbered fatty acids are generally a minor constituent of the individual brain lipids, although they are present in the polyphosphoinositides,[79] or cerebrosides,[80] in as high a concentration as the normal even-numbered carbon acids. The odd-numbered acids show the greatest increases with age in comparison to the other fatty acids of rat brain cerebrosides.[81] At least part of these fatty acids may be synthesized by chain elongation of propionic acid.[82,83] With labeled acetate, the odd-numbered fatty acids have specific activities that are similar to the adjacent even-numbered acids.[83] Thus a second route to the odd-numbered fatty acids has been described involving a one-carbon degradation of the even-numbered fatty acids. This latter pathway proceeds by hydroxylation of the non-hydroxy acids in the 2-position[84,85] followed by decarboxylation to the shorter acid.[86,87]

A brain microsomal system,[88,89] which may require ascorbic acid, decarboxylates the α-hydroxy and 2-keto fatty acids. Tentative evidence indicates that this one carbon degradation system also may utilize the odd-numbered acids as substrates.[90]

### 5. *Branched-Chain Fatty Acids*

Substantial amounts of a $C_{20}$ branched-chain fatty acid: 3,7,11,15-tetramethyl hexadecanoic acid (phytanic acid) were found by Klenk and Kahlke[91] in the brain and other tissues of patients afflicted with heredopathia atactica polyneuritiformis (Refsum's disease). The concentration of phytanic acid in normal individuals[92] has been shown to be smaller by several orders of magnitude than the concentrations in patients with Refsum's disease.[93-96]

In addition to the occurrence of the branched chain acids in brain, Hansen[97] has shown that in the sciatic nerve of these patients the acids are present in concentrations exceeded only by palmitate and oleate. There are significant changes in the amounts of these latter two acids. In both nerve and brain there is a significant increase in total lipid palmitate and an even greater decrease in oleate. The possibility that the biochemical lesion in Refsum's disease involves more than the metabolism of the branched-chain acids has been discussed.[98]

A major pathway for the degradation of phytanic acid has been shown to proceed by an α-decarboxylation to pristanic acid (2,6,10,14-tetramethylpentadecanoic acid).[99] The formation of α-hydroxyphytanic acid, a likely intermediate in the degradation of phytanic acid in mammalian tissues, has been demonstrated.[100]

The failure of this one-carbon degradation has been shown to be the enzymatic error in Refsum's disease by Steinberg and co-workers.[101]

A detailed discussion of the physiological and biochemical findings and the abnormalities which occur in the nervous system recently has been published.[102]

## C. Influences on Fatty Acid Biosynthesis

Fatty acid biosynthesis may be influenced by various metabolic conditions and chemical agents.

Starvation has a profound influence on lipid biosynthesis in peripheral nerve. Although the nerves appeared histologically normal, the rate of acetate incorporation was reduced to one-half the control value after 5–9 days of restricted food intake.[104] Smith[103] has studied the effect of fasting on lipid metabolism of the nervous system. Phospholipid synthesis of the brain is very little affected by a 3-day fast, but in the spinal cord the incorporation of the labeled precursors was decreased by 50%. The conclusion that lipid synthesis of the brain is little affected by fasting may not be warranted since these experiments were conducted with 8-week-old animals. More confidence might be placed on such a conclusion if younger animals were used, since restricted food intake by preweaned rats results in a significant decrease in the complex lipids that contain fatty acids in the central nervous system of these animals.[105]

The influence of insulin on fatty acids and total lipid synthesis in normal and diabetic animals has been studied by several workers. Eliasson and Hughes[106] have observed a constant depression of fatty acid synthesis from acetate in the spinal cord of diabetic rats but a synthesis 2–19 times higher than the normal control in the diabetic sciatic nerve. In contrast to this report, Eliasson[107] recently found a decrease of 50% or more in the conversion of radioactive acetate to triglyceride, phospholipids, and glycolipids and cerebroside fatty acids in the sciatic nerve of diabetic animals.

A similar depression of acetate incorporation into triglyceride, diglyceride, monoglyceride, and cholesterol esters compared to normal sciatic nerve was found in tissue from the diabetic rabbit.[108] Phospholipid radioactivity, however, was higher in the diabetic tissue.

Nerves from diabetic animals failed to respond to insulin in showing increased incorporation of acetate into lipid. The fatty acid synthesis in nerves from normal control animals, on the other hand, is profoundly stimulated by insulin.[108] For example, compared to the controls, the fatty acids of the cholesterol esters show an eightfold increase in radioactivity in the presence of insulin while those of phospholipid show a tenfold increase. In addition, insulin provides a stimulating effect on the incorporation of [1-$^{14}$C]acetate into fatty acids in brain slices from young and adult animals.[109] This effect is not seen when the precursors used are [2-$^{14}$C] mevalonic acid and [U-$^{14}$C] glucose.

Enhanced lipid biosynthesis was observed by the addition of a nerve growth factor from the submaxillary gland to ganglia preparation from chick embryos.[110] Contact radioautography of thin-layer chromatographs of the lipid mixture showed that the increased synthesis occurred in all lipid classes. Since the stimulatory effect is inhibited by the addition of actinomycin-D, it was concluded that the mechanism of action of the nerve growth factor is by induction of enzymes involved in lipid synthesis through a stimulation of DNA-primed RNA synthesis.

The conversion of labeled glucose into lipids and lipid classes in brain slices from anoxic and anoxic–ischemica rats is significantly lower than in the normal control.[111] Metabolic changes in rats exposed to an oxygen-enriched environment also have been examined.[112]

Homogenates of brain from mice with muscular dystrophy showed a twofold increase in fatty acid biosynthesis from acetate when compared to normal litter mates.[113] Increases in labeling above control values also were found in the phospholipid plus galactolipid fractions of the brainstem and spinal cord of animals with experimental allergic encephalomyelitis.[114] Diphtheritic polyneuritis, on the other hand, induces a profound decrease in acetate conversion to lipid in peripheral nerve, with a measureable decrease apparent as early as 24 hr after injection of the toxin.[103] Liver slices from normal and treated animals showed no such differences.

Difficulties involved in studying drug effects on brain lipid metabolism have been discussed by Ansell and Hawthorne.[115] In addition, the effects of psychotropic drugs on lipid biosynthesis in the brain and other tissues have been detailed by Holmes.[116] Several points are of concern in discussing drug effects: Different concentrations of the drug may have directly opposite results; the drug may have an effect on lipogenesis only at certain stages of development; metabolites of the drug may have a different time course of action; and neuropathological effects of the drug may differ in various parts of the nervous system.

Thus low concentrations of chlorpromazine ($10^{-4}$ to $10^{-5}$ M) increase the rate of incorporation of precursors into the fatty acids of the phospholipids in the brain slices of young rats; higher concentrations ($10^{-3}$ M) give an inhibition of incorporation. A similar concentration effect is also found *in vivo*.[117] Imipramine and desmethylimipramine have similar concentration effects.[118] These two drugs have different effects when given *in vivo*, probably because of the different time course of effective concentration within the tissue. With adult rat tissue, chlorpromazine has no effect.[117]

Mipafox (bis-mono-isopropyl aminofluorphosphine oxide), a choline esterase inhibitor, produces a peripheral neuropathy. In the proximal portion of the sciatic nerve, where no lesions are demonstrable, the drug causes a depression of acetate incorporation into total lipids.[119] In the degenerating branches, an increase in synthesis from radioactive acetate occurs.

## D. Physiological and Pharmacological Influences on Fatty Acid Composition of Lipids

The changes in fatty acid composition induced by dietary influences have been detailed in the section on transport of fatty acids into the brain. Compositional changes in the fatty acids of lipid classes and individual lipids are also dependent on the state of development of the nervous system, the anatomical region from which the lipid is isolated, and the neuropathological state, if any, of the tissues. These changes have been studied primarily on individual lipids and are discussed in the relevant sections. In addition, pharmacological and environmental influences also may effect alterations in the fatty acid patterns.

Although major neurological changes and effects on lipid synthesis are observed in Mipafox-treated animals, the polyunsaturated fatty acid patterns of nervous tissue remain remarkably constant except for a marked decrease in brain pentaenoic acid concentration.[120] Starvation produces a marked increase in tetraenoic acid and a decrease in pentaenoic acid concentration in the brain. Since these results bear no resemblance to those characteristic of essential fatty acid deficiency, it has been suggested that these results are due to a failure in the conversion of $\Delta^{(5,8,11,14)}$-eicosatetraenoic acid to $\Delta^{(4,7,10,13,16)}$ docosapentaenoic acid. Other changes in the fatty acid composition of peripheral nerves of organophosphorus-poisoned animals have been noted.[121] A similar decrease in docosahexaenoic acid is found in glial tumors. All the polyunsaturated fatty acids are not affected in the same fashion; linoleic acid concentration shows a tenfold increase over that found in normal tissues.[122] Deficiencies in the amounts of fatty acids, and also other fatty acid changes, are found in white matter in multiple sclerosis[123,124] and in diabetes.[125]

Concentrations of polyunsaturated fatty acids are significantly reduced compared to control animals in the brain of rats treated with high concentrations of vitamin A.[126] There are no effects on the saturated fatty acid patterns. In contrast, a cholesterol biosynthesis inhibitor, AY-9944, produces a doubling of brain sphingomyelin palmitate with a concomitant decrease in stearate; the unsaturated fatty acid patterns are also influenced by the drug.[127]

Of possible significance in the role of fatty acid composition to maintain a specific liquid–crystalline state of cellular membranes in the brain is the increase in the degree of unsaturation observed in changing the environmental temperature of goldfish down to 5°.[128]

## III. THE FREE FATTY ACIDS

### A. Occurrence

It has been reported that the free fatty acids account for little more than 1% of the dry weight of rat tissue that has been ground in a mortar and stored at −20°.[129] However, the levels of free fatty acids in nervous tissues is apparently greatly dependent on the isolation procedure since other workers report values of unesterified fatty acids one tenth of this in fresh rat tissue.[130] Brain slices incubated at 0 or 37° and fractionated to determine concentrations in subcellular fractions showed increases in free fatty acids of two- to threefold relative to the fresh tissue.

Increases in levels of free fatty acids have been found in different focuses of anemic softening of human brain that is independent of the age of the lesion.[131] Changes in phospholipids during 24 hr autolysis of rat brain, with a particular increase in lysolecithin, have been observed,[132] but no detectable changes in lecithin were found with much shorter incubation times.[133] The numerous enzymes for the release of free fatty acid from complex lipids will be described in the sections on the individual lipids.

### B. Biochemical Effects of Free Fatty Acids

Injection of short-chain fatty acids produces a narcotic-like effect in normal animals.[134] Aliphatic alcohols of similar chain length have been found to give an inhibition of respiration of electrically stimulated cortex slices.[135]

Initial stimulatory effects on the respiration of brain cortex slices by decanoate and other fatty acids are reversed to give an inhibition after a short period (45–60 min) of incubation.[136] This appears to be characterisitic of brain tissue since this effect of the fatty acid was not observed with either kidney or liver slices. A greatly stimulated aerobic glycolysis, as measured by $CO_2$ evolution, also was observed in the presence of decanoate. However, no effect on glucose utilization was found in the same time periods when cortex slices were incubated with palmitate.[137] The two experiments may not be precisely comparable since in the latter experiment the fatty acids were added as a complex with albumin whereas in the former experiment the salts of the fatty acids were used. In addition, the chain length of the fatty acid used also has a significant effect in other systems on such important metabolic factors as mitochondrial swelling and function.[138]

Differences in chain length of the fatty acid also may account for the different effects observed on anaerobic glycolysis. While decanoate slightly stimulates anaerobic glycolysis, as measured by $CO_2$ evolution,[136] longer chain fatty acids such as palmitate, stearate, and oleate give an inhibition of anaerobic glycolysis, as measured by lactate production.[139]

Uncoupling of oxidative phosphorylation is observed with isolated rat brain mitochondria incubated with fatty acids.[140] At similar concentrations, on the order of 0.4 mM, a 50% inhibition of the adenosine-triphosphate-$^{32}$Pi exchange also is observed.[136]

A potential role of unesterified fatty acids as carriers of cations through membranes also has been suggested.[141]

## IV. FATTY ACID OXIDATION

Since the respiratory quotient of the brain is close to 1, it has been concluded that the brain *in vivo* oxidizes primarily glucose. The ability of the brain to utilize lipid as an energy source has been inferred from glucose-free perfusion experiments of Geiger, Magnes, and Geiger.[142] Both fatty acids[17,143] and ketone bodies[137] have been shown to be oxidized by the brain.

The presence of the acyl-activating enzymes is masked in whole brain homogenate, since no activation of palmitate is found.[144] Subcellular fractionation, however, reveals that the activating enzyme is present and is localized primarily in the supernatant fraction, with significant amounts of activity in the mitochondrial and microsomal fractions. The enzyme preparation is capable of activating fatty acids, ranging from butyrate to oleate, with the highest activity found for stearate and palmitate.

Other enzymes involved in fatty acid oxidation are palmityl CoA dehydrogenase, which has been demonstrated in extracts of acetone powder of brain mitochondria,[(144)] and β-ketothiolase, which is present in low activity in brain.[(145)]

Beattie and Basford have studied the activation of fatty acids in bovine brain mitochondria.[(146)] They have presented evidence for two different activating systems, an ATP-linked reaction that is dependent primarily on respiration and a system that utilizes the high-energy intermediate of oxidative phosphorylation. In these mitochondria the incorporation of palmitate into phospholipids requires ATP and CoA; fatty acid oxidation, although proceeding at a rate 4 times that of phospholipid formation, does not require the addition of coenzyme A.[(147)] The capacity of brain mitochondria to oxidize fatty acids is comparable to that of mitochondria from liver or heart tissue[(148)] when prepared in a particular fashion;[(149)] other preparations give lower values.[(144)]

Carnitine enhances the rate of fatty acid oxidation in isolated mitochondria.[(148)] The concentration of free carnitine is the lowest in brain of the organs examined.[(150)] The organ and subcellular distribution of palmityl CoA:carnitine palmityl transferase has been studied in biopsy samples from man.[(151)] The brain samples had relatively low activities compared with most other tissues. The brain, like the other tissues examined, had most of the activity in the mitochondrial fraction, but significant activity was present in the microsomal and particle-free supernatant fraction.

The oxidation of fatty acids by brain mitochondria[(152)] is inhibited by sodium ions; this inhibition persists in the presence of oligomycin and can be reversed by increasing concentrations of fatty acid.

## V. TURNOVER OF FATTY ACIDS

The capacity of the brain to oxidize fatty acids indicates that turnover of the complex lipids occurs. Early experiments using deuterium oxide suggested that approximately 20% of the fatty acids of the brain were broken down and resynthesized.[(153)]

Radin and co-workers have pointed out that several factors must be evaluated in discussing turnover data.[(66,90,155)]

The time course of changes in radioactivity in the individual fatty acids of specific lipids is dependent on the age of the animal at the time of injection. In addition, the specific activity of a particular fatty acid will vary not only with time but also with the specific lipid under examination.[(155)]

The time intervals chosen for sampling are extremely important; profound changes in radioactivity of individual fatty acids may occur within 4 hr after injection. For example, the radioactivity of the cerebroside C20 : 0 fatty acid declines rapidly 4 hr after injection of precursor, begins to rise again within 12 hr, reaches a peak by 14 days after injection, and then gradually declines.[90] This has been interpreted as indicating that two or more pools of cerebroside exist that turn over at greatly different rates.

The turnover of the individual lipids of myelin is discussed in Chapter 21 of this volume.

## VI. CONCLUSION

Since the fatty acid composition of a lipid may exert a great influence on the characteristics of the membranes with which it is associated, the ultimate goal of studies on the transport and metabolism of the fatty acids of the nervous system is control over the fatty acid composition of individual lipids. By controlled variations in the fatty acid composition of these lipids, it is hoped that insights into the metabolic and physiological role of the individual lipids in normal and abnormal functional states of the nervous system can be gained.

## VII. ACKNOWLEDGMENT

The author wishes to thank Miss Eileen Allinson for her help in preparing this manuscript and to acknowledge the support of the U.S. Public Health Service under Grant NB-03356.

## VIII. REFERENCES

1. W. M. Sperry, H. Waelsch, and V. A. Stoyanoff, Lipid metabolism in brain and other tissues of the rat, *J. Biol. Chem.* **135**:281–290 (1940).
2. A. F. D'Adamo, Jr., unpublished observations.
3. B. K. Siesjo and W. O. B. Thompson, The uptake of inspired $^{14}CO_2$ into the acid-labile, the acid-soluble, the lipid, the protein and the nucleic acid fractions of rat brain tissue, *Acta Physiol. Scand.* **64**:182–192 (1965).
4. J. I. Kessler, M. Demeny, and H. Sobotka, Rates of tissue uptake of palmitic acid -1-$^{14}$C complexed with albumin by two different procedures, *J. Lipid Res.* **8**:185–190 (1967).
5. D. Engelmann, I. Stork, H. D. Eisenbarth, A. Naher, D. Heyse, and K. Schreier, Uber die Resorption und den Stoffwechsel einiger 1-$^{14}$C-markierter freier Fettsäuren bei jungen Kaninchen, *Clin. Chim. Acta* **9**:126–137 (1964).
6. B. L. Walker, Maternal diet and brain fatty acids in young rats, *Lipids* **2**:497–500 (1967).
7. H. Mohrhauer and R. T. Holman, Alteration of the fatty acid composition of brain lipids by varying levels of dietary essential fatty acids, *J. Neurochem.* **10**:523–530 (1963).

8. L. A. Biran, W. Bartley, C. W. Carter, and A. Renshaw, Studies on essential fatty acid deficiency. Effect of the deficiency on the lipids in various rat tissues and the influence of dietary supplementation with essential fatty acids on deficient rats, *Biochem. J.* **93**:492–498 (1964).
9. G. J. Marco, L. J. Machlin, E. Emery, and R. S. Gordon, Dietary effects of fats upon fatty acid composition of the mitochondria, *Arch. Biochem. Biophys.* **94**:115–120 (1961).
10. L. A. Witting, C. C. Harvey, B. Century, and M. K. Horwitt, Dietary alterations of fatty acids of erythrocytes and mitochondria of brain and liver, *J. Lipid Res.* **2**:412–418 (1961).
11. L. Rathbone, The effect of diet on the fatty acid compositions of serum, brain, brain mitochondria and myelin in the rat, *Biochem. J.* **97**:620–628 (1965).
12. J. Tomasch, Dietary factors and nerve myelination, *J. Anat.* **95**:180–190 (1961).
13. L. J. Machlin, G. J. Marco, and R. S. Gordon, Effect of diet and encephalomalacia on the fatty acid composition of the brain of young and old chickens, *J. Am. Oil Chem. Soc.* **39**:229–232 (1962).
14. B. Century and M. K. Horwitt, Role of arachidonic acid in nutritional encephalomalacia: Interrelationships of essential and nonessential polyunsaturated fatty acids, *Arch. Biochem. Biophys.* **104**:416–422 (1964).
15. K. K. Carroll, Levels of radioactivity in tissues and in expired carbon dioxide after administration of 1-$C^{14}$-labelled palmitic acid, C2-$^{14}$-labelled erucic acid, or 2-$C^{14}$-labelled nervonic acid to rats, *Can. J. Biochem. Physiol.* **40**:1229–1238 (1962).
16. S. Laurell, Distribution of $C^{14}$ in rats after intravenous injection of non-esterified palmitic acid-1-$C^{14}$, *Acta Physiol. Scand.* **46**:97–106 (1959).
17. C. Allweis, T. Landau, M. Abeles, and J. Magnes, The oxidation of uniformly labelled albumin-bound palmitic acid to $CO_2$ by the perfused cat brain, *J. Neurochem.* **13**:795–804 (1966).
18. E. T. Pritchard, The formation of phospholipids from $^{14}C$-labelled precursors in developing rat brain *in vivo*, *J. Neurochem.* **10**:495–502 (1963).
19. K. Miyamoto, L. M. Stephanides, and J. Bernsohn, Incorporation of [1-$^{14}$C] linoleate and linolenate into polyunsaturated fatty acids of phospholipids of the embryonic chick brain, *J. Neurochem.* **14**:227–237 (1967).
20. S. Gatt, Metabolism of [1-$^{14}$C] lignoceric acid in the rat, *Biochim. Biophys. Acta* **70**:370–380 (1963).
21. R. J. White, M. S. Albin, and J. Verdura, Isolation of the monkey brain; *In vito* preparation and maintenance, *Science* **141**:1060–1061 (1963).
22. E. T. Pritchard, The formation of phospholipids from $C^{14}$-labelled precursors in slices from immature rat brain, *Can. J. Biochem. Physiol.* **40**:353–361 (1962).
23. G. R. Webster, The incorporation of long-chain fatty acids into phospholipids of respiring slices of rat cerebrum, *Biochem. J.* **102**:373–380 (1967).
24. H. Keen and C. Chlouverakis, Metabolism of isolated rat retina. The role of non-esterified fatty acid, *Biochem. J.* **94**:488–493 (1965).
25. J. S. Andrews and T. Kuwabara, Net triglyceride synthesis by rabbit cornea *in vitro*, *Biochim. Biophys. Acta* **54**:315–321 (1961).
26. D. W. Clarke and L. Geiger, Effect of experimental allergic encephalomyelitis serum on fatty acid output of brain slices, *Nature* **201**:401 (1964).
27. C. H. Tator, J. R. Evans, and J. Olszewski, Tracers for the detection of brain tumors. Evaluation of radio-iodinated human serum albumin and radio-iodinated fatty acid, *Neurology* **16**:650–661 (1966).
28. S. Garattini, P. Paoletti, and R. Paoletti, Lipid biosynthesis *in vivo* from acetate-1-$C^{14}$ and 2-$C^{14}$ and mevalonic-2-$C^{14}$ acid, *Arch. Biochem. Biophys.* **80**:210–211 (1959).

29. A. F. D'Adamo, Jr., and A. P. D'Adamo, Acetyl transport mechanism in the nervous system. The ketoglutarate shunt and fatty acid synthesis in the developing rat brain, *J. Neurochem.*, **15**:315–323 (1968).
30. P. Hill, Incorporation of citrate 1, 5-$^{14}$C and $^{3}$H palmitic acid and the composition of the $\beta$-acyl chain in phosphatides of rat tissue, *Can. J. Biochem. Physiol.* **44**:1285–1289 (1966).
31. R. G. Gould, The Biosynthesis of Cholesterol, *in Cholesterol* (R. P. Cook, ed.), p. 212, Academic Press, New York, 1958.
32. C. J. van den Berg, P. Mela, and H. Waelsch, On the contribution of the tricarboxylic acid cycle to the synthesis of glutamate, glutamine and aspartate in brain, *Biochem. Biophys. Res. Commun.* **23**:479–484 (1966).
33. K. Miyamoto, L. M. Stephanides, and J. Bernsohn, Acetate-1-$^{14}$C incorporation into polysaturated fatty acids of phospholipids of developing chick brain, *J. Lipid Res.* **8**:191–195 (1967).
34. A. F. Spencer and J. M. Lowenstein, The supply of precursors for the synthesis of fatty acids, *J. Biol. Chem.* **237**:3640–3648 (1962).
35. I. B. Fritz and K. T. N. Yue, Effects of carnitine on acetyl-CoA oxidation by heart muscle mitochondria, *Am. J. Physiol.* **206**:531–535 (1962).
36. A. F. D'Adamo, Jr., L. I. Gidez, and F. M. Yatsu, Acetyl transport mechanisms. Involvement of *N*-acetyl aspartic acid in *de novo* fatty acid biosynthesis in the developing rat brain, *Exptl. Brain Res.*, **5**:267–273 (1968).
37. A. F. D'Adamo, Jr., and F. M. Yatsu, Acetate metabolism in the nervous system. *N*-acetyl-L-aspartic acid and the biosynthesis of brain lipids, *J. Neurochem.* **13**:961–965 (1966).
38. P. A. Srere and A. Bhaduri, Incorporation of radioactive citrate into fatty acids, *Biochim. Biophys. Acta.* **59**:487–489 (1962).
39. J. Schuberth, J. Sollenberg, A. Sundwall, and B. Sorbo, Determination of acetyl-coenzyme A in brain, *J. Neurochem.* **12**:451–454 (1965).
40. D. J. Pearson and P. K. Tubbs, Carnitine and derivatives in rat tissues, *Biochem. J.* **105**:953–963 (1967).
41. V. P. Wittaker, The application of subcellular fractionation techniques to the study of brain function, *Progr. Biophys. Mol. Biol.* **15**:39–96, 1965.
42. J. Schuberth, On the biosynthesis of acetyl coenzyme A in the brain, *Biochim. Biophys. Acta* **98**:1–7 (1965).
43. S. Tucek, Subcellular distribution of acetyl-CoA synthetase, ATP citrate lyase, citrate synthase, choline acetyltransferase, fumarate hydratase and lactate dehydrogenase in mammalian brain tissue, *J. Neurochem.* **14**:531–545 (1967).
44. R. W. Korff, The effects of alkali metal ions on the acetate activating enzyme system, *J. Biol. Chem.* **203**:265–271 (1953).
45. R. A. Deitrich and L. Hellerman, Pyruvate metabolism. V. Pyruvate utilization by mitochondria of rat brain, *J. Biol. Chem.* **239**:2735–2740 (1964).
46. R. E. Koeppe, R. M. O'Neal, and C. H. Hahn, Pyruvate decarboxylation in thiamine deficient brain, *J. Neurochem.* **11**:695–699 (1964).
47. R. E. Koeppe, G. A. Mourkides, and R. J. Hill, Some factors affecting routes of pyruvate metabolism in rats, *J. Biol. Chem.* **234**:2219–2222 (1959).
48. A. D. Friedman, P. Rumsey, and S. Graff, The metabolism of pyruvate in the tricarboxylic acid cycle, *J. Biol. Chem.* **235**:1854–1855 (1960).
49. S. Tucek, The use of choline acetyltransferase for measuring the synthesis of acetyl CoA and its release from brain mitochondria, *Biochem. J.* **104**:749–756 (1967).
50. R. E. McCaman, M. W. McCaman, and M. L. Stafford, Carnitine acetyltransferase in nervous tissue, *J. Biol. Chem.* **241**:930–934 (1966).

51. M. S. Kornacker and J. M. Lowenstein, Citrate and the conversion of carbohydrate into fat, *Biochem. J.* **94**:209–215 (1965).
52. R. O. Brady, Biosynthesis of fatty acids. II. Studies with enzymes obtained from brain, *J. Biol. Chem.* **235**:3099–3103 (1960).
53. J. D. Robinson, R. M. Bradley, and R. O. Brady, Biosynthesis of fatty acids. III. Utilization of substituted acetyl coenzyme A derivatives as intermediates, *J. Biol. Chem.* **238**:528–532 (1963).
54. C. Landriscina, V. Liso, M. N. Gadaleta, and A. Alifano, Sintesi *in vitro* di acidi grassi in varie frazioni di cellule di cervello di ratto, *Boll. Soc. Ital. Biol. Sper.* **42**:473–476 (1966).
55. A. H. Hughes and S. G. Eliasson, Synthesis of cholesterol and fatty acids in fractions of peripheral nerve, *J. Clin. Invest.* **39**:111–115 (1960).
56. J. D. Robinson, R. O. Brady, and R. M. Bradley, Biosynthesis of fatty acids: IV. Studies with inhibitors. *J. Lipid Res.* **4**:144–150 (1963).
57. R. O. Brady, Studies of inhibitors of fatty acid biosynthesis. III. Mechanism of action of tetrolyl-coenzyme A, *Biochim. Biophys. Acta* **70**:467–468 (1963).
58. R. Lindbohm and H. Wallgren, Oxidation of acetate by rat cerebral cortex *in vitro* and the effect of stimulation, *J. Neurochem.* **13**:573–577 (1966).
59. P. Reich, E. Henneman, and M. L. Karnovsky, Oxidative metabolism of glucose in resting and active sciatic nerve, *J. Neurochem.* **14**:447–456 (1967).
60. G. Majno, E. L. Gasteiger, M. LaGattuta, and M. L. Karnovsky, Lipid biosynthesis *in vitro* by electrically stimulated rat sciatic nerves, *J. Neurochem.* **3**:127–131 (1958).
61. H. J. Nicholas and B. E. Thomas, The metabolism of cholesterol and fatty acids in the central nervous system, *J. Neurochem.* **4**:42–49 (1959).
62. P. J. McMillan, G. W. Douglas, and R. A. Martensen, Incorporation of $C^{14}$ of acetate-1-$C^{14}$ and pyruvate-2-$C^{14}$ into brain cholesterol in the intact rat, *Proc. Soc. Exptl. Biol.* **96**:738–741 (1957).
63. P. A. Srere, I. L. Chaikoff, S. S. Treitman, and L. S. Burnstein, The extrahepatic synthesis of cholesterol, *J. Biol. Chem.* **182**:629–634 (1950).
64. G. Majno and M. L. Karnovsky A biochemical and morphological study of myelination and demyelination. 1. Lipide biosynthesis *in vitro* by normal nervous tissue, *J. Exptl. Med.* **107**:475–496 (1958).
65. C. E. Rowe, The occurrence and metabolism *in vitro* of unesterified fatty acid in mouse brain, *Biochim. Biophys. Acta* **84**:424–434 (1964).
66. Y. Kishimoto and N. S. Radin, Metabolism of brain glycolipid fatty acids, *The Lipids* **1**:47–61 (1966).
67. E. Lorch, S. Abraham, and I. L. Chaikoff, Fatty acid synthesis by complex systems. The possibility of regulation by microsomes, *Biochim. Biophys. Acta* **70**:627–641 (1963).
68. D. H. Nugteren, The enzymatic chain elongation of fatty acids by rat-liver microsomes. *Biochim. Biophys. Acta* **106**:280–290 (1965).
69. R. B. Guchhait, G. P. Putz, and J. W. Porter, Synthesis of long-chain fatty acid by microsomes of pigeon liver, *Arch. Biochem. Biophys.* **117**:541–549 (1966).
70. W. R. Harlan, Jr., and S. J. Wakil, The pathways of synthesis of fatty acids by mitochondria, *Biochem. Biophys. Res. Commun.* **8**:131–135 (1962).
71. W. R. Harlan, Jr., and S. J. Wakil, Synthesis of fatty acids in animal tissues, 1. Incorporation of $C^{14}$-acetyl coenzyme A into a variety of long chain fatty acids by subcellular particles, *J. Biol. Chem.* **238**:3216–3223 (1963).
72. E. Klenk, Incorporation of $^{14}C$-labelled acetate into some lipids of nervous tissue, in *Metabolism of the Nervous System* (D. Richter, ed.), pp. 396–398, Pergamon Press, Oxford, 1957.

73. A. J. Fulco and J. F. Mead, The biosynthesis of lignoceric, cerebronic, and nervonic acids, *J. Biol. Chem.* **236**:2416–2420 (1961).
74. Y. Kishimoto and N. S. Radin, Biosynthesis of Nervonic acid and its homologues from carboxyl-labeled oleic acid, *J. Lipid Res.* **4**:444–447 (1963).
75. J. C. Kirschman and J. G. Coniglio, Polyunsaturated fatty acids in tissues of growing male and female rats, *Arch. Biochem. Biophys.* **93**:297–301 (1961).
76. B. Gerstl, M. J. Kahnke, J. K. Smith, M. G. Tavaststjerna, and R. B. Hayman, Brain lipids in multiple sclerosis, *Brain* **84**:310–319 (1961).
77. R. R. Brenner and R. O. Peluffo, Effect of saturated and unsaturated fatty acids on the desaturation *in vitro* of palmitic, stearic, oleic, linoleic and linolenic acids, *J. Biol. Chem.* **241**:5213–5219 (1966).
78. R. R. Brenner and R. O. Peluffo, Inhibitory effect of docosa-4,7,10,13,16,19-hexaenoic acid upon the oxidative desaturation of linoleic into $\gamma$-linolenic acid and of $\alpha$-linolenic into octadeca-6,9,12,15-tetraenoic acid, *Biochim. Biophys. Acta* **137**:184–186 (1967).
79. S. E. Kerr and W. W. C. Read, The fatty acid components of polyphosphoinositide prepared from calf brain, *Biochim. Biophys. Acta* **70**:477–478 (1963).
80. Y. Kishimoto and N. S. Radin, Isolation and determination methods for brain cerebrosides, hydroxy fatty acids, and unsaturated and saturated fatty acids, *J. Lipid Res.* **1**:72–76 (1959).
81. Y. Kishimoto and N. S. Radin, Composition of cerebroside acids as a function of age, *J. Lipid Res.* **1**:79–82 (1959).
82. W. Pedersen, L. Hausheer, and K. Bernhard, Weitere beitrage zur neurochemie: Die inkorporation von [1-$^{14}$C]-propionate in die fettsauren der gehirn-cerebroside, *Helv. Chim. Acta* **46**:675–677 (1963).
83. A. K. Hajra and N. S. Radin, Biosynthesis of the cerebroside odd-numbered fatty acids, *J. Lipid Res.* **3**:327–332 (1963).
84. A. K. Hajra and N. S. Radin, *in vivo* conversion of labeled fatty acids to the sphingolipid fatty acids in rat brain, *J. Lipid Res.* **4**:448 (1963).
85. J. F. Mead and G. M. Levis, Alpha oxidation of the brain fatty acids, *Biochem. Biophys. Res. Commun.* **9**:231–234 (1962).
86. J. F. Mead and G. M. Levis, A I carbon degradation of the long chain fatty acids of brain sphingolipids, *J. Biol. Chem.* **238**:1634–1636 (1963).
87. J. F. Mead and G. M. Levis, Enzymatic decarboxylation of the alpha-hydroxy acids by brain microsomes, *Biochem. Biophys. Res. Commun.* **11**:319–324 (1963).
88. W. E. Davies, A. K. Hajra, S. S. Parmar, N. S. Radin, and J. F. Mead, Decarboxylation of 2-keto fatty acids by brain, *J. Lipid Res.* **7**:270–276 (1966).
89. G. M. Levis, The possible role of ascorbic acid in the $\alpha$-hydroxyacid decarboxylase of brain microsomes, *Biochim. Biophys. Acta* **99**:194–197 (1965).
90. A. K. Hajra and N. S. Radin, Isotopic studies of the biosynthesis of the cerebroside fatty acids in rats, *J. Lipid Res.* **4**:270–278 (1963).
91. E. Klenk and W. Kahlke, Uber das Vorkommen der 3.7.11.5-Tetramethyl-hexadecansäure (Phytansäure) in den Cholesterinestern und anderen Lipoidfraktionen der Organe bei einem Krankheitsfall unbekannter Genese (Verdacht auf Heredopathia atactica polyneuritiformis [Refsum-syndrome]), *Hoppe-Seylers Z. Physiol. Chem.* **333**:133–139 (1963).
92. J. Avigan, Pristanic acid (2,6,10,14-tetramethylpentadecanoic acid) and phytanic acid (3,7,11,15-tetramethylhexadecanoic acid) content of human and animal tissues, *Biochim. Biophys. Acta* **125**:607–610 (1966).
93. W. Kahlke, Refsum-Syndrom.-lipoidchemische Untersuchungen bei 9 Fällen, *Klin. Wochschr.* **42**:1011–1016 (1964).

94. W. Kahlke and R. Richterich, Refsum's disease (heredopathia atactica polyneuritiformis). An inborn error of lipid metabolism with storage of 3,7,11,15-tetramethyl hexadecanoic acid. II. Isolation and identification of the storage product, *Am. J. Med.* **39**:237–241 (1965).
95. S. Laurell, Separation and characterization of phytanic acid-containing plasma-triglycerides from a patient with Refsum's disease, *Biochim. Biophys. Acta* **152**:75–79 (1968).
96. W. S. Alexander, Phytanic acid in Refsum's syndrome, *J. Neurol. Neurosurg. Psychiat.* **29**:412–416 (1966).
97. R. P. Hansen, 3,7,11,15-Tetramethylhexadecanoic acid: Its occurrence in the tissues of human afflicted with Refsum's syndrome, *Biochim. Biophys. Acta* **106**:304–310 (1965).
98. D. Steinberg, C. Mize, J. Avigan, H. M. Fales, L. Eldjarn, K. Try, O. Stokke, and S. Refsum, On the metabolic error in Refsum's disease, *Trans. Am. Neurol. Assoc.* **91**:168–172 (1966).
99. J. Avigan, D. Steinberg, A. Gutman, C. E. Mize, and W. A. Milne, Alpha-decarboxylation, An important pathway for degradation of phytanic acid in animals, *Biochem. Biophys. Res. Commun.* **24**:838–844 (1966).
100. S. C. Tsai, J. H. Herndon, B. W. Uhlendorf, H. M. Fales, and C. E. Mize, The formation of alpha-hydroxy phytanic acid from phytanic acid in mammalian tissues, *Biochem. Biophys. Res. Commun.* **28**:571–577 (1967).
101. D. Steinberg, J. H. Herndon, Jr., B. W. Uhlendorf, C. E. Mize, J. Avigan, and G. W. A. Milne, Refsum's disease: Nature of the enzyme defect, *Science* **156**:1740–1742 (1967).
102. D. Steinberg, F. Q. Vroom, W. K. Engel, J. Cammermeyer, C. E. Mize, and J. Avigan, Refsum's disease—A recently characterized lipidosis involving the nervous system, *Annal. Intern. Med.* **66**:365–395 (1967).
103. G. Majno and M. L. Karnovsky, Experimental study of diphtheritic polyneuritis in the rabbit and guinea pig. II. The effect of diphtheria toxin on lipide biosynthesis by guinea pig nerve, *J. Neuropath. Exptl. Neurol.* **19**:7–24 (1960).
104. M. E. Smith, The effect of fasting on lipid metabolism of the central nervous system of the rat, *J. Neurochem.* **10**:531–536 (1963).
105. W. J. Culley and E. T. Mertz, Effect of restricted food intake on growth and composition of preweanling rat brain, *Proc. Soc. Exptl. Biol. Med.* **118**:233–235 (1965).
106. S. G. Eliasson and A. H. Hughes, Cholesterol and fatty acid synthesis in diabetic nerve and spinal cord, *Neurology* **10**:143–147 (1960).
107. S. G. Eliasson, Lipid synthesis in peripheral nerve from alloxan diabetic rats, *Lipids* **1**:237–240 (1966).
108. R. A. Field and L. C. Adams, Insulin response of peripheral nerve. II. Effects on lipid metabolism, *Biochim. Biophys. Acta* **106**:474–479 (1965).
109. E. Grossi, P. Paoletti, and M. Poggi, The effect of insulin on brain cholesterol and fatty acid biosynthesis, *World Neurol.* **3**:209–215 (1962).
110. A. Liuzzi, P. U. Angeletti, and R. Levi-Montalcini, Metabolic effects of a specific nerve growth factor (NGF) on sensory and sympathetic ganglia: Enhancement of lipid biosynthesis, *J. Neurochem.* **12**:705–708 (1965).
111. D. P. Kosow, H. P. Schwarz, and A. Marmolejo, Lipid biosynthesis in anoxic-ischemic rat brain, *J. Neurochem.* **13**:1139–1142 (1966).
112. A. D. Bond, J. P. Jordan, and J. B. Allred, Metabolic changes in rats exposed to an oxygen-enriched environment, *Am. J. Physiol.* **212**:526–529 (1967).
113. J. L. Rabinowitz, Enzymic studies on dystrophic mice and their littermates (lipogenesis and cholesterolgenesis), *Biochim. Biophys. Acta* **43**:337–338 (1960).

114. M. E. Smith, Lipid biosynthesis in the central nervous system in experimental allergic encephalomyelitis, *J. Neurochem.* **11**:29–37 (1964).
115. G. B. Ansell and J. N. Hawthorne, *The Phospholipids*, pp. 352–362, Elsevier, Amsterdam, 1964.
116. W. L. Holmes, Drugs affecting lipid synthesis, *in Lipid Pharmacology* (R. Paoletti, ed.), pp. 131–184, Academic Press, New York, 1964.
117. E. Grossi, P. Paoletti, and R. Paoletti, The *in vitro* and *in vivo* effects of chlorpromazine on brain lipid synthesis, *J. Neurochem.* **6**:73–78 (1960).
118. R. Fumagalli, E. Grossi, and P. Paoletti, The effect of imipramine and desmethylimipramine on lipid biosynthesis in brain and liver, *J. Neurochem.* **10**:213–217 (1963).
119. G. Manjo and M. L. Karnovsky, A biochemical and morphologic study of myelination and demyelination. III. Effect of an organo-phosphorus compound (Mipafox) on the biosynthesis of lipid by nervous tissue of rats and hens, *J. Neurochem.* **8**:1–16 (1961).
120. C. D. Joel, H. W. Moser, G. Majno, and M. L. Karnovsky, Effects of bis-(monoisopropylamino)-fluorophosphine oxide (Mipafox) and of starvation on the lipids in the nervous system of the hen, *J. Neurochem.* **14**:479–488 (1967).
121. J. F. Berry and W. H. Cevallos, Lipid class and fatty acid composition of peripheral nerve from normal and organophosphorus-poisoned chickens, *J. Neurochem.* **13**:117–124 (1966).
122. A. A. Stein, E. Opalka, and I. Rosenblum, Fatty acid analysis of two experimental transmissible glial tumors by gas–liquid chromatography, *Cancer Res.* **25**:201–205 (1965).
123. B. Gerstl, M. G. Tavaststjerna, R. B. Hayman, L. F. Eng, and J. K. Smith, Alterations in myelin fatty acids and plasmalogens in multiple sclerosis, *Ann. N.Y. Acad. Sci.* **122**:405–416 (1965).
124. J. N. Cumings, R. C. Shortman, and T. Skrbic, Lipid studies in the blood and brain in multiple sclerosis and motor neurone disease, *J. Clin. Pathol.* **18**:641–644 (1965).
125. B. Gerstl, M. G. Tavaststjerna, R. B. Hayman, J. K. Smith, and L. F. Eng, Lipid studies of white matter and thalamus of human brains, *J. Neurochem.* **10**:889–902 (1963).
126. U. K. Misra, Fatty acids of brain in hypervitaminosis A in rats, *Can. J. Biochem. Physiol.* **44**:1539–1542 (1966).
127. P. Hill, Effect of a cholesterol-biosynthesis inhibitor on the fatty acid composition of phospholipids in the serum and tissues of rats, *Biochem. J.* **98**:696–701 (1966).
128. P. V. Johnston and B. I. Roots, Brain lipid fatty acids and temperature acclimation, *Comp. Biochem. Physiol.* **11**:303–309 (1964).
129. C. Galli and D. Cecconi, Lipid changes in rat brain during maturation, *Lipids* **2**:76–82 (1967).
130. G. G. Lunt and C. E. Rowe, Unesterified fatty acid in brain and its release in subcellular fractions, *Biochem. J.* **104**:56P–57P (1967).
131. F. Lindlar and R. Guttler, Die Lipoide der weissen Hirnsubstanz während der Autolyse und bei der anämischenErweichung, *Acta Neuropathol.* **6**:349–358 (1966).
132. R. Niemiro and J. Przyjemski, Changes in phospholipids during autolysis of rat brain and lung, *Acta Biochim. Polon.* **10**:107–116 (1963).
133. G. B. Ansell and S. Spanner, The breakdown of endogenous ethanolamine and choline phospholipids in rat-brain homogenates, *Biochem. J.* **88**:26P–27P (1963); The magnesium ion dependent cleavage of the vinyl ether linkage of brain ethanolamine plasmalogen, *Biochem. J.* **94**:252–258 (1965).
134. F. E. Samson, Jr., and N. Dahl, Coma produced by injection of short chain fatty acids, *Federation Proc.* **14**:129 (1955).

135. R. Lindbohm and H. Wallgren, Changes in respiration of rat brain cortex slices induced by some aliphatic alcohols, *Acta Pharmacol. Toxicol.* **19**:53–58 (1962).
136. K. Ahmed and P. G. Scholefield, Studies on fatty acid oxidation. 8. The effects of fatty acids on metabolism of rat-brain cortex *in vitro*, *Biochem. J.* **81**:45–53 (1961).
137. F. S. Rollesteon and E. A. Newsholme, Effects of fatty acids, ketone bodies, lactate and pyruvate on glucose utilization by guinea-pig cerebral cortex slices, *Biochem. J.* **104**:519–523 (1967).
138. J. Zborowski and L. Wojtczak, Induction of swelling of liver mitochondria by fatty acids of various chain lengths, *Biochim. Biophys. Acta* **70**:596–598 (1963).
139. A. L. Luzzati, Effetto di acidi grassi e di isoottano sulla glicolisi anaerobia del tessuto nervoso, *Boll. Sel. Soc. Ital. Biol.* **36**:1893–1895 (1960).
140. P. G. Scholefield, Studies of fatty acid oxidation. 4. Effects of fatty acids on the oxidation of other metabolites, *Can. J. Biochem. Physiol.* **34**:1211–1225 (1956).
141. C. E. Rowe, The metabolism of unesterified fatty acid in mouse brain *in vitro*, *Biochem. J.* **88**:48P–49P (1963).
142. A. Geiger, J. Magnes, and R. S. Geiger, Survival of the perfused cat's brain in the absence of glucose, *Nature* **170**:754–755 (1952).
143. M. E. Volk, R. H. Millington, and S. Weinhouse, Oxidation of endogenous fatty acids of rat tissues *in vitro*, *J. Biol. Chem.* **195**:493–501 (1952).
144. P. M. Vignais, C. H. Gallagher, and I. Zabin, Activation and oxidation of long chain fatty acids by rat brain, *J. Neurochem.* **2**:283–287 (1958).
145. F. Lynen, Participation of coenzyme A in the oxidation of fat, *Nature* **174**:962–965 (1954).
146. D. S. Beattie and R. E. Basford, Brain mitochondria. IV. The activation of fatty acids in bovine brain mitochondria, *J. Biol. Chem.* **241**:1412–1418 (1966).
147. D. S. Beattie and R. E. Basford, Brain mitochondria. V. Incorporation of fatty acids into phospholipids in bovine brain mitochondria, *J. Biol. Chem.* **241**:1419–1423 (1966).
148. D. S. Beattie and R. E. Basford, Brain mitochondria. III. Fatty acid oxidation by bovine brain mitochondria, *J. Neurochem.* **12**:103–111 (1965).
149. W. L. Stahl, J. C. Smith, L. M. Napolitano, and R. E. Basford, Brain mitochondria, *J. Cell Biol.* **19**:293–307 (1963).
150. N. R. Marquis and I. B. Fritz, Enzymological determination of free carnitine concentration in rat tissues, *J. Lipid Res.* **5**:184–187 (1964).
151. K. R. Norum, The organ and the subcellular distribution of palmityl CoA:carnitine palmityltransferase in man, *Acta Physiol. Scand.* **66**:172–181 (1966).
152. D. S. Beattie and R. E. Basford, Sodium ion and fatty acid oxidation in bovine brain mitochondria, *Biochem. Biophys. Res. Commun.* **22**:419–424 (1966).
153. H. Waelsch, W. M. Sperry, and V. A. Stoyanoff, A study of the synthesis and deposition of lipids in brain and other tissues with denterium as an indicator, *J. Biol. Chem.* **135**:291–296 (1940).
154. Y. Kishimoto, W. E. Davies, and N. S. Radin, Turnover of the fatty acids of rat brain gangliosides, glycerophosphatides, cerebrosides and sulfatides as a function of age, *J. Lipid Res.* **6**:525–531 (1965).

*Chapter 20*

# CHOLESTEROL METABOLISM


## Alan N. Davison

*Department of Biochemistry*
*Charing Cross Hospital Medical School*
*London, England*


## I. INTRODUCTION

Cholesterol is found in high concentration in adult nervous tissue. Large amounts of the sterol are found in the white matter and histochemical studies show cholesterol to be a major constituent of the myelin sheath. The subcellular distribution of cholesterol has also been determined in fractions separated by differential and gradient centrifugation. All membrane fractions contain cholesterol but the highest proportion is found in the myelin layer (Table I). The localization of cholesterol in membranes and especially its presence in myelin has an overriding influence on the metabolism of brain sterols. Thus, cholesterol synthesis is highly active during myelination of the brain (see LeBaron, Chapter 21 of Volume III of this series) whereas in the adult it appears from isotope studies that there is relatively little synthesis and turnover of cholesterol.

**TABLE I**

**Distribution of Cholesterol in Subcellular Fractions of the Adult Brain**[a]

| | Rat[1] | Guinea pig forebrain[2] |
|---|---|---|
| Whole brain | 100% | 100% |
| Nuclei | 4.4 | 15.5 |
| Myelin | 35.6 | 36 |
| Nerve endings | 19.3 | 10.3 |
| Mitochondria | 2.3 | 1.1 |
| Microsomes | 11.3 | 5.7 |
| Supernatant | 8.1 | 1.1 |

[a] Brain homogenates in sucrose solution were separated by centrifugal procedures.

## II. CHANGES IN STEROL CONTENT OF THE DEVELOPING NERVOUS SYSTEM

In developing brain prior to the onset of myelination the lipid composition more closely resembles that of other tissues but as myelin is deposited the concentration of cholesterol increases—particularly in the white matter. The rate of deposition of cholesterol in the brain reaches a maximum shortly after the onset of myelination. For example, in rats, mice, and rabbits, the rate of deposition of cholesterol is greatest between about 10–20 days postpartum (Fig. 1); in the pig the rate is also at a maximum soon after birth;[3] however, in other species such as the guinea pig, peak accretion of brain cholesterol occurs before birth.[4] Deposition of cholesterol continues in the brain for a considerable time until adult concentration is reached. The adult concentration of cholesterol is then maintained with the increasing growth of the brain. Maximum incorporation of labeled precursors both *in vivo* and *in vitro* and deposition of cholesterol[5,6] coincides with the period of maximal myelination as judged by histochemical observations.[7]

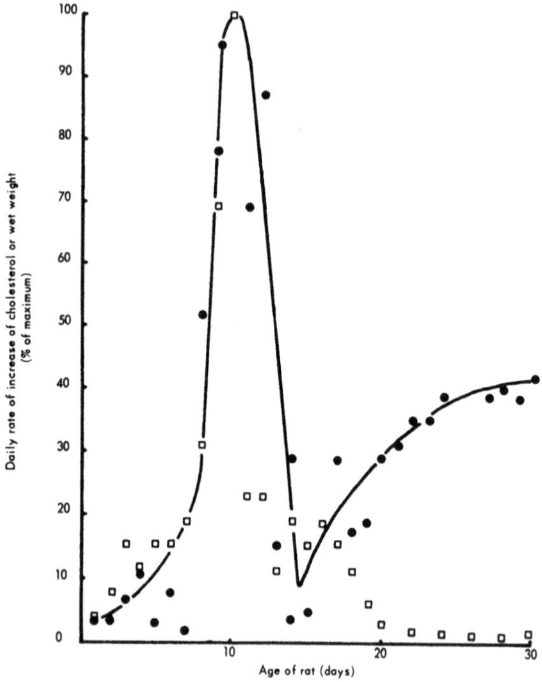


Fig. 1. Changes in the rate of cholesterol accumulation and increase in wet weight in the rat brain during development. The results are shown as percentages of the maximum rate of cholesterol deposition (●) and rate of increase of brain weight (□) per day.[1] (By courtesy of the *Biochemical Journal*.)

It is therefore most appropriate to employ actively myelinating nervous tissue preparations for the investigation of the biosynthetic steps leading to the synthesis of cholesterol.

## III. BIOSYNTHESIS OF CHOLESTEROL

Immediately before myelination commences in the brain Marchi-positive lipid droplets accumulate in the neuropil and small amounts of cholesterol esters, desmosterol ($\Delta^{5,24}$-cholestadien-3$\beta$-ol), and traces of many other sterol precursors are found (see p. 550). As myelination proceeds these droplets disappear and the sterol precursors are replaced by unesterified cholesterol. Cholesterol esterase activity is most pronounced in the brain at about this time[(8)] so that it seems possible that hydrolysis of cholesterol esters can supply both free sterol and fatty acids to the developing brain although this source of cholesterol is probably of only minor importance. Thus it seems unlikely that cholesterol esters are transported in large amounts from the liver to the central nervous system for, at least in the chick, the spectrum of cholesterol ester fatty acids is dissimilar in brain, spinal cord, and liver (Table II). In addition, following injection of labeled precursors, measurement of relative changes in brain and plasma cholesterol specific radioactivity make it clear that the labeled sterol found in the brain is largely the result of endogenous synthesis.[(9)] Small amounts of $^{14}C$ cholesterol from the blood can, however, be taken up by the brain[(10-12)] and this uptake parallels accretion of cholesterol in the organ.

While the metabolic pathways for the biosynthesis of cholesterol from acetate in the mammalian liver is well established (see Paoletti *et al.*, Chapter 10 of Volume I of this series) not all such steps have been recognized in neural tissue.[(13,14)] Substances such as amino acids, pyruvate, and glucose which on metabolism can give rise to acetate serve as efficient sterol precursors. Incorporation of radioactive glucose into brain cholesterol following injection into mice is markedly greater than that of acetate, leucine, or mevalonate,[(6,15)] but this may be ascribed to the preferential utilization of glucose as substrate by nervous tissue. Acetate is, however, the preferred substrate for synthesis of cholesterol by brain slices.[(16)] It has been unexpectedly found that both *in vivo* and *in vitro* acetate is a more efficient precursor of brain cholesterol than mevalonate[(15,17-20)] although in the liver mevalonate is a better precursor. Similar results have been obtained with cell-free preparations, thus eliminating the possibility that mevalonate penetrates less readily into brain cells than into liver cells. Paoletti and his colleagues have therefore suggested that there is a deficiency in mevalonic kinase and decarboxylase in the brain preparations from young animals although there is no direct evidence on this point. Mevalonic lactonase is lacking in the brain, in comparison with the liver in the same animal, and as a result potassium mevalonate is far better utilized than mevalonolactone by brain preparations. Fumagalli *et al.*[(21)] further suggest that the equili-

TABLE II

**Cholesterol and Its Esters in the Central Nervous System and Liver of the 21-Day-Old Chick Embryo**[a]

| | Brain (mg/organ) | Spinal cord (mg/organ) | Liver (mg/organ) |
|---|---|---|---|
| Cholesterol | 6.28 | 0.7 | 3.28 |
| Cholesterol ester | 0.25 | 0.06 | 21.70 |
| Major fatty acids[b] | | Percentage of total fatty acid | |
| 16 : 0 | 16.5 | 30.0 | 6.9 |
| 16 : 1 | 11.2 | 14.9 | 5.1 |
| 18 : 0 | 3.8 | 9.4 | 2.9 |
| 18 : 1 | 49.3 | 34.0 | 62.0 |
| 18 : 2 | 19.2 | 11.7 | 23.1 |

[a] Sterols were extracted with chloroform:methanol (2:1, v/v) and fatty acids released by alkaline hydrolysis of ester were determined by gas–liquid chromatography. Similar results have been obtained for samples obtained from 17- and 20-day-old embryos and from newly hatched chick (Davison and Oxberry, unpublished).

[b] We are grateful to Dr. I. MacDonald (Guy's Hospital Medical School, London) for fatty acid analyses.

brium between mevalonolactone and the acid form may regulate the rate of utilization of mevalonate for the synthesis of brain cholesterol.

## A. Control of Biosynthesis

Further information of the later stages of cholesterol biosynthesis in nervous tissue and its control has been provided by the work of Fish *et al.*,(22) Holstein *et al.*,(23) and Paoletti and his colleagues in Milan. It was suggested that the accumulation of desmosterol in the developing brain(16,24–27) can be ascribed to a rate-limiting step in the pathway of cholesterol syntheses due to impaired ability to reduce the $\Delta^{24}$ unsaturation of precursor sterols. Accumulation of desmosterol in the brain can be enhanced by administering triparanol,(27,28) for this drug inhibits the enzymic reduction of the 24,25 double bond. Desmosterol is found in small amounts in the myelin of the developing brain(29) and after treatment of rats with triparanol incorporation into myelin is increased(30) so that it seems that desmosterol may be biologically indistinguishable from cholesterol. Trace amounts of other intermediary sterols have also been found in the normal developing brain: zymosterol ($\Delta^{8,24}$-cholestadien-3$\beta$-ol), $\Delta^{7,24}$-cholestadien-3$\beta$-ol, and $\Delta^{5,7,24}$-cholestatrien-3$\beta$-ol have been detected.(23,24,28,31) Holstein and his colleagues have traced the pattern of labeling of brain sterol intermediates at different times following injection of [1-$^{14}$C] acetate into 5-day-old rats. Their studies showed a slow turnover of radioactivity (about 27 hr) in the zymosterol fraction which suggested a rate-controlling step along the pathway leading from the 8,9 double bond to the 5,6 double bond compounds (Fig. 2). The second slow step, detected in the same experiment, was found to resemble that previously observed in chick embryo brain and involved impaired ability to reduce the 24,25 double bond of desmosterol. This accounts for the accumulation and slow turnover (about 72 hr) found in the desmosterol fraction.

The pathway described above for the synthesis of cholesterol from lanosterol involved transformation of steroid intermediates with unsaturated lateral chains ($\Delta^{24}$ compounds). Lanosterol can also be reduced by the relatively nonspecific $\Delta^{24}$ reductase to give 24,25-dihydrolanosterol, which itself can lead through successive saturated side-chain compounds to 7-dehydrocholesterol ($\Delta^{5,7}$ cholestadien-3$\beta$-ol) and cholesterol (Fig. 2). Evidence for the presence of such a metabolic pathway comes from studies using the drug AY-9944 (trans-1,4-bis [2-dichlorobenzylaminomethyl] cyclohexane dihydrochloride).(31) This compound inhibits the enzyme catalyzed reduction of 7-dehydrocholesterol.(32) Following administration of AY-9944 to growing rats there is accumulation in the brain of $\Delta^{5,7}$-cholestadien-3$\beta$-ol, $\Delta^{7,24}$-cholestadien-3$\beta$-ol, and $\Delta^{5,7,24}$ cholestatrien-3$\beta$-ol(31) with a concomitant decrease in cholesterol content. If mature rats are treated with AY-9944, 7-dehydrocholesterol accumulates in the brain only after a very prolonged period (Table III).(31,33) When drug treatment is discontinued for 6 months no trace of 7-dehydrocholesterol can be found in the adult brain. These observations are consistent with metabolic experiments suggesting that very slow synthesis and turnover of cholesterol occurs in the adult rat brain.

Thus with increasing age it becomes more difficult to demonstrate net synthesis of cholesterol, and several workers have failed to find incorporation of labeled precursor into adult brain cholesterol.(34) However, by using suitably fortified minced brain preparations from 1-year-old animals,

**TABLE III**

**Effect of Chronic Treatment of Rats with AY-9944 on Rat Brain Sterols**

| | Cholesterol (mg/100 g) | 7-Dehydrocholesterol (mg/100 g) | Total sterol (mg/100 g) |
|---|---|---|---|
| Control | 1880 | 0 | 1880 |
| Treatment | | | |
| 2.32 mg/Kg/day for 4 days | 1880 | Trace | 1880 |
| 5 mg/Kg/day for 12 months | 886(33) | 850 | 1736 |
| Recovery for 6 months | 1936(33) | 0 | 1936 |

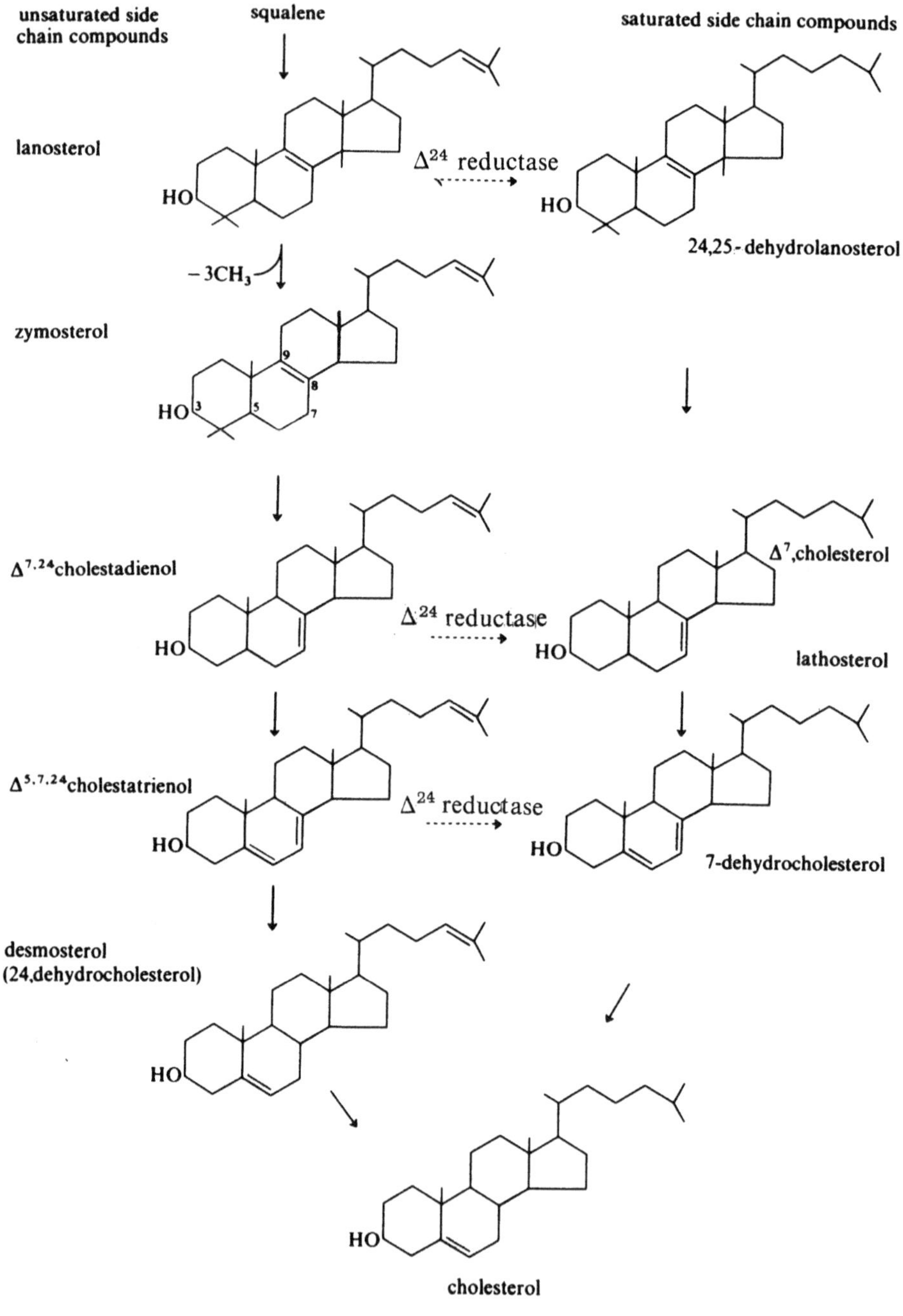


Fig. 2. Steps in the synthesis of cholesterol.

Nicholas[35] was able to show utilization of [2-$^{14}$C] acetate and mevalonate to form labeled nonsaponifiable material. After incubation $^{14}$C-squalene and $^{14}$C-cholesterol were isolated from the preparations, by precipitation as digitonide and conversion to the dibromide, and thus a definite conversion by adult rat brain was established. As expected, liver was much more effective in synthesizing cholesterol, but Nicholas[35] was able to show that both adult brain and liver slices utilized mevalonate more efficiently than acetate. This therefore indicates an interesting difference in the cerebral metabolism of young and mature rats, as in 10-day-old rat brain slices Garattini *et al.*[17] found that [1-$^{14}$C] acetate was a much better precursor of $^{14}$C labeled cholesterol than [2-$^{14}$C] mevalonic acid[21] (see also p. 549). These experiments were extended by Nicholas and Aexel,[36] who incubated adult rat brain homogenized in $MgCl_2$-nicotinamide buffer with added cofactors and various $^{14}$C-labeled substrates. A small but definite synthesis of cholesterol was obtained although most of the radioactivity was accumulated into squalene and other cholesterol intermediates. It may be concluded that earlier failures to demonstrate cholesterol biosynthesis by adult brain slices were due to the fact that the very small quantities of labeled cholesterol synthesized are not easily detected (see below).

## IV. CHOLESTEROL METABOLISM

The early work of Waelsch *et al.*[37,38] established that heavy water was readily incorporated into the unsaponifiable lipids of the brain when fed to newborn rats. The extent of such uptake was inversely related to the age of the rat at the time of dosage and directly related to their current rate of myelin deposition. Only traces of deuterium were found in the brain sterols of adult rats fed heavy water for a period of 4–7 days. Bloch *et al.*[39] fed cholesterol labeled with deuterium for 3 days to an adult dog; although at the end of 6 days the marked material could be recovered from the other organs examined, none was found in the brain or spinal cord. It was concluded that there was a lack or paucity of metabolic interchange between the sterol of the central nervous system and that of the blood. These observations and the restricted synthesis of cholesterol found in the adult brain both *in vivo* and *in vivo* led to the view that brain sterols were metabolically stable.

However, some investigators suggest an alternative explanation for failure to incorporate labeled cholesterol and to utilize precursors, for the synthesis of the sterol in the brain may be ascribed to a selective barrier between blood and brain or nerve. Thus, it is possible to isolate labeled cholesterol from the brain following intracerebral or intraventricular injections of $^{14}$C acetate or mevalonate although this is not possible with equivalent doses given parenterally. It was therefore argued that this procedure bypassed the "blood–brain barrier." Dobbing[40,41] has however pointed out that interpretation of these experiments is exceedingly difficult; for example, it is not possible to equate a dose given directly into the brain to that given by parenteral injection. Precursor of very high specific activity presented locally by injection into the brain will produce labeled cholesterol even if it is poorly utilized. The sum of current evidence therefore supports the view that synthesis of cholesterol in the adult brain is restricted. This in turn suggests that much of this cholesterol must be metabolically stable, for the neural sterol content remains essentially constant throughout adult life.

Isotopic experiments indicate that a small amount of cholesterol in the adult brain undergoes quite rapid exchange equivalent to a complete turnover time of 80 min.(9,15) In addition, a second pool of brain cholesterol undergoes rather slower turnover ($t_{0\cdot5}$ approximately 30 days) but a high proportion is metabolically relatively stable. This metabolic heterogeneity of cholesterol metabolism in the brain indicates the complexity of the system and underlines the importance of relating metabolism to anatomical structure.

Since cholesterol is readily incorporated into the nervous system during development, it is possible to label neural cholesterol by administration of suitable precursors during this period. Alternatively cholesterol may be specifically labeled by giving $^{14}C$ cholesterol. Thus, [4-$^{14}$ C]cholesterol(10) or $^{3}H$ labeled cholesterol(42,43) has been injected into the yolk sac of one-day-old chicks and the incorporation of a small amount of labeled cholesterol and its subsequent persistence have been studied. Animals were allowed to survive for various periods up to 420 days thereafter. Radioactive cholesterol taken up into the brain underwent only slow turnover, and little loss of $^{14}C$ occurs over at least 200 days after injection. Essentially similar results have been obtained in other species: in rats(44,45) and rabbits.(46) In rabbits slow turnover of [4-$^{14}$C] cholesterol was found to be associated with the gray matter sterol and little if any turnover was seen in the white matter of the brain or in the spinal cord. It seemed possible that this remarkable persistence of cholesterol in the central nervous system was due to the continual reutilization of $^{14}C$ released by degradation. However, radioactive cholesterol isolated from the brain of animals long after injection of specifically labeled [4-$^{14}$C] cholesterol was found to retain almost all the label in the 4-C position.(47) Moreover, it was found that all the radioactivity in the original crude extract of the central nervous system was due to its content of $^{14}C$ cholesterol and other lipids were not labeled. In long-term radioautographic experiments on young mice, Torvik and Sidman(48) also concluded that there was no evidence of recycling of labeled lipid and persistence was localized in the white matter. Nicholls and Rossiter(49) have found incorporation and persistence of [4-$^{14}$C] cholesterol for periods up to 30 weeks in the sciatic nerve of rats. Four times as much labeled sterol was taken up into the nerve after crushing, and again once incorporated there was only very slow catabolism. Long-term persistence of $^{14}C$-labeled cholesterol has also been demonstrated in the peripheral nerve and brain by Simon(50) after giving radioactive cholesterol to young rats.

These experiments therefore suggest that most of the cholesterol deposited during development and growth of the central and peripheral nervous system undergoes only very slow turnover. The difficulties of labeling brain or nerve cholesterol in adult animals can now be related to this large mass of cholesterol essentially removed from the normal dynamic exchange processes of the body.

## V. INTERPRETATION OF METABOLIC STUDIES

Since much of the cholesterol of mature nervous tissue is present in the myelin sheath it may be inferred that this structure has a high degree of metabolic stability. This same stability is shown by other myelin lipids[51] and also by the myelin proteins.[52] These observations have been amplified by the result of direct experiment. Thus months after injection of isotopically labeled precursors into young rats, persisting cholesterol can be isolated in the myelin fraction but some remaining labeled cholesterol is also present in other subcellular fractions, notably the mitochondria.[44,53] It has been suggested that the stable lipids may form a complex in myelin, for the molar ratio of cholesterol equals that of the other stable lipids (galactolipid + sphingomyelin + phosphatidyl ethanolamine).[54] In the brain mitochondrial membrane cholesterol, sphingomyelin, and cardiolipin seem to form a stable complex[44,53] whereas other phospholipids undergo turnover. The general metabolic stability of neural cholesterol, the absence of any active catabolic system, and the relative lack of synthetic ability in the mature brain is of some biological importance, for failure in sterol synthesis during development or loss of cholesterol from adult brain may result in irreparable damage.

## VI. BRAIN CHOLESTEROL—THE EFFECTS OF STRESS

The lipids of the nervous system have a predominantly structural role and even after prolonged starvation no alteration has been found in the lipid content of the brain.[55] For example, in the hen starvation for 15–16 days produces a 22% loss in body weight and while some changes in the polyunsaturated fatty acid content of the brain were noted, the overall lipid content of the organ remained unaffected.[56] In 8-week-old rats, after 3 days of starvation no reduction was observed by Smith[57] in total brain or spinal cord cholesterol. Similarly Dobbing[4] fed adult rats on a diet of cane sugar until the animals lost an average of one third of their body weight. No differences were seen in the lipid content of the brain of normal controls and the malnourished rats (Table IV).

However even mild malnutrition has an effect on the developing brain and its cholesterol content,[4,58,59] provided the stress occurs during the period of maximum myelination[60] (Table IV). In pigs, undernourished for one year after birth, cholesterol in the brain was about 60% of controls.[61] Even two years after such severe malnourishment it was not possible to restore brain cholesterol to normal concentration. When weanling rats (that is, after the vulnerable period of development) were underfed for 8 weeks the reduction in brain cholesterol could be reversed on rehabilitation.[62] Since much of the cholesterol in the developing brain is incorporated into myelin, the changes in cholesterol content seen in these experiments may be taken to represent effects on this structure—an interpretation supported by histological examination of the optic nerve myelin in undernourished rats.[63] The relative metabolic stability of myelin therefore accounts for the apparent resistance of the adult brain to the effects of undernutrition.

## TABLE IV

**Undernutrition and Brain Cholesterol[4]**

| Age of rat | Period of undernutrition | Body weight compared to controls | Whole brain cholesterol deficiency compared to controls | Whole brain cholesterol deficiency after re-habilitation |
|---|---|---|---|---|
| Newborn | 3 Weeks | 41% | 24% | 14% |
| 3 Weeks | 8 Weeks | 80 | 17 | 10 |
| Adult | 5 Weeks | 55 | 100 | 100 |

Hence, the period of myelination may be regarded as a vulnerable period in development[60] since restrictions at this time can limit the supply of precursors and reduce biosynthesis much more easily than in the adult where these processes are virtually quiescent.

The interpretation of experimental observation on the effects of undernutrition and its relationship to the development of intelligence—particularly in man—is a subject of considerable difficulty.[60,64] In rats there is some evidence that early malnutrition has an effect on subsequent memory,[65,66] in dogs cortical electrical activity is retarded,[67] and in pigs a long-lasting behavioral abnormality has been obtained by feeding a severely imbalanced diet to very young pigs.[68]

In certain diseases of the developing nervous system there is a delay in myelination—a condition accompanied by reduction in the accumulation of the myelin lipids: cerebrosides, some phospholipids, and cholesterol. Such amyelination is seen in Swayback and Border disease of sheep[69,70] and in phenylketonuria in man.[71,72] In the adult nervous system damage resulting in demyelination is followed by formation of esterified cholesterol and loss of cholesterol.[73]

## VII. CONCLUSION

Synthesis of cholesterol in nervous tissue is most pronounced during the period of myelination and in the adult brain only small amounts are formed. Although the pathway for biosynthesis of cholesterol is well established for the liver, not all the steps have so far been identified in other tissues such as those of the nervous system. Two rate-controlling stages have so far been recognized in sterol biosynthesis by young brain but other factors influencing the course of biosynthesis and the rate of accumulation of sterols in the developing brain before and during myelination are at present largely unknown.

Much of the brain cholesterol is localized in myelin and other membranous structures within the nervous system. Since myelin and possibly part of the mitochondrial membrane appear to be metabolically stable, the period of formation of these membranes may be regarded as a vulnerable period in the development of the nervous system.

## VIII. REFERENCES

1. M. L. Cuzner and A. N. Davison, The lipid composition or rat brain myelin and subcellular fractions during development, *Biochem. J.* **106**:29–34 (1968).
2. J. Eichberg, V. P. Whittaker, and R. M. C. Dawson, Distribution of lipids in subcellular particles in guinea pig brain, *Biochem. J.* **92**:91–100 (1964).
3. J. T. W. Dickerson and J. Dobbing, Prenatal and postnatal growth and development of the central nervous system of the pig, *Proc. Roy. Soc.* **166**:384–395 (1967).
4. J. Dobbing, *in Malnutrition, Learning and Behavior* (N. S. Schrimshaw, ed.), in press, MIT Press, Cambridge, Mass.
5. S. R. Korey and A. Stein, *in Regional Neurochemistry* (S. S. Kety and J. Elkes, eds.), pp. 175–189, C. C. Thomas, Springfield, Ill. (1961).
6. H. W. Moser and M. L. Karnovsky, Studies on the biosynthesis of glycolipids and other lipids of the brain, *J. Biol. Chem.* **234**:1990–1997 (1959).
7. S. Jacobson, Sequence of myelinization in the brain of the albino rat, *J. Comp. Neurol.* **121**:5–29 (1963).
8. E. T. Pritchard and N. E. Nichol, Cholesterol esterase activity in developing rat brain, *Biochim. Biophys. Acta* **84**:781–782 (1964).
9. J. J. Kabara and G. T. Okita, Brain cholesterol: biosynthesis with selected precursors *in vivo*, *J. Neurochem.* **7**:298–304 (1961).
10. A. N. Davison, J. Dobbing, R. S. Morgan, and G. Payling Wright, Metabolism of myelin: the persistence of [4-$^{14}$C] cholesterol in the mammalian central nervous system, *Lancet* **1**:658–660 (1959).
11. J. Dobbing, The entry of cholesterol into developing rat brain, *J. Neurochem.* **10**: 739–742 (1963).
12. R. Clarenburg, I. L. Chaikoff, and M. D. Morris, Incorporation of injected cholesterol into the myelinating brain of the 17 day old rabbit, *J. Neurochem.* **10**:135–143 (1963).
13. A. N. Davison, Brain sterol metabolism, *Advan. Lipid Res.* **3**:171–196 (1965).
14. A. N. Davison, *in Applied Neurochemistry* (A. N. Davison and J. Dobbing, eds.), pp. 178, Blackwell, Oxford (1968).

15. J. J. Kabara, Brain cholesterol. XI. A review of biosynthesis in adult mice, *J. Am. Oil Chemists' Soc.* **42**:1003–1008 (1965).
16. D. Kritchevsky, S. A. Tepper, N. W. DiTullio, and W. L. Holmes, Desmosterol in developing rat brain, *J. Am. Oil Chemists' Soc.* **42**:1024–1028 (1965).
17. S. Garattini, P. Paoletti, and R. Paoletti, The incorporation of 2-$^{14}$C mevalonic acid into cholesterol and fatty acids of brain and liver *in vitro*, *Arch. Biochem. Biophys.* **80**:210–211 (1959).
18. S. Garattini, P. Paoletti, and R. Paoletti, Lipid biosynthesis *in vivo* from acetate-1-$C^{14}$ and mevalonic-2-$C^{14}$ acid, *Arch. Biochem. Biophys.* **84**:253–255 (1959).
19. H. J. Nicholas and B. E. Thomas, Intracerebral incorporation of [2-$^{14}$C] mevalonic acid into adult rat brain squalene and cholesterol, *Biochim. Biophys. Acta* **36**:583–585 (1959).
20. H. J. Nicholas and B. E. Thomas, Cholesterol metabolism and the blood–brain barrier: An experimental study with [2-$C^{14}$]-sodium acetate, *Brain* **84**:320–328 (1961).
21. R. Fumagalli, E. Grossi, M. Poggi, P. Paoletti, and S. Garattini, Cholesterol synthesis in rat brain: Differential incorporation of mevalonolactone-2-$C^{14}$ and potassium mevalonate-2-$C^{14}$, *Arch. Biochem. Biophys.* **99**:529–533 (1962).
22. W. A. Fish, J. E. Boyd, and W. M. Stokes, Metabolism of cholesterol in the chick embryo. III. Localization and turnover of desmosterol (24-dehydrocholesterol), *J. Biol. Chem.* **237**:334–337 (1962).
23. T. J. Holstein, W. A. Fish, and W. M. Stokes, Pathway of cholesterol biosynthesis in the brain of the neonatal rat, *J. Lipid Res.* **7**:634–638 (1966).
24. D. Kritchevsky and W. L. Holmes, Occurrence of desmosterol in developing rat brain, *Biochem. Biophys. Res. Commun.* **7**:128–131 (1962).
25. R. Fumagalli and R. Paoletti, The identification and significance of desmosterol in the developing human and animal brain. *Life Sci.* **5**:291–295 (1963).
26. R. Fumagalli, E. Grossi, P. Paoletti, and R. Paoletti, Studies on lipids in brain tumours I, *J. Neurochem.* **11**:561–565 (1964).
27. T. G. Scott and V. C. Barber, An enzyme histochemical and biochemical study of the effect of an inhibitor of cholesterol synthesis on myelinating mouse brain, *J. Neurochem.* **11**:423–429 (1964).
28. T. J. Scallen, R. M. Condie, and J. Schroepfer, Inhibition by triparanol of cholesterol formation in the brain of the newborn mouse, *J. Neurochem.* **9**:99–103 (1962).
29. M. E. Smith, R. Fumagalli, and R. Paoletti, The occurrence of desmosterol in myelin of developing rats, *Life Sci.* **6**:1085–1091 (1967).
30. N. L. Banik and A. N. Davison, Desmosterol in rat brain myelin, *J. Neurochem.* **14**:594–595 (1967).
31. R. Fumagalli, R. Niemiro, and R. Paoletti, Inhibition of the biogenetic reaction sequence of cholesterol in rat tissues through inhibition with AY-9944, *J. Am. Oil Chemists' Soc.* **42**:1018–1023 (1965).
32. R. Niemiro and R. Fumagalli, Studies on the inhibitory mechanism of some hypocholesterolemic agents on 7-dehydrocholesterol $\Delta^7$-bond reductase activity, *Biochim. Biophys. Acta* **98**:624–631 (1965).
33. D. Dvornik, Inhibition of cholesterol biosynthesis and its usefulness, *Proc. Intern. Symp. Drug Res.* (*Cand.*) (June 1967).
34. R. J. Rossiter, *in Metabolism of the Nervous System* (D. Richter, ed.), pp. 355–380, Pergamon Press, London (1957).
35. H. J. Nicholas, Cholesterol, *J. Kansas Med. Soc.* **62**:358–361 (1961).
36. H. J. Nicholas and R. T. Aexel, Biosynthesis of cholesterol in cell-free extracts of adult rat brain, *Federation Proc.* **26**:342 (1967).

37. H. Waelsch, W. M. Sperry, and V. A. Stoyanoff, A study of the synthesis and deposition of lipids in brain and other tissues with deuterium as an indicator, *J. Biol. Chem.* **135**:291–296 (1940).
38. H. Waelsch, W. M. Sperry, and V. A. Stoyanoff, Lipid metabolism in brain during myelination, *J. Biol. Chem.* **135**:297–302 (1940).
39. K. Bloch, B. N. Berg, and D. Rittenberg, Biological conversion of cholesterol to cholic acid, *J. Biol. Chem.* **149**:511–517 (1943).
40. J. Dobbing, The blood–brain barrier, *Physiol. Rev.* **41**:130–188 (1961).
41. J. Dobbing, *in Applied Neurochemistry* (A. N. Davison and J. Dobbing, eds.), p. 317, Blackwell, Oxford (1968).
42. D. Kritchevsky and V. Defendi, Persistence of sterols other than cholesterol in chicken tissue, *Nature* (*Lond.*) **192**:71 (1961).
43. D. Kritchevsky and V. Defendi, Deposition of tritium labeled sterols (cholesterol sitosterol, lanosterol) in brain and other organs of the growing chicken, *J. Neurochem.* **9**:421–425 (1962).
44. M. L. Cuzner, A. N. Davison, and N. A. Gregson, Turnover of brain mitochondrial membrane lipids, *Biochem. J.* **101**:618–626 (1966).
45. M. E. Smith and L. F. Eng. The turnover of the lipid components of myelin, *J. Am. Oil Chemists Soc.* **42**:1013–1018 (1965).
46. A. N. Davison, R. S. Morgan, M. Wajda, and G. Payling Wright, Metabolism of myelin lipids: Incorporation of [3-$^{14}$C] serine in brain lipids of the developing rabbit and their persistence in the central nervous system, *J. Neurochem.* **4**:360–365 (1959).
47. A. N. Davison and M. Wajda, Persistence of cholesterol-4-$^{14}$C in the central nervous system, *Nature* (*Lond.*) **183**:1606–1607 (1959).
48. A. Torvik and R. L. Sidman, Autoradiographic studies on lipid synthesis in the mouse brain during postnatal development, *J. Neurochem.* **12**:555–565 (1965).
49. D. Nicholls and R. J. Rossiter, Metabolism of lipids of peripheral nerve regenerating after crush, *J. Neurochem.* **11**:813–818 (1964).
50. G. Simon, Cholesterol ester in degenerating nerve: Origin of cholesterol moiety, *Lipids* **1**:369–370 (1966).
51. A. N. Davison, *in Metabolism and Physiological Significance of Lipids* (R. M. C. Dawson and D. N. Rhodes, eds.), pp. 527–540, Wiley, London (1964).
52. A. N. Davison, Metabolically inert proteins of the central and peripheral nervous system, muscle and tendon, *Biochem. J.* **78**:272–282 (1961).
53. A. A. Khan and J. P. Folch, Cholesterol turnover in brain subcellular particles, *J. Neurochem.* **14**:1099–1105 (1967).
54. L. F. Eng and M. E. Smith, The cholesterol complex in the myelin membrane, *Lipids* **1**:296 (1966).
55. J. P. Folch, *in Psychiatric Research* (A. Drinker, ed.), p. 23, Harvard Univ. Press, Cambridge, Mass. (1947).
56. C. D. Joel, H. W. Moser, G. Majno, and M. L. Karnovsky, Effects of bis-(monoisopropylamino)-fluorophosphine oxide (mipafox) and of starvation on the lipids in the nervous system of the hen, *J. Neurochem.* **14**:479–488 (1967).
57. M. E. Smith, The effect of fasting on lipid metabolism of the central nervous system of the rat, *J. Neurochem.* **10**:531–536 (1963).
58. J. Dobbing, The effect of undernutrition on myelination in the central nervous system, *Biol. Neonatorum* **9**:132–147 (1966).
59. J. W. T. Dickerson, J. Dobbing, and R. A. McCance, The effect of undernutrition and subsequent rehabilitation on the growth and composition of the central nervous system of the rat, *Brain* **90**:897–906 (1967).

60. A. N. Davison and J. Dobbing, Myelination as a vulnerable period in brain development, *Brit. Med. Bull.* **22**:40–44 (1966).
61. J. T. W. Dickerson, J. Dobbing and R. A. McCance, The effect of undernutrition on the postnatal development of the brain and cord in pigs, *Proc. Roy. Soc. Ser. B* **166**:396–407 (1967).
62. J. Dobbing and E. M. Widdowson, The effect of undernutrition and subsequent rehabilitation on myelination of rat brain as measured by its composition, *Brain* **88**: 357–366 (1965).
63. A. R. Buchanan and J. E. Roberts, Relative lack of myelin in optic tracts as result of underfeeding in the young albino rat, *Proc. Soc. Exptl. Biol. Med.* (*N.Y.*) **69**:101–104 (1948).
64. J. Dobbing, *in Applied Neurochemistry* (A. N. Davison and J. Dobbing, eds.), pp. 287, Blackwell, Oxford (1968).
65. V. Novakova, J. Faltin, V. Flandera, P. Hahn, and O. Koldovsky, Effect of early and late weaning on learning in adult rats, *Nature* (*Lond.*) **193**:280 (1962).
66. R. H. Barnes, S. R. Cunnold, R. R. Zimmermann, H. Simmons, R. B. Macleod, and L. Krook, Influence of nutritional deprivations in early life on learning behavior of rats as measured by performance in a water maze, *J. Nutr.* **89**:399–410 (1966).
67. J. Myslivecek, M. W. Fox, and J. Zahlava, Maturation retardée de l'activité bio-éléctrique corticale provoquée par malnutrition, *J. Physiol.* (*Paris*) **58**:572–573 (1966).
68. W. G. Pond, R. H. Barnes, R. B. Bradfield, E. Kwong, and L. Krook, Effect of dietary energy intake or protein deficiency symptoms and body composition of baby pigs fed equalised but suboptimal amounts of protein, *J. Nutr.* **85**:57–66 (1965).
69. J. McC. Howell, A. N. Davison, and J. M. Oxberry, Biochemical and neuropathological changes in Swayback, *Res. Vet. Sci.* **5**:376–384 (1964).
70. A. N. Davison and J. M. Oxberry, A comparison of the composition of white matter lipids in Swayback and Border disease of lambs, *Res. Vet. Sci.* **7**:67–71 (1966).
71. L. Crome, V. Tymms, and L. I. Woolf, A chemical investigation of the defects of myelination in phenylketonuria, *J. Neurol. Neurosurg. Psychiat.* **25**:143–148 (1962).
72. B. Gerstl, N. Malamud, L. F. Eng, and R. B. Hayman, Lipid alterations in human brains in phenylketonuria, *Neurology* **17**:51–58 (1967).
73. J. N. Cumings, The cerebral lipids and mental retardation, *Proc. 2nd. Intern. Congr. Ment. Retard., Vienna, Part* 1, 111–122 (1963).

Chapter 21

# METABOLISM OF MYELIN CONSTITUENTS

Francis N. LeBaron

*Department of Biochemistry*
*The University of New Mexico School of Medicine*
*Albuquerque, New Mexico*

## I. INTRODUCTION

As noted in Chapter 9 of Vol. I, the material usually referred to as myelin is a lamellar structure produced by glial cells and extruded in concentric layers around nerve axons. The chemical constituents listed in Table I and discussed more fully in the chapter previously referred to are associated in large macromolecular arrays, the exact structure of which are yet to be determined in detail. This lamellar material still must be considered part of the living cells and, as such, it must have associated with it at least a minimal metabolic activity for maintenance of its integrity. This deduction is obvious from the known facts of demyelinative processes that can be triggered by a variety of stimuli. Nevertheless, there has been a continuous controversy as to the importance and degree of these metabolic processes. This controversy has been caused by many factors germane to the study of myelin specifically, and these problems had to be clarified first. For example, the relatively slow metabolism of myelin was overemphasized by limited passage of metabolic precursors through the "blood–brain barrier"; the role of glial cells in elaborating myelin had to be elucidated; interrelationship of the myelin constituents in complex macromolecules was not at first fully appreciated; and so on. It is only recently that some definitive evidence has become available by analysis of isolated myelin, but even these recent techniques have their drawbacks and the whole question of the relative rates of biosynthesis and degradation of the various constituents of myelin is only beginning to be examined in a definitive manner.

The detailed description of enzymatic pathways involved in synthesizing and catabolizing the various lipid and protein constituents of myelin are adequately given in other chapters of these volumes, and no discussion of these aspects of metabolism will be given in the present chapter. Instead the

TABLE I

**Lipid and Protein Constituents of Myelin**

| Lipids | Lipids found in isolated myelin[8,10] | Other lipids presumed associated with myelination[11] |
|---|---|---|
| Sterols | Cholesterol | |
| Galactolipids[a] | Cerebrosides | |
| | Sulfatides | Galactosyl diglycerides[12] |
| Phospholipids | Sphingomyelins | |
| | Phosphatidyl serines | |
| | Phosphatidyl cholines | |
| | | Choline plasmalogens |
| | Phosphatidyl ethanolamines | |
| | Ethanolamine plasmalogens | |
| | Phosphatidyl inositols | |
| | Diphosphoinositides | |
| | Triphosphoinositides | |
| | | Inositol plasmalogens |
| | Phosphatidic acids | |
| | Diphosphatidyl glycerols | |
| | | Plasmalogenic acids |
| *Proteins*[5] | | |
| Folch–Lees proteolipids | | |
| Wolfgram acid-soluble proteolipid | | |
| LeBaron acid-soluble proteolipid | | |
| Basic protein (EAE antigen) | | |
| Insoluble protein | | |

[a] Gangliosides have been reported in some preparation of isolated myelin. No metabolic information is yet available for this possible myelin lipid.

meager knowledge directly related to the formation and maintenance of myelin itself will be covered. By reference to other parts of these volumes the reader then can judge how best to plan future investigations.

To the author's knowledge a number of good reviews on various aspects of myelin chemistry and metabolism are currently in preparation or have recently appeared. Some of these are listed at the beginning of Section V.[1–9] One can only hope that this plethora of reviews will stimulate many new investigations to attain much needed definitive metabolic data on more specific aspects of general biological and chemical importance. It should be emphasized that the bibliography in this chapter includes only some of the most recent contributions from the many groups now working in this field. The reader is referred to these papers and to the reviews cited above for a more complete survey of the literature.

## II. METHODS FOR STUDYING METABOLISM SPECIFICALLY OF MYELIN

A continuing problem in studying myelin metabolism has been the need to ascertain whether the measurements one was making were indeed those of myelin per se and not some other brain constituent. Thus, e.g., the bulk of the cholesterol in brain or nerve is probably in myelin, but there are small amounts in other tissue constituents. For this reason interpretation of data must take into account the possibility that a small, very active metabolism of a minor tissue constituent could give the erroneous indication that the myelin itself was metabolizing more rapidly.

### A. Study of Myelin-Rich Tissue Fractions

The most obvious methods are those designed to study tissue samples in which myelin can be presumed to be concentrated. The crudest of these would be comparisons of metabolic processes in gray matter and white matter of the central nervous system. Such studies can be little more than indicative because of the obvious presence of other tissue constituents such as glial cells and axons in the white matter. A further refinement is the study of extremely small, highly localized samples.[13,14] More recently the technique of differential centrifugation has led to a considerable refinement of this type of analysis. By this technique several laboratories have been able to study metabolic processes in quite highly purified myelin.[8,10,15,16]

This technique can be used both for measuring enzyme concentrations and for incorporation of labeled precursors in biosynthetic processes. While these samples are certainly the purest yet obtainable, there is still the possibility of contamination of the myelin fractions by traces of other tissue constituents with very active metabolism. In addition one needs to evaluate the effects on the metabolic process to be studied of the various procedures and reagents used during the isolation of the myelin.

### B. Studies During Myelination or Demyelination

A more indirect method of studying the myelin constituents of brain tissue is to make parallel biochemical analysis and histological examination during the postnatal development of an animal that does not begin the process of myelination until after birth.[11,15,17] In this type of experiment accumulation of chemical constituents in parallel with histologically demonstrated deposition of myelin strongly suggests that the constituents are in the myelin. In addition, relatively rapid incorporation of labeled precursors also suggests the biosynthesis of myelin constituents.[18,19] The subsequent rate of disappearance of the labeled constituents then indicates their relative metabolic stability. In the reverse situation, the disappearance of chemical substances during demyelination helps to indicate which of these substances are myelin constituents. However, in this case the metabolic processes that would be measured would be more a reflection of the metabolism of macrophages or other scavenging cells than of processes inherent to myelin itself.

### C. Histochemical and Radioautographic and Other Methods

Other studies that could indicate metabolic processes of myelin are those in which histological sections are stained to indicate localization of enzyme activity. The big advantage of these methods is their potentially accurate localization. However, such studies are much more difficult to conduct and interpret and thus far have yielded little except negative information.[20] Nevertheless, continued refinement of these techniques should yield more definitive information when the apparently very low concentrations of enzymes in the myelin are no longer below the range of the methods. The use of radioautography also shows promise in localization of metabolic activity and movement of metabolic products within the tissue.[21]

Finally, a more direct method for studying metabolic processes involved in myelination is the use of tissue culture. It has been shown that glial cells will form concentric layers around axons in tissue culture, and the study of metabolic processes going on at this time should lead to definitive information about the myelination process.[22]

## III. METABOLISM OF INDIVIDUAL MYELIN CONSTITUENTS

### A. General Concepts

In discussing the metabolism of myelin one needs to address oneself to the problem by dividing the reactions involved into three main categories: the processes involved in the laying down of myelin, the processes involved in maintenance of the integrity of the myelin once it is produced and the processes involved in removal of the myelin after it or its nerve cell is damaged. In the first category some of the questions to be answered are as follows: Assuming one knows the complete chemical structure of myelin (a goal yet to be completely accomplished), is this structure formed as a unit or is the glial membrane layed down in a lamellar structure and then altered in some stepwise manner to produce the structure which is stable for a long time? Which, if any, of the constituents of the myelin ultrastructure are synthesized at the site of deposition and which are synthesized elsewhere in the cell or surrounding tissue and incorporated *in toto*? What are the precursor substances in limiting concentrations?

In the second category the most pertinent questions would be different. Can any of the small molecule constituents of the macromolecular complexes be renewed by simple exchange? If so, which? Must submolecular blocks be renewed instead? If so, what is the composition of such blocks? What are the critical biochemical reactions needed *in situ* for maintenance of myelin integrity? At what location within the lamellar ultrastructure do they take place? What points in these reactions are influenced by the various factors that trigger demyelination, and how can these be prevented?

As noted elsewhere, the actual process of demyelination is probably not strictly a function of the myelin tissue itself. Rather the metabolic processes measured during demyelination are a reflection of the activity of scavenging cells.[7,23,24] In discussing myelin metabolism per se, then, the critical questions are those asked above and the processes to be discussed take place before invasion of these secondary cells. Nevertheless, as seems probable from available information, if there are processes whereby many of the constituents of myelin are renewed at various rates depending upon the constituent in question, there must be both anabolic and catabolic processes involved. Perhaps these terms are inappropriate in that they imply synthesis and destruction where in some cases the processes really are those of supply, exchange, and removal of constituents. As with most of the questions posed here, even the best way to frame the questions cannot be decided without further data. The data currently available are summarized in the following sections.

## B. Lipids

To summarize the voluminous literature about the metabolism of myelin lipids we have resorted to an outline form to indicate available knowledge of metabolic activity of the individual lipids and their components during the three stages noted above.

### 1. *Sterols*

During myelination, there is rapid biosynthesis of cholesterol in brain, which has been demonstrated from a number of labeled precursors.[4] Exogenous labeled cholesterol is also incorporated directly without alteration of the label in the molecule.[3] The bulk of the cholesterol in myelin is then very stable after it is deposited. However, a very low rate of loss of radioactivity can be measured over long periods if the cholesterol is labeled during myelination. Also, a very slight incorporation of radioactivity from $^{14}C$-labeled acetate or glucose can be detected in the myelin of adult animals.[8,25] During demyelination the cholesterol is converted to cholesterol esters.[7]

### 2. *Sphingolipids*

*a. Cerebrosides.* The indications are that the metabolism of the cerebrosides of myelin is similar to that of cholesterol,[8] i.e., rapid synthesis occurs at the time of myelination, the lipids undergo only very slow turnover in adult animals, and in the process of demyelination they are destroyed by the scavenging cells. The problem is more complicated by the probable presence of cerebrosides with different fatty acids in other tissue constituents that metabolize more rapidly. The critical studies to determine whether or not turnover in adult myelin involves synthesis of complete cerebroside molecules or exchange of constituents are yet to be completed, but there is evidence[26] to suggest that the fatty acids are metabolized.

*b. Sulfatides.* The sulfate moiety of sulfatides has been known for some time to be one of the most stable myelin constituents. This is true whether one measures stability of the sulfate moiety or of $^{14}C$-labeling in the rest of the molecule.[2,8,27] Comparison of stability with that of myelin cerebrosides is consistent with the accepted pathway whereby sulfatides are formed by sulfation of cerebrosides. More recently, Davison and Gregson[28] have found that there also is a small pool of sulfatide in myelin that has an active turnover of sulfate even in the adult animal.

*c. Sphingomyelins.* When the metabolism of sphingomyelins of myelin is measured by uptake and turnover from $^{14}C$-labeled substances, the results parallel those obtained with cerebrosides and cholesterol.[8] By comparison of metabolism of the phosphorus moiety of sphingomyelin in gray and white matter Ansell and Spanner[29] reached the same conclusions for this constituent. In the adult animal $^{32}P$ incorporation is slowest into sphingomyelin among all myelin phospholipids yet studied.[30]

### 3. *Phospholipids*

*a. Phosphatidyl Serines.* As noted by Smith,[8] there is presently some disagreement in the reported results concerning metabolism of serine-containing diester phospholipids, depending upon the age of the animals studied and the radioactive precursor used. There is an indication that these lipids may change their metabolic processes more than other phospholipids in the transition from myelination to the maintenance processes. Nevertheless, some slow metabolism does occur in adult myelin.[30]

*b. Phosphatidyl Cholines.* There is also some discrepancy in the literature concerning the rate of metabolism of phosphatidyl cholines as related to other phospholipids (see Smith[8] for discussion). Most data indicate that these lipids have a more active metabolism than the other phosphatides that are present in relatively large concentrations in myelin. Nevertheless, they are considered stable lipids.[31] Much of the difficulty in interpretation of data is caused by the necessity to compare data from different investigations that have employed different radioactive precursors and different experimental conditions. It does seem apparent, however, that the fatty acids in the two different positions of the diester are renewed at parallel rates.[8]

*c. Choline Plasmalogens.* The choline plasmalogens, as opposed to the ethanolamine plasmalogens, represent a very small proportion of choline-containing phospholipids and of plasmalogens in myelin. Consequently little or no specific data on their metabolism is yet available. Mandel and Nussbaum[30] have reported, however, that the choline plasmalogens are metabolized somewhat more slowly than the diester form in adult myelin.

*d. Phosphatidyl Ethanolamines.* The reverse of the case with choline phospholipids occurs in the ethanolamine phospholipids, i.e., 80% of these lipids are plasmalogens and 20% are diesters. In this case, then, the data for the diesters is less exacting than for the plasmalogens, and they are frequently reported as a combined group. Nevertheless, Mandel and Nussbaum[30] have reported a slightly higher metabolism of the diesters in this case also. These diesters are metabolized nearly as rapidly as the phosphatidyl cholines in adult myelin.

*e. Ethanolamine Plasmalogens.* Most studies have indicated that the total group of ethanolamine phospholipids are metabolized somewhat more slowly than choline phospholipids. This may be a reflection of the fact that, in general, plasmalogens are metabolized more slowly than diester lipids and the proportion of plasmalogens is many times higher in the ethanolamine group. However, definite data is not completely available yet. Smith[8] has indicated that the vinyl ether side chain of ethanolamine plasmalogen is metabolized at the same rates as the adjoining ester-linked fatty acid, suggesting that the molecule turns over as a unit. On the other hand, Debuch,[32] working with whole brain, reports differences between these two moieties. The data concerning exact relationships of the choline, serine, and ethanolamine lipids in general are not yet consistent, however, either for metabolism of adult myelin in the maintenance phase or for the processes involved in myelination.

*f. Inositol Phosphatides.* The considerable amount of data recently reported concerning the metabolism of inositol phosphatides in nervous tissue is summarized in Chapter 17. This work has been somewhat specialized since these lipids occur in relatively low concentrations as compared to many others in myelin, but the di- and triphosphoinositides, at least, seem to be localized in myelin, and some of the constituents of these particular lipids have a much more active metabolic turnover in mature myelin than most other myelin constituents. The relative activities of the several phosphoinositides have been studied in various subcellular fractions including myelin,[33,34] but recent evidence that some of the enzymes associated with this metabolism are not present in myelin[35] seems to complicate the situation. It seems established that at least the 4 and 5 phosphates of triphosphoinositides have very rapid turnover *in vivo*, a rate much higher than that of the bulk of myelin phospholipids and one that appears to be maintained at a very high level in mature myelin. Whether this seeming exception to the general stability of mature myelin is related to processes involved in maintaining myelin or in processes related to the activity of the axon awaits the results of further work. The localization of the small amount of triphosphoinositide within the lamellar macromolecular structure of myelin needs to be determined, and then one can investigate the intriguing possibility that it is associated with loci of active metabolism of other constituents as well.

The metabolism of phosphatidyl inositol has been more carefully compared with other myelin phosphatides[8,30] and shown to be more active than most. The turnover of the various constituents of these monophosphoinositide molecules and the corresponding components of the di- and triphosphoinositides appears to be quite active but slower than the 4 and 5 phosphate positions of the latter. This has led to some speculation about possible involvement of these latter phosphate radicals, at least indirectly, in transport processes.

*g. Phosphatidic Acids and Polyglycerophosphatides.* In many tissues including nerve tissues these lipids seem to have higher metabolic activity than most other phospholipids. Little or no definitive data are available regarding myelin per se, however. Mandel and Nussbaum[30] do report that $^{32}P$ incorporation into polyglycerophosphatides of mature myelin is more rapid than any other phospholipids they studied. McCaman *et al.*,[36] however, report relatively low activity of phosphatidic acid phosphatase in white matter areas.

### 4. *Other Lipids*

The accumulation of three other lipids in brain during growth suggests that they are myelin constituents.[11] No definitive data concerning their metabolic activity in myelin is available yet, however. These lipids are galactosyl diglycerides, inositol plasmalogens, and plasmalogenic acids.

## C. Proteins

Information on the metabolism of myelin proteins is very scanty at present principally because of a relative lack of knowledge of the chemistry of the molecular species involved.[5] Initial studies on biosynthesis of several probable protein constituents of myelin have been made, however. Thus a preliminary report of biosynthesis of the "basic protein" that is responsible for induction of experimental allergic encephalomyelitis has been made.[37] The only other studies reported have been involved with proteolipid protein and total brain protein or insoluble brain protein residues. The latter are not pertinent to a discussion concerning myelin.

With regard to proteolipid protein, studies both *in vivo* and *in vitro* have shown that uptake of labeled amino acids into this constituent is much more rapid in young animals at the stage of rapid myelination.[38-42] The labeled precursor is retained for a relatively long time, indicating that the protein is not being as rapidly renewed as most tissue protein. Nevertheless, some slower incorporation does take place in adult animals. The amino acid composition of human adult proteolipid is slightly different from that from the immature brain also.[43]

Thus for this major protein constituent of myelin macromolecules, the long-term kinetics are similar to those of cerebrosides, cholesterol, and some other of the myelin lipids. However, the detailed sequence of assembly of all these lipid and protein constituents into the macromolecules and subsequently into the lamellar ultrastructure is yet to be determined. Sequential and parallel studies of lipid and protein turnover in the slow maintenance phase of myelin metabolism are also needed.

There undoubtedly are other protein species in myelin, but there is no information yet available about their nature or their metabolism. Undoubtedly the metabolism of myelin proteins is equally important to that of the lipid constituents, and it may be even more important in early stages of demyelination since an active proteinase is one of the few enzymes certainly present in myelin.[20] A great deal of work is needed in this field.

### D. Other Constituents

Certainly the lipids of the myelin macromolecular complexes are their most characteristic constituent and those which have been studied most extensively. The protein constituents are beginning to be investigated in more detail. Undoubtedly the inorganic constituents, water and salts, are of equal significance in maintenance of the structural integrity of the myelin ultrastructure. As yet there is no available information concerning the stability or the possible exchange of these constituents.

## IV. SUMMARY AND CONCLUSIONS

### A. Metabolic Processes During Myelination

Most evidence currently available is consistent with the hypothesis that myelin is not synthesized and deposited as a macromolecular unit that is never modified. Rather it would seem that the glial cells lay down lamellae of cell membrane and then, subsequently, the macromolecular composition of this structure is modified for a period of time before the mature, very stable structure is achieved.[8,10,11,15,16,44,45] These later changes involve an increase in the chain length and degree of unsaturation of some fatty acids in some lipid constituents, an increase in the proportion of cerebrosides, and a decrease in the proportion of the more labile phospholipids. Quantitatively the most significant decrease is in phosphatidyl cholines. Changes in the proportions of the several protein constituents also have been reported, and a substitution of cholesterol for desmosterol has been suggested as an early change. It also has been suggested that the establishment of a relatively stable structure involves the formation of a complex of the major lipids with cholesterol, and this structure requires the presence of relatively long-chain fatty acids in the lipids.[8,46]

For the most part, the evidence for the above hypothesis has been gained indirectly, and it has not yet been rigorously examined by experiments designed for that purpose. More detailed temporal correlation of the increase in the many biochemical constituents and morphological characteristics of myelin is necessary. Careful consideration of differences between peripheral and central nervous system myelin and between myelin of different species also needs to be made. On the other hand, in view of the advances that have been made in recent years in elucidating the metabolic pathways and enzyme catalyzed reactions involved in biosynthesis and degradation of the various lipids and proteins, some of the most fruitful investigations on myelin metabolism should be studies designed to answer the other questions posed earlier. Which of the constituents of the myelin ultrastructure are synthesized at the site of deposition and which must be supplied from other parts of the glial cell or other cells? What are the metabolic precursors that are in limiting concentrations?

## B. Metabolic Activity of Mature Myelin

It seems reasonably well established that the bulk of the quantitatively important chemical constituents of myelin are unusually stable in comparison to constituents of many other body structures. However, it also seems probable that a small metabolic activity of these constituents is present. The data of Davison and Gregson,[28] among others, suggests that in fact the turnover of the mature myelin is due to a relatively rapid metabolism of a quite small proportion of these constituents. This has led these authors to postulate[2] two separate pools of mature myelin constituents, a small active one and a large inert one. They suggest that the more active metabolism takes place in the outer layers of myelin.

One also might extend this hypothesis to include the constituents, such as triphosphoinositide, which occur in small concentration but are relatively labile metabolically and apparently do not have any stable pool. It could be suggested that these constituents occur exclusively in the outer layers where the active pools of the other constituents are localized. On the other hand, an equally logical hypothesis might be made that the more active metabolic sites occurred in the inner lamellae next to the axons. The need of a healthy axon for maintenance of its myelin is obvious from pathological studies, and Klenk,[47] among others, has suggested that the axon supplies necessary materials for maintenance of myelin integrity. In any event maintenance of healthy myelin is dependent upon healthy metabolism of glia and neurons. The reader is referred to other chapters for information on these processes. At present, when evidence on many points is only beginning to be available, many other theories can be postulated. There seem to be variations in the composition of myelin along the length of the axon,[48] and transport of materials along the axon has been claimed for some substances including phospholipids.[49]

It should be emphasized, however, that many basic critical considerations concerning the metabolism of mature myelin are yet to be completely evaluated. To be definitive all such studies of "turnover" of constituents and measurements of rates of incorporation of radioactively labeled substances require knowledge of, and measurement of, activity in immediate precursor pools as well as in the products being studied. This type of study is particularly difficult in studies *in vivo*. For the most part such information is not yet available for studies of myelin, and all theories must remain in question until it is available.

### C. Stimulus for Demyelination

Most of the questions posed earlier in this chapter regarding possible causes of demyelination are still largely unanswered, and perhaps the most important of these from a practical point of view relate to the cause and preventions of demyelination in human diseases. While some clues have been obtained in relation to possible defects in genetic diseases that result in inadequate supply of necessary myelin constituents, there is as yet no indication of any possible specific triggering mechanism to cause the disruption of healthy, mature myelin. This can be caused by so many seemingly unrelated agents that a single triggering mechanism may not be common to all diseases. However, since it seems probable that a small amount of quite active metabolism is continually taking place in mature myelin, active investigation to establish the localization of these processes should yield fruitful results. The exact nature and interrelationships of these processes needs careful elucidation. The reader is referred to appropriate chapters in Vol. III for further information on these matters.

## V. REFERENCES

1. R. D. Adams and E. P. Richardson, Jr., The Demyelinative Diseases of the Human Nervous System. A Classification; A Review of Salient Neuropathologic Findings; Comments on Recent Biochemical Studies, *in Chemical Pathology of the Nervous System* (J. Folch-Pi, ed.), pp. 162–194, Pergamon Press, London (1961).
2. A. N. Davison, Myelin Metabolism, *in Metabolism and Physiological Significance of Lipids* (R. M. C. Dawson and D. N. Rhodes, eds.), pp. 527–537, Wiley, New York (1964).
3. A. N. Davison, Brain sterol metabolism, *Advan. Lipid Res.* **3**:171–196 (1965).
4. J. J. Kabara, Brain cholesterol: The effect of chemical and physical agents, *Advan. Lipid Res.* **5**:279–327 (1967).
5. F. N. LeBaron, The Lipid-Protein Complexes of Myelin, *in Structural and Functional Aspects of Lipoproteins in Living Systems* (A. Scanu and E. Tria, eds.), Academic Press, London (in press).
6. G. Wright Payling, The metabolism of myelin, *Proc. Roy. Soc. Med.* **54**:26–30 (1961).
7. R. J. Rossiter, The Chemistry of Wallerian Degeneration, *in Chemical Pathology of the Nervous System* (J. Folch-Pi, ed.), pp. 207–227, Pergamon Press, London (1961).
8. M. E. Smith, The metabolism of myelin lipids, *Advan. Lipid Res.* **5**:241–278 (1967).
9. H. E. Whipple (ed.), Research in demyelinating diseases, *Ann. N.Y. Acad. Sci.* **122**:1–570 (1965).
10. W. T. Norton, S. E. Poduslo, and K. Suzuki, Rat brain myelin: Compositional changes during development, *Abstr. 1st Intern. Meet. Intern. Soc. Neurochem.* 161 (1967).

11. M. A. Wells and J. C. Dittmer, A comprehensive study of the postnatal changes in the concentration of the lipids of developing rat brain, *Biochemistry* **6**:3169–3175 (1967).
12. J. M. Steim, Monogalactosyl Diglyceride: A new neurolipid, *Biochim. Biophys. Acta* **144**:118–126 (1967).
13. A. Pope, H. H. Hess, and J. N. Allen, Quantitative Histochemistry of Proteolytic and Oxidative Enzymes in Human Cerebral Cortex and Brain Tumors, *in Ultrastructure and Cellular Chemistry of Neural Tissue* (H. Waelsch, ed.), pp. 182–191, Hoeber-Harper, New York (1957).
14. E. Robins, D. E. Smith, and M. K. Jen, The Quantitative Distribution of Eight Enzymes of Glucose Metabolism and Two Citric Acid Cycle Enzymes in the Cerebellar Cortex and Its Subjacent White Matter, *in Ultrastructure and Cellular Chemistry of Neural Tissue* (H. Waelsch, ed.), pp. 205–211, Hoeber-Harper, New York (1957).
15. M. L. Cuzner and A. N. Davison, The lipid composition of rat brain myelin and subcellular fractions during development, *Biochem. J.* **106**:29–34 (1968).
16. L. F. Eng, B. Gerstl, D. V. Pratt, and M. G. Tavaststjerna, The maturation of human CNS myelin, *Abstr. 1st Intern. Meet. Intern. Soc. Neurochem.* 62 (1967).
17. J. Folch-Pi, Composition of the Brain in Relation to Maturation, *in Biochemistry of the Developing Nervous System* (H. Waelsch, ed.), pp. 121–133, Academic Press, London (1955).
18. H. S. Maker and G. Hauser, Incorporation of glucose carbon into gangliosides and cerebrosides by slices of developing rat brain, *J. Neurochem.* **14**:457–464 (1967).
19. K. G. Manukyan, A. A. Smirnov, and E. V. Chirkovskaya, Metabolism of brain phospholipid phosphorus during ontogenesis, *Biokhimiya.* **28**:246–252 (1963).
20. C. W. M. Adams, A. N. Davison, and N. A. Gregson, Enzyme inactivity of myelin: Histochemical and biochemical evidence, *J. Neurochem.* **10**:383–395 (1963).
21. A. Torvik and R. L. Sidman, Autoradiographic studies on lipid synthesis in the mouse brain during postnatal development, *J. Neurochem.* **12**:555–565 (1965).
22. T. Yonezawa, M. B. Bornstein, E. R. Peterson, and M. R. Murray, A histochemical study of oxidative enzymes in myelinating cultures of central and peripheral nervous tissue, *J. Neuropathol. Exptl. Neurol.* **21**:479–487 (1962).
23. P. W. Lampert and M. W. Kies, Mechanism of demyelination in allergic encephalomyelitis of guinea pigs. An electron microscopic study, *Exptl. Neurol.* **18**:210–223 (1967).
24. G. Majno and M. L. Karnovsky, A biochemical and morphologic study of myelination and demyelination—III. Effect of an organo-phosphorus compound (mipafox) on the biosynthesis of lipid by nervous tissue of rats and hens, *J. Neurochem.* **8**:1–16 (1961).
25. H. J. Nicholas, Cholesterol turnover in the central nervous system, *J. Am. Oil Chem. Soc.* **42**:1008–1012 (1965).
26. Y. Kishimoto and N. S. Radin, Metabolism of brain glycolipid fatty acids, *Lipids* **1**:47–61 (1966).
27. G. M. McKhann and W. Ho, The *in vivo* and *in vitro* synthesis of sulphatides during development, *J. Neurochem.* **14**:717–724 (1967).
28. A. N. Davison and N. A. Gregson, The physiological role of cerebron sulphuric acid (sulphatide) in the brain, *Biochem. J.* **85**:558–568 (1962).
29. G. B. Ansell and S. Spanner, The metabolism of labelled ethanolamine in the brain of the rat *in vivo*, *J. Neurochem.* **14**:873–885 (1967).
30. P. Mandel and J. L. Nussbaum, Incorporation of $^{32}P$ into the phosphatides of myelin sheaths and of intracellular membranes, *J. Neurochem.* **13**:629–642 (1966).

31. M. L. Cuzner, A. N. Davison, and N. A. Gregson, Chemical and metabolic studies of rat myelin of the central nervous system, *Ann. N.Y. Acad. Sci.* **122**:86–92 (1965).
32. H. Debuch, Uber die bildung der plasmalogene zur zeit der myelinisierung bei der ratte, II. *Z. Physiol. Chem.* **344**:83–88 (1966).
33. M. Kai and J. N. Hawthorne, Incorporation of injected [$^{32}$P] phosphate into the phosphoinositides of subcellular fractions from young rat brain, *Biochem. J.* **98**:62–67 (1966).
34. J. G. Salway, M. Kai, and J. N. Hawthorne, Triphosphoinositide phosphomonoesterase activity in nerve cell bodies, neuroglia and subcellular fractions from whole rat brain, *J. Neurochem.* **14**:1013–1024 (1967).
35. G. Hauser, J. Eichberg, S. M. Gompertz, and M. Ross, Triphosphoinositide phosphohydrolases of rat brain, *Abstr. 1st Intern. Meet. Intern. Soc. Neurochem.* 93 (1967).
36. R. E. McCaman, M. Smith, and K. Cook, Intermediary metabolism of phospholipids in brain tissue. II. Phosphatidic acid phosphatase, *J. Biol. Chem.* **240**:3513–3517 (1965).
37. R. E. Martenson, M. K. Gaitonde, and D. Richter, The metabolism of the basic proteins of rat brain during development, *Abstr. 1st Intern. Meet. Intern. Soc. Neurochem.* 151 (1967).
38. M. K. Gaitonde, The rate of incorporation of [$^{35}$S] methionine and [$^{35}$S] cystine into proteolipids and proteins of rat brain, *Biochem. J.* **80**:277–284 (1961).
39. A. N. Davison, Metabolically inert proteins of the central and peripheral nervous system, muscle and tendon, *Biochem. J.* **78**:272–282 (1961).
40. L. C. Mokrasch, Incorporation of [$^{14}$C] amino acids into the proteolipid of subcellular preparations of rat brain *in vitro*, *J. Neurochem.* **13**:49–58 (1966).
41. C. B. Klee and L. Sokoloff, Amino acid incorporation into proteolipid of myelin *in vitro*, *Proc. Natl. Acad. Sci.* **53**:1014–1021 (1965).
42. A. A. Abdel-Latif and L. G. Abood, *in vivo* incorporation of L-[$^{14}$C] serine into phospholipids and proteins of the subcellular fractions of developing rat brain, *J. Neurochem.* **13**:1189–1196 (1966).
43. A. L. Prensky and H. W. Moser, Changes in the amino acid composition of proteolipids of white matter during maturation of the human nervous system, *J. Neurochem.* **14**:117–121 (1967).
44. N. L. Banik and A. N. Davison, Desmosterol in rat brain myelin, *J. Neurochem.* **14**:594–596 (1967).
45. L. A. Horrocks, R. J. Meckler, and R. L. Collins, Variations in the Lipid Composition of Mouse Brain Myelin as a Function of Age, *in Variations in Chemical Composition of the Nervous System as Determined by Developmental and Genetic Factors* (G. B. Ansell, ed.), p. 46, Pergamon Press, London (1966).
46. L. F. Eng and M. E. Smith, The cholesterol complex in the myelin membrane, *Lipids* **1**:296 (1966).
47. E. Klenk, Die Chemie der Markreisung und das Problem der Entmarkung, *Verh. Dtsch. Ges. Inn. Med.* **61**:331–339 (1955).
48. R. L. Friede and K. H. Hu, Increase in cholesterol along human optic nerve, *J. Neurochem.* **14**:307–315 (1967).
49. N. Miani, Analysis of the somato-axonal movement of phospholipids in the vagus and hypoglossal nerves, *J. Neurochem.* **10**:859–874 (1963).

# SUBJECT INDEX

# ERRATA

**HANDBOOK OF NEUROCHEMISTRY, VOL. 3**

**Chapter 8—SULFUR AMINO ACIDS—by M. K. Gaitonde**

A line was accidentally omitted on page 225. The text 13 lines from the bottom should read

> ...(2) the enzyme cysteine sulfinic acid decarboxylase is predominantly a particulate enzyme in the brain but a soluble enzyme in the liver, and (3) taurine is deaminated to form isethionic acid by rat brain and heart and not by liver....

In the equation on page 240 a subscript has been omitted and the second term on the right should read

$$CH_2{=}CHCH(NH_2)COOH$$

The curved arrows in the equation on page 250 are misdrawn. The equation should read

α-Oxoglutarate ↘↗ Cysteine sulfinic acid

Glutamate ↙↘ β-Sulfinylpyruvate $\rightleftarrows$ pyruvate + $SO_2 \rightarrow SO_3^{2-} \leftrightarrows SO_4^{2-}$

**Chapter 14—GANGLIOSIDES—by Lars Svennerholm**

In Table I, the structures in the last line on page 427 and the first line on page 428 have been accidentally interchanged. Thus, the last line on page 427 should read

| | | | |
|---|---|---|---|
| Gal(β,1 → 3)GalNAc(β,1 → 4)Gal(β,1 → 4)Glu(1 → 1)Cer<br>Gal: (3 ← 2)NAN; Gal (inner): (3 ← 2)NAN | Disialosyl-*N*-tetraglycosylceramide | $G_{D1a}$ | $G_{NT}2a$ |

and the first line on page 428 should read

| | | | |
|---|---|---|---|
| Gal(β,1 → 3)GalNAc(β,1 → 4)Gal(β,1 → 4)Glu(1 → 1)Cer<br>Gal (inner): (3 ← 2)NAN(8 ← 2)NAN | Disialosyl-*N*-tetraglycosylceramide | $G_{D1b}$ | $G_{NT}2b$ |

[The abbreviation Glu was used for glucose in this volume.] Note also that the correct code designation for the latter entry is $G_{D1b}$ and not $C_{D1b}$.

GPSR Compliance

*The European Union's (EU) General Product Safety Regulation (GPSR) is a set of rules that requires consumer products to be safe and our obligations to ensure this.*

*If you have any concerns about our products, you can contact us on ProductSafety@springernature.com*

In case Publisher is established outside the EU, the EU authorized representative is:

Springer Nature Customer Service Center GmbH
Europaplatz 3
69115 Heidelberg, Germany

**Batch number: 10837688**

Printed by Printforce, the Netherlands